Biochemical Adaptation

Response to Environmental Challenges
from Life's Origins to the Anthropocene

Biochemical Adaptation

Response to Environmental Challenges from Life's Origins to the Anthropocene

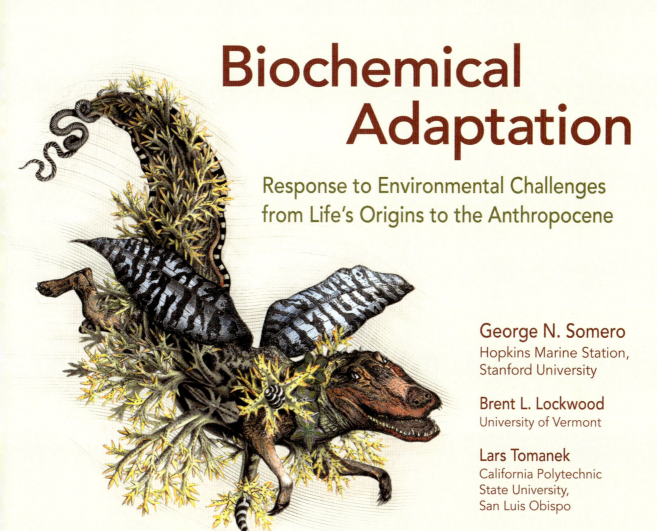

George N. Somero
Hopkins Marine Station,
Stanford University

Brent L. Lockwood
University of Vermont

Lars Tomanek
California Polytechnic
State University,
San Luis Obispo

Sinauer Associates, Inc. Publishers
Sunderland, Massachusetts U.S.A.

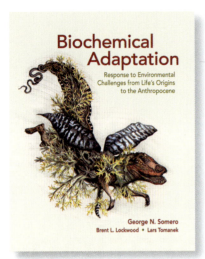

About the Cover

The cover is original artwork by Lissa Herschleb, who has a unique way of portraying—indeed, of creating—organisms. Here, she has assembled a collage comprising parts of many different types of organisms: microorganisms, plants, invertebrates, and vertebrates. Her multi-taxon creature is intended to emphasize the concept of "unity in diversity" that forms a central theme of the book. Thus, as different as organisms may be in appearance and in where and how they live, they are remarkably similar "under the skin"—at the biochemical level.

Address editorial correspondence and orders to:
Sinauer Associates, 23 Plumtree Road, Sunderland, MA 01375 USA
FAX: 413-549-1118
Email: publish@sinauer.com
Internet: www.sinauer.com

Library of Congress Cataloging-in-Publication Data

Names: Somero, George N., author. | Lockwood, Brent L., author. | Tomanek, Lars, author.
Title: Biochemical adaptation : response to environmental challenges, from life's origins to the Anthropocene / George N. Somero. Hopkins Marine Station, Stanford University, Brent L. Lockwood, University of Vermont, Lars Tomanek, California Polytechnic State University, San Luis Obispo.
Description: Sunderland, MA : Sinauer Associates, 2016. | Includes bibliographical references and index.
Identifiers: LCCN 2016040061 | ISBN 9781605355641 (casebound)
Subjects: LCSH: Adaptation (Physiology)
Classification: LCC QP82 .S66 2016 | DDC 578.4/6--dc23
LC record available at https://lccn.loc.gov/2016040061

Printed in China
5 4 3 2 1

We dedicate this book to Peter Hochachka, catalyst of our field.

Peter Hochachka aboard the R/V Alpha Helix in Alaska, summer of 1968

Contents

4   Water and Solutes: Evolution and Regulation of Biological Solutions   339

Preface

"Where is the knowledge we have lost in information?"

T. S. Elliot (1888-1965)

As we prepared to develop this volume, we discussed our intended project with a number of colleagues who work in the broad area of biochemical adaptation. The responses we received were at once highly encouraging, yet in one context also a bit discouraging. Although all felt that it would be a worthwhile project, they wondered how it would be possible to distill from the huge literature now available a clear "view of the forest," a perspective on the fundamental principles that characterize the types of adaptations organisms make at the biochemical level. One colleague presented us with an explicit and quantitative estimate of this challenge: Currently, there are close to 4,000 papers published *per day* in the biological sciences, broadly defined to include both medical and nonmedical research. The number of new journals increases daily, it seems, so tracking this literature becomes more challenging by the hour. As a point of reference, since *Strategies of Biochemical Adaptation*[1] was published in 1973, the number of publications per day in biology has risen by over an order of magnitude. Much of this material is "chaff" that can be neglected—but how does one find the "wheat" in all of these studies? And is it possible to glean overarching concepts from this vast wealth of information? In view of the information overload we all face, we think our readers will share the feeling expressed by T. S. Elliot back in 1934, to the effect that knowledge—here, an appreciation of key principles—can be obscured by an over-abundance of information.

To address this challenge, we have tried to review an adequate fraction of the relevant literature in order to provide an up-to-date perspective on some of the principal strategies of biochemical adaptation found in the three domains of life, the Archaea, Bacteria, and Eukarya. We have sought to identify strategies common to all domains of life as well as unique adaptations that are found in only certain taxa. It would clearly be impossible, at least in a book of this length, to cover a substantial fraction of the studies that have a relevant message to deliver. We hope that our choices of stories to tell strike the reader as appropriate for the task at hand. We especially hope that the new information given about archaea and bacteria will provide new insights for our colleagues who work on multicellular eukaryotes, which much of our past writing has focused on. In this go-round, we've tried to be much more ecumenical, in an effort to present generalizations that span all types of life.

---

[1] *Strategies of Biochemical Adaptation* (1973). P. W. Hochachka and G. N. Somero. Saunders, Philadelphia.

Our presentational strategy is as follows. We first examine how individual abiotic environmental factors—oxygen availability, temperature, water availability, osmolality, pH, and hydrostatic pressure—influence biochemical structure and function. This analysis comprises two fundamental lines of inquiry. First, how does variation in these factors perturb living systems? What is the nature of biochemical stress and how is this stress ameliorated through adaptation at the biochemical level? We address this question in a holistic manner by showing how adaptive changes in macromolecules are complemented by changes in the milieu in which these large molecules function. This first line of analysis is, in essence, a study of how the status quo is maintained through adaptive modification in biochemical structures and function. These adaptations help to explain the underlying biochemical unity found across all domains of life and in all environments. The second line of analysis focuses on the question of the new opportunities that environmental change affords an organism. Here, the essential question concerns not the conservation or restoration of the status quo, but rather the "invention" of new biochemical traits that provide the organism with better means for functioning in a changed environment. Thus, two themes—conservation and innovation—provide a framework for much of this volume's discussion.

The first four chapters focus principally on the effects of single types of environmental factors on diverse biochemical systems in a wide array of organisms. The discussions found in these chapters are intended to provide a foundation for the analyses of global change given in the final chapter—Adaptation in the Anthropocene. In this final chapter we touch on several issues. We show the importance of interactions among multiple stressors in determining the severity of the effects of global change. We consider the relative vulnerabilities of different organisms to our rapidly changing environment and examine why species differ so greatly in their susceptibility to environmental change. This analysis includes an examination of the role of genetic variation in establishing a species' likelihood of coping with anthropogenic environmental change. For the reader whose primary interest is in anthropogenic global change, we hope that our analysis will help to provide a perspective that allows the "why" of global change biology to be appreciated as well as the "what." Here's what we mean by that statement. There have been many excellent phenomenological analyses of the effects of anthropogenic global change in the biosphere, including studies of changes in timing (phenology) of natural events and in the distribution ranges of species in our changing world. These phenomena are what we refer to as the "what" aspect of global change: What's happening to the biosphere as anthropogenically driven changes to the planet's temperature, rainfall patterns, atmospheric and oceanic circulation, seawater pH, and oxygen levels continue—and continue at ever-increasing rates. Our treatment of global change will attempt to provide a mechanistic explanation for these phenomenological analyses, and show "why" these changes in the biosphere have occurred. This cause-effect approach is not only explanatory of what has already been observed, but it may help to facilitate predictions of what is likely to occur in the future as global change continues. If our presentation sparks an interest in biochemical adaptation and leads the reader to pursue this diverse subject in greater detail, we feel that we will have accomplished one of our goals. However, if the reader's analysis of the subject also leads to a sense of apprehension about what faces the biosphere because of anthropogenic activities, and if this apprehension catalyzes efforts to address and ameliorate these human-made challenges, then the authors will feel especially successful in their endeavor.

Acknowledgments

The authors are extremely grateful for the marvelous collaboration we have enjoyed with the folks at Sinauer Associates: Andy Sinauer, Laura Green, Liz Pierson, David McIntyre, Chris Small, Ann Chiara, Jan Troutt, and the entire production staff. Laura Green's editorial contributions were so important to the evolution of the book that we even feel she rightly merits being considered as a co-author! Evolution of the chapters was also facilitated by the help of the several reviewers who provided us with very important critical input, including Brad Buckley, Ronald Burton, Wes Dowd, John Duman, Peter Fields, Steve Hand, Kristy Kroeker, Dietmar Kültz, Jason Podrabsky, Raul Suarez, Pat Walsh, and Paul Yancey. Their broad knowledge of the topics we covered and their sense of how to organize and present this knowledge were of major help in refining the five chapters of this volume.

George wishes to acknowledge the intellectual companionship he has enjoyed over the years, especially that provided by Peter Hochachka, the mentor and friend to whom this volume is dedicated. George wishes to give special thanks to the graduate students, postdoctoral scholars, and sabbatical visitors who have populated his laboratory over the past half-century. He hopes that, when their material is presented in the different chapters, we have done justice to their intellectual efforts. He thanks Brent and Lars for joining this project. Lars brought to the effort his deep knowledge of metabolism and Brent brought his wide expertise in molecular biology, evolution, and genetics. He thanks Jim Childress for getting him into the study of deep-living species and Art DeVries for honing his skills in ice fishing and introducing him to the notothenioids, a group of fishes he's studied since 1963. During George's years at the Scripps Institution of Oceanography, he cherished the companionship and intellectual exchanges with Vic Vacquier, Paul Dayton, Farooq Azam, and Nick and Linda Holland. The late Dick Rosenblatt at Scripps played a major role in George's career by pointing him in the direction of the longjaw mudsucker, which become another favorite study species. At Oregon State University, George's frequent exchanges with Bruce Menge and Jane Lubchenco helped him put biochemistry into the context of the challenges that face intertidal organisms. George's Stanford colleagues Mark Denny, Jim Watanabe, and David Epel also merit special thanks for their friendship and intellectual stimulation over many decades. Most recently, George's close interactions with colleagues in China—Yunwei Dong, Xianliang Meng, and Cuiluan Yao—have opened up an exciting new chapter in his career. George's gratitude to Amy Anderson for coping with his obsessive preoccupation with the word processor during the writing of this book (and, for that matter, during most of our long and happy marriage) is deeper than can be put into words. Lastly, the roles of Gabe, Sophie, Logan, Lily, and Luka, canine officemates who've kept George's spirits high over the years must be recognized; one writes better with a smile on one's face.

Brent would like to acknowledge George Somero, Kristi Montooth, Colin Meiklejohn, Russ Lande, Stephen Palumbi, Dmitri Petrov, Thom Kaufman, Brian Calvi, Peter Fields, Lars Tomanek, and Ward Watt for their invaluable mentorship. He also thanks Jon Sanders for his grace and friendship inside and outside the lab, and Brandon Cooper and Luke Hoekstra for lively discussions of evolutionary genetics and thermal physiology. Brent would like to thank Aurelia Lockwood and Charlie Lockwood for providing balance and wonder in life. Lastly, Brent would like to thank his closest colleague and friend, Melissa Pespeni, who continues to inspire him as a scientist, mother, and wife.

Lars would like to thank his academic mentors George Somero and Dietmar Kültz for their invaluable guidance and the intellectual challenges they have provided. Lars would also like to express his appreciation for his colleagues and collaborators over the years, including Nikki Adams, Nann Fangue, Peter Fields, Kristin Hardy, Nishad Jayasundara, Sean Lema, Brent Lockwood, Frank Melzner, Don Mykles, Hans-Otto Pörtner, Jonathon Stillman, Inna Sokolova, and Anne Todgham, who have contributed greatly to his research efforts, some of which found their way into this book. He would also like to express his gratitude for the tremendous energy that his many students at Cal Poly San Luis Obispo have contributed to his research over the last ten years, who patiently waited for his attention while he juggled his many projects. All of Lars' research has required expensive instrumentation, which was acquired with the support of Deans Phil Bailey and Dean Wendt, whom he would like to acknowledge for seeding his research program. Lars would also like to thank the National Science Foundation for their support and the many reviewers who have evaluated so many of his proposals. Lars wants to give a special thanks to Christina Vasquez, who stepped in to deftly manage several research projects and his bustling lab while he was busy writing this book. Finally, he is grateful for the patience and understanding of his best friend Ruth Rominger, without whom he would have been unable to sustain this effort.

Lastly, the authors wish to compliment and thank Lissa Hershschleb for her cover artwork. Her wonderful composite critter is intended to emphasize the theme of "unity in diversity" that plays a major role in structuring the analyses found in this volume.

CHAPTER 1

A Conceptual Foundation for Understanding Biochemical Adaptation

"The ideal scientist thinks like a poet and only later works like a bookkeeper."

E. O. Wilson

In this opening chapter we sketch out the central themes that will serve to focus and structure the analyses of biochemical adaptation presented in this volume. These themes are applicable to all of the individual abiotic (chemical and physical) environmental factors we focus on in the specialized chapters that follow, whether the environmental issue in question is oxygen availability, temperature, or water and osmotic relationships. These same themes will prove helpful when, in the final chapter, we build on the foundation established by Chapters 2, 3, and 4, to examine the complex issues the biosphere faces in confronting the multifaceted challenges presented by anthropogenic global change. Thus, it is our goal to, first, examine in fine detail how different abiotic factors affect the structures and functions of different types of biochemical systems. Then, our analysis moves from this mechanistic or reductionist mode of investigation to an integrative analysis that addresses questions at higher levels of biological organization. By attaining a deep understanding of how abiotic factors influence biochemical systems, we will be in an advantageous position to analyze effects of the environment on such large-scale phenomena as biogeographic patterning and ecosystem structure. This multilevel analysis will also allow us to peer into the future and predict how anthropogenically driven global change is likely to alter the biosphere. Our analytical approach, then, is somewhat like using a camera with a zoom lens. We initially use this tool to focus deeply—submicroscopically—into the structures and functions of the fundamental biochemical systems common to all of life. And after completing this mechanistic analysis, we will "zoom out" and see how these biochemical responses to environmental change contribute to determining the properties of biological systems at the highest scales of complexity.

1.1 Conservation and Innovation: Two Sides of Biochemical Adaptation

The focus of this volume, in a nutshell, is this. In view of how sensitive biochemical systems are to the physical and chemical features of the environment, how can we account for the prevalence of life over the wide range of temperatures, salinities, oxygen concentrations, and hydrostatic pressures that characterize the diversity of habitats in which life occurs (Table 1.1)? How have organisms, during their evolutionary histories, met the challenges posed by the abiotic environment—and how successful might they be in meeting these challenges in the future, in a rapidly changing world? To address these central questions, we will employ a multifaceted analytical approach. First, to bring into focus the challenges organisms face from alterations in the abiotic environment, we will examine how a given environmental factor influences—perturbs—the structures and functions of core biochemical systems common to most if not all organisms. This mechanistic understanding will bring the challenges associated with coping with a changed environment into clear focus. Next, we will examine how organisms modify their biochemical systems to readjust their environmental optima and tolerance ranges over different timescales. This analysis will incorporate study of multigenerational evolutionary processes that involve acquisition of new genetic tools, and the processes involved in generating adaptive phenotypic modifications occurring during an organism's lifetime, which must rely on differential use of existing genetic information. This analysis will, in large measure, help us better understand how organisms can maintain a common set of biochemical processes in the face of changes in their environment. These adaptive changes in the phenotype can be viewed as largely conservative responses to environmental change—a retention of the biochemical status quo.

At the same time that natural selection leads to conservation of core biochemical structures and functions across most taxa, it also can lead to radically different and innovative types of biochemistries that reflect exploitation of new features of the environment. In some cases, the environmental features that offer opportunities for invention of novel forms of life are ones that have been generated by biological activities, notably metabolic activities that have altered the geochemical properties of the lands, waters, and atmosphere of the planet. Chapter 2 will illustrate how biological activities during the first approximately 2 billion years of evolution influenced the development of the atmosphere and how these atmospheric changes in turn led to new opportunities for life. The generation of dioxygen (O_2) by green-plant type (oxygenic) photosynthesis led to profound evolutionary changes: New, highly efficient modes of energy metabolism (aerobic generation of ATP) became possible, in part because an oxidizing atmosphere made elements like copper available for exploiting in oxidation-reduction reactions; more efficient metabolism, in turn, led to more active modes of life; transport systems for carrying O_2 through the body allowed for the evolution of larger-sized metazoans; and as O_2 entered the atmosphere in large amounts, an ozone (O_3) shield was generated that reduced penetration of ultraviolet (UV) light to Earth's surface, thus allowing life to emerge from the sea onto land. "Innovation" was piled on "innovation" to yield the diversity of life we see today, a diversity that is possible only because of the appearance of oxygen generated by oxygenic photosynthesis.

Apropos of E. O. Wilson's apt comment on how scientists work, this introductory chapter will present a general and somewhat qualitative presentation of the key themes that run through the volume. We certainly can't promise that this presentation will strike the reader as "poetic," but we do believe that this generalized, conceptual introduction will help prepare readers for the more detailed and mechanistic ("bookkeeping-level")

TABLE 1.1 ▮ ▮ ▮▮▮ ▮▮▮ ▮▮▮▮▮▮▮▮▮▮▮▮▮▮▮▮▮▮▮▮▮▮▮▮▮▮▮▮▮▮▮▮▮

Environmental ranges of abiotic conditions encountered by contemporary organisms

Examples of extremes of absolute values (lowest and highest) and breadth of ranges of values are shown. Examples of organisms living in these habitats are given. Many of these organisms will be discussed in specific contexts at relevant places in this volume.

TEMPERATURE	HYDROSTATIC PRESSURE
Absolute extremes	**Absolute extremes**
Lowest: ~–70°C	*Lowest*: 1 atm (0.101 MPa)
Arctic insects during winter dormancy	Shallow water species
Highest: ~130°C	*Highest*: 1100 atm (111 MPa)
Thermophilic archaea from deep-sea hydrothermal vents	Organisms in deep-sea trenches
Ranges	**Ranges**
Narrowest: –1.9°C to ~2°C (Southern Ocean and Antarctica); 2°C to 3°C (deep sea)	*Narrowest*: ~1 atm (0.101 MPa)
Ectotherms of McMurdo Sound, Antarctica (78°S) Deep-living fishes and invertebrates with minimal changes in depth during their life histories	Organisms with minimal changes in vertical distribution during their life histories
Widest: –70°C to ~25°C	*Widest*: ~250 atm (25.25 MPa)
High-latitude terrestrial species, e.g., Arctic insects	Vertically migrating species such as deep-sea rattail fish
SALINITY (OSMOLALITY)	**OXYGEN**
Absolute extremes	**Absolute extremes**
Lowest: ~0 mosmol l^{-1}	*Lowest*: 0 mmol l^{-1}
Freshwater invertebrates	Bacteria, archaea, and animals inhabiting oxygen-free sediments
Highest: 3500 mosmol l^{-1}	*Highest*: 24 mmol l^{-1}
Extremely halophilic archaea and brine shrimp (*Artemia franciscana*) in salt ponds	Organisms inhabiting tidepools with high photosynthetic activity
Ranges	**Ranges**
Narrowest: ~0 mosmol l^{-1}	*Narrowest*: ~0 mmol l^{-1}
Stenohaline marine species	Many terrestrial species and pelagic marine species
Widest: 0–3500 mosmol l^{-1}	*Widest (seasonal/daily)*: 0–24 mmol l^{-1}
Archaea and bacteria in shallow salt-flat ponds	Organisms inhabiting aquatic environments with high biological activity (respiration and photosynthesis) such as swamps, mudflats, tidepools, and surfaces of coral reefs

treatments given to the specific topics treated in the subsequent four, more specialized chapters. Each chapter is, indeed, a large ledger of information, whose comprehension will be facilitated by keeping the appropriate thematic perspective in place.

1.2 Life Prevails in the Face of a Remarkable Set of Environmental Conditions

One of the most striking aspects of life is its presence in such an extraordinary diversity of environments (see Table 1.1). Organisms thrive in niches whose temperatures span a range of approximately 200°C. Oxygen may be abundant or completely absent (anoxia). Salinity may vary from the dilute water of freshwater lakes to the hypersaline brines of desert salt ponds. Acidity (proton activity) varies by several orders of magnitude. Pressures may exceed 1100 atmospheres (atm) (111 megapascals [MPa]) in the depths of the ocean. Solar radiation may be intense or absent. Toxic chemicals may be abundant in the waters of a habitat. Suffice it to say, life prevails under circumstances that often seem incredibly extreme to us, from our perspective as a species that has a stable and quite moderate body temperature, lives under 1 atm pressure, and generally has access to abundant oxygen and water. An archaeal cell living at a seafloor hydrothermal vent (photo at left) at a depth of 3 km (pressure 30.3 MPa), where it is bathed by acidic, oxygen-free, sulfide-rich water with a temperature over 100°C, might have a different perspective on "extreme," of course. And if, as some have conjectured, life arose in deep-sea hot springs, then our salubrious (to us) terrestrial environment might most appropriately be perceived as being characterized by an "extreme" set of conditions for evolution to have mastered!

A black smoker chimney at a hydrothermal vent site.

Despite the relative nature and subjectivity of the adjective "extreme," it is fair to say that the prevalence of life at earthly extremes of temperature, water availability, salinity, acidity, oxygen concentration, and hydrostatic pressure should strike one as remarkable from a biochemical perspective. Why "remarkable"? The short answer to this question is that all of these abiotic factors, and, as we will emphasize in Chapter 5, combinations of these variables, have strong effects on the structural integrity and functions of the biochemical systems common to all organisms. The prevalence of life in so many diverse habitats is decidedly *not* a reflection of insensitivity of biochemical systems to the physical and chemical conditions in the environment. Rather, life's prevalence is a reflection of how effectively biochemical systems have undergone selection to overcome abiotic stress and, indeed, evolved in ways to exploit the abiotic environment, so as to thrive under a remarkable array of physical and chemical conditions. How organisms have achieved these abilities is the principal subject of this volume.

1.3 What Is "Stress"?

At this juncture, the concepts of stressor and stress must be examined and given precise definitions that will assist in our analyses of adaptation at the biochemical level. Our use of these two concepts differs from what might be the more familiar and customary uses of the words, as found, for example, in textbooks that focus on animal physiology. Here we will focus principally on biochemical manifestations of stress, rather than on higher-level physiological processes that may involve such phenomena as hormonal signaling via stress hormones, changes in rates of organ (e.g., cardiac) function, and other types of physiological responses that occur when an animal experiences stress. The definitions we present are applicable to all biochemical systems of all organisms—eukaryotes, archaea, and bacteria. These concepts thus focus on universal aspects of stress that are common among all forms of life. The effects of stress on the fundamental molecules of life provide an exceptionally good illustration of the principle of **unity in diversity**. As different as organisms may seem at "skin" level, "under the skin" there is a commonality of biochemical constituents that share, across species, a set of universal sensitivities to environmental factors (Somero 2000; Hochachka and Somero 2002).

An environmental factor that can perturb biochemical structures and functions is termed a **stressor**. The perturbation—the change induced in the system by the stressor—is termed a **stress**. The latter term normally has a negative connotation in common English usage. Indeed, in most analyses of stress found in the physiological literature, stress is a "bad thing." However, as we show in numerous contexts, the biochemical perturbations induced by stressors can yield *three* different types of outcomes. First, the change in the biochemical system (the stress) may be without significant influence on the organism's performance. There may be a range of values for a particular biochemical system in which stressor-induced changes are readily tolerated; organisms have evolved to cope well with a certain range of stressor effects. When this is the case, the stress we observe should not be viewed in the context of the conventional, negative usage of that word. We will attempt to provide quantitative criteria for evaluating whether a stressor-induced change in a biochemical system does, in fact, represent a stress that requires amelioration. Second, stressor effects may indeed prove threatening to homeostasis—"stress" in its typical usage—and these perturbations will require amelioration if the organism's performance is to be restored. These restorative processes often depend on a third influence of changes in a stressor. Here, the change induced in a biochemical system by a stressor can serve useful roles, for example, in signaling to the cell that an adaptive response to the stressor is needed. Stressor-induced changes in macromolecules can also convert them from inactive to active states, which may facilitate adaptive responses to the changed environment. Stress can thus serve as a key type of information for modulating the activities of cellular processes, notably those involved in restoration of homeostasis. Suffice it to say, the concept of "stress" is a challenging one for biologists because the word has colloquial—usually pejorative—connotations that may conflict with the proper uses of the term in the context of biological relationships.

1.4 Homeostasis: A Multitiered Phenomenon

In the preceding paragraph, we introduced another term—**homeostasis**—that also figures importantly in analysis of stressor effects. As this term implies (*homeo-* is Greek for "like" or "alike"), homeostasis refers to the maintenance (or restoration) of a state for a system that enables the organism to function optimally. As in the case of the concepts stressor and stress, our focus on homeostasis will be at the biochemical level. We will examine what particular structural states of proteins, lipids, and nucleic acids are optimal for function and must be conserved for molecular-level homeostasis to be maintained. In turn, we will provide examples of how homeostasis in molecular structure and function enables higher-order homeostasis in more complex physiological systems. The tiered nature of homeostasis thus begins with conserving the appropriate structures of biochemical systems to allow higher-level physiological processes to retain (or regain) their appropriate characteristics.

1.5 Life Is Fragile because It Also Must Be Efficient, Accurate, and Responsive

In order to appreciate why abiotic stressors can lead to perturbation of biochemical homeostasis, it is critical to understand why selection has made biochemical systems so vulnerable to perturbation by these stressors. In other words, we can ask, "Why is life so fragile?" There are of course multiple factors that render biochemical systems sensitive to perturbation by stressors. Each chapter will examine how a specific factor—ranging from oxygen and its complex chemistry, to temperature with its myriad effects on physiological structures and processes, to the incredibly diverse and critical roles of water and small solutes—affects biochemical systems. Although each environmental factor has unique effects on cellular chemistry, there are some general features of stress that pertain across biochemical systems.

One ubiquitous type of biochemical stress arising from physical factors like temperature and pressure is perturbation of higher orders of molecular organization in proteins, nucleic acids, and lipid-containing systems like membranes. To appreciate this widely occurring form of stress, we need to briefly review three important properties of these large molecular systems that are at once essential for their diverse biological roles and yet have the consequence of rendering these systems vulnerable to perturbation from stressors. This analysis will go a long way to answering the question raised above about the **ultimate cause**—the selective significance—of life's fragility.

The molecules that make biochemical systems so *efficient*—here, think of enzymatic enhancement of chemical reaction rates; *accurate*—here, think of copying, transcribing, and translating the genetically coded information in the cell; and *responsive*—here, think of the rapid and sophisticated signaling and subsequent effector responses that cells make to changes in the external or internal environment—all depend on *reversible changes in molecular shape during function.* Enzymatic proteins must alter their conformations during their binding and catalytic activities, and these changes in shape must occur rapidly if enzymes are to attain the high efficiencies for which they are noted. Regulation of gene expression at the levels of transcription and translation likewise involves alterations in molecular shape. DNA must open its structure to enable the assembly of transcriptional complexes. Translation of mRNAs is attendant on the ability of the ribosome-binding region of the mRNA to flex into the correct three-dimensional structure to recognize the appropriate, complementary site on the ribosome. Membrane-localized processes commonly require rapid conformational changes in proteins that are embedded in the lipid bilayer. Membrane-localized processes also frequently entail movement of proteins across the plane of the bilayer. Thus, none of these macromolecular systems can be overly rigid if they are to function well. In fact, the structures of all of these large systems tend to be poised in a state that is commonly termed **marginal stability** (Box 1.1 and Figure 1.1; see also Box 4.4). This expression denotes a structural state, which some authors discuss in terms of **fluidity**, in which the molecule has adequate stability to ensure its existence in a conformational state that allows it to recognize the entities with which it must interact, thus allowing the process in question to be initiated, yet not so great a stability that the molecule is handicapped in undergoing the subsequent changes in three-dimensional structure that are needed for it to complete its function. Sustaining this delicate "balancing act" will be seen to play central roles in biochemical adaptation in the face of physical and chemical stressors.

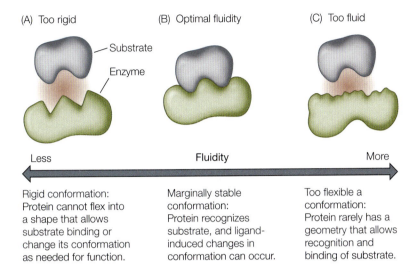

(A) Too rigid

Substrate
Enzyme

(B) Optimal fluidity

(C) Too fluid

Less   Fluidity   More

Rigid conformation:
Protein cannot flex into
a shape that allows
substrate binding or
change its conformation
as needed for function.

Marginally stable
conformation:
Protein recognizes
substrate, and ligand-
induced changes in
conformation can occur.

Too flexible a
conformation:
Protein rarely has a
geometry that allows
recognition and
binding of substrate.

Figure 1.1 Molecular fluidity, marginal stability, and biochemical function. Proteins, nucleic acids, and lipid-rich systems like membranes possess abilities to alternate between different three-dimensional structures (conformations), and this ability is influenced by their structural fluidities. When fluidity is optimal (B), the system possesses the correct marginal stability to (i) retain the structural geometry needed to recognize the ligand with which binding must occur, and (ii) alter the geometry of the structure to allow the conformational change needed for effective binding and function. If the system is too rigid (A), the binding site may lack the right geometry for binding, and/or binding might not be followed by the change in conformation needed for function to occur. If the system is too fluid (C), the geometry of the molecule is distorted into a conformation that does not permit effective binding. The green shape is the molecule—a protein (shown here), nucleic acid, or lipid—that interacts with a ligand (blue shape) at the binding (interaction) site. The ligand may be a small molecule—for example, a substrate of an enzymatic reaction—or another large molecule, as in the case of, for example, two proteins that form a multiprotein complex, a protein and a nucleic acid involved in gene regulation, or a protein and a membrane phospholipid that form part of a transmembrane channel. Although only three fluidity states are shown, it must be appreciated that fluidity is a continuum—as the double-headed arrow denotes.

BOX 1.1

The concepts of marginal stability and fluidity

One of the central unifying concepts that will appear in numerous contexts in this volume is *fluidity*, which is used to refer to the abilities of the molecular systems in question to fluctuate among different structural—and hence functional—states (see Figure 1.1). Fluidity thus refers to a molecule's ability to alter its structure in response to a chemical or physical factor, for example, a change in temperature. Recall that only at absolute zero (0 K) is there an absence of molecular motion. Thus, across the full spectrum of temperatures at which life occurs (see Table 1.1)—roughly –70°C to 130°C—the structures of

biomolecules like proteins, nucleic acids, and lipids will fluctuate among a suite of what are often termed *configurational microstates*. Even at temperatures to which an organism is evolutionarily adapted and currently acclimatized, its macromolecules and large molecular assemblages like membranes and nucleosomes will be "wiggling around" among a vast number of slightly different conformations. As temperatures rise to higher values, the diversity of microstates will accordingly rise. Proteins may unfold into a wider range of conformations, some of which may be dysfunctional (see Chapter 3). DNA may

BOX 1.1 *(continued)*

open up its structure and thereby alter its transcriptional activities. The membrane bilayer may become so fluid that its barrier function is compromised and the activities of membrane-associated proteins are impaired.

How are the concepts of fluidity and marginal stability related? Perhaps the most general answer to this question is that the defense of a marginally stable structure for a protein, membrane, or nucleic acid represents conservation of the appropriate capacity for changing shape as physiological or environmental circumstances dictate. The fluidity of the system thus reflects a capacity for altering shape. Other terms are also employed in this context. One is protein "softness," which represents the ease with which a protein's surface can alter its geometry (Isaksen et al. 2014).

This brief and qualitative treatment of the effects of fluidity should suggest to the reader that another central theme of this volume, conservation of key values for traits, is closely related to the phenomenon of fluidity. In fact, we will see that the structures of proteins, nucleic acids, and lipids (notably the lipids that form the membrane bilayer) reflect *pervasive adaptation to conserve the degree of fluidity that is appropriate for the conditions the organism is facing*, especially (but not exclusively) the temperature regimen it is experiencing. In all of these cases we can provide a quantitative index for measuring fluidity, as discussed in the specific contexts of protein-, nucleic acid-, and lipid-based structures. What will be seen in all cases is that conservation of fluidity is a central "target" of the processes of physiological acclimatization and evolutionary adaptation. Maintaining or regaining the optimal values for the fluidity of the system involves a broad array of mechanisms, including the rapid adjustment of the composition of the cellular solution, the activation of transcriptional and translational processes to modify the macromolecular complement of the cell, and multigenerational processes that include adaptive substitutions in the amino acid sequences of proteins. The degree to which different species can access these strategies contributes to determining how successfully they can adapt to the challenges of a changing world, an issue that has taken on increasing importance as organisms eke out survival in the rapidly changing world of the Anthropocene (see Chapter 5).

1.6 Why Are Biochemical Systems So Easily Perturbed by Environmental Stressors?

If we recognize that the ultimate cause (**Box 1.2**) of the marginal stabilities of large molecular systems relates to optimizing functions that require changes in molecular shape or organization, what are the **proximate causes**—the mechanisms that achieve this structural balancing act? The sensitivities of proteins, nucleic acids, and lipid-containing structures like membranes to stressor effects stem from the reliance of these large molecular systems on noncovalent interactions (hydrogen bonds, van der Waals interactions, ionic interactions, and hydrophobic effects) that support their higher orders of structure (the native folded states of proteins, the secondary and tertiary structures of nucleic acids, and the organization of large assemblages like membranes, nucleosomes, and ribosomes). The balancing act involved in modulating structural stability so as to ensure optimal function is seen to involve a judicious selection of the right types and numbers of noncovalent bonds to stabilize the system.

The study of protein evolution provides key insights into the importance of marginal stability and the manners in which it is attained in different environments. Even though thousands of weak chemical bonds contribute to stabilizing the native structure of a protein, the **net stabilization free energy**, which is what governs overall stability, lies in the range of only several weak bonds (see Chapter 3). Evolution thus has established a balancing act that leads to a protein with just the right level of fluidity of structure to support optimal

BOX 1.2

A critical philosophical perspective: Ultimate and proximate causes

The notion of "causation" runs throughout all of the sciences. In the biological sciences, it is important to recognize that there are two types of causes: ultimate cause and proximate cause. Understanding these two sorts of causation can help us develop another important component of the structure of our treatment of adaptation. Ernst Mayr (1982) has perhaps given the clearest distinctions between these two aspects of causation in biology. *Ultimate causes* refer to the selective advantages that led to the origin and retention of a trait. "Why" questions in evolutionary biology almost invariably demand analyses of ultimate causation. Many of the central points developed in this volume seek to answer "why" questions about ultimate causes. For example, we will ask "why" certain pH values are optimal across all taxa; and "why" certain types of small organic molecules and inorganic ions are preferred for adjusting osmotic balance of cells. Ultimate causation obviously pertains to the factors that led to the initial appearance of a trait—the restricted use of "adaptation" we discuss—and to the later exploitations of the system in question to generate exaptations.
Proximate causes refer to "mechanism," that is, the means by which the beneficial trait is achieved. Optimal states of protein stability provide a good illustration of what we mean here—and also show that a given ultimate cause may be realized by different types of proximate causes. In terms of ultimate cause, a certain intermediate level of protein structural stability—neither too rigid nor too flexible—is important for ensuring optimal protein function (see Box 1.1). The proximate causes of this optimal state of structural stability may be amino acid substitutions that modify the inherent structural stability of the protein. However, the same end-result might be achieved by modulating the intracellular fluid that bathes the protein. For example, in extremely thermophilic bacteria and archaea, protein-stabilizing low-molecular-mass organic solutes are produced in a temperature-dependent fashion to modify stabilities of proteins (see Chapter 4). In fact, both types of proximate causes are likely to be important in achieving the desired level of protein stability. Similarly, the benefits of freeze-avoidance pertain to a wide range of taxa exposed to cold temperatures (see Chapter 3). The ultimate cause—the benefit of "staying liquid," as it were—is achieved by a wide and fascinating variety of biochemical mechanisms (proximate causes) that comprise adaptive changes in proteins and in the suite of low-molecular-mass organic molecules in the cell. Freeze-resistance is one of several cases where we show how a common "cat" (ultimate cause) is "skinned" in many different and equally effective ways.

function. One might ask at this juncture, "Couldn't evolution have done a better job by generating proteins that are much stronger, that is, resistant to stressor effects, yet still able to do their functions well?" We seem close to being able to answer this question—with the answer being "probably not." By exploiting contemporary tools of molecular biology like site-directed mutagenesis, biochemists have successfully built much stronger proteins than those typically found in nature. However, the high stabilization energies of these laboratory-created proteins typically make them too rigid to be able to rapidly undergo the types of conformational changes needed for optimal catalysis and regulation. Suffice it to say, evolution "knows what it's up to" when selection for marginal stability occurs. Selection does not favor the toughest possible structures, despite their resistance to environmental stressors. Rather, selection favors the structure that confers optimal functional properties on the system—and these structures tend to be only marginally stable.

Marginally stable structures have another benefit: the capacity to sense and respond appropriately to environmental changes. This is the "good" type of stress we alluded to earlier. We will see that the intrinsic sensitivities of macromolecules to stressors like temperature provide them with the raw materials for evolving exquisitely sensitive antennae for detecting changes in the environment. These antennae transduce the information they've acquired to modulate downstream systems that bring about adaptive responses. For example, RNA thermometers, which are based on thermal disruption of base pairing in mRNAs, are capable of detecting changes in temperature of 1°C or less. The temperature-induced conformational changes in the RNA thermometer region of the mRNA allow binding to the ribosome, which is followed by rapid changes in translation of the mRNA into protein (see Chapter 3; Kortmann and Narberhaus 2012). As emphasized above, examples such as this require that we view environmental perturbation of macromolecular structures in a holistic manner, recognizing that such structural alterations can bode either good or evil, depending on the context in which the change in structure occurs.

In conclusion, sensitivity to perturbation by abiotic environmental factors is something that life "has to live with," because of the importance of maintaining biochemical systems in particular states of dynamic structural balance in which function is optimized. However, vulnerability to stressor-induced perturbation also is a consequence of this balance. In subsequent chapters we will analyze many specific examples that show why this balance is needed and what mechanisms are employed to sustain it.

1.7   Adaptation: A "Loaded" Concept

In the preceding sections we have used the word "adaptation" quite generically and have not yet provided a specific definition of what this term denotes and how it should be used in discussions of organisms' responses to the abiotic or biotic environment. "Adaptation" is, in fact, a "loaded" concept that has generated a great deal of discussion among evolutionary biologists. Entire volumes have been written on the use (and misuse) of this concept (Rose and Lauder 1996). A watershed publication on the problematic uses of "adaptation" was the 1979 paper by Stephen Jay Gould and Richard Lewontin that offered a "critique of the adaptationist programme" (Gould and Lewontin 1979). Gould and Lewontin rightly emphasized that it is too easy to fall into the practice of creating "just so" stories (in the sense of author Rudyard Kipling's *Just So Stories*) about traits that, to our intuitions, may appear to be adaptive but which, on closer and more logical analysis, either are not adaptive at all or play roles that differ from the one(s) we might think are obvious in terms of adaptive value.

It thus is incumbent upon a biologist to provide rigorous standards for interpreting whether or not, say, a difference between conspecifics (members of a common species) or between species is adaptive in the sense that this difference enhances the survival and, ultimately, the likelihood of propagation and persistence of the species or population in its changed habitat. We will illustrate procedures for identifying changes in biochemical systems that we feel are, without question, of adaptive significance.

Optimal values for traits: An empirical definition

One mode of analysis reflects the conservative perspective mentioned above: When an environmentally sensitive trait is conserved across species or populations living under widely different environmental conditions, the biochemical changes that offset this perturbation and facilitate conservation of the trait in question can almost certainly be regarded as adaptive. Part of the analysis of conservative adaptations of this general type requires that we identify **optimal values for traits**. Here we use a strictly empirical definition: *optimal values of a trait are those values that we observe to be strongly conserved across all species in the face of environmental differences that would be expected to perturb the trait's value*. We will provide quantitative criteria for identifying optimization of biochemical systems in several different contexts throughout this volume. Biochemical traits can, of course, only be optimized in the context of physical and evolutionary constraints, but given the set of physical laws and evolutionary forces that have influenced the evolution of all life, this analysis will help provide a sense of the "targets" that adaptive processes need to "hit."

Adaptation and exaptation

One of the most interesting—yet often extremely challenging—facets of the debate about adaptation concerns distinguishing between the initial adaptive significance of a trait and subsequently evolved, different contributions of the trait to the organism's well-being. Adopting the terminology found in some of the more philosophical literature on "adaptation," one restricts the term "adaptation" to the *initial* beneficial function of the trait. Subsequent exploitations of the trait for other functions are termed **exaptations**. We find this distinction between adaptation and exaptation to be a useful one in the context of the evolutionary "invention" of new biochemical and physiological capacities, because new traits are almost invariably fabricated using preexisting genetic information coding for what are usually quite different functions. Often, the new capacity the organism acquires from novel use of existing genetic information concerns a very different environmental challenge from that which led to the initial origin of the trait. For example, the heat-generating capacities of mammals in thermogenic brown adipose tissue (BAT) (see Chapter 3) reflect exploitation (exaptation) of a preexisting biochemical system for reducing the production of reactive oxygen species (see Chapter 2). Using contemporary evolutionary terminology, then, the adaptation for reducing damage from ROS developed into an exaptation that facilitates closely regulated heat production when mammals face challenges from low ambient temperature. BAT cells themselves may be an exaptation of a cell type that originally functioned as a precursor only to muscle cells (Trajkovski et al. 2012).

The evolution of BAT is a relatively clear instance of how exaptation occurs. In many other instances, the multiple functional roles of a trait make it challenging to determine which of many contemporary adaptive features of the trait came first—what is the adaptation and what are the exaptations? We will focus on several biochemical traits that illustrate the ways in which exploitation of a given type of molecule for solving different types of problems leads to novel evolutionary change. Teasing out the initial use of the molecule is often not possible, however, so distinguishing adaptation from exaptation often remains problematic for students of biochemical evolution. The multiple uses of the small carbohydrate molecule glycerol offer an excellent case study in this regard (Box 1.3).

BOX 1.3 ▮▮ ▬ ▬ ▬ ▬▬▬▬▬▬▬▬▬▬▬▬▬▬▬▬▬▬▬

Glycerol: A "multitasking" molecule par excellence

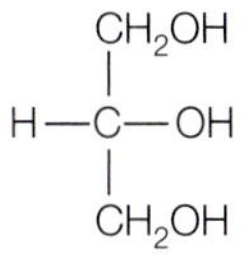

```
        CH₂OH
          |
   H —— C —— OH
          |
        CH₂OH
```

Glycerol

An excellent illustration of the broad range of functions a given biochemical constituent can serve—and an especially good example of exaptation—is given by glycerol. As shown in the structural diagram above, glycerol is a relatively simple molecule comprising 3 carbon, 8 hydrogen, and 3 oxygen atoms.

Important to most of glycerol's roles in biochemistry is the occurrence of three hydroxyl (–OH) groups. These support ester- and ether-forming chemical reactions, which are important in lipid biosynthesis; they allow critical interactions with water molecules, which affect properties like undercooling (supercooling) ability; and they influence glycerol's abilities to interact with surfaces of macromolecules, which is important in affecting protein stability. In view of this wide range of contributions that glycerol makes to cellular structure and function, can we deduce what glycerol's *initial* role in biochemistry was—its primor-

BOX 1.3 *(continued)*

dial adaptation—as distinct from subsequent exploitation in different contexts of exaptation?

It seems possible that glycerol's earliest role in the cell was as a substrate within pathways of carbohydrate metabolism. Glycerol is an energy-rich molecule that can enter the glycolytic pathway of ATP generation (see Chapter 2). However, one might argue instead for a different adaptive role. Thus, if the importance of fabricating a cell membrane was one of the very earliest requirements for establishing cellular life, and if glycerol's current role in lipid structure is indicative of the role that it served at the dawn of cellular evolution, then perhaps its membrane-building function is its true adaptation (see Chapter 3). The subsequent exploitation of glycerol as an osmotic solute, an antifreeze agent, a supercooling agent, and a protein stabilizer may reflect further exaptations (see Chapter 4). Suffice it to say, the most important point here is not what glycerol first contributed to the success of life, but rather the fact that evolution has been strikingly effective in

finding multiple uses for a single type of molecule. We will see that, whereas glycerol is something of an extreme example of this type of multiple-use strategy, it is not unique by any means. In particular, many other "micromolecules" have assumed a wide range of roles in cellular chemistry that at once tend to obscure their initial adaptation, but illustrate the widespread role of multifunctioning small molecules in adapting to all of the abiotic stresses we examine in this volume (see Chapter 4). A given type of macromolecular system, too, may perform multiple roles—one adaptive and others exaptive. We are increasingly aware of proteins, even enzymes of core metabolic pathways, that have multiple functions, which raises the issue of which of these functions represents the initial adaptation and which functions represent exaptations. Multitasking micromolecules and macromolecules contribute importantly to both the conservative and the innovative facets of biochemical adaptation that are a primary focus of this volume.

1.8   Adaptation, Acclimation, and Acclimatization: The Elements of Time and Complexity in Responding to Environmental Change

A further complexity—and another common source of semantic confusion—that surrounds the term "adaptation" concerns the time-course over which an organism responds to environmentally induced stress. As the discussion in the preceding section indicates, the *noun* "adaptation," strictly speaking, should be restricted to evolutionary processes that involve changes to the genome. Adaptation, in this restrictive sense of usage, is thus multigenerational. However, such strict adherence to a narrowly defined use of the word is not always found or, indeed, always practical. Changes that occur during an organism's lifetime are also spoken of as being "adaptive" when the resulting change in phenotype confers an improved ability for coping with or exploiting the environment. Indeed, it seems entirely fair (and certainly consistent with common usage) to employ the *adjective* "adaptive" in these cases, to denote improved likelihood of surviving and thriving in the changed environment. The semantic point here is to keep in mind the distinct meanings of the noun "adaptation" and the adjective "adaptive." The former refers to evolutionary change; the latter denotes the beneficial effects of change that occurs either across generations or during the lifetime of an individual.

Adaptive phenotypic changes that occur during an individual's lifetime are an example of **phenotypic plasticity**, that is, when the expression of the phenotype depends on the environmental conditions facing the organism. Studies of phenotypic plasticity have involved both field and laboratory investigations, and different terminology is used to distinguish between these two experimental approaches. **Acclimatization** refers to phenotypic changes that occur in an organism under natural (field) conditions during or subsequent to exposure to changes in the environment. The environmental changes in question can occur over a range of time periods, for example, from diel changes (on a 24-h timescale) to those that occur seasonally. Importantly, acclimatization almost always involves simultaneous changes in more than one abiotic factor. For example, in the rocky intertidal zone, tidal rhythms expose sessile organisms in particular to alterations in temperature, access to food, oxygen, and water (desiccation stress). Teasing apart the effects of an individual abiotic factor like temperature from the effects of other factors can be very difficult (Dowd et al. 2013).

Partly because of this experimental complexity, biologists often bring the organism into the laboratory and subject it to change in one or more abiotic factors of interest. This experimental approach is termed **acclimation**. Some users of this term restrict "acclimation" to the study of a single variable, for example, temperature, with all other variables being held constant. Some softening of this narrow definition has occurred in recent years, however, with laboratory studies focused on two or more variables now being referred to as acclimation studies as well. The key point is that acclimation studies diverge from real-world conditions in which a host of factors vary, many of which might not be amenable to laboratory study and some of which may not even rise to the attention of the investigator.

The greater share of the literature in the broad field of environmental and ecological physiology—the discipline that seeks to elucidate how physiological processes play into such ecological phenomena as distribution patterns and energy flows—has come from work in the laboratory using acclimation as the experimental paradigm. Although this more narrowly focused mode of analysis has taught us abundant lessons about the effects of single environmental factors on diverse biochemical and physiological systems, it is becoming increasingly clear that multifactor (multistressor) experiments are critical to provide realistic analyses of the effects of environmental change on organisms in their natural habitats. The requirement for multistressor analysis is especially crucial in several contexts related to global change, where simultaneous changes in several environmental variables confront organisms (see Chapter 5). In the seas, for example, rising temperatures, falling pH, and decreasing levels of dissolved oxygen are already presenting serious challenges to life (Somero et al. 2016).

The element of the time available for generating adaptive responses to environmental change is important in the contexts of both acclimatization and evolutionary adaptation. In the case of acclimatization, the speed with which an organism can mount an adaptive response may determine its ability to survive environmental change. In general, slow rates of change in the environment give an organism a better ability to adaptively restructure its phenotype, compared with the situation that pertains when changes are especially rapid. Acclimatization is commonly viewed as a "first line of physiological defense" in coping with global change. In fact, behavioral responses—moving away from damaging stress—may be the most effective first response, but this avenue of escape may be closed for many organisms (see Chapter 5). It is thus critical to understand how rapidly acclimatization can occur and how species differ in this ability to modify their phenotype (Somero 2010).

From an evolutionary perspective too, rates of environmental change are apt to be instrumental in determining the odds of successful adaptation. One important question that will be addressed at several junctures in this volume concerns the relationships among generation time, effective population size, existing levels of genetic variation, capacities for acquiring new genetic tools (for example, via de novo mutation and/or horizontal gene transfer), strength of selection, and the pace of global change. This question essentially asks, "Can organisms 'keep up with' the pace of environmental change?"

Whereas an organism's capacity for acclimatization of its phenotype may determine its short-term capacity for coping with environmental change, over the long run—that is, over multiple generations—success may be related to the available sources of genetic variation, as we discuss in the next section. Differences in amounts of existing genetic variation, in mechanisms by which genetic information can be generated and exchanged, and in capacities for epigenetic adaptation may be of pivotal importance in determining "winners" and "losers" in the evolutionary race to cope with global change.

1.9 Multiple Sources of Genetic "Raw Material" Provide Tools for Adaptation

As we will demonstrate throughout this volume, some of the most interesting and provocative insights into biochemical adaptation stem from efforts to answer the following question: "Just where did the genetic 'raw material' for this trait come from?" These stories of the origin of biochemical novelties—many of which can be viewed as some of the most important "innovations" in evolution—are being revealed by the rapidly growing volume of gene sequence information that is being generated for diverse taxa and by a deepening of our understanding of how a variety of epigenetic processes contribute to the generation of biological diversity. Genomic studies are providing more and more examples of how a given type of genetic information has been exploited to generate novel traits, often ones that bear no resemblance to the original traits encoded by the genes. The origin of glycoprotein "antifreezes" in Antarctic fishes from a fragment of a gene encoding a digestive enzyme is a case in point (see Chapter 3). We are also coming to understand some of the genetically based factors that explain why certain taxa are so effective in acquiring novel traits—and why others are strikingly limited in this regard, and therefore face an uncertain future in the face of global change (see Chapter 5).

In the broad arena of epigenetics (*epi* is Greek for "over" or "outside of"), heritable changes in genetic function that are not based on any changes in the nucleotide sequences of DNA are being shown to have significant potential for driving adaptive changes in biochemical function. It is well recognized that a variety of types of epigenetic changes in the chromosome, including methylation of DNA and covalent modifications of histone proteins, are critical during an individual's lifetime for controlling processes such as development and metabolic regulation (Storey 2015). Transgenerational effects of epigenetic changes may play a role in evolution (Burggren 2014, 2015); however, due to the context dependence of epigenetic modification, the degree to which epigenetics influences long-term evolutionary trajectories is uncertain (Jablonka and Raz 2009; Suter et al. 2013). In the context of biochemical adaptation, transgenerational inheritance of epigenetic changes could facilitate adaptation to environmental changes over shorter timescales of a few to several generations (Ho and Burggren 2012).

Below we provide brief overviews of some of the sources of genetic and phenotypic variation that provide the "raw material" for adaptive change. In later chapters we will illustrate the importance of many of these sources of novelty for assisting organisms to cope with—and at times exploit in new ways—the changes occurring in the world around them.

Horizontal gene transfer (HGT)

In this process, genes are exchanged either between different evolutionary lineages or between conspecifics, through mechanisms distinct from normal reproductive processes in which vertical transmission of genetic information takes place. HGT establishes an enormous potential for organisms to acquire the types of genes needed for local environmental conditions (Goldenfeld and Woese 2007). HGT has been shown to favor rapid and extensive rebuilding of genetic "tool kits" in marine microbes, for example, such that a complement of widely occurring "core" genes needed for basic functions like transcription and translation is complemented by acquisition of changing sets of genes that allow exploitation of the abiotic and biotic characteristics of the local environment. HGT can operate in a wholesale manner, bringing into a lineage a large set of genes that radically change the organism's environmental optima and tolerance limits. For example, there is evidence that the earliest archaea were thermophilic cells, and that the subsequent radiation of archaea into lower-temperature environments resulted from "grabbing" scores of genes from mesophilic bacteria through HGT (López-García et al. 2015; see Chapter 3). Thus, rather than fine-tuning protein thermal sensitivities on a protein-by-protein basis, as we will discuss in Chapter 3, a wholesale introduction of genes that encoded proteins with thermal optima at cooler temperatures sped the evolution of archaea in mesic habitats. HGT has been studied mostly in archaea and bacteria, which can gain and shed genes rapidly in the face of changing environmental conditions. However, more and more examples of HGT are being discovered in eukaryotes as well. For example, as we will discuss in Chapter 4, a desiccation-tolerant midge appears to have acquired bacterial genes that encode a set of proteins that permit cells to endure extreme desiccation (Gusev et al. 2014). A single-celled red alga has acquired genes from bacteria and archaea that enable it to be one of the most stress-tolerant eukaryotes ever discovered—thriving in acidic hot springs at temperatures near 56°C (Schönknecht et al. 2013; see Chapter 5).

Gene duplication

The duplication of genes, followed by sequence changes that lead to adaptive variations on different protein themes, can be viewed as a major mechanism for facilitating exaptation at the protein level. Gene duplication occurs on two scales. At one extreme, the entire genome is duplicated, resulting in a change in **ploidy** (number of copies of the organism's set of chromosomes). At the other extreme, an individual gene is duplicated, a process termed **tandem gene duplication**. Tandem gene duplication can involve expansion of entire blocks of genes that support specific biochemical functions. For example, in Antarctic notothenioid fishes, long-term evolution under stably cold temperatures has been marked by the duplication of at least 118 protein-coding genes that support functions that appear critical for life at near-freezing temperatures (Chen et al. 2008). In some cases, the copy number of a gene has been increased by about 300-fold. This type of expansion in gene copies not only provides raw material for evolution of novel protein variants, but also supports a capacity for strong upregulation of genes that are required for specific environmental conditions—extremely low temperature in this case. Here then, gene duplication can be seen to serve a regulatory function: providing the organism with an enhanced ability to upregulate the production of specific types of proteins. We will return later to this second, regulatory contribution of gene duplication when we discuss antifreeze proteins in Chapter 3.

Through increases in ploidy or through tandem gene duplication, **paralogous** genes encoding protein **paralogs** are produced. Subsequent to gene duplication, paralogous proteins can follow different evolutionary trajectories. The wealth of genetic raw material produced by gene duplication, especially by increases in ploidy, has been of central importance in the evolution of complex multicellular organisms (Ohno 1970). For example, a duplicated enzyme may acquire different regulatory properties or substrate affinities that suit it well for function in a particular type of cell or tissue. In multicellular species, tissue-specific paralogs of enzymes are more the rule than the exception. Thus, for example, in the case of the glycolytic enzyme lactate dehydrogenase (LDH) in bony fishes, different paralogs (or different ratios among paralogs) are generally found in white skeletal muscle, heart muscle, brain, and gonad—organs that differ widely in their relative reliance on aerobic versus anaerobic pathways of ATP generation (see Chapter 2).

Whereas the functional properties of different paralogs can often be interpreted in the context of tissue- or cell-type-specific roles of the protein variants, in other cases differential expression of paralogs occurs to provide the cell with the "right" form of the protein for the environment (or developmental stage) in question (Schulte 2004). Perhaps the most common example of life-stage-specific expression of paralogs is in the case of hemoglobins, where, for example, fetal and adult paralogs are found in mammals. There are increasing examples of environment-specific expression patterns as well. For example, in certain eurythermal fishes, multiple forms of myosin with different thermal sensitivities are encoded in the genome, and are expressed differentially during thermal acclimation (Watabe 2002). By virtue of there being "hot" and "cold" variants of a single type of protein, biochemical function can be optimized over a wide range of temperatures. The differential expression of paralogs has also been seen in osmotic adaptation. Expression of paralogous isoforms of the sodium pump (sodium-potassium ATPase) has been observed in gill tissue of rainbow trout following acclimation to different salinities (Richards et al. 2003).

A further consequence of the process of evolution by gene duplication is that molecules that have already proven their compatibility within the cellular system are used for generating a novel function, requiring only some minor modification to structure and function, but not the de novo reinvention of an entire new structure. For instance, for enzymes that have distinct binding sites for ATP and a specific substrate, novel function can be generated through retention of the existing ATP-binding site and a slight modification of the substrate-binding region. In some instances, it has been shown that a single amino acid substitution can convert the substrate preference of an enzyme. This makes gene duplication a process that can lead relatively easily to novel genes that can broaden the metabolic repertoire of the cell.

Allelic polymorphism

In diploid species, when nonsynonymous nucleotide changes in one allele for a protein-coding gene lead to alterations in amino acid sequence, allelic variants termed **allozymes** are generated (*allo-* means "other" or "different" in Greek). Assuming that both alleles are expressed, a heterozygous individual would have two forms of the protein (two allozymes) in the cell. In the context of evolutionary adaptation, a long-standing question about such allelic variation concerns the factors that support its persistence in the population. How can we determine if this variation is neutral (without functional consequences and, therefore, without selective importance) or adaptive in some way? In an adaptive scenario, the persistence of allelic variation in a population could result from different environmental optima among allozyme forms, such that selection would favor different allozymes in different environments. One of the best lines of evidence of this type of biochemical adaptation is the presence of clinal variation in a population across environmental gradients—that is, the relative frequencies of different allozymes change across the full environmental range of the population (see Chapter 2; Place and Powers 1984; Eanes 1999; Watt and Dean 2000; Zera 2011).

In cases where clinal variation exists within a population, retaining allozyme variants with different environmental optima may provide a species with a high potential for coping with environmental change.

Allelic variants may also differ in their expression levels. This type of effect is well understood in the context of gene dosage effects in female mammals, where one of the two X-chromosomes is silenced by epigenetic mechanisms (the **Lyon effect**). However, **allele-specific expression** (**ASE**) also occurs in a more restricted context, namely in regulation involving specific genes rather than full chromosomes (Knight 2004). Recent surveys of gene expression in humans and other biomedical model species have revealed large numbers of differentially expressed mRNAs for allelic variants, and these patterns of ASE can be heritable (Knight 2004; Pastinen 2010). The amount of ASE found in a species under particular developmental and environmental conditions remains unclear, however, for several reasons. The extent of ASE depends on cell type and the preexposure of cells to different stimuli. Inter-individual variation among conspecifics is also significant. Furthermore, the sensitivities of the molecular techniques used to detect ASE vary considerably, so the detection of ASE levels exhibits a strong technique dependence (see Pastinen 2010). The functional significance of this differential expression also remains to be established in almost all cases, albeit ASEs are known to characterize several disease states in humans (Pastinen 2010). It remains to be discovered how ASE varies among species and whether ASE contributes importantly to acclimatization to environmental change.

Local adaptation

Local adaptation of populations of a species refers to assembling the right types of genetic variation to enhance the population's success under its particular set of local habitat conditions, which may differ considerably from the conditions found by conspecific populations at other locations within the species' range (Sanford and Kelly 2011). Local adaptation comprises a wide range of processes, including selection for allelic variants of proteins that differ in thermal optima (Dahlhoff and Rank 2000) and altered patterns of gene expression (Gleason and Burton 2015). Locally adapted populations that possess greater tolerance to environmental extremes than other populations of the species may be of critical importance in fostering the species' survival in the face of global change (see Chapter 5). Such populations may replenish the species at sites where a previous population became extinct because of environmental changes that exceeded its capacities for acclimatization.

RNA editing: Splice variants

While the previous sources of genetic variation are centered on changes at the level of the gene or the genome, the messenger RNA (mRNA) molecules transcribed from genes typically are not yet prepared for function in the translational process. To gain competence for translation, most types of mRNA in eukaryotes must first undergo splice-editing: introns must be removed and exons must be joined. (An interesting exception to this general rule is found with intron-free mRNAs encoding some stress-induced proteins. These messages must be translated without delay if stress is to be ameliorated as quickly as possible.) One reason that cells "go to all of this trouble" of mRNA splice-editing is the potential for this type of editing to generate molecular variation. For a large fraction of genes, the initially transcribed mRNA can undergo variations in splicing patterns, such that a given unedited mRNA can yield several mature mRNAs capable of being translated into different proteins known as **splice variants**. The discovery of closely regulated generation of splice variants of a common type of mRNA in widely different species suggests a deep and important evolutionary history for this variation-generating process. A striking illustration of this phenomenon is the generation of functionally distinct splice variants of the muscle contractile protein troponin T, a phenomenon that is similar in insects and mammals (Marden et al. 2001; Schilder et al. 2011). This process provides a mechanism for generating different variants on a protein theme for employment in different cell types or in different ontological stages. However, the broader importance of splice variants in adaptation to the environment remains to be elucidated.

RNA editing: Changing base composition

There is a second way in which newly transcribed mRNA molecules can be modified to generate differences in proteins. Messenger RNA molecules may be chemically modified to alter the amino acid sequence of the proteins they encode. RNA editing of this type, which involves changing codons, thus can give rise to multiple protein isoforms. One type of base change is the conversion of adenine to inosine by the enzyme adenosine deaminase. There is good evidence that RNA editing can be adaptive to environmental factors like temperature (Garrett and Rosenthal 2012; Rosenthal 2015). Thus, the codon changes resulting from RNA editing may lead to temperature-adaptive changes in amino acid composition that modify protein thermal stability (see Chapter 3).

Small regulatory (noncoding) RNAs

A potentially important mechanism for generating phenotypic diversity involves genes that encode small RNA molecules that are transcribed but not translated (Baek et al. 2008). These small regulatory RNAs have been placed into several categories (microRNAs, sRNAs [bacterial], etc.). They are known to fulfill several regulatory functions, for example, determining whether a specific mRNA gets translated into protein and controlling the half-lives of mRNAs. How small regulatory RNAs fit into the broader picture of adaptation to different physical and chemical environments remains largely unknown. However, from what has been discovered about their functions in model systems, there would seem to be a strong likelihood of a widespread role for these small RNAs in biochemical adaptation.

Stress-induced unmasking of cryptic genetic variation

A recent discovery has opened up an exciting, controversial, and still largely unexplored source of phenotypic variation: Stress-regulated proteins such as the molecular chaperone heat-shock protein 90 (Hsp90) can suppress (mask) cryptic genetic variation until the organism experiences a particular type of stress (Rutherford and Lindquist 1998; Rohner et al. 2013). If the released (unmasked) genetic variation leads to adaptive changes in the phenotype, rapid evolutionary change may result, including major changes in morphology.

Here is how Hsp90-mediated effects are conjectured to alter development. Under nonstressful circumstances, defined here as conditions under which protein stability is not challenged, Hsp90 plays many important regulatory roles. In particular, it assists in the correct folding of marginally stable signal transduction proteins, including transcription factors and certain signaling proteins that can play critical roles in guiding development. If not properly folded, these proteins can lead to altered pathways of development, resulting in altered phenotypes. In its role as a chaperone for marginally stable signal transduction proteins, then, Hsp90 functions as a *canalization factor* by constraining development within a narrow "channel."

Environmental stressors can perturb canalization by altering the role that Hsp90 plays in the cell. When a stressor reduces protein stability, thereby increasing the need for molecular chaperones, Hsp90 dissociates from the signal transduction proteins with which it interacts under nonstressful conditions and begins to function in the chaperoning of other classes of proteins (see Chapter 3). Thus "abandoned" by Hsp90, the signal transduction proteins can fold into alternative structures and cause changes in developmental patterning. If a phenotype unmasked in this way is adaptive for the new stressor conditions, selection will favor an increase in the frequency of that phenotype. We would like to point out that the function of Hsp90 as a canalization factor probably evolved secondarily to its function as a molecular chaperone.

Researchers have discovered a potential role for stress-induced unmasking in the evolution of cavefish, which lost their eyes subsequent to entry into a continuously dark subterranean habitat (Rohner et al. 2013). Laboratory studies revealed that the change in water conductivity these fish would have experienced in moving from an ancestral stream into a cave triggers a cellular stress response. If the colonization of cave habitats induced a similar stress response in the evolutionary past, then the release of Hsp90 from signal transduction proteins in response to this stress may have led to changes in the developmental pathway involved in eye formation, resulting in increased phenotypic variation. Indeed, when non-cave-dwelling members of the same species were treated with a reagent that blocked Hsp90 function during eye development, there was a large increase in the variation of eye size—some individuals developed larger eyes and some developed smaller eyes than normal. Thus, the stress associated with entry into the cave environment may have unmasked cryptic variation in eye size. Because loss of eyes may be adaptive in permanently dark waters, selection then could have favored the individuals who possessed the genetic variants leading to the eyeless condition. Alternatively, phenotypic evolution could have been the result of random genetic drift, such that the eyeless phenotype was selectively neutral in a cave habitat and once this phenotype was revealed its frequency could have randomly fixed in the population. This alternative scenario is equally as plausible as that of adaptation, given that population bottlenecks are a likely consequence of colonizing a new habitat. In either case, this is the type of rapid, major morphological transformation that stress-induced alteration of Hsp90 function may support. The extent to which Hsp90-mediated effects on development have contributed to phenotypic evolution in other study systems remains to be determined.

Transgenerational inheritance of epigenetic changes (TGI)

We mentioned above that research in the field of epigenetics is providing important new insights into evolutionary processes, as well as into the mechanisms that govern cellular differentiation during ontogeny or control the activities of metabolic pathways. When epigenetic changes that modify gene transcription pass from one generation to the next—a process termed "transgenerational inheritance of epigenetic changes (TGI)"—these heritable changes could be important drivers of adaptation (Burggren 2014). This phenomenon is difficult to study because the organisms' environment must be controlled for multiple generations. For this reason, the extent to which TGI can help organisms cope with environmental stress is largely unknown, but the potential for achieving rapid shifts in phenotype that can be passed on to future generations may be widespread among taxa and categories of stress responses.

The possible persistence of TGI across multiple generations is an important consideration in the study of environmental and evolutionary epigenetics (Burggren 2015). Do heritable epigenetic changes to a chromosome "wash out" over a few generations—or are they long-lasting and perhaps permanent? This question remains unanswered, but some researchers have argued for a paradigm shift in evolutionary theory to accommodate these seemingly Lamarckian phenomena. TGI certainly expands our conventional understanding of the basis of inheritance; however, these phenomena do not necessarily pose a challenge to evolutionary theory. Epigenetic modifications, if inherited, exist on chromosomes that segregate in the same manner as all nuclear genetic material and thus are subject to the same evolutionary genetic processes that ultimately determine whether they will persist or be lost from a population (Suter et al. 2013). Moreover, epigenetic modification itself is caused by proteins, such as methyl-binding proteins, which themselves are encoded by genes that are subject to evolutionary forces, such as mutation, genetic drift, and natural selection. More empirical data are needed to address the relative importance of TGI in biochemical adaptation, as compared with other mechanisms of adaptation, but this will no doubt be a worthwhile pursuit in future studies.

Symbiosis: Partnering-up to solve a problem

Many major evolutionary advances have arisen from new genetic information originating from outside the species' own genomes. Acquisition of new traits through symbiotic partnerships has been one of the most important drivers of evolutionary novelty. All readers will be familiar with certain examples of symbioses. Mitochondria, which originated from an α-proteobacterium incorporated into an early progenitor of the eukaryotic cell, possibly an archaeon (Spang et al. 2015), are perhaps the classic instance of this type of evolutionary advance in biochemical function. This view of eukaryotic evolution is supported by the finding that the majority of genes encoding proteins of eukaryotic energy metabolism are of bacterial origin, whereas those of the translational machinery are of archaeal origin (Müller et al. 2012). Coral reefs (see photo at left) constructed by cnidarians that house photosynthetic algal symbionts termed zooxanthellae are another familiar example of symbiosis, one that we will examine in the context of global change (see Chapter 5).

The wide-ranging importance of symbioses in plant and animal life must be appreciated to understand the many different contexts in which symbiosis has led to evolutionary diversity (McFall-Ngai 2015). We will touch on a few examples of the benefits of symbiosis in different parts of our treatment of adaptation, including the potential ability of animals to reconfigure their set of symbionts when the existing symbionts are not well adapted for functioning in a changed environment (see Chapter 5). An especially intriguing example of environmentally mediated shifts in symbiont populations is found in small mammals undergoing thermal acclimation. Acclimation of mice to cold leads to major changes in the composition of the intestinal **microbiome**, the full spectrum of microbes in the gut. When the "cold" microbiome is transplanted into axenic (germ-free) mice, a broad suite of physiological changes is triggered that leads to a cold-acclimated phenotype: higher heat-generating potential and enhanced capacity for uptake and catabolism of nutrients (Chevalier et al. 2015; see Chapter 3).

1.10 "Cranes" and an Integrative Perspective on Biochemical Conservation and Innovation

The additions of variation to the genome of an organism, whether these occur through changes in DNA sequence or epigenetic modifications of DNA and histones, are the engines for evolutionary innovation. Some types of changes to the genome and to gene expression, like gene duplication and release of cryptic genetic variation, respectively, can generate major cell-, organ-, or organism-level novelties in organization and functional abilities. These novelties can be viewed as "cranes" in the way the philosopher Daniel Dennett used the analogy to describe the potential of these sources of novel genetic information to "lift" biological organization and function into a new "landscape" where evolution of new types of structures and modes of life becomes possible (Dennett 1995). Thus, the cranes that lift organisms into a new evolutionary landscape set the stage for major evolutionary innovations. However, these new innovations can only succeed if there is conservation of the critical adaptations that allow biochemical systems to function within the suite of the abiotic conditions found in the environment. Innovation and conservation must work hand in hand to allow organisms to realize the potential offered by crane effects, to evolve new abilities while remaining well adapted to the specific challenges posed by abiotic factors like temperature, salinity, and availability of oxygen. We will explore numerous types of evolutionary processes in which innovation and conservation have taken place concurrently.

A good example of how a specific type of biochemical function can support the generation of novelty concerns the relative speeds of translation noted in bacteria and eukaryotes. The substantially slower translation speed in eukaryotes versus bacteria (4 versus 20 amino acids s^{-1}) may permit the cotranslational folding of domains with slower folding kinetics, thereby enabling the evolution of large, multidomain proteins in eukaryotes (Kim et al. 2013). We may, therefore, say that slower cotranslational folding kinetics is able to maintain productive folding pathways for more complex proteins. Thus, the slower rate of translation in eukaryotes must not be viewed as a disadvantage, but rather as a foundation—a crane in Dennett's terminology—for allowing the evolution of large, multidomain proteins that can fold properly during the relatively sluggish translational process. The "biochemistry" here is the productive folding pathway leading to the native state, while the evolutionary novelty at the level of the cell is the translational machinery supporting slower speeds of synthesis. Thus, major evolutionary novelties will challenge the biochemistry to maintain function under new conditions, giving rise to biochemical adaptation.

1.11 The Steno- and the Eury-: Can We Explain Differences in Environmental Tolerance Ranges?

Organisms differ markedly not only in the absolute values of the environmental factors they tolerate, but also in the breadth of the range of values they can withstand (see Table 1.1). Thus, another theme that will be woven through this volume concerns the widely different tolerance ranges that organisms exhibit toward environmental variables like temperature, oxygen, salinity, and hydrostatic pressure. Some species are remarkably tolerant of wide ranges of one or more environmental factors: these are **eurytolerant** species (*eury* is Greek for "wide" or broad"). We thus speak of eurythermal, euryoxic, euryhaline, and eurybaric organisms. Other species have extremely narrow tolerance ranges and are termed **stenotolerant** species (*steno* is Greek for "narrow"). Here we speak of species that are stenothermal, stenohaline, and so on. One of the critical determinants of the degree of eurytolerance a species possesses is its capacity for acclimatization—its physiological or phenotypic plasticity. Acclimatization ability, and the genetic mechanisms that support it, will figure importantly in our discussion in each chapter, but most notably in Chapter 5, when we develop our analyses of what global change portends for diverse types of organisms. Acclimatization capacity may be a critical determinant of whether a species becomes a "winner" or a "loser" in a rapidly changing world (Somero 2010). To illustrate how species belonging to a single taxonomic group can differ in tolerance ranges and acclimatization abilities, we here briefly introduce some of the "cast of characters" that will figure in several of the analyses we present in later chapters.

The mudsucker and the emerald notothen

Gillichthys mirabilis

Two species of bony (teleost) fish that have been studied extensively at different levels of biological organization serve as good illustrations of the wide variation in environmental tolerance ranges that exist among aquatic species. One species is a goby, the longjaw mudsucker (*Gillichthys mirabilis*), shown at left. As its species name implies, this fish has rather "miraculous" abilities for coping with changes in its environment. The mudsucker occurs in shallow bays and estuaries along the eastern Pacific coastline, from just north of San Francisco to Baja California and the Sea of Cortez. The mudsucker is arguably one of the most eurytolerant vertebrates in the biosphere. It tolerates temperatures from approximately 0°C to 35°C–40°C (Jayasundara and Somero 2013), salinities from approximately 1.3 times that of seawater to that of freshwater (Kültz and Somero 1995), and oxygen concentrations from air-saturated levels to miniscule values near 5%–10% of saturation, a capacity assisted by the species' ability to air-breathe (Gracey et al. 2001). It can live out of water for periods of several hours by relying on cutaneous respiratory exchange. Not surprisingly, in view of its "eury" nature, the mudsucker has substantial ability to acclimatize to different temperatures, salinities, and oxygen tensions (Logan and Somero 2010, 2011; Jayasundara and Somero 2013).

Trematomus bernacchii

In contrast to the mudsucker, the Antarctic emerald notothen (*Trematomus bernacchii*), shown at left, is remarkably stenotolerant, as are most members of the teleost suborder to which it belongs, the Notothenioidei. This suborder is restricted largely to the Southern Ocean, although a few species occur in cold, high-latitude waters in South America and New Zealand (Eastman 1993). *T. bernacchii* normally lives at subzero temperatures; water temperatures in the Southern Ocean range from −1.86°C, the freezing point of seawater, to about 1°C–2°C in the warmest regions around the Antarctic Peninsula. It dies of heat death at temperatures of 4°C–6°C (Somero and DeVries 1967; Podrabsky and Somero 2006). Notothenioid fishes also are steno-oxic, especially members of the family Channichthyidae (white-blooded icefish), which lack hemoglobin (Beers and Jayasundara 2015). Notothenioid families that do contain hemoglobin have relatively low numbers of erythrocytes and limited hemoglobin-based oxygen transport. And importantly, polar notothenioids appear to lack a substantial capacity for acclimatization, at least relative to eurythermal species (Logan and Buckley 2015).

What accounts for these large differences in environmental tolerance ranges and abilities to acclimate? As we will see, to a large extent these differences reflect two characteristics of the contents of the genome. One is the set of protein-coding genes that are in the genomic "tool kit." Here, the example of Antarctic white-blooded ice-fishes, which have lost genes for hemoglobin, provides a good illustration of what the absence of a particular type of genetic information can mean. Without hemoglobin, oxygen transport is severely affected by rising temperatures, which at once reduce oxygen solubility and boost metabolic rates. Limitations in the ability to obtain needed amounts of oxygen thus may limit thermal tolerance (see Chapter 5). Because of their diminished genomes, stenotolerant species that have evolved for millions of years in highly stable environments are likely to be inadequately prepared for coping with the rapid global changes that characterize the Anthropocene, a conjecture we will develop in more detail in Chapter 5.

The second type of genomic tool that may differentiate "steno" and "eury" species is the set of regulatory mechanisms that allow differential use of genetic information in acclimatization processes. Thus, even if a species should happen to possess all of the protein-coding genes needed for developing the appropriate phenotype for a changed environment, if the regulatory circuitry for adjusting expression of these genes is inadequate, survival may be threatened. For example, unlike the mudsucker, which has a robust cellular stress response (Logan and Somero 2011), *T. bernacchii* has lost the ability to upregulate critical stress-related proteins whose activities are needed to repair heat-induced protein damage (Buckley and Somero 2009). Whereas the genes coding for the needed proteins may be present, the regulatory switches that orchestrate the expression of these genes may be lost (Hofmann et al. 2000). The contrasting abilities of "steno" and "eury" species to acclimatize to environmental change may play a major role in determining which species survive in the face of global change. Lacking this "first line of physiological defense" as well as colder waters to which they can retreat in the face of warming, Antarctic ectotherms in general may be especially vulnerable to the environment that lies in their future.

Even "eury" species face challenges: The stressful intertidal zone

The foregoing comparison of stenothermal ectotherms of the Southern Ocean and estuarine gobies in thermally variable environments must not be interpreted to mean that organisms that have evolved in habitats with wide ranges of temperature are necessarily less vulnerable to future changes in temperature than more stenothermal species. Although ectotherms that have evolved for long periods of time in stably cold waters are relatively limited in acclimatization ability and live close to their upper thermal limits, organisms in other habitats are not necessarily "safe" from rising temperatures either. This point is brought out clearly by recent studies of organisms found in rocky intertidal habitats (see photo at right), which will serve as an important focus for many of our analyses.

As several broadly comparative analyses have recently shown, organisms inhabiting thermally variable environments like the rocky intertidal zone may be highly vulnerable to rising temperatures, and this sensitivity may differ significantly among species distributed at different vertical positions (see Chapters 3 and 5). For several sets of congeneric invertebrates, including molluscs and crustaceans, species

Porcelain crabs, *Petrolisthes eriomerus* (left) and *P. cinctipes* (right).

occurring highest along the intertidal to shallow subtidal gradient and having the greatest absolute tolerances of high temperatures show the greatest vulnerability to further increases in temperature (Tomanek and Somero 1999; Stillman 2003; Stenseng et al. 2005). Among the organisms to show this pattern of thermal sensitivity are congeners of porcelain crabs (genus *Petrolisthes*) that occur at different vertical positions (photo at left). The higher-occurring species, *P. cinctipes*, is more tolerant of high temperatures than its lower-occurring congener, *P. eriomerus*, yet appears more vulnerable to rising temperatures associated with global warming.

This vulnerability derives from two factors. First, the most warm-adapted species live closer to their upper lethal temperatures than do lower-occurring, more cold-adapted species. Like the extreme stenotherms native to the Southern Ocean, warm-adapted species may be highly capable of thriving under current thermal conditions, but they are not "preadapted" for coping with higher temperatures than they've experienced during their long evolutionary histories. This generalization, in fact, applies across most taxa, including terrestrial species, where tropical species are found to be more vulnerable to rising temperatures than are mid-latitude species (see Tewksbury et al. 2008). Second, the most warm-adapted species may have the least ability to acclimatize to temperatures that exceed the current upper habitat temperatures. Because of this limited acclimatization capacity, many warm-adapted eurythermal ectotherms appear to be as vulnerable to global warming as are stenothermal organisms that live in thermally stable environments like the Southern Ocean or many tropical habitats (Tomanek 2010).

Studies of the effects of abiotic variables—temperature in particular—on intertidal species are informative about global change for another reason. Latitudinal differences in the timing of the tidal cycle may have stronger effects on the physiology of these species than do latitudinal gradients in average temperature (Helmuth et al. 2002; see Chapter 5). Thus, in predicting the consequences of anthropogenic global change, one must take into account fine-scale variation among habitats in the times at which low tides occur. The intensity of thermal stress is not a strict function of generalized latitudinal temperature gradients, but depends on timing of tides and, at a given locale, the precise height at which organisms occur as well as their orientation to the sun. Across such complex **mosaic environments**, a substantial potential for local adaptation may be present, and this genetic variation may be crucial for species survival in the face of environmental change.

1.12 Integrative Analysis: Biochemistry Is Not Just Macromolecules

Another theme running through this volume is the importance of conducting research in an integrative manner, one that contains within its focus as many (ideally, all) of the variables that contribute to the status of the system of interest. Below and in the next two sections we examine briefly three aspects of integration, the first involving the complex sets of interacting molecules within the cell; the second involving the complex suite of abiotic variables in the external environment, which may have strong interacting effects on biochemical systems; and the third focusing on the importance of conducting analyses of adaptation across multiple levels of biological organization.

As we will discuss in the opening section of Chapter 4, despite the fact that most biochemistry (and certainly most funding for biochemical work) focuses on "big" molecules, namely proteins and nucleic acids, the numerically dominant molecules of the cell are "small": water and a diverse set of low-molecular-mass solutes. These "lesser" molecules are often viewed as "background" players on the stage of life, which is dominated by the big actors, the genes and the gene products that fabricate the structures of cells and support all types of metabolic activity and information transfer. We attempt to redress this misbalance of emphasis that is characteristic of much of biochemistry. We will show that, in the contexts of many of the environmental challenges organisms face from alterations in abiotic factors, conserving the structures and functions of "big" biochemical entities requires simultaneous changes in "small" chemical species found in the diverse biological fluids. We commonly use the term **micromolecules** when referring to low-molecular-mass constituents of biological solutions, which comprise protons, inorganic ions, and numerous types of small organic molecules.

The adaptive interplay between large biochemical systems—macromolecules and complex systems like membranes—and micromolecules will be examined in two different time frames. In the first time frame, we will see how macromolecular and micromolecular systems have coevolved, such that the proper structures and functions of macromolecules, proteins in particular, have come to depend on being bathed in the right micromolecular milieu. As we will see, the conservation of critical traits of proteins is through complementary alterations in the protein itself, that is, in its amino acid sequence, and in the medium in which it is bathed. One of the key points we will emphasize in discussing micromolecular adaptations is that they can achieve a *global solution* to the problems facing macromolecules, because of the similarities in effects of most micromolecules on all classes of proteins. Thus, a small organic solute that stabilizes one type of protein is very apt to stabilize most if not all of the proteins in the cell. This is another facet of the "unity" among biological systems that is of great importance in evolutionary adaptation to extreme physical and chemical conditions (Somero 2000).

In the second time frame we will examine, acclimatization to the environment may also feature adjustments in the micromolecular constituents of cells, notably in proton activity (pH) and organic osmolyte concentrations. Changes in the availability of water, external ion concentrations, hydrostatic pressure, and temperature during an organism's lifetime all may elicit adaptive alterations in the types and the concentrations of cellular micromolecules, as will be shown in Chapter 4. As in the case of evolutionary changes in micromolecular constituents of cells, these acclimatizations foster a status quo condition for proteins and membranes through counteracting the perturbations that are caused by environmental change. Offsetting perturbation on a macromolecular system by a physical factor like hydrostatic pressure or temperature through modulating the levels of a micromolecule can allow a rapid and effective response to environmental change (Yancey and Siebenaller 2015). In the temporal contexts of both evolutionary adaptation and phenotypic acclimatization, then, an integrative perspective that incorporates all of the biochemical constituents of organisms that are relevant for the adaptive response in question must be achieved if a full account of biochemical adaptation is to be developed. Caveats for experimental design will naturally follow from some of our analysis of micromolecular influences; biochemists have too frequently been neglectful of the important effects of the "small" players in complex biochemical systems.

1.13 Integrative Analysis: Spanning All Levels of Biological Organization

Our presentational strategy in each chapter will be much the same: We initially will build a foundation that lays out the most basic and fundamental ways in which the particular environmental factor of interest affects biochemical systems or offers opportunities for novel types of exploitation. These introductory sections of each chapter will be basic to the biological analyses that follow, for they will outline the "ground rules," based on the laws of physics, that all organisms are forced to follow. The physical relationships we present will at once spell out the challenges organisms face when the variable in question changes or reaches certain extreme values, and suggest as well the adaptational options that might be available for countering the perturbation the organism is facing. Thus, the basic laws of physics can at once portend problems for life yet also offer avenues of adaptive escape to organisms. In some cases we will discover adaptive exploitations of physical relationships that, at first glance, would be expected to have negative and perhaps lethal consequences for the organism. For example, strong inhibition of metabolism by some environmental changes can actually work to the organism's advantage by allowing it to enter a quiescent period during unfavorable environmental conditions—and emerge unscathed and prepared for active life when more favorable conditions reappear (see examples in Chapters 2 and 4; Podrabsky and Hand 2015). Another example of exploiting a seemingly negative phenomenon involves induction of extracellular ice formation in freeze-tolerant species. The initial formation of ice in extracellular fluids can be used to protect water inside the cells from freezing—an event that is almost always lethal—and allow freeze-tolerant organisms to survive periods of extreme cold with ice-laden extracellular fluids but ice-free intracellular fluids (see Chapter 3; Duman 2015).

A good example of integration across diverse levels of biological organization is afforded by study of ocean acidification (see Chapter 5). Here we will begin with a molecular phenomenon, the biochemistry of pH regulation, but broaden out to achieve a much wider context that may involve, for example, behavioral and ecological effects. This type of analysis will serve us well in all contexts in which adaptation (or acclimatization) is studied, and may be especially helpful for gaining a new and deeper appreciation for some of the challenges posed to organisms by anthropogenic global change. Changes in the pH of seawater affect acid-base balance and ionic regulation in fishes (see Chapter 5). In bony fishes, decreases in pH are managed through an acid-base regulatory process that leads to an increase in plasma bicarbonate ion concentration, which is accompanied by a fall in plasma chloride ion concentration to maintain charge balance. These changes in plasma ion concentrations can have several downstream effects. For example, reduced plasma chloride concentrations can interfere with activities of ion channels in nerve cells and thereby impair important sensory functions like olfaction (Nilsson et al. 2012). In bony fishes, elevated bicarbonate concentration also favors the growth of larger otoliths (ear bones), which are composed largely of calcium carbonate (Checkley et al. 2009). The enlarged otoliths may affect the fish's auditory sensitivities and abilities to detect locomotory acceleration. Thus, a change in the pH of seawater can lead to alterations in fishes' sensory capacities and behavior, which in turn can have wide-ranging ecological effects in such contexts as discrimination of predators from prey and location of suitable habitats (Nilsson et al. 2012). Our analysis, then, will attempt to be highly integrative—as in this case, where changes in the tiniest of micromolecules, the proton, can be shown to influence the behavior of organisms and perhaps even ecosystem structure.

1.14 Integrative Analysis: Effects of Multiple Stressors

In the final chapter, "Adaptation in the Anthropocene," we will shift our focus from a largely single-variable analysis to examine the influences of multiple variables in the context of anthropogenic global change. Global change is confronting organisms with an array of changes in environmental factors. For marine organisms, these include changes in temperature, dissolved oxygen concentration, pH, carbonate saturation state, and osmolality (in concert with an intensifying hydrologic cycle). In most cases, the rapid rates at which these factors are changing are either completely unprecedented or have not been experienced for tens of millions of years or more. We believe that an excellent way to develop an understanding of the nature and magnitude of these multiple stressors is to first look in depth at the effects of individual stressors like temperature, oxygen availability, and osmolality—the focus of Chapters 2, 3, and 4—and, with this foundation beneath us, examine the true complexity of the real—and rapidly changing—world. We hope that by the time readers reach the final chapter, their understanding of the effects of individual factors will be sufficiently developed to allow, first, an appreciation of how all of these separate influences on biochemical and physiological systems can interact to affect organisms and ecosystems and, second, the development of soundly based predictions of what lies ahead for terrestrial and aquatic ecosystems. One must move from physics to biochemistry to ecology to appreciate the challenges that lie ahead for life in our rapidly changing world.

CHAPTER 2

Oxygen and Metabolism

"Of the four 'biogenic elements'—carbon, hydrogen, oxygen, and nitrogen—oxygen stands out as being a geochemical and biochemical anomaly. In a combined state it is a highly abundant, fundamental constituent of the inner planets, their satellites, and of all known living systems; yet of all the planets of the solar system, molecular oxygen is a major component only of the atmosphere of the Earth, and in this uncombined state, it is at the same time both lethal to many organisms and an absolute requirement for many others."

D. J. Chapman and J. W. Schopf (1983)

2.1 The Two-Sided Nature of Oxygen and the History of Life

In Chapter 1 we emphasized that biochemical adaptations commonly fall into two categories, conservative adaptations that enable organisms to maintain capacities for carrying out critical existing functions, and innovations that enable organisms to exploit in novel ways the potentials offered by changes in the abiotic environment. In this chapter we focus principally on the interactions between oxygen and living systems, where the capacities of organisms to exploit changing environmental conditions—here, the availability of molecular oxygen (O_2)—provide what we feel are some of the most striking examples of innovative adaptations. Perhaps the most concise summary of oxygen's role in evolution has been given by biochemist Nick Lane, who titled his book on the diverse impacts of oxygen on life *Oxygen: The Molecule that Made the World* (2002). What this title denotes is that oxygen not only has important and wide-reaching effects on biological systems, but also on our planet's geochemistry and atmospheric composition (the "World"). It is this interplay among geochemistry, atmospheric properties, and life that we will first examine, in order to build a strong foundation for understanding the oxygen relationships of contemporary organisms.

We then will examine how contemporary organisms exploit different pathways of metabolism, both aerobic and anaerobic, to conduct a wide variety of functions that are characterized by differences in the degree to which they require oxygen. We will show that adaptive variation, in the *types* of metabolic pathways used to generate ATP and in the *intensities of activities* of these pathways, helps organisms sustain key physiological activities under widely different oxygen availabilities. These adaptive variations in pathway type and activity are observed among species, between tissues and organs of a single species, and also in response to changes in oxygen availability. The orchestration of the activities of these pathways reflects a type of conservative adaptation: Organisms conserve the ability to generate needed quantities of ATP despite the widely varying access they have to molecular oxygen. However, the different modes of reducing the requirement for ATP, a process called metabolic depression (or hypometabolism), often in response to reduced levels of oxygen, represent a type of innovative adaptation specific to the life-history traits of certain species.

The earliest life evolved under oxygen-free conditions (anoxia)

One often reads that water is a pre-condition for life, a point that we develop in detail in Chapter 4. However, life's dependence on water is broader than may be commonly realized. Whereas water is essential for its role as a solvent and medium for the complex biochemistry of cells, water is also the source of the oxygen that has been essential for development of life's abundance, diversity, and complexity. When, more than 2.4 billion years ago, cyanobacteria evolved a photosystem that was able to scavenge electrons from water and produce oxygen as a consequence—a "waste product," as it were—the chemistry of Earth began a radical shift. Atmospheric and geochemical changes were extensive. Biological impacts were also substantial. Thus a microbial world adapted to—and only tolerant of—anoxic conditions needed either to retreat away from oxidizing habitats to other anoxic environments or evolve capacities to tolerate and/or exploit the arrival of O_2. As we discuss below, the earliest life did not depend on the presence of oxygen and ran its metabolism using **anaerobic** pathways. (It is important to note that the terms *anoxic* and *anaerobic* are not synonymous. **Anoxia** refers to the absence of oxygen, whereas anaerobic refers to living processes that are independent of oxygen.) The central position of glycolytic reactions in contemporary organisms is something of a "fossil" that reveals the ancient biochemistry of the earliest cells. Even today, some microbes are unable to survive in the presence of oxygen, a sensitivity that most likely reflects their origins under conditions where availability of oxygen was low to nonexistent. However, the abundance of life in oxygen-replete habitats shows that the challenges life faces in the presence of oxygen can be met, and that once this is done, evolution is able to craft a wide range of biochemical processes that take advantage of oxygen's unique properties for gaining and releasing electrons. The quote from Chapman and Schopf (1983) that heads this chapter is a concise summary of this dual-faceted nature of oxygen. We will explore both the "good" and "bad" sides of this "anomalous" element in the discussion that follows.

Oxygen and multicellular life

The complex multicellular organisms of today's world rely on oxygen to achieve a variety of functions. Thus the abundance of multicellular life that we see today is only possible because of the relatively high concentration of atmospheric oxygen. Without the efficient aerobic ATP-generating systems that operate when oxygen is present, the flux of energy through living systems would certainly be vastly smaller. Moreover, one of the essential components of the system for exploiting oxygen in energy metabolism, the **cytochrome c oxidase system**, not only supports an enhanced yield of ATP per substrate molecule, but also provides a mechanism for decreasing the amounts of toxic oxygen by-products produced during aerobic ATP generation (**Box 2.1**). Perhaps more than any

BOX 2.1

Evolution of an electron capacitor: The cytochrome *c* oxidase system

Perhaps no other biochemical entity in the cell better illustrates how evolution has dealt with the dual nature of oxygen's chemistry—its "good" and "bad" sides—than the cytochrome *c* oxidase system. Cytochrome *c* oxidase plays a central role in ATP production by the electron transport system (ETS), serving as an *electron capacitor* that stores four electrons until the system can be discharged to generate the nontoxic end product, water. This **tetravalent reduction** of oxygen (four electrons are added simultaneously to a single dioxygen molecule to form two molecules of water; see figure) enables completion of the energetically downhill flow of electrons through the ETS. Thus oxygen's "good" side—its role in generating biologically useful forms of energy (ATP)—is achieved.

The capacitor function of cytochrome *c* oxidase also reduces the risk of **univalent reduction** of dioxygen, which generates ROS (see text). The five metal ions of cytochrome *c* oxidase—two heme irons and two copper centers with two and one coppers each—hold on to their single electrons until the capacitor is fully charged and tetravalent reduction of O_2 can occur. Cytochrome *c* oxidase is able to hold on to partially reduced forms of oxygen until they can be fully reduced, and this greatly decreases the likelihood of ROS production.

The origin of cytochrome *c* oxidase was attendant on the rise of oxygen levels to concentrations at which transition metals like copper could be reversibly oxidized and reduced. A strongly reducing atmosphere, as existed early in Earth's history, would preclude this reversible gain and loss of electrons. As oxygen levels rose to values that allowed reversible oxidation and reduction of copper, innovative adaptations like the cytochrome *c* oxidase capacitor were able to evolve. This exploitation of copper is perhaps one of the most important "innovations" in the history of biochemical adaptation, for it has at once allowed the development of powerful aerobic metabolic systems with high ATP turnover and, simultaneously, minimized the dangers of ROS production in the face of high rates of oxygen use (reduction).

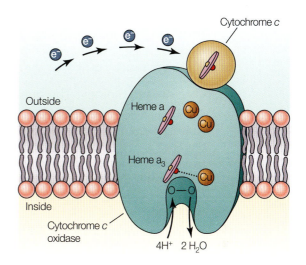

Reduction of oxygen by cytochrome *c* oxidase.

other biochemical system in the broad set of metabolic reactions involved in intermediary metabolism, cytochrome *c* oxidase shows the innovative nature of biochemical adaptation and the close interplay between geochemical change and evolution.

The essential role of oxygen in certain biosynthetic reactions is probably a less-appreciated aspect of oxygen's role in evolution, but it is certainly a critical one. For instance, oxygen is required for the synthesis of important structural elements in plants and animals. Large vascular plants would not be possible without **lignin**, and lignin would not be possible without a biosynthetic pathway that uses oxygen. In animals, one of the most important oxygen-dependent biosynthetic pathways produces **collagen**. One can ponder what animal life would be like without this vital structural protein. Oxygen is also required for the biosynthesis of steroid compounds such as **cholesterol** that play numerous roles in cellular structure and function. In addition to playing an essential role in biosynthesis, oxygen's entry into the biosphere also established a chemical environment that permitted the evolution of numerous other biochemical capacities, which underlie such major biological characteristics as rate of function, body height and mass, and locomotory capacity.

These adaptations reflect the positive side of oxygen's chemistry, but other consequences of reliance on oxygen-dependent biochemical systems are much less favorable to life. The presence of oxygen in the cell sets up a potentially dangerous situation in which oxygen-derived chemicals termed **reactive oxygen species** (**ROS**) can be generated. As we will illustrate when we discuss the atomic structure of oxygen, molecular oxygen, **dioxygen** (O_2), is able to accept either pairs of electrons, as occurs in the reactions of the electron transport system that generates ATP, or single electrons, which lead to the production of ROS. ROS pose numerous and serious threats to cells because of their high reactivity; they damage DNA, lipids, and proteins, threatening the integrity of the genome and the structures of existing biomolecules on which the cell depends. Damage from ROS has been implicated in ageing: Life span may be a direct reflection of the rate at which ROS are produced in aerobic metabolism, as we discuss in Section 2.8. ROS generation commonly increases as a result of cellular stress. Many of the downstream effects of exposure to extremes of temperature, osmotic concentrations, and oxygen availability are a consequence of the ROS produced as a result of perturbation of cellular structures such as the inner membrane of the mitochondrion.

Oxygen and biology: The questions we will address

To help us understand the importance of oxygen for the biochemistry of life, we will be reviewing the history of oxygen on Earth. This analysis will give us some insights into the origins of the biochemistry of oxygen-dependent life and will illustrate the innovative nature of biochemical evolution mentioned in Chapter 1. We will review some of the hypotheses concerning how oxygen influenced the evolution of life, including the rise of metabolically efficient eukaryotes, the appearance of large land plants, and the proliferation of highly mobile animals. With this evolutionary perspective as a foundation, we then will discuss some of the central properties of metabolic systems that vary in their dependence on oxygen. This analysis will lead us into a discussion of the distinct aerobic and anaerobic routes for ATP production, and what the opportunities and limitations of these different pathways of energy metabolism are for diverse types of organisms. In conjunction with examining aerobic ATP generating pathways, we will review the production and scavenging of ROS and their biological effects. What these analyses will reveal is the major role that oxygen plays throughout the biosphere, where its "good" and "bad" influences can be seen in much of what organisms are—their biochemical structures—and in what they do.

2.2 The History of Oxygen on Earth: How and When Did Oxygen Appear?

Almost all of the oxygen on Earth originated from **O_2-producing (oxygenic) photosynthesis**, which splits a water molecule to yield oxygen and provide hydrogens—reducing power—to drive ATP production and the photosynthetic fixation of carbon dioxide. However, the origin of oxygenic photosynthesis is not tightly correlated with the initial rise in atmospheric oxygen, which now accounts for approximately 20% of atmospheric gases. The oxygen generated during the earliest period of oxygenic photosynthesis reacted with minerals, and these oxidation reactions placed a damper on the development of an oxygen-rich atmosphere (Lalonde and Konhauser 2015). Thus the geochemical evidence that we will review here is mainly an account of the accumulation of oxygen to levels high enough to make the atmosphere and shallow waters oxygen-replete. The two pieces of evidence that help establish when oxygen began to be produced in high quantities are the disappearance of mass-independent fractionation patterns and banded-iron formations.

Geochemical evidence for the rise of atmospheric oxygen

Stable isotopes have enormous utility for revealing past chemical events. Most chemical reactions in biological and nonbiological systems manifest a preference for lighter isotope(s) of an element because of energetically more favorable kinetics of the reaction (**kinetic fractionation**). As a result of this preference, molecules produced by these chemical processes become enriched in the preferred (lighter) isotopes (isotope fractionation). In contrast, photochemical reactions do not exhibit such dependency on element mass—they are an example of **mass-independent fractionation** (**MIF**). Thus analysis of the stable isotope composition of rock formations of different ages can reveal important aspects of the radiation environment that elements experienced during different periods of Earth's history. Mass-independent fractionation of isotopes of certain elements is characteristic of long geological periods. In the case of sulfur, MIF patterns are found from between 4.0 and 2.5 billion years ago (Gya)(the Archean time period of Earth; Figure 2.1) (Farquhar et al. 2000). These patterns are likely based on a reaction between ultraviolet (UV) radiation and volcanic SO_2 (Farquhar et al. 2001). Today, extremely high levels of UV radiation don't penetrate Earth's atmosphere, because the radiation is blocked by the **ozone (O_3) layer**, and thus MIF is relatively infrequent. But in the distant past, an O_3 layer could not form until there was abundant oxygen in the atmosphere; before that, there was nothing to protect Earth from intense UV radiation. Therefore, when MIF patterns disappear in the geological record, one can reasonably conclude that a substantial amount of oxygen had been added to the atmosphere.

Based on the disappearance of the MIF pattern of sulfur isotopes ~2.4 Gya, we can infer that before that time O_2 levels in the atmosphere must have been very low, probably less than 10^{-5} % to 10^{-3} % of **present atmospheric levels** (**PAL**) (Holland 2006). At this time the

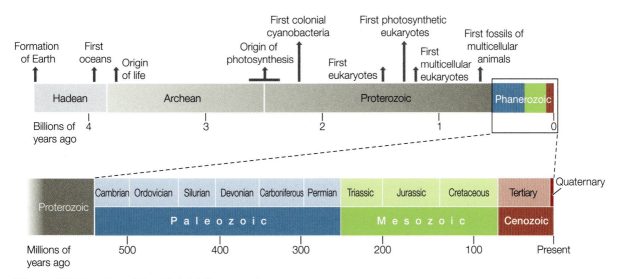

Figure 2.1 Timeline of Earth's 4.5-billion-year history.

deep ocean was almost certainly anoxic, but it is likely that O_2 levels were rising in some shallow oceans. Evidence for oxygenation of shallow waters comes from the appearance of what are called **banded-iron formations (BIFs)**, which are made up of alternating layers of dark rock and red rock that contain oxidized iron and occurred frequently before 2.4 Gya but became less frequent between 2.4 and 1.9 Gya (Canfield 2014). The bands are thought to result from periodic changes in the photosynthetic activity of cyanobacteria, which oxygenated shallow waters only periodically. As upwelling brought nutrients from deeper to shallower waters, cyanobacteria produced more oxygen, which would have oxidized elemental iron to iron oxides ("rust"). The seasonality of primary productivity—and thus of "rusting"—would have caused the red banding of BIFs. However, whereas cyanobacterial oxygenic photosynthesis likely contributed to the formation of BIFs early in the oxygenation of Earth's oceans, other reactions probably contributed as well. For example, elemental iron can also be oxidized as a result of anoxygenic photosynthesis by anaerobic purple bacteria, which use light as a source of energy to drive CO_2 fixation and obtain reducing power (electrons) from a source other than water (Widdel et al. 1993). Furthermore, the "disappearance" of the pattern of BIFs during subsequent periods indicates generally higher levels of oxygen, supposedly eliminating the alternation of oxic and anoxic conditions. Thus it is challenging to develop a consistent "detective story" for the time of oxygen's rise in Earth's atmosphere and waters.

Evidence for the accumulation of oxygen on land comes from the appearance of **red beds**, soils that retained iron because of it being oxidized by increasing concentrations of oxygen between 2.2 and 2.0 Gya (Rye and Holland 1998). Whereas BIFs suggest seasonally fluctuating levels of considerable concentrations of oxygen in shallow waters, red beds are seen as an indication of increasing levels of oxygen in the atmosphere, possibly to as high as 5%–18% of PAL.

Direct evidence for the appearance of cyanobacteria, the first O_2-producing organisms, dates to 2.7 Gya, but recent analyses of the data cast doubt on their validity (Brocks et al. 1999; Rasmussen et al. 2008). Indirect evidence of "whiffs" of atmospheric O_2 before 2.4 Gya comes from the appearance at that time of metals, for example molybdenum (Mo), that had previously been bound to sulfide minerals of Earth's crust. Oxidation released Mo from these minerals, and some of the element entered rivers and then the ocean. The element is now detected in organic-rich shales (Lyons et al. 2014). The existence of organic-rich, Mo-containing shale dating to the Archean age (4.0–2.5 Gya) indicates that oxygenic photosynthesis may have originated earlier than 2.7 Gya, as this bioenergetic pathway seems to be the only one that could elevate the production of organic matter to such high levels (Lyons et al. 2014).

Geochemical evidence thus presents a complex timeline for the appearance of oxygen in the atmosphere and seas. The disappearance of the MIF patterns of sulfur isotopes 2.4 Gya is consistent with increasing levels of atmospheric O_2 at that time. The lower frequency of BIFs between 2.45 and 2.08 Gya, indicating that oxygen levels were rising to higher levels across wider areas, and the subsequent appearance of red beds between 2.2 and 2.0 Gya, further suggests that rising oxygen levels in the atmosphere were preceded by increased oxygenic photosynthesis in the marine environment (Holland 2006).

Another way to establish the time O_2 began accumulating in the atmosphere is based on the fundamental balance between O_2 production through photosynthesis and O_2 consumption through aerobic (mainly microbial) respiration of organic matter. Excess O_2 can only be generated when organic matter is buried before O_2 can be consumed through aerobic respiration. Greater rates of burial of organic material during the late Proterozoic (2.5–2.0 Gya) allowed O_2 to accumulate to levels high enough to change the geochemical cycle of iron. One consequence was the disappearance of BIFs. The

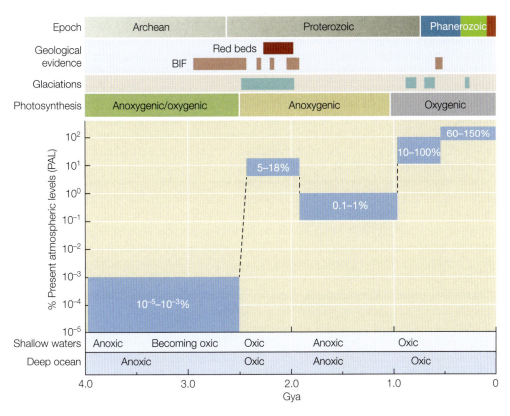

Figure 2.2 The history of oxygen on Earth. Approximate levels of oxygen are plotted as the percentage of present atmospheric levels (PAL) across major epochs. Dashed lines represent short time-periods when oxygen levels changed dramatically. Note the $\log_{10}$ scale of the y axis. Also shown are the occurrence of red beds and banded-iron formations (BIF) in the geological record, global glaciations, the types of photosynthesis that dominated (based on Johnston et al. 2009), and the oxygen contents of shallow and deep ocean habitats. The Great Oxidation Event (GOE) occurred ~2.4–2.0 Gya.

elevated oxygen levels also led to increased levels of sulfate (SO_4^{2-}) through oxidation of terrestrial sources of pyrite (FeS). Estimates for O_2 levels during the time period from 2.4 to 2.0 Gya range from 5% to 18% PAL (**Figure 2.2**) (Canfield 2005), an increase in O_2 relative to previous ages that justifies naming these changes the **Great Oxidation Event (GOE)** (Holland 2006). Although the increasing oxygen production of the surface oceans oxygenated the atmosphere at this time, the deep ocean is assumed to have stayed anoxic. Another early consequence of the rise of atmospheric O_2 might have been an increase in the consumption of methane (CH_4), which served as an important greenhouse gas while the sun's radiation was only 70%–80% as intense as it is today. CH_4 removal from the atmosphere through oxidation might have caused temperatures on Earth to plummet and led to the first occurrence of a "**Snowball Earth**" ice age (Huronian glaciation) from 2.4 to 2.0 Gya.

While the GOE marks the arrival of O_2 as a factor affecting Earth's atmospheric chemistry, the fact that extensive BIFs reappeared between 2.0 and 1.8 Gya suggests that there was a return to lower O_2 levels, possibly as low as those that existed before the GOE (10^{-3}% PAL; see Figure 2.2), for at least ~200 million years (Myr). Atmospheric O_2 generally stayed low during the mid-Proterozoic from 1.8 to 0.8 Gya, a period sometimes referred to as the "boring billion" (for a higher estimate see Holland 2006). In fact, recent studies suggest that the atmosphere may have been void of oxygen during this period (Planavsky et al. 2014). Because of the low levels of atmospheric O_2 during the mid-Proterozoic, the ocean surfaces stayed hypoxic and the ocean depths stayed globally anoxic. Oceans may have been locally productive in areas rich in hydrogen sulfide (H_2S) because of the contribution of anoxygenic photosynthesis by purple and green bacteria; the combination of hypoxia and high levels of hydrogen sulfide is referred to as a **euxinic** condition (Canfield 2014). Recent work also shows that deep-water conditions were iron rich, further indicating that little O_2 reached the deep ocean (Canfield et al. 2008; Planavsky et al. 2011).

Euxinic conditions, even if they did not expand globally, limited the availability of certain metals, such as Mo, which is insoluble under these conditions. Because Mo is required by two of the three classes of enzymes (nitrogenases) involved in the fixation of nitrogen (reduction of N_2 to NH_4), primary production can be limited when Mo concentrations are too low (Anbar and Knoll 2002). During the mid-Proterozoic, low Mo levels likely favored anoxygenic photoautotrophs, which have nitrogenases that only require iron (Johnston et al. 2009). A euxinic ocean also may have affected the incorporation of other metals into proteins (Dupont et al. 2010). As a consequence, the mid-Proterozoic was characterized by very low O_2 in the atmosphere and shallow oceans, and anoxic, H_2S- and iron-rich (or ferruginous) conditions in the deep oceans, as well as low primary production, except in some shallow locations due to anoxygenic photosynthesis by bacteria. Such conditions may have prevented eukaryotic life from thriving in certain habitats, even though eukaryotic life arose ~1.9–1.7 Gya. However, the geochemical situation changed dramatically in a period of ~300 Myr, the Neoproterozoic (0.85–0.55 Gya), and established more favorable conditions for the expansion of eukaryotic life.

The Neoproterozoic was characterized by a big step toward higher atmospheric O_2, with levels increasing to above 50% PAL and possibly close to 100% PAL (Canfield 2005; Holland 2006). Although ocean conditions were initially still mainly euxinic and ferruginous, trace metal enrichments suggest that the oceans were quickly becoming more oxygenated and that they were widely oxygenated between 630 and 550 Myr ago (Mya), a time that coincides with the emergence of Ediacaran animal forms (Lyons et al. 2014). Three major ice ages (Snowball Earth) during this time period might have made more nutrients available and led to an increase in primary production, possibly augmented by the rise of a new group of eukaryotic primary producers (green algae). A major increase in the burial of organic matter, possibly through changes in plate tectonics, might have been the proximate cause for an increase in O_2. The exact sequence of events and the forces that triggered this **Neoproterozoic Oxidation Event (NOE)** are unclear, but we do know that it loosely correlates with fossils indicating the radiation (not the origin) of animal phyla that directly followed the Ediacaran fauna.

Modeling of geochemical cycles of carbon and sulfur through the Phanerozoic age (550–0 Mya; Phanerozoic is from the Greek *phaneros*, "visible," and *zōion*, "animal") yields estimated O_2 levels of between 60% and 150% PAL (or 0.13–0.31 atm) (**Figure 2.3**; also see Figure 2.2) (Berner 2006). O_2 levels from the end of the Neoproterozoic into the Ordovician period (485–445 Mya) are estimated to have been in general slightly lower than PAL (0.15–0.2 atm). Oxygen levels were higher during the Silurian (445–420 Mya) and Devonian (420–360 Mya) periods, reaching levels higher than 100% PAL (0.2–0.25 atm). Levels then dropped again to pre-Devonian levels by the end of the Devonian. The highest O_2 levels were reached during the Carboniferous (360–300 Mya) and Permian (300–250 Mya) periods (see Figure 2.3). This large increase in O_2 levels is likely related to the proliferation of terrestrial green plants.

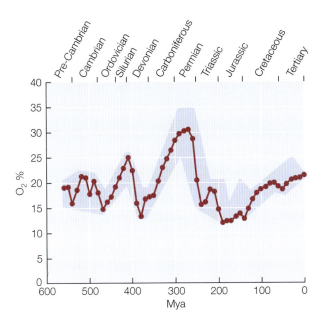

Figure 2.3 Estimates of oxygen levels during the Phanerozoic eon. The percentage of oxygen in the atmosphere was estimated based on the differential weathering rates of rocks, the effect of oxygen on carbon isotope fractionation and weathering of organic matter and pyrite, the exposure of land mass by sea level change, and the effect of CO_2 on the variability of volcanic rock weathering over time. The shaded area indicates the range of error in the estimates. (After Berner 2006.)

The success of terrestrial vascular plants, in turn, was strongly dependent on the evolutionary appearance of lignin, whose biosynthesis requires molecular oxygen. Lignin, which is composed of cross-linked phenolic polymers (see structure at right), is a key structural element of vascular plants' cell walls. The strength provided by lignin allows large-sized, tall plants to develop. Importantly, lignin is resistant to microbial degradation. Thus lignin-rich plant material would be relatively resistant to biological degradation, allowing it to accumulate and lock up organic carbon. Burial of organic matter reduces the amount of CO_2 returned to the atmosphere (or ocean) and favors a higher O_2/CO_2 ratio in the environment.

Following the increase in O_2 during the Permian, the end of the Permian (~250 Mya) was marked by a relatively sudden drop in O_2. A similar drop in O_2 occurred at the end of the Triassic (~200 Mya), and both of these events were accompanied by major extinction events in both marine and terrestrial environments. By the end of the Permian, O_2 levels were probably ~0.15 atm, and by the end of the Triassic they might have been as low as 0.1 atm (50% PAL). These lower levels of O_2 were probably the result of a combination of factors, including geological events, such as volcanism, and biological processes, such as greater rates of microbial respiration at

Lignin

the beginning of the Triassic (Meyer et al. 2011). Over the last 200 Myr since the end of the Triassic, O_2 levels increased again to today's levels of ~0.2 atm.

In summary, although the origin of O_2 is biological, geochemical reactions in Earth's atmosphere and crust served as sinks for removal of oxygen—for example, when O_2 reacted with reduced gases of volcanic origin and oxidized iron in the crust. These processes for removal of oxygen were especially prevalent during the Archean and Proterozoic. The rise of atmospheric and dissolved oxygen was driven by increased oxygenic photosynthetic activity that was enhanced by the increased levels of nutrients released by events like glaciation. Oxygen levels also rose as a result of the increased burial of organic carbon following evolution and proliferation of degradation-resistant, lignin-rich vascular plants. Tectonic processes also contributed to the burial of rich deposits of organic carbon (the source of the fossil fuels we are currently burning). This long and complex interplay between biology and geology largely accounts for the wide variation in atmospheric and dissolved oxygen levels throughout Earth's history. Changes in the availability of the "anomalous" molecule, O_2, have had profound impacts on the evolution of life, and as we show in the sections to follow, the "footprints" of oxygen's effects can be found at all levels of biological organization.

2.3 Atmospheric Oxygen and Animal Evolution

Whereas photosynthetically generated oxygen has influenced the evolution of taxa belonging to all three domains of life, it is among the diversity of animal species that we find some of the most striking examples of how oxygen has shaped the course of evolution at all levels of biological organization. Oxygen's imprint on evolution is evident at the anatomical, physiological, and biochemical levels, and these oxygen-driven traits profoundly affect ecological interactions as well, notably in predator-prey relationships and in the needs of highly active aerobic animals for an abundant food supply. Here, then, we focus strongly on the oxygen-related biochemical adaptations of animals, to illustrate the opportunities that oxygen-rich terrestrial and aquatic environments have afforded to animal evolution. The systems we examine below provide clear illustrations of the two-sided nature of O_2: its potential for allowing novel metabolic and physiological innovations and the challenges it presents through the production of toxic ROS.

Putting oxygen's role in evolution into a wider perspective

The conjecture that the rise in O_2 contributed to the diversification of animal phyla is based to a large extent on a correlation between increasing O_2 levels during the late Neoproterozoic and the appearance of fossils of the major animal phyla during the Cambrian. Molecular phylogenies tend to place the origin of major animal phyla deep into the early Neoproterozoic (Erwin et al. 2011). Because there is still uncertainty over the Neoproterozoic O_2 levels of shallow oceans (Lyons et al. 2014), the most likely environment for early animal life forms, the assumption that oxygen was the primary driver of radiation of animal phyla is somewhat problematic—or at least not the full story (Sperling et al. 2015). For example, little is known about how animals dealt with increasing levels of ROS and the damage these chemicals cause to cellular components. Other, indirect geochemical effects of oxygen might account for any correlation between a rise in O_2 and radiation of animal phyla. These effects include changes in the availability of metals, for example, molybdenum, that are needed for nitrogen fixation and primary production (Anbar and Knoll 2002) and the development of more complex food webs (Sperling et al. 2013).

Further complicating the connection between animal diversity and oxygen availability are factors such as changes in animal development that were occurring at the same time. The evolution of phylum-specific developmental programs had a major impact on animal radiation. Evolution of new developmental pathways may have been driven by selection resulting from the rise of more complex ecosystems, which were fueled by a rise in primary productivity. Thus, rising oxygen levels can be viewed as a "bottom-up" force that led first to enhanced metabolic capacities, and then to more robust and productive ecosystems that were able to support more and different forms of animal life. The importance of oxygen in allowing high levels of activity by animals is reflected in some of the major changes in animal form and function that occurred during periods when there were large increases in oxygen levels. The geological history of changing O_2 levels can help explain several major evolutionary transitions, including the diversification of animal phyla during the transition from the pre-Cambrian to the Cambrian (Canfield et al. 2007); the correlation of oxygen levels with an increase in body size of insects, reptiles, and mammals; and the evolution of respiratory structures and oxygen-transport systems (Berner et al. 2007; Falkowski et al. 2005). For this reason, one must view oxygen as a major "driver" of animal evolution. However, a more holistic or nuanced perspective is needed to more fully understand how oxygen worked with other "drivers" to facilitate the evolution of animal diversity.

Body size and oxygen

The challenges in accounting for animal traits strictly on the basis of levels of available oxygen are well illustrated by attempts to account for changes in animal body size during evolution. The overall idea behind the proposed effect of oxygen on animal size is based on the process of diffusion driving the delivery of oxygen along the gradient from the environment to the tissues and cells in the organism. As the rate of diffusion is positively dependent on the difference in oxygen concentration along this gradient, higher levels of oxygen could provide equal oxygen levels over greater distances and thus support larger animals. Oxygen-driven evolution of body size has been proposed to account for the increase in the size of placental mammals during the Tertiary (~65–2.5 Mya) (Falkowski et al. 2005). However, there are reasons to question this all-encompassing explanation. First, larger animals have lower mass-specific metabolic rates, so oxygen demands per unit of mass are reduced in larger species. Second, endothermic mammals, which would have had relatively high metabolic rates in comparison with their ectothermic counterparts, evolved during the Phanerozoic O_2 minimum, bringing into question the connection between metabolic requirements and O_2 availability, at least for mammals (Butterfield 2009).

   A connection between O_2 levels and body size was first proposed based on the appearance of giant insects during the Carboniferous and Permian periods, when O_2 levels were at a maximum of over 0.31 atm (or 150% PAL). Insects use a complex, branched tracheal system to deliver oxygen to the respiring tissues (see illustration at right). Airflow into the insect is regulated by the tightly controlled opening and closing of spiracles at the body surface. Smaller tubule systems, tracheoles, then carry oxygen to the cells. Insects likely are more directly dependent on

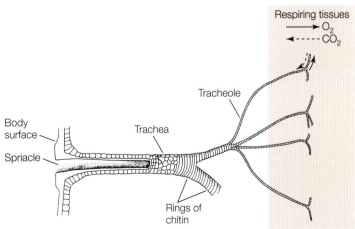

(After Barnes 1980.)

ambient oxygen concentrations than are animals with a circulatory system that is based on oxygen dissolving into an aqueous medium (hemolymph or blood), as the latter can be augmented with oxygen-binding proteins that facilitate increased oxygen transport.

The relationship between atmospheric oxygen levels and body size in insects is strikingly shown by dragonflies of the extinct order Protodonta. These giant insects had a wingspan of over 70 cm and twice the thoracic width of the largest extant dragonflies (see illustration at left). A survey of insect fossils showed that, whereas large insects were always rare, maximum insect body size correlated with atmospheric O_2 from the Carboniferous-late Permian O_2 maximum to the end of the Jurassic. Subsequently, this relationship became decoupled: Insect size stayed static during the Cretaceous and decreased during the Cenozoic (Clapham and Karr 2012). This breakdown of the body size-atmospheric O_2 relationship may have been caused by changes in biotic interactions, notably increased predation by birds, which evolved a more agile flight capability, and the appearance of insectivorous bats.

Several empirical studies with modern insects have attempted to evaluate the relationship between O_2 levels and insect body size, and thus substantiate the physiological basis of large insect body sizes during the Carboniferous-late Permian O_2 maximum (Harrison et al. 2010). The acute respiratory response of insects to changing O_2 levels involves the adjustment of airflow through the tracheal gas exchange system. As mentioned above, the opening and closing of spiracles regulate the ventilation of the tracheal system and thereby help maintain internal O_2 levels and control respiratory water loss. If O_2 levels fall below a certain point, insect performance begins to decline; the O_2 concentration at which this occurs is termed the **critical oxygen partial pressure (P_{crit})**. Higher P_{crit} values are therefore an indication of greater sensitivity toward oxygen limitation. It is not surprising that P_{crit} is highest for activities like flight, which are highly aerobic and energetically intensive. If larger body sizes were made possible by hyperoxia, one would see this reflected by an increase in P_{crit} with increasing body size, which would indicate that larger insects are more limited by oxygen and thus would benefit more from hyperoxia because the tracheal system is unable to match the greater needs of a greater tissue mass (Harrison et al. 2010). In contrast to this prediction, current evidence indicates that variation in P_{crit} is driven more by metabolic requirements of particular developmental stages or modes of activity, such as flight, than it is by variation in body size (Harrison et al. 2010). The lack of a consistent size-dependency of the effect of acute O_2 changes on performance suggests that the tracheal system generally matches O_2 demand across the range of body sizes of extant insect species. If these results can be extrapolated to much larger insects, the same would be assumed to have been the case during earlier time periods.

Nonetheless, other studies have found evidence for a mechanistic basis for oxygen-body size relationships in insects. Development under hypoxic conditions leads to a decrease in mean and maximum body size in most insects (Harrison et al. 2010). During development under hypoxia, insects modify their respiratory system by increasing the diameter of the trachea and the number of tracheoles. Hyperoxia, in contrast, leads to an increase in body size in some insects, for example, the giant mealworm (*Zophobas morio*), but not in the majority of insects studied. Thus, while there is a consistent response to hypoxia, the response to hyperoxia is less uniform. Fruit flies (*Drosophila melanogaster*) that were evolved for many generations under hyperoxia in the laboratory increased mean and maximum body size, whereas hypoxia led to the evolution of smaller body size (Klok et al. 2009). These differences were accompanied by a decrease (under hyperoxia) and increase (under hypoxia) in tracheal volume, respectively, suggesting that there is a decrease in the investment in the tracheal system with increasing oxygen levels (Henry and Harrison 2004). However, subsequent rearing under normoxia reversed these changes in the hypoxic but not in the hyperoxic group, suggesting that the reduction in body size under hypoxia was due to developmental plasticity rather than adaptation to different levels of oxygen.

The results of the fruit fly selection experiments suggest a clue for a possible mechanism for the role of oxygen in determining insect body size. In almost all animals other than insects, the structures supporting O_2 transport—the pulmonary and circulatory systems (including blood volume)—generally show an isometric relationship (direct correlation) with body size. This does not hold true for the tracheal systems of insects. Larger insects have to invest a relatively higher proportion of the body to their tracheal system to overcome the limitations of diffusion-based O_2 transport in long blind-ended tracheoles. This hypermetric relationship means that the tracheal system will at some point limit how much larger an insect can become. For example, in darkling beetles each leg is supplied with O_2 by a trachea that enters the leg from the rest of the body via an orifice that also contains nerves and other vital tissues. As the diameter of the trachea increases hypermetrically with body size, it takes up a greater proportion of the space in the orifice (Kaiser et al. 2007). When the empirical data were extrapolated to the largest extant beetles, the researchers found that the leg tracheae probably occupy 90% of the space in these beetles' orifices, leaving very little space for other tissues. Because of this, it may not be possible for the beetles to become any bigger (Kaiser et al. 2007). As a consequence, greater concentrations of oxygen may require less investment in trachael volume, thereby allowing for a greater body size to be achieved during evolution.

In summary, oxygen may have a direct effect on the evolution of body size in insects because of their tracheal system, but the patterns are far from unequivocal. Furthermore, given the lower mass-specific oxygen consumption rates of larger animals in general, the same causative relationship may not hold for any other group.

2.4 Energy Metabolism: Organismal Properties That Establish Animals' Metabolic Requirements

Oxygen has an immediate effect on an animal's ability to produce energy to maintain various functions, and it thus affects the energy metabolism of virtually all cells. One of the main functions of energy metabolism, and a major determinant of metabolic rate, is to provide an adequate supply of ATP (**Figure 2.4**). ATP turnover rates differ widely among species, among tissues within an individual, and among different activity levels. Metabolic rates also differ in relation to body mass (metabolic scaling) and the thermal biology of a species; ectotherms have metabolic rates that are only 10%–20% of those of endothermic homeotherms—birds and mammals (see Section 3.7). Because of the different sensitivities of ectotherms and endotherms to ambient temperature conditions, we must distinguish between **standard metabolic rates** (**SMRs**) in ectotherms, which vary with body temperature, and **basal metabolic rates** (**BMRs**) in endothermic homeotherms, which are measured at the typical body temperature of the species (~37°C for most

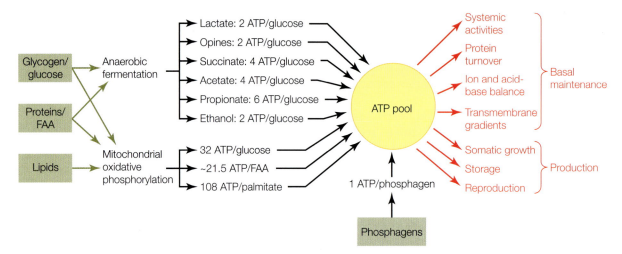

Figure 2.4 Simplified overview of the major ATP-producing and -consuming pathways. Organisms can oxidize carbohydrates (glycogen or glucose) and proteins (free amino acids) for ATP production through both aerobic and anaerobic pathways, while lipids are mainly oxidized through aerobic pathways (mitochondrial oxidative phosphorylation). See text for details. FAA, free amino acids. (After Sokolova et al. 2012.)

placental mammals). SMR and BMR are both measured in a state of fasting to eliminate interference from the temporary increase in metabolism that typically follows feeding. This postabsorptive increase in energy metabolism—referred to as **specific dynamic action (SDA)**—is thought to be caused by the production of proteins, an increased investment in mainly digestive tissues, and the breakdown of foodstuffs. This process is more pronounced in intermittently feeding animals (for example, pythons that feed irregularly), but SDA occurs in all animals to some extent (Secor 2009).

In this section we also look at how body size affects ATP turnover rates, as this relationship greatly affects the need for energy and places different constraints on small and large animals. Energy turnover also figures prominently in a wide range of ecological relationships. For instance, energy metabolism is intimately linked with the locomotory abilities of animals and with their capacities to remain active throughout the year. As we will see, major benefits can accrue from seasonal reduction in metabolism—hibernation and estivation—when food availability or severity of abiotic conditions in the environment favors reductions in activity.

What factors establish animals' metabolic rates?

To analyze the relationships between metabolic rate and biotic and abiotic factors, physiologists rely strongly on comparisons based on SMR and BMR. Using these rates as a foundation, it is possible to examine a wide range of factors that lead to either increases or decreases in rates of ATP turnover. One very important factor in governing needs for ATP is an organism's physical activity, in particular its locomotory activity driven by ATP turnover in muscle. For example, the ATP turnover rates of hummingbird breast muscle during hovering flight are among the highest known and are ~500 times basal (resting) metabolic rates (Suarez 1992). Tissues other than muscle can also have high turnover rates; for example, the ATP turnover rate of brain gray matter is comparable to that of leg muscles during a marathon, mainly because of the energy used to generate action potentials and synaptic transmission (Attwell and Laughlin 2001). Other factors affecting metabolism include digestion, reproduction (and the sex of the individual), time of day, and age. In addition to temperature, other abiotic environmental conditions can have strong effects on metabolism. For example, decreased availability of oxygen presents challenges for aerobic ATP production. Cellular stress, caused by a variety of environmental insults, can greatly increase the demand for ATP in order to maintain cellular homeostasis and repair tissue damage.

The variations in metabolic rate found among species, tissues, and environmental circumstances are all a consequence of two primary types of biochemical responses: changes in the *types* of metabolic pathways that are used and alterations in the *flux rates* through these pathways. The pathways that generate ATP may be strictly dependent on a continuous supply of oxygen as a terminal electron acceptor—**aerobic metabolism**—or they may be able to supply adequate amounts of ATP in an oxygen-independent manner—**anaerobic metabolism** (see Figure 2.4). Shifts between aerobic and anaerobic pathways of ATP generation may occur in response to the intensity of physiological activity, locomotion in particular, and as a consequence of changes in the availability of oxygen in the environment. We will begin our analysis of ATP-generating mechanisms with a review of these pathways, in order to provide a foundation for understanding why and how their activities are regulated in response to physiological and environmental circumstances.

Our initial focus will be on the use of different pathways to provide adequate levels of ATP for cellular activities, but we will also consider another important aspect of energy metabolism based on reduction of oxygen. As mentioned above, a major consequence of aerobic metabolism is the production of ROS. These compounds, which have in common the occurrence of an unpaired electron, present a fundamental challenge to cells. If unchecked, ROS can inflict significant damage to DNA, RNA, lipids, and proteins (Halliwell and Gutteridge 2015). This damage can add appreciably to the overall metabolic costs of life and introduce mutations into the genome. However, cells possess several biochemical reactions that reduce (and thereby scavenge) ROS, a process that requires NADPH as a cofactor (or a so-called reducing equivalent).

In summary, energy metabolism must be viewed in a broad context, one that takes into account not only the critical function of ATP supply to the cell, but also the threats posed by highly reactive by-products of aerobic metabolism, ROS. We will show how the evolution of the mitochondrial electron transport and proton translocation systems has been influenced by both of these key factors.

Energy metabolism: Mechanisms of ATP production

In animals, the main pathways of ATP production are **glycolysis** (also referred to as the Embden-Meyerhof-Parnas pathway), the reactions of the **tricarboxylic acid (TCA) cycle** (commonly known as the Krebs or citric acid cycle), and the mitochondrial **ATP synthase complex**, which uses the proton gradient generated by the **electron transport system (ETS)** across the inner mitochondrial membrane to synthesize ATP. Among these ATP-generating reactions are two fundamentally different mechanisms for adding a terminal phosphate to ADP. **Substrate-level phosphorylation** entails the transfer of a "high-energy" phosphate group from a metabolite, for example, phosphoenolpyruvate (PEP), to ADP, yielding ATP. Two reactions of glycolysis and one within the TCA cycle use substrate-level phosphorylation. The second mechanism of ATP production is **oxidative phosphorylation**, which involves the phosphorylation of ADP by the ATP synthase complex of the mitochondrial inner membrane. Electrons for this system are supplied by reducing equivalents in the form of NADH produced by glycolysis, the TCA cycle, and the β-oxidation of fatty acids.

Two other important reactions that can lead to short-term increases in ATP production or to buffering of ATP fluctuations during physiological activity involve the **phosphagens** creatine- and arginine-phosphate (see figure on p. 46). Creatine and arginine kinases catalyze the exchange of terminal phosphate groups between ADP and creatine-phosphate or arginine-phosphate. Below we discuss the importance of these reversible phosphorylation reactions in several metabolic contexts.

Metabolism: Production of reducing equivalents

As indicated above, an important role of metabolism is the production of **reducing equivalents**, mainly NADH, NADPH, and $FADH_2$. These reduced nucleotides serve several functions. First and foremost, they deliver electrons to the electron transport system and thereby provide a force to drive the transport of protons across the inner mitochondrial membrane, thereby establishing the proton gradient whose subsequent dissipation drives oxidative phosphorylation. The oxidized nucleotide is then able to serve again as a soluble carrier of the electrons and protons that are stripped from metabolites during their oxidation. It is important to realize that the oxidation of NADH (see figure at left) at the start of the ETS (complex I) is essential to maintain flux rates through the catabolic pathways that break down substrates like glucose. The oxidized form of this nucleotide (NAD^+) is an essential cofactor in one of the reactions of glycolysis and several of the reactions of the TCA cycle. Resupply of NAD^+ to these reactions is thus necessary to sustain flux through these pathways.

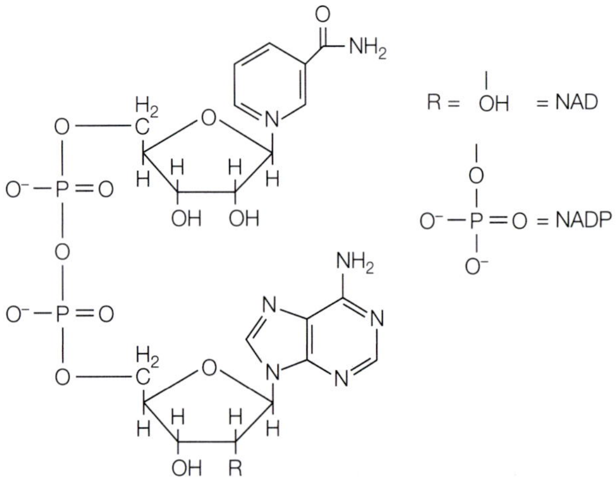

Among the array of catabolic pathways available to the cell, oxidative phosphorylation—involving reduction of oxygen through the cytochrome c oxidase reaction—has the highest yield of ATP per substrate molecule catabolized. However, sometimes there is not enough oxygen to support this pathway. Reduced oxygen availability may arise from two constraints. First, intense physiological activity can result in **physiological (functional) hypoxia** when the transport of oxygen from the environment to the respiring tissues cannot keep up with energy demands. When environmental oxygen levels are below those needed to support adequate levels of ATP production, **environmental hypoxia** occurs. Under either type of hypoxic condition, molecules other than oxygen serve as electron acceptors to allow regeneration of NAD^+ from NADH and continued catabolic activity. Reduction of an organic molecule to regenerate the oxidized cofactor is referred to as **fermentation**. A familiar NADH-requiring fermentation reaction is catalyzed by the enzyme **lactate dehydrogenase (LDH)**, which reduces the glycolytic metabolite pyruvate to lactate:

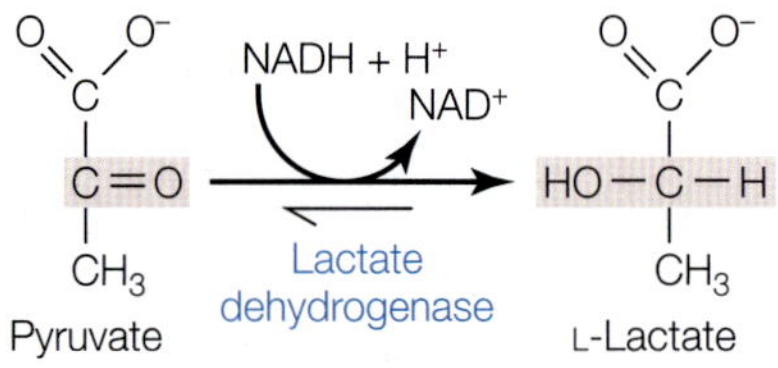

Continuous regeneration of NAD$^+$ is necessary to ensure continued degradation of glycogen and glucose for ATP production via substrate-level phosphorylation. Whereas the LDH reaction is of pivotal importance in vertebrates and many invertebrates, other metabolites can serve as electron acceptors to allow production of ATP to continue during hypoxia. Later we will discuss some of these alternative pathways of substrate-level phosphorylation and point out their benefits relative to the usual vertebrate anaerobic pathway of glycolysis and the LDH reaction.

In contrast to NADH, which is a central cofactor in ATP-generating reactions, NADPH is used in many anabolic reduction reactions in pathways that synthesize various cellular components. The pathways that generate NADPH differ from those that produce NADH. Reactions of the TCA cycle are the primary source of NADH, whereas the **pentose phosphate pathway** (**PPP**) is a major supplier of NADPH. The reaction catalyzed by the NADP-dependent isoform of isocitrate dehydrogenase (IDH), as well as several other reactions, which we detail below, can also make important contributions to NADPH production.

In addition to being a reductant in biosynthetic reactions, NADPH is used to reduce ROS. These reactions lead to the generation of water from a potentially toxic by-product of oxygen use (Pollak et al. 2007). As pointed out above, environmental stress from high temperatures, osmotic challenges, and transitions between high and low levels of oxygen frequently increases the production of ROS and consequently the demand for NADPH and its antioxidant activity (Tomanek 2015). The allocation of NADPH within the cell thus involves a trade-off between the cell's biosynthetic (anabolic) activities and its ability to eliminate toxic by-products of oxygen use.

Flavin-containing nucleotides (flavin adenine dinucleotide [FAD] is shown at right) are also important in reversible oxidation-reduction reactions and ATP generation. Reduced FAD (FADH$_2$) generated in the TCA cycle delivers electrons to the ETS, but unlike with NADH, the electrons are delivered to complex II through the succinate dehydrogenase (SDH) reaction. The β-oxidation of fatty acids also delivers FADH$_2$ to the ETS, specifically to coenzyme Q, through an electron carrier of the ETS called electron-transferring flavoprotein.

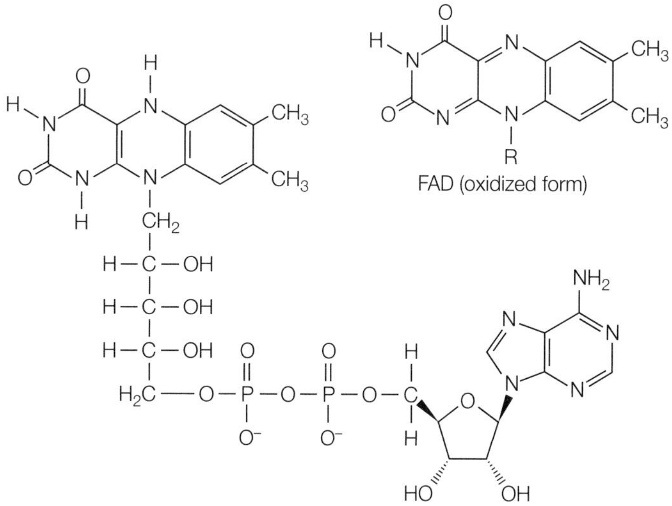

FAD (oxidized form)

FADH$_2$ (reduced form)

Metabolism: Providing building blocks for the cell

Another basic function of intermediary metabolism is to provide the various building blocks that cells require to grow and proliferate. For example, a metabolite of glycolysis (glyceraldehyde 3-phosphate) can be converted to glycerol 3-phosphate, which, when esterified to fatty acids (for different forms of fatty acids, see Section 3.6) yields triacylglycerides for fat storage and phosphoglyceride lipids for membranes. Many of the reactions of glycolysis are readily reversible thermodynamically—with important exceptions discussed below—and can take part in the resynthesis of glucose through a process called **gluconeogenesis**. The glucose thus formed may be deposited in glycogen (**glyconeogenesis**). **Glycogen** (see

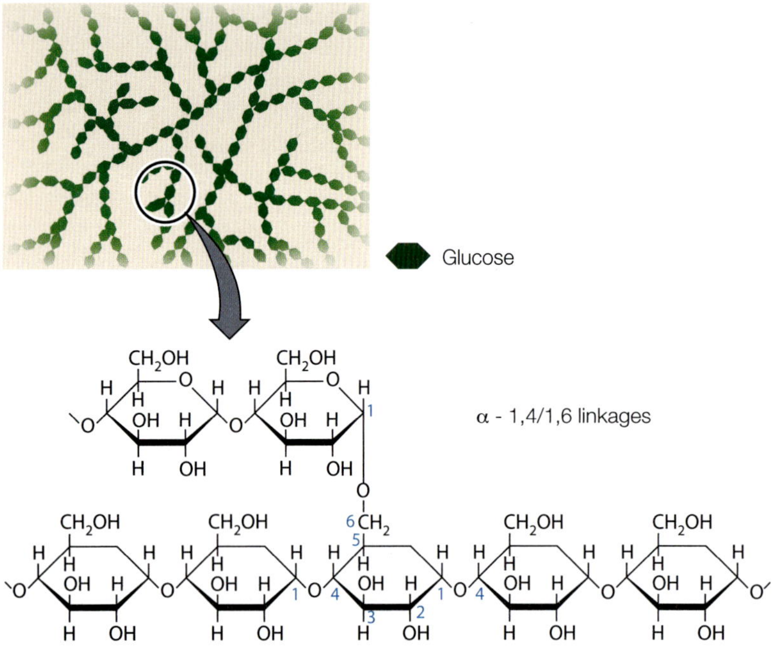

Glucose

α - 1,4/1,6 linkages

figure at left) is an energy storage molecule that can be exploited to sustain glucose levels in the blood and within cells. In animal locomotory muscle, glycogen is of particular importance in providing glucose when limited oxygen availability may demand an increase in substrate-level phosphorylation driven by glycolysis.

The TCA cycle provides building blocks for the synthesis of fatty acids and carboxylic acid skeletons for amino acid biosynthesis. The pentose phosphate pathway also provides important biosynthetic intermediates by diverting glucose 6-phosphate from glycolysis to the synthesis of pentose sugars needed for the synthesis of nucleotides (RNA and DNA).

In summary, the catabolic pathways of intermediary metabolism fulfill several functions and are tightly integrated with anabolic pathways of biosynthesis. A major—and perhaps most familiar—role of catabolic pathways of energy metabolism is to match ATP production with ATP demand, either through substrate-level phosphorylation (mainly glycolysis) or oxidative phosphorylation. In addition, these pathways can provide numerous intermediates for the biosynthesis of nucleotides, amino acids, and lipids. Because biosynthetic reactions commonly involve reduction of substrates, the reducing equivalents generated by catabolic pathways are of vital importance to the cell. Some reductants, notably the NADPH generated by the pentose phosphate pathway and the IDH reaction, are essential for scavenging ROS and thus defending the integrity of the cell. We now take a closer look at some of these pathways to better understand their roles in metabolism and how their activities are regulated in response to the shifting needs of the cell for energy and biosynthesis.

Temporal and spatial ATP buffering by phosphagens

For maintaining energy homeostasis in cells or tissues that are characterized by high and variable ATP turnover rates (e.g., muscle, neurons, and gill cells), the first line of defense is an increase in the reaction rates of creatine kinase (CK) and arginine kinase (AK). These reactions typically use intracellular stores of the phosphagens creatine-phosphate (CrP) and arginine-phosphate (ArP; see figure at left) to drive ATP production (Ellington 2001):

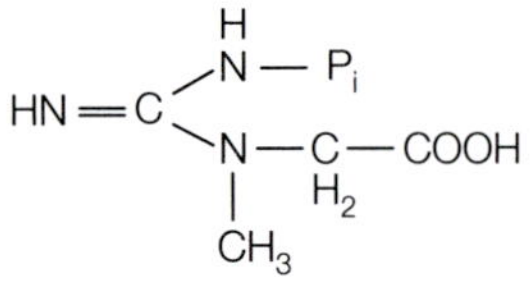

Creatine-phosphate

$$\text{Phosphagen-P} + \text{ADP} + \text{H}^+ \leftrightarrow \text{ATP} + \text{phosphagen}$$

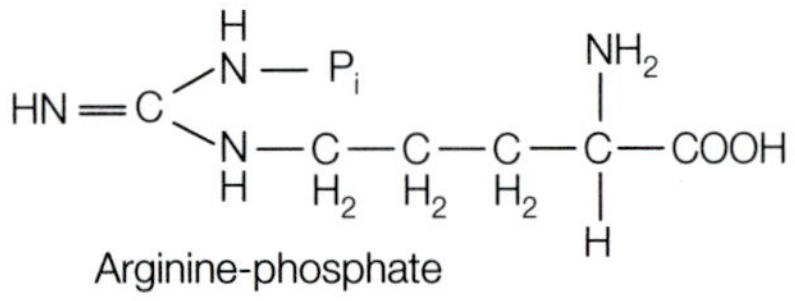

The type of phosphagen used varies among taxa: Arginine-phosphate is the main phosphagen in molluscs and arthropods, and creatine-phosphate is the major phosphagen in vertebrates. Most other phylogenetic groups have one of these phosphagens, but annelids also have several additional variants.

Arginine-phosphate

Although different phosphagen kinases play a common role in ATP generation, their kinetic properties differ and may reflect specific metabolic characteristics of the species in which they occur. Differences in the equilibrium constants (K_{eq} = [ATP][phosphagen]/[ADP][H$^+$][phosphagen-P]) of creatine kinase (K_{eqCK} = 100) and arginine kinase (K_{eqAK} = 13.2) reactions suggest that creatine kinase may be more able to regenerate ATP (buffer changes in ATP concentration) at higher ATP/ADP ratios, while arginine kinase may be more effective at buffering ATP during prolonged conditions of lower ATP/ADP ratios and low pH (higher [H$^+$] concentration) that are typical of prolonged hypoxia (Ellington 1989). This may explain why animals that possess both phosphagen systems generally express creatine kinase in spermatozoa, a cell type requiring high ATP turnover rates.

In addition to functioning as ATP buffers by maintaining ATP concentrations, phosphagens also bind a large fraction of inorganic phosphate (P$_i$). When released, the P$_i$ can serve as a substrate for glycogen phosphorylase and thus promote glycogenolysis. For example, exhaustive activity of the abdominal muscle of the shrimp *Crangon crangon* increased P$_i$ from 1 to 20 mM, leading to an increase in the activity of glycogen phosphorylase a (the phosphorylated form of glycogen phosphorylase) and thus stimulated glycogenolysis because P$_i$ is one of the substrates of the reaction (Kamp and Juretschke 1987). Furthermore, P$_i$ can assist in intracellular pH buffering, supplementing the pH buffering capacity of other compounds like bicarbonate ion and histidine imidazole groups (Griffiths 1981). By this mechanism, pH buffering by P$_i$ that is mediated by the activity of arginine kinase may be particularly important during heat stress in muscle and gill tissues of intertidal organisms, when intracellular pH decreases with increasing temperature during times of emersion (air exposure) and when CO_2 accumulation (hypercapnia) leads to increased concentrations of bicarbonate ions and protons (Burnett 1997; Garland et al. 2015).

Another major function of creatine kinase is to optimize ATP supply during episodes of high ATP turnover rates; under both anaerobic and aerobic conditions, the enzyme is a key part of the **phosphocreatine shuttle**, which channels and compartmentalizes the transfer of phosphoryl groups inside cells (**Figure 2.5**). The central contribution of this shuttle is that creatine and phosphocreatine, instead of ADP and ATP, are the molecules diffusing between ATP supply sites and sites of ATP use by major ATPases (Wallimann et al. 2007).

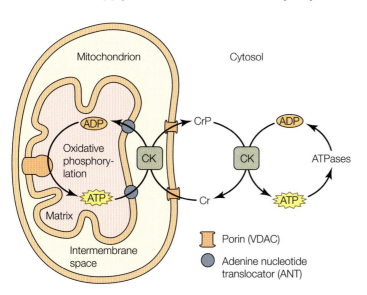

Figure 2.5 The phosphocreatine shuttle is based on the transfer of phosphoryl groups from ATP to creatine-phosphate (CrP) at the cellular site of ATP production (mitochondria) and the reverse transfer from CrP to ADP at the site of ATP demand (ATPases). Specific isoforms of creatine kinase are localized in the mitochondria and cytosol. The transport of ADP and ATP across the inner and outer mitochondrial membranes is regulated by adenine nucleotide translocator (ANT) and the voltage-dependent anion channel (VDAC; porin), respectively. CK, creatine kinase; Cr, free creatine. (After Guimarães-Ferreira 2014.)

This idea that ADP and ATP are localized for optimized function contrasts with the concept of the cell as a simple fluid-filled sac of chemicals whose reactions are determined by diffusion alone. Instead, the shuttle and compartmentation concepts propose that structural arrangements within the cell facilitate enzyme-substrate encounters and thus play a central role in the regulation and efficiency of metabolic pathways (Hochachka 2003).

One key to understanding the function of the phosphocreatine shuttle is that organisms express different isoforms of creatine kinase within a single cell (Guzun et al. 2012). Whereas the cytosolic isoforms are constitutive dimers, the mitochondrial isoforms are octamers when functional, and lose their role in the mitochondrion when converted into dimers through **posttranslational modification** (**PTM**; see Box 2.3) (phosphorylation). Furthermore, the cytosolic isoforms are associated with myofibrillar sarcomeres or membranes active in ion transport, whereas the mitochondrial isoforms are found in the mitochondrial intermembrane space in association with the **adenine nucleotide translocator** (**ANT**) and the **voltage-dependent anion channel** (**VDAC**), which regulate the transport of ADP and ATP across the inner and outer mitochondrial membrane, respectively. This locates the mitochondrial isoforms close to sites of ATP production, while cytosolic isoforms are associated with sites of ATP consumption (see Figure 2.5). By replacing ATP with CrP as the unit of energy transfer, this arrangement localizes the turnover of ATP to sites of production and consumption, thereby avoiding the mixture of ADP and ATP with the bulk phase of the cytosol. The advantage of using creatine-phosphate lies in (1) it being metabolically inert in comparison with adenine nucleotides, which serve as cofactors in many biochemical reactions; (2) the greater diffusion rates of both creatine and creatine-phosphate relative to adenylates due to their smaller molecular mass; and (3) the avoidance of ATP serving as an inhibitor of metabolic reactions. Furthermore, the local buffering of ATP supply through creatine kinase maximizes the change in Gibbs free energy of ATP hydrolysis by maintaining high ratios of ATP to ADP—based on the contribution of a higher ratio to $\Delta G_{ATP} = \Delta G°_{ATP} - RT \ln ([ATP]/[ADP][P_i])$—and thus enables reactions that require a high free energy change to function. Last, the close association of mitochondrial creatine kinase with ANT and VDAC facilitates the transphosphorylation from ATP to CrP and thus facilitates the recycling of matrix ADP to maintain favorable thermodynamic conditions of oxidative phosphorylation in the mitochondrion (**Figure 2.6**). The phosphocreatine shuttle thus facilitates both temporal and spatial ATP buffering.

Because the recycling of mitochondrial ADP by creatine kinase activates oxidative phosphorylation, the enzyme also regulates rates of mitochondrial respiration (Kay et al. 2000). In turn, by activating oxidative phosphorylation, creatine kinase activity dissipates the proton gradient across the inner mitochondrial membrane, thereby maintaining a depolarized mitochondrial membrane potential ($\Delta\Psi_m$), which in turn reduces the production of ROS (Meyer et al. 2006). Increasing levels of ROS can lead to PTMs of creatine kinase, which either protect the thiol (–SH) group of the active site cysteine through reversible binding of glutathione (S-glutathionylation) or irreversibly deactivate its activity through oxidation of cysteine by peroxynitrite (ONOO–). Peroxynitrite is strongly oxidizing and is one of a group of species called **reactive nitrogen species** (**RNS**), by analogy to reactive oxygen species (Wallimann et al. 2007).

These insights about the multiple functions of the phosphocreatine shuttle are largely based on studies of mammalian cardiomyocytes (heart muscle cells). Few studies have compared this shuttle in muscles of animals with greatly fluctuating performance, such as tunas and hummingbirds, or in hibernating organisms with pronounced metabolic depression but energetically costly periodic arousal episodes. However, a study of the effect of cold-induced hibernation (4°C versus 22°C) on thigh skeletal muscle of the

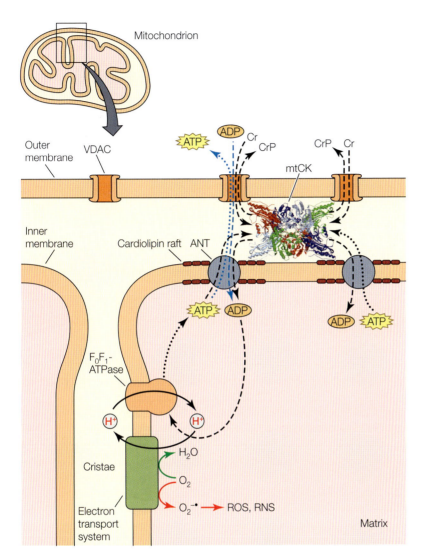

Mitochondrion

Outer membrane

VDAC

Inner membrane

Cardiolipin raft   ANT

mtCK

ADP   Cr

ATP

CrP

CrP   Cr

F_0F_1-ATPase

ATP   ADP

ADP   ATP

H⁺

H⁺

H_2O

O_2

$O_2^{-\bullet} \longrightarrow$ ROS, RNS

Cristae

Electron transport system

Matrix

Figure 2.6 Mitochondrial creatine kinase (mtCK) functions in the channeling of high-energy metabolites. During aerobic metabolism, respiration (green arrow), production of reactive oxygen and nitrogen species (ROS and RNS; red arrow), and synthesis of ATP are tightly coupled to the transphosphorylation of ATP to CrP by mtCK and the export of CrP across the outer membrane through the voltage-dependent anion channel (VDAC). The close proximity to the adenine nucleotide translocator (ANT) is facilitated by rafts of cardiolipin, leading to the assembly of a multiprotein complex of ANT, VDAC, and mtCK and the efficient channeling of ADP back to ATP synthase (F_0F_1-ATPase) for an energetically favorable oxidative phosphorylation. (After Guzun et al. 2012.)

ground squirrel *Spermophilus richardsonii* showed that creatine kinase activity was 20% lower during hibernation than during periods of normal body temperature (euthermia) (Abnous and Storey 2007). In addition, during hibernation the ATP affinity of creatine kinase was increased by PTM (phosphorylation). This PTM led to a reduced Michaelis-Menten constant (K_M; see below for a brief explanation) of the phosphorylated enzyme, which matched the lower levels of adenylates typically found in hibernating animals. By this PTM mechanism, creatine kinase maintained the ability to effectively regulate its reaction rate in response to fluctuating substrate levels under conditions of low substrate concentration. The creatine kinase of the hibernating individuals was completely phosphorylated, but euthermic individuals showed a mix of phosphorylated (low K_M) and dephosphorylated (high K_M) isoforms. This is a clear example of the role that the matching of K_M with substrate concentrations plays in regulating enzymatic activity (see Chapter 3).

As an added layer of complexity in the diversity of phosphagen systems, many species express different isoforms of creatine kinase in different tissues to support distinct metabolic requirements. In mammalian glycolytic fast-twitch muscle, which relies on anaerobic metabolism using glycolysis, the predominant isoform of creatine kinase is the cytosolic muscle-type variant (MM-CK), which is bound to the M band of sarcomeres, and quickly rephosphorylates ADP to supply the ATP demand of myosin-ATPase (Yamashita and Yoshioka 1991). By contrast, slow oxidative and cardiac muscle fibers have lower levels of the cytosolic isoform (MM-CK), but also express a mitochondrial isoform (sMtCK). Thus it seems that the cytosolic isoforms mainly contribute to energy homeostasis in anaerobic muscle, whereas oxidative muscle is characterized by a phosphotransfer shuttle that transports creatine-phosphate from sites of ATP production to sites of ATP use. However, not all oxidative muscles follow this arrangement. For example, cardiac muscle of several Antarctic notothenioid fishes that lack hemoglobin and myoglobin also lack the mitochondrial isoform of CK, despite the fact that the muscle relies heavily on β-oxidation of fatty acids and thus aerobic metabolism. This finding suggests that oxidative phosphorylation functions sufficiently without the phosphocreatine shuttle to quickly replenish the ADP in the mitochondrial matrix and that diffusion of ATP is fast enough to support myocardial function in these cold-adapted fishes (O'Brien et al. 2014).

2.5 Anaerobic Metabolism

As mentioned above, the need to rely on anaerobic pathways of ATP generation can arise from two conditions: limiting oxygen in the environment (environmental hypoxia), and high levels of physiological activity in which ATP needs cannot be sustained through oxygen-based ATP synthesis (physiological hypoxia). Many circumstances can lead to environmental hypoxia in the marine realm, including vertical migration into the oceanic oxygen minimum zone (OMZ) or exposure to air during low tides in intertidal zones, when many species restrict gas exchange with the environment to avoid desiccation. Physiological hypoxia most commonly arises from spikes in locomotory activity, as might occur during prey capture or predator avoidance. During the initial phase of a rise in muscular activity, muscle cells rely on their internal energy resources, namely oxygen-independent substrate-level phosphorylation and phosphagen-based ATP generation, at least until an increase in the delivery of oxygen through the circulatory system can be established. Phosphagens and glycolysis are therefore the major pathways providing ATP during critical periods of increased ATP demand.

Below we discuss how different species and different tissues within a species are adapted to have the correct balance between aerobic and anaerobic pathways of ATP production. This is a dynamic balance in the sense that rapid shifts from one type of ATP production to another must occur as changes in ambient oxygen or activity take place. Furthermore, during the transitions between aerobic and anaerobic metabolism, intertissue coupling of metabolites can be important. For example, the lactate generated by a rapidly working locomotory muscle may enter the bloodstream and then be taken up by the heart for use as a fuel through aerobic reactions, beginning with the conversion of lactate back to pyruvate. Pyruvate can then be directed toward the TCA cycle or be used to synthesize glycogen in preparation for the next episode of anaerobiosis. Thus energy metabolism must be viewed in a holistic and integrative manner.

Glycolysis is an ancient pathway of anaerobic ATP production

Under oxygen-replete conditions, aerobic production of ATP is highly favored in almost all circumstances. The energy of the electrons removed from reduced organic compounds ("food") is very effectively exploited as electrons flow along the ETS to the cytochrome *c* oxidase reaction, where tetravalent reduction of oxygen occurs (see Box 2.1). While recent estimates suggest that the ATP yield from aerobic metabolism is ~32 ATP per glucose (**Box 2.2**), glycolysis may yield only 2 net ATP, albeit some alternative anaerobic pathways such as those found in facultatively anaerobic invertebrates can produce up to 6 ATP per glucose (see Figure 2.4). Therefore, although the rates of anaerobic pathways can be upregulated in many animals in response to falling oxygen levels, this strategy is an inefficient way to produce ATP from existing energy stores (e.g., glycogen) unless the anaerobic end products can be reincorporated into metabolism. Thus, all things being equal, a continuous reliance on aerobic ATP generation would seem optimal.

The continued importance in most contemporary organisms of glycolysis in ATP production (and in the generation of substrates for biosynthetic processes) can perhaps best be understood through a historical analysis. The ten reactions of glycolysis represent

BOX 2.2 ■ ■ ■ ■ ■

ATP yield from different fuels and pathways

The oxidation of glucose, lipids, and amino acids results in different ATP yields (see figure). The first pathway of **glucose oxidation**, glycolysis, produces two NADH and two net ATP when the substrate is glucose and three net ATP when it is glycogen; glycogen phosphorylase does not require an ATP to generate glucose 1-phosphate, which can be converted to glucose 6-phosphate by phosphoglucomutase. The two NADH from glycolysis are transported into the mitochondrial matrix through the

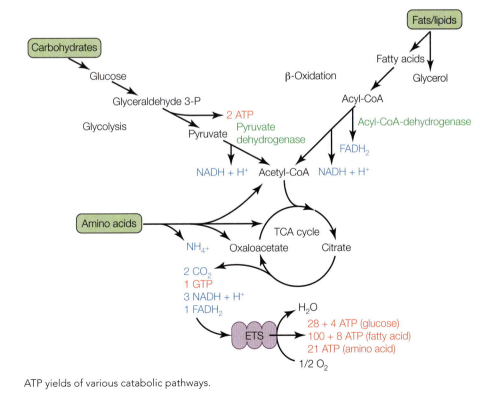

ATP yields of various catabolic pathways.

BOX 2.2 (continued)

malate-aspartate shuttle (see Figure 2.13).

For each pyruvate formed in glycolysis, the pyruvate dehydrogenase reaction generates one NADH and one acetyl-CoA, which can enter the TCA cycle. The TCA cycle generates six NADH, two $FADH_2$, and two GTP, which can be converted to two ATP. The complete reaction is therefore (Chandel 2015):

$$Glucose + 2 H_2O + 10 NAD^+ + 2 FAD + 4 ADP + 4 P_i$$

$$\rightarrow 6 CO_2 + 10 NADH + 6 H^+ + 2 FADH_2 + 4 ATP$$

As we elaborate in the main text, each NADH generates 2.5 ATP and each $FADH_2$ 1.5 ATP. Together with the 4 ATP from substrate-level phosphorylation (glycolysis and the succinyl-CoA synthetase reaction of the TCA cycle), the oxidation of glucose therefore yields 32 ATP.

The **β-oxidation of fatty acids** starts with the transport of the fatty acyl carnitine conjugate into the mitochondrial matrix (see Figure 2.16). For every two carbons of the fatty acid, the four-reaction sequence of β-oxidation generates one NADH and one $FADH_2$. The resulting acetyl-CoA enters the TCA cycle and generates an additional three NADH, one $FADH_2$, and one GTP. Complete β-oxidation of 16-carbon palmitoyl-CoA yields 8 acetyl-CoA, generating 24 NADH, 8 $FADH_2$, and 8 GTP. In addition, the seven rounds of β-oxidation generate seven NADH and seven $FADH_2$:

$$Palmitoyl\text{-}CoA + 7 CoA + 7 FAD + 7 NAD^+ + 7 H_2O$$

$$\rightarrow 8 \ acetyl\text{-}CoA + 7 FADH_2 + 7 NADH \ and \ 7 H^+$$

This generates 31 NADH and 15 $FADH_2$, which can yield 77.5 ATP (31 NADH × 2.5 ATP = 77.5 ATP) and 22.5 ATP (15 × 1.5 ATP = 22.5 ATP), respectively, for a subtotal of 100 ATP. By adding the 8 GTP we get another subtotal of 108 ATP. However, the initial activation of the fatty acid by conjugation to coenzyme A required 2 ATP that we need to subtract, resulting in a grand total of 106 ATP per 16-carbon palmitic acid or about three times the ATP yield from oxidation of one 6-carbon glucose (Chandel 2015).

The ATP yield of **amino acid oxidation** depends on the particular amino acid and its specific entry point into the main catabolic pathways. After deamination or transamination of the amino group (NH_2), the carbon skeleton of the amino acid enters these pathways as pyruvate, acetyl-CoA, α-ketoglutarate, succinyl-CoA, fumarate, or oxaloacetate. The net ATP yield also depends on the excretion pathway of the amino group, with the end product being either ammonia, urea, or uric acid. For example, the oxidation of alanine yields 12.5 ATP, but the excretion of the amino group as urea requires 2 ATP, resulting in 10.5 net ATP. The average ATP yield for animal muscle ("meat") is 21.5 ATP per amino acid (Jungas et al. 1992).

---

an ancient pathway that evolved long before the rise of oxygen allowed ATP to be generated aerobically (**Figure 2.7**). Thus ATP generation in the earliest life forms was of necessity based on the reactions of glycolysis (or other oxygen-independent reactions). Bear with us while we discuss some of the details of this pathway, as this analysis will provide a foundation for the discussion of other pathways, which we will cover in a more abridged fashion.

THE STAGES OF GLYCOLYSIS: INVESTMENT PHASE In the first reaction of glycolysis, **hexokinase** phosphorylates glucose using one molecule of ATP. In animals, cellular glucose may derive from glycogen stores or from uptake from the blood, mediated by several isoforms of glucose transporters. By converting glucose into a charged metabolite, glucose 6-phosphate (G6P), the cell traps the glucose because G6P cannot diffuse out of the cell across the cell membrane. G6P is next converted into fructose 6-phosphate (F6P) by an **isomerase** enzyme. Then another phosphorylation event occurs: F6P is converted to fructose 1,6-bisphosphate (FBP) by **phosphofructokinase 1 (PFK1)**. Thus, up to this point, glycolysis has used two ATP molecules to generate FBP, but no new ATP has been formed. Appropriately, this initial phase of glycolysis is termed the **investment phase**.

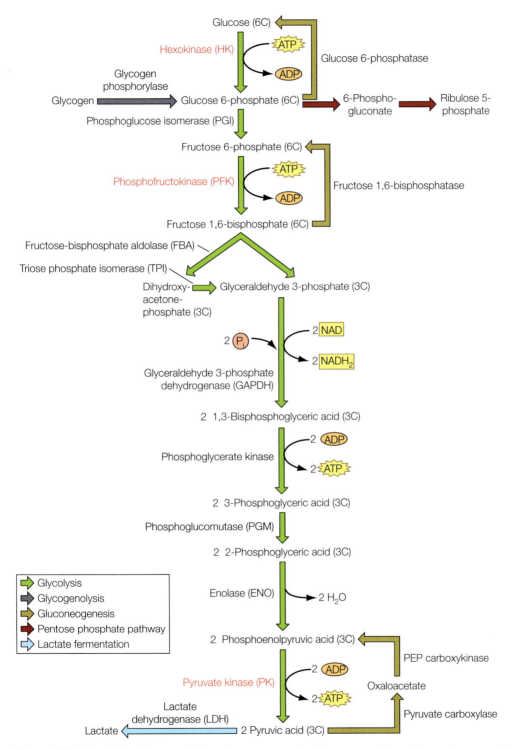

Figure 2.7 Glycolysis, glycogenolysis, gluconeogenesis, the pentose phosphate pathway, and the reaction of lactate dehydrogenase. The reactions of enzymes in red are characterized by a large drop in free energy that makes them irreversible. PEP, phosphoenolpyruvate.

THE STAGES OF GLYCOLYSIS: PAYOFF PHASE   The next reaction, catalyzed by **fructose 1,6-bisphosphate aldolase**, splits the six-carbon FBP into two three-carbon metabolites. One of them, glyceraldehyde 3-phosphate (GAP), is oxidized to 1,2-bisphosphoglycerate by **glyceraldehyde 3-phosphate dehydrogenase (GAPDH)**. The other three-carbon metabolite, dihydroxyacetone-phosphate (DHAP), can be converted to GAP by **triose phosphate isomerase (TPI)**, thus allowing the metabolite to contribute to glycolysis. Alternatively, DHAP can be converted to glycerol 3-P by **glycerol 3-phosphate dehydrogenase** and serve as a backbone for triacylglyceride biosynthesis. The reaction by GAPDH links the oxidation of GAP to another phosphorylation with an inorganic phosphate and the production of NADH, forming a metabolite with two phosphoryl groups, 1,3-diphosphoglyceric acid. In the following reactions, two glycolytic enzymes, **phosphoglycerate kinase (PGK)** and **pyruvate kinase (PK)**, will use phosphorylated metabolites to produce two ATP. In this way, two substrate-level phosphorylations generate a total of four ATP molecules per glucose—a *net yield* of two ATP—in what is termed the **payoff phase** of glycolysis.

Thus the phosphorylations of the investment phase of glycolysis set the stage for the subsequent payoff phase to generate two ATP from each of the two three-carbon metabolites. Importantly, when glycogen is the source of glucose, the cell invests only one ATP, because glycogen phosphorylase requires inorganic phosphate and not ATP to phosphorylate the glucose. Thus, when starting with glycogen, a total of three net ATP is achieved. The end product of these reactions is pyruvate, which can enter the mitochondria and the TCA cycle or be used in fermentative pathways.

GLUCONEOGENESIS AND REACTION REVERSIBILITY   Seven of the ten glycolytic reactions are near-equilibrium reactions: These reactions have small changes in free energy (ΔG) and thus can proceed in either the forward or the reverse direction, depending on the concentration of the metabolites. In contrast to these reversible reactions, those of hexokinase, phosphofructokinase, and pyruvate kinase are characterized by a steep drop in free energy and are thus favored to proceed in the forward reaction. These reactions are essentially irreversible under cellular conditions. During the process of gluconeogenesis, the reversible reactions of glycolysis are readily able to support production of glucose. However, the irreversible reactions require the use of other enzymes that are able to catalyze the glycolytic reaction in reverse through thermodynamically favorable bypasses of the catabolic glycolytic sequence.

The reverse reactions are catalyzed by **glucose 6-phosphatase, fructose 1,6-bisphosphatase, PEP carboxykinase**, and **pyruvate carboxylase** (see Figure 2.7). These reactions are typically activated under energy-replete conditions, when ATP supplies are adequate to allow the bypass reactions to work at a high level. Gluconeogenesis occurs in many tissues and generally relies heavily on the lactate produced in muscle during anaerobic glycolysis. Some of this lactate remains in the muscle, where gluconeogenesis converts it into depot glycogen. Some lactate is released into the blood, which transports it to the liver for uptake and entry into the gluconeogenic and glyconeogenic pathways. However, some fraction of blood-borne lactate is taken up by aerobic tissues like cardiac muscle for use as an energy-generating substrate.

Gluconeogenic reactions must be appreciated as constituting a pathway that is complementary to the catabolic reactions of glycolysis, in that gluconeogenesis uses glycolytic end products like lactate that, following return of oxygen-replete conditions, are present in excess of what is needed to fuel aerobic ATP production. This coupling of the thermodynamically "downhill" and "uphill" reactions of glucose degradation and resynthesis, respectively, ensures that little glucose carbon goes to waste.

Regulation of glycolytic reactions

The regulation of glycolysis—or of any metabolic pathway, for that matter—involves a suite of controls that range from direct alteration of enzyme activity by changing substrate levels to more complex mechanisms like posttranslational modification of proteins, changes in protein subunit assembly, and altered binding of glycolytic enzymes to contractile elements (actin). We begin with what can be regarded as the simplest and most direct type of regulation: changes in enzyme activity in response to alterations in substrate concentrations.

Velocity depends on substrate concentration: The important concept of enzyme-substrate affinity

Here it may be useful to briefly review how the rate (or velocity, V) of a biochemical reaction responds to changes in available substrate concentrations [S]: the $\Delta V / \Delta$[S] relationship (**Figure 2.8**). Rate versus [S] effects are important in a wide variety of contexts in environmental relationships, as discussed in Chapters 3 and 4. Thus the analysis we present in the following paragraphs will serve as a foundation for many later discussions of temperature- and solute-driven changes in metabolic function. Preserving the

(A)

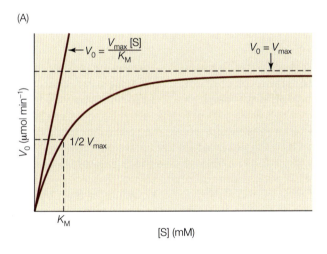

(B)

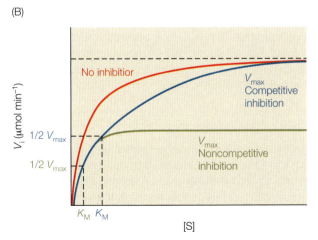

Figure 2.8 Relationship between (A) the velocity of an enzyme-catalyzed reaction and substrate concentration [S], and (B) the effect of inhibitors on this relationship. (A) At very low substrate concentrations (high K_M relative to [S]), the change in V is near-linear with changes in [S]. The slope approximates V_{max}/K_M. With increasing substrate concentration, the contribution of K_M to rate decreases and the velocity converges on V_{max}. (B) Competitive versus noncompetitive inhibition of enzyme reactions. Competitive inhibitors compete with the substrate for binding of the catalytic site and increase K_M without changing V_{max}. Noncompetitive inhibitors bind allosterically and reduce V_{max} without changing K_M.

appropriate $\Delta V/\Delta$[S] relationship is a key conservative type of biochemical adaptation in the face of a wide variety of environmental stressors.

The change in rate with rising substrate concentration is complex. During the initial increase in [S], the reaction rate increases in an approximately linear fashion. However, $\Delta V/\Delta$[S] levels off with increasing substrate concentrations and exhibits an asymptote as substrate concentrations reach their highest value. In terms of geometry, the V versus [S] curve has the form of a rectangular hyperbola. The asymptote of the $\Delta V/\Delta$[S] function is termed the **maximum velocity (V_{max})** of the reaction. Note that the highest velocity measured in an experiment is likely not the true V_{max} of the reaction, because various factors such as product inhibition may preclude the enzyme from attaining its highest possible (in theory) activity. The V_{max} measured or estimated in an experiment is the product of the concentration of enzyme [E] in the reaction medium and the inherent catalytic efficiency of the enzyme, the **k_{cat}** value (measured in units of sec^{-1}). k_{cat} values may be expressed on a per-enzyme basis, albeit for a typical multisubunit enzyme, a per-active-site rate is a more meaningful index of catalytic efficiency. As discussed in the context of temperature adaptation in Chapter 3, adaptive variations in V_{max} may be functions of k_{cat} and/or enzyme concentration.

The relationship between rate (V) and substrate [S] is described by the **Michaelis-Menten equation** (see Figure 2.8A):

$$V = \frac{V_{max}[S]}{K_M + [S]}$$

Here K_M is the **Michaelis-Menten constant**, the substrate concentration at which $V = 1/2\ V_{max}$. A low K_M value denotes a high **affinity**—more effective binding—between the enzyme and substrate. In the context of regulation of enzymatic activities, it is important to realize that K_M values generally are close to or slightly above the concentrations of substrate in the cellular compartment in which the enzyme functions. This pairing of K_M and [S] facilitates a strong regulatory role for changes in [S] because the changes in available substrate levels are in the low-velocity, near-linear range of the Michaelis-Menten curve. As discussed at several junctures in this volume, maintaining K_M values in the optimal range for regulation is a critical feature of biochemical adaptation, one that is achieved through protein evolution (see Chapter 3) and through appropriate adjustments in the milieu in which the enzyme functions in the cell (see Chapter 4).

Whereas enzymes have intrinsic K_M values that reflect their evolutionary histories (e.g., adaptation temperature), acute changes in physical and chemical factors can change K_M as well. Some of these effects are important in regulating enzymatic activity in the cell. For instance, inhibitors that compete with the substrate of the reaction for binding to the active site of an enzyme can increase the K_M value of an enzyme, thereby shifting the rate to lower values (**competitive inhibition**; see Figure 2.8B). Inhibitors that bind to an allosteric site of an enzyme, a site distant from the substrate-binding site (*allo-* is a Greek prefix meaning "different"), can lead to a change in the conformation of the enzyme and a reduction in V_{max} (**noncompetitive inhibition**).

Returning now to the specific roles of different regulatory mechanisms in governing flux through the glycolytic pathway, an increase in intracellular [glucose] will increase the rate of the first reaction of glycolysis, catalyzed by hexokinase, in accord with Michaelis-Menten relationships. The increase in intracellular [glucose] through the uptake of glucose from blood is controlled by three processes: (1) delivery of glucose to the tissue, (2) transport of glucose into the cell by glucose transporters (GLUT4 in skeletal muscle), and (3) phosphorylation of glucose by hexokinase (HK II in skeletal muscle), with the distribution of the control depending on the physiological conditions. For example, whereas overexpression of HK II in skeletal muscle of mice had no effect on glucose uptake during resting conditions, it increased uptake during both exercise and insulin stimulation (Wasserman et al. 2011). Any subsequent increases in substrate concentrations for the other reactions of the pathway will also increase their rates, to the extent that these enzymes also obey Michaelis-Menten kinetics. However, the reactions that are characterized by a large decrease in free energy—namely those catalyzed by hexokinase, phosphofructokinase, and pyruvate kinase—also can be modulated through allosteric inhibition and activation.

Control is differentially distributed among enzymes of glycolysis

Whereas all enzymes in a pathway make some contribution to the pathway's regulation, enzymes that catalyze reactions with large changes in free energy are commonly the most important regulatory sites. These "valve" enzymes monitor the status of the cell—for example, its needs for ATP—and alter their activities accordingly. For example, hexokinase, in common with most enzymes, can be inhibited by its product. Thus, as G6P levels rise, feedback inhibition indicates to hexokinase that a decrease in its activity is warranted. Phosphofructokinase and pyruvate kinase can be inhibited by high concentrations of ATP, acetyl-CoA (a common metabolite of carbohydrate, fatty acid, and amino acid catabolism), and citrate (a metabolite of the TCA cycle that signals to glycolysis that a reduction in pyruvate supply is appropriate). PFK is also regulated by changes in intracellular pH, which leads to alterations in the enzyme's compartmentation in the cell (its binding to actin filaments) and subunit assembly (see Chapter 4). By possessing the ability to "read" the levels of regulatory signals like citrate and pH, PFK is well poised to vary its activity—through a variety of mechanisms—in response to changes in the cell's need for ATP.

All three adenylates—ATP, ADP, and AMP—play important roles in regulating pathways of energy metabolism, including glycolysis. High rates of ATP consumption temporarily increase concentrations of ADP. The enzyme **adenylate kinase** converts two ADP to one ATP and one AMP, which provides some enhancement in ATP availability to the cell. Furthermore, the increase in AMP serves as an activating signal for PFK, which in turn can lead to an increase in the overall rate of glycolysis.

Another biochemical mechanism of regulating the rates of glycolytic reactions is called **feed-forward activation**. The reaction catalyzed by **phosphofructokinase 2**, which is not part of glycolysis (unlike PFK1), converts the glycolytic metabolite fructose 6-phosphate into fructose 2,6-bisphosphate, which activates PFK1. This activation accelerates the conversion of fructose 6-P to fructose 1,6-bisphosphate and thus increases glycolytic flux. Fructose 1,6-bisphosphate activates pyruvate kinase—a feed-forward effect—helping coordinate the early and later steps of the glycolytic sequence.

Another level of regulation is provided by several posttranslational modifications (PTMs) of glycolytic proteins (**Box 2.3**). While it has been known for a long time that the activity of glycolytic enzymes that catalyze irreversible reactions, namely phosphofructokinase and pyruvate kinase, are modified by phosphorylation, several additional PTMs of glycolytic enzymes have been discovered more recently. These include PTMs of enzymes that catalyze near-equilibrium reactions. Thus even enzymes involved in thermodynamically reversible reactions share a role in regulating glycolytic flux (see below). For example, the acetylation of the ε-amino group of internal lysine residues affects almost every glycolytic protein (Guan and Xiong 2011). For the muscle form of pyruvate kinase,

BOX 2.3

Posttranslational modifications and signaling by metabolic pathways

Several posttranslational modifications (PTMs) change the activity, location, or turnover of proteins in response to shifts in metabolism. In addition to phosphorylation, acetylation, methylation, and glycosylation, numerous other PTMs can affect the function of proteins (Ryslava et al. 2013; Walsh 2006; Walsh et al. 2005). Here we illustrate the principal role of metabolites in affecting proteins by focusing on acetyl-CoA, a central metabolite of both carbohydrate and fatty acid catabolism and a key building block for the synthesis of lipids and cholesterol.

High levels of acetyl-CoA in the mitochondria can be transferred to the cytosol and the nucleus through the transport of the TCA metabolite citrate into the cytosol, where the cytosolic **ATP citrate lyase (ACL)** converts citrate into oxaloacetate and acetyl-CoA (Hatzivassiliou et al. 2005). Acetylases use acetyl-CoA as a substrate to acetylate proteins, and deacetylases remove the acetyl group from proteins. Acetylation affects almost all metabolic enzymes and numerous nuclear regulatory proteins, including transcription factors and histones (Zhao et al. 2010). Increasing levels of acetyl-CoA indicate a high energy charge, signaling to the cell the potential for enhancing activities of anabolic pathways, cell proliferation, and differentiation. High levels

of acetyl-CoA in the nucleus trigger acetylation of lysine residues on histones, which leads to loosening of chromatin structure and activation of gene expression. The effect of acetylation on gene expression, a prerequisite for cell proliferation, is thus a way for cells to grow only when there is enough cellular energy available to support the growth of the cell (Metallo and Vander Heiden 2010). High levels of acetyl-CoA also lead to acetylation of many metabolic enzymes, lowering their activity; acetylation indicates a high energy charge and thus no need for a further increase in catabolism and ATP production. Histone deacetylases (HDACs) reverse the activation of gene expression through the deacetylation of histones and the formation of heterochromatin. This process plays an important regulatory role in turtles during the hypometabolic state and in hibernators during the response to cold anoxia and seasonal shortages of food (see main text). **Sirtuins**, another set of proteins in the HDAC family, deacetylate many nuclear, cytosolic, and mitochondrial proteins other than histones. Sirtuins use NAD$^+$ as a substrate to remove the acetyl group, generating O-acetyl-ADP-ribose and nicotinamide. By deacetylating metabolic enzymes, sirtuins generally increase the activity of catabolism for the generation of ATP (Zhao et al. 2010).

it has been shown that lysine acetylation decreases its activity (Xiong et al. 2011). Some of the same lysine residues can be modified by attaching the small peptide **ubiquitin**, which affects the degradation and thus the turnover of the protein (Tripodi et al. 2015). Attachment of the **small ubiquitin-related modifier 1 (SUMO-1)** to glycolytic proteins has been shown to promote glycolysis, possibly by reorganizing the spatial arrangement of the enzymes in the pathway (Agbor et al. 2011). Furthermore, the addition of the antioxidant tripeptide **glutathione** inactivates GAPDH by forming a mixed disulfide bond (Cotgreave et al. 2002). Other glycolytic enzymes, including aldolase, triose phosphate isomerase, phosphoglycerate kinase, enolase, and pyruvate kinase are glutathionylated in response to exposure to pro-oxidants in mammalian liver cells, which affects the activity of at least enolase (Fratelli et al. 2003). As if the complexity of multiple PTMs were not enough, glycolytic enzymes have multiple acetylation and ubiquitylation sites, some of which may be activating whereas others are inhibiting of the enzyme's activity; at some of the residues (e.g., lysines), competition between acetylation and ubiquitylation may occur (Tripodi et al. 2015).

Several glycolytic enzymes change activity through the modification of their subunit composition. The number of subunits may change, as in the case of phosphofructokinase and pyruvate kinase, or the composition of the enzyme may change, such that tissue-specific isoforms are expressed (Al Hasawi et al. 2014; Gupta and Bamezai 2010). In the case of PFK, tetramers are active whereas the monomeric form is inactive (see Chapter 4). Humans have muscle-, plasma-, and liver-specific isoforms of PFK, with the last one being most highly expressed in tumors with a high glycolytic efficiency (Al Hasawi et al. 2014).

Like PFK, pyruvate kinase is also most active as a homotetramer. Of the four different known isoforms of pyruvate kinase (L, R, M1, and M2), two originate from the same gene locus through alternative splicing (M1 and M2). The M2 isoform lacks part of the sequence involved in subunit association and tends to dissociate into dimers when phosphotyrosine signaling pathways are activated downstream of several growth factors. Dissociation of the M2 isoform leads to partial inhibition of glycolysis and accumulation of metabolites for anabolic purposes in proliferating cells (Gupta and Bamezai 2010).

In summary, the pattern that emerges from these studies is that reaction rates of individual glycolytic enzymes are regulated by a variety of mechanisms, including Michaelis-Menten kinetics, product feedback or feed-forward mechanisms, several inhibitory or stimulatory PTMs, subunit composition, and changes in cellular compartmentation and metabolon composition. Now we can address a more complex question: How do changes in rate of a single reaction affect the flux of the entire pathway? Which enzymes are changing in which way to increase flux?

From single glycolytic reactions to overall pathway flux

Given that the three glycolytic reactions of hexokinase, PFK1, and PK are far from equilibrium and essentially irreversible, it has often been hypothesized that changes in the activity or the abundance of these enzymes are key to a change in the overall flux of metabolites through the pathway (J_{max}). In contrast, glycolytic enzymes that operate near equilibrium have been assumed to simply move metabolites in the direction of a pathway's substrate-to-product gradient. Based on this model, changes in the activities of the former enzymes (hexokinase, PFK1, and PK) are often used as indicators for a shift in the overall flux of the pathway. However, the focus on these "rate-limiting steps" contrasts with theoretical considerations developed as part of **metabolic control theory**. This theoretical analysis suggests that all enzyme-catalyzed steps in a pathway can potentially contribute to the regulation of flux. In addition, changes in the activity of a single enzyme have only a small overall effect on the flux of a pathway; that is, their specific contributions to a change in flux—their **flux control coefficients** (the change in pathway flux as a particular enzyme changes activity)—are small (Fell 1997). This leads to the prediction that multiple enzymes have to change their capacity (V_{max}) simultaneously to account for the observed variation in the flux of metabolic pathways (Fell 2000).

In contrast to these theoretical predictions, there is empirical evidence in support of the hypothesis that far-from-equilibrium, essentially irreversible reactions contribute the most to controlling pathway flux. Previously in this chapter we discussed the role of phosphofructokinase (PFK1) in controlling glycolytic activity. We described how PFK's responses in vitro to allosteric modulators—both activators and inhibitors—can lead to severalfold changes in activity of the enzyme. It turns out that PFK represents a paradigmatic instance of how a far-from-equilibrium enzyme can control pathway flux. Evidence that other far-from-equilibrium enzymes play pivotal roles in governing pathway flux has come from many other studies that examined a variety of species, tissues, and cell types. Some studies have looked at metabolic regulation in a single tissue from a range of species that differ in pathway flux rates to infer which steps in a pathway contribute the most to controlling flux. For example, tropical orchid bees (Apidae; Euglossini), which vary 20-fold in body mass among species, show a close correlation between flight properties, mass-specific metabolic rates, and body mass (Darveau et al. 2005b). Due to their higher wing-beat frequencies, smaller orchid bees have higher mass-specific metabolic rates, which have to be sustained by higher rates of glycolysis, the TCA cycle, and the ETS in the thoracic flight muscle (Darveau et al. 2005a). Orchid bees thus provide an ideal opportunity to correlate pathway flux rates to the V_{max} of several enzymes of glycolysis, including those of irreversible (hexokinase and PFK) and reversible (phosphoglucose isomerase) reactions. The enzyme with the closest correlation between enzyme activity and body mass would be expected to have the greatest influence on overall flux rates. After accounting for the phylogenetic relationships among bee species, hexokinase was the only enzyme whose V_{max} values closely tracked mass-specific metabolic rates during flight, suggesting an important role of the first irreversible reaction of glycolysis in determining the overall rate of the pathway flux (**Figure 2.9**).

The important role played by hexokinase in flux limitation in insect flight muscles is further supported by a study that measured the effects of in vivo knockouts of glycolytic enzyme activities on flight performance in *Drosophila melanogaster* (Eanes et al. 2006). Among six enzymes (three nonequilibrium and three near-equilibrium) for which activity was reduced in vivo, only two enzymes (hexokinase and glycogen phosphorylase) showed significant effects on flight performance, and both of these enzymes catalyzed far-from-equilibrium reactions. Insect flight muscle energetics requires glycolysis as part of the obligate aerobic system of ATP production, which is also the case for other aerobically poised muscle tissue types in other species. In fact, in many cases it has been shown that in tissues that are highly dependent on glucose as a fuel, such as cardiac and red skeletal muscle in mammals (Kashiwaya et al. 1994; Wasserman et al. 2011), glycolytic rates are mainly controlled by hexokinase and glucose transport.

However, the pattern of differential contribution between far-from-equilibrium and near-equilibrium reactions to pathway flux goes beyond insect flight muscle and includes a variety of tissues and cell types. In a survey of 27 studies comparing enzyme activities with pathway fluxes in mostly mammalian tissues, principally liver and muscle, activities of most enzymes changed in response to hypoxia and diet-induced shifts in metabolism by a similar factor (to within the experimental error) when each study was looked at in isolation (Fell 2000). However, when initial differences in enzyme activity (more than 100-fold) and subsequent fold changes were large enough (5-fold or more), it emerged that shifts in the overall rates of pathway flux were accompanied by greater changes in activity among the least active far-from-equilibrium enzymes than in the most active near-equilibrium

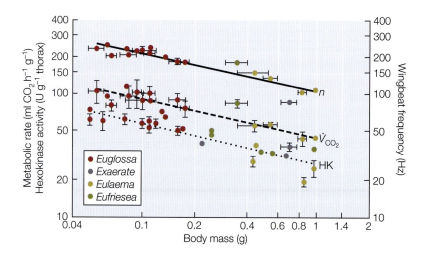

Figure 2.9 Relationship between body mass and orchid bee (Apidae; Euglossini) hovering flight wing-beat frequency (solid line: $n = 106M_b^{-0.31}$, $r^2 = 0.86$), mass-specific metabolic rate (dashed line: $\dot{V}_{CO_2} = 44M_b^{-0.31}$, $r^2 = 0.80$), and hexokinase activity (dotted line: HK $= 27M_b^{-0.33}$, $r^2 = 0.82$) on a log-log scale. (After Darveau et al. 2005a.)

enzymes. Moreover, in comparisons of glycolytic enzyme activities in cancer versus non-cancer cell lines from both rats and humans, it was found that the far-from-equilibrium enzymes hexokinase and PFK1 showed the largest increases in activity in cancer cells and contributed the most to the greater metabolic fluxes in highly proliferating cancer cells (Marín-Hernandez et al. 2006).

In summary, empirical data suggest that far-from-equilibrium "rate-limiting" reactions may contribute more to controlling pathway flux than do near-equilibrium reactions, but this pattern might be context-dependent. The contribution (expressed as the flux control coefficient) of each reaction may be tissue- or species-specific. Within a species and in a given cell type, the relative contributions of various enzyme-catalyzed steps are likely to change under different physiological conditions. We would like to point out that flux control is clearly determined by more complex processes than just changes in transcript and protein abundances, or even enzyme activities as measured by V_{max}. It is therefore necessary to confirm the relative roles of enzymes in controlling flux rates through pathways with metabolic control analysis (Fell 1997) by measuring the response of actual pathway flux to changes in enzyme activities.

The anabolic (biosynthetic) contribution of glycolysis

Energy homeostasis includes the ability of the organism to maintain tissues and individual tissue components that may turn over quickly or may need to be modified in response to stressors. Organisms that are synthesizing new structures—such as pinnipeds (seals) replacing skin and fur during their annual molt, snakes growing digestive tissue in response to episodic prey capture, and all organisms investing in tissue growth for reproduction—must obtain the precursors necessary for cell proliferation, which range from nucleotides to lipids. Changing metabolic demands may require the redirection of pathways to produce these precursors. The building blocks for the nucleotides, amino acids, and lipids needed to support such resource reallocation and tissue growth originate from intermediary metabolites of pathways such as glycolysis, the pentose phosphate pathway, and the TCA cycle (Lunt and Vander Heiden 2011). Exploring the anabolic role of glycolysis will thus help us appreciate how different life-history stages, for example molting, nursing, reproduction, and premigratory fat deposition, affect pathways that we often associate solely with ATP production.

The first such pathway for redirecting glycolytic metabolites starts with the conversion of glucose 6-P to 6-P-gluconolactone by glucose 6-P dehydrogenase (G6PD), which directs glucose toward the pentose phosphate pathway (PPP; see Figure 2.7). The major product of the PPP is D-ribulose 5-P, which serves as a precursor for all five types of nucleotides that form RNA and DNA. The early reactions of the PPP (the oxidative part of the pathway) also produce two NADPH, which can be used in the biosynthetic reactions that convert ribose (RNA) to deoxyribose (DNA) nucleotides (Lunt and Vander Heiden 2011). The NADPH can also be used in other biosynthetic reactions, especially if the D-ribulose 5-P is converted to the glycolytic metabolites fructose 6-P and glyceraldehyde 3-P by transketolase and transaldolase, respectively, instead of being used to synthesize nucleotides. The full recycling to glucose 6-P through the reverse glycolytic reaction of phosphoglucose isomerase generates a cycle that leads to the conversion of glucose 6-P to 6 CO_2 and 12 NADPH. Thus the PPP has long been viewed as a major source for NADPH; however, as we will see, other pathways also contribute to the production of this reducing equivalent.

The next metabolite that contributes to the biosynthetic capacity of glycolysis is dihydroxyacetone-phosphate (DHAP), which can be converted into glycerol 3-P, the precursor of membrane phospholipids and triacylglycerides for fat storage in adipocytes.

Other glycolytic metabolites (3-phosphoglycerate and pyruvate) can be converted to the amino acids serine, cysteine, glycine, and alanine. Glycolysis is therefore an important source for some of the amino acids that are needed for protein synthesis. Cysteine and glycine are also precursors for the antioxidant tripeptide glutathione, which additionally requires glutamate.

The role of glycolysis during hypoxia: Interspecific variation

Whereas the reduction of pyruvate to lactate is the common, "textbook" illustration of how the glycolytic pathway contributes to ATP generation under conditions of limiting oxygen, there are several enhancements to the glycolytic pathway that increase its ATP yield. Whereas vertebrates, as well as some invertebrates, rely on LDH as the terminal enzyme in glucose breakdown under conditions of limiting oxygen, many other invertebrates have significantly embellished the core glycolytic pathway, with the result that the yield of ATP per glucose is considerably higher (**Figure 2.10**). These alternative and more efficient pathways are prevalent in organisms that experience hypoxic or anoxic conditions in their environments for considerable periods of time. These pathways are associated with long-term environmental hypoxia rather than physiological (functional) hypoxia, which is commonly a short-term (timescale of minutes) phenomenon associated with bursts of muscular activity.

Long-term environmental hypoxia: ATP needs and buffering of pH

Environmental hypoxia that persists for many hours is common during periods of low tide in intertidal organisms that are either sessile or have limited abilities to escape hypoxia. Under environmental hypoxia, many invertebrate species, including molluscs, cnidarians, and annelids, condense pyruvate with an amino acid, thereby converting NADH to NAD^+, in a reaction that forms **imino acids** as end products (see Figure 2.10). These condensation reactions are catalyzed by amino acid-specific dehydrogenases. **Octopine dehydrogenase** (ODH) (see reaction at left) forms the imino acid octopine from pyruvate and arginine; the reaction between pyruvate and alanine catalyzed by **alanopine dehydrogenase** yields alanopine; and the reaction of pyruvate and glycine, catalyzed by **strombine dehydrogenase**, yields **strombine**.

Pyruvate + Arginine → Octopine (catalyzed by ODH, NADH/H^+ → NAD^+ + H_2O)

These reactions of the so-called **opine pathways** serve one function in common with LDH by maintaining the supply of NAD^+ in support of anaerobic glycolysis during the early phase of hypoxia. Whereas the opine pathways produce only two ATP per glucose—in common with the LDH-terminated pathway—the imino acids produced are less acidic than lactate, making it easier for the organism to accumulate high concentrations of these metabolic end products without perturbing intracellular pH (Grieshaber et al. 1994).

Although opine production does not lead to increased yields of ATP relative to lactate production, some facultatively anaerobic invertebrates have further embellished glycolysis to allow several additional ATP to be generated per glucose molecule catabolized

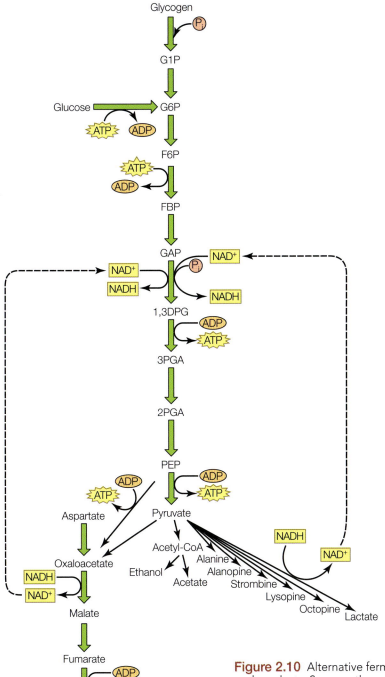

Figure 2.10 Alternative fermentative pathways that yield different end products. Some pathways—for example, that leading to propionate formation—may increase the yield of ATP per glucose molecule catabolized. Some marine invertebrates that regularly experience hypoxic or anoxic conditions condense pyruvate with an amino acid, thereby forming imino acids such as alanopine, strombine, lysopine, and octopine that are less acidic than lactate and thus can accumulate to higher levels.

(see Figure 2.10). We will examine these higher-yield anaerobic pathways after we review the reactions of the TCA cycle. Before we move on, however, it is important to examine another facet of anaerobiosis in vertebrates, to illustrate that fermentation to lactate is not the only route for sustaining ATP production under conditions of environmental hypoxia.

Anaerobic ATP production in oxygen minimum zone fishes

One of the most common types of environmental hypoxia is experienced by marine animals that migrate between oxygen-replete shallow waters and the deeper waters of **oxygen minimum zones** (**OMZs**), which generally occur at depths of ~200 m to ~1000 m in coastal areas with high productivity (see Chapter 5). OMZs originate from the microbial decomposition of sinking organic matter, which leads to major depletion of dissolved oxygen in these regions of the water column. Movements of animals into and out of the OMZ may occur ontogenetically, but the larger share of these movements occurs on a daily basis: Many species seek the deeper, darker, and thus safer waters of the OMZ during the daylight hours, but return to shallower, food-rich waters at night to feed. These marine vertical migrators—which comprise what is known as the **deep-scattering layer**—account for the largest migratory movements in the biosphere. These vertical migrations are of critical importance in transfer of organic materials through the marine water column, and the enormous number of animals that are able to function across a wide range of oxygen concentrations indicates a widespread ability to cope with environmental hypoxia.

Lanternfish

Lanternfishes (family Myctophidae; see photo at left) are the most abundant fishes in the sea and likely are the most numerous group of vertebrates in the biosphere. They occur worldwide and form a significant portion of the deep-scattering layer (Robinson et al. 2010). As vertical migrators, lanternfishes encounter different oxygen versus depth relationships in different regions of the ocean. Low oxygen does not preclude the presence of these fishes, for they enter some of the most hypoxic conditions found in the midwater regions. For example, in the Arabian Sea water becomes anoxic near 200 m depth and remains so to a depth of more than 1000 m (**Figure 2.11A**). In contrast, oxygen levels in the Gulf of Mexico drop to half-saturation levels over the same depth profile (**Figure 2.11B**).

Studies of activities of enzymes associated with anaerobic and aerobic pathways of ATP generation have provided insights into the biochemical adaptations of lanternfishes that allow their diurnal migrations into regions of low oxygen. A comparison of activities of lactate dehydrogenase, an indicator of anaerobic capacity; citrate synthase (CS), an indicator of aerobic capacity; malate dehydrogenase (MDH), an indicator of both aerobic and anaerobic potential; and alcohol dehydrogenase (ADH), an indicator of capacity for ethanol production, showed that lanternfishes have a pronounced ability for a fermentative metabolism that yields ethanol as an end product. Arabian Sea lanternfishes had a greater capacity for glycolysis and for the conversion of pyruvate to ethanol, but a lower aerobic capacity than fishes from the Gulf of Mexico (**Figure 2.11C**) (Torres et al. 2012). In addition, the ability to exchange reducing equivalents between the cytosol and mitochondria, as indexed by MDH activity, was greater in lanternfishes from the Arabian Sea than in those from the Gulf of Mexico.

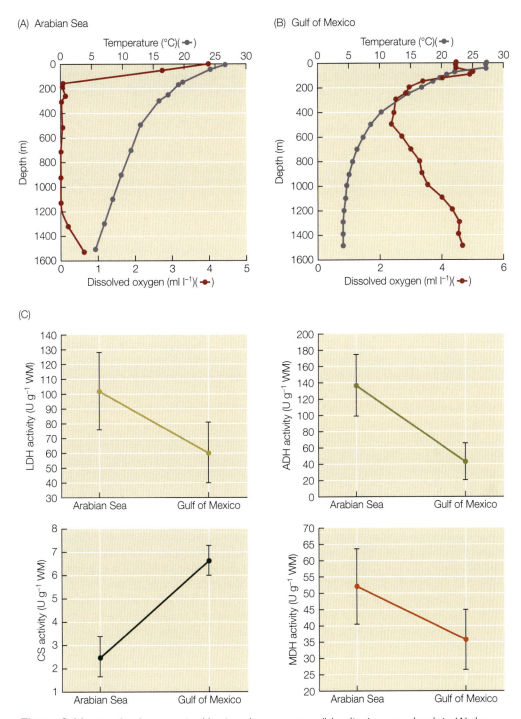

Figure 2.11 Dissolved oxygen (red line) and temperature (blue line) versus depth in (A) the Arabian Sea and (B) the Gulf of Mexico. (C) Activities of lactate dehydrogenase (LDH), alcohol dehydrogenase (ADH), citrate synthase (CS), and malate dehydrogenase (MDH) in epaxial white muscles of lanternfishes from the Arabian Sea and the Gulf of Mexico. Shown are mean values ± 95% confidence limits. All means were significantly different based on an analysis of covariance (ANCOVA; $P < 0.05$). WM, wet mass. (After Torres et al. 2012.)

The conversion of glycolytically generated pyruvate to ethanol is initiated by the decarboxylation of pyruvate to acetaldehyde by the enzyme pyruvate decarboxylase (see reaction scheme below) (Shoubridge and Hochachka 1980). ADH then catalyzes the conversion of acetaldehyde to ethanol, using glycolytically generated NADH in the process. This reaction thus restores the levels of NAD$^+$, allowing further glycolytic activity to occur. The end product of the ADH reaction, ethanol, can be released into the blood and is readily excreted across the fishes' gills. Interestingly, the capacity for ethanol formation is concentrated in the fishes' skeletal muscle even though several tissues, including brain, liver, and heart, also produce lactate. Inter-tissue shuttling of lactate thus plays an important role in the metabolic balance of these fishes, much as we discussed earlier for liver-based gluconeogenic reactions fueled by lactate released from working muscle.

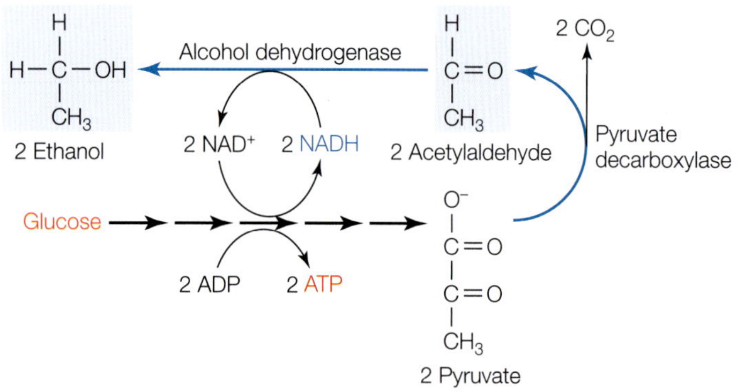

Reaction scheme: H–C–OH (CH$_3$), 2 Ethanol ← Alcohol dehydrogenase ← 2 NAD$^+$ / 2 NADH ← C=O (CH$_3$), 2 Acetylaldehyde ← Pyruvate decarboxylase → 2 CO$_2$; Glucose → (2 ADP / 2 ATP) → 2 Pyruvate (C=O, C=O, CH$_3$).

The importance of ADH in conferring anoxia tolerance was initially described for two freshwater fish species, carp and goldfish, that can survive for months during the winter in ice-covered ponds with limited or no oxygen (Shoubridge and Hochachka 1980). The study of marine lanternfishes highlights the broader ecological importance of this alternative anaerobic pathway in vertebrates. A capacity for ethanol-generating fermentation enables this group of fishes to escape during daylight hours to hypoxic waters that cannot be tolerated by most of their predators. This ability to exploit the dark waters of the OMZ no doubt contributes importantly to the extraordinary abundances of these vertically migrating fishes.

2.6 Aerobic Metabolism

We now return to our examination of aerobic metabolism, in which the use of molecular oxygen permits a vastly higher yield of ATP per substrate molecule and leads to the production of the nontoxic end product, water. These higher yields of ATP are, however, coupled with the challenges posed by the generation of ROS. As a consequence, enhanced capacities for aerobic metabolism must be paired with parallel increases in capacities for dealing with the "bad" side of oxygen, ROS formation. We begin this analysis by reviewing some of the basic structural and functional characteristics of the mitochondrion, the eukaryotic cell's "power plant," where the largest share of oxygen use occurs.

Mitochondrial metabolism

Mitochondria originated as a symbiosis between an α-proteobacterium and an archaeal host (Spang et al. 2015). This conclusion is supported by the finding that the majority of genes encoding proteins of eukaryotic energy metabolism are more closely related to bacterial genes than to those of archaea; the opposite pattern is found for genes encoding proteins involved in translation (Müller et al. 2012). Thus the consequences of this symbiosis are deeply engrained in eukaryotic metabolism.

Mitochondria are characterized by an intermembrane space between an outer and inner membrane, and by a matrix region lying inside the inner membrane. Infoldings of the inner membrane form **cristae** (*crista* is a Latin term for "tuft" or "plume") that reach into the matrix and serve to increase the surface area of the inner membrane. As the major metabolic pathways of mitochondria are localized either in the matrix or are associated with the inner membrane, the substrates of these pathways—for example, pyruvate, several amino acids, and fatty acids—and the metabolites that are being exported from these pathways need to be transported across two membranes. The exchange of metabolites between the cytosol and the intermembrane space is controlled by the voltage-dependent anion channel (VDAC), a **porin** with low selectivity, and the exchange across the inner membrane into the matrix is regulated by the adenine nucleotide translocator (ANT), which mainly controls the exchange of ATP and ADP, and various other metabolite-specific transporters (see Figure 2.6). Importantly, the reducing equivalent produced in the cytosol by glycolysis, NADH, is transported into the matrix through the **malate-aspartate shuttle (Figure 2.12)**.

The other end product of glycolysis, pyruvate, is oxidized (decarboxylated) by the pyruvate dehydrogenase (PDH) reaction, forming NADH and acetyl-CoA. Acetyl-CoA feeds its two carbons into the TCA cycle by condensing with oxaloacetate through the

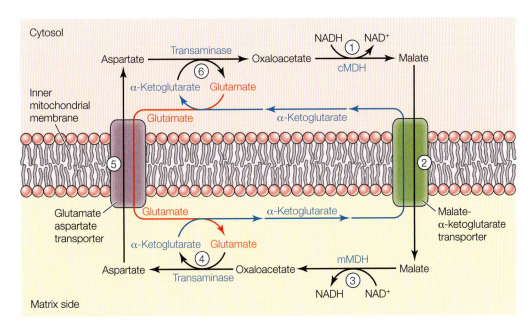

Figure 2.12 The malate-aspartate shuttle transports reducing equivalents (NADH) between the cytosol and mitochondrion. NADH enters the intermembrane space of mitochondria through the VDAC (not shown). The cytosolic isoform of malate dehydrogenase (cMDH) uses NADH to reduce oxaloacetate to malate [1], which enters the mitochondrial matrix in exchange for α-ketoglutarate (through the malate-α-ketoglutarate transporter; [2]). The mitochondrial isoform of MDH (mMDH) reverses the previous reaction [3], and aspartate aminotransferase converts glutamate and oxaloacetate through a transamination reaction [4] to aspartate and α-ketoglutarate. Aspartate can in turn be transported into the cytosol by the glutamate-aspartate transporter [5]; there, aspartate is converted back into oxaloacetate [6], completing the cycle.

Figure 2.13 The pyruvate dehydrogenase reaction and the eight reactions of the TCA cycle. Decarboxylations, yielding CO_2 and NADH, occur in three reactions (highlighted in red): pyruvate dehydrogenase, NAD^+-isocitrate dehydrogenase, and α-ketoglutarate dehydrogenase.

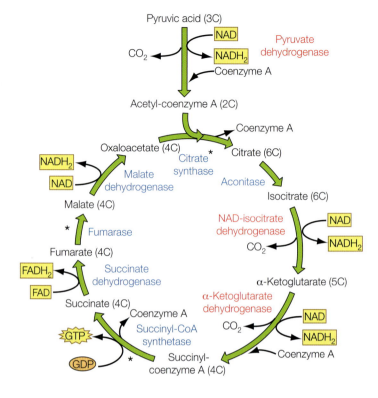

citrate synthase reaction to form citrate (**Figure 2.13**). This reaction can only function if any TCA cycle metabolite that is removed from the TCA cycle for anabolic processes is subsequently replaced. The cycle can be replenished using the carbon skeletons of amino acids such as glutamine, which, after conversion to glutamate, enters the cycle as α-ketoglutarate, in a process called glutaminolysis. Despite the common emphasis on the role of the TCA cycle in aerobic ATP production, it is likely that the reactions of the TCA cycle evolved before aerobic metabolism existed and initially served mainly anabolic functions, such as supplying α-keto acids for amino acid biosynthesis (Hochachka and Somero 2002).

The reducing equivalents NADH and $FADH_2$ enter the ETS at complex I (NADH dehydrogenase) and complex II (succinate dehydrogenase), respectively (**Figure 2.14**). Both complexes donate their electrons to ubiquinone (Q). Complex III (cytochrome *c* reductase) moves electrons from Q to cytochrome *c*. Finally, complex IV (cytochrome *c* oxidase) transfers electrons to oxygen to form water. The movement of electrons through the ETS is accompanied by the transport of H^+ by complexes I, III, and IV across the inner membrane into the intermembrane space, building a chemical and electrical gradient called the **proton motive force** (Δp), which is used by the F_0F_1-ATP synthase in a process called oxidative phosphorylation to generate ATP.

In addition to glycolysis, the other metabolic pathway that generates acetyl-CoA for the TCA cycle is the β-**oxidation of fatty acids** (see Box 2.2). After lipases hydrolyze triacylglycerides into glycerol and fatty acids, fatty acyl-CoA synthetase activates the fatty acids to fatty acyl-CoAs. The latter are transported into the mitochondrial matrix as carnitine conjugates by carnitine acetyl transferase (CPT1). The conjugate passes across the inner membrane through a carnitine-acylcarnitine translocase, an

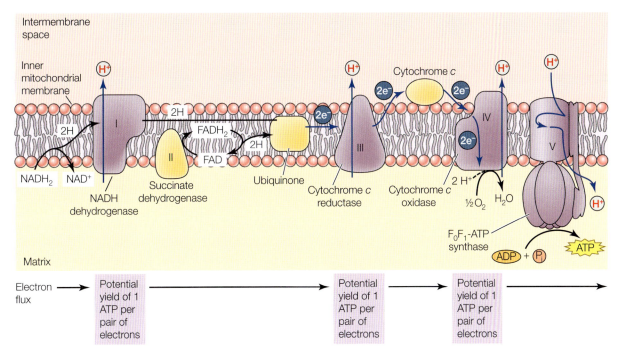

Figure 2.14 The electron transport system (ETS) starts with oxidation of NADH to NAD$^+$ through the reaction catalyzed by NADH dehydrogenase, complex I of the ETS. Because NADH delivers its electrons to complex I, the transport of its electrons leads to the pumping of protons across the inner membrane by complexes I, III (both four H$^+$), and IV (two H$^+$), for a total of ten H$^+$. The reaction of succinate dehydrogenase (SDH) generates FADH$_2$. SDH is also part of the ETS (complex II) and delivers electrons (from FADH$_2$) to ubiquinone and subsequently to complex III, skipping complex I. It causes the transport of six protons across the inner membrane. The transport of these H$^+$ is subsequently used to generate ATP by the F$_0$F$_1$-ATP synthase, which requires the transport of four H$^+$ back into the matrix to generate one ATP from ADP + P$_i$. Thus each NADH can generate 2.5 ATP, and each FADH$_2$ 1.5 ATP at the ATP synthase. The reaction of succinyl-CoA synthetase directly generates guanosine triphosphate (GTP), which can be converted to ATP. A very steep proton gradient can inhibit electron transport and lead to a univalent transfer of electrons to oxygen instead, leading to the production of ROS at complexes I and III.

antiporter, where the carnitine is replaced with coenzyme A by CPT2. The carnitine can then be exchanged with another conjugate by the translocase. Fatty acyl-CoAs are subsequently degraded to acetyl-CoA by the β-oxidation reactions. After each round of the sequence of four reactions making up β-oxidation, the fatty acid is two carbons shorter; thus the oxidation of an entire fatty acid requires multiple rounds of β-oxidation. Each round also produces one FADH$_2$ and NADH. The acetyl-CoA produced by β-oxidation can subsequently enter the TCA cycle and produce additional NADH and FADH$_2$.

The acetyl-CoA from β-oxidation can also be used to produce the ketones acetone, acetoacetate, and β-hydroxybutyrate. In vertebrates, the production of ketones occurs mainly in the liver during starvation, when metabolites from the TCA cycle are used for gluconeogenesis to provide glucose to a variety of tissues. If blood glucose levels decrease to critical levels, the brain and other tissues that cannot use fatty acids but can use ketones rely on the latter substrates to produce acetyl-CoA, which can enter the TCA cycle to produce ATP via oxidative phosphorylation.

Amino acids from the digestion of dietary proteins or from the degradation of cellular proteins produce ammonium (NH_4^+) and a carbon skeleton (see Box 2.2). The ammonium is either incorporated into amino acids or nucleotides or is excreted as urea or another nitrogenous end product, depending on the species in question. The carbon skeleton produced enters catabolic pathways either as pyruvate or one of several TCA cycle intermediates, such as α-ketoglutarate, succinyl-CoA, fumarate, or oxaloacetate, illustrating the central role of the TCA cycle in the catabolism of amino acids (see Figure 2.13). Importantly, during starvation the conversion of amino acids to TCA cycle intermediates can give rise to the production of glucose through gluconeogenesis and of fatty acids through the export of citrate to the cytosol and its conversion to acetyl-CoA and oxaloacetate by ATP-citrate lyase (**Figure 2.15**). While the oxaloacetate can be used for the synthesis of glucose, the acetyl-CoA can be used for the biosynthesis of fatty acids.

Metabolic strategies that exploit the reactions of the TCA cycle during prolonged exposure to hypoxia

As mentioned above, the ability of facultative anaerobes to survive long periods of environmental hypoxia is commonly based on metabolic pathways other than the well-known pathway leading to lactate. These alternative pathways comprise the core of the glycolytic reactions plus additional biochemical reactions that lead not only to higher yields of ATP per glucose but also to end products that, unlike lactate, do not create problems for pH regulation. Below we examine some of these extended anaerobic pathways, with a focus on intertidal animals.

Intertidal invertebrates, for example, mussels of the genus *Mytilus*, may face prolonged hypoxia during low tide, because they close their shells to reduce water loss. To maintain energy homeostasis, these animals need to employ strategies that generate adequate amounts of ATP but do not lead to accumulation of metabolic end products like lactate that can lead to pronounced acidosis. We saw earlier that marine invertebrates can use

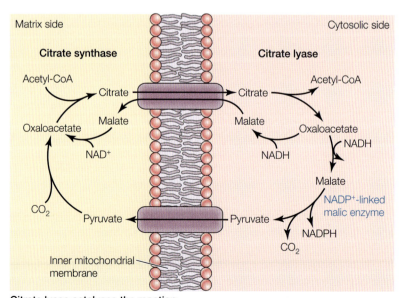

Figure 2.15 Transfer of acetyl-CoA from the mitochondria to the cytosol via citrate.

Citrate lyase catalyses the reaction
Citrate + ATP + CoA-SH + H_2O → acetyl-CoA + ADP + P_i + oxaloacetate

opine pathways to generate ATP during hypoxic conditions (see Figure 2.10), generating end products, opines (imino acids), that are less acidic than lactate.

Some invertebrates enhance their glucose-derived glycolytic activity by exploiting other substrates that can be metabolized through glycolysis. In particular, these animals use a pathway that links the anaerobic production of pyruvate by glycolysis to the amino acid pool, specifically aspartate, producing alanine and succinate as end products (Müller et al. 2012). In this way, the large glycogen reserves that characterize certain marine invertebrates and their abundant intracellular amino acid pools (see Chapter 4) can both be used to help support ATP production during hypoxic and anoxic conditions.

To illustrate how these alternative anaerobic pathways operate during hypoxia, let's look more closely at the anaerobic pathways used by a species of blue mussel (*Mytilus edulis*; Figure 2.16). To begin, two linked amino transferase reactions convert pyruvate into alanine, and aspartate into oxaloacetate. Oxaloacetate is converted to malate by the cytosolic form of malate dehydrogenase and the malate produced is transported into the mitochondrion by the malate-aspartate shuttle. Within the mitochondrion, malate is converted to succinate through the reverse reactions of the TCA cycle enzymes fumarase and fumarate reductase. During prolonged hypoxia or anoxia, propionate may be produced (see Figure 2.16). However, a portion of the malate is oxidized to acetyl-CoA, producing two NADH from the reactions catalyzed by malic enzyme and pyruvate dehydrogenase. The reaction catalyzed by acetate:succinate-CoA transferase transfers coenzyme A to succinate, producing acetate and succinyl-CoA. The latter can be converted back to succinate by succinyl-CoA synthetase, an abundant TCA cycle enzyme, generating one ATP.

The electrons for the reduction of fumarate come from complex I and rhodoquinone, which is analogous to but has a much lower redox potential than ubiquinone; the reducing equivalents (NADH) are provided by the acetate pathway. Thus malate provides both the reducing equivalents for complex I as well as the acceptor for the electrons (fumarate; see Figure 2.16). As a consequence, the simultaneous oxidation (to acetate) and reduction (to succinate) of malate, referred to as malate **dismutation**, can maintain the redox balance of the tissue, if acetate and succinate (propionate) are produced in a ratio of 1:2. Malate dismutation also generates a proton gradient through complex I, which can be used by the F_0F_1-ATP synthase to produce additional ATP. Furthermore, during prolonged anaerobiosis, succinate is first converted to methylmalonyl-CoA and then decarboxylated by propionyl-CoA carboxylase, yielding an additional ATP. Together these reactions yield two ATP from glycolysis, one each (for each pyruvate separately) from the formation of acetate and propionate, and an additional one from the synthesis of ATP by the F_0F_1-ATP synthase, generating a total of six ATP per glucose. Thus blue mussels and a variety of other eukaryotic facultative anaerobes, including flatworms (*Fasciola*), roundworms (*Ascaris*), lugworms (*Arenicola*), and peanut worms (*Sipunculus*), can exploit part of the metabolic machinery of the mitochondrion to produce additional ATP, an important biochemical adaptation to long-term environmental hypoxia (Müller et al. 2012).

In summary, organisms that are capable of coping with long-term environmental hypoxia typically embellish the reactions of the glycolytic sequence with additional biochemical systems that can increase the yield of ATP per glucose molecule catabolized. Intermediates of the TCA cycle play critical roles in adaptation to prolonged environmental hypoxia, which illustrates the multiple functions of the TCA cycle, that is, its roles in addition to the "conventional" contributions it makes to aerobic metabolism. Finally, it is important to realize that, in stark contrast to physiological hypoxia, which generally is a short-term (timescale of minutes) phenomenon that involves very high outputs of

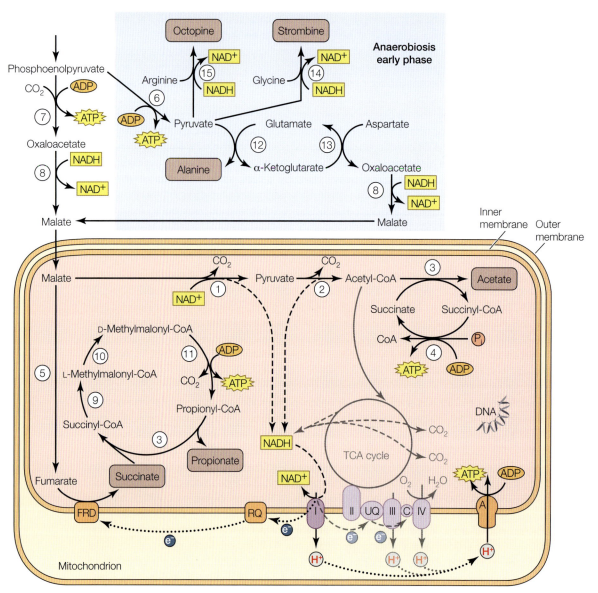

Figure 2.16 Alternative anaerobic metabolic pathways in the cytosol and mitochondria of a sessile intertidal bivalve, the blue mussel (*Mytilus edulis*). This mussel experiences hypoxia and, at times, anoxia during low tide when it is emersed and exposed to aerial conditions. The blue area depicts biochemical reactions during the early phase of hypoxia; these are mainly cytosolic reactions. During prolonged anaerobiosis, both succinate and propionate are produced. The electrons for the reduction of fumarate by fumarate reductase come from rhodoquinone (analogous to ubiquinone) and reducing equivalents (NADH) produced in the reactions of malic enzyme [1] and pyruvate dehydrogenase [2], producing acetate. The protons pumped by complex I (NADH dehydrogenase) can be used by F_0F_1-ATP synthase, which produces one ATP for four protons. I–IV, respiratory complexes I–IV; UQ, ubiquinone; RQ, rhodoquinone; C, cytochrome *c*; A, F_0F_1-ATP synthase; FRD, fumarate reductase; [3], acetate:succinate-CoA transferase; [4], succinyl-CoA synthetase; [1], malic enzyme; [2], pyruvate dehydrogenase complex; [5], fumarase; [6], pyruvate kinase; [7], phosphoenolpyruvate carboxykinase (ATP-dependent); [8], malate dehydrogenase; [9], methylmalonyl-CoA mutase; [10], methylmalonyl-CoA epimerase; [11], propionyl-CoA carboxylase; [12], alanine aminotransferase; [13], aspartate aminotransferase; [14], strombine dehydrogenase; [15], octopine dehydrogenase. (After Müller et al. 2012.)

work, environmental hypoxia typically is characterized by low work output that can be sustained over time periods of many hours and, in some cases, many days.

2.7 Evolution of Metabolic Systems: Genetic Aspects

We now turn to examples of how the core metabolic machinery used by cells to produce ATP and biosynthetic intermediates evolves to enable organisms to function well under different constellations of environmental conditions. Studies of a variety of species shed light on the intricacies of metabolic evolution, where capacities for both catabolic and anabolic activities are conserved by a wide array of genetic mechanisms (Zera 2011).

Intergenomic coadaptation of metabolic enzymes in the copepod *Tigriopus californicus*

Genetically differentiated populations can be the first step in the evolution of a new species. For the genetic differentiation to last, it is often necessary for genetic variation between populations to be selectively advantageous and that any introgression between populations results in reduced fitness. This is expressed as **hybrid breakdown**, a breakdown in the fitness of the F_2 hybrid generation in crosses between populations.

Populations of the intertidal harpacticoid copepod *Tigriopus californicus* (shown at right) that inhabit neighboring rocks yet exhibit genetic differentiation that is stable over time offer an ideal system to study the molecular mechanisms of hybrid breakdown. In particular, studies on *Tigriopus* point to a role for metabolic enzymes and their interactions as a potential driver of population divergence and as contributors to the postzygotic isolation of populations adapted to different environments (Burton et al. 2006).

Tigriopus californicus

The mitochondrial genome encodes 13 proteins that are subunits of several of the protein complexes of the ETS. These subunits interact with other subunits that are encoded by the nuclear genome, and thus any amino acid variation in one of the two genomes might affect a critical subunit interaction of one of the ETS complexes. Because of the maternal, essentially clonal, inheritance of the mitochondrial genome, the need to conserve interactions between proteins encoded by the two genomes can become an important reproductive isolation factor if mitochondrial genomes differentiate quickly relative to nuclear genomes. In the case of *Tigriopus*, mitochondrial sequences of the cytochrome oxidase I (COXI) subunit of cytochrome *c* oxidase (complex IV) diverged by more than 18% between populations on the north and south side of Santa Cruz Island in California (Burton et al. 2006). Furthermore, amino acid replacement rates were about tenfold higher in mitochondrially encoded compared with nuclear-encoded proteins. In experimental crosses, interpopulation crosses (as compared with intrapopulation control crosses) showed reduced survivorship in response to hyperosmotic stress, slower development, reduced fecundity, and lower viability in the F_2 generation, suggesting hybrid breakdown. Based on these results, the researchers hypothesized that **intergenomic (nuclear and mitochondrial) coadaptation** had occurred in the genetically distinct populations of *Tigriopus*. As a consequence, hybridization between populations would disrupt function and trigger hybrid breakdown.

This hypothesis was tested by looking into the interaction between nuclear-encoded cytochrome *c* (CYC) and the three mitochondria-encoded subunits COXI, COXII, and COXIII, which are critical to the function of cytochrome *c* oxidase. Together the COX genes differ by dozens of amino acid substitutions between populations ranging from San

Diego (SD) to Santa Cruz (SC) in California. The nuclear-encoded CYC, a small protein of 104 residues, showed up to five amino acid substitutions among natural populations of *Tigriopus*. This variation raised the question of whether the CYC proteins would perform better when matched with a partner from the same population than they would when matched with variants from other populations. When different isoforms of COX and CYC were coexpressed in *E. coli*, population-matched pairs indeed led to better performance of cytochrome *c* oxidase than did pairs of variants from different populations (Rawson and Burton 2002).

The variation of the CYC gene between the SD and SC populations is limited to three amino acid substitutions. Therefore the coadaptation between the COX and CYC gene products must be due to one or a combination of these three substitutions. By using site-directed mutagenesis and expressing the variants in *E. coli*, it was possible to evaluate the effect of each combination of substitutions on the activity of cytochrome *c* oxidase (Harrison and Burton 2006). These experiments identified one substitution that was mainly responsible for the higher COX activities when population-matching combinations of COX were used: the substitution of a lysine with a glutamine between the SD and SC CYC variants. The associated change in the charge state of the residue presumably modifies the interaction of CYC with the COXII subunit. However, other combinations of substitutions showed an effect on cytochrome *c* oxidase activity that was dependent on the assay temperature, suggesting a gene-by-environment interaction. Furthermore, the coadaptation of mitochondrial and nuclear genomes within a population was associated with higher rates of ATP production, higher larval survivorship, and faster developmental rate in comparison with values of populations that carried hybrid features of these genomes (Ellison and Burton 2006). Finally, it is possible that this kind of coadaptation contributes to lower rates of oxidative DNA damage among experimental lines of native in comparison with hybrid populations, as any disruptions of the electron flow due to a mismatch of genes can easily increase the production of ROS (Barreto and Burton 2013).

In summary, studies of *Tigriopus californicus* have provided important insights into the interplay between proteins encoded by different genomes—nuclear genes and mitochondrial genes—in adaptation to the environment. These studies reflect the complexities associated with the processes of local adaptation and hybridization. In Chapter 5 we will examine this species in additional contexts to show how interpopulation differences in their genomes provide insights into adaptive potentials for coping with global change.

Moonlighting functions of proteins

In addition to affecting metabolic flux, metabolic enzymes and their metabolites are also known for their roles in several so-called moonlighting functions that are different from

their roles in intermediary metabolism (Kim and Dang 2005; Marden 2013). In other words, metabolic enzymes can have pleiotropic effects that influence fitness in multiple ways.

One example of a key interaction of a metabolic protein, its metabolites, and a signaling pathway has emerged from the study of lowland populations of the Glanville fritillary butterfly (shown at left) in Finland (Marden et al. 2013). Initial studies showed that polymorphisms for phosphoglucose isomerase (PGI) and succinate dehydrogenase (SDH) affected the maximum metabolic rate during flight and the dispersal potential in these butterflies. Furthermore, the polymorphism in these two enzymes was also able to explain 60% of the variation in growth

Glanville fritillary butterfly

rate among local populations of this species. By what mechanisms can these enzymes achieve such wide-ranging effects?

At least in the case of SDH, it is known that the variants of the D subunit (SDHD) differ in an insertion-deletion sequence polymorphism 492 nucleotides from the stop codon of the 3′ UTR of the mRNA, with associated effects on SDH activity, peak metabolic rate, and mitochondrial integrity following peak metabolic activity (Marden et al. 2013). These effects of the polymorphism may be caused by variation in binding of a microRNA to the sequence; a 25% reduction in SDH activity is associated with the sequence of one of the variants (M) that allows the microRNA full access. Butterflies with this allele also have twice the cross-sectional area of trachea in their flight muscle and show less oxidative damage in their mitochondria in response to peak flight exercise. Also, greater cross-sectional area correlated positively with peak flight metabolic rate and ability for maintaining peak rates during low atmospheric oxygen. Given the reduction in SDH activity, it seems unlikely that an increase in pathway flux is responsible for these changes. Instead, low SDH activity may lead to a subtly higher level of succinate. And here is the possible surprise answer for how this might affect metabolic rate during exercise: Because succinate is known to inhibit the hypoxia-inducible factor (HIF)-prolyl hydroxylase enzyme (PHD), which tags the α subunit of HIF-1 for degradation by the ubiquitin-proteasome pathway during normoxic conditions, higher levels of succinate can lead to greater inhibition of PHD (**Figure 2.17**). This inhibition could prolong the half-life of HIF-1α, allowing it to form increased numbers of dimers with the other type of HIF subunit, HIF-1β. HIF-1α:HIF-1β dimers activate the expression of hypoxia-responsive genes. Some of these genes are known to promote the growth of tubular networks (such as trachea in insects) to enhance the delivery of oxygen (Selak et al. 2005). Supporting the hypothesis that higher levels of succinate (caused by low SDH activity) promote enhanced development of tracheae, addition of a cell-permeable form of succinate to butterfly pupae led to an accumulation of HIF-1α and activation of hypoxia-responsive and tracheal development genes (Marden et al. 2013). This is not a singular occurrence. Selection studies on *Drosophila* subjected to hyperoxic and hypoxic conditions also showed an effect on SDH activity (Ali et al. 2012).

In summary, these studies suggest that higher succinate concentrations due to reduced SDH activity inhibit the degradation of HIF-1α, inducing the HIF-1 signaling pathway and

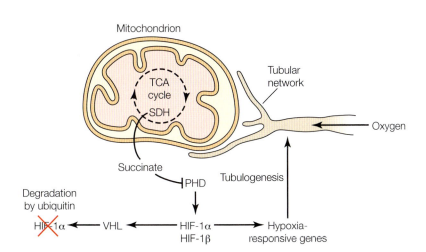

Figure 2.17 Role of the TCA cycle metabolite succinate in regulating the degradation of hypoxia-inducible factor 1 (HIF-1). Higher levels of succinate due to lower levels of succinate dehydrogenase (SDH) activity exert a greater inhibitory effect on HIF-prolyl hydroxylase enzyme (PHD), which tags HIF-1α for degradation by the von Hippel-Lindau tumor suppressor (VHL) via the ubiquitin-proteasome pathway. As a consequence, HIF dimers (HIF-1α:HIF-1β) accumulate to higher levels and can activate hypoxia-responsive and tracheal development genes, leading to tubulogenesis. (After Marden et al. 2013.)

activating the expression of several hypoxia-responsive genes. Although this may be the first study presenting direct evidence that a moonlighting function of a metabolic enzyme plays an important role in the adaptation of organisms to their environment, given the known multifaceted roles of metabolic enzymes, it is likely that they play a broader role than we have suspected.

2.8 Oxidative Stress: Production and Scavenging of ROS

Aerobic metabolism relies on the coordinated transfer of four electrons from the ETS to oxygen via cytochrome *c* oxidase (COX) (see Box 2.1). COX contains two iron (heme) clusters and three copper atoms arranged in two clusters, each of which can accept one electron. In effect, COX functions as an **electron capacitor**, holding a four-electron charge until the system can discharge the electrons through the sequentially **univalent reduction** of molecular oxygen (dioxygen) that leads to the production of water:

$$O_2 + 4\,H^+ + 4\,e^- \rightarrow 2\,H_2O$$

The transfer of the four electrons from COX to oxygen has to occur sequentially, without releasing any intermediates that are only partly reduced, which would form ROS. However, limited oxygen availability or interruptions of the ETS can disrupt the constant flow of electrons to oxygen, causing electrons along several sites of the ETS to transfer to oxygen in univalent reductions that generate ROS.

The chemical nature of reactive oxygen species

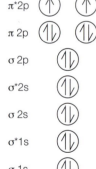

Oxygen has a unique arrangement of its outer shell electrons, which are involved in reacting (bonding) with other molecules. The two electrons of the outer orbital are unpaired and have parallel (the same) spins (spin direction is indicated by the arrows shown in the figure at left). Oxygen is termed paramagnetic for this reason. Importantly, electron pairs generated by intermediary metabolism, for example, the electron pairs carried by NADH, have opposite spins. In order for such molecules to react directly with oxygen—to donate their pair of opposite-spin electrons to O_2—one electron would have to be paired up with an electron of the same spin, an energetically unfavorable process termed **spin restriction**. Spin restriction causes a relative chemical inertness of molecular oxygen and thereby contributes to the accumulation of oxygen in the atmosphere. The spin restriction barrier can be overcome by adding the electrons to O_2 one at a time—univalent reduction of O_2—but this can lead to ROS generation, the "bad" side of oxygen (Fridovich 1998; Grivennikova and Vinogradov 2013). A note on terminology is worth introducing here: It is important to distinguish between **reactive** species, which commonly have a strong tendency to either provide (reduce) or extract (oxidize) electrons to or from other molecules, respectively, from **radicals**, which have an unpaired electron on the outer shell and are therefore even more strongly oxidizing.

Steps in the univalent reduction of dioxygen

The initial step in the univalent reduction of dioxygen generates the **superoxide anion** ($O_2^{-\bullet}$). Additional single electron transfers generate **hydrogen peroxide (H_2O_2)**, **hydroxyl radical (OH$^\bullet$)**, and finally water (H_2O) (**Figure 2.18**). Superoxide anion has a short life span, probably due to both its tendency to react with other cellular molecules and the scavenging activity of mitochondrial **manganese superoxide dismutase** (**Mn-SOD**) (Halliwell and Gutteridge 2015). Because hydrogen peroxide is not a radical, it is less reactive

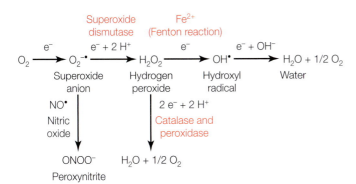

Figure 2.18 Generation of reactive oxygen and nitrogen species (ROS and RNS). This overview shows the major reactions (not fully balanced) that generate ROS and RNS (peroxynitrite only), along with the antioxidant (scavenging) proteins that catalyze the reduction of superoxide anion ($O_2^{-\bullet}$) and hydrogen peroxide (H_2O_2). Also shown is the contribution that ferrous iron (Fe^{2+}) makes to the formation of hydroxyl radicals ($OH^{\bullet}$) (Fenton reaction).

than other ROS, for example, the hydroxyl radical. However, this relatively low reactivity makes H_2O_2 more stable and long-lived than $O_2^{-\bullet}$ and allows it to diffuse easily across the inner mitochondrial, endoplasmic reticulum, and peroxisomal membranes, a process that is facilitated by aquaporins (Bienert and Chaumont 2014). H_2O_2 can therefore affect biomolecules at sites far from its origin. Importantly, when it reacts with ferrous iron (Fe^{2+}), it is converted to the hydroxyl radical, through what is known as the **Fenton reaction**. $OH^{\bullet}$ is the most reactive ROS; it reacts immediately with biomolecules around the site of its cellular origin. It is therefore not surprising that the primary function of the major ROS scavenging systems, the glutathione and thioredoxin-peroxiredoxin systems, as well as catalase, is to scavenge hydrogen peroxide before it can diffuse across membranes into other cellular compartments and generate hydroxyl radicals. In addition, the supply of ferrous iron can be regulated to reduce the probability that the Fenton reaction will occur; ferrous iron can be chelated by **ferritin**, which prevents Fe^{2+} from reacting with hydrogen peroxide (Finazzi and Arosio 2014).

While we focus here on ROS, they often give rise to other radicals or reactive species. For example, when nitric oxide ($NO^{\bullet}$), itself a radical, reacts with $O_2^{-\bullet}$, they form peroxynitrite ($ONOO^-$), a highly reactive nitrogen species, which is mainly produced in peroxisomes. Also, hydroxyl radicals reacting with bicarbonate (HCO_3^-) can form carbonate radicals, strong oxidizing agents that damage proteins by extracting protons from cysteines. ROS are therefore just the beginning of many chemical chain reactions that can be damaging to cellular structures (Halliwell and Gutteridge 2015).

Cellular damage by reactive oxygen species

ROS can damage DNA, RNA, lipids, and proteins. DNA oxidation can be caused by almost all ROS, but under in vivo conditions it seems most likely that the hydroxyl radical formed through the reaction between H_2O_2 and DNA-associated iron and copper ions (Fenton reaction) may be the major route of DNA damage (Halliwell and Gutteridge 2015). Metals may either be bound continuously to DNA or be released from other molecules, for example, iron and copper-containing proteins, in response to oxidative stress. While both purine and pyrimidine bases can be oxidized by ROS, the most common oxidation product is 8-hydroxyguanine (8OHG), possibly because of guanine's high potential for oxidation (see figure at right). Cells can respond to the formation of 8OHG by removing it from the pool of base

Guanine → Oxidative damage → 8-Hydroxyguanine (8OHG)

precursors, to prevent its incorporation into DNA. If 8OHG is incorporated into DNA, the defective base can be removed through a **base excision repair** or a **nucleotide excision repair** mechanism, involving several DNA damage recognition and repair enzymes. The conversion of a guanine to 8OHG in existing DNA can lead to incorporation of adenine instead of cytosine into the opposite strand of the DNA molecule when the DNA is repaired or replicated. If this occurs, **mismatch repair** is activated to remove the base (adenine) from the pair.

While damage to DNA—and the potential for mutagenesis—is a critical aspect of ROS relationships, damage of other biomolecules by ROS also poses major challenges to organisms. The oxidation of fatty acids and cholesterol is of great biological importance. ROS-driven modification of fatty acids can cause leakiness of membranes and alterations of lipid-protein interactions within membranes. In addition, ROS damage to lipids can lead to production of many reactive aldehydes, alkanes, isoprostanes, and other molecules that can affect DNA and proteins as well (Lushchak et al. 2012). The oxidation of polyunsaturated fatty acids (PUFAs, e.g., linoleic or arachidonic acid; see Chapter 3) by the hydroxyl radical (OH$^{\bullet}$) involves the removal of a proton from a methylene (–CH$_2$–) group, which results in the formation of a carbon radical (FA$^{\bullet}$) (see figure at left). FA$^{\bullet}$ in turn reacts with oxygen, which is highly soluble within membranes, forming a peroxyl radical (FAOO$^{\bullet}$). Removal of a proton from another methylene group forms a hydroperoxide (FAOOH) and another carbon radical (FA$^{\bullet}$). Both lipid peroxyl radicals and hydroperoxides can undergo further oxidation to products that readily decompose to reactive aldehydes, alkanes, and isoprostanes. Two of the most common of these are malondialdehyde (MDA) and 4-hydroxynonenal (HNE), which react with both DNA and proteins. MDA can be metabolized by aldehyde dehydrogenases, and HNE can be absorbed by fatty acid binding protein, proteins whose abundances change in response to cellular stress (Wang et al. 2009). HNE can also activate signaling pathways and enzymes at very low concentrations, illustrating its potential beneficial function as a messenger to induce greater resistance to higher levels of ROS (Halliwell and Gutteridge 2015).

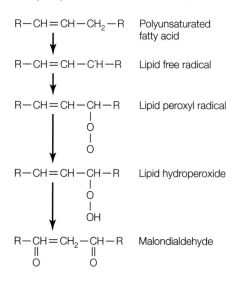

The oxidation of proteins begins with the removal of a hydrogen from the α-carbon atom of an amino acid residue, triggering several reactions that lead to the formation of an amino acid radical. The reaction of the radical with oxygen leads to an amino acid peroxide radical, which can be converted into more stable carbonyl compounds, whose concentration is often used as a global measure of protein oxidation. More specifically, the thiol groups of cysteines and methionines are readily oxidized by ROS, forming sulfenic acid. The subsequent reaction with other intra- or intermolecular thiol groups, including those of the non-enzymatic tripeptide glutathione, leads to the formation of disulfides, which can interfere with protein function, or to further irreversible oxidations that will lead to the degradation of proteins (**Figure 2.19**). The sensitivity of thiol groups to the redox state of the cell and their formation in the endoplasmic reticulum (ER) are probably the main reasons why ER chaperones are frequently activated in response to environmental stress (Tomanek 2015).

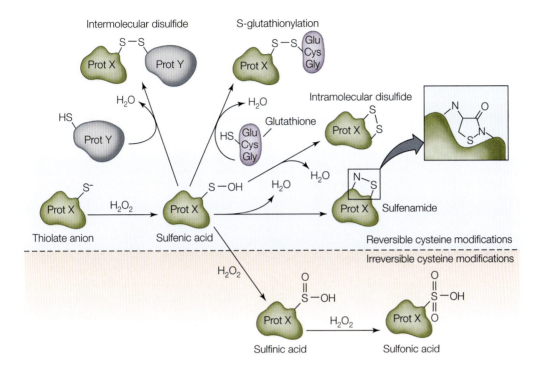

Figure 2.19 Oxidation of thiol groups (–SH) and formation of intra- and intermolecular disulfide bonds (–S–S–). Oxidation of cysteines leads to the formation of sulfenic acid, which can form either intramolecular disulfide bonds or intermolecular disulfide bonds with other proteins or the non-enzymatic tripeptide antioxidant glutathione (Glu-Cys-Gly), a reaction termed S-glutathionylation. Further oxidation of sulfenic acids leads to sulfinic and sulfonic acids, which form irreversible modifications of the cysteine.

Production of ROS in mitochondria

Oxidative stress, here defined as an increase in amounts of ROS above an established baseline, is the net result of several cellular processes that contribute to the production of ROS and their scavenging. ROS generation and scavenging are both potential targets of regulation for reducing ROS during stress. Whereas an overall reduction of mitochondrial activity (e.g., metabolic depression) in response to environmental stress would tend to reduce rates of ROS generation (Podrabsky and Hand 2015), not many studies have focused on the responses to stress of the specific biochemical reactions producing ROS. In contrast, the ROS scavenging systems of mitochondria—both enzymatic and non-enzymatic—have been the focus of numerous studies investigating the effects of stress (Abele et al. 2012; Halliwell and Gutteridge 2007). Before looking at how organisms adapt to challenges from ROS, we will first review some of the basic reactions involved in ROS generation and scavenging.

Sites of ROS production in mitochondria

Insights into the rate of ROS production are based mainly on isolated mitochondria and the measurement of H_2O_2 efflux across the mitochondrial membranes (Andreyev et al. 2005; Murphy 2009). A key element in governing rates of ROS generation by the mitochondrion is the maintenance of a steady flow of electrons from their entry point in the

Figure 2.20 Sources of ROS in the mitochondrion. Stress-induced interruption of the ETS and high NADH/NAD$^+$ ratios (as well as low rates of ATP production) lead to the generation of superoxide anion ($O_2^{-\bullet}$) at complex I (mode 1). A high CoQH$_2$/CoQ ratio generates $O_2^{-\bullet}$ at complexes I and III (mode 2). The E3 subunit of α-ketoglutarate dehydrogenase (dihydrolipoamide dehydrogenase; DLDH) also contributes to the generation of $O_2^{-\bullet}$. Blue arrows indicate the normal flow of electrons; the red arrow indicates the flow of electrons during reverse electron transfer. FMN, flavin mononucleotide; αKGDH, α-ketoglutarate dehydrogenase; CoQ, oxidized coenzyme Q; CoQH$_2$, reduced coenzyme Q. (After Murphy 2009.)

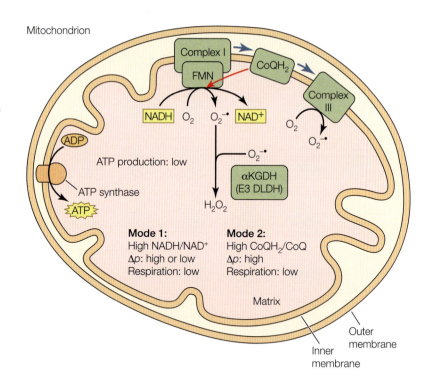

ETS to the cytochrome *c* oxidase complex. Any retardation or blockage of electron flow can lead to a buildup of electrons, with the concomitant danger that these electrons will leave the ETS and cause univalent reduction of O_2.

Several conditions affect the flow of electrons along the ETS and the production of ROS at complex I (NADH dehydrogenase) and complex III (cytochrome *c* reductase). For example, a high NADH/NAD$^+$ ratio in the matrix contributes to an increase in ROS production by complex I (mode 1; **Figure 2.20**). Normally, complex I accepts the electrons of NADH at the site of its cofactor flavin mononucleotide (FMN), from where the electrons are passed across several iron-sulfur centers to coenzyme Q (CoQ), starting the transfer of electrons along the ETS. However, any damage to the ETS or low ATP demand (and low respiration rates) inhibits the transfer of electrons and causes an increase in NADH/NAD$^+$. This leads to a backup of electrons and the direct transfer of electrons from NADH via FMN to O_2 and the formation of $O_2^{-\bullet}$ (Murphy 2009). Thus, high ratios of NADH/NAD$^+$ are an indication of a slow transfer of electrons along the ETS, but also can be seen as establishing conditions that favor the univalent reduction of oxygen to superoxide anion.

Other conditions, such as a large pool of reduced coenzyme Q (high CoQH$_2$/CoQ ratio), a high proton gradient (Δp), and a low rate of ATP synthesis, all contribute to the flow of electrons back from CoQH$_2$ to complex I (mode 2; see Figure 2.20), a process referred to as **reverse electron transfer** (**RET**). High CoQH$_2$/CoQ ratios can be formed when succinate, α-glycerophosphate, and fatty acid oxidation provide electrons to the ETS through complex II, causing RET toward complex I (for important exceptions, see below). This process is highly sensitive to the proton gradient across the inner mitochondrial membrane, with a higher gradient causing higher rates of $O_2^{-\bullet}$ production (Murphy 2009). Although the exact site of ROS production (FMN or the CoQ-binding site on complex I) is still debated, mode 2 generates the highest rates of ROS production.

Mode 3, defined by high rates of respiration combined with high rates of ATP synthesis and a small proton gradient (not illustrated in Figure 2.20), prevails during normal conditions. These conditions generate much lower rates of H_2O_2 efflux than modes 1 and 2. Complex III (*bc*1 complex or cytochrome *c* reductase) produces $O_2^{-\bullet}$ at a high rate when the exchange of electrons between two different sites (Q_i and Q_o) of its structure is inhibited, for example by antimycin (Andreyev et al. 2005). This stabilizes the electron-carrying quinone bound to one of the sites for a sufficient period to allow interaction with O_2 to form $O_2^{-\bullet}$; the latter is released toward both the intermembrane space and the matrix sides of the inner membrane. Although mode 3 of ROS production prevails through most of life, modes 1 and 2 are likely to occur under environmental stress.

Other sources of ROS in mitochondria are linked either to the NADH pool of the mitochondrial matrix or to CoQ in the inner membrane. For example, when perturbations of the ETS generate higher $NADH/NAD^+$ ratios (i.e., mode 1), the high levels of reduced NADH and low levels of NAD^+ can lead to the reversal of reactions that normally use NAD^+ as an electron receptor, for example, those catalyzed by α-ketoglutarate (2-oxoglutarate) dehydrogenase (αKGDH) or pyruvate dehydrogenase (PDH), leading to the transfer of electrons to oxygen instead and resulting in the formation of $O_2^{-\bullet}$ (Starkov 2013). Decreasing the activity of dehydrogenases such as αKGDH and PDH in response to stressful conditions that cause higher ratios of $NADH/NAD^+$ could thus decrease the production of ROS directly, as well as reduce production of $O_2^{-\bullet}$ at complex I during mode 1, by decreasing the production of NADH. Such an interpretation was proposed as an explanation for the decrease in PDH and malate dehydrogenase in gill tissue of blue mussels (genus *Mytilus*) in response to acute heat stress (Tomanek and Zuzow 2010).

A comprehensive picture of the sources of ROS in mitochondria also needs to consider the role of fatty acids in (1) uncoupling electron flow within the inner mitochondrial membrane, (2) interacting with ETS complexes, and (3) transferring electrons from fatty acid β-oxidation via FADH to the ETS (Schönfeld and Wojtczak 2008; Tahara et al. 2009). Importantly, recent evidence suggests that the transfer of electrons that originate from β-oxidation (in the form of FADH) does not cause reverse electron transfer and therefore ROS production from complex I. This is opposite to the action of succinate dehydrogenase (SDH), which transfers electrons (as FADH) from succinate, emphasizing that the substrate-specific entry of electrons to the ETS can affect ROS production (Chouchani et al. 2014; Schönfeld et al. 2010). Interestingly, the major flight muscle (*pectoralis*) in birds, which relies on fatty acid oxidation, appears to have lower rates of ROS production than the major locomotory muscle (*quadriceps*) in mammals, which relies on carbohydrate-based metabolism. This suggests that fatty acid oxidation may be associated with lower rates of ROS production (Kuzmiak et al. 2012).

Closely linked to any possible changes in enzyme activity in response to oxidative stress are possible changes in metabolites that act as antioxidants. There is evidence that pyruvate protects neurons from H_2O_2, and this might also be the case for other α-ketoacids (Desagher et al. 1997), including α-ketoglutarate (αKG) (Mailloux et al. 2007). A decrease in the activity of PDH and αKGDH might not only lower the production of ROS through the reduced production of NADH from the TCA cycle, but also facilitate scavenging of ROS through the accumulation of metabolites like pyruvate and αKG, which function as antioxidants (Mailloux et al. 2007). However, although α-ketoacids may serve as antioxidants, their relative importance to the other cellular ROS scavenging systems is considered to be small.

In summary, multiple enzyme reactions—including complexes I and III of the ETS, PDH, and αKGDH of the TCA cycle—can generate ROS, especially during conditions of high $CoQH_2/CoQ$ and $NADH/NAD^+$ ratios and a large proton gradient. These conditions are most likely to be a consequence of a disruption of the ETS caused by protein or membrane damage during stress. Is there any evidence that organisms can change these reactions in response to environmental stress? And if so, is it possible to infer the cause of these changes? Proteomics studies are beginning to shed some light on these questions and point to novel hypotheses for how organisms respond to environmental stress.

Abundance changes in ETS and TCA cycle enzymes during environmental stress in *Mytilus*

Several recent proteomic studies have found simultaneous changes in the abundance of several ETS and TCA cycle enzymes in marine invertebrates in response to acute and chronic temperature and hyposalinity stresses (Dilly et al. 2012; Fields et al. 2012; Tomanek and Zuzow 2010; Tomanek et al. 2012). These findings suggest that decreases in abundance, changes in subunit composition, and posttranslational modifications of these metabolic enzymes are potential mechanisms for regulation.

One study measured the proteomic response of gill tissue to acute heat stress (+6 degrees Celsius h^{-1}) in marine mussels of the genus *Mytilus* and discovered coordinated changes in the abundance of proteins involved in the ETS and TCA cycle, as well as proteins that function in antioxidant systems (Tomanek and Zuzow 2010). Strikingly, while heat stress led to decreases in abundances of proteins in complexes I and III of the ETS and of NADH-producing enzymes of the TCA cycle, proteins involved in the production of NADPH increased (Figure 2.21). This suggests a downregulation of metabolism and possibly ROS production, coupled with a switch to the production of NADPH, which can serve as a reducing equivalent for ROS scavenging.

Figure 2.21 Changes in the proteome (arrow up indicates an increase, arrow down a decrease in abundance) in gill of the mussel *Mytilus trossulus* detected in response to acute heat stress. While pro-oxidative NADH-producing enzymes (PDH and MDH) and subunits of complexes I and III decreased in response to heat stress, antioxidant NADPH-producing enzymes (citrate synthase and NADP-dependent isocitrate dehydrogenase [NADP-IDH]) increased in *M. trossulus*. The NADPH produced is most likely fed into the one of the two hydrogen peroxide scavenging systems, the thioredoxin-peroxiredoxin system and/or the glutathione system. GSH, glutathione; SOD, superoxide dismutase; Trx, thioredoxin. (Data from Tomanek and Zuzow 2010.)

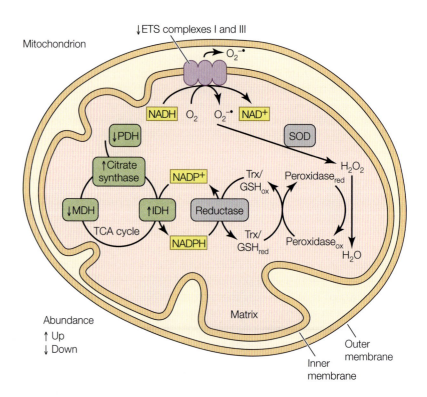

Together these changes indicate that when the gill cells of *Mytilus* experienced an increased challenge from ROS, they switched from mitochondrial pathways that produce pro-oxidant NADH to those that produce antioxidant NADPH to supply the reducing equivalents they needed for the scavenging of ROS or reactive molecules. The gill cells simultaneously downregulated pathways that could produce ROS, most likely as an effort of last resort to control oxidative stress near the upper limit of the mussel's thermal range (Tomanek and Zuzow 2010). The gram-negative bacterium *Pseudomonas fluorescens* showed a similar upregulation of NADP-dependent isocitrate dehydrogenase in response to oxidative stress, suggesting that the TCA cycle plays a role in the production of both NADH and NADPH (Mailloux et al. 2007).

Uncoupling proteins: "Safety valves" for reducing ROS production

The amount of ROS production that results from activity of the ETS depends strongly on the mitochondrion's ability to maintain a close balance between the flow of electrons entering the ETS and their subsequent addition to oxygen by cytochrome *c* oxidase. When the system is in perfect balance, the only end product of the ETS is water. However, as discussed above, if the cytochrome *c* oxidase electron capacitor is unable to discharge the electrons it receives from earlier steps in the ETS, a backup in electron transport will occur, leading to production of ROS from complexes I and III of the ETS.

When is a "safety valve" needed?

What factors determine whether the electrons entering the ETS can be rapidly and quantitatively added to molecular oxygen at the cytochrome *c* oxidase step? One very important factor is the supply of ADP to the ATP synthase reaction. As shown in **Figure 2.22A**, a steady flow of ADP must be provided to the ATP synthase reaction to allow reentry of the protons that have been moved outward across the mitochondrial inner membrane by the activity of the ETS. Protons reentering the mitochondrial matrix through the ATP synthase system can be viewed as supplying one of the two substrates for the cytochrome *c* oxidase reaction. Without these protons, electrons cannot combine with O_2, the second substrate, to generate H_2O. The supply of ADP, in turn, depends on the use of ATP to drive various types of work in the cell (ion transport, biosynthesis, etc.). If the rate of ATP use is too low, electrons will build up in the ETS and leakage of electrons will produce superoxide ($O_2^{-\bullet}$). $O_2^{-\bullet}$ production is strongly correlated with the magnitude of the proton motive force (Δp) across the inner mitochondrial membrane and the state of mitochondrial respiratory activity. When ATP use and production rates are low, the resulting large Δp is ideal for a high level of production of superoxide. Low ADP levels can also lead to a low rate of oxygen consumption that, in turn, favors increases in the oxygen concentration in the matrix. The highly reduced state of the ETS leads to leakage of electrons into the matrix, and $O_2^{-\bullet}$ production increases. When demands for ATP increase, rising levels of ADP increase respiration, causing a decrease in the Δp and a lowering of ROS production. This transition is marked by a rise in ATP production and a fall in the rate of $O_2^{-\bullet}$ generation.

Because of variations in demands for ATP as cellular work intensities change, there generally is imperfect coupling between entry of electrons into the ETS and their addition to oxygen at the end of the respiratory protein chain. It thus would seem advantageous to provide a "safety valve" of some sort that could allow reentry of protons into the mitochondrial matrix that is not coupled to ATP production (i.e., not dependent on availability of ADP). This safety valve is a family of **uncoupling proteins** (**UCPs**) localized in the inner mitochondrial membrane that allow proton reentry without ATP synthesis having to occur (Echtay 2007; Krauss et al. 2005; **Figure 2.22B**). UCPs are part of a large anion transporter

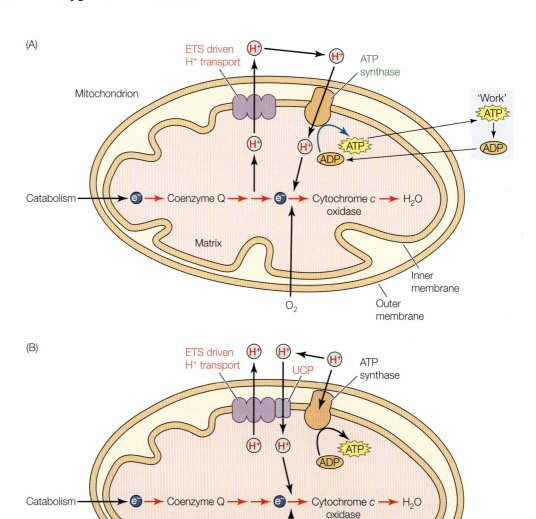

Figure 2.22 A simplified diagram of mitochondrial ETS activity and ATP synthesis shows the importance of a continuous supply of ADP for ATP production. (A) A balanced system in which enough ADP is generated through various types of work to ensure an ample supply of ADP for ATP synthesis coupled with proton entry through the ATP synthase. (B) An unbalanced system in which inadequate ADP is available to support the potential for ATP synthesis due to the transmembrane proton gradient. Under these conditions, proton movement occurs (in part) through the uncoupling protein (UCP) system. Movement through the UCP allows continued electron transport through the ETS and thus reduced risks of ROS formation (see text).

family, which may contain 35–50 members in eukaryotes (Echtay 2007). Proton-transporting UCPs thus dissipate ("waste") the energy that was used to drive protons from the mitochondrial matrix into the intermembrane space, but the payback for this lost energy is less ROS generation and concomitant reductions in damage to cellular structures and processes.

REGULATION OF UCP FUNCTION   Because of UCPs' potential to waste large amounts of energy, their activities are tightly regulated so that activity is upregulated only when the danger of ROS generation rises. (We will see in Chapter 3 that *all* of the energy of the ETS proton pumping system may be "wasted" by UCPs when they function in a thermogenic mode in brown adipose tissue [BAT]. In BAT, the UCP in question is UCP1, the first paralog of UCP to be discovered. Defense against ROS formation involves two other paralogs of UCP, UCP2 and UCP3 [Divakaruni and Brand 2011; Echtay 2007; Krauss et al. 2005].) This regulatory scheme is based on the ability of small amounts of one specific ROS, the highly reactive hydroxyl radical OH•, to cause a chain reaction that leads to UCP-mediated dissipation of the transmembrane proton gradient.

The steps in this regulatory process are shown in **Figure 2.23**. When there is a large proton motive force ($\Delta p\uparrow$), which is composed of both an electrical gradient ($\Delta\psi_m$) and a chemical gradient (ΔpH), in the mitochondrion and the supply of ADP to the ATP synthase reaction is inadequate to support quantitative flux of electrons to the cytochrome *c* oxidase system, electrons leak from the ETS and lead to formation of $O_2^{-\bullet}$. A portion of the $O_2^{-\bullet}$ produced in this univalent reduction of O_2 is rapidly converted to hydrogen peroxide (H_2O_2) by superoxide dismutase. However, some of the $O_2^{-\bullet}$ interacts with the iron (Fe)-containing enzyme aconitase. This interaction leads to release of Fe^{2+} (Fe[II]) into the matrix. As discussed earlier, the interaction of Fe^{2+} with $O_2^{-\bullet}$ can generate an extremely potent form of ROS, the hydroxyl radical (OH•). OH• reacts rapidly with polyunsaturated fatty acids (PUFAs) in the membrane, triggering a free radical chain reaction that forms aldehydic lipid-peroxidation intermediates. This chain reaction ultimately produces the aldehyde **4-hydroxynonenal** (**HNE**; structure shown above), which is the regulatory molecule that triggers the expression of UCP2 and UCP3 in a tissue-specific manner. HNE can activate the transcription factor Nrf2 (nuclear factor erythroid 2-related factor 2), which binds to an antioxidant response element (ARE) upstream of the *UCP2* and *UCP3* genes, thereby

OH (4-Hydroxynonenal structure)

4-Hydroxynonenal

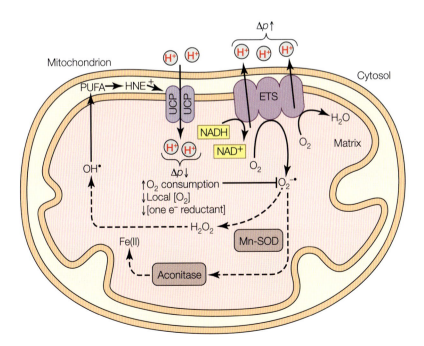

Figure 2.23 Regulation of UCP activity by ROS. Δp, proton motive force; ETS, electron transport system; HNE, 4-hydroxynonenal; Mn-SOD, manganese superoxide dismutase; PUFA, polyunsaturated fatty acids. (After Echtay 2007.)

inducing their expression (Lopez-Bernardo 2015). The resulting production of UCPs leads to a dissipation of some of the proton motive force ($\Delta p\downarrow$). This is accompanied by a rise in oxygen consumption by cytochrome *c* oxidase, a decrease of O_2 levels in the matrix, and a fall in production of ROS. Perhaps the key feature of this mode of regulating UCP activity is the feedback mechanism through which an initial spike in ROS generation leads to rapid downregulation of further ROS production.

It must be appreciated that proton leakage through the inner mitochondrial membrane can represent a substantial percentage of oxygen consumption—as much as 20% of basal metabolism in mammals (Echtay 2007)—and can be mediated by mechanisms other than UCPs. However, UCP2 and UCP3 provide a tightly regulated mechanism for adjusting the rates of uncoupled proton flux into the mitochondrial matrix in accord with threats of ROS production. UCPs, then, are important components of the broader arsenal of tools that cells have available to either reduce ROS generation or scavenge ROS once they are formed. We now turn to two of these ROS-destroying systems: a battery of different types of enzymes that degrade ROS, and a suite of small organic molecules that function as antioxidants.

ROS scavenging in mitochondria

Superoxide anion is produced in the mitochondrial matrix as well as the intermembrane space of mitochondria. As a first line of defense, $O_2^{-\bullet}$ is converted to H_2O_2 by the mitochondrial manganese superoxide dismutase (Mn-SOD) or the intermembrane space copper-zinc SOD (Cu-Zn-SOD) (**Figure 2.24**), which also occurs in the cytosol. Both SOD isoforms are distant orthologs (Fridovich 1998) and can occur in other cellular compartments, for example in peroxisomes (Bonekamp et al. 2009).

The second line of defense includes the dismutation of H_2O_2 by several enzymes, including catalase, glutathione peroxidase (GPx), and peroxiredoxin (Prx). Catalase is

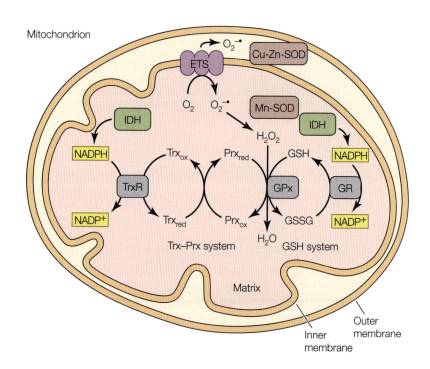

Figure 2.24 ROS scavenging reactions in mitochondria. Cu-Zn-SOD, copper-zinc superoxide dismutase; GPx, glutathione peroxidase; GR, glutathione reductase; GSH, reduced glutathione; GSSG, oxidized glutathione; IDH, NADP-dependent isocitrate dehydrogenases; Mn-SOD, manganese superoxide dismutase; Prx, peroxiredoxin; Trx, thioredoxin, TrxR, thioredoxin reductase. (After Mailloux et al. 2013; Murphy 2012.)

mainly localized in the peroxisome; mitochondria and the ER have low catalase activity in most tissues (Halliwell and Gutteridge 2015). GPx and Prx catalyze major scavenging reactions in the mitochondrion (see Figure 2.24) (Mailloux et al. 2013; Murphy 2012). GPx couples the reduction of H_2O_2 to the oxidation of glutathione, a tripeptide consisting of glutamine, cysteine, and glycine (see structures at right) and which is the main cellular non-enzymatic antioxidant. The oxidized form of glutathione (GSSG) is then reduced (GSH) by glutathione reductase (GR), a reaction that requires NADPH as a reducing equivalent. Two GPx isoforms play an important role in scavenging peroxides in mitochondria: GPx1 degrades hydrogen peroxides, and GPx4 catalyzes the conversion of lipid hydroperoxides to alcohols, thereby slowing down lipid peroxidation. Two other enzymes use GSH as an electron donor: glutathione S-transferase (GST) and glutaredoxin (Grx) (**Figure 2.25**) (Murphy 2012). GSTs are historically associated with detoxification reactions in which GSH is conjugated to a compound through a thioester linkage (Tew and Townsend 2012). These conjugates increase the solubility of the resulting compound and facilitate excretion from cells through efflux transporters. However, some GST isoforms can also catalyze the formation of mixed disulfides between proteins and GSH, a process called S-glutathionylation, which protects thiol groups from ROS (Townsend et al. 2009). There is also evidence that S-glutathionylation affects the activity of various metabolic proteins in addition to protecting them from ROS (Dalle-Donne et al. 2009; Fratelli et al. 2003). The various reactions involving glutathione are known as the **glutathione system**.

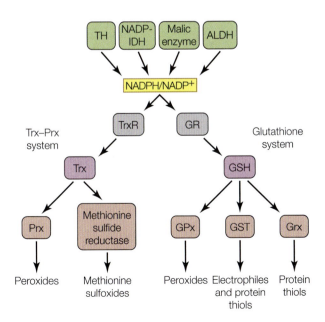

Figure 2.25 Overview of major elements of the cellular ROS scavenging system. A schematic overview of the glutathione and thioredoxin-peroxiredoxin (Trx-Prx) systems and the enzymatic reactions affecting various thiol modifications. ALDH, aldehyde dehydrogenase; GPx, glutathione peroxidase; GR, glutathione reductase; Grx, glutaredoxin; GSH, glutathione; GST, glutathione S-transferase; NADP-IDH, NADP-dependent isocitrate dehydrogenase; Prx, peroxiredoxin; TH, transhydrogenase; Trx, thioredoxin; TrxR, thioredoxin reductase. (After Murphy 2012; Mailloux et al. 2013.)

The **thioredoxin-peroxiredoxin system** is the other major hydrogen peroxide scavenging system (see Figures 2.24 and 2.25) (Mailloux et al. 2013; Murphy 2012). Thioredoxin (Trx) catalyzes the reduction of disulfide bonds on other proteins, specifically those of peroxiredoxins (Prx), which reside not only in mitochondria (Prx3 and 5), but also in the cytoplasm (Prx1, 2, and 6), the endoplasmic reticulum (Prx4) and peroxisomes (Prx5) (Cox et al. 2010). Whereas all peroxiredoxins scavenge hydrogen peroxide, Prx5 targets organic peroxides and peroxynitrite, a highly reactive RNS that forms from the reaction between $O_2^{-\bullet}$ and $NO^\bullet$ (Trujillo et al. 2008). A kinetic analysis of ROS metabolism in mice, which took into account rate constants for the ROS scavenging enzymes and published protein concentrations, estimated that 90% of the H_2O_2 produced in the mitochondrion is scavenged by Prx3, the main mitochondrial peroxiredoxin; this estimate will likely vary with tissue and cell type, however (Cox et al. 2010).

ANTIOXIDANT RESPONSES TO ACUTE TEMPERATURE STRESS   As we discussed previously in the example of *Mytilus* mussels, acute heat stress may induce oxidative stress. Accordingly, heat stress has been associated with increased production of ROS, increased oxidative damage, increased abundances of several antioxidant enzymes, and production of a more oxidized cellular redox potential in several marine organisms (Abele et al. 2002; Heise et al. 2003; Keller et al. 2004). Recent systems-level investigations have confirmed the role of established oxidative stress response pathways as well as discovered how these pathways link up with simultaneous shifts in energy metabolism and proteostasis.

Paralvinella sulfincola

Dilly and colleagues compared changes in the gill proteomes of two species of hydrothermal vent polychaetes in response to acute heat stress and discovered correlated changes in the abundances of antioxidant enzymes and constituents of the glutathione system (Dilly et al. 2012). They showed that the *Paralvinella* congeners differed in changes in abundance of two SOD isoforms, enzymes of the glutathione system, and the glutathione biosynthesis pathway. Specifically, the more heat-tolerant *P. sulfincola* (see photo at left) did not change abundance of either SOD isoform with increasing heat stress (**Figure 2.26**). In contrast, the less heat-tolerant *P. palmiformis* increased the abundance of the cytosolic Cu-Zn- and the mitochondrial Mn-SOD isoform. Two enzymes that are part of the cysteine biosynthesis pathway and therefore necessary for glutathione synthesis, cystathione β-synthase and γ-glutamylcysteine synthetase (see Figure 2.26), increased in abundance with heat stress in both congeners (Dilly et al. 2012). However, *P. sulfincola* increased the abundance of both enzymes of the glutathione scavenging system, glutathione peroxidase (GPx3; cytosolic) and glutathione reductase (GR), while *P. palmiformis* showed an increase in GPx3 and a decrease in GR. As a result, *P. sulfincola* was able to maintain total levels of GSH as well as the ratio of GSH/GSSG, in contrast to the less tolerant *P. palmiformis*, which decreased by about half its total GSH as well as decreased its GSH/GSSG ratio to one-third compared with levels and ratios at temperatures that were not stressful (Dilly et al. 2012). Using this example, we want to emphasize how many antioxidant pathways are activated during acute heat stress, more than the specific interspecific differences, because the latter may depend on the specific range of temperatures or the heating rate chosen. However, the differences in the ratio of GSH/GSSG indicate

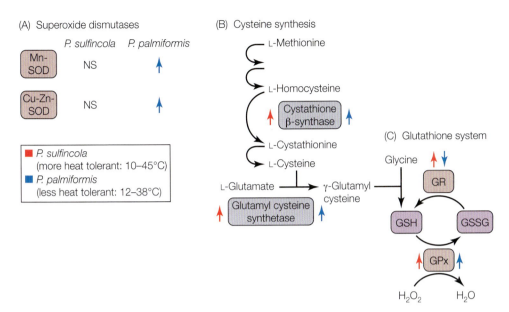

(A) Superoxide dismutases

(B) Cysteine synthesis

(C) Glutathione system

Figure 2.26 Changes in abundance of several antioxidant proteins in response to acute heat stress in gill tissue of two vent polychaete species, the more heat-tolerant *Paralvinella sulfincola* (red) and the less heat-tolerant *P. palmiformis* (blue). Upward arrows indicate an increase and downward arrows a decrease relative to the control; NS indicates no significant difference. The differences shown are a summary of the interspecific differences in abundance changes of (A) two isoforms of superoxide dismutase (SOD) and (B) proteins representing cysteine (cystathione β-synthase) and glutathione (glutamyl-cysteine synthase) synthesis pathways, and (C) the glutathione (peroxidase and reductase) ROS scavenging system. (After Dilly et al. 2012.)

that the differences in how the proteomes of the two species responded to heat stress may indeed affect their antioxidant capacities.

ANTIOXIDANT RESPONSES TO LOW pH CONDITIONS Estuaries are characterized by great fluctuations in CO_2 production due to the respiratory activity of the resident biota and temporarily limited exchange with open ocean water. CO_2 levels in estuaries can reach 1.3–4.7 kPa, and pH values frequently reach as low as 7.5–6.0 (Cochran and Burnett 1996; Ringwood and Keppler 2002). Mantle tissue of a common estuarine inhabitant, the eastern oyster (*Crassostrea virginica*; see photo at right), responded to a 2-week exposure to ~357 Pa P_{CO_2} (~39 Pa P_{CO_2} control), equivalent to a pH of 7.5 (8.3 control), by increasing the abundance of Cu-Zn-SOD, Prx2 (cytoplasm), Prx4 (endoplasmic reticulum), and Prx5 (mitochondria and peroxisome) (Tomanek et al. 2011). In addition, the thioredoxin homolog

Crassostrea virginica

nucleoredoxin also increased abundance in response to low pH, suggesting that the SOD-thioredoxin-peroxiredoxin system was activated to ameliorate the intracellular effects of increasing levels of ROS. Low pH may cause increased ROS production either through the release of heavy metals such as Fe^{2+} from intracellular stores, triggering the Fenton reaction (Dean 2010), or through the interaction of CO_2 with peroxynitrite, resulting in reactive carbonate, oxygen, and nitrogen species (Trujillo et al. 2008). These results were recently confirmed on the congeneric species *Crassostrea gigas* (Harney et al. 2016).

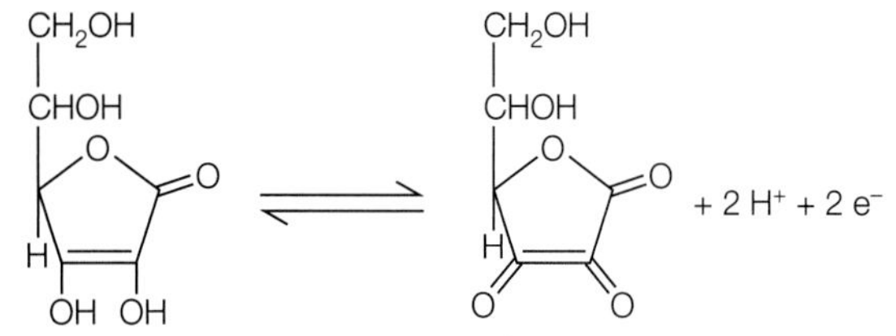

Non-enzymatic antioxidant responses

In addition to glutathione—the main cellular non-enzymatic antioxidant—three additional groups of compounds have been shown to contribute greatly to the antioxidant capacity of an organism in response to environmental change: **ascorbic acid** (vitamin C; see structure at left), **α-tocopherol** (vitamin E), and **carotenoids**. However, many more compounds are synthesized by photosynthetic organisms for protection against the effect of radiation and are found in various animal tissues to protect against ROS (Abele et al. 2012; Halliwell and Gutteridge 2015).

Ascorbic acid   Dehydroascorbic acid

Plants and most animals can synthesize ascorbic acid, but primates, fruit bats, and teleosts have lost this ability. Ascorbate is a cofactor for several enzymes, including prolyl and lysosyl hydroxylases, which are involved in oxygen sensing and the synthesis of collagen; dopamine-β-hydroxylase, which produces norepinephrine; enzymes of the biosynthesis pathway of carnitine, which is responsible for the transfer of fatty acids into the mitochondrion; and enzymes of tyrosine metabolism, which is important for an effective immune response. Ascorbic acid can scavenge superoxide anions, hydrogen peroxide, and water-soluble peroxyl ($RO_2^\bullet$) radicals. The oxidized form of ascorbic acid, dehydroascorbate, can be reduced by enzymes with dehydroascorbate reductase activity, similar to the glutathione reductase system. These enzymes include protein disulfide isomerase, glutathione S-transferase, and glutaredoxin (Halliwell and Gutteridge 2015). The ascorbate ROS scavenging system also cooperates with other antioxidant systems, for example, by regenerating the hydrophobic α-tocopherol from the α-tocopherol radical.

The hydrophobic molecule α-tocopherol (see structure below) is the most important antioxidant for inhibiting lipid peroxidation. It reacts faster with lipid peroxyl radicals than they interact with each other, and it is thereby able to terminate the radical chain reaction. Tocopherols enter the circulation as chylomicrons, and the liver selectively adds α-tocopherol into the very low density lipoproteins. Animals being fed large amounts of PUFAs have an increased need for α-tocopherol, otherwise they show signs of increased levels of lipid peroxidation.

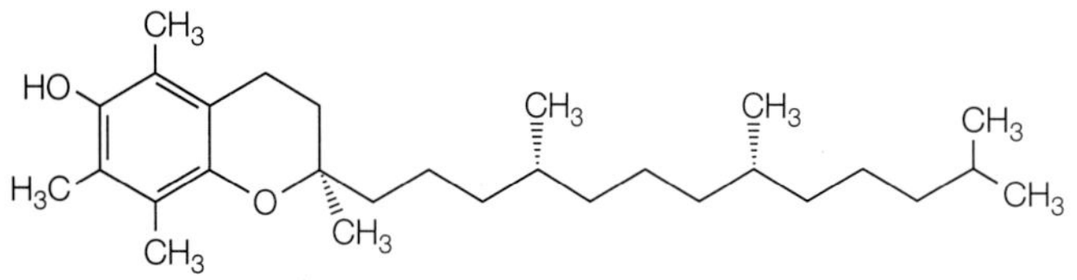

α-Tocopherol (vitamin E)

The long-chained carotenoids (e.g., β-carotene; see structure below) are capable of delocalizing electrons when they interact with radicals. They are fat-soluble and inhibit lipid peroxidation. Dietary sources of carotenoids are brightly colored vegetables and certain animal tissues, for example salmon skin. In humans, most carotenoids are found in adipose and liver tissue.

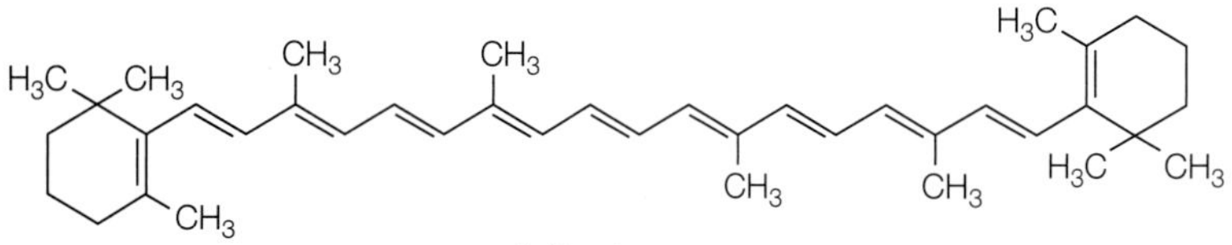

Non-enzymatic antioxidants may play an important role in adaptation to temperature extremes. For example, polar fishes, despite having similar antioxidant enzyme activities as temperate fishes, may

β-Carotene

have a greater capacity to scavenge peroxyl ($ROO^•$) and hydroxyl radicals ($OH^•$). Recent work has suggested that Antarctic fishes may have higher levels of glutathione, ascorbic acid, α-tocopherol, and carotenoids in their liver and blood plasma than temperate fishes (Heise et al. 2007; Benedetti et al. 2010).

Sources of reducing power (NADPH)

Glutathione reductase (GR) and thioredoxin reductase (TR) both require reducing equivalents in the form of NADPH to maintain levels of reduced glutathione and thioredoxin, respectively (see Figures 2.24 and 2.25). Thus the reactions producing NADPH are potentially important in contributing to an effective antioxidant response. Despite this, these reactions have received only scant attention as part of investigations into the antioxidant response to environmental stress.

Whereas the cytosolic pentose phosphate pathway (PPP), specifically the reaction of glucose 6-phosphate dehydrogenase, was long thought to be the major source of NADPH, other cytosolic and mitochondrial NADP-dependent dehydrogenase reactions have emerged as sources of reducing power during oxidative stress (Pollak et al. 2007). Specifically, isoforms of NADP-dependent isocitrate dehydrogenase (NADP-IDH) are found in the mitochondria, cytosol, and peroxisome (Jo et al. 2001; Margittai and Banhegyi 2008; Yoshihara et al. 2001). Cytosolic and mitochondrial malic enzyme produces NADPH and pyruvate, using malate as a substrate (Michal and Schomburg 2012). In addition, a cytosolic isoform of aldehyde dehydrogenase (ALDH6P) catalyzes the following reaction:

$$RCHO + NAD(P)^+ + H_2O \leftrightarrow RCOOH + NAD(P)H + H^+$$

and thereby contributes to the NADPH pool (Grabowska and Chelstowska 2003). While each of these enzymes contributes to the level of NADPH, in yeast none of these enzymes is essential on its own (Pollak et al. 2007). Furthermore, a mitochondrial membrane-based transhydrogenase uses the proton motive force across the inner membrane to reduce $NADP^+$ to NADPH (Rydstrom 2006). Finally, in the ER, reducing equivalents in the form of NADPH are provided by hexose 6-phosphate dehydrogenase (White et al. 2007).

ABUNDANCE CHANGES OF NADPH-GENERATING ENZYMES DURING ENVIRONMENTAL STRESS   Of the biochemical processes providing NADPH during oxidative stress (PPP, several NADP-dependent dehydrogenases, and transhydrogenase), enzymes of the oxidative part of the PPP (6-phosphogluconate dehydrogenase and lactonase) increased in abundance in response to acute heat stress in *Mytilus* (Tomanek and Zuzow 2010) and hypoxic stress in *Geukensia* (Fields et al. 2014). Both stressors also increased enzyme abundances in the non-oxidative part of the PPP (transketolase) in both genera. This is important because the non-oxidative part of the PPP recycles ribulose 5-phosphate to glucose 6-phosphate, generating a cycle that leads to the conversion of glucose 6-phosphate to 6 CO_2 and 12 NADPH (Nelson and Cox 2008). Transketolase was also more abundant in larvae of the barnacle *Balanus amphitrite* after exposure to CO_2-enriched acidic (pH 7.6) seawater (Wong et al. 2011). The land snail *Otala lactea* showed higher activity and substrate affinity of glucose 6-phosphate dehydrogenase, the first reaction of the PPP during estivation, presumably to increase the production of NADPH (Ramnanan and Storey 2006). Acute and chronic heat stress, as well as hyposaline stress, also led to an increase in abundance in additional proteins generating NADPH, specifically NADP-IDH and ALDH in *Mytilus* congeners (Fields et al. 2012; Tomanek and Zuzow 2010; Tomanek et al. 2012).

Together these studies establish that the production of NADPH is part of the cellular response to oxidative stress that contributes to the scavenging of ROS during environmental stress. Since multiple reactions contribute to the production of NADPH, several questions remain largely unexplored, among them: Is there a tiered sequence of which pathway dominates during acute and chronic stress? And what trade-offs exist between the production of NADPH as reducing equivalents for the biosynthesis of lipids, DNA, and RNA and for scavenging ROS?

Oxidative stress in the endoplasmic reticulum

The endoplasmic reticulum (ER) is a single-membrane enclosed compartment separate from the cytosol in which secreted and membrane proteins fold and mature. These processes include the folding of newly translated proteins, formation and proofreading of N-linked glycosylation, and importantly, disulfide bond formation (Braakman and Hebert 2013). The last process affects protein structure and function and is closely linked to the production of ROS (Bulleid 2013; Csala et al. 2010).

When newly translated proteins carry an ER signal sequence, they enter into the lumen of the ER where molecular chaperones assist their folding (**Figure 2.27**). The ER chaperones glucose-regulated protein 78 (Grp78, or BiP) and 94 (Grp94) are members of the Hsp70 and Hsp90 families, respectively, and are highly abundant and conserved (Braakman and Bulleid 2011). In addition, following the glycosylation of proteins, the carbohydrate-binding chaperones malectin, calnexin, and calreticulin recognize and promote the correct folding of glycoproteins (Pearse and Hebert 2010). The initial folding process brings cysteines into close proximity and continues by catalyzing the formation of disulfide bonds by members of the protein disulfide isomerase (PDI) family (Bulleid and Ellgaard 2011). The formation of disulfide bonds in the target protein (which involves the oxidation of two thiol groups) reduces PDI, which in turn requires an oxidase to recycle PDI back to its oxidized form (through the formation of a disulfide bond). This oxidase

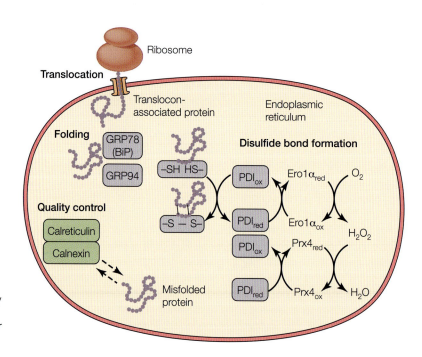

Figure 2.27 Chaperones of the endoplasmic reticulum and the generation of hydrogen peroxide by Ero1α link protein folding to oxidative stress. See text for details. (After Tomanek 2015.)

is ER oxidase 1α (Ero1α), which oxidizes PDI by mediating the transfer of electrons to the oxidase's FAD^+ group. Once FAD^+ is reduced to FADH, Ero1α reacts with oxygen to form H_2O_2 (Araki and Nagata 2012; Bulleid 2013). Thus disulfide formation requires an oxidizing environment, reflected by a much lower ratio of GSH/GSSG in the ER in comparison with the cytosol. However, the formation of non-native disulfides, which must be removed, also requires at the same time the reverse reaction, the oxidation of PDI (i.e., the reduction of protein disulfide bonds), in the same compartment. This suggests that the redox state of PDI is tightly regulated to facilitate both the formation (oxidation) and reduction of disulfides. Importantly, an enzyme that can reduce oxidized PDI has not been discovered, but the process is known to be either directly or indirectly dependent on GSH (Appenzeller-Herzog 2011).

Whereas Ero1α is essential in lower eukaryotes, higher eukaryotes have additional enzymes for oxidizing the reduced form of PDI. These include peroxiredoxin 4 (Prx4) and glutathione peroxidase 7 and 8 (Bulleid and Ellgaard 2011). Prx4 can serve as an additional electron acceptor to PDI by using the H_2O_2 from the Ero1α reaction as the terminal electron acceptor (Tavender et al. 2010). This reaction results in two PDIs being oxidized, one by Ero1α and the other by Prx4. Similarly, Gpx7 and Gpx8 show high PDI peroxidase activity and can also use the H_2O_2 generated by the Ero1α reaction as a terminal electron acceptor (Nguyen et al. 2011).

The maturation of proteins in the ER, which after glycosylation are exported (e.g., as glycoproteins in mucus), may be a source of ROS production under conditions of environmental stress. But the relevance of this protein maturation pathway as a source for ROS has only begun to be appreciated. Several proteomic studies have suggested that the ER deserves special attention as a compartment that is highly sensitive to environmental challenges that change the redox balance of the cell (Tomanek 2015). For example, recent evidence suggests that sudden reductions in salinity (hyposmotic stress) that are common during episodes of freshwater runnoff in intertidal habitats might induce oxidative stress in the ER among intertidal invertebrates, like mussels. In a study that measured proteomic responses to acute hyposmotic stress in blue mussels (*Mytilus* spp.), it was revealed that ER molecular chaperones, such as PDI, Grp78, and Grp94, increased in abundance immediately following a drop in salinity but subsequently decreased in abundance after mussels were allowed to recover in full-strength seawater (Tomanek et al. 2012). In conjuction with these responses, antioxidant enzymes, such as SOD and peroxyredoxin 5, also increased in abundance during acute hyposmotic stress and decreased in abundance after recovery.

These coordinated responses may be interpreted in the context of ER ROS production in the following way. If the expression of ER chaperones is a proxy for the maturation of proteins in the ER, then rates of H_2O_2 production from disulfide bond formation would track increases and decreases in ER chaperone abundance. This effect would then be mirrored by changes in antioxidant protein expression in response to changes in H_2O_2 concentrations. Thus these proteomic data may indicate that hyposmotic stress caused oxidative stress in the ER in marine mussels. Alternatively, these parallel shifts in ER chaperone and antioxidant enzyme expression might instead suggest that low salinity stress caused mussels to increase metabolic rates to meet higher energy demands, perhaps due to higher levels of protein synthesis, and thereby induced ROS in the mitochondria. However, the mitochondria may not have been a significant source of ROS production during acute hyposmotic stress in this study because proteins critical for aerobic metabolism, including electron transfer flavoprotein-α and mitochondrial ATP synthase, actually decreased in abundance during acute salinity stress (Tomanek et al. 2012), suggesting that aerobic respiration was repressed. Ultimately, further study will be required to sort out the context dependence of various sources of oxidative stress and specifically the contributions made to it by protein maturation in the ER.

2.9 Metabolic Rates and Oxygen

Basal metabolic rate (BMR) and standard metabolic rate (SMR) are expressions used for endotherms and ectotherms, respectively, to describe a metabolic state that supports the adult organism's basic needs while it is resting, fasting, not regulating its body temperature, or reproducing (Rolfe and Brown 1997). States in which the metabolic rate of an organism falls well below the BMR are referred to as **metabolic depression**, or in the more recent literature as **hypometabolism**. **Maximum metabolic rates (MMRs)**, which are achieved during exhaustive physical activity, are referred to as **hypermetabolism**. Organisms going about their lives sustain a widely varying metabolic rate that is referred to as **field metabolic rate**. It is important to realize that the relative contributions of organs to the BMR do not stay the same during conditions of hypo- and hypermetabolism. For example, kidney and digestive system function may decrease during hypometabolic conditions and thus contribute little to the metabolic rate. Thus conditions of hypo- and hypermetabolism are also characterized by major shifts in the relative contributions of the different metabolic processes conducted by different organs.

How metabolic rates are measured: Direct and indirect calorimetry

About 95% of the chemical energy that is released through the oxidation of food under standard conditions is turned into heat. Therefore, measuring the production of heat is the most direct way of quantifying metabolic rate (Rolfe and Brown 1997). This method is termed **direct calorimetry**. However, measurements of oxygen consumption, a method called **indirect calorimetry**, can be used as a close approximation of heat production if an **oxycaloric equivalent** is used (joules [J] of heat released per oxygen consumed). The amount of heat generated per oxygen varies with the type of fuel used by the cell, but ranges only between 18.7 J per ml of O_2 for proteins and 21.1 J per ml of O_2 for carbohydrates, with an average of 20.1 J (Hill et al. 2016). Most studies of BMR and SMR have used oxygen consumption as the index of metabolism, so most of the studies discussed below represent indirect calorimetry. Whereas most energy released through metabolic activities is in the form of heat, some of the energy is converted into excretion products, reproductive output, body growth, and external work (**Figure 2.28**).

Up to this point in our discussion of metabolism, we have mostly portrayed metabolism as ATP production and consumption, linked through intermediate metabolites (i.e., phosphogens and reducing equivalents). However, several other phenomena represent temporary storage of free energy released through metabolism. These include the proton gradient across the inner mitochondrial membrane and the Na^+ gradient across the cell membrane of animal cells. Maintaining these ion gradients requires considerable energy, but when the gradients are dissipated to conduct work, this initial investment of energy has important payoffs, notably in ATP production and active transport.

Contributions of individual organs to metabolic rate

The contribution of individual organs to BMR or SMR is based on estimating their oxygen consumption during in vivo or in vitro conditions. In vivo measurements are made on perfused organs, involving processes that are key to the function of an organ, such as the transport of ions in the kidney. In vitro conditions may involve isolated organs or

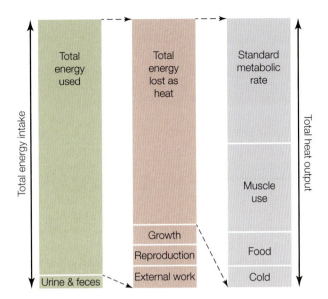

Figure 2.28 Distribution of assimilated energy in humans. Total energy intake (first column) is distributed to urine and feces and the actual energy used by the body. The energy being used is either lost as heat or invested into growth, reproduction, and external work (middle column). The heat that is produced is generated through the standard metabolic rate, muscle activity, food processing (specific dynamic action), and regulation of body temperature (third column). (After Rolfe and Brown 1997.)

tissues that are not carrying out these key processes and thus are likely to underestimate metabolic rate. Measuring metabolic rate under either condition is challenging, and the sum of all rates from all measurements rarely adds up to a perfect 100% of the metabolic rate of the whole organism.

The most general finding from studies of vertebrates is that some organs make a disproportionately large contribution to overall SMR or BMR (**Figure 2.29**). These organs include liver, kidney, heart, and brain and represent a relatively small fraction of the mass of the organism relative to skeletal muscle. However, an organ with low mass-specific metabolic rate, such as skeletal muscle, may make a major contribution to total metabolism under conditions of high activity, when metabolic rate greatly exceeds BMR or SMR. Adipose tissue is metabolically relatively inert and thus contributes minimally to BMR or SMR. The relative contributions of organs and tissues shift during development; for example, juvenile animals are often not fully capable of all locomotory activities because their skeletal muscle system is still developing.

The metabolic capacity of mammalian organs depends on the organism's lifestyle: Diving mammals

Animals with different lifestyles may have the same set of organs, but do these organs perform in similar ways across species? For example, are the same organs in terrestrial and marine mammals contributing in the same way to the resting metabolic rate (similar to the ideal BMR but distinguished as resting, as it is often impossible

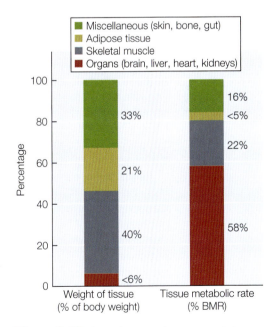

Figure 2.29 Contribution of organs and tissues to the basal metabolic rate (BMR) of a nonobese man. Note that several organs account for less than 6% of body weight but make a disproportionately high contribution (> 50%) to BMR. (After Elia 1992.)

to control such factors as food intake or state of reproduction in animals, especially in free-ranging ones), or are their relative and absolute contributions different? Terrestrial mammals increase their ventilation and heart rates when they exercise. They also use peripheral vasodilation to redirect more blood flow to the muscle and skin for heat dissipation. In contrast, marine mammals hold their breath (apnea), reduce their heart rate (bradycardia), and employ peripheral vasoconstriction to direct blood flow to the brain, lung, and heart during diving.

In marine mammals, such as the Weddell seal (*Leptonychotes weddellii*) shown at left, the ability for prolonged diving depends in large part on the animal's ability to perform efficient aerobic metabolism for as long as oxygen stores are available. Thus the animal attempts to retain an efficient aerobic means of ATP production for as much of the dive time as possible. Seals (pinnipeds) are able to do this in part by having a greater density of mitochondria in several of their organs, including skeletal muscle, liver, kidney, and stomach, in comparison with terrestrial mammals such as rat and dog (**Figure 2.30A**) (Kanatous et al. 1999; Fuson et al. 2003). The increase in mitochondrial density is conjectured to facilitate aerobic metabolism under the hypoxic conditions that prevail during a dive, by decreasing the diffusion distance between the mitochondria and the intracellular stores of oxygen bound to myoglobin. Oxygen is also more soluble in lipids than in aqueous solutions, so high densities of mitochondrial membranes facilitate oxygen transport (see Chapter 3). However, the biochemical adaptations that facilitate breath-hold diving involve considerably more changes in the ATP-generating apparatus. Especially in the heart and liver, there are higher levels of **hydroxyacyl-CoA dehydrogenase** (**HOAD**), even when standardized against the resting metabolic rate (RMR), in comparison with terrestrial mammals (**Figure 2.30B**) (Fuson et al. 2003). HOAD is indicative of capacity for aerobic fat metabolism. Compared with terrestrial mammals, seals turn out to have higher levels of HOAD relative to another aerobic enzyme, citrate synthase (CS), leading to a CS/HOAD ratio of less than 1. These findings suggest that marine mammals are capable of relying on fatty acid oxidation, which is more energy efficient than carbohydrate oxidation, during a dive to extract more ATP per oxygen, while they are still within their aerobic dive limit (ADL; the duration before they start relying mainly on anaerobic metabolism).

Weddell seal (*Leptonychotes weddellii*)

Relative to terrestrial mammals (rat and dog), seals also show greater levels of lactate dehydrogenase (LDH) activity—at least in liver, stomach, and intestine—when standardized to RMR, indicating a greater capacity for anaerobic metabolism once the animal has reached its ADL (**Figure 2.30C**). Thus the organs of marine mammals are characterized by a greater capacity to maintain aerobic metabolism during breath-hold dives, but once the ADL is reached, these organs are capable of long durations of anaerobic metabolism (Fuson et al. 2003). The ability for prolonged anaerobic metabolism is in part facilitated by the greater buffering capacity that characterizes the muscle tissue of marine versus terrestrial mammals (Castellini and Somero 1981). This greater buffering capacity attenuates the decreases in pH that occur in cells during prolonged reliance on anaerobic glycolysis and therefore is an important adaptation to enable diving mammals to maintain adequate levels of ATP production if their dive exceeds the ADL. The limited data available on buffering compounds in marine mammals suggest that histidine-containing dipeptide buffers (see Section 4.6) play an especially important role (Abe 2000).

(A)

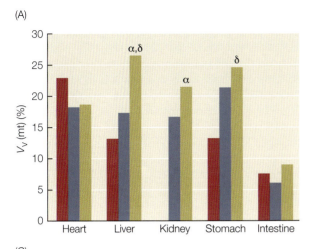

(B)

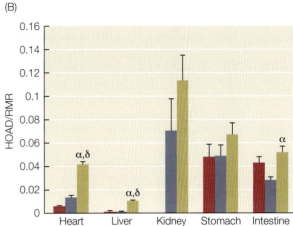

(C)

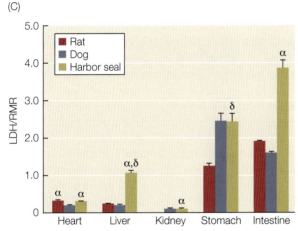

Figure 2.30 (A) Mean mitochondrial density, (B) hydroxyacyl-CoA dehydrogenase (HOAD) activity, and (C) lactate dehydrogenase activity standardized against tissue-specific resting metabolic rates (RMR) for rats, dogs, and harbor seals (*Phoca vitulina*). α significantly different from dog, δ significantly different from rat (ANOVA, $P < 0.05$). (After Fuson et al. 2003.)

Processes contributing to oxygen consumption under BMR and SMR

Oxygen-using reactions in addition to the cytochrome *c* oxidase reaction contribute to the aerobic metabolic rate of the organism. For example, several oxidases in peroxisomes that are involved in the oxidation of fatty acids, purines, and amino acids transfer one or two electrons to oxygen, generating $O_2^{\bullet-}$ and H_2O_2 (Nordgren and Fransen 2014). Cytochrome P450 residing in the ER and other organelles uses oxygen in the metabolism of xenobiotics, drugs, steroids, arachidonic acid, eicosanoids, and retinoic acid (Halliwell and Gutteridge 2015). Non-mitochondrial oxygen consumption is estimated to represent approximately 10% of total respiration (Rolfe and Brown 1997).

Oxygen consumption in the mitochondria is determined mainly by oxidative phosphorylation, that is, ATP synthesis. However, when this process is inhibited, there is a residual consumption of oxygen of about 20%, which is based on **proton leak** across the inner mitochondrial membrane. This leakage partially dissipates the proton gradient built by ETS activity that terminates in the reduction of oxygen to water. Thus even in the absence of ATP synthesis, a substantial amount of oxygen consumption by mitochondria is possible. As the proton leak is dependent on the potential of the proton gradient across the inner mitochondrial membrane, it has to be estimated at an in vivo membrane potential. As discussed earlier, proton leak may occur through membrane channels termed uncoupling proteins (UCPs), whose activities are tightly regulated in concert with needs to reduce ROS production or requirements for heat production (see Section 3.7).

Most estimates of ATP production rely on the measurement of oxygen consumption as a proxy. It is therefore important to establish how tight this link is and identify the processes that contribute to it. Given that part of the oxygen consumed by an organism is used by non-mitochondrial processes and the mitochondrial proton leak, it is important to first estimate the *theoretical* ratio of ATP synthesized to oxygen consumed—the P/O ratio, which is the H^+/O ratio (the number of H^+ transported for each O) multiplied by the ATP/H^+ ratio (the ATP synthesized to the number of protons used)—and compare it with the *effective* ratio of ATP produced to oxygen consumed at the whole-tissue level. These estimates provide a value for how much ATP can be synthesized for each reducing equivalent (NADH and FADH) produced through the various metabolic pathways. Older estimates of the theoretical ratio were 3 and 2 for mitochondria respiring on NADH or succinate (providing FADH), respectively, but current reliable estimates are 2.5 and 1.5 (Rolfe and Brown 1997). These numbers are in agreement with the estimates of the H^+/O ratios (10 for mitochondria oxidizing NADH and 6 for mitochondria respiring succinate) and a H^+/ATP ratio of 3 for mitochondria plus 1 H^+ for the transport of ADP and P_i into the organelle, leading to a total of 4 H^+/ATP ($ATP/H^+ = 0.25$). Now, if we take the non-mitochondrial consumption of oxygen and the proton leak into account, we get estimates of P/O = 1.8 for whole-body respiration and P/O = 2.0 for whole-mitochondria respiration.

Several ATP-consuming processes contribute differentially to the BMR or SMR. In order of the most likely percentage contribution (based on studies on humans and rats), with the highest first, these are: protein synthesis and degradation (~28%), Na^+-K^+-ATPase (19%–28%), gluconeogenesis (7%–10%), urea synthesis (nitrogen excretion; 7%–8%), Ca^{2+}-ATPase (4%–8%), and actomyosin ATPase (2%–8%), plus other processes that may include RNA turnover (Rolfe and Brown 1997). The contributions of these processes vary from organ to organ, and some change greatly with exercise, for example, Ca^{2+}-ATPase (Figure 2.31).

Another perspective for describing BMR and SMR is the relative contributions made by oxidation of different metabolic fuels, that is, fat, carbohydrate, and protein. At least for humans, the contributions are approximately 60%, 30%, and 10%, respectively, under BMR conditions. During fasting, the percentage contribution of fat metabolism increases (up to 90%), while amino acid metabolism (most likely to contribute glucose via gluconeogenesis) stays at 10%, and carbohydrate metabolism becomes minimal. The contribution of anaerobic glycolysis to ATP production during BMR conditions is estimated to be 2.4% and 3.1% in humans and rats, respectively. With the contribution of glycolysis during aerobic (mitochondrial) metabolism, the total percentage of glycolysis to BMR is likely between 4.6% and 7.8% (Rolfe and Brown 1997).

The effect of body mass on metabolic rate

Body mass has a fundamental effect on basal and standard as well as maximum metabolic rates. For example, an elephant may weigh ~4,000,000 g, about 400,000 times more than a mouse (~10 g); but even though the elephant requires more overall energy to maintain its body mass, its total metabolic requirement is a lot less than the difference in mass would suggest (only ~1600 times more than the BMR of a mouse weighing 10 g). Per gram of body mass, the metabolic rate of the elephant is about 1/25th the value for the mouse (Schmidt-Nielsen 1984). This has several important consequences, such as the requirement for more frequent food provisioning in smaller animals and a higher

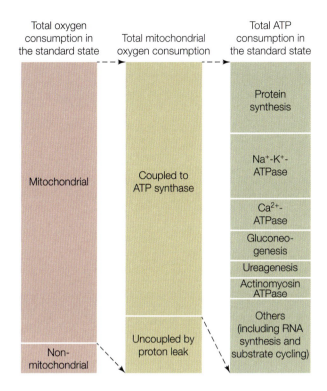

Total oxygen consumption in the standard state

Total mitochondrial oxygen consumption

Total ATP consumption in the standard state

Mitochondrial

Non-mitochondrial

Coupled to ATP synthase

Uncoupled by proton leak

Protein synthesis

Na$^+$-K$^+$-ATPase

Ca^{2+}-ATPase

Gluconeo-genesis

Ureagenesis

Actinomyosin ATPase

Others (including RNA synthesis and substrate cycling)

Figure 2.31 Estimated contributions of different cellular processes to total oxygen consumption, mitochondrial oxygen consumption, and ATP consumption to basal and standard metabolic rate (standard state). Mitochondrial and non-mitochondrial contributions to oxygen consumption are estimated to be 90% and 10%, respectively. ATP synthesis and proton leak contribute approximately 80% and 20% to mitochondrial respiration, respectively. The estimates in the first two columns are based on measurements from liver, heart, and skeletal muscle of humans and rats. The third column shows the contribution of various ATP-consuming processes to total ATP consumption. (After Rolfe and Brown 1997.)

foraging intensity. The relationship also leads to lower costs of transport and therefore a greater ability to migrate for larger animals. Furthermore, a given percentage of stored fuel (such as depot lipids) relative to body mass will last a larger animal longer, again making it easier for larger animals to migrate.

The relationship between metabolic rate (oxygen consumption) and body mass is described by the equation

$$\dot{V}_{O_2} = a\, M^b \quad (2.1)$$

The logarithmic form of the equation is

$$\log \dot{V}_{O_2} = \log a + b \log M \quad (2.2)$$

and describes a linear relationship, with $\dot{V}_{O_2}$ being the oxygen consumption, M the body mass in grams, a the intersection with the y axis, and b the slope of the line (the change in metabolic rate with each gram of body mass).

Much has been written about the exponent b, which is lower than 1 and therefore denotes an **allometric** rather than an isometric (proportional) relationship; metabolic rate increases with body mass at a slope less than 1 (Schmidt-Nielsen 1984). Early measurements by Rubner (1883) suggested a value of b near 2/3 (0.67). This is the ratio of the increase in surface area to increase in volume for an expanding sphere. A logical extension was to infer that smaller endothermic animals lose more heat across a relatively larger surface area than larger animals do across a smaller surface area relative to body volume. Thus scaling was initially thought of in terms of exhausting heat to the environment by mammals. However, later studies showed that the relationship also holds for ectotherms, including even unicellular organisms (Hemmingsen 1960). These broader findings showed that the scaling of aerobic metabolism is not just a consequence of

Figure 2.32 Relationship between mass-specific metabolic rate ($\dot{V}_{O_2}$) and body weight (M) in four vertebrate groups. The lines for birds (avian reptiles) and placental mammals show basal metabolic rates. The line for lizards shows standard metabolic rates at 37°C, the body temperature of placental mammals. The SMRs of amphibians were determined at 25°C. Notice that the exponents represent ($b - 1$), so the values for b are between 0.65 and 0.70. (After Hill et al. 2016; equations from McKechnie and Wolf 2004 [birds], Hayssen and Lacy 1985 [mammals], Templeton 1970 [lizards], and Whitford 1973 [amphibians].)

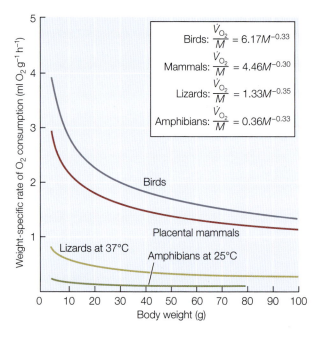

heat transfer relationships. Also, additional measurements by Max Kleiber (1932) suggested that the value of b is closer to 0.75 and therefore different from the ratio of the presumed change of surface area with an increase in volume. The question of whether the average value of b is closer to 0.67 or 0.75 is still being debated (Savage 2004; White and Seymour 2003).

In order to get mass-specific metabolic rate (MSMR), we have to divide Equation 2.1 by the mass (**Figure 2.32**):

$$\dot{V}_{O_2}/M = a\,M^b/M = a\,M^{(b-1)} \quad (2.3)$$

The logarithmic form of this equation forms a linear relationship with $b - 1$ being the slope of the relationship of change in MSMR with body mass:

$$\log \dot{V}_{O_2}/M = \log a + (b-1)\log M$$

While the overall relationship (shape of the curve) of MSMRs generally holds true among organisms, it is also important to point out that the constant a (y-intercept) varies among phylogenetic groups and between organisms from drier versus more humid environments. For example, a varies in the following order: birds > mammals > lizards and amphibians (see Figure 2.32; Hill et al. 2016). These differences may be caused by a larger proton leak and thus higher rates of non-phosphorylating respiration in birds (avian reptiles) and mammals in comparison with non-avian reptiles (Brand et al. 1991).

CAUSES OF METABOLIC SCALING   The relationship between mass-specific metabolic rate (MSMR) and body mass raises questions of causation in both ultimate and proximate contexts (see Chapter 1). Why is this scaling pattern found so commonly, and what mechanisms account for it? We first consider mechanistic issues. Here, a possible approach is to consider how different organs, tissues, and cellular processes contribute to the metabolic rate of the whole organism. Earlier in vitro measurements of organ oxygen consumption rates of mammals suggested that the MSMRs of the kidney, brain, liver, spleen, and lung follow approximately the same allometric relationship as the MSMR of the whole organism (Krebs 1950). More recent in vivo experiments have estimated organ oxygen consumption rates by measuring oxygen concentrations in blood flowing into and out of the organ and the rate of blood flow. Results of such measurements in five mammals ranging in mass from 0.48 to 70 kg showed values for b between 0.60 and 0.86 for liver, brain, kidney, heart, and all remaining tissues (Wang et al. 2001). In addition, the masses of the liver, brain, and kidney also vary allometrically with body mass, meaning that metabolically active tissues such as these represent a decreasing proportion of the whole body mass in larger organisms (by contrast, heart and remaining tissues vary isometrically). The combination of the resting metabolic rates of the various organs with their allometric contribution to body mass yielded an equation with values similar to the ones obtained by Kleiber, with an exponent of 0.76. The relationship indicates that the four most active organs account for 68% of the RMR of a mammal weighing 0.1 kg, but for only 34% of the RMR of a mammal weighing 1000 kg (Wang et al. 2001). Thus metabolic scaling is a reflection of the allometric relationship of metabolic rates of highly active organs to body mass times their decreasing contribution to body mass.

This leads us to ask if metabolic rates of single cells also decrease with increasing body mass. Indeed, measurements of single cells from mammals, birds, and non-avian reptiles confirm the same trend (Else et al. 2004; Hulbert et al. 2002; Porter 2001). Mitochondrial densities and the surface area of mitochondrial cristae also decrease with increasing body mass in brain, liver, kidney, heart, and skeletal muscle (Mathieu et al. 1981), and so does the proton leak across the inner mitochondrial membrane (Brand et al. 2003; Porter 2001). Other cellular ATP-consuming processes, such as the Na^+-K^+-ATPase (Couture and Hulbert 1995), Ca^{2+}-ATPase (Hamilton and Ianuzzo 1991), and protein synthesis (Hawkins 1991), also decline with increasing mass. Finally, mass-specific activities of oxidative enzymes decline while those of glycolytic enzymes increase with increasing mass in white muscle tissue in pelagic fishes (Childress and Somero 1990). The ultimate cause of the increase in glycolytic activity with increasing body mass is conjectured to reflect selection for conservation of burst-swimming ability in fishes of different mass (Somero and Childress 1980). Thus metabolic systems like glycolysis that contribute strongly to short-term ("burst") spikes in work, during which physiological hypoxia occurs, scale differently from aerobic systems that contribute to ATP generation under conditions where oxygen supply is adequate for supporting catabolism.

So far we have considered the underlying processes that contribute to the scaling of metabolic rates during conditions of basal or standard metabolic rates. During strenuous physical exercise, these contributions shift dramatically, even if physiological hypoxia is avoided. For example, oxygen consumption rates of skeletal muscle tissues involved in locomotion account for about 90% of the maximum metabolic rate (MMR), while the relative contributions of such organs as brain, liver, and kidney are greatly diminished. However, unlike mitochondrial respiration rates under BMR, during MMR mitochondrial respiration rates of skeletal muscle display mass-independent respiration rates under in vivo conditions, suggesting that the delivery of oxygen by the cardiovascular system is closely matched with the oxygen demands of mitochondria (Weibel et al. 2004).

In addition, under MMR the exponent b is also significantly greater than under BMR, lying closer to 0.86 than to 0.75 (Weibel et al. 2004). This suggests that the exponent itself is not as constant as suggested by some of the theories that attempt to explain the allometric scaling of metabolic rates with a single mathematical equation based on the supply of oxygen and nutrients (Suarez et al. 2004).

THEORIES TO EXPLAIN THE SCALING OF AEROBIC METABOLIC RATES   Given the high consistency among taxa of the allometric relationship between aerobic metabolic rate and body mass, the lack of a theory that provides a general causal (mechanistic) basis for this relationship has been a vexing challenge to physiologists (Schmidt-Nielsen 1984). Recent attempts to provide a mechanism behind this relationship can be divided into single-cause versus multiple-cause theories. The single-cause theories focus on the supply of oxygen and nutrients to cells—the supply side of metabolism—as the single main cause of the allometric scaling of metabolism. Several attempts at a single-cause theory focus on the efficiency of transport systems that have a fractal architecture (a replication of similar structures across different scales), not unlike the cardiovascular system; such analyses lead to a scaling factor of 0.75 based on the properties of a branching tubular delivery system. These approaches commonly lack specific physiological details and are presented instead as a general model based on mathematical and fractal geometry relationships (Banavar et al. 2002; West et al. 1997; West et al. 1999; West et al. 2002). In contrast, a multiple-cause theory emphasizes that multiple processes together establish the scaling exponent and that these causal factors shift in their contribution between BMR/SMR and MMR (Darveau et al. 2003).

To evaluate the merits of these theories we need to consider some of the foundations of metabolic control theory. There is widespread evidence that any flux through biological systems is controlled by more than one of the processes involved, with those exerting greater control being characterized by greater metabolic control coefficients—the change in pathway flux as a particular enzyme changes concentration or activity (Fell 1997). There is also an acknowledgement that these coefficients can shift under different circumstances. For example, Brand and colleagues (1993) used a top-down metabolic control analysis that clustered multiple reactions into blocks connected by a shared intermediate, in this case proton motive force (Δp) across the inner mitochondrial membrane. This analysis showed that the control coefficients of the Δp-generating reactions (substrate oxidation) and those of reactions dissipating Δp—the proton leak and ATP synthesis (or the phosphorylating system)—change when respiration shifts from state 4 (non-phosphorylating) to state 3 (phosphorylating) conditions (**Figure 2.33**). These and other studies indicate that metabolic rates are indeed controlled by multiple processes whose contributions can change. These insights from metabolic control analysis cast doubt on a single-cause explanation for the allometric scaling of metabolic rates. In addition, any theory based on the importance of the supply system as a single cause

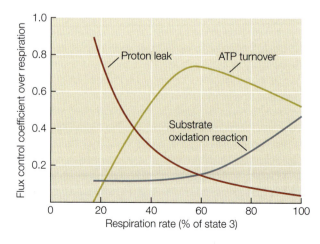

Figure 2.33 Flux control coefficients of blocks of reactions that either contribute to building the proton motive force (substrate oxidation) or dissipate the proton motive force (ATP turnover and proton leak) change when respiration shifts from non-phosphorylating (state 4) to 100% phosphorylating (state 3). (After Brand et al. 1993.)

must acknowledge that there is an almost tenfold increase in supply during the shift from standard to maximum metabolic rate. This difference suggests that a limitation to supply during BMR or SMR is hardly an issue, even though the supply of oxygen and nutrients does play a more controlling role during maximum metabolic rates (Suarez and Darveau 2005). Furthermore, the scaling exponent b changes from closer to 0.75 at basal or standard levels to 0.86 during MMR. Thus a theory that focuses exclusively on one value for b must be viewed as incomplete.

The multiple-cause theory illustrates its heuristic power by comparing the processes that contribute to maximum and basal metabolic rates (Darveau et al. 2002; Darveau et al. 2003). The idea behind this theory is that multiple processes have their own specific exponent b and differ in contribution and regulatory control (metabolic control coefficient) to the overall metabolic rate. Processes such as the activity of the Na^+-K^+-ATPase and protein synthesis contribute the most to the BMR or SMR, while Ca^{2+}-ATPase, ureagenesis, and gluconeogenesis contribute much less. Given that oxygen supply capacity is in large excess in relation to the oxygen requirements of organisms' organs during BMR or SMR, the contributions of alveolar ventilation, pulmonary diffusion, cardiac output, and capillary-mitochondria tissue diffusion to BMR or SMR are minimal. Thus metabolic rates are set in large part by the ATP turnover and not by any limitations in the oxygen supply system. The approach taken here is similar to the idea that each organ contributes to the overall metabolic rate with its unique allometric relationship (or exponent b), weighted by its allometric relationship to body mass, as discussed earlier.

During strenuous physical exercise or the metabolic response to extreme cold—conditions that require greatly elevated metabolic rates—the contributions of control coefficients of the various oxygen supply and ATP turnover processes change; that is, the coefficients for the oxygen supply components become much greater than the ones for the ATP turnover processes.

The ultimate aim of the quest to understand the mechanisms behind aerobic metabolic scaling is a general explanation that encompasses different phylogenetic groups as well as BMR/SMR and MMR. Such an explanation has to be grounded in the details of the actual biology we are trying to explain. Mathematical theories that neglect physiological complexity may fail to describe reality in an adequate manner. Further work on scaling seems warranted, however, because it is likely that any progress on this topic will also contribute to a better understanding of the control of metabolism overall.

Rates of living and length of life: Metabolic scaling and the ROS-lipid connection

Given the different life spans of smaller versus larger animals and the similar total energy expenditure over the lifetime of organisms across a wide range of body masses, physiologists have sought to establish a link between the intensity of life—mass-specific rate of oxygen consumption—and its duration (Hulbert et al. 2007). The realization that the underlying mechanism for this link could include the rate of ROS generation by mitochondrial oxygen use prepares us for addressing two intriguing questions that have long fascinated physiologists and biochemists. First, what biochemical mechanisms account for the higher mass-specific metabolic rates of small animals relative to larger ones? A second and closely related issue—the *causes of aging*—has perhaps an even deeper fascination for biologists and a longer history of discussion. Why do smaller species have shorter life spans than larger species? Interest in this issue goes back at least to Aristotle (384–322 BCE), who called for a research program on this point: "The reasons for some animals being long-lived and others short-lived, and, in a word, causes of the length and brevity of life call for investigation" (Grimm 2015). Why doesn't a mouse live as long as an elephant? Does the higher mass-specific metabolism of small animals doom them to shorter life spans? Can we gain insights into human life spans by seeing what differences exist between small rodents and large pachyderms?

We address these questions here in the contexts of birds and mammals—endothermic homeotherms, which have metabolic rates severalfold higher than those of ectotherms (see Chapter 3 and Figure 2.32). Comparisons of the metabolic rates of different-sized species within these two groups have shed important light on the underlying determinants of oxygen consumption rates and, by extension, rates of ROS production. In turn, insights into size-related generation of ROS have provided a unique perspective on the determinants of life span. Given the damaging effects of ROS, greater production of ROS is associated with metabolic decline—caused, for example, by damage to mitochondrial membranes and DNA—implying that higher mass-specific metabolic rates should bring an earlier onset of damage and thus aging. The answer to these closely related questions concerning the underlying biochemical determinants of metabolic scaling and life span may derive to a considerable extent—though certainly not entirely—from differences in membrane lipid chemistry that have recently been discovered in mammals and birds of different body masses (Hulbert et al. 2007). These differences in lipid composition may function as **metabolic pacemakers** and serve as a causal factor in both metabolic scaling relationships and longevity.

MITOCHONDRIAL ENZYME ACTIVITIES AND SCALING   If we narrow down the analysis of aerobic metabolic scaling to focus on the set of proteins that is responsible for oxidative phosphorylation and the reduction of molecular oxygen to water, we can then ask how the mitochondria of small mammals or birds succeed in carrying out more rapid rates of oxygen use than the mitochondria of their larger-mass kin. There would seem to be three distinct options for such size-dependent modulations of mitochondrial enzymatic activity: (1) more mitochondria per unit of mass of tissue, (2) more enzymes per unit of mass of mitochondria, or (3) enzymes with different specific activities (rate of function per enzyme active site). Broad comparisons of mitochondrial abundance and function in diverse mammals and birds have shown that the third type of adaptation is of primary importance in scaling: Different mass-specific rates of oxygen consumption are supported by approximately the same number of mitochondrial enzyme molecules per unit of mass of mitochondria in birds and mammals of different mass (Hulbert et al. 2007).

What makes the specific activities of mitochondrial enzymes higher in smaller species? Once again, three options would seem to be available. First, the intrinsic catalytic power of an enzyme, measured as the rate at which an active site can convert substrate to product—the catalytic rate constant (k_{cat})—might be evolutionarily adjusted such that small species have enzymes with higher k_{cat} values than the orthologous enzymes of larger species. In Chapter 3 we will see that species adapted to different temperatures have orthologous enzymes with adaptively different k_{cat} values, such that cold-adapted enzymes can work faster at a given temperature of measurement than orthologs of warm-adapted species. However, there appear not to be differences in the intrinsic catalytic power, the k_{cat} values, of orthologous mitochondrial enzymes of different-sized mammals and birds. Evolutionary temperature, not body mass, drives selection for k_{cat}. A second option might involve posttranslational modifications that modify the k_{cat} value of the proteins. This option, too, seems not to be exploited in metabolic scaling. A third option, one that we will encounter at several junctures in this volume, is to change the milieu in which the enzymes work to alter the specific activity of the enzyme. In the case of respiration, the key feature of the milieu is the lipid composition of the inner mitochondrial membrane. The lipid environment surrounding the proteins involved in respiration modulates their activity, an influence termed a **viscotropic effect**.

VISCOTROPIC EFFECTS: LIPID-ENZYME INTERACTIONS   Changes in membrane phospholipid composition can change enzyme specific activity (viscotropic effects) because of lipid-modulated effects on the energy barriers the enzymes face in changing their structural organizations during function. In a highly viscous lipid microenvironment, proteins are likely to encounter more resistance when they change their three-dimensional structures (conformations) or move laterally in the membrane. These protein movements are of key importance in governing reaction rates. Decreasing lipid viscosity thus would be predicted to increase the rate of protein function. Viscotropic effects have wide-ranging consequences in thermal biology, as we will see in Chapter 3 when we examine how ectothermic species adjust their membrane lipid compositions to cope with changes in body temperature; cold adaptation leads to reduction in the intrinsic viscosity of membranes, which in turn leads to some offsetting of the decelerating effects of reduced temperature on activities of membrane-localized enzymes. In endothermic homeotherms, where body temperatures are tightly regulated, viscotropic effects are concerned not with compensation to temperature, but with adjusting the viscosity of the "working environment" of mitochondrial enzymes to give them body-size-specific catalytic power.

The pattern of variation in membrane lipid composition—specifically, in the acyl chain composition of phosphoglyceride lipids—among birds and mammals of different body size is shown in **Figure 2.34**. Three trends, common to both birds and mammals, are

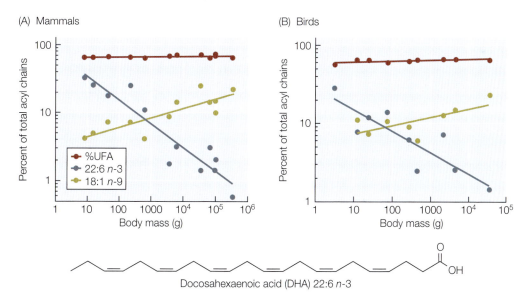

Docosahexaenoic acid (DHA) 22:6 *n*-3

Figure 2.34 Changes in membrane lipid composition with body mass in (A) mammals and (B) birds. The percentage contributions of different types of acyl chains vary regularly in both groups, even though the total amount of unsaturated fatty acids (UFAs) is essentially size-invariant. Nomenclature: The first number is the total carbon chain length; the second is the total number of double bonds (22:6 means a 22-carbon acyl chain with 6 double bonds); the "*n*" number is the carbon atom at which the first double bond is found, counting in from the terminal methyl group (left end of DHA molecule as shown below the graphs). (After Hulbert et al. 2007.)

evident. First, the total percentage of **unsaturated fatty acids** (**UFAs**), those containing one or more double bonds, does not vary significantly with body mass in either group of animals. Thus membrane viscosity is not modulated by an overall change in percentage of acyl chains containing double bonds. Second, a common acyl chain, the 18-carbon monounsaturated stearic acid, increases as a percentage of acyl chains with increasing body size. Third, and of critical importance for the mechanistic analysis that follows, the percentage of the polyunsaturated fatty acid **docosahexaenoic acid** (**DHA**) shows a regular decrease with increasing body mass. As a consequence, DHA has the same scaling pattern as metabolic rate.

DOCOSAHEXAENOIC ACID (DHA): AN ACYL CHAIN WITH GOOD AND BAD EFFECTS   The similar scaling patterns found for DHA and mass-specific metabolic rate have been proposed to reflect a causal linkage mediated by viscotropic effects. DHA has a strong fluidizing (viscosity-reducing) influence on the interior of the membrane bilayer. Thus high percentages of DHA would be expected to reduce the energy costs of protein conformational changes or lateral movements during mitochondrial respiration. Small birds and mammals, with their substantially higher levels of DHA, could thereby obtain higher specific activities of proteins involved in oxidative phosphorylation. In other words, the same number of enzyme molecules in the mitochondrial membranes of small birds and mammals can process more substrates per unit of time than can the orthologous enzymes in larger species because they work in a less viscous lipid milieu.

The benefits that accrue from high percentages of DHA are paired with negative consequences for the cell as well. These negative effects derive from the production of ROS during oxidative phosphorylation. First, a higher rate of electron transport, and thus use of oxygen, might be expected to lead, in and of itself, to greater ROS production. Second, DHA is unusually sensitive to damage from ROS (**Figure 2.35**), and the by-products that are formed when ROS interact with DHA include lipid peroxyl radicals that cause further downstream damage. As pointed out earlier, lipid peroxyl radicals foster chain reactions of free radical injury to lipids, proteins, and perhaps most critically, DNA. Thus, not only are existing cellular structures damaged, but lipid peroxidation can also lead to genotoxic effects that alter genome content, which in turn could disrupt gene expression or protein function. These effects on DNA integrity may have an especially critical impact on cellular survival—and life span.

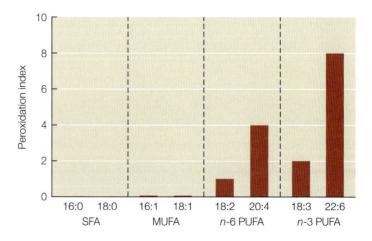

Figure 2.35 Relative susceptibilities of different fatty acids to peroxidation. Values are expressed relative to the rate of peroxidation for linoleic acid (18:2 *n*-6), which is assigned a value of 1.0. SFA, saturated fatty acids; MUFA, monounsaturated fatty acids; *n*-6 PUFA, omega-6 polyunsaturated fatty acids; *n*-3 PUFA, omega-3 polyunsaturated fatty acids. (After Hulbert et al. 2007.)

The close linkage between the propensity for ROS-induced damage—that is, for lipid peroxyl radical formation—and life span is shown by **Figure 2.36**. In birds and mammals, short-lived smaller species have a much higher potential for generation of lipid peroxyl radicals (see Figure 2.36A), as a consequence of their membrane phosphoglyceride acyl chain compositions (see Figure 2.34). Correspondingly, the scaling of life span for birds and mammals shows an allometric mass-specific scaling coefficient less than 1.0 (*b* values near 0.2 for both groups; see Figure 2.36B).

In summary, the proximate cause of aerobic metabolic scaling of BMR in birds and mammals appears to involve adjustments in membrane phosphoglyceride acyl chain composition that lead to modulations of rates of activities of proteins involved in electron transport and oxidative phosphorylation—viscotropic effects. The primary type of acyl chain involved in adjusting membrane viscosity, DHA, is highly sensitive to ROS damage, and when this damage occurs, lipid peroxyl radicals are formed with large downstream consequences, especially damage to DNA. Damage from ROS thus can be viewed as a size-dependent contributor to the aging process and an important illustration of oxygen's "bad" side. This analysis puts a mechanistic foundation under the expression, "Live hard, die young." Or as the musician Neil Young expressed this point, "The same thing [that "thing" being oxygen] that makes you live can kill you in the end."

DHA-ROS EFFECTS ARE ONLY PART OF THE LONGEVITY STORY We do not want to leave the reader with a sense that the correlations discussed above between body size, DHA content of mitochondrial membranes, and susceptibility of mitochondria to damage from ROS are the full story in terms of what causes aging and determines longevity. Indeed, longevity is a complex result of a variety of processes occurring at different levels of biological organization, from the molecular level—where ROS damage is likely important—to the ecological level, where predation pressures may be more instrumental in setting life span than molecular or cellular factors. Thus elephants, unlike small rodents, are seldom preyed on in nature. In a more apples-to-apples comparison, long-lived naked mole rats may owe their greater longevity relative to most other rodents to their fossorial life style and relatively predator-free habitat. Thus the fascinating correlations among size, aging, and membrane lipid composition discussed above are but one facet of the broad suite of factors that influence life span.

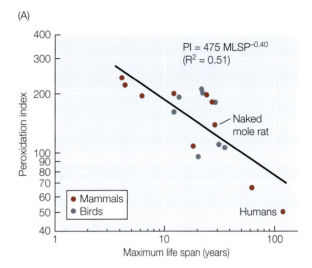

(A)

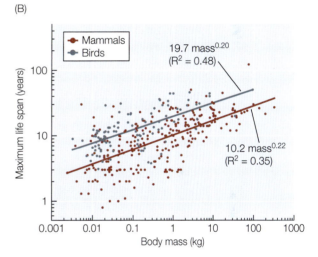

(B)

Figure 2.36 Propensity for lipid peroxide radical formation correlates with life span. (A) Size dependence of propensity for generation of lipid peroxides (peroxidation index, PI) in different-sized mammals and birds. (B) Scaling of maximum life span (MLSP) for birds and mammals. (After Hulbert et al. 2007.)

The interest in aging that goes back at least to Aristotle is manifested in a vast number of ongoing biomedical research programs, whose scope and findings cannot be adequately summarized in this short discussion. Suffice it to say that current research into the mechanisms that cause aging focuses on a wide range of cellular systems. One mechanism underlying aging is decrease in telomere length, which can result from many factors (Blackburn et al. 2015). Interestingly, telomeric DNA is more susceptible to oxidative damage than is the genome as a whole, which suggests another potential causal link between ROS production and aging. Other studies have focused on such disparate processes as changes in the gut microbiome (O'Toole and Jeffery 2015), ontogeny of stem cells (Goodell and Rando 2015), and NAD^+ metabolism (Verdin 2015). There is good evidence that NAD^+ levels fall with age and lead to loss of cellular function. In this case, cellular function may decrease because there is less NAD^+ to support the activities of sirtuins, which are important in retarding the cellular aging process (Verdin 2015). Another major focus of aging studies is mitochondrial dysfunction caused by several mechanisms other than DHA-related effects. However, despite repeated demonstrations that loss of mitochondrial function is a hallmark of aging (Wallace 2005), the exact mechanisms remain unresolved (Wang and Hekimi 2015). It remains to be seen how many of the determinants of aging will be traceable to ROS effects, but the role of ROS in size- and age-related physiological phenomena seems likely to be widespread and substantial—a significant part of the story, if not the whole story, of why we grow old.

2.10 Metabolic Depression and Hypometabolism

Adverse environmental conditions challenge an organism's ability to maintain ATP turnover rates, especially when it is unable to escape to more favorable conditions through migration. One of the most common responses to such conditions is a downregulation of metabolism or overall ATP turnover. Metabolic depression (or hypometabolism) can occur in response to reduced levels (hypoxia) or absence (anoxia) of oxygen in many types of animals, including intertidal mussels, goldfish, and the brine shrimps *Artemia*; a combination of starvation and cold temperatures in hibernating mammals, frogs, and turtles; and dehydration in tardigrades, snails, and diapaused insects and embryos of the annual killifish. Often a combination of external factors triggers metabolic depression, but in other instances—for example, estivating land snails—organisms enter metabolic depression actively, as part of their annual life cycle, while oxygen and hydration levels are still normal. Metabolic depression is also used during periods of programmed starvation, for example, the periods that marine mammals spend on land without access to food. Some life-history strategies employ metabolic depression over time periods of decades (e.g., *Artemia* cysts), while others use it as a short-term response to preserve energy (e.g., during diving or cold nighttime temperatures [torpor]).

Because hypometabolism is a downregulated state in comparison with the BMR and SMR, it is characterized by a reduction in mitochondrial respiration and ATP synthesis. Not surprisingly, hypometabolism also is marked by downregulation of ATP-consuming processes that contribute the most to the BMR and SMR. This downregulation leads to a reduction in ion regulation (for example, the activity of Na^+-K^+-ATPase), in protein synthesis, and in biosynthesis of other macromolecules and cellular assemblages. Hypometabolism is also characterized by increased stabilization of the existing cellular structures, including proteins and mRNA, to prolong their life span.

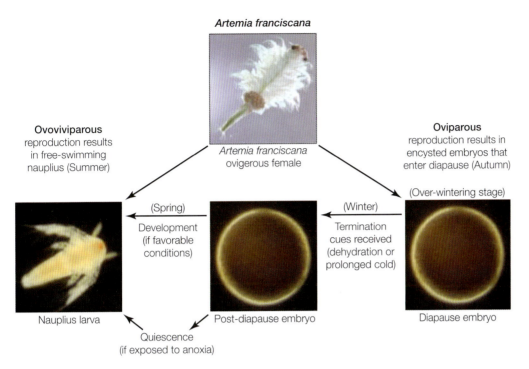

Artemia franciscana

Ovoviviparous
reproduction results
in free-swimming
nauplius (Summer)

Artemia franciscana
ovigerous female

Oviparous
reproduction results in
encysted embryos that
enter diapause (Autumn)

(Over-wintering stage)

(Spring)
Development
(if favorable
conditions)

(Winter)
Termination
cues received
(dehydration or
prolonged cold)

Nauplius larva

Post-diapause embryo

Diapause embryo

Quiescence
(if exposed to anoxia)

Figure 2.37 Life cycle of the brine shrimp *Artemia franciscana*. Females carry eggs that can give rise either to free-swimming nauplius larvae during ovoviviparous reproduction in the summer or to encysted embryos that enter a programmed arrest in development (diapause) in the fall. Repeated cycles of dehydration and cold exposure terminate diapause, and the post-diapause embryo can either develop into a nauplius larva or enter a quiescent period when exposed to anoxic conditions. (From Podrabsky and Hand 2015.)

Metabolic depression and diapause in the brine shrimp *Artemia franciscana*

Brine shrimps inhabit saline inland waters that lack predators because of the extreme salinities, which can reach 5–10 times the salt concentration of seawater, which is ~30–35 practical salinity units (psu). Females release free-swimming nauplius larvae from their brood pouch into the water during the summer months (**Figure 2.37**). With the onset of shorter days in the fall, females start laying encysted eggs containing embryos that develop only up to the gastrula stage before they enter diapause, a programmed delay in development (Podrabsky and Hand 2015). The cysts survive the winter on the shore in a dormant state and hatch with the onset of longer days, after repeated cycles of de- and rehydration in the spring (Drinkwater and Crowe 1987). These cysts are extremely hardy life forms. In addition to extreme salinities, these cysts survive laboratory temperatures from −200°C to 100°C, at least for a short time, and prolonged anoxia; yet these cysts can hatch within hours when returned to more benign conditions.

Following the embyos' entry into diapause, their respiration rates decline rapidly and by 3–5 days after release from the female fall to levels close to 1%–2% of the initial rate (**Figure 2.38**). This decline represents one of the steepest known metabolic depressions, and thus brine shrimps provide an ideal study system to investigate how such a steep decline is established (Clegg et al. 1996). One of the key questions is whether the decline is initiated through a decrease in pathways contributing to the synthesis of ATP, by a reduction in ATP-consuming processes, or by both simultaneously, a question we will address shortly.

Figure 2.38 Percentage of respiration rate of diapausing embryos of *Artemia franciscana* relative to day 1 after release of the embryo from the female. Data are based on samples obtained from Great Salt Lake and an independent sample from San Francisco Bay. (After Patil et al. 2013.)

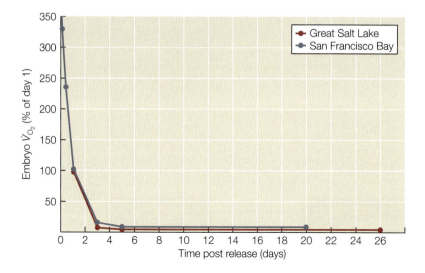

The main source of energy for the embryos is the disaccharide trehalose, and down-regulation of this sugar's catabolism is key to decreasing the supply of reducing equivalents to the ETS (Patil et al. 2013). Patil and colleagues (2013) measured the product-to-substrate ratios of several reactions of the glycolytic pathway and the reaction by trehalase that feeds glucose into glycolysis from trehalose (see reaction at left). A decrease in this ratio indicates an inhibition of the reaction, and the researchers found that three reactions were downregulated during entry into diapause: the reactions of trehalase, hexokinase, and pyruvate kinase (**Figure 2.39**). In addition, there was an increase in the phosphorylation of pyruvate dehydrogenase (PDH) at a serine of one of its subunits (E1α)—a modification that is known to inhibit the activity of PDH—that paralleled the time-course of metabolic depression (Patil et al. 2013). The inhibition of all four of these reactions would greatly inhibit the entry of acetyl-CoA into the TCA cycle and thus the production of reducing equivalents for the ETS, leading to an overall reduction in mitochondrial respiration.

$$\text{Trehalose} + H_2O \xrightarrow{\text{Trehalase}} 2 \text{ D-glucose}$$

Given that the proton leak across the inner mitochondrial membrane contributes about 20% to mitochondrial respiration, a reduction of metabolism to 1%–2% implies that the proton gradient is largely dissipated and thus that the proton motive force (Δp) is minimal or absent. Since the proton leak is dependent on the membrane potential ($\Delta\psi_m$) across the inner membrane, it is necessary to quantify the contribution of the leak to mitochondrial respiration at a range of potentials while ATP synthesis (ATP synthase) is inhibited. Such measurements of mitochondria from *Artemia* showed no difference between diapausing and non-diapausing embryos in the contribution of the proton leak to metabolic rate. This suggests that the proton conductance is the same in both embryonic states. Therefore a lack of ETS activity caused by a reduction in trehalose catabolism likely lowers the membrane potential across the inner membrane to minimize Δp and thus the loss of protons (Patil et al. 2013). Although the reduction of Δp saves energy, it potentially increases the likelihood that the mitochondrial transition pore complex will open and thereby trigger apoptosis; however, none of the typical inducers of apoptosis act in *Artemia* during diapause (Hand and Menze 2008).

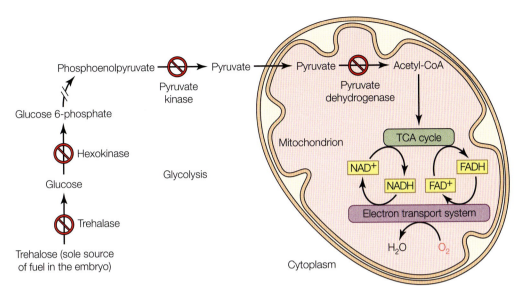

Figure 2.39 Putative metabolic reactions inhibited during diapause in the brine shrimp *Artemia franciscana*, in which the disaccharide trehalose serves as the main energy source for the embryo. Inhibition was detected by measuring the product-to-substrate ratios of several glycolytic reactions. (After Patil et al. 2013.)

Given the extreme metabolic depression in diapausing embryos, it is not surprising that their ATP/ADP ratios fall to about one-fifth of those found in post-diapause embryos, and that AMP levels increase greatly. The latter change—most likely mediated by the reaction of adenylate kinase, which converts two ADP into ATP and AMP—probably serves as the signal for regulating ATP-consuming processes in *Artemia* through the signaling pathway of AMP kinase (Hand et al. 2011). As a consequence of the reduced ATP/ADP ratios, ATP-consuming processes that normally contribute to the standard metabolic rate have to be reduced. One of the major costs under standard conditions is the maintenance of ion gradients across cell membranes by the Na^+-K^+-ATPase (>20%–30% of BMR or SMR). Given the degree of metabolic depression in diapausing *Artemia* embryos, it comes as little surprise that their Na^+-K^+-ATPase activity levels are very low. In addition, encysted embryos show greatly reduced ion and proton exchange rates with the environment, further reducing the costs for maintaining ion balance during diapause. Another major contributor to BMR and SMR, protein synthesis, which can account for up to 30% of the ATP consumption rate, is also greatly reduced during diapause (Clegg et al. 1996). Rates of protein degradation, which is also energetically costly, are reduced as well; these counterbalancing effects on protein turnover help maintain the proteome and cellular homeostasis (Anchordoguy and Hand 1994). The cell's existing macromolecular structures are further protected by the accumulation of compatible osmolytes (see Chapter 4) and molecular chaperones, such as small heat-shock proteins (Clegg 2007; Crowe et al. 1997).

The metabolic depression of diapausing *Artemia* embryos is not accompanied by any changes in the cytosolic pH (pH_i) of the cell. In contrast, during anoxia, metabolic depression is activated by a sharp drop in the cytosolic pH_i, suggesting that the regulation of these two hypometabolic states is distinct. During anoxia, the cytosolic pH_i drops from

>7.9 to 6.3, one of the largest drops in pH$_i$ ever reported, and this leads to a complete cessation of metabolism, the ametabolic state (Clegg 2007). This drop in pH$_i$ occurs within 1 h of the onset of anoxia and is accompanied by a sharp decline in ATP/ADP ratios. The intracellular source for all these protons has been an enigma for some time. Only a portion of the protons can be accounted for by the hydrolysis of ATP and the production of organic acids as products of anaerobic metabolism. A large part of the acidification most likely depends on the activity of the **vacuolar proton pump** or **V-ATPase** (Figure 2.40). The V-ATPase is proposed to acidify cellular compartments, such as lysosomal vesicles, that are separated from the cytoplasm by membranes (Covi et al. 2005). Exposure of embryos to anoxia releases protons from these acidified compartments into the cytosol; this release is proposed to contribute about half of the protons needed for the observed decrease in pH$_i$. Evidence for this mechanism is based on the failure of embryos to fully return to their normal pH$_i$ after anoxia when the activity of the V-ATPase is blocked. Thus part of the recovery of the cytosol to more alkaline pH$_i$ occurs through the transport of protons from the cytosol into compartments such as lysosomal vesicles by the V-ATPase (Covi et al. 2005). Consequently, it is now proposed that the hydrolysis of ATP and the dissipation of proton gradients both contribute to intracellular acidification (Covi et al. 2005). The decrease in the ATP/ADP ratio triggers a decrease in the activity of V-ATPase,

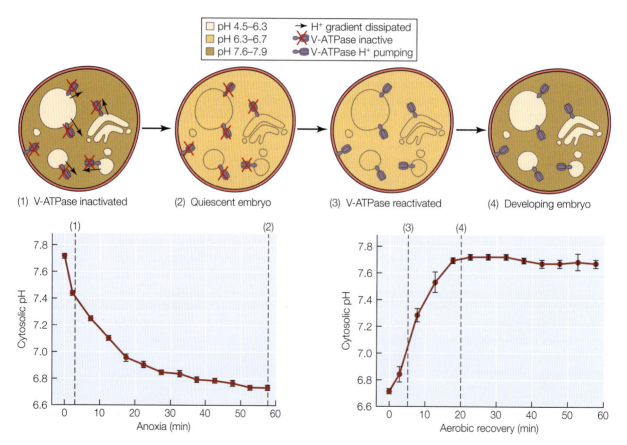

Figure 2.40 Exposure to anoxia induces a drop in cytosolic pH (pH$_i$) in *Artemia* embryos. See text for details. (After Covi et al. 2005.)

which normally consumes up to 31% of the aerobic energy budget of the embryos. This effect on V-ATPase activity further decreases pH_i, and facilitates the extreme degree of metabolic depression that occurs in the embryos.

Diapause in the annual killifish

Among vertebrates, some of the most extreme and longest metabolic depressions known occur in annual killifish found in regions of Central America and Africa that experience extreme dry and wet seasons (Podrabsky and Hand 2015; Wagner and Podrabsky 2015). Adults of *Austrofundulus limnaeus* inhabit ephemeral waters in the coastal deserts and savannas of the eastern shore of Lake Maracaibo in Venezuela. Individual adults survive for a few weeks to several months, depending on the water levels in these habitats, but their diapausing embryos can survive for months to years in the dry mud (**Figure 2.41**). Developing embryos can enter metabolic depression at three distinct stages, diapauses I, II, and III. Diapause I occurs prior to gastrulation and segmentation of the embryo, when future embryonic cells distribute across the surface of the yolk sac (Wourms 1972). Diapause II occurs midway through development and is characterized by the development of the central nervous system, sensory systems, and heart and the establishment of segmentation; however, it precedes organogenesis of organs other than the heart. Entry into diapause II is not a completely preprogrammed developmental step, but instead is dependent on egg provisions from the mother and the physical environment

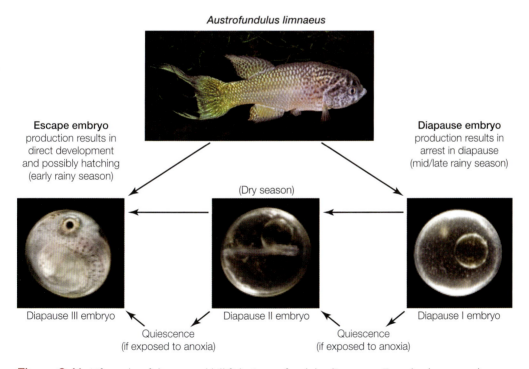

Figure 2.41 Life cycle of the annual killifish *Austrofundulus limnaeus*. Females lay eggs that develop into embryos that go through development directly without diapause at the beginning of the rainy season. Late in the rainy season, before ponds dry out, females lay eggs and the embryos enter immediately into diapause. Embryos move through several stages of diapause during development and can enter a quiescent state if they experience anoxia. (From Podrabsky and Hand 2015.)

the embryos experience (Podrabsky et al. 2010). Diapause III occurs after development is completed but before the embryo hatches. All three stages have unique features, and the stimulus to enter any of the three is to some extent dependent on environmental factors, specifically cold temperatures that are typical for the native habitats of the species in winter. However, diapause II is the stage with the greatest tolerance toward environmental stressors.

Respiration rates start to decrease even before entry into diapause II and III, and can fall by 80% to 90% (Podrabsky and Hand 1999). It is remarkable that these changes occur under normoxic and optimal temperature conditions and in fully hydrated embryos.

In contrast to *Artemia franciscana*, *Austrofundulus limnaeus* maintains mitochondrial respiration rates during diapause at a level that would still be able to support the proton leak across the inner mitochondrial membrane. However, oxidative phosphorylation is depressed through a reduction in the activity of complexes II, IV, and V of the ETS in diapausing embryos (**Figure 2.42**) (Duerr and Podrabsky 2010). This depression in activity of several ETS complexes is reversible within 1 h and is most likely due to the removal of inhibitory posttranslational modifications of subunits of these complexes.

Diapausing embryos of *A. limnaeus* have extremely low Na$^+$-K$^+$-ATPase activity, and this extends to all other ATP-consuming processes, including protein synthesis (Podrabsky and Hand 2015). This low activity is possible in large part because these embryos reduce to a minimum any exchanges with the aquatic environment that would involve

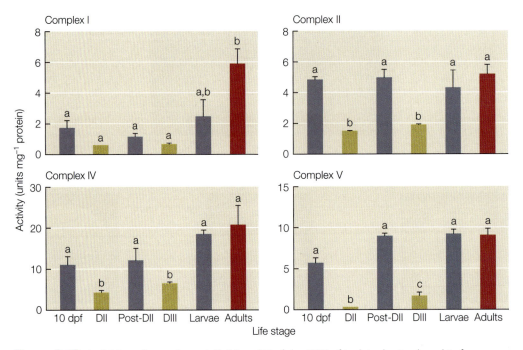

Figure 2.42 Activities of complexes I, II, IV, and V of the ETS of isolated mitochondria from different life stages of the annual killifish *Austrofundulus limnaeus*. For a description of the diapause stages, see main text and Figure 2.41. Larvae were 1–2 days posthatch, and mitochondria from adults were isolated from liver tissue. Bars represent mean ± S.E.M. Different letters indicate that means are different based on an ANOVA and a post-hoc Tukey's analysis, *P* < 0.05. DII, diapause II; DIII, diapause III; dpf, days postfertilization; 10 dpf is considered to be pre-diapause. (After Duerr and Podrabsky 2010.)

the activity of Na^+-K^+-ATPase. Similar to the brine shrimps, diapausing embryos have ATP/ADP ratios that are 40% lower than those of post-diapause embryos, but their ATP levels are still three times higher than their ADP levels despite the fact that ATP synthesis is essentially nonexistent. The ATP levels that stay unused during diapause are thought to provide the embryo with an energy charge that can be used to restart ATP-consuming processes during the escape from metabolic depression.

Hypoxia-induced hypometabolism

The previous two examples of hypometabolism are, at least partly, programmed stages in the life cycle of the organisms and are induced even under normoxic conditions. However, it is common for a hypometabolic state to be activated under conditions of limited or no oxygen availability (hypoxia and anoxia) and cold temperatures. Organisms exposed to anoxic conditions often experience additional changes in other environmental factors simultaneously, such as freezing temperatures and food shortages, all of which can contribute to a hypometabolic state. In some cases the hypometabolic state is accompanied by **anhydrobiosis**, a life stage with limited intra- and extracellular hydration (see Chapter 4). The vertebrates we will look at here have shown remarkable adaptations to prolonged periods of hypoxia and even anoxia, often in combination with cold temperatures. However, other organisms, for example, several terrestrial snails, survive drought conditions in the summer by going through a hypometabolic life stage (estivation) under fully normoxic conditions.

ANOXIA RESPONSE OF OVERWINTERING FRESHWATER TURTLES: REGULATION OF ION TRANSPORT AND NEURONAL ACTIVITY   Several temperate turtle species enter a hypometabolic state in the fall when temperatures drop and oxygen levels decline in their aquatic environments, first because of a decrease in primary productivity, then because of oxygen depletion by microbial respiration. In addition, food is limited during winter when primary productivity is very low. Submersion itself poses a hypoxia challenge to turtles because they can't employ their lungs to breath underwater; however, cutaneous uptake of oxygen can at least partially compensate for this. However, after prolonged periods of submersion during overwintering, turtles may even experience anoxic conditions, especially if they are embedded in mud that becomes progressively depleted of oxygen. Amazingly, some turtle species—for example, the painted turtle (*Chrysemys picta*) and red-eared slider (*Trachemys scripta*)—can survive for several months during cold anoxia. Thus turtles are, together with the crucian carp and annual killifish, among the most anoxia-tolerant freshwater vertebrates known (Ultsch 2006). Turtles survive these long periods of anoxia in a near-comatose state that is accompanied by a matching reduction of most physiological functions.

The turtles' extreme tolerance of anoxia is strikingly manifested in the observation that both cardiac muscle and brain emerge from overwintering conditions unharmed. This contrasts with the high sensitivity of these tissues to anoxic conditions in other vertebrates. Here we explore the biochemical mechanisms behind this extreme tolerance.

First, hypoxic conditions induce a reduction in heart rate and blood flow in turtles to about 10% of normoxic levels, yet systemic blood pressure is maintained through vasoconstriction (Overgaard et al. 2007). The reduction in blood flow is accompanied by a redistribution of blood flow among the various organs of the turtle (Stecyk et al. 2004). For example, cold anoxic turtles reduce blood flow to the digestive and urogenital organs, but increase flow to the liver, whose glycogen stores play an important role during anoxia, and to the shell, which provides calcium and magnesium carbonates that help buffer the acidosis caused by anaerobic production of lactate. Flow rates to the brain and heart stay constant.

The turtles' cardiac muscle cells undergo intracellular changes that accompany the shift from normoxic to anoxic conditions and the reduction in heart rate. These changes include a shift from aerobic to anaerobic pathways of ATP production and thus a reduction in overall ATP synthesis, which is matched by a reduction in ATP-consuming processes (Figure 2.43). The largest contribution to the reduction in energy consumption comes from a complete suppression of protein synthesis, but other contributions come from a reduction in the activity of the contractile apparatus as well as ion-transport processes involving the Na^+-K^+-ATPase and Ca^{2+}-ATPase (Fraser et al. 2001). However, the heart still has to produce enough ATP, relying only on lactate fermentation,

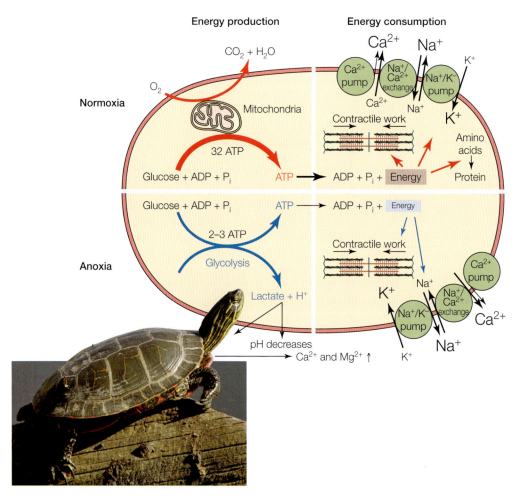

Figure 2.43 Major effects of anoxia on the production and consumption of ATP in the cardiac muscle cell of a turtle. The top half shows the relative magnitude of ATP production and consumption under normoxic conditions, including the production of 32 ATP and the consumption of ATP by protein synthesis, contractile work, and ion-transport processes. The Na^+ gradient built by the Na^+-K^+-ATPase is used to exchange Ca^{2+} from the intra- to the extracellular compartment. The bottom half shows the same cellular processes under anoxic conditions, including the major reduction in ATP yield from glucose (2 ATP) and glycogen (3 ATP). (After Overgaard et al. 2007.)

to maintain the contractile activity required to distribute blood glucose from the liver to other organs and to deliver lactate to the shell where it can be buffered by carbonates. This buffering in turn increases the extracellular concentrations of Ca^{2+} and Mg^{2+} (**Figure 2.44**). In addition, the reduced activity of the Na^+-K^+-ATPase leads to the accumulation of extracellular K^+, which depolarizes the cardiac cell membrane. These changes are not limited to the heart, and thus prolonged anoxia leads to the extracellular accumulation of H^+, Ca^{2+}, Mg^{2+}, and K^+ (see Figure 2.44) (Overgaard et al. 2007). To maintain adequate output during anoxia, the turtle heart defends its energy status by sustaining a high anaerobic poise to metabolism, as indicated by high activities of creatine kinase and pyruvate kinase relative to cytochrome oxidase activity (Christensen et al. 1994). The former may be important in the temporal and spatial buffering of ATP and in maintaining a high Gibbs free energy for the remaining ATPase activities, while pyruvate kinase is indicative of a high capacity for lactate fermentation.

Another challenge the cardiac muscle faces is to maintain contractility (Overgaard et al. 2007). Increased acidosis, especially in conjunction with anoxia and cold, reduces contractility by about one-third. The elevated levels of extracellular K^+ (hyperkalemia) slowly depolarize the membrane and shorten the typically long (hundreds of milliseconds) action potential of the cardiac muscle, causing reduced influx of Ca^{2+} and contractility, more so at low temperatures. However, it is likely that the increasing levels of extracellular Ca^{2+} (hypercalcaemia) and catecholamines (epinephrine), which increase twitch force in isolated ventricular muscle fibers, compensate for the effect of hyperkalemia and increased acidosis (Overgaard et al. 2005). For example, combined acidosis and hyperkalemia reduced the contractile force to 5%, while addition of calcium and epinephrine increased the force to 20%, partially compensating for the effect of the former treatments.

The turtle brain is one of the most anoxia-tolerant vertebrate brains (Lutz and Milton 2004). Depriving the typical vertebrate brain of oxygen for only a few minutes leads to a drop of ATP to about two-thirds of normoxic levels, which compromises ATP-dependent ion-transport processes and dissipates ion gradients across membranes. This depolarizes the cell membrane and increases influx of Ca^{2+}, causing the uncontrolled release of neurotransmitters, such as glutamine, to levels that are toxic and cause neuronal death (Lutz et al. 2003). In contrast, the turtle brain can tolerate anoxia for days, even at room temperatures, and this tolerance is achieved through a combination of adaptations at the biochemical, cellular, and organismal levels (**Figure 2.45**).

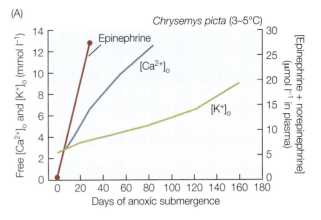

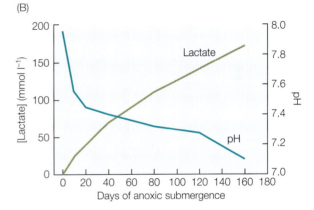

Figure 2.44 Changes in the extracellular concentration of (A) calcium $[Ca^{2+}]_o$, potassium $[K^+]_o$, and epinephrine and (B) lactate, as well as changes in blood pH over a 160-day time period under cold anoxic conditions in the turtle *Chrysemys picta*. (After Overgaard et al. 2007.)

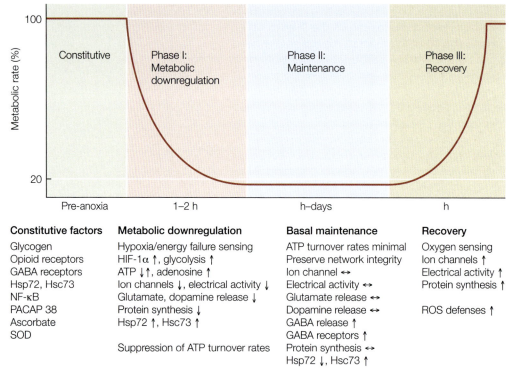

Constitutive factors	Metabolic downregulation	Basal maintenance	Recovery
Glycogen	Hypoxia/energy failure sensing	ATP turnover rates minimal	Oxygen sensing
Opioid receptors	HIF-1α ↑, glycolysis ↑	Preserve network integrity	Ion channels ↑
GABA receptors	ATP ↓↑, adenosine ↑	Ion channel ↔	Electrical activity ↑
Hsp72, Hsc73	Ion channels ↓, electrical activity ↓	Electrical activity ↔	Protein synthesis ↑
NF-κB	Glutamate, dopamine release ↓	Glutamate release ↔	
PACAP 38	Protein synthesis ↓	Dopamine release ↔	ROS defenses ↑
Ascorbate	Hsp72 ↑, Hsc73 ↑	GABA release ↑	
SOD		GABA receptors ↑	
	Suppression of ATP turnover rates	Protein synthesis ↔	
		Hsp72 ↓, Hsc73 ↑	

Figure 2.45 The various cellular processes involved in the four general stages of response to cold anoxia in the turtle brain. Phase I (1–2 h) of the anoxia response involves the downregulation of ATP-consuming processes to 20% of normoxic levels. This includes a reduction in ion channel and electrical activity, neurotransmitter release, and protein synthesis, and an increase in molecular chaperones. Phase II (hours to days) involves maintenance of the basal requirements at a reduced level. During phase III, which takes hours, the brain recovers to operate again at normoxic levels by increasing the activity of ion channels and neurons. See text for additional details. (After Lutz and Milton 2004.)

Given that turtles commonly experience short periods of anoxia during routine diving, turtle brains are constitutively primed to deal with prolonged anoxic conditions in winter (Lutz and Milton 2004). For example, on a per unit of mass basis the turtle brain stores five times as much glycogen as a rat or a trout, to provide enough fuel for glycolysis. The turtle brain has lower activities of aerobic enzymes and Na^+-K^+-ATPase, and lower densities of voltage-dependent Na^+ channels, indicating an overall lower rate of metabolism, as well as lower maintenance requirements of ion gradients and neuronal activity. Turtle brains also have greater densities of inhibitory gamma-aminobutyric acid (GABA) receptors and greater binding capacities of opioid receptors, which can protect against greater excitatory activity caused by the uncontrolled release of glutamate that is typically observed in anoxia-intolerant vertebrate brains. Additional constitutive measures to protect turtle brains from anoxia include higher levels of molecular chaperones and enzymatic (superoxide dismutase, peroxiredoxin, glutathione peroxidase, and glutathione S-transferase) as well as non-enzymatic (ascorbic acid, vitamin E, and glutathione) antioxidants to protect the brain against cellular stress (Storey 2007).

The reduced rate of ATP production in the brain under anoxia has to be matched by a quick reduction in ATP-consuming processes. This involves the downregulation of ion fluxes across neuronal membranes, which consume about 50% of the cellular energy in a normoxic neuron. Reduced ion flux is observed for K^+, which is responsible for setting the resting potential, rate of repolarization, and neuronal firing rates (Meir et al. 1999). K^+ flux is reduced to 35% of normal rates within 2 h of anoxia, at least in part through the downregulation of voltage-gated K^+ channels (K_v channels). In addition, the density of voltage-gated Na^+ channels, which are responsible for generating action potentials, is also lowered in the turtle brain with the onset of anoxia. The arrest of voltage-gated K^+ and Na^+ channels (**channel arrest**) helps reduce ion fluxes and neuronal activity (**spike arrest**), respectively, protecting neurons from dissipating ion gradients while ATP-dependent ion-transport processes are greatly reduced during anoxia (Lutz and Milton 2004).

Postsynaptic ligand-gated ion channels are also downregulated in turtle brains during anoxia. For example, the activity of N-methyl-D-aspartate (NMDA)-modulated glutamate receptors is downregulated within minutes in response to anoxic conditions; during prolonged anoxia, neurons further downregulate the activity by retrieving the receptors from the cell membrane (Bickler et al. 2000). The downregulation is presumed to protect neurons from the effects of uncontrolled glutamate-induced influx of extracellular Ca^{2+}. The neurons are further protected by a moderate increase in cytosolic Ca^{2+} levels, caused by the release of Ca^{2+} from mitochondria (**Figure 2.46**). Experimental evidence

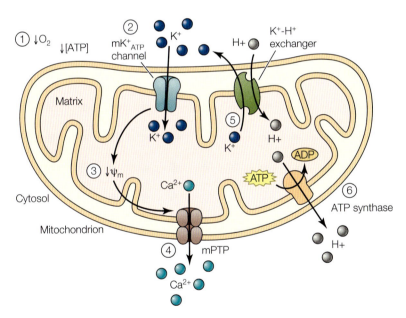

Figure 2.46 Processes involved in the release of calcium from mitochondria in response to hypoxia or anoxia in the brain of the western painted turtle (*Chrysemys picta bellii*). Reduced levels of oxygen lead to a reduction in ATP synthesis (1). Reduced levels of ATP lead to the opening of ATP-sensitive mitochondrial mK^+_{ATP} channels, and an increase in K^+ influx into the mitochondrial matrix (2). This leads to the decrease of the mitochondrial membrane potential (Ψ_m; 3), which in turn triggers the formation of the mitochondrial permeability transition pore (mPTP) and an increase in Ca^{2+} release (4). While the mK^+_{ATP} channels are constitutively open, mitochondrial $[K^+]$ is balanced by a K^+-H^+ exchanger (5). The increase in $[H^+]$ is balanced by a reversal of the F_0F_1-ATPase (ATP synthase; 6) to prevent the dissipation of the proton gradient while maintaining Ψ_m at a new depolarized set point. Oxygenation reverses these processes, leading to an increase in ATP production, mK^+_{ATP} channel closure, reestablishment of the original Ψ_m, and closure of mPTP and therefore cessation of Ca^{2+} release. (After Hawrysh and Buck 2013.)

suggests that the release of Ca^{2+} from mitochondria is triggered by the opening of an ATP-sensitive mitochondrial K^+_{ATP} channel; the resulting influx of K^+ leads to depolarization of the mitochondrial matrix and is followed by the formation of the mitochondrial permeability transition pore (mPTP), which functions as a Ca^{2+} channel (Hawrysh and Buck 2013).

Turtle brains are protected from hypoxia in yet another way. Vertebrate brains that are less hypoxia tolerant typically respond to hypoxia with an increase in the release of glutamate (and dopamine). This response is limited in the turtle brain through an immediate reduction in the release of glutamate at the presynaptic membrane and stable uptake rates through glutamate transporters in response to acute hypoxia (Milton et al. 2002). The reduction in glutamate release by presynaptic membranes in the turtle brain in response to anoxia is modulated in the early phase by adenosine receptors—which belong to the family of P2 purinoreceptors (or nucleotide receptors) that regulate neurotransmission and inflammation—and K^+_{ATP} channels; however, during prolonged anoxia, adenosine and GABA receptors play a more dominant role. The synthesis, release, and uptake of neurotransmitters are all ATP-dependent processes that have to be defended while ATP production is reduced. Thus, the fact that the turtle brain maintains the release and uptake of glutamate and dopamine, even if at lower levels than under normoxic conditions, suggests that this investment is serving a function important for survival during anoxic conditions. Although the electrical activity of the brain is greatly reduced during anoxia, it is possible that the continued release and uptake of neurotransmitters play an important role in the occasional burst activity of mixed frequency of the brain that may represent periodic checks for an arousal signal (Lutz and Milton 2004).

ANOXIA RESPONSE OF OVERWINTERING FRESHWATER TURTLES: REGULATION OF TRANSCRIPTIONAL ACTIVITY   While the energetic costs of transcription are relatively small in comparison with the costs of protein synthesis, the downregulation of gene expression may play an important regulatory role during the entry into a hypometabolic state. The mechanisms that have been studied in detail include the modifications of histones, the posttranscriptional sequestration of mRNA, and the regulatory role of microRNAs (miRNA) (Storey 2015).

Epigenetic modifications can control access of gene-regulatory factors to the genome and thereby regulate gene expression patterns without changing the primary DNA sequence. These modifications can occur within a generation (intragenerational) and can, for example, affect future generations' (intergenerational) responses to hypoxic and cold conditions if they affect the germline cells (Burggren 2014). Epigenetic modifications change the chromatin structure of DNA by modifying the fundamental structural unit of chromatin, the **nucleosome**. The nucleosome winds a 147-bp DNA segment around a histone octamer, which consists of one $(H3-H4)_2$ heterotetramer and two H2A-H2B heterodimers. While the C termini of the histones make up the core of the nucleosome with the DNA wrapped around it, the N-terminal end of each histone protrudes away from the core and is modified by several posttranslational modifications that affect the chromatin scaffold (Li et al. 2012). For example, the acetylation of lysine residues and the phosphorylation of serine/threonine residues of histones tend to favor the open structure of chromatin, the so-called euchromatin, which allows transcription factors to access the DNA. In contrast, the methylation of lysine and arginine residues condenses the chromatin (heterochromatin) and inhibits the binding of transcription factors.

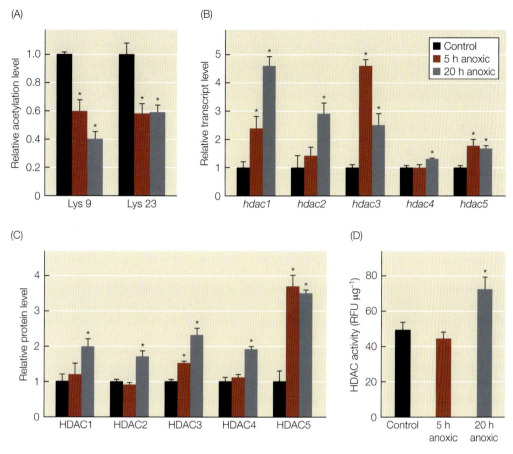

Figure 2.47 Responses of histone deacetylases (HDACs) and histone lysine acetylation in white muscle of the turtle *Trachemys scripta elegans* experiencing 5 h and 20 h of anoxia. (A) Acetylation of lysines (Lys) 9 and 23 of histone H3 decreased within 5 h and stayed low at 20 h of exposure. (B) Transcript and (C) protein levels of five different HDACs increased with increasing anoxic submergence time. (D) Total HDAC activity increased at 20 h of anoxia exposure. Data are means ± S.E.M.; $n = 3–4$; * = significantly different from control group ($P < 0.05$). (After Krivoruchko and Storey 2010.)

In turtles, exposure to anoxia during submergence in nitrogen-bubbled water led to a reduction in the acetylation of histones (H3) in white muscle by more than 50% over a 5-h time period, and remained low at 20 h (**Figure 2.47**) (Krivoruchko and Storey 2010). This reduction in lysine acetylation correlated with an increase in gene and protein expression of five different histone deacetylases (HDACs 1–5) after 20 h. This resulted in a 1.5-fold increase in HDAC activity after 20 h of anoxia. Thus it is likely that HDACs are involved in the condensation of chromatin, leading to reduced gene expression in response to hypoxia in the turtle's white muscle. Similar results were obtained for liver, but the heart showed no changes in the acetylation of H3 (Krivoruchko and Storey 2010). Thus the changes in HDAC activity and histone acetylation in response to anoxia are organ-specific and may reflect distinct roles of the different organs in maintaining function during anoxia.

Over the past decade, **microRNAs** (**miRNAs**) and their functions have been revealed to be an important mechanism for regulating the expression, storage, and degradation of mRNA (Qureshi and Mehler 2012). MicroRNAs are short sequences of 18–23 nucleotide base pairs whose coding sequences make up only 1%–2% of the genome, but they may regulate the expression of up to 60% of all the protein-coding genes. Furthermore, one miRNA can affect multiple mRNAs, and a specific mRNA can be affected by several different miRNAs, giving rise to myriad combinations that can affect the fate of an mRNA (Ebert and Sharp 2012).

MicroRNAs also play a role in the response of turtles to hypoxia, especially in suppressing the cell cycle. Given the costs associated with building new cells and the possibility of energetic constraints during replication, it would make sense to delay cell proliferation until environmental conditions become more benign. The transcription factor **p53** plays an important role in controlling the cell cycle, responding to DNA damage, and shifting mitochondrial metabolism. Turtles exposed to anoxic submergence showed elevated levels of p53 in several tissues, an accumulation of p53 in the nucleus, and an upregulation of mRNA transcripts that are under the control of p53 (Zhang et al. 2013). One of the genes transcribed by p53 encodes the miRNA miR-34a, whose transcript levels increased two- to threefold in turtle liver and muscle in response to hypoxia and whose overexpression in other organisms leads to cell cycle suppression through the downregulation of several important cell cycle regulatory proteins. Two additional miRNAs, miR-16-1 and miR-15a, are also upregulated in turtles responding to hypoxia and are known to prevent entry into the G1 phase of the cell cycle by inhibiting cyclin D1. The role of these miRNAs in regulating the cell cycle was further supported by the fact that the protein, but not the transcript levels of cyclin D1, decreased with hypoxia (Biggar and Storey 2012). Thus several miRNAs are an essential part of the suppression of the cell cycle and cell proliferation, helping animals save energy when it is limited because of low levels of oxygen.

MAMMALIAN HIBERNATORS: SURVIVING PROLONGED COLD EXPOSURE THROUGH DEEP TORPOR   **Torpor** is the controlled reduction of metabolism, body temperature, and a suite of physiological processes in endotherms. This strategy is employed in order to survive times of cold exposure while food sources are limited. When this hypometabolic state occurs on a daily basis we refer to the animals as **daily heterotherms**; when it occurs annually during winter we refer to the animals as **hibernators** (Ruf and Geiser 2015). What distinguishes these two groups is the maximum or mean torpor bout duration, which is generally shorter than 24 h in daily heterotherms and 2–32 days in hibernators. Some physiologists have pointed out that these two groups are only the extremes along a continuum (van Breukelen and Martin 2015), but a recent meta-analysis of studies on more than 200 species supports a bimodal distribution and the distinction between daily heterotherms and hibernators (Ruf and Geiser 2015). Hibernation, which can last up to 9 months, is characterized by long periods of torpor, interrupted by periodic arousals that can last for 48 h, during which the animals return to a euthermic state (interbout euthermia) before reentry into torpor (**Figure 2.48**). Daily heterotherms are characterized by a shallower reduction in body temperature (T_B), from 37°C to 10°C–35°C, while some hibernators reduce their T_B to temperatures from below 0°C–10°C. Furthermore, daily heterotherms reduce their metabolic rate during torpor (TMR) to about 35% of BMR on average, while hibernators reduce their rate to about 6% of BMR. The two groups also vary in body size, as daily heterotherms tend to be smaller than hibernators (Ruf and Geiser 2015). Here we are mainly interested in the regulation of the hypometabolic state of hibernators and the biochemical adaptations that accompany the increase in metabolism and T_B during periods of interbout arousals (interbout euthermia).

Entrance into torpor is characterized by depressed breathing and heart rates and an overall reduction in oxygen consumption by 95% or more compared with the euthermic animal during summer; a reduction in T_B then follows (Storey et al. 2010). The energy homeostasis network of the hypothalamus controls this process, including the setting of T_B, but the specific endocrine and neural signals that regulate the behavior and the metabolic depression during torpor are still largely unknown (Drew et al. 2007). However, recent work showed that the activation of the adenosine A1 receptor (A_1AR) in the central nervous system is sufficient to induce torpor during the mid-hibernation season (Jinka et al. 2011). A_1AR is a G protein-coupled receptor, which produces the second messenger diacylglycerol through the activation of phospholipase C. In turn, diacylglycerol activates protein kinase C ε (PKCε), which leads to the opening of ATP-sensitive potassium channels. Although this is an important finding in elucidating the neural pathway, how the activation of A_1AR exactly leads to entrance into torpor has yet to be determined.

During hibernation we can distinguish cellular processes, such as mitochondrial respiration, protein synthesis, and transcription, that simply—one could say passively—follow the drop in T_B; these processes display reduced activities and thus almost cease to function during torpor. The decreased rates of these processes are reversible upon an increase in T_B during interbout euthermia. Other processes, however, continue to function during torpor, albeit at greatly reduced levels, including the activities of the central and peripheral nervous systems, the cardiovascular system, the pulmonary system, and white and brown adipose tissue (Carey et al. 2003). The majority of the former, passively responding cellular processes change reaction rates according to changes in T_B, that is, according to the Q_{10} relationships (see Chapter 3). Overall there is

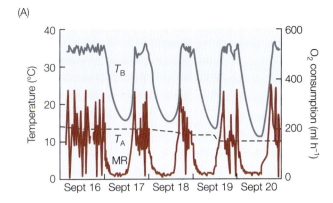

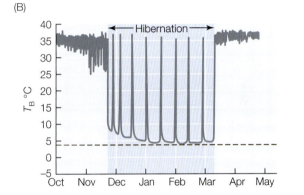

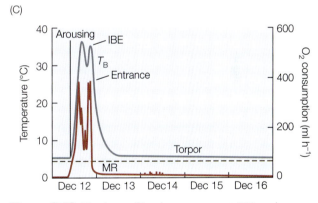

Figure 2.48 Rhythms of body temperature (T_B) and metabolic rate during daily torpor and hibernation. (A) Daily torpor in the dormouse *Glis glis* showing body temperature (T_B), ambient temperature (T_A), and metabolic rate (MR) over a 5-day period. (B) Homeothermic and heterothermic time periods lasting several months during the hibernation of a ground squirrel. (C) Body temperature and metabolic rate (MR) in *Glis glis* during different phases of hibernation, including a day of interbout euthermia (IBE) and several days of torpor. Blue lines represent body temperature, red lines represent metabolic rate, and dashed black lines represent ambient temperature. (After van Breukelen and Martin 2015.)

limited evidence that the inhibition and restoration of these processes are driving the entry into or exit out of torpor. For example, mitochondrial respiration in the fully coupled state 3 is depressed in the liver of the golden-mantle ground squirrel *after* entry into the torpid state and reactivated *after* the beginning of the interbout arousal, suggesting that these changes do not regulate these transitions (Martin et al. 1999). In addition, mitochondria did not show any differences in the contribution of the proton leak (state 4) to respiration between these hibernation stages, with the exception that state 3 and 4 respiration both decrease in liver but not skeletal muscle mitochondria of the arctic ground squirrel (Barger et al. 2003).

Similarly, protein synthesis is almost completely downregulated during the entry into torpor in heart, liver, spleen, pancreas, and kidney, and upregulated during entry into the interbout arousal in ground squirrels (Zhegunov et al. 1988). However, brown adipose tissue may be an exception to this pattern (Hittel and Storey 2002). Curiously, protein synthesis, as indicated by the association of actin mRNA with polysomes, is not fully downregulated until torpid ground squirrels reach a T_B of 18°C, but resumes during interbout arousals before the animal reaches 18°C, suggesting that a drop in temperature alone is not the sole factor controlling the rates of translation (van Breukelen and Martin 2001). The phosphorylation of both the translation initiation factor eIF2α and the elongation factor eEF2 likely contributes to the reduced rate of protein synthesis (Frerichs et al. 1998). Phosphorylation of specific proteins also controls several other cellular processes, for example, energy metabolic pathways, during torpor (Storey 1997). However, proteomic studies showed that more proteins are modified by acetylation than by phosphorylation during the heterothermic periodicity of torpor and interbout arousals, at least in comparison with animals collected during summer that are characterized by strict homeothermy (Hindle et al. 2014).

Transcription initiation and elongation rates showed a reduction that was similarly steep to the reduction seen for translation during hibernation (van Breukelen and Martin 2002). However, despite a global reduction in transcriptional activity, specific transcription factors are activated (see below), and the expression of their downstream targets is an important element of the hibernator's ability to deal with the metabolic, redox, and proteostasis challenges encountered during torpor.

Posttranscriptional processes also contribute to the regulation and timing of mRNA translation (Storey 2015). For example, even with transcription and translation being suppressed during hibernation, large pools of mRNA exist and are ready for translation upon entrance into interbout arousal. In some cases, a message may have been initially transcribed during entrance into torpor and then sequestered into nuclear loci for later use by the animal during arousal. Like proteins, mRNA transcripts are processed, chaperoned, stored, and degraded, in this case through the activities of RNA-binding proteins and RNA chaperones (Moore 2005). For example, the RNA-binding proteins T-cell intracellular antigen 1 (TIA-1), TIA-1 related (TIAR), and poly A-binding protein (PABP-1) affect the transcription, splicing, localization, degradation (stability), and translation of mRNA (Suswam et al. 2005). Liver sections of torpid (5 days at T_B of 5°C–8°C) thirteen-lined ground squirrels were compared with euthermic controls and showed a much greater presence of large subnuclear bodies containing TIA-1/TIAR and PABP-1 (Tessier et al. 2014). Subsequent quantification of protein levels showed that the two splice variants TIA-1a and TIA-1b were elevated four- to sevenfold in nuclear extracts of the liver of torpid animals. An increase in the abundance of subnuclear bodies containing RNA-binding proteins may be a way to suppress protein synthesis during torpor and yet recruit these messages for the immediate resumption of translation upon entrance into interbout arousal.

Given the overall reduction in metabolism and the reduced rates of transcription and translation during hibernation, it is not surprising that both cell division and proliferation are also greatly reduced. What is surprising is that cell proliferation of the enterocytes in the gastrointestinal tract recovers during interbout arousals, despite the fact that the animal does not take in any food (Carey et al. 2003). Ion- and nutrient-transport processes are also greatly suppressed in the gastrointestinal tract, but they are able to recover quickly during interbout euthermia. However, the overall gut mass is reduced. Although ion-transport processes overall are reduced during torpor, hibernators are generally better at maintaining ion gradients because of less leakage of Na^+ and K^+ across their cell membranes—for example, across membranes of red blood cells—and an overall greater Na^+-K^+-ATPase transport activity in comparison with nonhibernators. Somewhat puzzling is the observation that hibernators do not show signs of homeoviscous adaptation—the adjustment of the membrane composition to maintain membrane fluidity in response to cold exposure—a typical response of other organisms to the same temperature challenge (see Chapter 3) (Aloia and Raison 1989). Perhaps this "failure" to alter membrane chemistry is in fact a sign that a reduction in the activities of certain membrane-localized proteins during hibernation is advantageous. Viscotropic effects thus could serve as contributors to downregulation of membrane-localized enzymes.

One of the hallmarks of hibernation is a switch of metabolism from using carbohydrates to using fats as the main fuel. This is indicated by a decrease in the **respiratory quotient** (**RQ**), the ratio of CO_2 produced to oxygen consumed, from 1.0 to closer to 0.7, suggesting a switch toward lipid oxidation. However, even while the animal is in a state of torpor, the *RQ* value may rise if T_B increases to 20°C or decreases to near-freezing temperatures (Buck and Barnes 2000). This suggests that sole reliance on lipid oxidation is limited to a narrow range of temperatures, for example, 4°C–8°C in the arctic ground squirrel. Additional evidence for the switch comes from measurements of the activities of glycolytic, gluconeogenic, lipogenic, and ketogenic enzymes, which revealed that hibernators have, overall, a greater glycolytic capacity than nonhibernators, but that carbohydrate and amino acid use are reduced during hibernation (Williams et al. 2011). The analysis of the liver proteome also detected a shift from a carbohydrate-based catabolism and lipid anabolism during summer to lipid and ketone body catabolism as well as lipid mobilization during winter (Grabek et al. 2015). There was also a shift in protein metabolism, from elevated activities of amino acid and nitrogen catabolism and the urea cycle during summer to a focus on maintaining proteostasis and conserving amino acids during hibernation (**Figure 2.49**). Levels of proteins of oxidative phosphorylation and the ETS were elevated during summer in comparison with winter. Finally, hibernators showed greater abundances of proteins involved in the cellular stress response, including antioxidant proteins that scavenge ROS. These changes may increase the animals' abilities to survive variation in T_B, blood flow, and oxygen availability.

The shift from carbohydrate to lipid metabolism associated with the transition from summer homeothermy to winter heterothermy is controlled at the enzymatic level in part through the phosphorylation of pyruvate dehydrogenase (PDH) by PDH kinase 4 (PDK4) and an increase in the activity of both hormone-sensitive lipase (HSL) and pancreatic triacylglyceride lipase (PTL). The latter two enzymes mobilize non-esterified

(A)

(B)

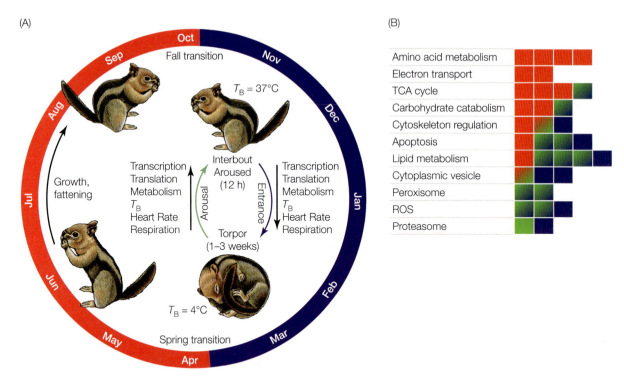

Figure 2.49 Hibernation phases and the accompanying proteomic changes. (A) Hibernation consists of two cycles, an annual cycle between summer homeothermy and winter heterothermy, and another cycle during winter heterothermy between prolonged torpor and short interbout arousals or euthermia. Several metabolic processes change during these transitions (see text). (B) General changes in functional categories of the proteome accompanying the different hibernation phases. More squares indicate a greater contribution to the overall proteome. The color of each square indicates the hibernation phase with the highest expression of that pathway's or cellular process's components: red, the homeothermic phase of the annual cycle; blue, the low T_B phase of winter heterothermy; green, the high T_B phase of winter heterothermy (see A). Overall increases during winter heterothermy are indicated by a green-to-blue transition square. The red-to-green transition indicates that proteins of that functional category showed similar changes in animals with warm T_B (homeothermic-phase animals and aroused hibernators). Some categories were merged: proteasome with ubiquitylation; cytoplasmic vesicle with plasma proteins and endocytosis; lipid metabolism with fatty acid metabolism; and ROS with redox homeostasis. The figure is based on a compilation of results from several comparative proteomic studies. (After Grabek et al. 2015.)

fatty acids from the fat deposits of white adipose tissue (WAT) to be reesterified to triacylglycerides in other tissues (**Figure 2.50**) (Bauer et al. 2001; Frank et al. 1998). At the transcriptional level, peroxisome proliferator-activated receptor α (PPARα) activates the expression of PDK4 and several genes involved in the intra- and extracellular mobilization of lipids and mitochondrial β-oxidation of fatty acids (Wu et al. 2001). Interestingly, PPARα is activated by long-chain fatty acids, such as linoleic and arachidonic acid, that occur at elevated levels in serum before and during hibernation. The protein levels of both PPARγ, a closely related isoform of PPARα, and its coactivator PGC-1α are also upregulated in hibernating bats, and induce the expression of different isoforms of fatty acid binding protein (FABP), which facilitates the transport of

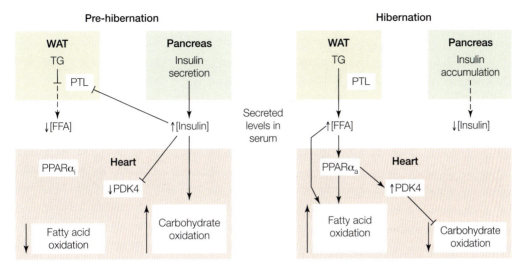

Figure 2.50 Model of the controlling steps in the shift from carbohydrate to fatty acid oxidation during hibernation. Effects of serum levels of free fatty acids (FFA) originating from triacylglycerides (TG) and insulin secreted from the pancreas on the activity of pancreatic triacylglyceride lipase (PTL) in white adipose tissue (WAT), and of peroxisome proliferator-activated receptor α (PPARα) and pyruvate dehydrogenase kinase 4 (PDK4) on the levels of fatty acid and carbohydrate oxidation in the heart of a typical hibernator. Upward arrows indicate upregulation and downward arrows indicate downregulation of the rate of a process or a concentration. Blunt arrows indicate inhibition. Solid and dashed lines indicate major and minor regulatory pathways, respectively. (After Carey et al. 2003.)

fatty acids across the cytosol in brown adipose tissue and the heart (**Figure 2.51A**) (Eddy and Storey 2003).

Several other transcription factors are activated with entry into torpor and affect processes other than the shift toward oxidation of fatty acids. For example, hypoxia-inducible factor 1α (HIF-1α) is activated in brown adipose tissue and skeletal muscle, the two thermogenic organs of hibernators (Morin and Storey 2005). HIF-1α normally recruits genes that help the organism deal with hypoxic conditions, for example by elevating ATP production through anaerobic pathways or by improving delivery of oxygen to the organs by stimulating capillary growth and accelerating the proliferation of red blood cells (Semenza 2010). However, the exact target genes of HIF-1α in these tissues are not known.

Nuclear factor erythroid 2-related factor 2 (Nrf2) is a transcription factor that increases expression of several antioxidant proteins (Ma 2013). In ground squirrels, high expression of both Nrf2 transcript and protein coincide with elevated levels of aflatoxin aldehyde reductase, Cu-Zn-SOD, heme oxygenase, and several peroxiredoxin isoforms in the heart during entry into torpor and long-term torpor (**Figure 2.51B**) (Storey et al. 2010). Superoxide dismutase scavenges superoxide anions, and peroxiredoxins are thought to be responsible for the bulk of the scavenging of hydrogen and lipid peroxides (Cox et al. 2010). Nuclear factor-κB (NF-κB) is another transcription factor that is induced by oxidative stress and activated upon entry into and during torpor, but it also affects the immune response (Carey et al. 2000). These proteins can protect the heart from ROS and lipid peroxides throughout the torpor period as well as during arousal periods when high metabolic rates and thermogenesis lead to high rates of ROS generation. Together with

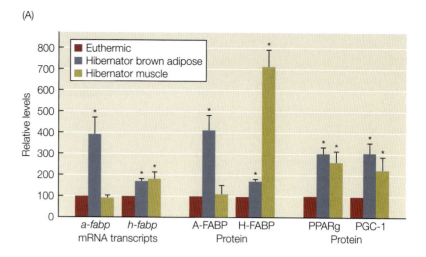

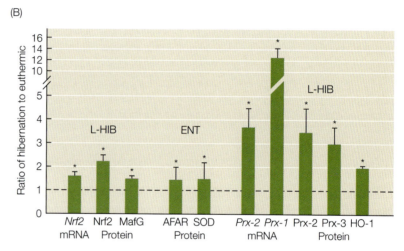

Figure 2.51 Transcription factor responses in various tissues of hibernating bats and ground squirrels and the effect on the expression of several downstream genes. (A) Protein levels of peroxisome proliferator-activated receptor γ (PPARγ) and its coactivator, PGC-1, were upregulated two- to threefold in brown adipose and muscle tissue of hibernating bats (*Myotis lucifugus*). This coincided with upregulation of the transcript and protein of adipose- and heart-specific isoforms of fatty acid binding protein (A-FABP and H-FABP), whose expression is regulated upstream by PPARγ. (B) Nrf2 activates the expression of antioxidant proteins. Both transcript and protein of Nrf2 and its binding partner MafG were upregulated in the heart of hibernating thirteen-lined ground squirrels (*Spermophilus tridecemlineatus*). Levels of protein and transcript of several genes involved in the antioxidant defense and under the control of Nrf2 were upregulated during entrance into torpor (ENT) and long-term torpor (L-HIB). AFAR, aflatoxin aldehyde reductase; SOD, Cu-Zn-superoxide dismutase; Prx, several isoforms of peroxiredoxin; HO-1, heme oxygenase. Data are means ± S.E.M; $n = 3–7$; * = significantly higher than corresponding value for the euthermic animal ($P < 0.05$). (After Storey et al. 2010.)

the elevated levels of molecular chaperones, specifically those localized to the endoplasmic reticulum (Mamady and Storey 2008), and enzymatic and non-enzymatic antioxidants, such as ascorbate, these proteins form a bulwark against the cellular stress that hibernators experience because of variation in T_B and changes in redox balance (Carey et al. 2003).

Although the physiological processes we covered in the previous paragraphs almost cease to function during deep hibernation, with few exceptions the central nervous system, sensory processes of the peripheral nervous system, cardiac muscle, respiratory system, and adipose tissue still function under these conditions, albeit at a much lower level. For example, even during torpor, hibernators are able to respond to progressively lower ambient temperatures by increasing metabolism and generating heat through shivering and nonshivering thermogenesis to maintain a minimum T_B. Thus hibernators are able to sense changes in ambient temperature and respond with an increase in metabolism to control T_B even during torpor. Furthermore, the cardiac muscle of hibernators continues to function at temperatures at which nonhibernators display patterns of arrhythmia, atrial fibrillation, and increased ventricular excitability with subsequent asystole, a lack of neuronal activity of the cardiac muscle (the flat line), and finally, complete cardiac arrest (Johansson 1996). The underlying biochemical and neural mechanisms may entail a greater reliance of cardiac tissue on fatty acid oxidation at colder temperatures, an increase in gap junctions that facilitate low-resistance electrical conduction in the muscle and thus greater synchronicity of muscle contraction, and an enhanced capacity for calcium uptake (Andrews 2007).

One of the many consequences of prolonged exposure to cold during hibernation is a loss of dendritic branches and synaptic profiles (spine densities) in the central nervous system (**Figure 2.52**). However, this effect is minimized in hibernators in comparison with nonhibernators, mainly because hibernators are more resistant to the effect of temperature on synaptic profiles and undergo rapid synaptogenesis upon arousal (Magarinos et al. 2006; von der Ohe et al. 2006). For example, synaptogenesis may contribute to cognitive enhancement in arctic ground squirrels within 24 h after arousal (Weltzin et al. 2006).

Hydrogen sulfide as a potential regulator of metabolism

Hydrogen sulfide (H_2S) is best known for its toxic effect on cytochrome c oxidase (Szabo et al. 2014). It is therefore not suprising that organisms that live in sulfide-rich environments have developed mechanisms to break sulfide down to avoid poisoning of their ETS and

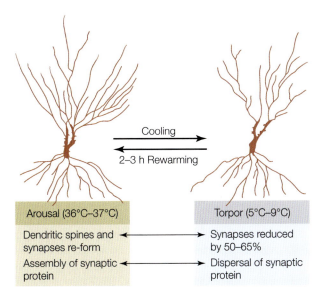

Figure 2.52 Neuronal plasticity in hibernating animals. Two drawings of the same brain neuron during arousal and during torpor, based on findings by Magarinos et al. (2006) and von der Ohe et al. (2006) in golden-mantle ground squirrels and by Magarinos et al. (2006) in European hamsters. (After Andrews 2007 and Magarinos et al. 2006.)

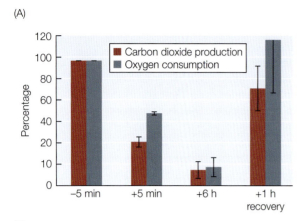

(A)

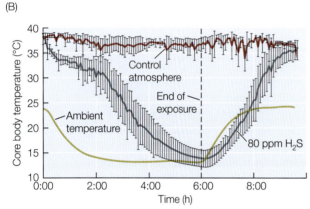

(B)

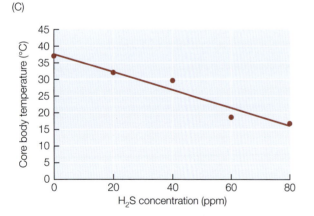

(C)

Figure 2.53 Body temperatures (T_B) and metabolic rates of mice exposed to H_2S. (A) Relative O_2 consumption and CO_2 production of mice exposed to 80 ppm H_2S for 5 min and 6 h and after 1 h of recovery. (B) Body temperatures of mice over a 6-h exposure to 80 ppm of H_2S or a control atmosphere. Data in (A) and (B) are means ± 1 S.D. (C) Linear relationship between H_2S concentration and T_B after 6 h of exposure ($R^2 = 0.95$). (After Blackstone et al. 2005.)

inhibition of oxidative phosphorylation (Powell and Somero 1986; Doeller et al. 1999). At low concentrations, however, H_2S can function as an electron donor to cytochrome c, leading to an increase in oxygen consumption rates and thus contributing to the production of ATP.

Over the last decade the role of H_2S as an endogenously produced gaseous signaling molecule similar to NO has taken physiologists by surprise (Szabo et al. 2014). Dissolved H_2S is lipophilic and easily diffuses across cell membranes. Although the enzymes that contribute to the synthesis of H_2S are known, the regulation of their activity is not understood (Olson and Straub 2016). The first evidence for a potential role of H_2S as a regulator of hypometabolism came from studies on the effects of relatively low levels of H_2S on mice (Blackstone et al. 2005). When mice were exposed to 80 ppm of H_2S in air, oxygen consumption and carbon dioxide production dropped by 50% and 60%, respectively, within 5 min (**Figure 2.53A**). Treatment under the same exposure conditions for 6 h caused a further drop of oxygen consumption to 10% of the control levels, a drop that was accompanied by a decrease in breathing rates from ~150 breaths per minute (BPM) to ~10 BPM. The reduction in metabolism was followed by a drop in body temperature to within 2 degrees Celsius of the ambient temperature (**Figure 2.53B**). Following exposure to H_2S for 6 h, animals returned to air without H_2S and room temperature quickly recovered their metabolic rate and returned to their previous T_B. The effect of H_2S was dose-dependent (**Figure 2.53C**), and at the H_2S levels used in this experiment, animals showed no signs of H_2S-induced tissue damage. Together, these results showed that H_2S affects mice, which are not hibernators, in a way that resembles the metabolic depression and subsequent reduction in T_B of hibernators.

Another recent study showed that the endogenous formation of H_2S may play a role in the remodeling of lung tissue in hibernating Syrian hamsters (Talaei et al. 2012). Although it is not known why the remodeling of lung tissue is necessary, lungs undergo structural changes during torpor that are rapidly reversed during arousal. In combination with cold exposure, Syrian hamsters also increase the production of H_2S through the increased activity of the enzyme cystathione β-synthase (CBS). The specific remodeling of lung tissue during hibernation was confirmed through the increase of the collagenous and noncollagenous hydroxyproline content, a change that modifies the twisting of the collagen helix and thereby the stability of collagen (Talaei et al. 2012). The remodeling of the extracellular matrix coincides with inhibition of Zn^{2+}-dependent matrix metalloproteinases (MMP-2 and MMP-9; also called gelatinases), which bind to collagen. Inhibition of CBS led to an increase of gelatinase activity in torpid animals, suggesting that H_2S may normally suppress its activity. Adding NaHS, an H_2S source, decreased the activity of gelatinase, whereas supplementing lung tissue with Zn^{2+} increased it. These results suggested that the sulfide ion of endogenously produced H_2S may quench Zn^{2+} and thereby decrease gelatinase activity, leading to the remodeling of lung tissue (**Figure 2.54**). Although much needs to be clarified about the role of H_2S during hibernation, the emerging evidence suggests that its role as a regulator of metabolism is more comprehensive than has been appreciated so far. H_2S also protected the viability of murine macrophages, and the effect was greater with increasing levels of hydrogen peroxide, illustrating an antioxidant role for hydrogen sulfide (Szabo et al. 2014). Clearly, we will be discovering more about the role of hydrogen sulfide in regulating the physiology of animals.

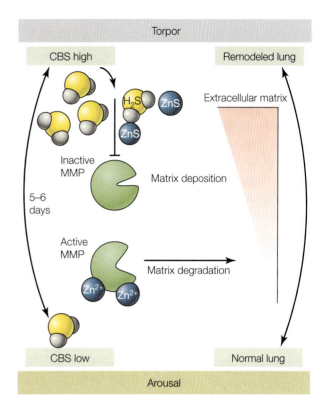

Figure 2.54 Putative effects of H_2S on the remodeling of lung tissue during torpor in the Syrian hamster (*Mesocricetus auratus*). The activity of matrix metalloproteinase (MMP) is regulated by zinc (Zn^{2+}). Increased expression of cystathionine β-synthase (CBS) increases the production of H_2S, which sequesters the Zn^{2+}, decreasing the activity of MMP. This leads to deposition of extracellular matrix during torpor. In contrast, during arousal the expression of CBS and levels of H_2S are low, with high MMP activity degrading the extracellular matrix as a consequence. (After Talaei et al. 2012.)

Biochemical strategies for predictable times of food shortage: Marine mammals

Hypometabolic states are generally accompanied by food shortages and greatly reduced food intake. When these time periods are a normal part of the life cycle of the organism, we must distinguish between times of predictable food deprivation, called fasting, and times of starvation that arise due to unpredictable food shortages. We already covered the biochemical strategies that accompany hibernation, which are characterized by a shift from carbohydrate-based metabolism during summer euthermia to a metabolism dominated by the oxidation of fatty acids during winter heterothermia. Hibernators are also distinguished by their great capacity for metabolic depression and their behaviorally inactive state. In contrast to hibernation, periods of fasting in marine mammals coincide with activities that are energetically costly, such as lactation and territorial and mating behavior in phocid and otariid pinnipeds (earless and eared seals, respectively) and long-distance migration in whales (Rosen and Hindle 2016). Similar to hibernators, marine mammals rely heavily on lipid resources for fuel during these periods, and they too are capable of lowering their metabolic rates in order to extend the time they can subsist on their onboard lipid fuel, depending on the stage of their life cycle.

During nursing, seals generally pursue two different fasting strategies, depending on their phylogeny, the stability of the nursing site (i.e., the physical stability of pack ice), and their body reserves. Phocid seals—for example, gray and northern elephant seals—tend to fast continously during lactation, whereas otariid seals—for example, sea lions and fur seals—intermittently leave their pups to forage. The seals that use these two strategies are referred to as *capital breeders*, which essentially rely on their capital of fat, and *income breeders*, which rely on intermittent food provision (**Figure 2.55**).

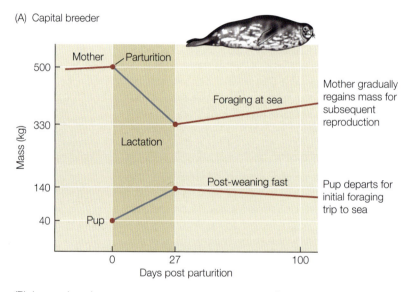

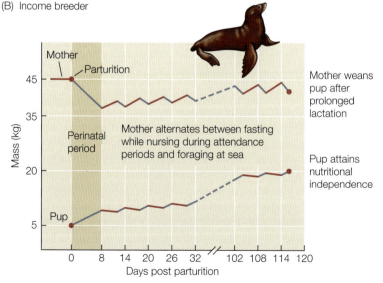

Figure 2.55 Lactating, fasting, and foraging strategies during the breeding season of (A) capital breeding and (B) income breeding pinnipeds. Each graph shows the approximate duration and change in body mass in mothers and pups (top and bottom line in each graph). Blue lines indicate periods of fasting and nursing for the mother, while red lines indicate periods of foraging trips of the mother out to sea and fasting for the pup. Phocid seals (e.g., elephant seals) generally exhibit a capital breeding strategy during which females fast for the duration of the nursing (lactation) period and pups endure a prolonged fast after weaning. In contrast, most otariid seals exhibit an income breeding strategy in which females alternate between brief periods of attendance to nurse the pup and foraging trips to sea. Thus otariid pups alternate between periods of feeding and fasting throughout their development. (After Champagne et al. 2012a.)

Phocid seals experience fasts during their haul-outs on land, which coincide with the breeding season and the annual molt. In the case of the northern elephant seal (*Mirounga angustirostris*), older adult males arrive in December at their breeding grounds along the California coast and defend their territories until the end of February. During this 3-month period they rely mainly on their fat reserves while defending their territories. Females arrive mid-December to mid-January and spend the next 4–5 weeks giving birth and nursing their single pup. During this time females may lose more than 50% of their initial energy reserves, the majority going toward the production of milk and the rest toward the maintenance of the female's metabolism (Crocker et al. 2001b). The juveniles increase in weight from about 30–35 kg to 120–150 kg after 4 weeks of nursing, relying on their mother's milk, which consists of more than 50% fat and about 10% protein but only trace amounts of carbohydrate (Champagne et al. 2012a). Post-weaning, juvenile elephant seals fast for about 8 weeks on land, while also undergoing their first molt, before they enter the sea for their first extended journey (Le Boef and Laws 1994). During this time phocid seals also develop the physiology that enables them to face the challenges of foraging at sea, including locomotion in an aquatic environment, increasing demand for breath-holding, and an increased capacity for thermoregulation (Bennett et al. 2010; Burns et al. 2014). The diving capacity of seals also requires that they have a greater oxygen storage capacity, which they achieve through a large blood volume, high hematocrit, and higher concentrations of both hemglobin and myoglobin relative to nondiving mammals (Kooyman and Ponganis 1998).

The general metabolic strategy used by seals during fasting is highlighted by changes in the predominant fuel they use. Initially, seals preferentially use glucose or glycogen. Because glycogen stores are depleted within hours to days, the animals start to rely on their lipid reserves during the longest and most stable period of fasting, which can last for weeks to months for phocid seals. Lipid reserves provide between 80% and 95% of the energy needed by fasting phocid seals. Although otariids undergo the same shift in the preferred metabolic fuel, they generally fast for a much shorter time. For example, female sea lions will fast for a few days after they return from a foraging trip in order to continue nursing their pup; the fasting period is relatively short due in part to the sea lions' more limited fat reserves (see Figure 2.55). In either group of seals, blood glucose levels have to be maintained at some minimum level to fuel the metabolic needs of the nervous tissue (i.e., the brain) and red blood cells, which rely heavily on glucose as their major fuel. Toward the end of a fast, the animal depletes more and more of the fat reserves located in its hypodermal (or subcutaneous) blubber layer. Because blubber also serves as an insulation against heat loss, its reduced thickness during fasting thus further challenges the animal to maintain its energetic balance. Although catabolism of proteins can help maintain blood glucose levels, proteins are a fuel of last resort because their catabolism requires the breakdown of functionally important cellular components, for example in skeletal muscle. Once a seal starts to degrade its proteins at higher rates, it has to start feeding soon to compensate for the degradation and limit further breakdown of proteins. Protein breakdown also requires the energetically costly production and excretion of urea under conditions when the animals may have no access to external water (when they rely solely on metabolic water). The most obvious reason for the prominent role of fats as a fuel during long fasts is their relatively high energy density. In addition, the oxidation of fats yields a much higher amount of water per gram (although glycogen yields higher levels of water per ATP produced). While fat deposits meet general energetic needs during fasting, they are also used by fasting, lactating females to provision the milk produced in the mammary gland, which contains high levels of non-esterized fatty acids (Champagne et al. 2012a).

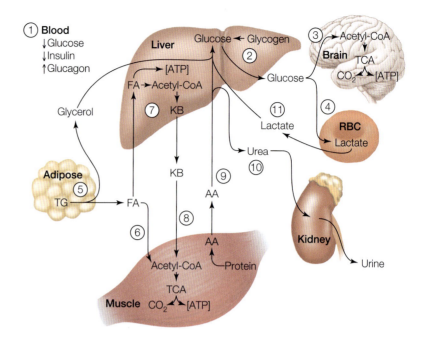

Figure 2.56 Metabolic interactions among organs during fasting. (1) Blood glucose and insulin levels go down, while glucagon levels go up to generate blood glucose through glyconeogenesis. (2) Glycogenolysis contributes to the production of blood glucose as long as glycogen reserves exist. (3) Blood glucose is used by the brain to generate acetyl-CoA and ATP. (4) Glucose is also used by red blood cells (RBC), but because RBCs lack mitochondria, they produce lactate to maintain the redox balance in the cell and to maintain high rates of glycolysis. (5) Triacylglycerides (TG) are converted to glycerol and fatty acids (FA) during lipolysis. (6) The fatty acids are oxidized to acetyl-CoA through β-oxidation, which fuels the TCA cycle and produces ATP. (7) In the liver, acetyl-CoA can be turned into ketone bodies (KB). (8) Ketone bodies are released into the blood and subsequently used by other tissues, for example muscles. (9) During prolonged fasting (or high cortisol levels), protein catabolism generates amino acids (AA) that are used to produce glucose through gluconeogenesis. (10) A by-product of amino acid catabolism is urea. (11) The lactate produced in the RBCs can be converted to glucose through gluconeogenesis. See text for more detail. (After Rosen and Hindle 2016.)

The major metabolic pathways that are used by fasting seals are shown in **Figure 2.56**. The liver takes a central role in providing the seal with a constant, although lower, level of blood glucose during fasting. To do this, the liver degrades its own reserves of glycogen, and it converts lactate produced by red blood cells, amino acids from muscle tissue, and glycerol from the lipolysis of adipose tissue to glucose through gluconeogenesis. Lipolysis of triacylglycerides in adipose tissue (i.e., blubber) by adipose triacylglyceride lipase generates fatty acids and glycerol (Fowler et al. 2015). The fatty acids are released into the blood as non-esterified (free) fatty acids and can be used by other organs, including skeletal muscle and liver, to produce acetyl-CoA. Because nervous tissue does not use fatty acids, the liver may produce and release ketone bodies (acetoacetate and β-hydroxybutyrate) to provide the brain with metabolic fuel, although the ketone bodies' contribution to ATP production may be smaller in species adapted to fasting than in nonfasting mammals (Champagne et al. 2005).

Earlier we stated that proteins are a fuel of last resort, but the extent to which this statement applies depends on the life-history stage of the fasting seal (Houser and

Crocker 2004; Rea et al. 2009). For example, all fasting seals mobilize proteins to maintain a certain level of blood glucose through gluconeogenesis, but the overall contribution of proteins to maintain blood glucose levels is rather small in comparison with that of glycerol and lactate—as low as 2% in weaned elephant seal pups and less than 14% in females (Champagne et al. 2012a). In addition, fasting and lactating females provide proteins for the synthesis of milk, losing some 25% of their total protein stores. Fasting males use proteins for wound healing and the production of sperm, and fasting pups use them for the increased synthesis of respiratory pigments and erythrocytes. Seals that simultaneously fast and molt use proteins to replace their pelage. Importantly, the degree to which protein stores may be tapped by the seal depends in large part on its fat reserves at the beginning of the fast. Furthermore, the fat reserves determine the time when the nursing mother will have to start mobilizing more of her protein reserves and thus the duration of lactation, as the mother resumes foraging soon after tapping into her protein reserves (Champagne et al. 2012a).

Despite surprisingly high levels of blood glucose (5–10 mmol l^{-1}) in fasting phocid and otariid seals in comparison with other mammals of their size, respiratory quotient measurements indicate that glucose oxidation actually contributes only minimally to the overall energy demands of the animals. The measurement of endogenous glucose production (EGP) using isotopic tracers has revealed that weaned elephant seal pups decrease EGP by 25% during their 8-week fast, while females do not reduce the rate at all during their fast. Furthermore, the contributions of both amino acids (2% in weaned pups and 14% in lactating females) and glycerol (~5% in weaned pups and adult females) to gluconeogenesis are small (Champagne et al. 2012a). Instead, the majority of glucose is derived from phosphoenolpyruvate (PEP) and coincides with high flux rates through the TCA cycle, PEP carboxykinase, and pyruvate cycling. In this scenario, the recycling of pyruvate may compensate for the gluconeogenic production of PEP from oxaloacetate, given the limited anaplerotic flow (resupply of intermediates) into the TCA cycle in fasting seals (Champagne et al. 2012b).

A reduction in metabolism during periods of fasting is generally constrained by the fact that seals are simultaneously engaged in activities that increase their metabolic rate, namely lactation, mating, molting, and development. For example, lactating female seals may increase their metabolic rate five times over BMR. However, some seals haul out on land to molt and thereby decrease their activity levels, save on the costs for thermoregulation, and lower their resting metabolic rates. The decrease in resting metabolic rates can be achieved through the downregulation of mass-specific metabolic rates of the most energy-demanding organs, for example, the digestive organs and the kidney, while the animal is on land and fasting (Rosen and Hindle 2016). Little is known about how seals might reduce their metabolic rates by reducing the rates of subcellular processes. For example, when juvenile elephant seals decrease their metabolic rate during the post-weaning phase (Tift et al. 2013), there is a parallel decrease in the activities of metabolic enzymes and mitochondrial respiration rates in skeletal muscles (Chicco et al. 2014; Somo et al. 2015).

The shifts in metabolic fuel preferences during the life cycle of elephant seals illustrate that the metabolic needs of an organism must be analyzed in the context of the role of various organs, their functions and requirements for a certain type of metabolic fuel, and their capacities to store metabolic fuel and to metabolize end products of the metabolic activity of other organs. It is this integration across levels of biological organization—from metabolic pathways to organs to the whole organism—that provides a comprehensive understanding of how biochemical adaptation enables organisms to thrive in various environments with different life cycles.

2.11 Energy Homeostasis and Environmental Stress: A Conceptual Framework

Given the greater energy yield that is achieved with oxygen as the final electron acceptor, organisms that evolved the capacities to exploit oxygen were able to increase the flux of energy through their cells. Perhaps this effect directly enabled the evolution of multicellular complexity. Alternatively, oxygen may have facilitated the development of complexity as a by-product of the cellular compartmentalization events that first led to eukaryotic mitochondrial aerobic respiration. Either way, oxygen has been a major factor influencing the evolutionary paths of multicellular organisms (Decker and van Holde 2011; Lane 2002). With oxygen's role as life's most efficient oxidant at center stage, we can place many of the metabolic pathways and biochemical strategies that we have been discussing into a conceptual framework. Such conceptual models are helpful to generalize the changing energetic demands of an organism in response to its life cycle and the environmental changes it experiences. However, a future goal should be to identify in greater detail the subcellular processes that accompany these changes and the molecular signals that trigger them. These models also offer a potential bridge to integrate changes at the biochemical level with those at the level of the organ and the organism, in order to understand the organism as a whole across all levels of biological organization.

Energy flow plays an important part in an organisms' ability to meet its own metabolic needs, and to invest in growth and reproduction (Sokolova et al. 2012). The flow of energy through an organism is inherently limited by the availability of food, the ability of the organism to acquire it, the absorption and conversion of its chemical bond energy into biologically useful forms like ATP, and the organism's ability to store energy-rich molecules to buffer fluctuations in food availability. Importantly, the conversion of food into biologically useful energy is in part dependent on the maximum metabolic capacity, which is in turn set by the supply of oxygen to cells through the cardiovascular system, as well as the aerobic metabolic capacity of mitochondria (Guderley and Pörtner 2010). It is therefore paramount for an organism to balance the limited energy acquired between the energetic demands of its life and the competing demands of responding to and maintaining homeostasis in the face of environmental stressors.

These trade-offs have been incorporated into a framework called the **dynamic energy budget (DEB) model**, which balances energy acquisition and allocation among competing physiological processes within an individual. The model also provides a way to integrate various physiological processes in order to estimate the costs imposed by single or multiple environmental stressors (Kooijman 2010). In addition, the physiological model of **oxygen- and capacity- limited thermal tolerance (OCLTT)** provides a framework for how oxygen supply sets limits to stress tolerance (Pörtner 2010). Both models incorporate biochemical indicators to a limited extent, but they provide a path toward understanding the organism as a whole system.

Organisms are essentially nonequilibrium systems that require energy consumption to meet their biological needs. In order to deal with environmental stressors, animals rely in part on energy reserves, as shown by the examples of wintering turtles, hibernators, and fasting seals. These reserves are also used for other situations that require temporarily high energy flow that cannot be met by regular food intake alone—for example, during reproduction and fluctuating food availability—thus creating a trade-off among the various energy demands (Sokolova et al. 2012). The female northern elephant seal illustrates this trade-off well; it must simultaneously provision for its own metabolic maintenence

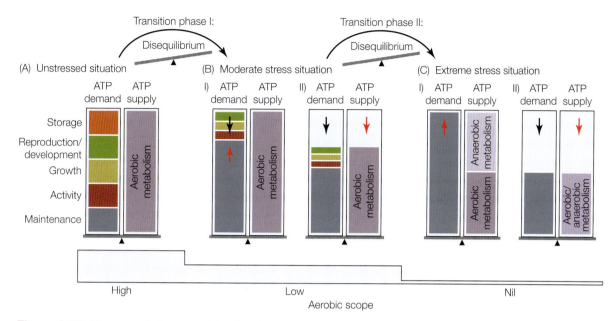

Figure 2.57 Bioenergetic framework for the shift in energy allocation in response to various levels of environmental stress. Red arrows indicate the impact of stress on ATP demand or supply, black arrows indicate the resulting shift in energy allocation. The squares representing the ATP demand of various cellular processes in the unstressed condition are drawn at equal size for clarity only; in an actual animal, these processes would use different proportions of the animal's ATP. (After Sokolova et al. 2012.)

and for the demands of nursing, during a period when it is not foraging and is relying mainly on its fat reserves (Crocker et al. 2001a).

The available energy flux through any organism will have to prioritize the maintenence of basal (or standard) metabolism. This occurs even to the detriment of growth, storage, and reproduction, if the energy flux is limited. Thus the latter processes are supported by the energy flux that remains after the organism has taken care of its basal (standard) metabolism. The surplus metabolic capacity—the difference between the basal (standard) and maximum metabolic rate of an organism—is also referred to as the organism's **aerobic scope**, as it can only be maintained over the long term through aerobic metabolism (**Figure 2.57**). The degree of aerobic scope available to an organism also plays a critical role in setting an organism's capacity to cope with environmental stress.

As illustrated in Figure 2.57, an organism is able to maintain a high aerobic scope under unstressed or optimal conditions. This means that the ATP supply from aerobic metabolism is maximal and that the organism can support activity, growth, reproduction, and development in addition to its own maintenance. It can also store a reserve supply of energy to buffer any future fluctuations in food availability. Such a balance of energy demands maintains a steady state, which is fully supported by aerobic ATP production, that allows the individual organism to thrive and reproduce.

During moderate stress, the energy demand for maintenance increases to respond to abiotic and biotic (pathogens and predators) stressors and to maintain homeostasis. The aerobic scope, although dimished due to the increase in maintenance, is still sufficient to support some activity, growth, and reproduction (see A versus B in Figure 2.57). In addition, the stressor can also impair aerobic metabolism and the acquisition of food, lowering the overall supply of ATP and diminishing the aerobic scope further (see Figure 2.57B, part II). The transition from the unstressed to the moderately stressed condition is characterized by an imblance between ATP supply and demand. Although acclimation or acclimatization can reestablish a balance (homeostasis), it is through trade-offs between the costs for maintenance and activity, growth, and reproduction.

During the transition to extreme stress, ATP demand for maintenance goes up and ATP supply through aerobic metabolism is impaired, making it necessary to provide additional ATP through anaerobic pathways (see Figure 2.57C). The increase in anaerobic pathway activity can extend the time of survival, but the duration of this period depends in part on the fuel reserves available—for example, glycogen and fat deposits—and the accumulation and excretion of anaerobic end products. Extreme stress thus all but eliminates any aerobic scope and instead may require metabolic depression as a longer-term survival strategy. The examples of brine shrimps and killifish illustrate that a hypometabolic state can enable organisms to survive periods of their life cycle that they could not otherwise survive. Furthermore, anaerobic metabolism may be of little help if stressors, such as low pH and high temperatures, impair protein function regardless of the dependency of a pathway on oxygen. In such cases, it is the stress proteome that can protect the organism temporarily until conditions and energy flux improve.

Stress may also cause shifts in metabolic activities that involve reallocation of limited energy resources. For example, because most stressors increase the production of ROS, a greater portion of the NADPH that is normally used for anabolic pathways associated with growth and reproduction is directed to coping with oxygen's "bad" side—the generation of damaging by-products of oxygen use. Regulation of the cell's metabolic activities thus is seen to require appropriate adjustments in the total metabolic flux and the reallocation of metabolic products like ATP and NADPH among different pathways whose needs vary with such factors as the ability to acquire oxygen, the organism's activity level, and the challenges posed by stressors. In Chapter 3 we will see how alterations in a physical environmental variable—temperature—affect the structures and processes of all biochemical systems and thereby have strong impacts on the metabolic rates and energy allocation patterns of organisms.

CHAPTER 3

Temperature

"…it is clear that nature has learned so to exploit the biochemical situation as to escape from the tyranny of a simple application of the Arrhenius equation. She can manipulate living processes in such a way as to rule, and not to be ruled by, the obvious chemical situation."

Sir Joseph Barcroft (1934)

3.1 The Pervasiveness of Temperature's Effects on Life

A substantial fraction of the many papers and reviews that focus on the effects of temperature on living systems begin by emphasizing the pervasive effects of temperature on biological structures and processes. In fact, the states of all biological structures and the rates of all biochemical reactions are affected by changes in temperature as a consequence of basic thermodynamic relationships that are universal—and thus inescapable—across all of life. In essence, everything an organism is and does is affected by the temperature of its cells. However, despite the inherent sensitivities of living systems to temperature, common types of biochemical processes and structures are found in organisms living across the full range of temperatures at which *active* life is observed, a span of almost 200 degrees Celsius (**Figure 3.1**). An even wider range of temperatures can be tolerated by some organisms when they enter a dormant state and temporarily suspend essentially all biochemical activities. Conservation of common types of biochemical capacities in all organisms is definitely not a reflection of insensitivity to temperature effects, but rather a manifestation of how effectively adaptation to temperature has occurred. Our two principal tasks in this chapter, then, are to gain an understanding of how changes in temperature perturb the structures and rates of activity of living systems and, with this foundation, investigate how a wide variety of adaptive changes have made it possible for life to overcome these thermal challenges and thrive in habitats ranging from the deep cold of polar habitats to the extremely high temperatures of deep-sea hot springs (see Figure 3.1). This coverage of the basic effects of temperature also is intended to provide a useful mechanistic foundation for the analyses in Chapter 5, where we examine the known and potential effects of global change.

FIGURE 3.1 Ranges of thermal tolerance in active and inactive states of organisms. Metabolically active life requires liquid water. The thermal tolerance ranges of active organisms lie between ~125°C—archaea living in deep-sea geothermal vents at elevated pressures where water can remain in the liquid state even at very high temperatures (top photo)—and approximately –60°C, where the presence of undercooling (supercooling) agents in some Arctic insects (e.g., the beetle *Cucujus clavipes puniceus* shown in the second photo from the bottom) keeps cellular water liquid. The thermal tolerance ranges of fully desiccated, metabolically inactive organisms are very broad. Fully dried organisms have been held near absolute zero (0 K) and later revived successfully (a tardigrade [bottom photo] is shown as an example of survival near 0 K). Fully desiccated organisms may withstand extremely high temperatures as well (not shown). Species also differ in the ranges of temperature they can tolerate while remaining metabolically active. The longjaw mudsucker (*Gillichthys mirabilis*; fourth photo from bottom) tolerates temperatures between ~2°C and 40°C. The Antarctic notothenioid fish *Trematomus bernacchii* (third photo from bottom) and many Antarctic invertebrates have temperature tolerance ranges of only several degrees Celsius (Peck et al. 2014). Among the most heat-tolerant eukaryotes are snails of the genus *Echinolittorina* (fifth photo from bottom), which withstand temperatures near 55°C.

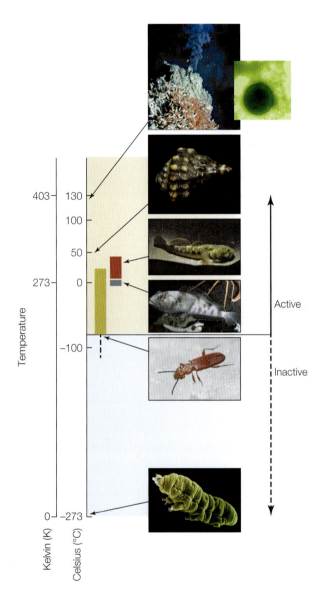

3.2 Why Have Organisms Remained So Vulnerable to Thermal Stress?

Several central questions and themes will serve as the framework for our analysis of the pervasive influences of temperature on living systems. A first set of questions addresses the fundamental bases of temperature sensitivity seen in biochemical structures and rates of chemical reactions. Here we will identify the sources of thermal perturbation of macromolecular structure and discover why even seemingly minor changes in cell temperature can have functionally significant effects on the structures of proteins, nucleic acids, and large molecular assemblages like cellular membranes. Our analysis of the bases of thermal sensitivity of structure will also suggest potential mechanisms of adaptation to temperature change, for example, means for altering the thermal stabilities of these large molecular systems.

The critical importance of molecular fluidity and marginal stability of structure

As we investigate the "tools" in organisms' "tool kits" for adapting to temperature, we will almost immediately come upon what may seem like a paradox. We will learn that it should be possible, at least in principle, to evolve much more thermally resistant macromolecular structures and membranes so as to reduce the potential for perturbation by temperature. Thus, organisms would seem to have the potential raw materials available to evolve much more temperature-resistant molecular systems than those we find in nature. Yet organisms have not done this. Why? The answer to this question involves one of the most important themes of this volume: molecular **fluidity** (see Chapter 1). This term refers to the requirement that proteins and nucleic acids not be overly rigid; instead, they must retain sufficient lability (fluidity) of structure to permit them to undergo the changes in three-dimensional configuration that are essential for function. A similar argument applies in the context of lipid-rich structures like membranes: An intermediate state of structural stability is the state that is optimal for function. The critical requirement that macromolecules and large molecular assemblages have a state of structural stability that allows them to alter their three-dimensional configurations during function is often discussed in terms of the retention of **marginal stability**. *Marginal stability denotes a structural state in which the molecular system in question is poised at the right fluidity to support optimal function.* The systems are stable enough to retain the right geometries for allowing the initial step in a process to occur—for example, ligand binding by enzymes or transcription factor binding to DNA—but the systems are not so stable that they cannot then alter their shape to carry out the subsequent events in the process, for example, catalysis or regulation of transcription. In summary, marginal stability denotes a structural balancing act where the net free energy of stabilization of the molecular system in question is, in the sense of the Goldilocks fable, not too stable, not too labile, but "just right." This stability-lability balancing act underlies both the remarkable specificities and efficiencies of biochemical systems and their vulnerability to disruption by temperature. As we examine this free energy balancing act in species adapted to widely different temperatures, we will see how evolution has finely tuned macromolecular fluidity to ensure optimal function, and we will also gain insights into mechanisms that set ultimate thermal limits to life.

Thermal perturbation can be "put to good use"

When we initially examined the concepts of *stressor* and *stress* in Chapter 1, we pointed out that perturbation of molecular structures has the potential to serve the interests of the cell as well as to disrupt biochemical function. As a major part of our examination of molecular fluidity and marginal stability of macromolecules, we will study an important emerging theme in thermal biology: Perturbation by temperature of macromolecular structure has been exploited during evolution to generate temperature-sensing mechanisms that allow organisms to mount adaptive responses to thermal damage. In effect, some macromolecules—RNA thermometers are a prime example—have evolved to have structures that are readily perturbed by even slight (~1 degree Celsius) changes in temperature, and this structural "damage" allows these temperature-sensing molecules to regulate a suite of adaptive responses to the new temperatures facing the organism. Evolution of regulatory mechanisms that rely on a high level of macromolecular thermal sensitivity, particularly mechanisms based on thermally labile nucleic acid structures, is a fascinating frontier area in evolutionary biology and biomedicine. It is a domain of study that is extremely broad, providing deep insights into the evolution of regulatory mechanisms that govern such widely occurring processes as the heat-shock response and medically important processes like the activation of pathogenic systems when microbes invade warm-bodied species. Suffice it to say, we must keep in mind that even though much of the analysis of macromolecular perturbation by changes in temperature has been discussed in the context of the challenge this damage poses to organisms—stress in the negative sense—there is a positive side to this type of temperature sensitivity as well: Macromolecules that have evolved to have especially high sensitivities to temperature may be crucial for regulating processes of thermal adaptation and for setting thermal limits and thermal optima of life. As we emphasized in Chapter 1, not all stress is bad!

3.3 Thermal Relationships: Basic Concepts and Definitions

To lay the groundwork for our wide-ranging analysis of thermal relationships of diverse species, it is important to begin by presenting some key definitions and concepts. These are terms that we will employ repeatedly in categorizing organisms' thermal relationships and in drawing distinctions among species that have widely different responses to the challenges and opportunities posed by temperature.

Thermal optima and thermal tolerance limits

Thermal optima and thermal tolerance limits are centrally important themes of this chapter. They obviously are key defining features of an organism's overall thermal relationships. Thermal optima and tolerance limits can be investigated at all levels of biological organization, from whole organism performance to activities of individual biochemical reactions. One goal of this part of our analysis will be to link higher-level phenomena, such as biogeographic distributions and whole organism performance, to underlying—and causal—temperature effects on biochemical systems. If we can provide an analysis of the factors that establish thermal optima and limits at the biochemical level, we are then likely to be better prepared to do *explanatory* and *predictive* analyses of thermal relationships at all levels of biological organization. Thus, as will be seen in the analyses in Chapter 5, biochemical studies may allow us not only to explain contemporary patterns of organismal distributions, but also to formulate well-founded predictions of the effects of global change on such large-scale phenomena as biogeographic distributions, ecosystem structures, and extinctions.

Defining "optimal" values of traits: Comparative analysis reveals patterns of conservation

To initiate a biochemically focused investigation of thermal optima and limits, it is necessary to begin with a comparative analysis that, in essence, asks, What properties of temperature-sensitive biochemical systems must be conserved across species, regardless of taxa or temperature? By comparing organisms that evolved at different temperatures, we will learn that certain values for temperature-sensitive biochemical traits—for example, ligand binding by proteins, transcriptional competence of DNA, regulatory sensitivity of switching devices like RNA thermometers, and static order of membranes—are strongly conserved in all species at their normal temperatures of function. This pattern of conservative adaptation, which may characterize differently evolved and differently acclimatized organisms, is illustrated in a highly generalized and schematic manner in **Figure 3.2**. The temperatures at which close conservation of specific values of temperature-sensitive biochemical traits is observed across all species, from all thermal environments, are here defined as **optimal temperatures** for the different organisms. As we emphasized in Chapter 1, when we initially presented the definition of optimal values for

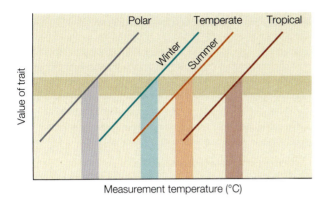

Figure 3.2 Defining optimal values for biochemical systems using comparative analysis. In this generalized presentation, the value of a trait is plotted as a function of measurement temperature for four differently adapted species or differently acclimated conspecifics. Despite the temperature sensitivity of the trait, there is a strong conservation of a particular range of values (horizontal shaded box) for the trait at each species' normal body temperatures (indicated by color-coded shaded boxes that project down to the x axis). This conservative pattern is sometimes referred to as maintenance of *corresponding states* at different temperatures.

environmentally sensitive traits, this is an empirically determined relationship, not one based on theoretical considerations. Here, "optimal" does not refer to "the best of all possible worlds." Rather, the term refers to the values of traits that have been selected by evolution under a wide range of temperatures, using the biochemical raw material available to the organism. When changes in temperature drive the values for these traits out of this highly conserved (optimal) range, sublethal effects are likely to ensue; physiological processes are no longer working at their best. **Thermal tolerance limits** are reached when values for these traits depart too far from the conserved (optimal) range and biochemical systems cease to be able to provide the resources necessary for life. We will have much more to say about determinants of these limits, including their temporal aspects. Rates of change in temperature and duration of exposure to potentially lethal temperatures are two critical factors involved in setting thermal tolerance limits. Acute thermal death, which may occur within seconds or minutes of exposure to a thermal extreme, is almost invariably caused by mechanisms different from those that cause death over periods of days to weeks. And it must be remembered that even if sublethal thermal stress does not kill an individual, it may so hamper its biochemical capacities as to preclude reproduction. In this context, thermal stress may eliminate a population found at a particular site—**local extinction**—even though no individuals of the population are direct casualties of the stress.

Thermal tolerance limits: Active versus dormant stages of life

The complete thermal tolerance range of a species may include temperatures over which normal physiological activities—feeding, growth, locomotory activity, and reproduction—take place, as well as an extended range of temperatures, especially at the lower end of the temperature scale, when periods of quiescence or dormancy are required for survival during harsh conditions. Thus, the span of temperatures over which active life is possible is much narrower than the range of temperatures that can be tolerated by quiescent life stages. Figure 3.1 illustrates these two different ranges of temperature and introduces several phenomena that will serve as important topics in the discussions that follow. Note, first, that the lower limits of temperature that can be tolerated by certain species approach absolute zero (0 K). (**Box 3.1** provides definitions of temperature scales and units for measuring heat and temperature.) Rotifers, tardigrades, and the spores of certain microorganisms can survive, seemingly indefinitely, in a deeply cold "solid state." Taking guidance from the "tricks" used by these organisms, biotechnologists have developed effective means for cryopreservation of cells, tissues, organs, and in some instances, whole organisms in solid state. These efforts have been guided in large measure by

BOX 3.1

Thermal energy, heat, and temperature

Physical relationships: Definitions and units

Thermal energy: The internal energy present in a system due to its temperature. Thermal energy is the statistical mean of the microscopic fluctuations of the kinetic energy of the system's particles. The thermal energy of a system scales with the system's size and is therefore an *extensive* property, not an *intensive* property like temperature. Units: joules (J) or calories (cal); 1 calorie = 4.18 joules. (Note that we employ both calories and joules in this chapter, using the particular unit that was employed by the authors of the studies we are analyzing.)

Heat: In physics, the term *heat* refers to the thermal energy transferred across a boundary between two regions. In biology, the terms *thermal energy* and *heat* are often used interchangeably. Units: joules or calories.

Temperature: The absolute temperature of a system is proportional to the mean kinetic energy of the motions of the particles that constitute the system (electrons, atoms, and molecules). Temperature is an *intensive* property of a system. Three scales (units) for temperature are in common use:

- **Absolute scale:** Measured in kelvins (K) with absolute zero defined as 0 K ($-273.15°C$ on the Celsius scale, and $-459.67°F$ on the Fahrenheit scale).
- **Celsius (or centigrade) scale:** Measured in °C (1°C = 1 K), with 0°C taken as the freezing temperature of pure water at 1 atmosphere pressure and 100°C taken as the boiling point of pure water at 1 atmosphere pressure.
- **Fahrenheit scale:** Measured in °F (1°C = 1.8°F). This scale is not generally used in science but is commonly used in meteorological measurements in the United States.

Heat capacity: The ratio of the heat added to (or removed from) a system to the resulting change in the system's temperature. Units (SI): joules per kelvin.

Specific heat: The amount of heat (joules) needed to raise the temperature of a mass by 1 K.

Biological relationships

Thermal optima: Temperature ranges over which function, including growth, reproduction, movement, and so on, are optimal (see text for more specific treatments of what constitutes "optimal"). For bacteria and archaea, the following terms are commonly used:

- *Psychrophiles*[1]: cold-loving species, thermal optima below ~10°C.
- *Mesophiles*[2]: species that prefer moderate temperatures, 20°C–45°C
- *Thermophiles*: heat-loving species, ~40°C–80°C
- *Hyperthermophiles*: extreme heat-loving species, > 80°C (upper limit of life ~125°C)

Thermal tolerance range: Generally, the range of temperatures over which the organism can remain active. In certain states, notably under extreme desiccation, thermal tolerance ranges can be much greater than those characteristic of active stages (see text and Figure 3.1). When experiments are designed to determine the thermal tolerance range of a species, it is critical to keep in mind that the rate of temperature change and the duration of exposure to extremes of temperature will affect tolerance. Thus, to define the thermal tolerance range, three temperature-related variables need to be considered: absolute temperature, rate of change in body temperature, and duration of exposure to extreme temperatures. Other factors, notably the acclimatization history of the individuals being studied and,

knowledge about the adaptations made by certain cold-tolerant species under natural conditions, as we discuss later in this chapter. The most important consideration in both naturally occurring and laboratory-based tolerance of subfreezing temperatures is the nature of the water present in cells. The physical state of water and the total amount of water in the cell commonly determine tolerance of thermal extremes. At low temperatures, formation of intracellular ice must be prevented in almost all cases; if solid-state existence is to occur, water must be solidified under conditions that foster formation of a glasslike state—vitreous water. One way to avoid freezing is to rid the cell of most of its water and generate a state of extreme desiccation or anhydrobiosis. Chapter 4 will present information on the transitions between hydrated and desiccated states of living

BOX 3.1 (continued)

for terrestrial or intertidal species, relative humidity and wind velocity, may also affect the range of temperatures that can be tolerated. It is also important to keep in mind that temperatures that are acutely lethal (effects seen in minutes to a few hours) may provide a misleading estimate of thermal tolerance; some lethal effects may manifest only after days or weeks (e.g., Dowd and Somero 2013).

Stenotherms[3] have relatively narrow thermal tolerance ranges. There is no strict definition of the precise range of temperatures that qualify an organism as a stenotherm, however. The most stenothermal species known are Antarctic marine ectotherms such as notothenioid fishes, which may have thermal

tolerance ranges as narrow as 4–6 degrees Celsius (–1.9°C to 4°C–5°C; Peck et al. 2014; Beers and Jayasundara 2015).

Eurytherms[4] tolerate relatively wide ranges of body temperature. Again, there is no precise range of temperatures that qualifies an organism as a eurytherm. Extremely eurythermal species may remain active over ranges of body (cell) temperature that exceed 50°C.

---

[1] *Psychro-* is Greek for "cold"; "-phile" is Greek for "love."
[2] *Meso-* is Greek for "middle."
[3] *Steno-* is Greek for "narrow."
[4] *Eury-* is Greek for "wide."

---

systems and show why large-scale reductions in water content can greatly increase cells' capacities to deal with environmental extremes, be they low humidity or extremes of temperature.

The upper thermal limits for life also depend strongly on the properties of water. Liquid water is essential for metabolic activity to occur, and this would imply that the upper temperature limit of life—at least metabolically active life—should lie near the thermodynamic boiling point of water, assuming that macromolecular stability can be retained at such high temperatures. Because the boiling point of water is increased by elevated hydrostatic pressure, the most heat-tolerant organisms known are microorganisms found in deep-sea hydrothermal vent ecosystems or deep in Earth's crust, where temperatures in excess of 100°C are common (see Figure 3.1). To date, the most heat-tolerant organisms that have been discovered are **hyperthermophilic** (sometimes called ultrathermophilic) members of the domain Archaea. Some hyperthermophiles are capable of growth (cell division) at temperatures close to 125°C and have minimum doubling times (= fastest growth) near 121°C (Kashefi and Lovely 2003; Takai et al. 2008). As in the case of tolerance of extreme cold, tolerance of extreme heat can also be increased when organisms undergo complete desiccation. Thus, fully desiccated tardigrades can tolerate temperatures from ~0 K to >373 K (Jönsson and Bertolani 2001). Fully desiccated larvae of the insect *Polypedilum vanderplanki* withstood temperatures between –270°and 102°C (Hinton 1960). Suffice it to say, temperature-water relationships are a critical element in thermal stress at both high and low extremes of temperature.

Eurytherms and stenotherms: Why do thermal tolerance ranges differ so widely?

Among the most intriguing facets of thermal relationships are the wide differences found among organisms in their **thermal tolerance ranges**, as distinct from differences in the absolute temperatures they withstand. Some species—**eurytherms**—are capable of remaining active across wide ranges of temperature, perhaps as great as ~60 degrees Celsius in the case of high-latitude terrestrial insects (Duman 2015). Other species—**stenotherms**—tolerate much narrower ranges of temperature, perhaps as small as 3–8 degrees Celsius in the case of many ectotherms of the Southern Ocean (Peck et al. 2014; Beers and Jayasundara 2015). Does this variation in thermal tolerance range derive from evolved differences in the inherent thermal sensitivities of the phenotype or from different abilities to modify the phenotype through temperature acclimatization—or both? In addressing this central question, one is immediately led to ask what fundamental differences in genetic information and capacities for regulation of gene expression might account for interspecific differences in thermal tolerance. We will find that the breadth of a species' thermal tolerance range is likely to depend far less on the absolute temperatures at which the species evolved than on the variation in temperature that characterized its evolutionary history. Organisms that have evolved in relatively variable thermal conditions may exhibit a higher degree of **phenotypic plasticity**, an ability to modify the phenotype through acclimatization, than stenothermal species. Phenotypic plasticity is likely to be a first line of physiological defense in coping with climate change; thus, the capacity of an organism to acclimatize to rising temperature may determine how successful it will be in dealing with a world that is warming too rapidly to allow sufficient time for evolutionary adaptation in the intrinsic thermal sensitivities of biochemical systems (see Chapter 5). A key factor we will focus on in the analysis of acclimatization to rising temperatures is the proximity of current maximum habitat temperatures to the species' upper thermal tolerance limits. We will see that many warm-adapted eurythermal species are likely to face especially severe threats from global warming because they currently are living at temperatures very close to their upper tolerance limits. Thus, being relatively heat-tolerant and eurythermal does not necessarily reduce a species' vulnerability to climate change (see Chapter 5).

Ectothermy and endothermy: Determinants of body temperature

Another organizing principle in this chapter is based on organisms' abilities to regulate cell or body temperature. We distinguish between organisms whose cell or body temperatures covary closely with (are largely determined by) the ambient temperature—**ectotherms** (*ecto-* is Greek for "outside")—and those organisms that possess abilities to control the temperatures of parts or all of their bodies by regulating the production and dissipation of heat—**endotherms** (*endo-*, Greek for "inner" or "within"). Ectotherms, which are also known as **poikilotherms** (*poikilo-*, Greek for "varied"), comprise the vast majority of species. The prevalence of ectotherms is a reflection of several principles. They generally have much lower metabolic rates—lower energy requirements—than endotherms of similar mass, so a given level of primary production can support a vastly larger number of ectotherms than endotherms. In aquatic habitats, organisms that extract oxygen from the surrounding water (water breathers) face severe challenges in retaining metabolic heat; transfer of heat from the body to the environment at respiratory surfaces is more rapid than the exchange of dissolved gasses. Thus, aquatic water-breathing species, with rare exceptions, are ectotherms. Body size is another factor that influences the occurrence of ectothermy and endothermy. Because of surface-to-volume relationships (see Chapter 2), endotherms must be of sufficient mass to enable effective retention of metabolic heat. Ectotherms thus can be much smaller than endotherms, and because of this, they can occupy niches that endotherms cannot penetrate. For these and other reasons, ectotherms constitute more than 99.9% of the organisms in the biosphere, both in terms of numbers of species and total individuals, and occupy a vastly larger number of niches than endotherms.

Despite having limited abilities to maintain body temperatures that differ from ambient temperatures—especially in the case of aquatic organisms—ectotherms are not without resources for regulating their body temperatures. **Behavioral thermoregulation** is common, especially in terrestrial organisms that can regulate their exposure to sunlight during the day or migrate seasonally to seek more suitable thermal conditions. Vertical movement in the water column may allow aquatic species to seek out temperature conditions that are more favorable than those found at other depths. Thus, ectotherms may respond to changes in ambient temperature through rapid behavioral shifts, as well as through slower acclimatization processes that rebuild their biochemistries. Furthermore, some ectotherms—for example, tunas with muscle-localized countercurrent heat exchangers—can trap the heat generated in the swimming musculature to elevate and stabilize muscle temperatures, enabling enhanced locomotory performance. This is termed **regional endothermy**. Ectotherms, while lacking the refined thermoregulatory abilities of systemic endothermic homeotherms nonetheless have certain capacities for maintaining their cell and/or body temperatures within ranges where their biochemical systems can function at or near optimal levels.

Systemic endothermic homeothermy, the capacity to maintain a stable body temperature through regulation of heat production and heat dissipation, is an energetically costly strategy relative to ectothermy because of the amount of food that must be obtained to fuel the high metabolic rates needed to achieve stable and generally relatively high body temperatures. However, expending energy in thermal regulation may preclude having to spend energy rebuilding the cell's biochemical systems through acclimatization processes, in order to regain optimal states for these systems when ambient temperature changes. Furthermore, acclimatization may require days to weeks to be complete; endothermic homeotherms thus can function at their optimal biochemical states continuously, despite changes in ambient temperature. Retaining stability of function in the face of wide ranges of ambient temperature is clearly a major advantage of systemic endothermic homeothermy. Endothermy can allow stability in behaviors that may be essential, for example, in predator-prey relationships—the acquisition of food or avoidance of becoming someone else's food. This stability in behavioral performance is paired with—and causally related to—conservation of optimal values for numerous biochemical traits in the face of variation in ambient temperature (see Figure 3.2). We now turn to an analysis of these conservative patterns and examine their widespread occurrence among taxa and biochemical systems.

3.4 Proteins

We begin our analysis with proteins, cellular workhorses that provide the cell with most of its structural elements and the machinery for generating energy and doing varied types of work, including locomotion, transport, and biosynthesis. A focus on proteins is an appropriate way to begin our wide-ranging study of thermal relationships, because of proteins' central roles in all physiological processes. Furthermore, some of the basic concepts we will develop through our analysis of temperature-protein interactions will apply as well to other large molecular systems, including nucleic acids and lipid-rich structures like cellular membranes.

Intrinsic versus extrinsic adaptations: A holistic view of biochemical adaptation

At the start of this analysis of temperature effects on specific types of macromolecules, an important caveat must be raised. Our initial mechanistic analysis will, by intent, tell only part of the story. It will focus on what we term **intrinsic adaptations**, defined as chemical changes in the large molecules themselves, for example, changes in amino acid sequences of proteins, alterations in nucleic acid base sequence, and acyl chain compositions of membrane lipids. In Chapter 4 we will return to many of these topics and offer a more complete and holistic analysis, one that expands our treatment by considering **extrinsic adaptations**—alterations in the medium that bathes the large molecules of the cell that adaptively modify the stabilities and functional properties of proteins and other macromolecules. The low-molecular-mass constituents (**micromolecules**) of the aqueous phases of the cell—notably the inorganic ions and small organic molecules that comprise the majority of the solutes present in cellular water—will be seen to undergo adaptive changes that conserve or restore optimal functional states for macromolecules. To reiterate a central theme of this volume (see Chapter 1), macromolecules and micro-molecules evolve together in a complementary way; many of the challenges posed by environmental change to macromolecules are solved through modification of the micromolecular compositions of the cellular and extracellular fluids rather than through changes in the macromolecules themselves.

A holistic perspective on what it means to be a protein

In view of what we have just said about the need to examine both the intrinsic proper-ties of a large molecule like a protein and the ways in which its structure and function are influenced by the chemistry of the milieu in which it occurs, it seems advantageous to define, in as biologically realistic a manner as possible, just what a protein *is*. The textbook definition of a protein emphasizes that it is a long string of amino acids linked together by peptide bonds. Each protein has a unique amino acid sequence, which is known as its **primary structure**. To gain function, however, the protein must fold into a three-dimensional shape (**conformation**) that provides the protein with the correct geometry to achieve its function, whether as a catalyst, a regulator of gene expression, a structural element of the cell, or a contributor to some other function. This function-ally competent, folded state of the protein is often referred to as its **native structure**. In fact, as we will shortly see, proteins occur in an ensemble of different conformational microstates that vary in their functional competence. Thus, a realistic portrayal of a pro-tein is not as one single, static structure—as a textbook illustration might suggest—but as a population of structures that differ subtly in their geometries and, consequently, in their functional capacities.

Another defining feature of a protein derives from the nature of the interactions it has with other molecules, both large and small. Most important, perhaps, are a protein's interactions with water, which govern the thermodynamics of protein folding and protein stability. For a protein to fold into its native, biologically active conformation, it must be bathed in a solution that provides the right conditions to facilitate proper folding—that is, to allow the information in the primary structure to guide the folding of the protein along what is termed a **productive folding pathway** that leads to its native structure. To become and remain functional, a protein must be hydrated. Thus, part of what a protein "is" includes the vast number of water molecules in its immediate vicinity. A similar argument, one developed more fully in Chapter 4, can be made for the inclusion of small organic molecules and inorganic ions found in the aqueous phase immediately surrounding the protein. These solutes, like water, can play critical roles in fostering the correct folding, stability, and function of a protein.

Solvent conditions also figure importantly in determining the effects of temperature on a protein. Many of the properties of the aqueous solution bathing a protein are affected by changes in temperature, which makes protein-temperature interactions even more complex. For instance, solution viscosity is highly dependent on temperature, and vis-cosity has a profound effect on protein stability and function (see Chapter 4; Beece et al. 1980). Thus, to more fully define what we will study when we examine temperature-protein interactions, we need to take an integrative and holistic approach and look beyond the boundaries of the amino acid sequence and three-dimensional shape of the protein and consider the influences on the protein coming from the local—and temperature-sensitive—milieu in which it carries out its function(s).

Last, in defining what a protein is, we must not neglect two other phenomena that help a protein achieve its characteristic structure and role in the cell. First, a protein may interact with other large molecules, and through these interactions, its structure and function may change in biologically important ways. Second, a protein may be subject to posttranslational modifications that alter its function. These chemical alterations of the protein may play vital roles in regulating its activity, for example, in the context of intermediary metabolism (discussed in Chapter 2).

In summary, we have tried to define what a protein is in a way that takes into account the broad array of factors—both intrinsic and extrinsic—that influence the structure(s) that a protein assumes and the function(s) that ensue from gaining these structures. By taking this holistic perspective on what defines a protein, we also lay the groundwork for our analysis of how changes in temperature influence a protein. These influences stem in part from direct effects on the protein itself, that is to say, effects arising from the particular amino acid sequence of the protein. However, of even greater importance in most cases are effects that are mediated through thermal influences on the solution that bathes the protein and on the interactions of that solution with the protein surface.

We will organize our initial analysis of temperature effects on protein structure by examining the different classes of noncovalent (weak) chemical bonds that are the primary drivers of protein folding and protein assembly. The different types of weak bonds exhibit different thermal sensitivities: Some are strengthened by rising temperature, and others are weakened. We will see in our later analysis of protein evolution that selection has exploited these different temperature sensitivities to finely adjust the intrinsic thermal stabilities of proteins.

Thermodynamics of protein folding: Water and the hydrophobic effect

Almost all protein folding requires the presence of abundant water (exceptions are proteins like Late Embryogenesis Abundant [LEA] proteins which are involved in tolerance of desiccation and assume their native structures only under conditions of low water activity; see Chapter 4). The requirement for an aqueous folding environment for proteins reflects the importance of what is termed the **hydrophobic effect** in protein folding and protein-protein interactions. The way this relationship is often presented is by emphasizing that "oil and water do not mix" (or at least not in the ways we customarily think about compounds going into solution). Here, in the context of protein folding and assembly, the "oil" in question is the set of amino acid side chains that are nonpolar. These nonpolar side chains, notably those of leucine, isoleucine, valine, tryptophan, and phenylalanine, tend to bury themselves in the interior of the protein, away from contact with water. This is a primary thermodynamic driving force for protein folding and assembly (Creighton 1993). The structural changes arising from the hydrophobic effect lead to a protein that is folded in such a way that charged and polar side chains (plus many of the peptide backbone linkages between residues) are mostly located near or on the protein surface, in contact with water, whereas nonpolar side chains are largely in the interior, "oily" phase. On the protein surface, networks or webs of hydrogen-bonding interactions occur among exposed polar and charged groups of the protein and with the water coating the protein surface (sometimes termed **vicinal water**). As discussed below, this web of hydrogen bonds is important in stabilizing the mature conformation of the protein, and is also partially responsible for determining the effects of temperature on protein structure. The inorganic ions and small organic molecules in the surrounding solution also play important roles in setting protein stability, in part because of their interactions with vicinal water (see Chapter 4).

Changes in temperature affect hydrogen-bonding interactions and the side chain burial processes driven by the hydrophobic effect. Burial of nonpolar side chains is endothermic (**Box 3.2** defines terms used in thermodynamic functions). This requirement for an input of heat energy arises from the need to break up the organized shell of water (termed a **clathrate**, derived from the Latin word *clatratus*, referring to a latticework or cage) that surrounds a nonpolar group when it is present in an aqueous phase. Clathrates of water must form around a nonpolar group to allow its accommodation in the aqueous solution. When energy is sufficient to break up the clathrates, the nonpolar groups can then move into the oily phase, the interior of the folding protein, and largely cease their interactions with solvent. In thermodynamic terms, what "pays" for this removal of organized water is the increase in disorder of water that results from breaking up the organized clathrates and releasing their water molecules to the **bulk water** phase. Thus, although the enthalpy change (ΔH) associated with transfer of hydrophobic groups from contact with water to the oily interior of the protein is positive (thermodynamically unfavorable in terms of the effect on ΔG of folding), the entropy change is favorable (ΔS is also positive). $T\Delta S$ is of larger absolute value than ΔH, so the net change in free energy ($\Delta G = \Delta H - T\Delta S$) is negative. As temperature is reduced, the thermodynamics of burial of hydrophobic groups changes, as does the efficacy of the hydrophobic effect as a stabilizing force. As temperature falls, $T\Delta S$ decreases and reductions in thermal energy also reduce the probabilities that water cages will be disrupted. Low temperatures thus can lead to protein unfolding because of a weakened hydrophobic effect. Thus, both high and low temperatures can lead to loss of native folded structure—**denaturation**.

BOX 3.2 ▌▬ ▬ ▬ ▬▬▬▬▬▬▬▬▬▬▬▬▬

Thermodynamic functions

There are three thermodynamic variables:
- **Free Energy** (ΔG): Change in the amount of energy available to perform work. Units: kilojoules (kJ) per mole (mol) or kilocalories (kcal) per mole (1 kcal = 4.18 kJ).
- **Enthalpy** (ΔH): Change in heat content. Units: kJ/mol or kcal/mol.
- **Entropy** (ΔS): Change in energy not capable of performing work. Units: kJ mol^{-1} °C^{-1}.

The interplay among these three variables is given below in the famous **Gibbs-Helmholtz equation**:

$$\Delta G = \Delta H - T\Delta S$$

This formulation relates the energy changes that accompany a chemical reaction as it reaches thermodynamic equilibrium. A similar equation describes the energetics of activation processes in enzyme-catalyzed reactions that characterize attaining the activated state(s) of the reactants (the double-dagger symbol is used to denote activation energy changes):

$$\Delta G^{\ddagger} = \Delta H^{\ddagger} - T\Delta S^{\ddagger}$$

Four terms are commonly used to describe the thermodynamics of chemical reactions:
- **Exothermic**: Heat is released during a reaction. ΔH is negative.
- **Endothermic**: Heat is required to drive reaction. ΔH is positive.
- **Exergonic**: ΔG is negative—reaction is favored thermodynamically.
- **Endergonic**: ΔG is positive—reaction is not favored thermodynamically.

Thermodynamics of protein folding: Hydrogen bonds and ionic interactions

Hydrogen bonds and ionic interactions exhibit a different temperature dependence from what is seen for hydrophobic interactions. These two types of weak bonds form exothermically: Heat is released during their formation. The negative ΔH term contributes more to the free energy of bond formation than the (positive) $T\Delta S$ term. Thus, rising temperatures, which may help stabilize hydrophobic interactions, tend to disrupt hydrogen bonds and ionic interactions. The different effects of temperature on these different classes of weak bonds (hydrophobic effect versus hydrogen bonding and ionic interactions) should suggest a means by which proteins could alter their stability during evolution at different temperatures—a point we return to below.

The thermodynamic rules that apply in the case of folding of an individual protein also come into play during equilibria involving protein-protein interactions. For example, hydrophobic patches on the surface of a protein subunit may bind to complementary hydrophobic regions of another subunit (or another type of protein). Protein-protein binding sites are typically rich in hydrophobic residues. Proteins that intercalate into membranes typically have strongly hydrophobic surface regions that favor burial of the membrane-intercalated portion of the protein into the lipid bilayer. In contrast, proteins binding to the surface of a membrane may rely on charged or polar interactions to stabilize the protein's attachment to charged or polar lipid head groups.

Net stabilization free energies (ΔG_{stab}) of proteins are low

To a student new to protein chemistry, the hundreds to thousands of noncovalent (weak) chemical bonds that are involved in stabilizing the three-dimensional structures and aggregation states of proteins might seem to ensure that proteins attain a maximum degree of rigidity of structure, a toughness that would seem to ensure a long lifetime in the cell and a resistance to thermal perturbation. In fact, this view is naïve: The **net stabilization free energy** (ΔG_{stab}) of proteins is only of the order of several noncovalent bonds, despite the large number of such bonds found within the protein structure and between the protein and the surrounding solvent (Jaenicke 2000). Why should this be the case? The strictly mechanistic answer, phrased in thermodynamic terms, is that the net changes in enthalpy and entropy that occur during protein folding tend largely to cancel each other out in terms of establishing the overall free energy change of folding, the net stabilization free energy. However, looking at the issue from the standpoint of evolution and natural selection, the biological or selective reason for this state of affairs is that a protein functions best when it is neither too rigid nor too flexible. As discussed earlier, this intermediate state of stability is termed marginal stability, to indicate the delicate balance between stabilizing and destabilizing forces at work in determining the overall stability of a protein—its fluidity or flexibility (Box 3.3). Proteins are marginally stable because of the selective advantage of this structural state; proteins could be much more stable than they are in nature, were it to the advantage of the organism to possess more rigid, inflexible proteins. Genetic engineering technology has, in fact, led to the production in the laboratory of proteins with greatly enhanced stabilities relative to the natural counterparts of the proteins (Wintrode and Arnold 2001). Under conditions expected in vivo, however, some of these more rigid, thermally stable proteins may not work as effectively as naturally evolved proteins do. For instance, even though highly heat-resistant laboratory-evolved proteins may have high activities at low temperature, they also may have much lower K_M values than naturally evolved proteins (Wintrode and Arnold 2001). Thus, despite having high stability and maximum rates of function, under the low substrate concentrations found in cells, these laboratory-evolved proteins likely

BOX 3.3

What is meant by protein flexibility?

Biochemists frequently use the term *flexibility* to refer to the ease with which a protein's conformation can be changed. We have relied on the term *fluidity* up to this point, to indicate how "flexible" a protein is. This qualitative—and intuitive—way of speaking about a protein's resistance to forces that can change its conformation is useful in discussing how proteins adapt to temperature. However, a more rigorous definition of flexibility—in fact, of flexibilities—is needed to gain a deeper sense of what fine-scale motions in a protein are and how these motions are measured. Thus, depending on the method used, one can measure either *static flexibility* or *dynamic flexibility* (Siddiqui and Cavicchioli 2006; Marx et al. 2007).

Static flexibility is defined as the number and structural varieties of all possible protein conformations. This aspect of flexibility is indexed by techniques that provide an average value for the mobility of amino acid side chains. The B-factor (temperature factor) data provided by X-ray crystallography and data from studies of kinetics of hydrogen/deuterium exchange can be used to estimate static flexibility.

Dynamic flexibility is defined as a measure of how rapidly structural elements in an enzyme interconvert between alternate conformations. Dynamic fluorescence quenching is one method used to

estimate this type of flexibility. In this procedure, a quencher of tryptophan fluorescence such as acrylamide is used to assess the permeability of the protein, that is, the ease with which the quenching agent can access its normally buried target (tryptophan). The side chain of tryptophan emits a strong fluorescence signal when stimulated at the proper wavelength. Because tryptophan residues are normally buried in the core of the protein, this method provides a quantitative measure of the rapid opening and closing—the *breathing*—of the protein's structure. Cold-adapted proteins breathe more rapidly than orthologs of warm-adapted species at a common temperature (Siddiqui and Cavicchioli 2006).

In our discussion of adaptation of protein structure to temperature, we generally will use *flexibility* in a generic sense and not refer to the specific form of flexibility that has been measured. It should be kept in mind that not all regions of a protein have the same static or dynamic flexibility; regions involved in catalysis that undergo large changes in conformation are likely to have a higher level of both static and dynamic flexibility. The conformational microstate model (see Figure 3.3) incorporates both types of flexibility in its analysis of interconversion of conformational states.

---

would have impaired regulatory capacities—a poor ability to vary their rates of function as substrate levels changed—because of the possibility that the low K_M values would lead to continuous saturation of the binding sites (see discussion of K_M relationships in Chapter 2). The excellent review by Wintrode and Arnold (2001) describes the wide variety of changes in stability and function that have been engineered into proteins, and provides deep insights into the mechanisms by which amino acid substitutions affect binding, rate of function, and structural stability, relationships that we return to in naturally evolved proteins later in this chapter.

The conformational microstate perspective of protein structure

One of the most important consequences of the marginally stable state of most proteins is that proteins exist not as single three-dimensional forms (as textbook illustrations might misleadingly suggest) but rather as ensembles of what are termed **conformational microstates** (**Figure 3.3**; Wrabl et al. 2011). The relatively low net stabilization free energies of proteins enable protein conformation to "flicker" among a wide number of slightly different geometries. These changes in shape may involve fluctuations in positions of side chains, shifts in position of structural elements like loops involved in binding (see below), or rearrangement of protein folds. These rapid and continuous variations in three-dimensional shape are extremely important in the contexts of protein function and

(A)

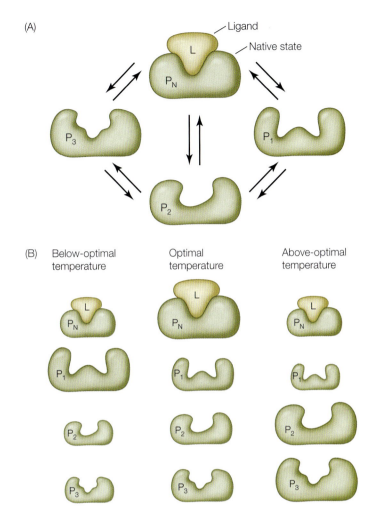

Ligand

Native state

Figure 3.3 Proteins exist as an ensemble of conformational microstates. (A) A given type of protein "flickers" among a diversity of microstates because of the effects of thermal energy on the marginally stable protein structure. Only a subset of these microstates possesses the correct geometry to recognize and bind the ligand (L). Here, P_N designates the native state that is competent to bind ligands, and P_1, P_2, and P_3 represent conformations that are not able to recognize and bind the ligand. The figure is highly diagrammatic; thousands of microstates may coexist at a given instant. (B) Changes in temperature may shift the distribution of microstates within the population of proteins present in the cell, with the consequence that the fraction of P_N may decrease. Here, the relative frequency of a microstate is indicated by the size of the individual image. (After Tokuriki and Tawfik 2009.)

(B)   Below-optimal   Optimal   Above-optimal
       temperature   temperature   temperature

evolution (Wrabl et al. 2011), as well as in the relationships of proteins to environmental factors like temperature.

Binding relationships provide one example of the important roles of rapidly interconverting microstates in mediating protein function. As shown in Figure 3.3, only a subset of all possible conformational microstates possesses the geometry needed for the protein to recognize and bind the ligand(s) with which it must interact to achieve its function as a catalyst, transporter, or gene regulatory molecule. Because of this, it would seem appropriate for natural selection to select for proteins having a propensity to exist, by and large, in microstates that are functionally competent. A protein with too flexible a structure might drift into too many nonfunctional microstates, such that, at any given moment, only a small fraction of the population of protein molecules would be able to recognize and bind the ligands needed for achieving its function. This conjecture leads directly to a question we've raised earlier: In view of the need to have the correct shape for function, why hasn't natural selection led to highly rigid proteins—those that would essentially always have the right shape to recognize the partners with which they need to interact? The short answer to this question, which we have given earlier, is that proteins not only need to have the correct geometry to recognize their ligands,

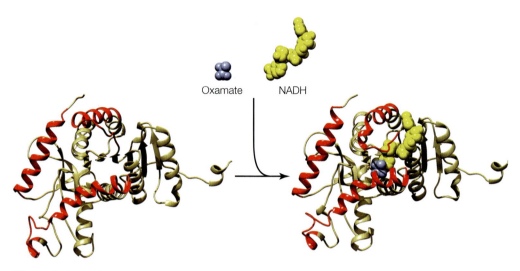

Figure 3.4 Conformational changes in a subunit of lactate dehydrogenase (LDH) during binding of substrate analog (oxamate) and cofactor (NADH). When the two ligands interact with an LDH binding site that possesses the right conformation to permit binding, they enter the active site pocket and trigger a conformational change that leads to formation of the catalytic vacuole, where the transformation of substrate to product occurs. Conformational changes in regions of the enzyme near the binding region, highlighted in red on the protein structure, are as large as ~17 Å for LDH and are the rate-limiting step for this reaction. A highly detailed picture of the conformational changes that occur during binding and catalysis in LDHs is given by Peng et al. (2015).

but they also need to be able to change their conformations to carry out their activities, whether these involve catalysis, transport, gene regulation, or some other function. In an enzymatic reaction, the formation of the enzyme-ligand complex is accompanied by a reorganization of the protein's conformation that brings the ligand(s) into contact with the active site residues essential for catalysis. Biochemists describe this structural change as a process of forming the **catalytic vacuole**, in which the geometrical arrangement of reactants (ligands and essential amino acid residues) allows conversion of substrate to product (**Figure 3.4**). Were the protein too rigid to allow this change in conformation to occur quickly, its function would be reduced, or perhaps blocked entirely. For many enzymes, the **rate-limiting step** in the reaction is formation of the catalytic vacuole. Thus, once the catalytic vacuole is formed, conversion of bound substrate to product may be extremely rapid. These dual requirements for having the correct geometry to allow binding and the needed flexibility of structure to allow subsequent catalysis go a long way toward explaining why natural selection has led to protein marginal stability and a conservation of a particular fluidity or flexibility of structure. This selective process can be seen as a type of compromise, where effective function trumps intrinsic stability and reduced sensitivity to perturbation by temperature. **Box 3.4** provides a short history of the views held by enzymologists about the importance of flexibility in protein structure for binding events.

A BOOST FROM QUANTUM MECHANICAL EFFECTS?   The roles of rapidly interconverting microstates in enzymatic catalysis continue to be elucidated. One fascinating aspect of the contribution of conformational flexibility to catalysis concerns the existence of microstates that may facilitate **hydrogen tunneling**. This quantum mechanical phenomenon enhances rates to an extraordinary degree because the particle in question tunnels through an energy barrier that, by the rules of classic physics, it should not be able to surmount (Benkovic and Hammes-Schiffer 2006). The picosecond timescale fluctuations in conformation that occur after ligands are bound enable the enzyme to sample configurations that provide a favorable electrostatic environment for hydrogen tunneling. This mechanism would likely be of greatest importance for enzymatic reactions in which the "organic chemistry" that occurs after the formation of the catalytic vacuole represents the rate-limiting step.

ALLOSTERIC REGULATION AND PROTEIN MICROSTATES   Conformational microstates also are important in regulation of protein activity by **allosteric** phenomena. Allosteric regulation involves binding of effectors, positive or negative modulators of protein activity, at sites on the protein distant from the site where the functional activity of the protein, for example, catalysis, takes place (*allo-* is a Greek prefix meaning "other" or "different") (Yang et al. 2010; Wrabl et al. 2011). Thus, effector binding may lead to shifts in the microstate distribution that, in turn, cause increases or decreases in protein function. For instance, binding of a negative modulator might lead to a shift in microstate distribution that disfavors substrate binding.

MICROSTATES AND INTRINSICALLY DISORDERED PROTEINS   In the context of regulation, the microstate model is extended in an interesting way by what are termed **intrinsically disordered proteins (IDPs)**, which may comprise more than 30% of the **proteome** (the full set of proteins in the cell) (Hilser 2013). IDPs are often critically important in regulating cellular activities such as gene expression. They belong to diverse functional classes and have in common a relatively low structural stability, and therefore generally limited amounts of tertiary structure when not interacting with their ligands. Formation of the protein-ligand complex leads to a large increase in protein order and the assumption of a functional conformation. The structural transformations that IDPs undergo are essentially at the extreme end of the continuum of protein microstate phenomena. During allosteric regulation, an effector may bind to a disordered region of an IDP and promote the formation of an organized effector-binding domain. This event, in turn, leads to structural organization of the functional domain of the protein, for example, a catalytic site.

BOX 3.4

Locks and keys, hands and gloves, and cows in tents

A brief historical aside seems appropriate here. In the history of enzymology, the first concept of enzyme-ligand interaction, developed by Emil Fischer early in the 20th century, was phrased in terms of rigid "locks" (the binding site) and "keys" (the ligand). This lock-and-key model was discarded when the intrinsic flexibility of proteins came to be appreciated; "locks and keys" became "hands and gloves." Daniel Koshland (2004) has given a concise and entertaining account of our changing perspective on enzyme-ligand interactions during this conceptual transition. Beece and colleagues (1980) provide what may be a better analogy for protein-ligand binding: "A protein is not like a solid house into which the visitor (the ligand) enters by opening doors without changing the structure. Rather it is like a tent into which a cow strays." Whichever analogy you prefer, the key point is that protein structure must be malleable (fluid) and able to accommodate the entry of a ligand into the binding pocket where the "action" (catalysis, transport, or regulation) takes place.

This new perspective on allostery emphasizes the roles of the protein structure as a whole—as well as of the protein-solvent system that influences protein stability (see Chapter 4)—in achieving regulation, and deemphasizes the older view that allostery relied on a network of structural perturbations involving contiguous residues within a fully-folded protein (Wrabl et al. 2011; Hilser 2013). The new conformational microstate perspective also highlights the sensitivities that allosteric regulation may have to changes in temperature, which not only shift microstate distributions (see Figure 3.3) but also influence the capacities of IDPs to acquire the conformations needed to carry out their physiological activities. The manner in which IDPs evolve to maintain the appropriate level of disorder—and capacities for gaining order—to allow optimal function is a poorly understood aspect of protein evolution, and one that merits study in light of the diverse roles of IDPs.

DO DIFFERENT CONFORMATIONAL MICROSTATES OF AN ENZYME HAVE DIFFERENT KINETIC PROPERTIES?   One final point about the significance of protein microstates bears emphasizing in the context of understanding just what we are measuring when we determine a functional characteristic of a protein. When we measure the properties of a protein—for example, the catalytic rate constant (k_{cat}) of an enzyme—we are measuring an average of a range of values. The values vary because a protein does not need to have one specific conformational form to be functional and because the protein's different functional conformations are likely to differ slightly in performance (Xie 2013). Thus, while all functional microstates are likely to be highly similar in conformation, there will nonetheless be some variation in performance among functionally competent microstates. Because the techniques used to determine enzyme properties like k_{cat} typically involve examining a huge number of protein molecules, the value of k_{cat} that one obtains will reflect an average of a great many, slightly different enzyme conformations and performance levels.

In addition to microstates that rapidly flicker in and out of existence, there is evidence that longer-lasting conformational variants of an enzyme with different functional properties may exist. Xue and Yeung (1995) used a sophisticated technique for measuring the activity of single molecules of lactate dehydrogenase. They found an approximately fourfold difference in activity among LDH molecules with identical primary structure but with potential differences in higher-order structures that influence catalytic activity. The differences in catalytic activity persisted over a period of hours, consistent with the view that these putative conformational variants of the enzyme were relatively stable. In a related study, Fields and colleagues (Fields and Somero 1997; Fields et al. 2002) examined the functional and structural properties of LDH-A_4 orthologs of two goby congeners, *Gillichthys mirabilis* and *Gillichthys seta*, that differed in kinetic properties and thermal stability but had identical deduced amino acid sequences and showed no evidence of posttranslational modifications (Fields and Somero 1997). LDH-A_4 purified from the more warm-adapted congener, *G. seta*, displayed a less temperature-sensitive K_M^{pyr} than the ortholog of *G. mirabilis* (see Figure 3.6). However, when the *G. seta* LDH-A_4 was denatured with urea and then renatured by removal of the denaturant, its kinetic properties became indistinguishable from the ortholog of *G. mirabilis*. A variety of physical techniques, including hydrogen/deuterium exchange and infrared spectroscopy, pointed to differences between the orthologs in conformational flexibility in certain α-helical regions of the enzyme. These putative differences in flexibility of structure could underlie the observed differences in kinetics and thermal stability (Fields et al. 2002). The factors that might lead to folding into different conformations during protein synthesis in the two species are not known. However, the differences observed between these two species' LDH-A_4 orthologs are intriguing in that they suggest that a single primary structure can follow different folding pathways and lead to conformational variants with different—and apparently temperature-adaptive—functional properties. These studies of LDHs suggest that the sampling of multiple conformations of proteins can lead to conformational variants with very short half-lives—the types of rapid fluctuations associated with the conformational microstate model shown in Figure 3.3—or much longer half-lives, as suggested by the work of Xue and Yeung (1995) and Fields and colleagues.

Last, in an evolutionary context, the existence of conformational microstates might permit enzymes to adopt a conformation that allows them to bind a ligand different from the normal ligand they have evolved to bind. This "promiscuous" binding and its role in generating evolutionary novelty in protein function are discussed in **Box 3.5** (Tokuriki and Tawfik 2009).

Protein stability and function reflect evolutionary thermal history

In view of the acute thermal sensitivities of proteins due to their marginal stabilities, adaptive responses of proteins to thermal stress are critical under all time-courses of temperature change, ranging from sudden shifts in cell temperature during the course of a day, to seasonal changes in temperature, to long-term temperature changes occurring on the timescale of evolutionary processes. In fact, adaptive responses of proteins

BOX 3.5

Microstate ensembles and "promiscuous" protein function

The evolutionary roles of microstates may be significant in the context of "invention" of new types of proteins. Tokuriki and Tawfik (2009) propose that proteins with a relatively high degree of disorder in their three-dimensional structures are likely to exhibit functional "promiscuity," defined as an ability to bind ligands (L*) in addition to the one that the protein has evolved specifically to bind (L; see the figure). If the promiscuous binding of L* leads to a new chemical reaction having favorable effects on cellular function, then mutations in the protein that lead to conformations that favor binding of the new ligand may be selected. If only one allele at the protein-coding gene locus acquires mutations that increase probability of binding of L*, then each of the two allelic forms of

the enzyme may have at least slightly different functional roles in the cell. The original allelic variant continues to perform its long-standing metabolic task; the new allele with promiscuous ability can continue to perform the original function and a new function as well. Gene duplication events can be appreciated as a means for enhancing this type of promiscuity and evolution of new function. Thus, duplicated genes, whether they arise from tandem gene duplication or from increases in ploidy, provide the organism with potentially important sets of new genetic raw material for development of novel metabolic potentials. Selection based on promiscuity could be an important component of the acquisition of new biochemical capacities via gene duplication (see Chapter 1).

Native conformer/function

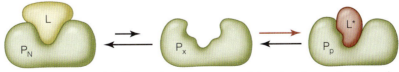

Conformational microstates, promiscuous binding, and protein evolvability. A protein's (P) conformation interconverts among a large number of microstates, three of which (P_N, P_x, P_p) are shown here. The most probable microstate is the so-called native state (P_N), which binds the ligand L, the binding event that has been selected during evolution. One

or more of the minor (less probable) conformations (P_p in the figure) may bind a different ligand, L*. If this promiscuous binding has a favorable effect on the organism, selection may favor mutations that increase the probability of formation of the promiscuous conformer. (After Tokuriki and Tawfik 2009.)

to these various temporal patterns of thermal stress have been well documented and shown to involve several underlying mechanisms, ranging from adaptive change in protein sequence to rapid modulation of the working environment—the intracellular solution or membrane microenvironment—in which the protein conducts its functions. In this chapter and the next, we will examine the variety of mechanisms by which conservation of protein function is achieved in the face of temperature change.

We first examine temperature-adaptive differences in function, structural stability, and amino acid sequence among proteins that have evolved at different temperatures. **Orthologous proteins** (Table 3.1) of species that have evolved at different temperatures typically

---

TABLE 3.1

Terminology associated with variation within a functional class of proteins

Ortholog. Orthologous proteins are proteins that are encoded by a common gene in different species and generally, but not invariably, carry out the same function in all species. For example, orthologs of the A-type lactate dehydrogenase (LDH-A_4), which is the predominant form of LDH found in vertebrate skeletal muscle cells poised for glycolytic function, are encoded by the same gene locus across species. Adaptive variation in gene (protein) sequence allows the common function to be sustained across widely different conditions of temperature, as discussed in the text. LDH-A_4 has functional (kinetic) properties that favor the reduction of pyruvate to lactate (hence, in terms of function, the enzyme could be termed a pyruvate reductase).

Paralog. Paralogous proteins are variants on a common protein theme that are encoded by different gene loci. Lactate dehydrogenase provides a good—indeed, a classic—example of this phenomenon. Whereas LDH-A_4 is critical for sustaining glycolytic flux under conditions of limiting oxygen by regenerating NAD^+ through reduction of pyruvate to lactate (see Chapter 2), the B-paralog of LDH (LDH-B_4) functions under aerobic conditions to convert lactate to pyruvate. The pyruvate that is generated can be shunted into aerobic pathways of ATP generation by entry into the TCA cycle, or converted back to glycogen through the multistep process of gluconeogenesis. The mechanisms that control the direction of LDH flux, in accord with oxygen availability, include pH modulation of substrate binding (see Chapter 4). Gene duplication events commonly lead to divergence in functions of paralogs (Ohno 1970).

Allelic variants. Within an individual organism, a given protein may exhibit sequence variation. For example, in diploid species the ortholog of a protein encoded by the gene contributed by one parent may have a sequence that differs from that of the protein encoded by the gene from the other parent. This is classic allelic variation. The term *allozyme* is sometimes used to refer to these variants. Similarly, among the individuals of a population of a species, significant amounts of genetic variation at a locus may exist, leading (if the variation is nonsynonymous) to a variety of amino acid sequences for a protein. There has been a long-running debate in the literature of population genetics about the functional relevance of allelic variation. The question of how much variation is neutral and how much is adaptive or selective has generated hundreds of theoretical papers, but a much smaller number of explicit empirical tests of protein function.

Splice variants. A eukaryotic protein-coding gene typically is a mixture of protein-coding exons and intervening noncoding introns. Splicing of the mRNA transcribed from a protein-coding gene entails the removal of introns and the joining of exons to generate a mature mRNA molecule that can be translated into a functional protein. Multi-exon genes can be assembled into more than one final protein product because of the opportunities for multiple patterns of splicing of mRNA. It is becoming increasingly clear that for a majority of protein-coding genes, alternate splicing of mRNA takes place. The functional significance of such splice variants is discussed in the text.

Posttranslational modification (PTM) variants. In Chapter 2 we emphasized that metabolic regulation often involves varying the posttranslational states of proteins—for example, through reversible phosphorylation or acetylation. Less is known about the role of PTM in responses to thermal stress, but this means of modifying the functions of proteins, especially enzymatic and contractile proteins, is likely to be of some (although as yet unclear) importance in facilitating protein function under conditions of varying temperature. In concert with regulating overall metabolic rates at values appropriate for existing thermal conditions, PTM also may help ensure that temperature-specific relative activities occur in different pathways.

exhibit differences in thermal responses that reflect the proteins' evolutionary thermal heritage. At one extreme are proteins of hyperthermophilic archaea and bacteria, which may be stable at temperatures near the boiling point of water. These are temperatures at which proteins of mesophiles and psychrophiles would literally be "cooked" (fully denatured) (Feller 2010). At the other extreme are proteins of psychrophilic (cold-loving; see Box 3.1) species that may function best at temperatures near the freezing point of water (Fields and Somero 1998; Fields 2001; Siddiqui and Cavicchioli 2006). At the low temperatures favored by psychrophiles, proteins of thermophiles would function poorly, in part because of the high rigidity of structure that would result from exposure of these heat-adapted proteins to extreme cold. As we show below, conservation of the appropriate marginal stability for function at normal cellular temperatures is the end-result—the primary selective advantage—of adaptive alterations in amino acid sequence arising from evolution at different temperatures. The number of such adaptive substitutions, their location(s) in the protein, and the mechanisms through which these substitutions cause adaptive changes in stability and function will be the focus of the following sections of our analysis.

Orthologous proteins of differently thermally adapted congeners

We begin this analysis with a detailed look at orthologous proteins from congeneric species that have evolved under different temperature conditions. Comparison of orthologous proteins of congeners is an effective way of discerning the amino acid substitutions that underlie adaptation of stability and function. Very few differences in amino acid sequence may exist between orthologs, and the changes that are observed may largely be adaptive, rather than neutral background noise in the sequence that has no effect on function or stability. In fact, analysis of orthologs of congeners calls into question how much amino acid change can, in fact, be without effect on the protein's stability or function. The finding that amino acid substitutions distant from an enzyme's catalytic vacuole have significant effects on catalytic function suggests that much of the protein, notably its surface, may be involved in establishing the energy changes accompanying binding and catalysis—in a sense, the entire enzyme may be an "active site" (Holland et al. 1997).

Once this analysis of closely related species is completed, we will broaden our perspective and compare proteins from all domains of life. Studies focused on archaea, bacteria, and eukaryotes that have evolved at temperatures across the approximately 200-degree-Celsius range of temperatures that are permissive of active life will lead to new insights into temperature-protein interactions. We will investigate whether different domains of life follow different rules in adaptation, whether the properties of proteins in contemporary species can provide insights into the temperatures at which life arose, and whether there are general patterns in amino acid use in proteins that have evolved under widely different temperatures. We will also examine recent literature that offers insights into the acquisition of temperature adaptations through horizontal gene transfer (HGT), a mechanism for attaining adaptive genetic variation that is especially important in bacterial and archaeal species.

A logical starting point for our analysis is to focus on essential protein traits that depend on maintenance of an appropriate level of marginal stability. These are traits that are especially sensitive to temperature and likely to be highly conserved among orthologous proteins of species with different evolutionary thermal histories. These patterns of trait conservation within optimal ranges, as portrayed schematically in Figure 3.2, are in fact a major outcome of the amino acid substitutions that occur during evolution at different temperatures. A clear demonstration of this pattern is shown by comparisons of ligand (substrate and cofactor) binding capacity and thermal stability in orthologous proteins from congeneric species that have evolved at different temperatures.

Figure 3.5 Adaptive variation in thermal stability (right panels) and ligand (NADH) binding ability (left panels) among cytosolic malate dehydrogenase (cMDH) orthologs of marine molluscs adapted to different temperatures. Rate of loss of cMDH activity during incubation at high temperature is used as an index of protein structural stability (see text). Function is indexed by the Michaelis-Menten constant of NADH (K_M^{NADH}). (A) Congeners of the genera *Chlorostoma* and *Littorina*. The two littorine species occur high in the intertidal zone. Congeners of *Chlorostoma* include Pacific Coast species with different vertical positions (*C. funebralis* > *C. brunnea* > *C. montereyi*), and a Gulf of California intertidal species, *C. rugosa*. (B) Congeners of limpets of the genus *Lottia*. *L. digitalis* is a northern species; *L. austrodigitalis* occurs at lower latitudes. (C) Abalone of the genus *Haliotis*. Species names (common names): *H. kamtschatkana* (pinto), *H. rufescens* (red), *H. cracheroidii* (black), *H. corrugata* (pink), and *H. fulgens* (green). Error bars indicate standard errors. (A, unpublished data of X. Meng and G. Somero; B after Dong and Somero 2008; C after Dahlhoff and Somero 1993a.)

CONSERVATIVE PATTERNS IN TEMPERATURE ADAPTATION   **Figure 3.5** illustrates how enzyme thermal stability (indexed by rate of loss of enzymatic activity during incubation at a high temperature) and thermal effects on functional capacity (indexed by the effects of temperature on the Michaelis-Menten constant of NADH [K_M^{NADH}]) differ among orthologs of cytosolic malate dehydrogenase (cMDH) from four sets of congeneric species adapted to different temperatures. All genera belong to the phylum Mollusca, and each genus comprises two or more species adapted to a different range of temperatures. Figure 3.5A presents data for congeners of turban snails of the genus *Chlorostoma* (formerly *Tegula*) and littorine snails of the genus *Littorina*. *Chlorostoma* congeners are found in habitats with temperatures ranging from near 0°C (for high-latitude populations of the mid-intertidal snail *C. funebralis*) to over 40°C (for low latitude populations of *C. funebralis* and the Gulf of California species, *C. rugosa*). As discussed later in the context of interspecific differences in the heat-shock response, congeners of *Chlorostoma* provide an excellent study system for examining protein evolution in a common genus that has evolved over an exceptionally wide range of adaptation temperatures. Littorine snails are commonly found in the highest reaches of the intertidal zone, where they encounter temperatures greater than those experienced by lower-occurring species, including turban snails. Figure 3.5B shows data for cMDH orthologs of two limpets of the genus *Lottia* with different latitudinal distributions but similar vertical distributions. The biogeographic ranges of both species are shifting in parallel with global warming: The southern limit of distribution of *L. digitalis*, the more cold-adapted species, is moving to higher latitudes, as is the northern limit of distribution of *L. austrodigitalis*, the more warm-adapted species (Dong and Somero 2008). These two species are what are known as **cryptic congeners**, because they are essentially indistinguishable on morphological grounds. Cryptic congeners are very common among marine invertebrates. Often, the discovery that one is examining two (or perhaps more) rather than one species is attendant on application of biochemical and molecular methods that reveal differences that are not apparent to the naked eye. Figure 3.5C presents data for congeners of *Haliotis* (abalone) that differ in their biogeographic and vertical distributions. Two species, *H. kamtschatkana* (pinto) and *H. rufescens* (red), have northern distributions; three are found in more southerly habitats: *H. cracheroidii* (black), *H. corrugata* (pink), and *H. fulgens* (green) (for distribution data, see Dahlhoff and Somero 1993a).

   In all comparisons of these sets of differently thermally adapted congeners, consistent trends are noted between the adaptation temperatures of the species and thermal stability of cMDH function and structure. This trend is seen across wide latitudinal ranges and, within a single habitat, along a vertical gradient from the subtidal zone to the mid- to high intertidal zone. For congeners of *Chlorostoma*, the two mid- to

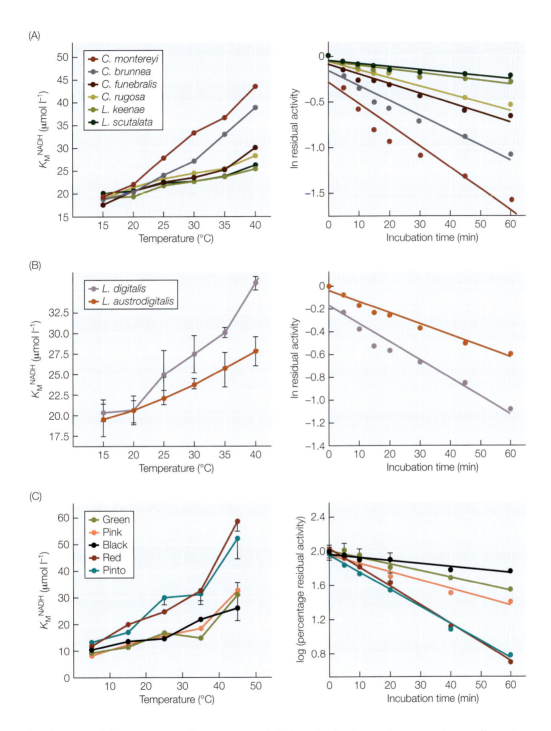

(A)

(B)

(C)

high-intertidal congeners, *C. rugosa* and *C. funebralis*, have the most thermally resistant cMDHs. These orthologs have longer half-lives when heated at high temperatures and more thermally stable K_M^{NADH} values than the orthologs of the low-intertidal and subtidal congeners, *C. brunnea* and *C. montereyi*, respectively. The loss of ligand (NADH) binding ability at temperatures above the normal habitat temperatures of the latter two species (> ~20°C–25°C), which is indicated by the relatively sharp increases in K_M^{NADH} for these orthologs, is a sign of heat-induced reduction in capacity to initiate the catalytic cycle by binding NADH, the initial and rate-limiting step in the MDH reaction. Similar differences in thermal stability of function and structure among cMDH orthologs are seen for the congeners of *Haliotis*. For the littorine snails, thermal stabilities are similar between species and are greater than those found for the cMDHs of the other genera, in keeping with the high vertical distribution of the littorine snails. The trends noted for cMDHs of marine invertebrates reflect a pattern seen in comparisons of MDHs of psychrophilic and thermophilic bacteria: Improved efficiency of function at low temperatures by the MDH of the psychrophile *Aquaspirillum arcticum* is accompanied by a reduced intrinsic stability of structure (Kim et al. 1999).

Comparisons of orthologs of another dehydrogenase enzyme, the A paralog of lactate dehydrogenase (LDH-A$_4$; **Figure 3.6**) from vertebrates whose body temperatures span a range of approximately 45 degrees Celsius, reveal patterns similar to those found for cMDHs of marine invertebrates. At normal body temperatures, K_M values of pyruvate (K_M^{pyr}) are highly conserved among species. The cold-adapted Antarctic notothenioids, whose body temperatures are near the freezing point of seawater (−1.9°C), and the thermophilic desert iguana (*Dipsosaurus dorsalis*), whose core temperature may exceed 40°C, have identical K_M^{pyr} values at their normal body temperatures. Above the normal range of body temperatures, however, sharp upswings in K_M^{pyr} may occur. The observed differences among congeners of *Sphyraena* (barracuda fishes) reflect the types of inter-congener differences seen for cMDH. The three species of barracudas compared in this

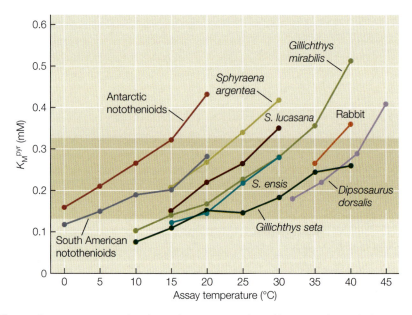

Figure 3.6 The effects of temperature on binding of pyruvate (indexed by K_M^{pyr}) by orthologs of LDH-A$_4$ from differently thermally adapted vertebrates. Congeners of barracuda fishes (genus *Sphyraena*) provide another illustration of adaptive differences among similar species evolved at different temperatures. Congeners of *Gillichthys* have LDH-A$_4$s that have identical deduced amino acid sequences, but may differ in folded structure (see text). The desert iguana (*Dipsosaurus dorsalis*) has the highest maximum body temperature of the species studied, approximately 43°C. The darkly shaded area indicates the range of values for K_M^{pyr} that are conserved among species across their range of normal body temperatures. Sources of data: Notothenioid fishes (Fields and Somero 1998); barracuda fishes (Graves and Somero 1982; Holland et al. 1997); *Gillichthys* congeners (Field and Somero 1997); *Dipsosaurus dorsalis* (Somero 1995).

study, the temperate congener *S. argentea*, the subtropical congener *S. lucasana*, and the tropical congener *S. ensis*, differ in maximum habitat temperature by 3–8 degrees Celsius (Graves and Somero 1982; Holland et al. 1997). This difference in evolutionary temperature is sufficient to favor selection for temperature-adaptive differences in binding properties. The LDH-A$_4$s of the two congeners of *Gillichthys* reflect the differences in evolutionary thermal history of the extremely heat-tolerant Gulf of California species, *G. seta*, and the more widely distributed (from central California to the Baja California peninsula) *G. mirabilis*. Here, unlike in the other sets of congeneric LDH-A$_4$s, no difference in amino acid sequence between orthologs of congeners was found (see above).

   Another component of temperature adaptation of proteins is illustrated by the comparisons of LDH-A$_4$ orthologs of three congeners of damselfishes (genus *Chromis*; **Figure 3.7**). Along with the conservation of K_M^{pyr} in the tropical (*C. caudalis* and *C. xanthochira*) and temperate (*C. punctipinnis*) congeners, there is a temperature-adaptive difference in the k_{cat} values of the orthologs. The ortholog of the temperate congener has a higher rate of conversion of substrate to product (per active site) than the orthologs of the two tropical congeners. This difference in k_{cat} can be regarded as temperature compensatory, for it enables the cold-adapted species to sustain a rate of LDH function at its normal habitat temperatures (15°C–20°C) that equals or exceeds the rates found in the tropical species at their substantially higher habitat (body) temperatures (26°C–28°C) (Johns and Somero 2004). The trend in k_{cat} values exhibited by LDH-A$_4$s of damselfishes has been observed in several other comparisons of orthologous proteins (Graves and Somero 1982; Fields and Somero 1998). In most comparisons, however, k_{cat} values are not fully compensated for temperature (see the Section 3.8, p. 297). In all of these comparisons of orthologs of differently thermally adapted species, there is a consistent correlation between the magnitudes of K_M and k_{cat}, such that, at a common temperature of measurement, orthologs with high values for K_M also have high k_{cat} values. We now turn to an analysis of the basis of this correlation, which will provide a link between structural stability and kinetic properties.

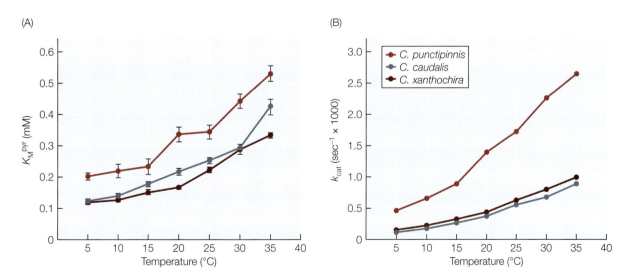

Figure 3.7 Comparisons of temperature effects on (A) binding (K_M^{pyr}) and (B) rate of function (k_{cat}) for LDH-A$_4$ orthologs of three congeners of damselfishes, two tropical species (*Chromis caudalis* and *C. xanthochira*) and a temperate species (*C. punctipinnis*). Error bars indicate standard errors. (After Johns and Somero 2006.)

STRUCTURAL BASES OF ADAPTATION IN KINETIC PROPERTIES   The relationships between temperature, K_M, and k_{cat} shown in Figures 3.5, 3.6, and 3.7 can be explained in terms of the evolved marginal stabilities of proteins and the occurrence of a range of conformational microstates whose distribution pattern is temperature-dependent (see Figure 3.3). Conservation of K_M across orthologs at normal body temperatures is conjectured to reflect the occurrence of a large fraction of the enzyme population in conformational microstates that are competent to recognize and bind ligands. The upswings that occur in K_M as temperature increases are viewed as a manifestation of a shift in the distribution of conformational microstates that leads to a decreased fraction of the population of enzyme molecules having the requisite geometry needed for recognizing and binding cofactor or substrate. Thus, through evolutionary modifications of structural stability, the microstate distribution curve is shifted so as to enable a large fraction of the protein population to have the correct geometry for ligand binding at normal cell temperatures in differently thermally adapted species.

This conservative trend is a manifestation of the maintenance of a **corresponding state** of protein structure at normal temperatures of cellular function (see Závodszky et al. 1998). One of the most direct demonstrations of such conservation of structural state comes from hydrogen/deuterium exchange experiments, which provide a quantitative index of protein structural flexibility. Using this method, Závodszky and colleagues (1998) showed that orthologs of 3-isopropylmalate dehydrogenase (IPMDH) from thermophilic and mesophilic bacteria differed greatly in thermal optima and intrinsic stability. However, at the respective optimal temperatures for the bacteria, the rates of hydrogen/deuterium exchange, that is, protein flexibility, were nearly identical.

The interspecific differences in k_{cat} likewise can be explained in terms of evolved differences in stability. Because the rate-limiting events in an enzymatic reaction may be the changes in conformation that occur during either binding of substrates or release of products, the more flexible proteins of cold-adapted species may be able to function more quickly than those of warm-adapted species (when comparisons are made at a common temperature of measurement). Flexibility in loop regions, as seen for LDH-A$_4$ (see Figure 3.4), may be especially important in adjusting thermal sensitivities. This has been conjectured for the orthologous dehydrogenases discussed above and has been demonstrated as well for other enzymes in which loop movement is rate-determining, for example, uracil DNA glycosylase (Olufsen et al. 2005).

Linking structure to function: How do amino acid substitutions lead to adaptive changes in stability, K_M, and k_{cat}?

To elucidate how differences in protein primary structure lead to adaptive changes in thermal stability, K_M, and k_{cat}, we must address three related questions. First, *how many* amino acid substitutions are required to adaptively modify stability and function? Second, *where* in the protein's sequence and three-dimensional structure do these substitutions occur? Are particular regions of the protein optimal sites for adaptive change—or can adaptation occur at a great many sites? And third, *how* does a particular change in sequence lead to changes in the protein's thermal sensitivities? Can we account, *mechanistically*, for the coupled changes in thermal stability, K_M, and k_{cat}?

To address these questions we will conduct a closer examination of orthologous dehydrogenases from congeneric animals adapted to different temperatures, for which consistent patterns of adaptive variation in temperature-sensitive traits have been discovered. By comparing the sequences and three-dimensional structures of these orthologs of cMDH and LDH-A$_4$, researchers have discovered several key mechanistic links between changes in sequence and thermal sensitivity of structure and function. These studies allow us to answer all three questions posed in the preceding paragraph.

ONE SUBSTITUTION OUTSIDE THE CATALYTIC VACUOLE CAN ACHIEVE ADAP-
TATION IN STABILITY AND KINETIC PROPERTIES   A good example of this
type of linkage between changes in sequence and alterations in structure
and function is shown for the cMDHs of two cryptic congeners of *Lottia* (see
Figure 3.5B; *Lottia digitalis* is shown at right). These enzymes' differences
in temperature sensitivity of NADH binding and structural stability have
been traced to a single-residue difference in their 333-residue amino acid
sequences (Dong and Somero 2008; **Figure 3.8**). Thus, modification of less
than 1% of the cMDH sequence is sufficient to shift the enzyme's thermal
characteristics. The single difference is at position 291 of the amino acid

Lottia digitalis

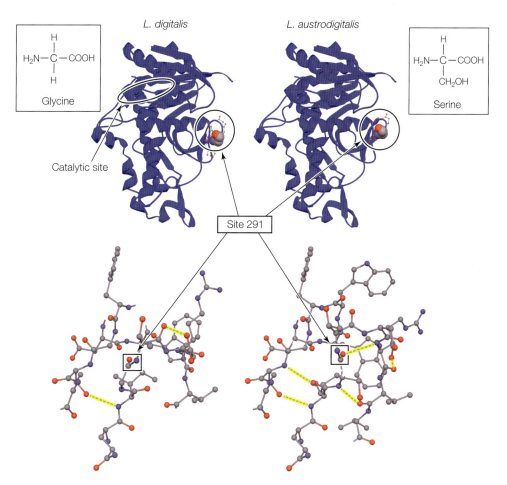

Figure 3.8 Orthologs of cMDH of two congeners of *Lottia*. The upper drawings show a complete subunit of cMDH and identify the site of the adaptive substitution; the ball-and-stick drawings below show an enlargement of the area around this site. A single amino acid substitution
at site 291 (black circles and boxes) in the 333-residue sequence distinguishes the two orthologs. The substitution in the ortholog of *L. austrodigitalis* of a serine for the glycine found in its
congener's cMDH leads to formation of three additional hydrogen bonds (all hydrogen bonds
are highlighted in yellow), which stabilize the warm-adapted protein. Site 291 is distant from the
catalytic site (black oval) yet influences ligand binding through a stabilizing effect transmitted
through the molecule. Active site (catalytic vacuole) residues are fully conserved between the
species. (After Dong and Somero 2008.)

sequence, a location that is distant from the ligand-binding site of the enzyme (see the upper structures in Figure 3.8). In fact, for all types of enzymes so examined, no comparisons of orthologs from differently thermally adapted species have revealed changes in sequence within the catalytic vacuole region (Isaksen et al. 2014; Fields et al. 2015). This is not surprising. Alterations in the amino acid residues with which substrates and cofactors interact could eliminate the catalytic activity of the protein (or shift its substrate preferences, as noted in comparisons of MDH and LDH, "sister" enzymes with highly similar catalytic vacuoles but different substrate preferences; see Fields et al. 2015).

How, mechanistically, can this minor change in cMDH primary structure lead to adaptive alterations in binding and stability? Molecular modeling suggests that the shift from a glycine residue in the cold-adapted species (*L. digitalis*) to a serine residue in the warm-adapted species (*L. austrodigitalis*) enables three additional hydrogen bonds to form (see Figure 3.8). These additional bonds confer an increase in net structural stability. Furthermore, glycine generally disfavors structural stability because its small side-chain (–H) enables a relatively high amount of structural flexibility compared with the side chains of other amino acids (Matthews 1987). Importantly, the enhanced stability conferred by serine is not restricted to the immediate site of the substitution—the region near residue 291—but is transmitted to other regions of the cMDH molecule, including the region encompassing the ligand-binding site. The increased stability is interpreted as favoring an increase in the fraction of the conformational microstates of cMDH molecules that allow recognition and binding of ligands at elevated temperatures. Thermal stability of the protein as a whole is also enhanced by the three additional noncovalent bonds.

The adaptive patterns observed in the studies discussed above, all of which were done with dehydrogenase enzymes isolated from animals, reflect the types of adaptive variation found in studies of other enzymes from a wide range of taxa (**Box 3.6**). Studies of bacterial enzymes have surveyed a substantial number of different classes of proteins in a wide range of species, included psychrophiles, mesophiles, and thermophiles. For

BOX 3.6

Activity-stability relationships in enzymes of psychrophiles, mesophiles, and thermophiles

Activity-stability relationships differ among psychrophiles, mesophiles, and thermophiles. **Figure A** shows temperature-adaptive differences among three orthologs of α-amylase in the net Gibbs free energy of stabilization (ΔG_{stab}) of the proteins: net ΔG_{stab} values (in kcal mol^{-1}) are lowest in the psychrophile (AHA), intermediate in the mesophile (PPA), and greatest in the thermophile (BAA), in keeping with many sets of similar observations. At either high or low extremes of temperature for each ortholog, ΔG_{stab} becomes zero and the protein becomes highly unstable. These temperatures vary among orthologs, such that the upper denaturation temperature is positively correlated with adaptation temperature. However, the temperature of cold denaturation (determined by extrapolation of the curves to subzero temperatures) is highest in the cold-adapted (AHA) ortholog (see Feller and Gerday 2003). The cold-adapted protein is thus the least heat- and cold-stable of the set of orthologs. Note that, despite the differences in magnitude of ΔG_{stab} and temperatures of denaturation, the temperatures of maximum stability are relatively similar; all orthologs exhibit a maximum value of ΔG_{stab} between ~20°C and 30°C. The similarity in temperatures of maximum stability is thought to reflect the major role played by hydrophobic interactions in stabilizing proteins (D'Amico et al. 2003). Hydrophobic stabilization is maximum at ~20°C (see Feller and Gerday 2003). Additional stabilization in the mesophilic and thermophilic proteins is provided by other classes of weak bonds (see text).

BOX 3.6 *(continued)*

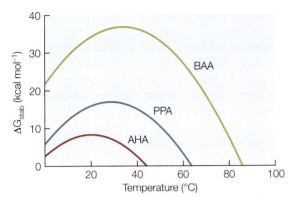

Figure A Temperatures of maximum stability (ΔG_{stab}) of α-amylases from psychrophilic (*Pseudoalteromonas haloplanktis*; AHA), mesophilic (pig; PPA), and thermophilic (*Bacillus amyloliquefaciens*; BAA) species. (After D'Amico et al. 2003.)

Figure B illustrates another important difference in thermal stability among the orthologs. For the orthologs of the mesophile (PPA) and thermophile (BAA), the temperature at which unfolding of the protein occurs, as shown by a rapid increase in relative fluorescence with rising temperature of incubation of the protein, is essentially the same as the temperature at which activity begins to fall. However, for the psychrophilic enzyme (AHA), activity begins to fall off at temperatures that are at least 10 degrees Celsius lower than the temperature at which gross unfolding of the protein occurs. D'Amico and colleagues (2003) interpret this discrepancy to reflect the extreme flexibility of the binding-site region of the psychrophilic ortholog: Local disruption of the geometry needed for ligand binding occurs even when the overall structure of the protein is not being denatured (see text).

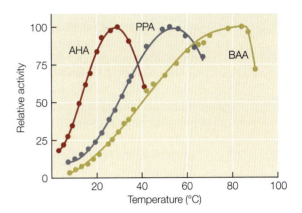

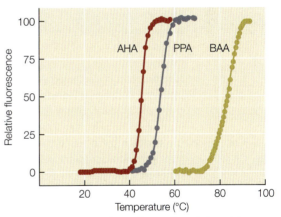

Figure B Temperature dependence of catalytic activity (top panel) and structural unfolding (bottom panel) for the α-amylases shown in Figure A. (After D'Amico et al. 2003.)

example, work done with proteins of cold-adapted bacteria, including Antarctic species living at subzero temperatures, provides support for the conclusions reached in studies of animal dehydrogenases, namely the occurrence of a close coupling of changes in K_M, k_{cat}, and stability during adaptation to different temperatures. Studies by D'Amico and colleagues (D'Amico et al. 2001) of an α-amylase from the Antarctic bacterium *Pseudoalteromonas haloplanktis* are a case in point. These investigators used site-directed mutagenesis to test the effects of different amino acid substitutions on kinetic properties and structural stability. Mutations in the cold-adapted protein that increased its thermal stability typically reduced k_{cat} and decreased K_M, in common with the patterns observed in dehydrogenases. Furthermore, mutations causing these coupled effects were located far from the catalytic vacuole, as has been seen in most studies of orthologous enzymes of animals (Isaksen et al. 2014). This, then, is further evidence that a considerable fraction of the protein, or at least the protein surface, is in play when selection favors temperature-adaptive change in stability and function.

LESSONS FROM STUDIES OF TRYPSINS: EVOLVING "SOFTNESS" OF SURFACE STRUCTURE   We have been discussing how amino acid substitutions at regions on the protein surface distant from the catalytic vacuole site can change the protein's kinetic properties. This discovery has received further support from computational studies that have modeled changes in protein surface properties in orthologous trypsins from a fish (salmon) and a mammal (cow) (Isaksen et al. 2014). The salmon trypsin has a higher specific activity at low temperatures than the bovine enzyme, in keeping with the dehydrogenases and other enzymes discussed above. The ligand binding sites of the orthologous trypsins are completely conserved; the key differences in structure that account for the large differences in kinetic properties are located at regions distant from the binding site where substitutions alter the "softness" of the protein surface. "Softness," the term used by these authors, is the inverse of the "rigidity" that other investigators commonly speak of when discussing protein structural stability. Here, the concept of "fluidity" introduced in Chapter 1 may be useful in appreciating the softness-rigidity balance that is so important in protein function.

Isaksen and colleagues showed that several factors are responsible for the adjustment in softness (or fluidity) of the trypsins. One important factor is the lower propensity of the salmon trypsin to form extensive, structure-stabilizing hydrogen-bonding networks among polar residues (e.g., serine, threonine, asparagine, and glutamine). Such networks involve interactions among the polar residues themselves as well as interactions among a network of water molecules near the protein's surface. A less extensive web of hydrogen-bonding interactions "softens" the protein's surface and lowers the energy barriers to conformational changes associated with catalysis. Because lower temperatures stabilize hydrogen bonds, amino acid substitutions that favor "softening" may be a key element of adaptation to cold, and substitutions that favor "hardening" of heat-labile hydrogen bonds may likewise be critical in adaptation to high temperatures. Although the protein surfaces of the bovine and salmon orthologs exhibit different "softnesses," their internal cores are relatively rigid and likely do not contribute to the inter-ortholog differences in catalytic function.

What is a biochemically meaningful definition of an "active site"?

These findings, in concert with discoveries made with orthologous dehydrogenases, raise an interesting question: What exactly comprises an "active site" of an enzyme? The catalytic vacuole, in which the actual conversion of substrate(s) to product(s) takes place, has commonly been treated synonymously with "active site." However, one could argue that any region of a protein that influences the energy changes that occur during binding, catalysis, and release of products could be included in a broader definition of "active site." Thus, if one broadens the definition of "active site" to include regions of the protein that influence the rate-determining activation free energy changes that accompany the reaction, then a substantial part of the protein molecule, especially relatively mobile regions of the protein surface, may be incorporated into an expanded concept of what comprises an active site (Holland et al. 1997).

Are there many ways to "skin the cat" in adapting proteins to temperature?

Discovery that a single amino acid substitution is sufficient to achieve adaptive change in structural stability and function raises two broad questions about protein evolution. First, how commonly is protein adaptation to temperature achieved by a single amino acid substitution? And, second, when a single substitution suffices, is the same region of the protein always the site of adaptation?

As biochemists compare more sequences of orthologous proteins from differently adapted congeners, more instances of single amino acid substitution-based adaptations are being discovered (reviewed in Fields et al. 2015). **Figure 3.9** summarizes the patterns that have been observed to date in analyses of cMDH and LDH-A$_4$. The three structures in the figure show sites at which a single change in amino acid sequence has been correlated with adaptive changes in substrate-binding and stability (and for some comparisons, in k_{cat} as well). In each of these six cases of adaptation, a *different* site in the primary structure has been modified. Thus, there are multiple sites where a single amino acid substitution suffices to adjust thermal sensitivities. The "cat" of adaptation can be "skinned" in many ways.

What these sites have in common is that they all lie on the surface of the protein, where they may be involved in long-range interactions, for example, through hydrogen-bonding networks among residues and water molecules near the protein surface. These interactions may couple the stabilizing (or destabilizing) effects noted at the site of the substitution to broader (de)stabilization throughout the molecule, as discussed above in our consideration of evolution of protein "softness." Single substitutions may also affect subunit-subunit interactions that are important in governing the energetics of ligand binding. This effect has been noted in LDH-A$_4$ orthologs of barracudas (genus *Sphyraena*). The orthologs of *Sphyraena idiastes* (a south-temperate species) and *S. lucasana* (a subtropical species) have a single amino acid difference (at sequence position 8, labeled "N8D" in Figure 3.9B) that alters the interactions between the N terminus of one subunit and regions on an adjoining subunit that affect the mobility of α-helix-H,

(A) cMDH

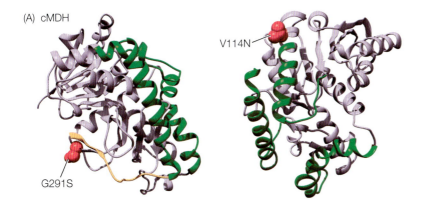

V114N

G291S

(B) LDH-A$_4$

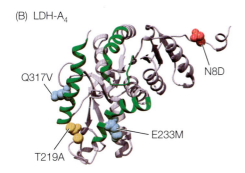

Q317V

N8D

E233M

T219A

Figure 3.9 Six different sites (positions) in the sequences of cMDH and LDH-A$_4$ subunits at which single amino acid substitutions have been shown to account for adaptive variation in stability and function. Individual subunits are shown; cMDH functions as a dimer and LDH as a tetramer. (A) Left: cMDH of *Lottia digitalis* and *L. austrodigitalis* (see Figure 3.8). Right: cMDH of *Mytilus trossulus* (V) and *M. galloprovincialis* (N) (Fields et al. 2006). (B) LDH-A$_4$. Shown are changes responsible for cold adaptation in Antarctic notothenioids (blue residues, positions 233 and 317; Fields and Houseman 2004), the sequence difference between *Chromis* congeners (yellow, position 219), and the sequence difference between *Sphyraena idiastes* and *S. lucasana* (pink, position 8; Holland et al. 1997). (After Fields et al. 2015.)

which undergoes large changes in orientation during function. Thus, this one substitution is responsible for changes in K_M^{pyr} as well as in K_M^{NADH}, thermal stability, and k_{cat} (Holland et al. 1997).

How much of the proteome requires adaptive change in sequence?

The discovery that one or a very few amino acid substitutions can be sufficient to alter the stability and function of a protein suggests that the rate at which proteins might evolve in the face of global warming could be relatively rapid. That is, adapation might occur more quickly than it would if several amino acid changes were necessary to "tune up" a protein to function adequately at temperatures a few degrees greater than the maximum temperatures that characterize the evolutionary history of a species. However, another question related to rates of adaptation looms: What fraction of the proteome (i.e., how many of an organism's proteins) may need to adapt according to the patterns seen for cMDH and LDH-A$_4$? Are some proteins more perturbed by changes in temperature than others?

Here, there are insufficient data to provide anything like a general answer. One initial approach to this question involved comparing orthologous enzymes of two congeneric blue mussels, *Mytilus galloprovincialis* (warm-adapted) and *M. trossulus* (cold-adapted), the cMDHs of which are distinguished by a single amino acid substitution (Fields et al. 2006; see Figure 3.9A). Five additional enzymes involved in ATP-generating pathways were compared by measuring K_M values at a single high temperature, where adaptive differences would be most apparent (see Figures 3.5 and 3.6) (Lockwood and Somero 2012). Of these five enzymes, only one showed a significant difference in K_M between the congeners that might indicate temperature adaptation of the enzyme. This enzyme is the NADP-specific paralog of isocitrate dehydrogenase (IDH), which in *Mytilus* plays important roles in redox balance under anaerobic conditions and in providing reducing equivalents for reactive oxygen species (ROS) scavenging (see Chapter 2). **Figure 3.10** illustrates the kinetic and structural differences found between the IDH orthologs of the blue mussel congeners. Two amino acid substitutions in a loop on the protein involved in subunit-subunit interactions distinguish the warm- and cold-adapted orthologs. In the warm-adapted IDH of *M. galloprovincialis*, two additional noncovalent bonds are formed that stabilize the subunit interaction site involved in forming the functionally active dimer (see Figure 3.10C,D). At least one of these additional hydrogen bonds also stabilizes the conformation of the protein in the region of the active site (Lockwood and Somero 2012). Thus, across the different types of dehydrogenases that have been examined to date, there appears to be a common pattern of adaptation through which a small number of amino acid substitutions influences the conformation and stability of the region of the protein where binding and then catalysis occur.

The fact that four of the six blue mussel enzymes compared did *not* differ in their responses to temperature suggests that only a small fraction of the proteome may require modification in order for an organism as a whole to adapt to temperature changes of 1 to a few degrees Celsius. A priori, proteins might be expected to be differentially sensitive to temperature. Large, multisubunit proteins that undergo complex changes in subunit-subunit interactions and proteins that have large conformational changes as part of their mechanism—both properties of dehydrogenases—would be expected to be more sensitive to changes in temperature than small, single-subunit proteins that might undergo relatively minimal changes in structure during function. However, at present there seems to be no good basis for predicting the relative sensitivities of proteins to the amount of change in environmental temperatures expected over the next several centuries. The question of how much of the proteome needs to adapt is, however, clearly a fundamental one in the context of global change biology. The larger the share of the proteome that requires adaptive change, the greater will be the challenges for organisms to cope with

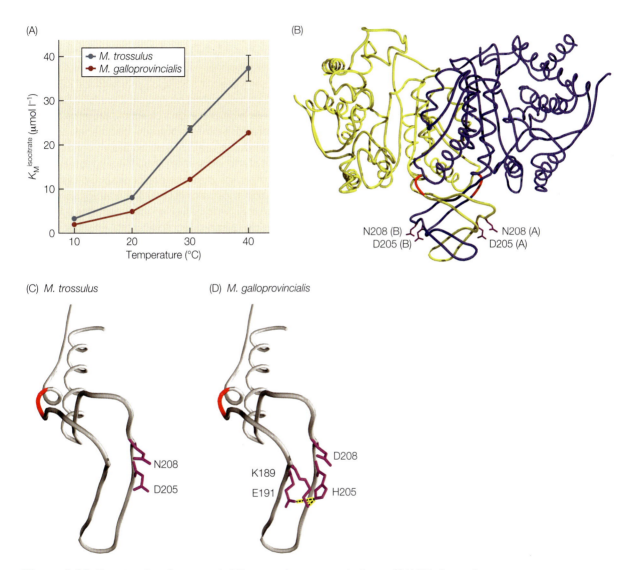

Figure 3.10 Functional and structural differences between orthologs of NADP-dependent isocitrate dehydrogenase (IDH) of the blue mussels *Mytilus galloprovincialis* and *M. trossulus*. (A) Effects of measurement temperature on the K_M of isocitrate. Error bars indicate S.E.M. (B) Model of the IDH dimer of *M. trossulus* (the two identical subunits, designated A and B, are shown in different colors). Red color denotes the highly mobile region of the active site. The two sites of sequence differences between species, positions 205 and 208, are indicated. (C) Magnified image of *M. trossulus* IDH monomer A, illustrating residues D205 and N208 in the intermonomer interaction region and the highly mobile region of the active site (red). (D) Monomer A of the IDH of *M. galloprovincialis* showing the two sequence differences (H205 and D208) and mobile region of the active site (red). Note that residues K189 and E191 on the adjacent loop can freely form hydrogen bonds (highlighted in yellow) with H205 in the *M. galloprovincialis* ortholog but not with D205 in the *M. trossulus* ortholog. (After Lockwood and Somero 2012.)

rising temperatures (see Chapter 5). The development of bioinformatic approaches to screen sequence databases and identify temperature-adaptive amino acid changes may contribute to this "detective" effort (see Fields et al. 2015). We will return to this issue shortly, in the section "The larger picture: The search for genome-wide patterns and rules of adaptation."

Activation parameters: Effects of activation enthalpy and entropy on catalytic function

Further insights into the mechanisms whereby changes in protein fluidity modify enzyme function can be obtained through analysis of the thermodynamic parameters of an enzyme-catalyzed reaction (see Box 3.2). The rate constant of a chemical reaction (k_{rxn})—for enzymatic reactions, k_{cat} is commonly used as the term for the rate constant—is determined by the free energy of activation of the reaction ($\Delta G^{\ddagger}$), the increase in system free energy that must occur to reach the transition state of the reactants. (The double-dagger symbol is used to distinguish an activation thermodynamic parameter from an energy change associated with the equilibrium position of a chemical reaction [ΔG]). $\Delta G^{\ddagger}$ is the algebraic difference between two additional thermodynamic parameters, the enthalpy of activation ($\Delta H^{\ddagger}$) and the entropy of activation ($\Delta S^{\ddagger}$):

$$\Delta G^{\ddagger} = \Delta H^{\ddagger} - T\Delta S^{\ddagger}$$

The equation below summarizes these effects in terms of how temperature (T) exerts its exponential effect on reaction rate (k_{rxn}).

$$k_{rxn} = \kappa \left(\frac{kT}{h}\right) e^{-\frac{\Delta G^{\ddagger}}{kT}} = \kappa \left(\frac{kT}{h}\right) e^{\frac{\Delta S^{\ddagger}}{kT}} \cdot e^{\frac{-\Delta H^{\ddagger}}{kT}}$$

Here, κ and h are Boltzmann's and Planck's constants, respectively. It is clear from this equation that the smaller the value of $\Delta G^{\ddagger}$, the greater the velocity of the reaction. A reduction in $\Delta G^{\ddagger}$ of 1 kcal mol^{-1} leads to an approximately fivefold increase in reaction velocity (Isaksen et al. 2014). From this relationship between $\Delta G^{\ddagger}$ and k_{cat}, it is apparent that a relatively small change in $\Delta G^{\ddagger}$, perhaps only a few hundred cal mol^{-1}, would be sufficient to cause a temperature-compensating change in enzymatic activity, as seen for the LDH-A$_4$s of the damselfishes (see Figure 3.7).

Where the three thermodynamic activation parameters ($\Delta G^{\ddagger}$, $\Delta H^{\ddagger}$, and $\Delta S^{\ddagger}$) for orthologous enzymatic reactions have been determined, a consistent relationship has been found: Reactions catalyzed by cold-adapted orthologs have lower values for $\Delta G^{\ddagger}$ and $\Delta H^{\ddagger}$, but higher values for $\Delta S^{\ddagger}$ (Table 3.2; Low et al. 1973; Coquelle et al. 2007; Isaksen et al. 2014). This regular inter-ortholog variation in thermodynamic activation parameters can be explained by the differences in enzyme fluidity found in comparisons of cold- and warm-adapted orthologs. The higher $\Delta H^{\ddagger}$ values associated with warm-adapted enzymes reflect the increased structural rigidity of these orthologs. That is, a higher energy (enthalpy) cost is entailed in achieving the conformational changes associated with catalysis in more thermally stable homologs. The higher $\Delta S^{\ddagger}$ for reactions catalyzed by cold-adapted orthologs reflects a relatively large increase in protein order during function. Thus, at a common temperature of measurement, the more flexible protein of a cold-adapted species will be less ordered than the orthologs of more warm-adapted species. To form the enzyme-ligand complex, then, a larger increase in order—a greater reduction in entropy—is needed for a cold-adapted enzyme.

TABLE 3.2

Variations in thermodynamic activation parameters among reactions catalyzed by orthologous lactate dehydrogenases, trypsins, and α-amylases from species adapted to different temperatures

	$\Delta G^{\ddagger}$ (kcal mol^{-1})	$\Delta H^{\ddagger}$ (kcal mol^{-1})	$\Delta S^{\ddagger}$ (kcal mol^{-1} °C^{-1})
Lactate dehydrogenase[a]			
Antarctic fishes (*Pagothenia*)	14.00	10.47	–12.7
Mammal (rabbit)	14.34	12.55	–6.4
Trypsin[b]			
Atlantic salmon	18.2	9.9	–8.3
Bovine	19.0	20.4	1.4
α-amylase[c]			$T\Delta S^{\ddagger}$ (kcal mol^{-1})
Psychrophile	13.8	8.3	–5.5
Mesophile	14.0	11.1	–2.9
Thermophile	15.0	16.8	1.8

[a] Data from Low et al. 1973.

[b] Data from Isaksen et al. 2014.

[c] Data from Feller and Gerday 2003. Three bacterial species from different thermal habitats (cold, mesic, hot) were compared.

It can be said that the coupled changes in $\Delta H^{\ddagger}$ and $T\Delta S^{\ddagger}$ reflect an **enthalpy-entropy "compensation"** that leads to a change in $\Delta G^{\ddagger}$ that is much smaller than would be predicted on the basis of either the change in $\Delta H^{\ddagger}$ or in $T\Delta S^{\ddagger}$ alone (Low and Somero 1974). This result is just what would be expected if the inter-ortholog differences in function arise from different levels of fluidity of the protein surface among species. The entropy-enthalpy compensation relationship is a common outcome of adaptation to temperature in different classes of enzymes in different taxa (Feller 2008). This observation argues for the generality of this mechanism of thermal adaptation—that is, modifications of fluidity that exert their effects through influencing the energetics of conformational changes.

Are numbers of covalent (disulfide) bonds adjusted to modify thermal stability?

The above discussion of intrinsic protein stability has centered almost exclusively on the contributions of different types of noncovalent interactions in governing protein stability. However, there is an even more potent mechanism for modulating structural stability: differential use of strong covalent bonds—disulfide (–S–S–) linkages—between two sulfur atoms of cysteine. One might reasonably ask whether proteins have been so "naïve" as to have missed exploiting this mechanism. The use of disulfide bonds for stabilization has been generally thought to occur primarily in extracellular or compartmentalized (e.g., periplasmic) proteins, which often face more challenging chemical conditions than proteins localized within the cytoplasm. Moreover, the reducing environment of the cytoplasm has been viewed as rendering disulfide bonds only marginally stable (Fahey et al. 1977). Finally, as we discuss in the next section, high temperatures can break disulfide bonds, and attempts to genetically engineer increased thermal stability in proteins by increasing numbers of disulfide bonds have not met with great success. For these reasons, the conventional wisdom has been that adjusting the numbers of disulfide bonds in proteins is not a common mechanism of thermal adaptation.

This view has been called into question by an analysis of genomes of 199 bacteria and archaea with different optimal growth temperatures (OGTs). Beeby and colleagues (2005) examined genome sequences with an algorithm that predicts the proximities of cysteine residues in a protein, a key determinant of whether disulfide bonds can be formed. A significantly greater potential for forming disulfide bonds was found in thermophilic and hyperthermophilic bacteria and archaea. Interestingly, many of these same microbes had low percentages of cysteine in their proteomes. Beeby and collaborators suggest that this might indicate selection against free thiol (–SH) groups in these species. In addition, variation in propensity for disulfide bond formation was noted among microbes with different metabolic characteristics. Strict anaerobes had low propensities for disulfide bond formation, which is likely a reflection of the strongly reducing conditions in their cells. Other environmental factors that may favor increased stabilization through elevated numbers of disulfide bonds are osmolality, pH, and radiation: Halophiles, acidophiles, alkalophiles, and radiation-tolerant microbes all showed evidence for increased numbers of disulfide bonds. Adjusting disulfide content thus may be a relatively widespread and general strategy for modulating protein stability, at least in bacteria and archaea facing extreme physical and chemical environments.

What types of biochemical systems are required to support enhanced numbers of disulfide linkages? Beeby and colleagues addressed this question by searching for proteins that are unique to those microbes with high amounts of disulfide bonds. After screening many archaeal and bacterial genomes, they identified only a single protein that met this criterion, **protein disulfide oxidoreductase (PDO)**. This protein's catalytic functions are not fully understood but likely include serving as a disulfide oxidase, reductase, and isomerase. A protein in the endoplasmic reticulum of eukaryotic cells, protein disulfide isomerase (PDI), carries out a comparable function by facilitating the formation of correct disulfide bonds in mature proteins. Beeby and colleagues conjecture that the PDO of a thermophilic microbe might have been ancestral to the PDI of eukaryotes. Even in cells with the appropriate PDO, the capacity to regulate stability of disulfide bonds may also depend on providing the necessary redox microenvironment in the cytoplasm. How extremophilic microbes maintain a redox state that stabilizes disulfide bonds remains to be discovered. Clues to how this might be achieved can be found in Chapter 2, where we review mechanisms for modulating redox state in the cell.

What is perhaps most provocative about this new perspective on protein stabilization through enhanced use of disulfide bonds is that it suggests that different lineages of thermophiles and hyperthermophiles have alternate strategies for protein stabilization. In species lacking PDO, "conventional" stabilization mechanisms that involve noncovalent interactions may be the primary and possibly the sole mechanism for increasing intrinsic thermal stability. In those lineages possessing PDO, a different tack may have been taken, one that relies strongly on increased covalent stabilization through employment of relatively large numbers of disulfide bonds in diverse intracellular proteins. However, not all proteins in PDO-containing species rely on increased numbers of disulfide bonds. Many base their stabilization on the conventional mechanism employing exclusively noncovalent bonds. In these species, therefore, a mosaic of stabilization strategies is present in the proteome. And, as we have mentioned earlier, extrinsic stabilizers (thermoprotectants) provide a complementary strategy for modulating protein thermal stability and may be extremely important for maintaining native protein structures in hyperthermophilic bacteria and archaea (see Chapter 4).

Is damage to covalent bonds a threat at high temperatures?

Although the covalent bonds that establish peptide backbone linkages and the structures of amino acid side chains are strong relative to noncovalent interactions (which are fittingly referred to as *weak bonds*), even some types of covalent bonds can break at the high temperatures experienced by thermophiles and hyperthermophiles (Ahern and Klibanov 1985). Breakage (hydrolysis) of peptide bonds can occur at high temperatures, notably at sites where an aspartyl (D) residue is present. An adjacent serine (S) or threonine (T) residue, both of which contain an –OH group that can react with a peptide bond, can greatly increase the sensitivity of a D-containing peptide bond to hydrolysis (Ahern and Klibanov 1985; Haney et al. 1999). The reduced numbers of S and T residues in proteins of extreme thermophiles (Haney et al. 1999) may be explained, at least in part, as a way to decrease the likelihood of peptide bond hydrolysis at high temperatures.

High temperatures can also lead to thermal destruction of amino acid side chains. Deamidation (loss of NH_3) from asparagine (N) and glutamine (Q) residues occurs at high temperatures (Ahern and Klibanov 1985). Damage happens more readily at N than at Q residues, so reduction in N content could stabilize proteins during adaptation to high temperature. In fact, Haney et al. (1999) reported that percentages of both N and Q were reduced in proteins of thermophiles. Disulfide bonds between cysteines (forming cystine) can also be broken at high temperatures, and cysteine sulfhydryl groups can be oxidized as well (Tomazic and Klibanov 1988). Thus, while disulfide bridges can stabilize proteins at high temperatures, they are not without their own intrinsic vulnerability to thermal damage.

Despite the trend toward decreased amounts of serine, threonine, asparagine, and glutamine found in some analyses of the proteins of thermophiles (Haney et al. 1999), it remains to be determined how universal these changes are in adaptation to high temperatures and how the role of substitutions that reduce the risk of covalent damage compares with that of amino acid substitutions that affect noncovalent bonding. A broad analysis of bacterial and archaeal sequences (see Table 3.3) seems to suggest that some of the amino acids that would be expected to increase the potential for heat-induced damage to the primary structure (covalent bonds) are not selected against during evolution at high temperatures. For example, asparagine (N) and aspartate (D) are both predictors of a high OGT in bacteria and archaea (Ma et al. 2010), even though these amino acids can lead to damage to covalently bonded structures at high temperatures. This seeming contradiction might be resolved by closer examination of sequences to determine, for example, if selection disfavors S and T occurring adjacent to D residues. In any case, it is clear that amino acid substitutions can enhance the stabilities of proteins of thermophiles by two mechanisms: increasing stability through enhancing noncovalent bonding and reducing the risk of damage to covalent bonds.

The larger picture: The search for genome-wide patterns and rules of adaptation

Current efforts to more fully characterize the processes of temperature adaptation in proteins across all taxa and temperatures of life have begun to exploit the information contained in the large and growing sequence databases for DNA and proteins. The combination of these in silico analyses and in vitro biochemical characterization of protein function and stability is yielding new insights into some of the most general patterns of adaptation to temperature. Below we examine several questions that are being addressed by these new approaches.

IN VITRO PROTEIN CHEMISTRY: HOW COMMON IS TEMPERATURE-ADAPTIVE VARIATION BETWEEN ORTHOLOGS?   We introduced this question above, but in an extremely limited context: the frequency of adaptive differences between orthologous proteins of cold- and warm-adapted blue mussels (Lockwood and Somero 2012). In these comparisons only six different proteins were studied, only two of which gave indications of adaptive differences. In view of the number of different proteins and splice variants that have been estimated from genomic and proteomic studies, the in vitro examination of proteins done to date has barely reached the tip of a large iceberg. Much of what we know has come from studies of well-characterized enzymes of two central pathways of energy metabolism, glycolysis and the tricarboxylic acid (TCA) cycle. The focus on these two sets of enzymes reflects their abundances in the cell—there is a lot of material to work with—and the ease with which many of these abundant proteins can be studied in vitro. In addition to the adaptations seen in dehydrogenases (e.g., LDH, cMDH, and IDH), adaptive patterning similar to that discussed above has been found in the glycolytic enzymes phosphoglucose isomerase (PGI) (Dahlhoff and Rank 2000; Wheat et al. 2005) and pyruvate kinase (Somero and Hochachka 1968).

A variety of enzymes not embedded in pathways of energy metabolism, including digestive enzymes and enzymes involved in various biosynthetic reactions, also exhibit the patterns of structural and functional adaptations noted for enzymes of glycolysis and the TCA cycle (Feller 2008, 2010). Studies of stability-function relationships in non-enzymatic proteins have also revealed consistent patterns of adaptation to temperature. For example, thermal stabilities of eye lens proteins (crystallins) are directly proportional to adaptation temperature across a wide range of species—from Antarctic fishes to thermophilic reptiles (McFall-Ngai and Horwitz 1990; Kiss et al. 2004). Tubulins, components of cellular microtubules, likewise exhibit temperature adaptations that ensure correct assembly over a wide range of adaptation temperatures (Detrich et al. 2000). Affinity differences among homologs of the calcium-binding proteins parvalbumin (Erickson et al. 2005) and troponin C (Gillis et al. 2007) stabilize binding abilities at the species' respective body temperatures. Oxygen-binding proteins also exhibit temperature-adaptive differences in structural stability and oxygen-binding ability (Madden et al. 2004).

Given the consistent patterns of temperature adaptations among dehydrogenases and the other proteins discussed above, many and perhaps most classes of proteins might be expected to follow a similar evolutionary trajectory. However, classes of proteins may differ in their intrinsic sensitivity to perturbation by temperature. As conjectured earlier, in the context of the blue mussel study, proteins that undergo extensive changes in conformation or assembly state during function may be more sensitive to temperature change than, for example, a small, single-subunit protein that is not embedded in a biochemical pathway or compartmentalized with other proteins in a metabolome (an assemblage of proteins dedicated to a common function; see Chapter 4, Figure 4.24). Another determinant of how large a fraction of the proteome will exhibit adaptation to a new temperature is the magnitude of the change in absolute temperature that characterizes the evolutionary process. To take an extreme example, we conjecture that *all* orthologous proteins of cryophilic and hyperthermophilic archaea—proteins evolved at temperatures of ~0°C and ~120°C, respectively—will exhibit temperature-adaptive differences. Tests of this conjecture should be possible through comparisons of genomes of these and other differently thermally adapted species. We now turn to the potential of these in silico investigations.

FROM THE IN VITRO TO THE IN SILICO: ARE THERE QUICKER WAYS TO SCREEN PROTEOMES FOR ADAPTIVE VARIATION?   Whereas comparative analyses involving in vitro biochemical methods could, in principle, be extended to a substantial fraction of the proteome, such an approach would be daunting in terms of effort and expense. A more tractable way to screen proteomes for evidence of adaptation to temperature might involve sophisticated computational analyses of sequence databases. Such in silico analyses could use existing knowledge about what types of amino acid substitutions modify proteins' structural and functional responses to temperature to uncover the extent of adaptation across the proteome and elucidate differences in extent of adaptation among different functional classes of proteins. Computational analyses can also exploit the abundance of sequence data from bacteria and archaea, which would allow examination of organisms adapted to the widest possible range of temperatures. Analyses across this wide range of temperatures might be necessary for elucidating the full set of mechanisms used to adjust protein stability, in view of the finding that "conventional" noncovalent bonding and covalent (disulfide) bonding may both contribute to adjusting stability, depending on taxa and temperature.

Gu and Hilser (2009) examined differences in thermal adaptation across the proteomes of 24 species of extremophilic (thermophilic and psychrophilic) bacteria and archaea. They used an analytical algorithm termed *eScape* to determine the contribution of each residue of a protein to the stability of the native structure (Gu and Hilser 2008). The eScape algorithm also enables quantification of the roles of amino acid substitutions in modulating the global stability of the entire protein and the local flexibility of specific regions of the protein, for example, ligand-binding sites. As suggested by the studies of cMDH, LDH, IDH, and other enzymes (Feller 2008), modifications of flexibilities of regions of enzymes that undergo conformational changes during activity may be of central importance in adaptation to temperature, whether or not they change the global stability of the protein.

As expected, the eScape analysis found that, overall, the proteomes of thermophiles were significantly more thermally stable than those of psychrophiles, in keeping with a large body of evidence from in vitro biochemical studies on the correlation between adaptation temperature and protein stability. However, when different types of proteins were compared in the different microbes, it was found that selection for increased stability in thermophiles was not uniform across the proteome. Rather, stability showed the greatest level of adaptation in certain classes of proteins, notably those having catalytic activity and a role in regulating catalysis. In particular, enzymes that affect the composition of the intracellular milieu were found to be especially likely to show thermal adaptation. Gu and Hilser (2009) conjectured that some of these enzymes are responsible for regulating the concentrations of low-molecular-mass solutes that are protein stabilizers (thermoprotectants), thus facilitating wide-scale ("global") stabilization of other classes of proteins (see Chapter 4 for a detailed analysis of the roles and mechanisms of action of thermoprotectant solutes). Other proteins, notably those involved in transcription and translation (e.g., ribosomal proteins), showed lower selection for stability. The finding that ribosomal proteins did not exhibit as high a selection for thermal stability could reflect the fact that these proteins function in a tightly organized multiprotein unit, such that proteins stabilize their neighbors in the compact ribosomal structure. Protein-rRNA interactions might also influence stability. Furthermore, because of the importance of these macromolecular interactions for ribosomal function, there may be constraints on amino acid substitutions that occur at sites on the protein surface that are involved in stabilizing these interactions.

Different types of proteins also differed in terms of whether selection for thermostability involved modulation of global- or local-scale flexibility. For some proteins, global stability was not different between thermophiles and psychrophiles, whereas local flexibility did vary in an adaptive manner; the opposite relationship was found for other proteins. As the generation of sequence data continues its rapid expansion, additional analyses of the sort conducted by Gu and Hilser (2009), but using a variety of newly developed stability-predicting algorithms, may reveal additional patterns of adaptation to temperature across the proteomes of eukaryotes as well as those of archaea and bacteria (see Fields et al. 2015).

NEUTRON SCATTERING AND PROTEIN RESILIENCE   Another novel way of examining the stability of the proteome as a whole involves the technique of neutron scattering (Tehei et al. 2004). The data from this type of analysis reveal the "resilience" (rigidity) of the proteome. Using cellular extracts (~70% protein) of three species of bacteria adapted to different temperatures, Tehei and colleagues found that the resilience of the proteomes increased regularly with adaptation temperature, supporting the view that adaptation is widespread across the proteome. At room temperature, resilience values ranged from 0.2 Newtons m^{-1} (N m^{-1}) in a psychrophile to 0.6 N m^{-1} in a hyperthermophile. However, when measurements were made at the three species' normal physiological temperatures, the measured resilience values were extremely similar, indicating the conservation of a corresponding state of fluidity in the proteomes.

It will be fascinating to see how the neutron scattering technique is employed in future studies of temperature adaptation of proteins. Neutron scattering data might be useful in determining the accuracy of in silico methods for computing protein stability. Several other unresolved questions seem amenable to analysis by neutron scattering. Can this technique detect significant differences between species whose evolutionary temperatures are only slightly different? Data of this sort could help elucidate how much change in evolutionary temperature may be required to favor strong selection on thermal stability. And at a deeper level of analysis, could neutron scattering reveal differences among varied classes of proteins? For example, if proteins were to be separated into different classes—for instance, structural proteins versus soluble enzymes—before being subjected to neutron scattering, might we discover different evolutionary trajectories among different functional categories of proteins?

Choosing the right amino acid for the job: Trends in amino acid composition in differently thermally adapted organisms

The repeated discovery of temperature-correlated differences in thermal stability among proteins of differently thermally adapted species raises an important question: Are there consistent rules for choice of amino acid substitutions in adaptation to temperature? Analyses of bacterial and archaeal proteins have shed considerable light on this issue, and have identified the types of amino acid replacements most commonly used to either increase or reduce structural stability. These studies also have revealed how shifts between different types of noncovalent bonds and how changes in different elements of secondary structure and protein folding contribute to adaptation to temperature. All told, we are gaining deep insights into the diverse "tools" that organisms possess for modulating the intrinsic stabilities of their proteins in order to attain comparable fluidities—corresponding states—at their widely different temperatures of adaptation.

THE BASIC RULES FOR ATTAINING HEAT-RESISTANCE   We begin our analysis with a short review of what we have already discussed about the mechanisms available for adjusting the intrinsic stabilities of proteins. When viewed from the perspective of gaining increased tolerance of high temperatures, four types of adjustments in noncovalent bonding are observed: increased van der Waals interactions, increased hydrophobicity of the protein core, increased networks of hydrogen bonding, and increased ionic interactions. In terms of three-dimensional structure, thermophilic proteins tend to have increased propensities for formation of secondary structure, increased packing density, and decreased lengths of surface loops. Increased packing density per se will favor increases in noncovalent bonding among the amino acid side chains of the protein.

Although similar rules for modifying protein stability are found in all domains of life, extremely thermophilic archaea and bacteria may differ in the degree to which different strategies are followed. Contemporary thermophilic archaea that are thought to be descended from thermophilic ancestors have proteins with greater compactness than proteins of thermophilic bacteria, many of which had mesophilic ancestors. In thermophilic bacteria, proteins are stabilized by a broad suite of stabilizing amino acid substitutions and do not follow the trend toward extremely compact structures noted in proteins that originated (evolved de novo) in thermophilic archaea. Explanations as to why thermophilic bacteria have not taken the same route as thermophilic archaea center on issues of the structural "options" available to de novo evolving proteins relative to options available to proteins secondarily adapting to higher temperatures (see Berezovsky and Shakhnovich 2005). Stabilizing amino acid substitutions—for example, those substitutions that lead to increased numbers of ionic interactions—may occur more easily than wholesale reorganization of the structure to increase compactness.

The types of amino acid substitutions that underlie differences in thermal stability have been analyzed in a large number of studies that have used bioinformatic techniques to extract general trends from protein and DNA databases. Some of these studies compared congeneric species adapted to different temperatures (Haney et al. 1999). The archaeal genus *Methanococcus*, for example, includes congeners adapted to temperatures differing by approximately 50 degrees Celsius (OGTs from 35°C to 85°C). Comparisons of amino acid sequences of these congeners provided a study system in which high sequence similarity minimized the uncertainties that might result from amino acid substitutions not associated with adaptive change. Furthermore, the large amount of sequence data available for the *Methanococcus* congeners allowed 115 proteins to be compared for either their full or partial sequences. The key trends noted in this study were that proteins of thermophiles have increased hydrophobicity, higher residue volumes, more charged amino acids (notably glutamate, arginine, and lysine), and fewer uncharged polar residues (serine, threonine, asparagine, and glutamine). These trends were robust and applied to between 83% and 92% of the proteins for which complete sequences could be obtained.

The mining of sequence databases has also involved comparisons of taxa that are highly diverged evolutionarily. Table 3.3 illustrates, for proteins of microbes adapted to different temperatures, how monomeric proteins and those involved in protein-protein complexes differ in amino acid composition overall and in different regions of the folded protein: interface (in complex-forming proteins only), surface, intermediate, and core (Ma et al. 2010). Here the correlations are between OGT of the microbes and sets of amino acids that serve as "predictors" of OGT (as indicated by the correlation coefficient, R). The trends for both monomeric and complex-forming proteins are clear: Adaptation to

TABLE 3.3 ▮ ▬ ▬ ▬ ▬▬▬▬▬▬▬▬

Amino acid groups that serve as predictors of optimal growth temperature (OGT) for bacteria and archaea adapted to different temperatures

The higher the R value, the greater the likelihood that the group of amino acids shown will be favored during adaptation to high temperature. Comparisons include sets of proteins that form protein-protein complexes and monomeric proteins. Different "depths" of the proteins are considered: surface, intermediate, and core, as well as interface regions for complex-forming proteins.

Class of protein	Structural region	Predictor amino acids[a]	R
Complex-forming	Interface	I, L, P, V, W, <u>N</u>, <u>Y</u>, *D, E, K, R*	0.91
	Surface	W <u>Y</u>, *E, K, R*	0.92
	Intermediate	I, L, P, V, <u>Y</u>, *E, K, R*	0.92
	Core	I, V, <u>A</u>, <u>Y</u>, *K*	0.83
	OVERALL	I, V, W, <u>Y</u>, *E, K, R*	0.93
Monomeric	Surface	W, <u>Y</u>, *E, K, R*	0.92
	Intermediate	I, L, P, V, <u>Y</u>, *E*	0.90
	Core	I, V	0.76
	OVERALL	I, P, V, <u>Y</u>, *E, K, R*	0.92

Source: Ma et al. 2010.
[a] Bold, hydrophobic; underlined, polar; and italic, charged residues. A, alanine; C, cysteine; D, aspartic acid; E, glutamic acid; F, phenylalanine; G, glycine; H, histidine; I, isoleucine; K, lysine; L, leucine; M, methionine; N, asparagine; P, proline; Q, glutamine; R, arginine; S, serine; T, threonine; V, valine; W, tryptophan; Y, tyrosine.

high temperatures favors increases in charged residues (glutamate [E], lysine [K], and arginine [R]) and some hydrophobic residues (isoleucine [I], valine [V], and tyrosine [Y]). Polar residues (alanine [A], histidine [H], asparagine [N], glutamine [Q], serine [S], and threonine [T]) are underrepresented in proteins of thermophiles.

NEGATIVE DESIGN AND POSITIVE DESIGN   The general trends presented in Table 3.3 do not tell the full mechanistic story about how some of the temperature-related alterations in amino acid composition serve to enhance stability and foster accuracy in formation of protein-protein complexes. Although increases in hydrophobic amino acids can easily be explained as a means for enhancing thermal stability—the hydrophobic effect is favored by high temperatures—it is harder to explain the increase in charged amino acids, notably positively charged K and R. Looking first at the protein core, the increase in buried positively charged amino acids is interpreted in large measure as a sign of what is termed **negative design** (Ma et al. 2010). When buried in the protein core, the positively charged side chains are positioned so as not to interact and repel each other. If unfolding begins, however, the likelihood of equivalently charged side chains interacting and repelling each other increases. This repulsion "pushes back" on the unfolding process, favoring return to the native state of the protein. Core-positioned positively charged amino acids are considered to be elements of "negative design" because they disfavor unfolding.

In contrast, other properties that tend to stabilize proteins are elements of **positive design**. These stabilizing forces include interactions among hydrophobic residues in the protein core and **salt bridges** on the protein surface. A salt bridge is a combination of two noncovalent interactions: hydrogen bonding and electrostatic interactions between ionized groups of different charge. The illustration at right shows these two elements of a salt bridge between lysine and glutamic acid. In proteins of thermophiles, formation of salt bridges between surface-exposed residues is thought to be of major importance in conferring a high thermal stability on the proteins (Georlette et al. 2003). To summarize the distinction between these two forms of protein "design," positive design decreases the free energy of the native state of the protein through the set

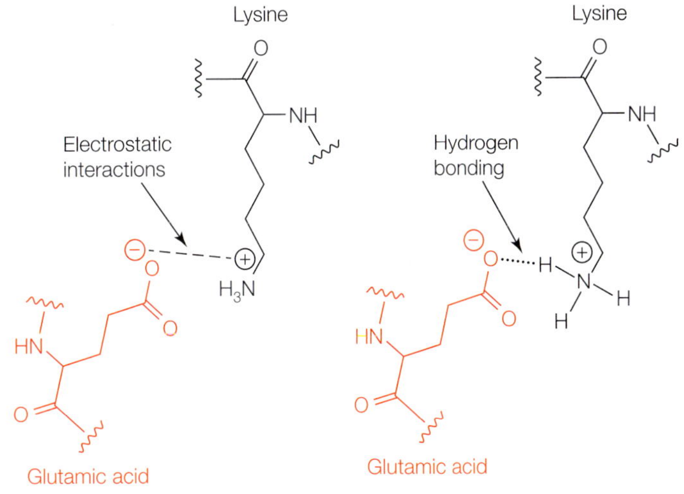

of amino acid changes discussed above. In contrast, negative design increases the free energy of the unfolded protein through addition of amino acids that, in the unfolded state, establish conditions like charge repulsion that favor their reburial and return to the native state.

Similar principles apply to the formation of protein complexes. With rising OGT, the propensities for forming protein-protein complexes and for doing so correctly were found to increase (Ma et al. 2010). Increased hydrophobicity of protein interfaces in thermophilic species enhances protein-protein stabilization; thus with increasing OGT there is a rise in hydrophobicity of sites involved in complex formation. An increase in the number of charged residues at interaction sites enhances the specificity of binding (properly positioned negative and positive charges attract each other and stabilize the complex) and contribute to negative design as well; like charges repel each other and favor reorientation of the interacting surfaces in a way that fosters native protein-protein interactions.

In conclusion, similar types of changes in amino acid composition characterize adaptation to increased temperature in individual protein subunits and at the contact surfaces that stabilize protein-protein interactions. Stability changes involve two different types of "designs": one, positive design, lowers the free energy of the folded protein subunit or the assembled complex; the other, negative design, increases the free energy of the unfolded subunit (monomer) or disassembled protein complex. Together, these two mechanisms achieve an appropriate stability for proteins at their adaptation temperatures.

SUMMARY: HOW MULTIPLE TOOLS ARE EXPLOITED TO ADJUST STABILITY AND ATTAIN CONSISTENT CORRESPONDING STATES The observed variations in amino acid composition and primary structure among proteins from species evolved at different temperatures allow us to draw several conclusions. First, several types of adaptive change are available for the evolutionary adjustment of protein stability (Table 3.4) (see Feller and Gerday 2003; Siddiqui and Cavicchioli 2006). Different proteins make varied use of these options in adjusting their stabilities—their net free energies of stabilization (ΔG_{stab})—in a temperature-adaptive manner. Second, differences in ΔG_{stab} are small, lying in the range of 40–50 kJ mol^{-1} for orthologous proteins of psychrophilic versus mesophilic species

TABLE 3.4

Mechanisms for adjusting net stabilization free energies (ΔG_{stab}) of proteins

Amino acid change	Mechanism of effect
Adaptation to cold	
Reduce numbers of large hydrophobic residues (e.g., isoleucine) in the protein core	Increases distance between side chains and peptide backbone linkages, weakening van der Waals interactions
Increase hydrophobicity of protein surface	Increases water organization in clathrates around hydrophobic groups, making protein more flexible
Introduce glycine residues	Increases flexibility of structure
Reduce proline content	Allows freer movement of protein backbone
Increase surface charge	Enhances interaction with solvent, leading to more flexible structure
Increase surface negative charge	Increases charge repulsion, leading to less rigid surface
Reduce number of salt bridges	Reduces protein stability
Adaptation to heat	
Increase hydrophobicity of core (large hydrophobic side chains)	Strengthens van der Waals interactions, increasing stability
Increase hydrogen-bonding networks on protein surface	Leads to hardening of surface
Increase numbers of ionic interactions (salt bridges)	Leads to more rigid structure
Increase disulfide bridges	Leads to more rigid structure
Reduce glycine content	Reduces conformational flexibility

(Siddiqui and Cavicchioli 2006). This difference in ΔG_{stab} among orthologs should be viewed in the context of the bond energies of hydrogen bonds (4–10 kJ mol^{-1}) and salt bridges (12–21 kJ mol^{-1}) (Vieille and Zeikus 2001). Thus, only one to a few changes in the numbers, types, and strengths of noncovalent bonds are sufficient to cause the shifts in ΔG_{stab} seen among orthologs from species adapted to different temperatures. This conclusion seems fully consistent with the findings that a single amino acid substitution that adds or subtracts one to a few weak bonds in the protein's structure is sufficient to alter structural stability and kinetic properties. Third, not all parts of a protein may adapt similarly. Selection for structural flexibility around the catalytic vacuole site may favor a localized change in stability, but global stability, as indexed by the denaturation of the entire protein, may not be changed. This has been observed in multidomain enzymes in which clear adaptive change is found near the catalytic site—higher flexibility occurs in the cold-adapted enzymes—whereas the remainder of the molecule has similar stability in cold- and warm-adapted orthologs (Feller and Gerday 2003). This discovery has been used to counter the conjecture that the more flexible structures of cold-adapted proteins are merely a consequence of genetic drift, which results from reduced selection for stability. If this were the case, one would not expect this drift to be confined to the catalytic site regions of proteins (Feller 2008). Fourth, and very important, a primary outcome of selection for protein stability is the retention of a corresponding state of structure—a conserved balance between flexibility and rigidity that enables orthologs to have a consistent fluidity at their different evolutionary temperatures (Závodszky et al. 1998; Feller 2008). A further, strong argument for the importance of conserving corresponding states will be presented in Chapter 4, where we examine the roles of low-molecular-mass organic solutes in establishing and defending an optimal state of structural fluidity for protein function.

Stabilizing protein sequences in thermophiles: Codon usage in adaptation to temperature

Whereas the patterns of codon usage in genomes of differently thermally adapted species not surprisingly reflect the temperature-adaptive trends in amino acid composition summarized in Table 3.4, studies of thermophiles have revealed an additional criterion that seems to influence which codons are preferred for a given amino acid. The particular codon(s) used can also ensure that the right types of amino acids—for example, those fostering increased protein stability in thermophiles—remain in a protein's amino acid sequence even if mutations lead to change in nucleotide sequence. Thus, codon choice may be subject to selection based in part on whether a change in one of the codon's three bases will allow the protein to retain at that position the particular type of amino acid that has been selected during adaptation to temperature. This is a type of "bet hedging" that could be of widespread importance in temperature adaptation.

An example of this phenomenon was discovered in studies of arginine codon usage in 173 genomes of bacteria and archaea with widely different OGTs: mesophiles (OGT < 50°C), thermophiles (50°C < OGT < 80°C), and hyperthermophiles (OGT > 80°C) (Van der Linden and Farias 2006). Arginine is specified by six codons: CGU, CGC, CGA, CGG, AGA, and AGG. In mesophilic species the preferred codon for arginine is CGN (where "N" stands for any base). In thermophiles and hyperthermophiles, the preferred codon for arginine is AGR (where "R" stands for A or G). The latter choice of codon reflects what is referred to as **positive error minimization**; mutation in a base is likely to lead to retention of an amino acid with properties similar to arginine. Thus, mutations from G to A in the second base of the codon would lead to a substitution of lysine for arginine, that is, to retention of a positive side chain at this site in the sequence and, as a consequence, retention of a thermally stable structure. Conversely, the codon CGN is one mutation away from histidine and glutamine, two amino acids that are avoided in hyperthermophiles. Avoidance of the CGN codon in thermophiles and hyperthermophiles thus can be viewed as **negative error minimization**, a codon choice that reduces the likelihood that a mutation could lead to substitution in the sequence of an inappropriate amino acid.

Another potential influence on codon use in thermal adaptation has been proposed. Some investigators have conjectured that elevated G+C percentages in genomes help determine what codons are used; that is, codons with G and/or C would be favored over those with A and/or T. This prediction is not consistent with the codon usage trends found by Van der Linden and Fairas (2006). In fact, they found that use of the arginine codon AGR did not correlate with—was not driven by—G+C content. However, AGR codon use was significantly correlated with an amino acid composition ratio shown to reflect protein thermal stability: (glutamate + lysine)/(histidine + glutamine) (Farias and Bonato 2003). The (E + K)/(H + Q) ratio increases with adaptation temperature in bacteria and archaea, in parallel with increased preference for the positive error minimization codon for arginine, AGR, and decreased occurrence of CGN, the codon that is one base change away from codons for the destabilizing amino acids H and Q.

In summary, these two strategies of error minimization show that temperature adaptation of codon usage involves two important considerations. First, selection of course favors codons for particular amino acids that confer the appropriate stability on the protein. Second, selection favors codons that, if a nonsynonymous mutation in the codon were to occur, would still be likely to encode a similar amino acid, resulting in a conservative "like for like" substitution. As a result, mutations would be less likely to significantly alter protein stability. Whether this two-faceted choice of codon usage extends beyond the case of arginine—and if so, how widespread it is among other amino acids and their codons—merits further investigation in all domains of life.

How do paralogous and allelic protein variants contribute to temperature adaptation?

Our discussion of adaptive changes in protein sequence has focused principally on orthologous proteins of species evolved at different temperatures. Whereas we used examples of proteins from all domains of life, we said little about differences among these domains that might contribute to the genetic diversity accessible to a species, such as by providing raw material for adaptation. Thus, our discussion might have led the reader to conclude that an individual organism has only a single version of a given type of protein to rely on when challenges from changing temperature arise. Whereas this is largely true within the domains Archaea and Bacteria, where the **haploid** condition pertains, in Eukarya, whose members are **diploid**, the "tool kit" of proteins available for coping with a change in temperature can be broader than this.

As discussed briefly in Chapter 1, eukaryotes commonly have two or more genes encoding a given type of protein, and the products of these genes may differ in their thermal sensitivities as well as in their specialized metabolic functions. Furthermore, in diploids the two copies of each gene locus may encode proteins with different primary structures that confer different responses to temperature. In the former case we speak of **paralogous** genes and proteins; in the latter case we are dealing with **allelic variants**. Both types of protein variants can contribute to an organism's ability to function at different temperatures.

PARALOGOUS VARIANTS   When genes are duplicated, either through changes in ploidy or through tandem duplication of individual genes, the organism has new genetic resources to work with. Mutations in paralogous genes can engender a family of proteins with different functional characteristics, as discussed at several junctures in this volume. Paralogous proteins can differ not only in their functional adaptations, but also in the way their expression is regulated in response to changes in environmental conditions. In the context of adaptation to temperature, paralogous proteins with different thermal optima and expression patterns have been discovered in ectotherms. For example, a species of carp (*Cyprinus carpio*) has three paralogs of myosin heavy-chain proteins with different thermal stabilities, kinetic properties, and temperature-dependent patterns of expression (Watabe 2002). The myosins expressed in 8°C-acclimated carp were less thermally stable and supported a higher rate of filament sliding than the paralogs found in 30°C-acclimated specimens. These contractile proteins thus manifest the same types of temperature-adaptive differences found among orthologous enzymes from differently thermally adapted species: Cold-adapted variants have enhanced activity, which supports temperature-compensation of function, and these higher activities stem from a reduced structural rigidity.

An especially interesting example of the roles of paralogous proteins in thermal relationships has been discovered in studies of temperature-sensing mechanisms. Thermosensing is critically important to animals: Temporal changes or spatial gradients in environmental temperature may elicit thermoregulatory behaviors in a motile animal that have wide-ranging benefits on its performance and, ultimately, its fitness. The molecular mechanisms of temperature sensing appear to be widely conserved across the animal kingdom (Garrity et al. 2010). Strikingly similar mechanisms are shared by organisms as evolutionarily divergent as two model ectotherms, the fruit fly *Drosophila melanogaster* and the roundworm *Caenorhabditis elegans*, and endotherms such as mammals and birds. Central to these mechanisms are sensory neurons with membrane-localized proteins that modulate transmembrane movement of ions in response to temperature changes. These neurons initiate downstream neural signaling that terminates in muscular activities that facilitate thermoregulatory behavior.

Here we focus on a large family of proteins known as **transient receptor potential (TRP) proteins** that occur throughout the animal kingdom and play important roles in a variety of sensory functions, including temperature sensing (Dhaka et al. 2006; Venkatachalam and Montell 2007). TRP proteins are cation channels localized in the membranes of neurons that respond to a variety of stimuli, including chemicals, light, gravity, and temperature (Venkatachalam and Montell 2007). When they act as thermosensors, TRPs are commonly referred to as **thermoTRPs**. It should be noted, however, that the particular sensory role played by a given type of TRP is determined by the cell type in which it is expressed. A TRP that functions in temperature sensing in one type of cell may function as a chemosensor when expressed in a different cell type that is specialized for sensing noxious chemicals (Kang et al. 2012).

In a given species of animal, TRP proteins are encoded by numerous gene loci; the human genome encodes 28 TRPs, and the *Drosophila melanogaster* genome encodes 16 TRPs (Rosenzweig et al. 2008). The thermal response of a particular thermoTRP variant reflects three characteristics of an organism's thermal relationships. First, the temperature at which an ortholog of a given type of thermoTRP is activated—the temperature at which the ion channel opens and, thereby, neuronal firing can occur—reflects the evolutionary thermal history of the species; activation temperatures of TRP channels reflect differences among species in optimal body temperatures (Myers et al. 2009). Second, thermoTRPs differ in whether they function to detect increases or decreases in temperature. Their distribution among different classes of temperature-sensing neurons also differs because neurons may be specialized for detection of either increases or decreases in temperature (Rosenzweig et al. 2008). Third, the responses of different paralogs of heat-detecting thermoTRPs to increases in temperature differ with respect to the severity of thermal stress to which they respond. As a broad generalization, thermosensing systems are tuned to respond either to relatively innocuous temperatures that lie just beyond the optimal range, or to noxious and potentially damaging temperatures that lie well outside the optimal range (McKemy 2007; Garrity et al. 2010). The downstream neurological and biochemical changes triggered by different paralogs of heat-sensing thermoTRPs, have yet to be well characterized but likely include locomotory effects at moderate temperatures and activation of cellular stress processes at more extreme temperatures.

ThermoTRPs commonly function in conjunction with other temperature-sensing ion channel systems. For example, an alternative temperature-sensing system relies on cyclic GMP (cGMP) as a regulatory signal. The concentration of cGMP in the cytoplasm varies with temperature—as, therefore, do the activities of cGMP-gated ion channels that modulate activities of thermosensing neurons (Dhaka et al. 2006; McKemy 2007; Garrity et al. 2010; Glauser et al. 2011). Whatever the battery of molecular mechanisms found in a thermosensing neuron, these neurons' capacities for detecting changes in temperature are truly remarkable. In fruit flies and roundworms, thermotaxis (movement to or away from a region with a different temperature) occurs in response to temperature fluctuations of less than 0.005 degrees Celsius per second (Garrity et al. 2010). In thermosensing neurons, the ion flux rates through temperature-sensing ion channels are extraordinarily responsive to slight changes in temperature; flux rates have Q_{10} values that can exceed 10^{21}—most impressive in view of the "normal" Q_{10} values near 2–3 found for most biochemical processes (Garrity et al. 2010; Glauser et al. 2011).

As mentioned above, within a single species, thermoTRPs may be specialized for sensing temperatures that lie either above or below the preferred temperature, T_p. For example, in wild-type *D. melanogaster*, in which an adult has a T_p of 24°C–26°C, different TRP paralogs are involved in avoidance of temperatures that are either too warm or too cold (Hamada et al. 2008; Rosenzweig et al. 2008). (Note that *D. melanogaster* and its many congeners exhibit a wide range of T_p values, which vary among species, across life stages of a given species, and with acclimatization conditions; see Dillon et al. 2009 for a comprehensive review of the thermal biology of fruit flies.) At least three variants of thermoTRPs are involved in regulating responses to temperatures that lie above the preferred range, and these paralogs differ in the threshold temperatures at which they become active. dTRPA1 responds to mild heating; it is activated by warming to 25°C–27°C. Activation of this TRP would, in conjunction with neuromuscular activity, enable a fly to remain near its T_p, perhaps through behavioral thermoregulation. Next in the thermal threshold hierarchy is the TRP PYREXIA, which is activated near 35°C and is thought to protect flies from heat-induced paralysis. A third TRP, termed PAINLESS, is activated at temperatures greater than 42°C. These three thermoTRPs may be paralogs originating from a common gene (Dillon et al. 2009). Different neurons are involved in sensing these different types of thermal stress, and this functional specialization is reflected in the types of thermoTRPs that are expressed (McKemy 2007). It is likely that as body temperature crosses each TRP activation threshold, different physiological and biochemical responses are triggered. The full suite of these temperature threshold-specific responses to rising temperature remains to be characterized.

Responses to suboptimal temperatures in *D. melanogaster* entail activation of two different paralogs of TRPs (Rosenzweig et al. 2008). In photosensitive cells of fruit flies, these TRP paralogs are involved in phototransduction. However, in the cold-sensing nerves in which these TRPs function as thermoTRPs, the systems required for phototransduction are absent. Interestingly—and perhaps as a result of the types of differences discussed below for orthologous TRPs of vertebrates—the *D. melanogaster* TRPs involved in cold sensing are not involved in heat sensing, and vice versa. Thus, the heat- and cold-detecting TRP paralogs are thermal specialists, much as seen for the myosin paralogs of carp. Through the activities of heat- and cold-sensitive TRPs, flies are able to respond to very slight changes in ambient temperature, so as to keep their body temperatures within the preferred range.

Insects also use TRP-based thermosensors for functions other than thermoregulation. In the case of blood-sucking insects such as lice and mosquitoes, these sensors allow identification of animal hosts (see Hamada et al. 2008). Thus, the yellow fever- and dengue fever-transmitting mosquito, *Aedes aegypti*, uses thermosensory neurons to detect the presence of warm-bodied mammals.

There is also evidence of temperature adaption among TRP orthologs of different vertebrate species that are adapted to different temperatures (Myers et al. 2009). In fact, temperature-adaptive differences in kinetic properties among thermoTRP orthologs reflect the patterns discussed earlier for orthologous enzymes. That is, similar to how the half-saturation constant of substrate, K_M, is conserved at normal body temperatures, so is the midpoint of thermoTRP channel activation (**Figure 3.11**). Myers and colleagues (2009) studied the thermal responses of orthologs of a cold-activated TRP

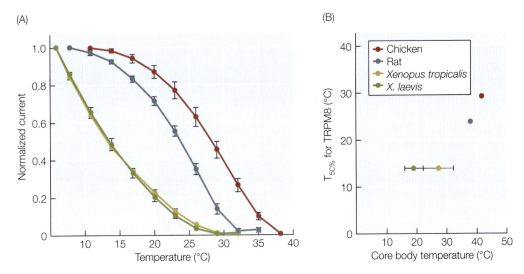

Figure 3.11 Temperature sensitivity of cold-activated TRP channel function in vertebrates adapted to different temperatures. (A) Current versus temperature relationships for TRPM8 channels for chicken, rat, and two congeners of frog, *Xenopus tropicalis*, and *X. laevis*. (B) Relationship between core body temperature and temperature of half-maximum activation of the TRPM8 channel ($T_{50\%}$). (After Myers et al. 2009.)

channel (TRPM8) in four vertebrates adapted to different temperatures: the chicken (core temperature ~41°C), a rat (core temperature ~39°C), and two congeners of the frog *Xenopus*, whose body temperatures span a wide thermal range (thermal tolerance range ~14°C – ~32°C). Recall that frogs are ectothermic, so their preferred temperature equals their core body temperature. In contrast, mammals and birds select preferred temperatures based on their ability to maintain their body temperatures through endothermic heat production and dissipation of heat to their surroundings. The rat has a preferred ambient temperature near 28°C, *X. laevis* has a preferred temperature near 22.4°C, and *X. tropicalis* has a preferred temperature somewhat higher than that of *X. laevis*.

Thermosensitivities of the cold-activated TRPM8 channels are distinctly different among these four species. For all four orthologs, falling temperature activates the channel (current flow increases), but the temperatures at which activation commences and reaches its maximum value differs among the orthologs. These differences reflect differences in the ranges of temperatures that would be perceived as suboptimal by the four animals. The relationship between core body temperature and the temperature at which 50% activation occurs for TRPM8 ($T_{50\%}$; see Figure 3.11B) shows the same pattern of trait conservation discussed earlier for ligand binding by enzymes. In other words, there is a positititve correlation between the core body temperatures of the species and the temperatures that induce activation of their TRPM8 orthologs.

Another exciting line of research on thermoTRPs concerns plasticity in the thermal activation threshold for a given type of TRP. In mammals, signaling systems that operate in conjunction with TRP channels, including tyrosine kinases and G protein-coupled receptors, can alter the temperature thresholds of thermoTRPs by several degrees (Huang et al. 2006). Effects of this type could be instrumental in modulating the behaviors of TRPs during acclimatization.

In summary, the importance of precise thermosensing and thermoregulatory behavior is manifested in the complex families of proteins, notably thermoTRPs, that sense even extremely small changes in temperature and rapidly initiate neurological activity that leads to appropriate changes in behavior. The widely varying thermal responses of orthologous and paralogous variants of TRPs offer an exciting arena for the study of molecular evolution and adaptation to temperature.

The occurrence of temperature-adaptive variation among paralogs of TRPs and myosins (Watabe 2002) leads one to ask: How common is this type of protein adaptation in ectotherms? It is important to emphasize that most paralogous proteins differ in function not in terms of their thermal responses, but rather in terms of the particular metabolic requirements of the cells or organs in which they are found (see Chapter 2). Thus, a temperature-driven change in relative expression levels of enzyme paralogs may reflect temperature-related shifts in metabolic demands rather than differences in the enzymes' thermal sensitivities. For example, in the case of aquatic ectotherms, changes in water temperature will alter metabolic rate and the amount of oxygen dissolved in the surrounding water. These effects of temperature may favor a shift in metabolic poise, for example an increased reliance on anaerobic ATP generation at higher temperatures, particularly when high levels of locomotory activity are required. Paralogs of LDH, an enzyme that is involved in both anaerobic and aerobic metabolism of glucose, are known to have temperature-dependent patterns of expression, a phenomenon first reported by Hochachka (1965). Changes in the relative expression levels of LDH-A_4, which supports anaerobic glycolysis (lactate generation), and LDH-B_4, which converts lactate to pyruvate under oxygen-replete conditions, thus might be driven more by shifts in metabolism related to activity levels and oxygen availability than to any differences in the temperature responses of the two LDH paralogs. Metazoans have multiple paralogs of most enzymes of ATP-generating pathways, which is a testimony to the importance of gene duplication and paralog evolution in the development of cell-, tissue-, and organ-specific metabolic potentials. One must be cautious, therefore, in interpreting temperature-driven changes in paralog expression: Metabolic reorganization rather than protein thermal sensitivities may be the principal (or only) basis for such changes in protein expression.

ALLELIC VARIANTS   Within a species, multiple variants (alleles) of a gene locus may encode proteins with different amino acid sequences, leading to a potential for adaptive allelic variation (for review see Watt and Dean 2000). If cold- and warm-adapted allelic variants evolve, then selection may favor a differential balance of genotypes across thermal gradients. This type of balancing selection, which we will discuss in more detail in Chapter 5, has been observed in a variety of terrestrial and aquatic ectotherms. A classic example is the fish *Fundulus heteroclitus*, which exhibits clines in allele frequencies along the Atlantic coast of North America that reflect the latitudinal thermal gradient (Powers et al. 1993). For one of the enzymes that exhibits an especially clear latitudinal pattern, LDH-B_4, the kinetic differences between allelic variants reflect a pattern of adaptation to temperature similar to that shown for LDH-A_4 orthologs (see Figure 3.6) (Place and Powers 1984).

Among other classic studies of allelic variation related to thermal gradients are those of Ward Watt and colleagues, who studied populations of *Colias* butterflies across altitudinal gradients (Watt 1983; Watt and Dean 2000). As in the case of the latitudinal patterning of gene frequencies, altitudinal temperature gradients led to selection for different frequencies of cold- and warm-adapted allelic variants. Finer-scale patterns of allele frequency gradients have also been observed, at times over distances measured in meters. For example, in the intertidal gastropod *Littorina saxatilis*, two allelic variants of aspartate amino transferase (AAT) are found, Aat^{100} and Aat^{120} (Panova and Johannesson 2004). In upper-shore individuals Aat^{120} is more common, whereas in lower-shore individuals Aat^{100} predominates. Correlated with this difference in allele frequency distribution is a difference in AAT activity: Significantly higher activities are found in lower-shore individuals, suggesting an adaptation to lower temperatures. Thus, even over distances of meters, temperature-correlated variation in gene frequencies of potential adaptive significance is found. In all cases, having alternate forms of a given enzyme may help the species tolerate a range of temperatures that might not be permissive without "cold" and "hot" variants of the protein in question.

One additional question about the roles of allelic variation in adaptation to temperature merits our attention. This concerns the possibility of differential expression of the two alleles at a given gene locus. Although this mechanism for temperature acclimatization remains to be demonstrated, a priori the potential importance of allele-specific expression could be great. For example, if an individual that possessed cold- and warm-adapted allelic variants were to find itself in a relatively cool habitat, could it selectively express only the more cold-adapted protein variant encoded in its genome? Such ability could provide an important mechanism of rapid acclimatization to temperature change. Although we currently have no basis for evaluating how allele-specific expression contributes to such acclimatization processes, there is evidence that allele-specific expression may be common under normal (non-stress-related) conditions (see Chapter 1). Thus, allele-specific differences in abundance of transcripts may occur in up to 30% of loci in a single individual, and across a population of a species, even wider variation in relative transcript levels of allelic genes is found (Pastinen 2010). How differential expression of allelic variants within an individual might play out during adaptation to temperature remains unknown, but this is clearly an area where future research is needed.

In summary, the generation of protein variants through changes in DNA sequence has long been recognized as a critically important mechanism in adaptation to temperature. These genetic changes have been well characterized in comparisons of orthologous proteins of differently thermally adapted species, and the role of allelic variants with different thermal responses is evident in some cases where populations of a given species thrive in widely different thermal conditions. There are, however, other ways to generate protein variants that do not involve changes in the DNA sequence that encodes the protein. These are the processes of **splice variant generation** and **RNA editing**. These posttranscriptional processes for generating protein variants may prove to be of widespread importance in adaptation to temperature.

Making the most of your exons: Splice variants in adaptation to temperature

One of the major advances in our understanding of how protein variants are generated came with the discovery that genes comprise exons and introns. The former contain the protein-coding sections of the gene; the later lack protein-coding function but play important roles in gene regulation. As shown in **Figure 3.12A**, introns may be the larger share of a gene's total sequence. Exons and introns are both transcribed, but the introns are removed (spliced out) as the gene transcript is processed; only the remaining exons are translated into a protein by the ribosome. If, during processing of the transcript, a gene's exons are joined in different combinations—producing **splice variants**—this can generate an assembly of different protein variants. It is likely that a majority of genes in eukaryotes undergo alternative splicing.

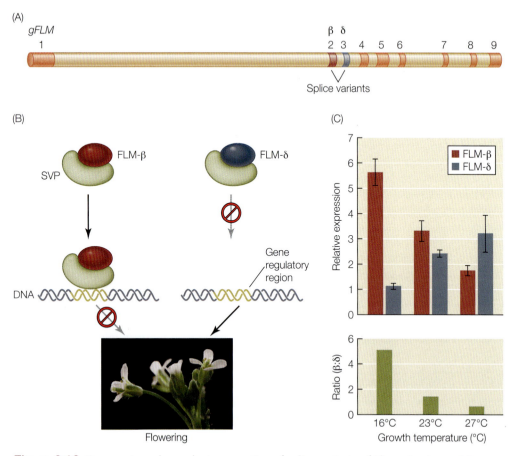

Figure 3.12 Temperature-dependent generation of splice variants of Flowering Locus M (FLM), a transcription factor that regulates flowering. (A) Structure of the gene that encodes FLM (*gFLM*), showing its nine exons (numbered segments colored orange, dark red, or blue) and eight introns (cream-colored segments). The two splice variants of FLM, FLM-β and FLM-δ, are formed from different combinations of these exons: Exon 2 (dark red) is used in translation of FLM-β, and exon 3 (blue) is used in FLM-δ. (B) A model for the regulation of flowering through interactions between FLM splice variants and a second transcription factor, SVP. The SVP-FLM-β complex, which predominates at low temperatures, binds to the gene regulatory site that inhibits flowering and prevents the development of flowers. The SVP-FLM-δ complex, which predominates at high temperatures, cannot bind to this regulatory site and flowering is not suppressed. (C) Upper graph: Relative expression levels of FLM-β and FLM-δ as a function of temperature (16°C, 23°C, and 27°C). Error bars denote standard deviations of three biological replicates, each with three technical replicates. Lower graph: The resulting ratios of FLM-β to FLM-δ. (After Posé et al. 2013.)

The role of alternative splicing in adaptation to temperature is now being elucidated. We consider but one example here, illustrating how this process helps control temperature-responsive gene expression related to a phenomenon that has a strong seasonal link: flowering in plants. The effect of temperature on the timing of biological processes—**phenology**—is a fascinating issue that clearly has great relevance to climate change. Many examples of shifts in timing of natural events have accumulated as the planet warms. Understanding the molecular basis of these temperature-phenology relationships may help us better understand and predict the effects of ongoing global change, as discussed in Chapter 5.

CONTROL OF FLOWERING THROUGH GENERATION OF TEMPERATURE-DEPENDENT SPLICE VARIANTS OF PROTEINS   Flowering in vascular plants is governed by a variety of seasonal conditions, including photoperiod and temperature. A study of flowering in the model plant *Arabidopsis thaliana* is helping reveal how temperature regulates the timing of flower development by influencing the alternate splicing of RNAs that encode transcription factors active in the flowering process (Posé et al. 2013). Exposure of *A. thaliana* to different temperatures leads to changes in expression of genes that encode enzymes involved in mRNA processing (Balasubramanian et al. 2006), so temperature-dependent production of splice variants may be of wide importance in this (and other) species. As illustrated in **Figure 3.12B**, binding of a transcription factor complex comprising two proteins, Flowering Locus M (FLM) and Short Vegetative Phase (SVP), to a gene regulatory region that helps control the onset of flowering can lead to suppression of flower development. However, the ability of the SVP-FLM complex to bind the regulatory site depends on which variant of FLM is present in the cell. The two primary splice variants of FLM present in the strain of *A. thaliana* used in this study both can bind to SVP, but only one of them, FLM-β, generates an SVP-FLM complex that can bind to a regulatory site that inhibits flowering. The other splice variant, FLM-δ, forms a complex with SVP that cannot bind to the site. Thus, flowering can commence only when the SVP-FLM complexes in the cell are formed largely with FLM-δ.

Because flowering tends to be favored by higher temperatures in *A. thaliana*, one would predict that the ratio of the two splice variants (FLM-β:FLM-δ) would show a temperature-dependent pattern. This is precisely what Posé and colleagues observed (**Figure 3.12C**). Increased temperature favored a reduction in levels of FLM-β and an increase in FLM-δ. As a result of this change in splicing pattern, the ratio (β:δ) of the two splice variants showed a striking change, decreasing from approximately 5.6 at 16°C to less than 1 at 27°C. Thus, with rising temperature the fraction of SVP bound to FLM-δ increased, removing suppression of flowering.

The control of flowering involves a variety of other mechanisms, so these changes in FLM splice variants are only part of the story. Nonetheless, this study shows clearly that alternative splicing of exons can lead to temperature-adaptive changes in protein isoforms. The extent to which this form of temperature-adaptive regulation occurs with other proteins is a question that merits a great deal of additional study. Another important feature of the *A. thaliana* FLM system is that different strains of the species have genetic variants of the FLM system, including deletion of the FLM-coding gene (Balasubramanian et al. 2006), that lead to different flowering responses. These genetic variants might be important in counteracting negative effects of climate change on flowering time. In plants, these effects include a decoupling of the timing of flowering from the arrival of pollinating insects. Selection on genetic variation in timing of flowering might lead to the reestablishment of optimal flowering times in plants that face this type of reproductive decoupling.

RNA editing: Posttranscriptional changes in mRNA can lead to changes in thermal responses of the encoded proteins

What if temperature were able to modify ("edit") the sequence of a messenger RNA molecule in a way that led to a temperature-adaptive change in the protein that is translated from this edited message? As recently as a few years ago, if one had been asked this question, one's answer likely would have been, Something like that is almost certainly too good to be true! However, one should never sell evolution short. In 2012 a mechanism of this type was discovered in studies of sequence-function relationships of potassium (K^+) channels in octopuses (Garrett and Rosenthal 2012a,b; Rosenthal 2015).

These K+ channels are expressed in neurons and are involved in regulating membrane potential and the nerves' rates of firing. This work demonstrated that the most common form of RNA editing, the conversion of adenosine to inosine (A-to-I conversion; **Figure 3.13A**), was frequent in these proteins and that the amount of editing that occurred was greater at low than at high temperatures (**Figure 3.13B**). The enzyme governing this editing is **adenosine deaminase acting on RNA** (**ADAR**), and its activity and structure may themselves be temperature-dependent. Here we consider what is currently known about RNA editing by ADAR and what the broader role of this type of editing might be in adaptation and acclimatization to changes in temperature.

One of the most important consequences of RNA editing catalyzed by ADAR is that A-to-I conversion can lead to a change in the amino acid specified by a codon, because I is read as guanosine (G) during translation. Importantly in the context of adaptation to temperature, the codon changes that result from A-to-I editing tend to favor shifts toward less stabilizing amino acids. Specifically, the A-to-I editing process that occurs at reduced temperatures increases the amounts of glycine, valine, and arginine and decreases the amounts of lysine, isoleucine, and glutamate. These shifts in amino acid composition reflect those found in broad comparisons of thermophilic versus psychrophilic microbes (see Table 3.3). To review these findings briefly, increased glycine content makes sense as an adaptation to cold, because glycine allows the greatest conformational flexibility of any amino acid. The shift from isoleucine to valine (from larger to smaller nonpolar

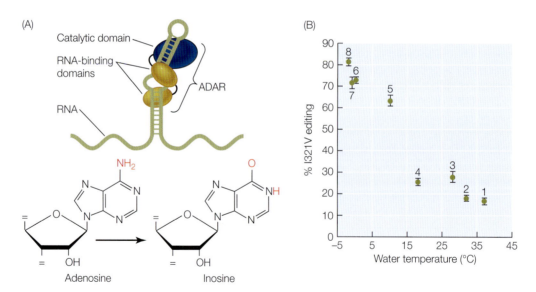

Figure 3.13 RNA editing by adenosine deaminase acting on RNA (ADAR). (A) The biochemistry of adenosine deamidation. At top is the ADAR enzyme interacting with an RNA substrate. Double-stranded RNA (dsRNA) is needed for binding to the enzyme. The lower panel shows the chemical structures of adenosine and inosine. (B) Temperature-related variation in the amount of RNA editing at the codon for residue 321 in the K+ channel protein from octopuses adapted to different temperatures. Editing leads to replacement of an isoleucine by a valine, which favors destabilization of this region of the protein's structure and enhances its rate of activity (see text). Error bars represent standard errors of the mean. The eight species studied and the water temperatures at the sites of collection are: 1, *Octopus digueti* (37°C); 2, *O. defilippi* (32°C); 3, *O. vulgaris* (28°C); 4, *O. bimaculatus* (18°C); 5, *O. rubescens* (10°C); 6, *Benthoctopus piscatorum* (0°C); 7, *Bathypolypus arcticus* (–1°C); and 8, *Pareledone* sp. (–1.9°C). (After Garrett and Rosenthal 2012a.)

side chain) reduces packing density in the protein's hydrophobic core. The decreases in content of lysine and glutamate also reflect basic principles of protein stabilization. The charged side chains of lysine and glutamate are instrumental in forming stabilizing salt bridges on the protein surface. In summary, the codon changes that result from enhanced A-to-I editing at lower temperatures lead to precisely the types of changes in amino acid composition that evolutionary adaptation achieves, but do so on a timescale of minutes to hours rather than over many generations.

The adaptive changes in ADAR-driven A-to-I editing in octopus K^+ channels illustrate the potential of RNA editing for generating temperature-adaptive protein variants (Garrett and Rosenthal 2012a; Rosenthal 2015). The sequences of the genomically encoded (nonedited) K^+ channels of an Antarctic octopus (*Pareledone* sp.) collected in McMurdo Sound (water temperature about −1.9°C) and a tropical octopus (*Octopus vulgaris*) collected in tropical waters (~28°C) were extremely similar, differing at only four sites. However, cDNAs prepared from isolated mRNAs for the K^+ channel protein from the two species revealed A-to-I conversions that would cause several functionally significant differences in the protein's amino acid sequence. Of particular interest was site 321, where editing was shown to lead to a temperature-adaptive change in the nerve's rate of firing. The frequency of editing at site 321, where an isoleucine is replaced by a valine in the edited sequence, was significantly higher in the Antarctic species. Functional studies showed that the isoleucine-to-valine substitution destabilized the ion channel's open state, which led to an increase in the rate of channel closing. Therefore, the edited K^+ channel prevalent in the Antarctic species has a temperature-compensated rate of activity due to a shortening of the refractory period between spikes, which leads to a higher repetitive firing rate. Comparisons of K^+ channels in octopuses from polar, temperate, and tropical zones revealed that the frequency of editing at site 321 is inversely related to environmental temperature (see Figure 3.13B). As a result of RNA editing, there is a regular and temperature-compensatory change in the rate of channel closing across species of octopuses that encounter habitat temperatures differing over a range of almost 40 degrees Celsius.

There are several additional characteristics of ADAR-mediated RNA editing that merit attention in the context of thermal relationships. One is that editing is not complete, but rather mRNAs are present in the cell with different levels of editing. This variation in mRNA sequence could lead to production of multiple isoforms of the protein in question, if all of the differently edited mRNAs are translated. This type of mRNA and protein variation might be of particular advantage to eurythermal ectotherms that experience wide changes in body temperature on a short-term (e.g., daily) basis. Having "warm" and "cold" variants of a protein might be possible if this type of editing process allows two (or more) different sequences for the protein to be generated at the same time.

SELF-EDITING OF THE EDITOR   A second aspect of ADAR function would seem to open the door for an even greater potential for generating temperature-specific isoforms of proteins: ADAR is self-editing (auto-editing) (see Garrett and Rosenthal 2012b). For example, in *Drosophila*, auto-editing by ADAR produces a new isoform of ADAR that contains a single change in amino acid sequence (Palladino et al. 2000). This substitution leads to a reduction in editing efficiency relative to the genome-encoded enzyme. Up- and down-regulation of ADAR activity thus might be mediated by auto-editing. Potentially, such auto-editing could be modulated by changes in temperature. Another way of generating ADAR isoforms is through alternative splicing of the ADAR mRNA (Garrett and Rosenthal 2012b). Many instances of splice variants of ADAR have been discovered in a variety of animals (mammals, squid, and *Drosophila*), and these variants have been shown to differ in activity in several cases. If the splicing process is temperature sensitive, then generation of ADAR variants with the appropriate function for the new thermal regime —for example, enhanced editing capacity at low temperatures—might be possible.

RNA CONFORMATION INFLUENCES RNA EDITING   Another important feature of ADAR function concerns the geometry of its RNA substrate. A single type of RNA can assume a variety of temperature-dependent conformations. This is a potentially important factor in RNA editing because the ability of ADAR to carry out its editing functions depends on the three-dimensional configuration of its substrate. ADAR works only on double-stranded RNA (see Figure 3.13A), and the double-stranded structure of an RNA molecule is likely to vary with temperature. In general, reductions in temperature favor increased amounts of double-stranded structure. If temperature-induced changes in RNA secondary or tertiary structure alter ADAR's access to editing sites, then a further mechanism for temperature-dependent editing would appear to be present. For example, exposure of an editing site when an RNA molecule changes structure during a fall in temperature might allow ADAR to edit the mRNA in a way that favors a more cold-adapted protein. Suffice it to say, RNA editing based on ADAR-mediated A-to-I conversion, coupled with temperature-dependent changes in ADAR sequence and RNA conformation, opens up a wealth of regulatory possibilities for generating temperature-adaptive variants of proteins. This type of mechanism for generating protein variants encoded by a common gene could have a "rheostat" type of function, by providing the cell with the most advantageous population of proteins for the thermal conditions at hand. How these mechanisms play out in the real world should prove to be a fascinating arena of study for investigators interested in biochemical adaptation. Likewise, the roles played by RNA editing in different taxa remain largely unknown. Endothermic homeotherms like mammals appear to have relatively low amounts of RNA editing compared with the few ectothermic species that have been examined. Perhaps temperature-adaptive RNA editing is not needed much, if at all, in organisms with stable body temperatures. In contrast, ectotherms that experience wide ranges of temperature might be major beneficiaries of this strategy for modulating protein sequence and function in the face of temperature change.

Before leaving the topic of RNA editing, it is important to point out that ADAR-mediated A-to-I conversion occurs in all classes of RNA molecules, including tRNAs and microRNAs (miRNAs) as well as mRNAs (Rosenthal 2015). A-to-I editing of mRNAs not only can foster changes in protein sequence, but also can modify the half-lives of the mRNAs and affect the splicing patterns that occur during RNA maturation. Editing of miRNAs by ADAR has been shown to regulate their posttranscriptional maturation and capacities for recognizing sites on mRNA molecules whose activities they regulate. How these additional functions of A-to-I editing might contribute to an organism's responses to temperature remains to be discovered.

Horizontal gene transfer: Gaining new genetic information for protein adaptation to temperature

Up to this point in our discussion of thermal adaptation of proteins, we have focused strongly on changes in primary structure that alter proteins' thermal sensitivities. Sources of genetic variation to support protein adaptation include point mutations, gene duplication, alternate splicing, and RNA editing. We now look at a source of genetic variation that has received considerable attention in the context of adaptation to temperatures in bacteria and archaea—the acquisition via **horizontal gene transfer** (**HGT**) of genes that encode proteins with adaptive thermal characteristics. The DNA elements involved in HGT have been shown to encode proteins that conduct a variety of functions, which range from numerous "housekeeping" activities in intermediary metabolism to repair of DNA damage induced by high temperatures.

HGT has been a major contributor to the evolution of bacteria and archaea (Goldenfeld and Woese 2007), and because of its prevalence in these domains, a given microbe may have a **mosaic genome** that reflects contributions of genes from multiple domains of life. It is estimated that at least 20% of bacterial genes and ~40% of archaeal genes have been transferred between these two domains during evolution (van Wolferen et al. 2013 and references therein). In Chapter 5 we will return to HGT and examine it in the broader context of global change, where alterations in temperature, pH, osmolality, and oxygen availability may all confront an organism. As we will show in that analysis, access to HGT may give bacteria, archaea, and unicellular eukaryotes a uniquely high potential for adapting rapidly to multiple stressors. Here we restrict our focus to the roles that HGT is known to play in adapting bacteria and archaea to new ranges of temperature. Box 3.7 provides a brief overview of the basic mechanisms by which HGT can take place and considers its broad importance in evolution.

BOX 3.7 ▌▌ ▬▬ ▬▬▬ ▬▬▬▬▬▬▬▬▬▬▬▬▬▬▬▬

Horizontal gene transfer (HGT): Mechanisms and evolutionary implications

How is HGT mediated in bacteria and archaea?

Several mechanisms for HGT are known in bacteria and archaea. **Natural transformation** is the process by which a cell takes up cell-free DNA—likely released into the surrounding medium from other, dying cells—and incorporates the new DNA into its genome. This type of transformation is largely under the control of the DNA recipient. In contrast, **conjugation** is a process through which a donor cell governs unidirectional transfer of DNA into a recipient cell, which may or may not be closely related to the donor cell. Direct contact between cells is needed for DNA transfer. The DNA moves through channels, such as pili, that protect the DNA from degradation, thereby facilitating the transfer of intact genes from donor to recipient. Most DNA transferred via conjugation is carried by **plasmids**. Conjugation is thought to be the most important mechanism of HGT, in part because linked sets of genes related to a common function may be carried by a single plasmid (van Wolferen et al. 2013). **Transduction** is another mechanism for DNA movement from one cell to another. Here, a viral carrier is involved in the movement of genes. HGT among viruses also occurs, leading to major diversification of viral genomes.

Another mechanism for DNA transfer involves membrane vesicles (**MVs**), which bacteria and archaea release into the surrounding medium (Deatherage and Cookson 2012). Some bacteria and archaea even have devices termed *nanopods* that project MVs several micrometers from the cell (Shetty et al. 2011). Although DNA may be present (and likely well protected from degradation) within MVs and may be taken up by cells, the roles of MV-mediated transfer of DNA remain unclear. Materials contained within MVs are known to be transferred between cells of a single species, but it is not clear whether this mechanism transfers DNA between cells of different species, as would be required for HGT (Deatherage and Cookson 2012). It is pertinent to point out that HGT, while of wide occurrence in archaea and bacteria, is not symmetrical in these two groups; most interdomain gene transfers appear to involve HGT from bacteria to archaea (Nelson-Sathi et al. 2015).

The importance of HGT in evolution— and the question of "What is a microbial species?"

In their short essay published in *Nature* in 2007, Goldenfeld and Woese presented a thought-provoking analysis of the phenomenon of HGT and its profound implications for evolution. They empha-

BOX 3.7 *(continued)*

sized the high degree of novelty that can be generated when genes pass among organisms of different evolutionary lineages. They stated that "microbes absorb and discard genes as needed, in response to their environment. Rather than discrete genomes, we see a continuum of genomic possibilities, which casts doubt on the validity of the concept of a 'species' when extended into the microbial realm." An excellent illustration of the challenges the species concept faces in the microbial realm is the discovery that in the "garden variety" bacterium *Escherichia coli*, only about one-third of all genes are common to all lineages of this bacterium (Fuhrman 2009). These are the "core" genes that are complemented

by a second set of genes—the "flexible" genes—that confer on the lineages a set of functional capacities that reflect the adaptational challenges offered by the lineage's particular environment. The term *species* has been replaced in much microbial analysis by the **operational taxonomic unit** (**OTU**), which is defined using the sequences of 16S ribosomal RNAs (16S rRNA) as the taxonomic yardstick. The criterion for assigning two microbial lineages to the same OTU is 97% sequence similarity. Note that, from the perspective of an animal taxonomist, this would group together in a single OTU all primates, from chimpanzees to lemurs to humans.

HGT AND ADAPTATION TO HIGH TEMPERATURE: DNA REPAIR SYSTEMS   Researchers have characterized the role of transferred DNA in adaptation to high temperatures in several archaea and bacteria (van Wolferen et al. 2013), and have identified several important contributions of HGT to thermal tolerance and adjustment of thermal optima. One role of transferred DNA is in DNA repair. Double-strand breaks (DSBs) in DNA can only be repaired through homologous recombination. For homologous recombination to succeed in repairing DSBs, a second, nondamaged strand of highly similar DNA is needed. This "healthy" DNA can be obtained through uptake of DNA from a closely related species. It is noteworthy in this context that DNA uptake often is strongly biased in favor of DNA from the same or a closely related species (Seitz and Blokesch 2012). This preference would of course favor the likelihood of obtaining the necessary strand of DNA—one with a similar sequence—for repair through homologous recombination. To increase the odds of obtaining DNA from a suitable donor—that is, a closely related OTU (see Box 3.7)—a microbe may induce cell death of its neighbors ("microbial fratricide") (van Wolferen et al. 2013).

Uptake of DNA for DSB repair can be viewed as a mechanism for regaining the status quo—no new genes are acquired, but existing damage to the genome is repaired. In contrast, HGT that brings new types of genes into a cell can potentially lead to adaptation to a new habitat, for example, a higher temperature. The role of HGT in facilitating bacterial adaptation to high temperatures is thought to be substantial. Hyperthermophilic bacteria have been found to contain a higher percentage of archaeal-derived genes than do mesophilic bacteria (van Wolferen et al. 2013). In some bacteria, a large percentage (up to ~50%) of the archaeal-derived genes are encoded in a single plasmid. This discovery illustrates the efficiency of HGT in providing the recipient cell with an appropriate "package" of useful genetic information—for example, genes encoding proteins whose function is critical at high temperatures. Among the types of genes acquired by bacteria from hyperthermophilic archaea are genes that encode DNA repair enzymes, including a **reverse gyrase** that is found in all hyperthermophiles. DNA repair is likely of critical importance to maintenance of genome integrity at high temperatures, so the occurrence of reverse gyrase at high levels may be one hallmark of tolerance of extreme heat.

The differences in protein stabilization mechanisms noted above between proteins of thermophilic archaea and bacteria are also seen among proteins of a single type of thermophilic bacterium that are encoded either by bacteria-derived or archaea-derived genes. Proteins obtained by HGT from an archaeal source have highly compact structures that are stabilized by strong hydrophobic interactions—a signature feature of proteins of thermophilic archaea. Proteins encoded by genes with bacterial ancestry show the characteristic bacterial structure: a less compact protein that is stabilized by a host of different types of enhanced weak bonds (Berezovsky and Shakhnovich 2005). Thus, the mosaic genomes of thermophilic bacteria reinforce the finding that different adaptive routes can lead to the same result: heat-resistant protein structures.

HGT: EVOLVING MESOPHILIC ARCHAEA   We now turn to a quite different role of HGT in the evolution of thermal optima: the provision to thermophilic archaea of sets of genes needed to thrive in environments characterized by moderate to low temperatures. These mesic environments also may present archaeal cells with conditions of pH and ion concentrations that differ from those faced by their ancestors. The need for evolving these new mesic-adapted biochemical systems stems from the environmental conditions that are thought by some to have been present at the origin of the domain Archaea. There is still some controversy about the environment in which this domain arose, but there is strong evidence that it was a hot environment, one that selected for hyperthermophiles. Hyperthermophiles are species that perform optimally at temperatures greater than 80°C and, conversely, perform poorly at cooler (mesic) temperatures. When Archaea initially began to be studied by microbiologists, the species that received almost all of the attention were extremophiles found in environments characterized by such conditions as extremely high temperatures, salinities, and acidities. This initial focus on extremophilic archaea tended to create the impression that archaea, as a group, are largely restricted to extreme environments. Subsequently, a large variety of archaeal species have been found in mesic environments. The discovery of high archaeal biodiversity in mesic habitats has led to genomic studies designed to determine where the genetic information for encoding mesic-adapted proteins and lipids might have originated. The most recent analyses of this question suggest that a substantial fraction of the genes needed for function in mesic environments has come from mesophilic bacteria through HGT (López-García et al. 2015). Whereas we focus here on the role of HGT in adaptation to temperature, these new bacterial genes seem also to have adapted archaea to two other abiotic variables that are being altered by global change: oxygen availability and pH.

Analyses of archaeal and bacterial genomes have led to the conclusion that the origins of the major archaeal clades correspond to the acquisition of a wide variety of genes from bacteria (Nelson-Sathi et al. 2015). Nelson-Sathi and colleagues undertook a detailed, comparative analysis of HGT events between archaea and bacteria, supported by the availability of large numbers of sequenced genomes from the two domains. Their comparison of 134 archaeal genomes and 1847 bacterial genomes revealed more than 2000 transfers from bacteria to archaea. There was an approximately fivefold higher rate of transfer from bacteria to archaea than vice versa (see Box 3.7). The proportion of bacterial genes in archaeal genomes differed severalfold among archaeal taxa, and there was a significant negative correlation between average OGT and the fraction of the archaeal genome that was bacterial in origin (Figure 3.14A; López-García et al. 2015). This correlation supports the conjecture that the ability of archaea to function well in mesic environments, especially environments of moderate to low temperature, was attendant on acquisition of bacterial genes that had evolved in such environments. This relationship also shows that the most thermophilic archaea—those with OGTs over

(A)

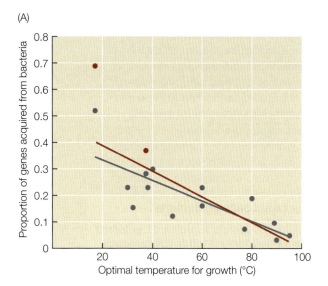

(B)

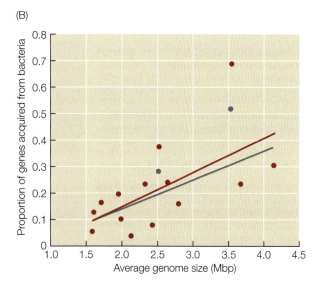

Figure 3.14 Proportion of bacterial genes passed by horizontal gene transfer (HGT) to the ancestors of archaeal groups as a function of (A) average optimal temperature for growth and (B) genome size. Blue data points indicate the proportions of HGT events that are viewed as ancestral to the major archaeal taxa. Red data points include all of the HGT events that occurred in the major archaeal taxa but include both ancestral and recent transfers. (After López-García et al. 2015.)

80°C—rely only minimally on genes obtained from bacteria. Again, this may reflect a high-temperature origin for the domain Archaea; archaea achieved their high thermal tolerance early in the domain's evolutionary history and did this "on their own," without significant help from the domain Bacteria.

The acquisition of new genetic material through HGT generally entails an increase in genome size, and this trend is seen during the evolution of mesophilic adaptation by archaea. Based on the relationship between genome size and proportion of bacterially derived genes in the genome, it seems clear that the genes acquired by archaea through HGT were, in fact, added on to the existing genome (**Figure 3.14B**).

What types of genes were included in these additions to the archaeal genome? Some of the newly acquired genes enabled lineages whose metabolisms were strictly based on anaerobic processes, for example, chemolithotrophic methanogens, to acquire capacities for aerobic metabolism in habitats where oxygen was abundant, as is the case in many moderate- to low-temperature habitats. Other genes facilitated biochemical activities at lower temperatures. Good examples of these types of temperature adaptations are seen in genes that encode enzymes involved in conformational changes in DNA. For example, acquisition of **DNA gyrase** (a bacterial type II topoisomerase) provided the archaeal recipients with an ability to unwind DNA effectively at low temperatures through introduction of negative supercoils during replication and transcription. As we saw above in the context of HGT from archaea to bacteria, hyperthermophiles have reverse gyrases, which have the opposite effect on DNA compaction. Thus, maintaining DNA secondary structure at the right balance between flexibility and rigidity is critical for adaptation to temperature. Genes encoding different classes of heat-shock proteins were also obtained by archaea through HGT. The common 70-kDa heat-shock protein Hsp70 appears to be absent in hyperthermophilic archaea but is found in mesophilic archaea, likely as a result of HGT. Bacterially derived genes encoding **cold-shock protein A (CspA)** are found in mesophilic archaea but not in hyperthermophiles. Another important class of genes acquired through HGT encodes proteins responsible for modulating membrane lipid composition in temperature-adaptive manners.

A remaining question about biochemical adaptation of hyperthermophiles and thermophiles to mesic conditions concerns what becomes of the genes that arose during evolution in high-temperature habitats. Assuming those genes have been retained (an assumption supported by the larger genome sizes of mesophiles), have the encoded proteins undergone the types of adaptive changes that would be required to ensure optimal fluidity (marginal stability) of structure, ligand-binding ability, and catalytic rate? Although the genes acquired by HGT may have provided an "entry ticket" for life in mesic environments, many—and perhaps most—of the genes carried over from long times spent under extreme thermal conditions may have had to evolve considerably to alter their environmental optima as well.

We leave the story of HGT at this point. In Chapter 5 we will return to this topic and examine the general utility of HGT in adapting to new environments, especially those in which multiple stressors are changing simultaneously—a hallmark of the Anthropocene. HGT will be seen to offer a potentially good "escape route" from the effects of global change, at least for unicellular organisms that have a high ability to take up DNA from other species.

Molecular thermometers: Exploiting thermal perturbation of protein and nucleic acid structures to regulate transcription and translation

Most of our analysis of thermal effects on protein structure has focused on negative consequences of structural perturbation, for example, thermal disruption of enzyme-ligand interactions or heat-induced unfolding of native conformations. However, as we emphasized in Chapter 1, the word *perturbation* (or *stress*) should not be viewed in a strictly pejorative sense: Many temperature-induced alterations in protein and nucleic acid structures serve important functions in regulating cellular processes like the heat-shock response, as we discuss later in this chapter. In other words, evolution has taken advantage of thermal perturbation of macromolecular structures to fabricate regulatory networks with remarkably high thermal sensitivities and short response times. Detection of differences in temperature in the range of ~1 degree Celsius has been observed, and temperature-mediated activation of gene expression can occur within minutes. The high temperature sensitivity of macromolecular higher-order structures thus can enable these molecules to serve as remarkably sensitive **molecular thermometers**. Later (see Section 3.5), we provide several striking illustrations of "useful" thermal perturbation of structure in the context of temperature-nucleic acid interactions. First, however, we focus on the role of temperature changes on hydrophobic interactions between proteins and show how thermal effects on one particular class of protein, those containing a structural motif termed a **leucine zipper**, provide a rapid mechanism for upregulation of gene expression when cell temperature rises.

LEUCINE ZIPPERS AND TEMPERATURE SENSING   Leucine zippers are a structural motif found in a variety of classes of proteins that are present in all domains of life. Although research on leucine zipper proteins has emphasized their roles in control of gene expression (DNA binding activities), zipper domains occur in proteins other than transcription factors and are regarded as one of the most important structural features for facilitating reversible protein-protein interactions. The N-terminal portion of leucine zipper-containing transcription factors is rich in arginine and lysine residues that function in binding to the major groove of DNA (**Figure 3.15A**). The C-terminal zipper region, which is about 80 amino acid residues in length, is rich in the strongly hydrophobic amino acid leucine, hence the name given to this type of structure. Leucine residues are present at every seventh

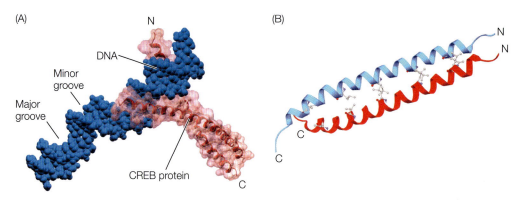

Figure 3.15 Leucine zipper proteins. (A) A leucine zipper transcription factor (CREB) interacting with a molecule of DNA. (B) A close-up of the leucine zipper region of a leucine zipper protein. N and C refer to the amino- and carboxyl-terminal amino acids of the zipper.

position in the α-helical zipper sequence (a heptad repeat structure; **Figure 3.15B**). The zipper helix is amphipathic, with a polar, water-preferring surface and a strongly hydrophobic surface that contains the leucine side chains. A functional complex is formed when the leucine side chains of one zipper α-helix interdigitate with those of the zipper sequence of another zipper protein. Leucine zipper domains in the oligomerized proteins have a type of structure referred to as a **coiled coil**: These structures are formed of two or more α-helices that are coiled around each other to generate what is termed a **supercoil**. This assembly process yields an oligomer that, in the case of transcription factors, is able to bind to DNA and modify gene expression. Most studies of leucine zipper proteins have focused on the class of eukaryotic transcription factors known as bZIP (basic-region leucine zipper) proteins, which form dimers as shown in Figure 3.15 (Krylov and Vinson 2001). Larger complexes can form as well. For example, as we discuss in the context of the heat-shock response in the next section of this chapter, the assembly of the active form of heat-shock factor 1 (HSF1) involves trimerization of HSF1 monomers, forming a three-strand α-helical coiled coil (Zhong et al. 1998).

Because formation of hydrophobic interactions involves a positive change in enthalpy, one would expect the interactions of leucine zipper α-helices to be favored by an increase in temperature. Direct heat effects on hydrophobic interactions may, in fact, explain some or all of the heat-induced trimerization of HSF1 (Zhong et al. 1998). Using purified HSF1 from *Drosophila* in controlled in vitro experiments, Zhong and colleagues showed that an increase in assay temperature from 20°C to 40°C led to a sigmoidal increase in the percentage of HSF1 that occurred in a trimer form. Reducing the assay temperature back to 20°C demonstrated full reversibility of the trimerization reaction. Whereas there may have been high-temperature-induced changes in HSF1 monomer conformation that altered exposure of the leucine zipper regions, the subsequent polymerization of these hydrophobic regions to form a coiled-coil structure would be enhanced by rising temperature. This type of direct temperature effect has the benefit of assisting in a rapid response to a change in temperature, a type of rapidity that is also characteristic of the regulatory responses of RNA thermometers involved in regulation of temperature-dependent gene expression (see Section 3.5).

The broader impacts of leucine zipper proteins in thermal biology largely remain unexplored. For example, to what extent do subunit-subunit interactions of enzymatic proteins rely on leucine zippers (or other types of hydrophobic structural elements) to modulate assembly states in a temperature-adaptive manner? Can the assembly thermodynamics of leucine zipper proteins evolve such that the thermal sensitivity of the proteins' coiled-coil assembly is modulated to "fit" the thermal conditions experienced by the organism? For example, do cold-adapted species have leucine zipper proteins that can reversibly assemble effectively at low temperatures despite the destabilizing effects of cold on hydrophobic interactions? In cold-adapted species, might the involvement of exothermic bonding between polar or charged residues be needed to supplement the effects due to hydrophobic stabilization of the oligomerized zipper proteins? The involvement of polar residues in determining the specificity of oligomerization between leucine zipper proteins suggests that such "fine-tuning" of supercoiling thermodynamics is possible (Hakoshima 2014). There would seem to be a vast field of interesting phenomena to explore in connection with these regulatory proteins that function both as thermometers and as transducers of a temperature-induced change in higher-order structure into alterations in gene expression.

We turn now to a detailed examination of temperature-driven changes in protein structure that initiate and regulate what is termed the heat-shock response (HSR). We will see that the negative effects of rising temperature on protein structure—denaturation—are paired with favorable effects of the sort just discussed: rising temperature not only causes widespread damage to the proteome but also activates the regulatory circuitry that is involved in addressing the consequences of this damage.

From the basic principles of thermodynamics to the reality of protein folding in the cell

Up to this point in the chapter, we have focused on proteins to illustrate the effects of changes in temperature on biochemical structures and processes. Beginning with this reductionist approach is appropriate in the context of biochemical adaptation because it can build a foundation for a multilevel analysis of how perturbations of individual types of macromolecules can affect higher levels of biological organization. Indeed, protein perturbation has widespread and complex consequences at the cellular and organismal levels, which in turn can affect an organism's fitness and, thereby, biochemical and physiological evolution.

Thermal perturbation of protein structures poses challenges to cellular physiology, and numerous cellular processes respond to these challenges by working to restore **protein homeostasis** (proteostasis). These repair mechanisms are present in virtually all types of cells, which is an indication of their importance and their early origins in cellular evolution. This ancient and conserved set of cellular responses is collectively referred to as the **heat-shock response** (**HSR**). In fact, perturbation of proteins occurs during all forms of abiotic stress, so a more generic term—the **cellular stress response** (**CSR**)—is often used to refer to the processes that achieve proteostasis (Kültz 2005). Here, with our focus on temperature, we will use HSR, but it is important for the reader to keep in mind that any type of environmental change that disrupts protein structure will trigger this same set of restorative processes. As with many questions about the origin of life, we can't be certain that temperature was the original selective agent that led to the evolution of the HSR. For example, heavy metal stress and UV stress may have been recurrent environmental challenges to cells living in the "primordial soup" that life inhabited billions of years ago. Nonetheless, as we will describe in subsequent sections, temperature has been, and continues to be, a selective pressure that leads to evolutionary adaptations in the HSR.

An historical perspective on the discovery of the HSR

The HSR has been studied extensively for more than 50 years, and consequently there is a vast literature on the subject, including many good reviews (Lindquist 1986; Richter et al. 2010; de Nadal et al. 2011). Therefore, our goal is not to review the details of the molecular and cellular biology of the HSR, but rather to give a focused account of this cellular process from an ecological and evolutionary perspective: How does the HSR help organisms cope with thermal stress, and how has the HSR evolved to enable the appropriate responses in species adapted to widely different absolute temperatures and ranges of temperature? From this perspective, it will become apparent that the HSR is an essential evolutionary adaptation that evolved in response to the environmental stressors that perturbed protein structural stability during the early evolution of life on Earth. Many of the core proteins of the HSR occur in all taxa, but the regulatory switches that govern expression of the HSR exhibit a fascinating evolutionary diversity that illustrates how a given type of biochemical system can be fine-tuned for addressing the specific challenges that a given species might meet.

The history of the discovery of the HSR provides a fascinating example of the role of serendipity in the advancement of science (**Box 3.8**). The "accidental" discovery by Ritossa (1962) that heat stress in *Drosophila busckii* induces specific types of chromosomal puffing—changes in chromosomal architecture that subsequently were shown to reflect activation of transcription of specific sets of genes—led to further study of this type of response in other organisms. Subsequent studies confirmed the existence of this phenomenon in other species of *Drosophila* (Ritossa 1963; Ashburner 1970). As analytical techniques advanced and patterns of protein expression could be elucidated using procedures such as sodium dodecyl sulfate polyacrylamide gel electrophoresis (SDS-PAGE), which separates proteins by molecular mass, more and more examples of

BOX 3.8

Ferruccio Ritossa and his exploitation of serendipity

As is often the case in science, the discovery of the heat-shock response (HSR) was unintentional and serendipitous. We present this interesting story in some detail, for it carries a key message: "Keep your eyes open and exploit the unexpected!" In the early 1960s, a graduate student named Ferruccio Ritossa (**Figure A**) attended a small course in genetics at the Genetics Institute in Pavia, Italy, where he pursued a project to identify the type of nucleotide (i.e., RNA versus DNA) that was being synthesized during gene expression (Ritossa 1996). At this early stage in the study of gene expression, messenger RNA had not yet been discovered, and analytical approaches generally relied less on molecular techniques than on other, ostensibly more "primitive" approaches, including microscopic examination of chromosomes' appearances. An ideal study system for this purpose involved the **polytene chromosomes** of *Drosophila*, a genus of fruit fly that had been chosen as an ap-

propriate model system to investigate the mechanism of gene expression during development. In the salivary glands of larval and adult *Drosophila*, polytene chromosomes—tightly organized assemblages of many copies of a given chromosome—

Figure A Ferruccio Ritossa (From De Maio et al. 2012.)

BOX 3.8 *(continued)* ▮ ▮ ▮ ▮▮▮▮

provided a straightforward method for imaging changes in chromosomal architecture that signify loci where gene expression is occurring. In this context, loci of gene expression can be visualized as protrusions, or "puffs," along the length of a chromosome (**Figure B**).

During his experiments, Ritossa was working to characterize the changes in puffing patterns across the genome during a developmental time series in two species of fruit flies, *D. busckii* and *D. melanogaster*. Unbeknownst to him, a fellow lab member accidentally increased the temperature of Ritossa's incubator from 25°C to 30°C. Upon examining his chromosome preparations, Ritossa noticed a puffing pattern he had never seen before: A few new loci suddenly showed huge puffs, and other puffs along the genome expected for a given stage of development had entirely disappeared. The micrograph in the figure clearly illustrates the changes in chromosomal puffing patterns that are induced by heat stress in *Drosophila* larvae. It was later reported that the degree of chromosomal puffing increased with the duration and severity of the heat stress (Ashburner 1970).

Ritossa published his pioneering results in 1962, as a previously uncharacterized thermally induced chromosomal puffing pattern (Ritossa 1962). What these chromosomal puffs led to in terms of specific types of protein function was only clarified many years later. Yet had Ritossa not "kept his eyes open," the history of discovery in this field might read differently.

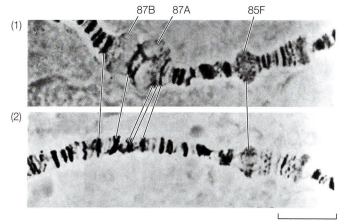

Figure B Changes in chromosomal architecture in polytene chromosomes isolated from salivary glands in *Drosophila simulans* prepupal larvae, (1) after heat stress (40 min at 37°C) and (2) control (25°C). Note the regions of the chromosome labeled at 87A and 87B that show major changes in puffing versus other regions like 85F that show little or no change. We now know that the regions at 87A and 87B correspond to the location of genes encoding Hsp70s (Attrill et al. 2016), and thus the changes in puffing patterns induced by heat stress signify induction of the HSR. (From Ashburner 1970.)

heat-induced changes in protein expression were revealed. The heat-induced proteins commonly were grouped into several molecular weight classes; for example, proteins with masses near 70 kDa were frequently found to be strongly upregulated following heat stress (Lewis et al. 1975; Koninkx 1976; Mirault et al. 1978). These proteins were named **heat-shock proteins** (**HSPs**), and the nomenclature for identifying specific types of HSPs commonly used their approximate molecular weights; for example, HSPs with masses near 70 kDa became known as Hsp70s. (In bacteria, the independent discovery of heat-induced proteins was followed by a different set of naming rules, ones that often referred not to the HSR but rather to the phenotypes of mutants that exhibited temperature-sensitive defects in DNA synthesis [Saito and Uchida 1977]. For example, *dnaK* and *dnaJ* were genes that were later shown to encode homologs of Hsp70 and Hsp40, respectively [Bardwell and Craig 1984; Blumberg and Silver 1991].)

Once HSPs had been discovered and classified into families with different masses, the next step was to determine just what functions HSPs carry out in heat-stressed cells. The observation that synthesis of HSPs was strongly upregulated by heat stress, and that synthesis of other types of proteins was strongly suppressed and often ceased entirely, suggested an important function. We now know that HSPs serve to protect and restore the functions of other proteins during heat stress, a role that has earned them the name **molecular chaperones**. The name *chaperone* was used by biochemist John Ellis (1993) to stress that these proteins function in a "Victorian" sense to prevent inappropriate interactions between mutually attractive entities with geometrically complementary surfaces. We will see below why this analogy is so apt.

The problem of thermal unfolding of proteins

Up to this point in our discussion of protein structure, we have focused on the underlying thermodynamic principles that determine the stability of a mature, folded protein. We saw that much of the driving force for protein stability involves protein-water interactions that determine the net stabilization free energy of the protein. However, we have not yet discussed the complex route that is traversed as a nascent (newly synthesized) protein folds into its native conformation in the cell, in accord with the relationships of thermodynamics. Whereas the amino acid sequence of a protein contains much of the information needed to determine the final conformational state, folding within the cell does not occur "in a vacuum." Instead, folding is governed by interactions with numerous factors in addition to water. These include other proteins, notably molecular chaperones; low-molecular-mass solutes that influence protein stability (see Chapter 4); and lipids, in the case of proteins that are integrated into cellular membranes. Likewise, when stress perturbs the conformation of a protein, restoration of its native three-dimensional shape may require help from other proteins and the solution microenvironment in which refolding occurs.

The marginal stability of protein structures ensures optimal function but also makes proteins extraordinarily sensitive to changes in temperature. We have seen, for example, how changes in temperature can disrupt protein-ligand interactions by influencing the distribution of proteins among different microstates. As thermal stress intensifies, proteins increasingly lose their native structures and become denatured, thereby losing function. Our comparisons of orthologous proteins showed how primary structure can be modified over evolutionary timescales to ensure the appropriate level of marginal stability for the adaptation temperature in question. However, we have not yet said much about the effects of *sudden* disruptions of protein structure that lead to denaturation and cessation of function. Can cells restore the native structures of heat-denatured proteins, allowing the proteome to resume normal function? The short answer to this question is yes, and with only rare exceptions, all organisms possess the HSR as a molecular mechanism for the management of thermal unfolding of proteins.

Protein unfolding, like protein folding, follows the rules of thermodynamics, as described by the Gibbs-Helmholz equation (see Box 3.2). Both folding and unfolding are influenced by temperature through thermal effects on weak bond interactions. In terms of protein configurational microstates, extreme temperature causes the unfolded (denatured) fraction to dominate the population of protein microstates. In other words, temperatures outside the normal thermal range for which proteins are adapted can cause unfolding and loss of function. This thermal unfolding of the proteome, which can occur in conditions of both extreme heat and extreme cold, not only causes loss of protein function but also can lead to partially unfolded proteins whose exposed hydrophobic amino acid side chains interact with other unfolded proteins to form aggregates. **Protein aggregation** can be cytotoxic for several reasons (for review see Stefani and Dobson 2003). For example, many neurodegenerative diseases may have their basis in protein aggregation. Membrane damage may occur as well because the exposed hydrophobic groups in large aggregates readily interact with membranous structures of the cell (e.g., cell membrane, nuclear membrane, and endoplasmic reticulum). These interactions can affect membrane permeability and lead to defects in cellular function and, where damage is extreme, to apoptosis (programmed cell death). Thus, sudden changes in temperature can disrupt proteostasis and trigger a cascade of events that results in increased levels of macromolecular damage and severely impedes cellular physiology.

We now turn to an examination of folding processes during protein synthesis and recovery from stress. This analysis will establish the foundation for our subsequent discussion of proteostasis.

Molecular chaperones: Protein folding assistants and regulators of proteome homeostasis

The pathbreaking work by Christian Anfinsen on the folding process of small proteins in vitro paved the way for the investigation of protein folding in vivo (Anfinsen 1973). Anfinsen was the first biochemist to convincingly show that the information encoded by the amino acid sequence of a protein—the thermodynamics set up by the sequence—could be sufficient to guide folding toward the native conformation. Anfinsen demonstrated this key point by first denaturing (and thus inactivating) small enzymes with a chemical denaturant and then removing the denaturant and following the recovery of the enzyme's activity. However, Anfinsen warned that his Nobel Prize-winning work, which was limited to small proteins, didn't take into account the complexity of folding in the cell and might not pertain to complex, multisubunit proteins. In the crowded intracellular environment, for example, interactions among unfolded or partially folded proteins could lead to dead-end complexes: aggregates of nonnative proteins (Anfinsen 1973).

Folding into the native, functional conformation does not wait until the entire protein sequence has been translated on the ribosomal apparatus. Rather, it is during the process of translation that proteins begin to fold into their native conformations. While some proteins can fold into their native state without assistance from other proteins, the majority of proteins depend on molecular chaperones, a diverse suite of proteins that generally function as folding catalysts. Some chaperones are able to bind partially folded or denatured proteins and "hold" these proteins in their current, non-native shape. These so-called **holdase** chaperones do not catalyze refolding to the native state in stress-damaged proteins, but rather keep a bad situation (partial unfolding) from becoming worse (full denaturation). Molecular chaperones that catalyze refolding iteratively bind and release a nascent protein so it can fold along what is termed a productive folding pathway towards its native state. Below we review the basic characteristics of this chaperoning activity and then show how molecular chaperones assist cells in reversing the effects of thermal stress on protein structure.

BASIC ROLES OF MOLECULAR CHAPERONES   In our analysis of the thermodynamics of protein folding, we emphasized the critical role played by hydrophobic interactions in stabilizing protein structure. However, hydrophobic interactions also pose challenges during the folding of a newly synthesized protein or refolding of a protein whose structure has been perturbed by stress: Hydrophobic patches on *different* proteins may form interactions, leading to protein aggregation. These aggregates, if not broken up, prevent a protein from progressing along the productive folding pathway and prevent restoration of native conformations in stress-damaged proteins. As mentioned above, protein aggregates may themselves be cytotoxic, so preventing their formation (or facilitating their removal, as discussed later in this chapter) is a key element in proteostasis.

The catalysis of protein folding by molecular chaperones is necessary for the initial de novo folding of the majority of newly translated proteins; for the transfer of proteins into membrane-bound organelles, for example, the mitochondrion and the endoplasmic reticulum; and for the refolding of partially denatured proteins following cellular stress. The importance of chaperoning differs among proteins. Protein folding is especially challenging for large proteins. These contain multiple domains that expose numerous hydrophobic amino acid sequence motifs during folding and require assistance to stay on a productive folding pathway and not fall into a thermodynamic trap that leads to and maintains nonnative states (Morimoto et al. 2012; Kim et al. 2013). Furthermore, molecular chaperones play an important role during disaggregation of protein aggregates and the transfer of irreversibly denatured proteins toward protein degradation pathways (Tyedmers et al. 2010). Molecular chaperones thus play multiple roles in ensuring homeostasis of the proteome during normal protein biosynthesis and during exposure to and recovery from protein-damaging types of stress, notably thermal stress.

The chaperone model of the HSR

The HSR is a complex cellular process, but at its core is a coordinated set of responses that work to restore proteostasis. The central players in this response are molecular chaperones that mitigate the effects of thermally induced protein unfolding by binding and sequestering unfolded proteins and helping disaggregate unfolded protein aggregates (Figure 3.16). The induction of the HSR is incredibly fast (see below) and involves two substantial changes in protein synthesis: (1) The expression of molecular chaperones increases by orders of magnitude, and (2) the synthesis of most other classes of proteins

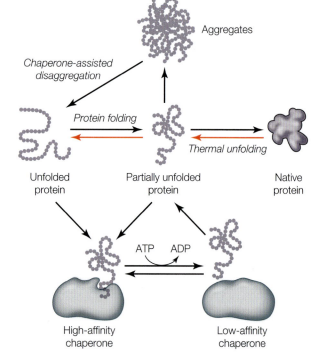

Figure 3.16 Chaperone model of the HSR. A sudden change in temperature disrupts native structures of proteins, leading to protein unfolding. Partially and fully unfolded proteins are prone to aggregation caused by the interaction of exposed hydrophobic side chains. Molecular chaperones exist in high-affinity and low-affinity conformations that bind and release unfolded proteins, respectively, to sequester unfolded proteins and help proteins proceed through productive folding pathways. Chaperones also act to disaggregate unfolded protein aggregates. Many molecular chaperone complexes, such as the Hsp70/Hsp40 complex, are dependent on ATP hydrolysis for release of bound proteins. (After Richter et al. 2010.)

ceases. This switch in protein synthesis is achieved by the fast and robust induction of transcription of heat-shock genes that encode HSPs and the preferential translation of heat-shock gene mRNAs by ribosomes. By increasing the concentration of molecular chaperones inside the cell, proteostasis can be regained by the mechanism of assisted refolding of denatured proteins and by the sequestration of partially and completely unfolded proteins to prevent protein aggregation. The concomitant decrease in the synthesis of most other proteins signifies an adaptation to minimize production of new proteins, whose partially unfolding structures would only exacerbate the unfolded protein load inside the cell. Preferential synthesis of HSPs also allows the ribosomal machinery and the ATP supply directed toward protein synthesis to be focused on the critical task of generating chaperones.

It is important to note that although the HSR is an effective mechanism for dealing with the problem of thermally induced protein unfolding, the induction of the HSR can impose significant physiological costs. Despite the speed with which the HSR is induced (on the order of minutes), the process of recovering proteostasis can take much longer (hours or days) following a significant heat stress event (Tomanek and Somero 2000). Moreover, the HSR is energetically expensive because heat-shock gene expression, HSP synthesis, and the function of many molecular chaperones require ATP (Box 3.9). Thus, as discussed later, several genes encoding proteins that govern rates of energy metabolism (ATP production) may also be induced during heat stress (Han et al. 2013). In the next section we examine the mechanisms by which the HSR is regulated to be induced when necessary but repressed when proteostasis is regained.

BOX 3.9

Metabolic costs of the HSR

Throughout the vast literature on the HSR there are countless references to the substantial metabolic costs associated with it. Such metabolic costs would pose serious constraints on the physiology and ecology of organisms that routinely face heat stress of sufficient magnitude to induce the HSR. Thus, it is pertinent to ask: What are the metabolic costs of inducing the HSR, in terms of ATP demand?

When one starts adding up all of the ATP that would be required to support the various elements of the HSR and related pathways, it quickly becomes clear that induction of the HSR would impose a high demand for ATP. First, there are the energetic requirements of initiating the transcription of heat-shock genes. Such a large increase in gene expression requires chromatin remodeling, release of paused polymerases, nucleosome remodeling, and nucleotide polymerization—all processes that require substantial amounts of ATP (Weake and Workman 2010). Second, the multiple steps of protein translation require energy, including amino acid activation

by tRNA synthetases, peptide bond formation, and proofreading. Third, HSPs and other protein products of the HSR require ATP for their function and/or regulation. For example, Hsp70s require ATP to release bound proteins (see Figure 3.16), and multiple steps of the ubiquitin-proteasome protein degradation pathway also require ATP. In addition, numerous components of the HSR are regulated by posttranslational modifications that require ATP, such as the hyperphosphorylation of HSF1 to initiate heat-shock gene transcription (see Figure 3.17).

What do these ATP-dependent processes amount to, in terms of the overall metabolic costs associated with the induction of the HSR? Despite decades of work in this field, very few studies have attempted to directly address this fundamental question. In fact, to our knowledge only *one study* has empirically measured the metabolic costs of the HSR directly. Hoekstra and Montooth (2013) quantified metabolic rates (MRs) during acute heat stress (1 h at 36°C) and recovery (22°C) in *D. melanogaster* larvae by measur-

BOX 3.9 *(continued)*

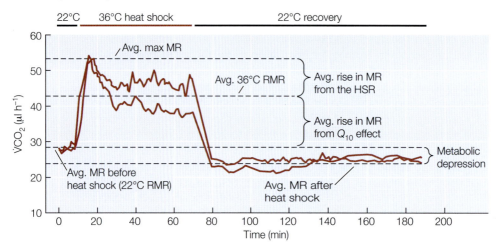

Figure A Representative metabolic rate (MR) tracings, plotted as volume of CO_2 (VCO_2) exhaled over 3 h, of *D. melanogaster* larvae before (10 min at 22°C), during (1 h at 36°C), and after heat shock (2 h at 22°C). MR before heat shock is the routine metabolic rate at 22°C (22°C RMR). Transient rise in MR during heat shock is the maximum MR (max MR). Sustained rise in MR during heat shock is the routine metabolic rate at 36°C (36°C RMR). (After Hoekstra and Montooth 2013.)

ing CO_2 production (VCO_2) via flow-through respirometry. **Figure A** shows two representative traces of VCO_2 of larvae exposed to acute heat stress. Note the transient increase in MR (max MR) followed by a more constant elevated MR (36°C routine metabolic rate [RMR]) during heat shock. Also note the drop in MR (metabolic depression) following heat stress.

To decouple the increase in MR due to Q_{10} effects from the metabolic demands that were specifically associated with the HSR, these authors used a fly strain that was genetically engineered to possess 12 extra copies of the *Hsp70* gene (total of 24 copies) in its genome (Welte et al. 1993), as well as another fly strain that was engineered to possess fewer copies (total of 6) of this gene (Gong and Golic 2004). As we discuss later in this chapter, the insertion of extra copies of *Hsp70* genes in this fly strain led to higher levels of both *Hsp70* mRNAs (Bettencourt et al. 2008; Hoekstra and Montooth 2013) and Hsp70 protein (Welte et al. 1993; Feder et al. 1996). As a control, Hoekstra and Montooth used a fly strain that possessed the same genetic background as the altered *Hsp70* copy number strains but that retained the wild-type state (12 copies of *Hsp70*). They then compared the transient rise in MR during heat shock (max MR – 36°C RMR), as well as the metabolic depression following heat shock (22°C RMR – MR after heat shock) among these three fly strains. A striking pattern emerged (**Figure B**). The number of copies

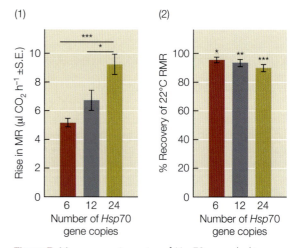

Figure B More genomic copies of *Hsp70* genes led to greater metabolic costs during and after heat stress (1 h at 36°C) in *D. melanogaster* larvae. Three genetically engineered fly strains have different copy numbers of *Hsp70* genes. Note that the copy numbers reflect the total diploid gene counts. (1) Rise in MR (max MR – 36°C RMR) among the three fly strains. Asterisks show statistical significance among strains (* = $P < 0.05$; *** = $P < 0.001$; ANOVA and Tukey post-hoc test). (2) Percent recovery of 22°C RMR (lower percentage means greater metabolic depression) among the three fly strains. Asterisks show that the difference in MR before and after heat shock is significantly different from zero (paired t-test; * = $P < 0.05$; ** = $P < 0.01$; *** = $P < 0.001$). Error bars represent standard error. (After Hoekstra and Montooth 2013.)

BOX 3.9 *(continued)* ▌■ ■ ■ ■

of the *Hsp70* gene—and thus the degree to which Hsp70 expression was induced—had a significant effect on the rise in MR in response to heat shock. In addition, more copies of the *Hsp70* gene resulted in greater metabolic depression (i.e., lower percentage recovery of the 22°C RMR) following heat shock.

These results are noteworthy for several reasons. First, they substantiate the claim that the HSR imposes significant metabolic costs to the organism. Previous studies showed indirect evidence of this effect by comparing the fitness consequences (i.e., survival and reproduction) of the HSR (Krebs and Loeschcke 1994; Krebs and Feder 1997). But the work of Hoekstra and Montooth (2013) is the only study to elucidate the physiological mechanism underlying these costs. Second, the results provide a mechanistic characterization of the major sources of energetic demand during heat stress. Thus it appears that the largest energetic cost of the

HSR is likely to be the immediate ATP demands of initiating heat-shock gene transcription, hence the greater transient rise in MR in the extra-copy strain. The greater metabolic depression in the extra-copy strain likely represents a response to recoup depleted energy stores. The ATP-dependent function of Hsp70s may incur longer-term costs that accumulate many hours or days after heat stress, because there was no evidence of a rise in MR during the 2-h recovery period of this experiment.

Future studies should use similar approaches to further characterize the metabolic costs of the HSR. For example, by measuring MR at later time points and/or in response to varying degrees of heat stress, we will be able to assess the energetics of long-term recovery from heat stress, as well as the effects of more severe heat stress events that induce pathways of proteolysis, cell cycle arrest, and apoptosis.

Regulation of the HSR: The cellular thermostat model

Figure 3.17 presents a simplified model of how the HSR is regulated in eukaryotes; prokaryotes have a related but simpler mechanism of HSR regulation. This model is called the **cellular thermostat model** and is characterized by multiple sensors that regulate the induction or repression of heat-shock gene transcription. Two molecular sensors of heat stress mediate the induction of the HSR: (1) thermally induced protein unfolding, and (2) thermally induced changes in the conformation, assembly state, and intracellular localization of the transcription factor **heat-shock factor 1** (**HSF1**). Thermally induced protein unfolding favors dissociation of a protein complex made up of HSF1 and several chaperones. Thus, as denatured proteins appear in the cell, the readily reversible equilibrium involving Hsp70, Hsp40, Hsp90, and HSF1 shifts in the direction of disassembly (see Figure 3.17). The "freed" chaperones then bind to denatured proteins and initiate refolding. The released HSF1 moves to the nucleus and assumes its role as a transcription factor. Because HSF1 is a leucine zipper protein, increases in temperature cause it to change conformation such that it trimerizes and attains the active conformation that binds to conserved motifs in the promoter regions of heat-shock genes, so-called **heat-shock elements** (**HSEs**). It is important to note that trimerized HSF1 must be hyperphosphorylated and sumoylated (modified by addition of SUMOs, small ubiquitin-like modifiers) before it can initiate gene transcription. The newly transcribed mRNAs for heat-shock proteins are rapidly and preferentially translated, leading to a large increase in the HSP pool of the cell.

The HSR is repressed via a simple negative feedback mechanism that involves resequestration of HSF1 into the multi-chaperone complex. The newly enlarged pool of HSPs binds unfolded proteins and catalyzes refolding. When the majority of the unfolded protein pool has been bound up by HSPs, excess HSPs bind to HSF1 and dissociate it from HSEs, thus halting the transcription of heat-shock genes. A return to nonstressful conditions (lower temperatures) also favors the dissociation of HSF1 trimers based on the thermodynamics of leucine zipper conformational and assembly state changes.

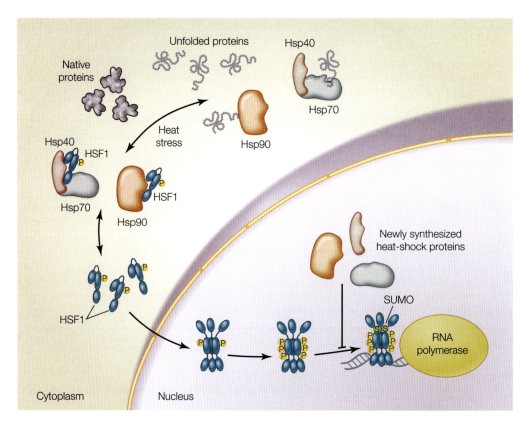

Figure 3.17 Cellular thermostat model of the HSR. In nonstressful conditions, heat-shock factor 1 (HSF1) monomers are bound to chaperone complexes comprising Hsp70, Hsp40, and Hsp90. The equilibrium between complex-bound and dissociated (free) chaperones is influenced by changes in temperature. Heat stress causes native proteins to unfold, and this favors a shift in the equilibrium toward dissociated HSPs, which can begin functioning as chaperones. HSF1, a leucine zipper protein, enters the nucleus, trimerizes, becomes hyperphosphorylated and sumoylated, and binds to HSEs in promoter regions of the chromosome. This activates heat-shock gene transcription via paused RNA polymerases. (After Richter et al. 2010.)

Depending on the organism and cell type in question, the details of the HSR may be more complex than what is presented in the chaperone model (see Figure 3.16) or cellular thermostat model (see Figure 3.17). Nonetheless, one of the most important lessons that we can glean from these models is that the HSR has evolved to be responsive and tightly regulated. Therefore, one would predict that evolution of the HSR has involved adaptations to multiple aspects of this cellular process in order to maintain the appropriate balance of induction and regulation.

THE HSR IS INCREDIBLY FAST AND ROBUST   Among the remarkable aspects of the HSR is the speed with which it can be induced. For example, changes in chromosomal puffing patterns of *Drosophila* larvae happen within *1 min* (Ashburner 1970), and HSP expression can be detected just *8 min* after heat stress (Mirault et al. 1978). That's less than 10 min to go from gene to protein! Thus, in eukaryotes the HSR is perhaps the fastest known gene expression response to an environmental stimulus. Bacteria show an even faster response to heat stress. For example, when cells of *Escherichia coli* are shifted from 30°C to 42°C, the induction of HSP expression happens almost immediately and peaks in less than 10 min (Lemaux et al. 1978; Yamamori et al. 1978; Yura et al. 1993).

There are many mechanisms that promote the speed of the HSR. First and foremost, the thermodynamics of HSF1 conformational states make this transcription factor highly sensitive to changes in temperature. The direct effect of temperature on HSF1 structure is an excellent illustration of the "good" effects of stress on macromolecules discussed in Chapter 1. Second, genome evolution has caused HSEs to be conveniently located in between nucleosomes so that they are readily accessible to HSF1. Third, at the chromosomal location of many heat-shock genes, such as *Hsp70*, RNA polymerases are bound to DNA in a paused state (so-called **paused polymerases**) and are ready to begin heat-shock gene transcription immediately upon binding of HSF1 to HSEs. For good reviews of these and other mechanisms of gene expression control in the context of the HSR, see the papers by Weake and Workman (2010) and de Nadal and colleagues (2011).

Another key aspect of the HSR is the robustness (amplitude) of the response in terms of the level of induced expression of heat-shock genes and the numbers of genes in the transcriptome that change in their expression. Transcriptomics studies in a diverse array of taxa, from yeast to marine invertebrates, have shown that as much as one-third of the transcriptome changes in expression in response to heat stress (Gasch et al. 2000). Moreover, the induction of the genes that encode heat-shock proteins is an extremely strong response. For example, in marine mussels of the genus *Mytilus*, the induction of a gene encoding Hsp70 (*Hsp70*) was upregulated more than 250-fold in response to acute heat stress (Lockwood et al. 2010). In *Drosophila* the change in chromosomal architecture shown in the figure in Box 3.8 corresponds to a marked increase in the expression of heat-shock genes. In *Drosophila* adults exposed to 36°C for 1 h, the gene encoding Hsp23 (*Hsp23*) was induced more than 90-fold (Brown et al. 2014). We note that there is not a 1:1 correlation between the increase in heat-shock mRNAs and the downstream translation of HSPs; however, increased levels of heat-shock mRNAs generally result in increased abundances of HSPs, albeit with different kinetics and different relative levels of induction (Buckley et al. 2006).

Adaptive benefits of the HSR

In the preceding sections we have gained an appreciation for the complexity and evolutionary origin of the HSR. Let's now go beyond the reductionist perspective and ask, How does the HSR provide an adaptive benefit at the whole-organism level? Several classic studies have used functional genetic techniques to manipulate the HSR in vivo, and these studies provide a valuable means by which we can evaluate the adaptive benefit of the HSR. In addition, experimental evolution (laboratory selection) experiments have been particularly informative in this realm. Finally, physiological studies that have compared the HSR among closely related species that inhabit distinct thermal niches have been instrumental in allowing us to evaluate the adaptive benefit of the HSR and to learn how evolution has fine-tuned the HSR's properties. We will now examine these various experimental approaches for exploring the evolution and physiology of the HSR in an adaptive context.

MOLECULAR GENETICS CONNECTS GENOTYPE TO PHENOTYPE   Here we address the question of the genotype-to-phenotype map: Does variation in the expression and/or function of heat-shock genes and their encoded proteins correspond to variation in whole-organism thermal tolerance? To answer this question requires the appropriate model organism in which to manipulate the expression of heat-shock genes and proteins in vivo. In a classic study, Welte and colleagues (1993) produced transgenic lines of *Drosophila melanogaster* in which 12 extra copies of the *Hsp70* gene (6 extra copies per homologous chromosome) were inserted into the genome. Because *D. melanogaster* naturally possesses 12 nearly identical *Hsp70* genes that are all heat-inducible, the insertion of 12 extra copies of one of these Hsp70-coding genes led to an accumulation of approximately twice the usual level of Hsp70 following heat shock, at double the rate of induction. It is important to note that this genetic manipulation did not change the basal levels of expression seen under nonstressful conditions. Moreover, the genetically engineered increase in the capacity of the HSR, in terms of inducible Hsp70 protein expression, led to a marked increase in whole-organism acute heat tolerance in embryos of the extra-copy strain compared with the control strain (Welte et al. 1993). This effect was later confirmed by a study by Feder and colleagues (1996) in which the extra-copy transgenic strain was assessed for thermal tolerance in later developmental stages (larvae and pupae) and in a more ecologically relevant heat stress scenario. Similar to the earlier embryonic stages, larvae and pupae of the extra-copy strain produced twice as much, or more, of the Hsp70 protein following sublethal exposure to heat stress (36°C for 1 h; **Figure 3.18A,B**). Importantly, this additional expression of Hsp70 led to a marked increase in heat tolerance at the whole-organism level (**Figure 3.18C**).

Let's consider in more detail the striking pattern revealed by these data and their broader implications. Not only did increasing the gene copy number elevate the levels of inducible Hsp70 expression, but also there was a tight correspondence between the level of induced Hsp70 and whole-organism heat tolerance. As can be seen in Figure 3.18B and C, at the time point that corresponds to the biggest difference between the two strains and highest levels of Hsp70 induction (1 h after heat shock), the extra-copy strain had 150% of the levels of Hsp70 that were induced in the control strain. This marked increase in Hsp70 levels corresponds to an approximately 133% increase in LT_{50} in the extra-copy strain versus the control strain. Effectively, these data demonstrate a direct link between Hsp70 expression and whole-organism thermal tolerance.

These studies (Welte et al. 1993; Feder et al. 1996) are two of the best empirical demonstrations of the phenotypic benefits of the HSR in the context of acute heat stress because they clearly demonstrate the genotype-to-phenotype connection of the HSR to whole-organism heat tolerance. In other words, in a controlled genetic background and environmental context, simply increasing the inducible expression of a single protein that is a major constituent of the HSR (i.e., Hsp70) substantially increases the inducible heat tolerance at the whole-organism level. Box 3.9 examines this same experimental system in the context of metabolic costs associated with the HSR. Next we consider the adaptive benefit and potential of the HSR in more realistic evolutionary and ecological contexts. Despite the well-established connection between HSP expression and heat tolerance, these more realistic experimental contexts complicate the picture of adaptation of the HSR, such that in many cases induction of the HSR must be viewed in the context of the particular environmental challenges that are posed to a given population or species.

EXPERIMENTAL EVOLUTION IN HIGH-TEMPERATURE ENVIRONMENTS Experimental evolution studies, in which model organisms are evolved over many generations in the laboratory in conditions of high temperatures, have been particularly informative for discerning the adaptive benefit of the HSR. In general, these studies have attempted to address the question, How do populations respond at the molecular level to selection for increased heat tolerance? Here we highlight four such studies, two in bacteria and two in fruit flies. We focus on these model organisms because they have historically

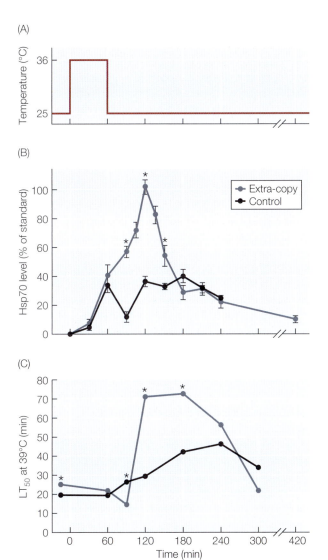

(A)

(B)

(C)

Figure 3.18 Extra copies of *Hsp70* (12 additional copies of the *Hsp70* gene compared with the wild type) led to higher induced levels of Hsp70 protein and higher heat tolerance in third-instar larvae of transgenic *Drosophila melanogaster*. (A) Experimental time-course of heat shock pretreatment. (B) Hsp70 expression over time measured in whole larvae using an ELISA assay. Hsp70 levels are expressed as a percentage of a positive loading control obtained from heat-shocked *D. melanogaster* tissue culture cells. * = $P < 0.05$ (Mann-Whitney U test comparing Hsp70 levels of the extra-copy strain versus the control strain). Error bars indicate standard error. (C) Heat tolerance measured as the time of exposure to a higher temperature (39°C) that induced 50% mortality (LT_{50}). Thus, higher LT_{50} values correspond to longer survival times at this acutely lethal temperature. * = $P < 0.05$ (log rank test comparing the survival times of the extra-copy strain versus the control strain). (After Feder et al. 1996.)

been the most experimentally tractable not only for studies focused on genetics and gene function but also for studies in experimental evolution. This is because these organisms can be maintained in the laboratory indefinitely in large numbers and have relatively short generation times; *E. coli* has an approximately 20-min doubling time at its preferred temperature of 37°C, and *D. melanogaster* has a 10-day generation time at the common rearing temperature of 25°C. As a consequence of these characteristics, these organisms can be used for experiments that create evolutionary microcosms that are designed to recreate the evolutionary dynamics that exist in the wild, but under controlled laboratory conditions of the experimenter's choosing. In the context of temperature adaptation, this allows the researcher to run an artificial selection experiment in which the individuals in a population that have the greatest thermal tolerance are propagated over many generations. Depending on the genetic makeup of the starting population in such an experiment—that is, whether the population is outbred or inbred—the response to selection regimes depends on the level of standing genetic variation and/or new mutations.

The outcomes of experimental evolution to conditions of changing temperature are, perhaps not surprisingly, dependent on the particular conditions under which the selective environment is manipulated. This dependence is interesting in the context of the HSR because it illustrates the fact that despite the high conservation of the HSR among most species, the details of how the HSR is induced and/or regulated can change depending on evolutionary history and ecological context. Rudolph and colleagues (2010) explored the evolution of heat tolerance in *E. coli* under conditions of rapidly increasing environmental temperatures. In this study, *E. coli* strains were selected for viability at increasingly higher temperatures through evolutionary time, beginning with the standard growth temperature of 37°C. These authors were able to select for strains that were viable at temperatures up to 48.5°C, 11.5 degrees Celsius higher than the starting temperature. This substantial evolutionary response to selection for increased heat tolerance was achieved over 620 generations—equivalent to ~10,000 years in human generation time. This evolution of increased heat tolerance was underpinned by striking changes in the HSR. In particular, the strains that evolved to survive at 48.5°C acquired through evolutionary time a permanently induced (constitutively expressed) HSR, as determined by two-dimensional gel electrophoresis; the proteins of greatest change in basal (steady-state) levels in the high-temperature strains were HSPs. Underlying these changes in protein expression were gene duplication events that led to increased copy numbers of heat-shock genes. Furthermore, no mutations were found in either the coding regions or the promoters of heat-shock genes, which indicates the important role of gene duplication (Rudolph et al. 2010). This is a similar mechanism of increasing expression of heat-shock genes as was demonstrated in the genetically engineered flies in the study by Welte and colleagues (1993) described above. However, there is a key difference here: The increased HSR in the *E. coli* experiment was induced *basally* under nonacute heat stress conditions, as opposed to being strictly inducible as in the case of *D. melanogaster*. Another difference was that *acute* heat stress led to similar relative increases in levels of HSPs in both high-temperature-evolved *E. coli* strains and control strains (Rudolph et al. 2010), whereas acute stress led to a much higher relative increase in Hsp70 in the extra-copy strain of *D. melanogaster* than in the control strain. So here we begin to see how the outcomes of temperature adaptation are likely to depend on the evolutionary context. In this study, bacterial strains evolved a marked increase in heat tolerance that was made possible only by the evolution of gene transcription to a state of permanent HSR induction. We will see evidence for other instances of this "preparative defense" strategy in other species below.

Another group of researchers conducted a laboratory evolution experiment on high temperature in *E. coli* and produced a different outcome in terms of the HSR (Riehle et al. 2003). In this study, *E. coli* strains were propagated at 41.5°C for 2,000 generations, which led to the evolution of increased survival after an acute heat shock of 50°C for 4 h. This study used DNA microarrays to measure changes in gene expression in the high-temperature strains versus the parental strains. Riehle and colleagues discovered that underlying this temperature adaptation of improved heat tolerance were evolved changes in the regulation of key heat-inducible genes. But in contrast to the study by Rudolph and colleagues (2010) discussed above, molecular chaperones showed very little change in transcriptional regulation. Instead, the majority of the changes in heat-inducible gene expression were among genes encoding proteins that localize to the periplasm—the space between the inner cytoplasmic membrane and the outer membrane of the bacterium—suggesting that periplasmic functions, like transport, play key roles in long-term evolution to higher temperatures in *E. coli*.

Why did these two studies of experimental temperature adaptation in *E. coli* produce such distinct evolutionary responses at the molecular level, despite both producing increases in heat tolerance at the whole-organism level? The differences reported in the molecular outcomes of these evolution experiments are perhaps due to the differences in the experimental designs of the two studies rather than in the genetic characteristics of the parental (starting) populations of bacteria. In the study by Rudolph and colleagues (2010), comparatively drastic increases in temperature were applied in several increments that occurred over short evolutionary timescales, whereas in the study by Riehle and colleagues (2003) there was only a single increase in temperature that was imposed on the evolving populations of cells, followed by long evolutionary adaptation under static conditions. It seems plausible, and perhaps likely, that in the former study (Rudolph et al. 2010) the populations that survived the abrupt and extreme shifts in temperature required more of the classic elements of the HSR to endure such acute thermal stress, whereas the more subtle and long-term adaptation that took place in the latter experiment (Riehle et al. 2003) likely involved more of an acclimatory response that was sustained over evolutionary time via adaptations that did not involve the HSR. This distinction is a good example of a general phenomenon—the HSR is a short-term physiological response that allows an organism to cope with sudden changes in temperature. The fact that the *E. coli* strains in the study by Rudolph and colleagues (2010) remained adapted at a permanently HSR-induced state is an interesting outcome that suggests that the HSR is central to these strains' abilities to remain viable at the extreme temperature of 48.5°C. One would predict that such a high temperature would cause significant thermal unfolding or denaturation of proteins, and in the absence of temperature adaptation of thermal stability across the proteomes of these cells, molecular chaperones would be necessary to help mitigate thermal protein unfolding at 48.5°C. Given enough evolutionary time, these strains may have been able to accumulate mutations that increased proteomic thermal stability, but this would have required protein-stabilizing mutations in a vast number of genes—likely too many adaptive mutations to be acquired in only 600 generations. Another missing piece of the puzzle involves the synthesis of small organic molecules that are strong protein stabilizers (see Chapter 4). Many bacteria increase the intracellular concentrations of these stabilizing (thermoprotectant) solutes under heat stress. Constitutively high levels of these stabilizing molecules could be favored by selection. To our knowledge, this possible mechanism of adaptation to elevated temperatures has not been examined in laboratory evolution studies.

Another set of classic studies in experimental evolution of temperature adaptation in *D. melanogaster* provides a good basis for considering the HSR in the broader context of how the HSR evolves in natural populations. A long-term experimental evolution experiment was conducted by Cavicchi and colleagues (1989) in which replicate populations of *D. melanogaster* were propagated from the same parental stock in three different constant temperature environments (18°C, 25°C, and 28°C) for more than 20 years. Subsequently, Bettencourt and colleagues (1999) assessed the degree to which these populations diverged in the HSR, namely in the expression of Hsp70 protein in response to acute heat stress. Surprisingly, the populations propagated at 28°C evolved to express significantly lower induced levels of Hsp70 than any of the populations evolved at either 18°C or 25°C. This pattern matched the acute heat tolerances of these populations, with

Figure 3.19 Experimental evolution for more than 20 years at constant high temperature results in reduced Hsp70 protein expression and reduced acute heat tolerance in larvae of *Drosophila melanogaster*. Hsp70 expression was induced by sublethal exposure to 36°C for 1 h and measured via ELISA. Hsp70 expression is plotted as a percentage of a loading standard. Acute heat tolerance was measured as the percentage (normalized to 25°C control animals) of larvae surviving to adulthood after a brief (1-hour) exposure to 39°C followed by recovery and development at 25°C. "A" and "B" represent replicate populations evolved at each indicated temperature. Error bars indicate standard error. (After Bettencourt et al. 1999.)

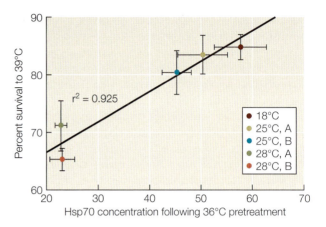

the 28°C populations evolving lower whole-organism acute heat tolerances than the populations evolved at 18°C or 25°C (**Figure 3.19**). This seeming paradox of evolution of lower acute heat tolerance, accompanied by lower levels of induction of Hsp70—despite evolution at higher temperatures—may be explained by the fact that all of these populations evolved in constant temperatures and not in acutely variable environments. In other words, even though 28°C is hotter than 18°C or 25°C, there is little evidence that 28°C is acutely stressful for *D. melanogaster*. In fact, the HSR has been shown to not be induced following short-term exposures at 28°C in *D. melanogaster* (Ashburner 1970). Therefore, the presence of highly inducible levels of HSPs may not be required for survival at 28°C and may even be deleterious. Interestingly, increased Hsp70 expression has been shown to lead to lower survival in constant environments in *D. melanogaster* (**Figure 3.20**) (Krebs and Feder 1997), and it is plausible that evolution at 28°C presented a significant evolutionary trade-off wherein individuals with higher Hsp70 expression had lower survival in this constant high-temperature environment than individuals with lower Hsp70 expression. This trade-off appears to not have been present, or at least wasn't a strong enough factor, to lead to reductions in Hsp70 expression after evolution at 18°C or 25°C. Box 3.9 discusses the potential mechanisms for the deleterious effects of increased induction of the HSR, but suffice it to say that it appears that increased expression of molecular chaperones like Hsp70 is adaptive in acutely stressful thermal environments, but not necessarily in more benign conditions.

CONNECTING EVOLUTION IN THE LAB WITH NATURAL VARIATION   This classic work in experimental evolution provides a framework for connecting what we have learned about the molecular biology and physiology of the HSR to evolutionary contexts. Moreover, in the case of *D. melanogaster* we can go one step further to connect these patterns observed in the laboratory to an ecological setting among natural populations. Bettencourt and colleagues (2002) characterized the alleles that were responsible for variation in Hsp70 expression following experimental evolution to constant environments of 18°C, 25°C, and 28°C. They discovered fixed differences among these populations in the presence of an insertion/deletion (indel) in the noncoding region of the genome at a cluster of *Hsp70* genes. A transposable element insertion at this locus was found to be fixed in the 28°C populations but absent in the 18°C and 25°C populations, and thus the presence of the insertion corresponded to the lower levels of Hsp70 induction and lower heat tolerance of the 28°C populations. The finding that experimental evolution led to divergence at a regulatory locus, and not in coding regions, is consistent with (1)

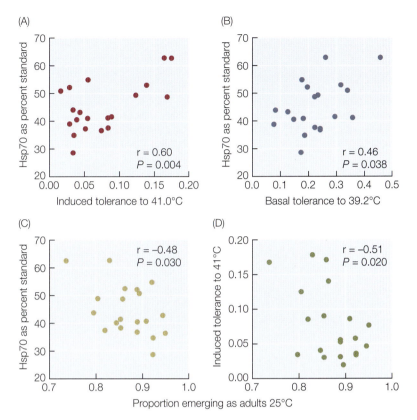

Figure 3.20 Hsp70 protein expression was positively correlated with acute heat tolerance but negatively correlated with survival in benign conditions (constant 25°C) in *D. melanogaster* larvae. (A) Relationship of induced heat tolerance to induced Hsp70 protein expression. First-instar larvae were exposed to 36°C for 1 h, allowed to recover at 25°C for 1 h, and then exposed to 41°C for 1 h; specimens were then scored for survival to adulthood (cultured at 25°C) or processed for measurement of Hsp70 expression via ELISA. Hsp70 levels are expressed as the percentage of a purified Hsp70 standard. (B) Relationship of basal heat tolerance to Hsp70 protein expression. First-instar larvae were exposed to 39.2°C for 1 h and scored for survival to adulthood (cultured at 25°C) or processed for measurement of Hsp70, as in part (A). (C) Relationship of viability (proportion of larvae emerging as adults) at 25°C to Hsp70 protein expression, as measured in parts (A) and (B). (D) Relationship of viability at 25°C, as in part (C), to induced heat tolerance, as measured in part (A). Correlation coefficients (r) and significance values (P) are indicated in each plot. (After Krebs and Feder 1997.)

evolutionary relationships among eukaryotic *Hsp70* genes (Krenek et al. 2013), whose copy numbers have waxed and waned throughout evolutionary time but which have shown little change in coding sequences, and (2) temperature adaptive differences that have evolved in regulatory regions that determine the inducibility of these genes through modifications of HSE promoter sequences (Tian et al. 2010; Nguyen et al. 2016).

Bettencourt and colleagues (2002) took their lab findings out into nature and sampled 10 populations of *D. melanogaster* from the wild, along the broad latitudinal range of this species on the coast of eastern Australia. They found natural allelic variation in the same indel that was differentially fixed in the lab-evolved strains. Interestingly, the frequency of the indel exhibited a clinal pattern in which the allele carrying the insertion was segregating at a higher frequency in high-latitude populations (**Figure 3.21**). Recall that the presence of this insertion in the lab-evolved 28°C populations corresponded to

Figure 3.21 Clinal variation in the frequency of alleles associated with Hsp70 induction among 10 natural populations of *Drosophila melanogaster* in eastern Australia. Pie charts show the relative frequency of a transposable element insertion at a cluster of *Hsp70* genes that is associated with lower Hsp70 induction (blue) versus the frequency of alleles that lack this insertion (red). Also reported here is the frequency of another allele (orange) that was amplified in an initial screen but later deemed by the authors to be a PCR artifact. (After Bettencourt et al. 2002.)

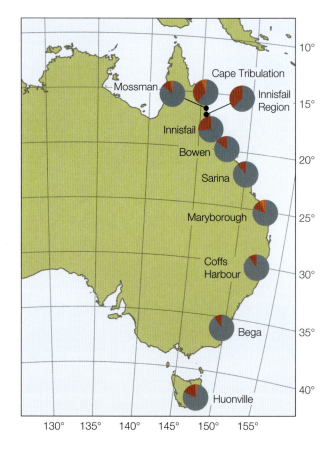

lower levels of Hsp70 induction and lower heat tolerances. Thus, the frequency of the indel alleles in the natural populations matched the thermal environment, suggesting that thermal selection maintained the clinal variation in allele frequencies. There is a strong thermal gradient across these latitudes in eastern Australia, mostly correlated with the mean daily maximum temperature during the hottest month of the year. It is important to note, however, that this insertion polymorphism was present in all 10 populations, despite exhibiting clinal variation—the frequency of alleles lacking this insertion increased from 17% in high-latitude populations to 34% in low-latitude populations. This is perhaps a reflection of the complex thermal environment in the natural setting of eastern Australia—that is, there are still hot days at higher latitudes, even though they are less frequent than at low latitudes, and these hot days likely select for individuals that lack the insertion. In addition, the insertion was found at relatively high frequencies in all 10 natural populations, perhaps a reflection of the costs associated with the induction of the HSR under more benign conditions. More recent work has characterized latitudinal variation in heat tolerance at the whole-organism level among *D. melanogaster* populations in eastern Australia; these studies have shown this temperature adaptive story to be even more complex than previously thought. In many cases, there is only weak evidence for clinal variation in whole-organism heat tolerance among these populations (Hoffmann 2010; Buckley and Huey 2016), a pattern that is probably driven by the complexities of the thermal environments that flies experience across this thermal gradient, and the fruit fly's ability to behaviorally thermally regulate (i.e., choose to avoid thermal extremes; Dillon et al. 2009).

Thus, complexities of the thermal environment and behavioral thermal regulation make it difficult to predict how thermal selection will lead to the evolution of the HSR and acute heat tolerance. This is particularly true in terrestrial environments, including intertidal regions during low tide, where ambient temperature can exhibit microenvironmental variability that far exceeds that which is characteristic of aquatic environments (Dowd et al. 2015). Nonetheless, the body of work discussed in the preceding paragraphs on thermal physiology and ecology of *D. melanogaster* has established clear connections between the molecular physiology of the HSR, whole-organism heat tolerance, evolutionary thermal history, and thermal ecology.

Next we consider several examples of variation among populations of a species, or divergence among closely related species, in the HSR and how these differences in molecular physiology correlate with thermal evolutionary history and large-scale patterns of biogeography. We now move our focus from the terrestrial to the marine realm, focusing mostly on various species of intertidal molluscs.

Adaptations to thermal environments in the intertidal zone

As we pointed out in Chapter 1, the intertidal zone has been one of the most powerful study systems for assessing the evolutionary and ecological links between the abiotic environment and organismal physiology (Denny 2016). This habitat, where many organisms face alternating exposure to seawater and air during the tidal cycle, is extremely dynamic and has high spatial structure for a variety of environmental factors, including temperature. Because of the dynamic nature of the intertidal, the body temperature of organisms, especially sessile invertebrates and algae, can change quickly—within minutes—and dramatically: Temperature shifts during the day can exceed ±20 degrees Celsius. Therefore, the HSR is a crucial physiological response in this environment, as it enables many organisms to survive sudden changes in temperature by conferring the ability to cope with the protein damage that results from this environmental perturbation.

Let's consider the intertidal zone's high degree of microspatial variation in temperature—on the order of several degrees Celsius over spatial ranges of less than 1 m—and how this spatial structure leads to structuring of the intertidal species community. Here we define the low, mid, and high tidal zones (also referred to as lower littoral, lower mid-littoral, and upper mid-littoral, respectively) generally based on tidal height. Low tidal zones are mostly submerged and only exposed during low tides. Mid tidal zones are submerged and exposed to air for approximately equal times. High tidal zones are exposed to air most of the time and only submerged during high tides. Perhaps not surprisingly, organisms that live higher in the intertidal tend to experience the greatest range of temperatures and more commonly are exposed to temperature extremes (Tomanek and Somero 1999). It is important to note that body temperature also depends on the orientation of surface substratum, solar irradiance, wave exposure, and wind speed (Denny and Harley 2006; Dowd et al. 2013). Nevertheless, for many species comparisons, tidal height closely matches the severity of the thermal environment because higher intertidal organisms are frequently exposed to aerial conditions, whereas the low intertidal is most often submerged. This condition leads to a steep thermal gradient across the intertidal zone, and accordingly, the composition of assemblages of intertidal species depends on tidal height—driven by the differential ability of different species to tolerate the conditions of the mid to high intertidal.

Intertidal ecologists have long appreciated the specific tidal heights at which species live as a key determinant of interspecies interactions in the perpetual struggle for living space on the rocks of the intertidal zone. But more recently, ecological physiologists have examined this relationship from the perspective of wanting to understand the physiological factors that influence where species live, with the hypothesis that perhaps interspecific differences in tolerances to abiotic stressors like temperature drive microspatial biogeographic patterns across the thermal gradient of the intertidal. One of the first studies to examine such a relationship from the perspective of the physiological consequences of heat stress was done with congeners of turban snails (genus *Chlorostoma*, formerly *Tegula*) (Tomanek and Somero 1999). This study examined the degree to which variation in the HSR could explain observed differences in preferred tidal heights among turban snail congeners.

As shown in **Figure 3.22A**, the three congeneric species of turban snails that live on the coast of central California have distinct vertical distributions in the intertidal zone: *Chlorostoma funebralis* lives in the mid intertidal, *C. brunnea* lives in the low intertidal, and *C. montereyi* has a mostly subtidal distribution. A fourth species that does not

Figure 3.22 Turban snail species (genus *Chlorostoma*) on the coast of central California live at disparate tidal heights and experience distinct thermal environments. (A) Cross section of the intertidal zone and the distinct vertical distributions of *C. funebralis* (mid intertidal), *C. brunnea* (low intertidal), and *C. montereyi* (subtidal). (B) Upper panel shows fluctuations in the tidal height; day and night are indicated by white and black bars, respectively. Lower panel shows body temperatures of snail-mimic temperature loggers, over the same tidal cycle, placed in the disparate habitats of *C. funebralis* versus *C. brunnea*. Seawater temperature is also shown, to indicate the expected body temperature of the subtidal *C. montereyi*. (C) The warm-adapted mid-intertidal *C. funebralis* has higher heat tolerance than the low-intertidal *C. brunnea* and subtidal *C. montereyi*. Snails were submerged and exposed to a heat ramp of +5°C h⁻¹. Heat death was assayed by examining the retraction response of the foot organ. (After Tomanek and Somero 1999.)

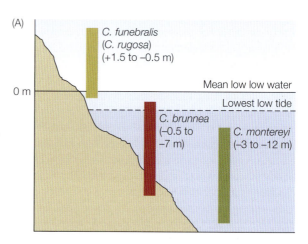

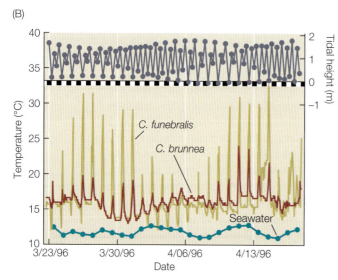

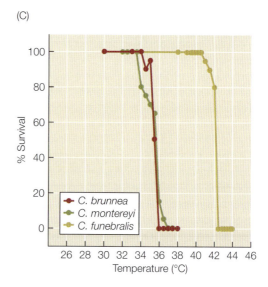

co-occur with the other three congeners, *C. rugosa*, has a subtropical range in the intertidal coastal waters of the Gulf of California, Mexico, and has a vertical distribution in the mid to high intertidal similar to that of *C. funebralis*. Perhaps not surprisingly, the thermal conditions to which these species are exposed correspond to the disparate tidal heights they inhabit (**Figure 3.22B**). In central California, the species that sees the greatest variation in its thermal environment is *C. funebralis*, which often experiences brief episodes of temperatures that exceed 30°C when low tides coincide with exposure to the sun. *C. brunnea* and *C. montereyi* experience thermal environments that are much less extreme, with *C. montereyi* experiencing the most stable thermal environment because it is submerged most of the time. Corresponding to the disparate thermal environments in which these three congeners live, these species exhibited differential survival after acute heat stress, with the most heat-tolerant species being *C. funebralis* (**Figure 3.22C**).

On further investigation of the molecular responses to acute heat stress in these three temperate-zone congeners, as well as in the subtropical congener *C. rugosa*, it was discovered that they exhibited divergent patterns in the HSR that reflected their adaptation temperatures (Tomanek and Somero 1999, 2000). In all species, heat stress caused protein expression—as measured by incorporation of ^{35}S-labeled amino acids (cysteine and methionine) into newly synthesized proteins—to shift in a canonical way that indicated the induction of the HSR. Synthesis of several size classes of HSPs increased with rising temperature, whereas synthesis of other types of proteins decreased and eventually ceased (**Figure 3.23**). However, distinct differences were found among species in the temperatures at which different aspects of the HSR were manifested. The temperature at which a significant increase in HSP levels was first observed (T_{on}) correlated with adaptation temperature, as did the temperature at which maximum HSP synthesis

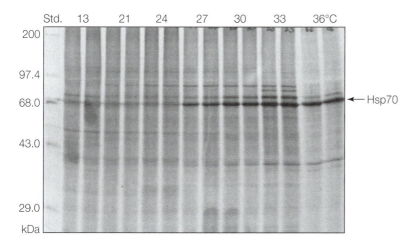

Figure 3.23 Effects of exposure temperature on protein synthesis patterns of isolated gill tissue from specimens of 13°C-acclimated *C. funebralis*. The images are autoradiographs of proteins labeled by ^{35}S during their synthesis at the indicated temperature (2.5 h incubation). Radiolabeled molecular mass standards (Std.) are shown in the left lane. Two specimens from each exposure temperature are illustrated. Temperatures above 24°C induce synthesis of HSPs in the molecular mass ranges of 38, 70, 77, and 90 kDa. With increasing exposure temperature, HSPs become an increasingly large fraction of the protein synthesized. (After Tomanek and Somero 1999.)

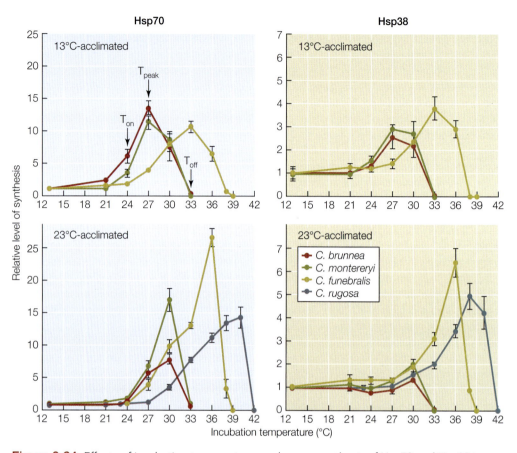

Figure 3.24 Effects of incubation temperature on de novo synthesis of Hsp70 and Hsp38 in isolated gill tissue of four congeners of *Chlorostoma*. Specimens were acclimated to either 13°C or 23°C for 1 month prior to the experiment. The amounts of newly synthesized proteins during 2.5 h at the indicated temperature were estimated from densitometric analysis of autoradiographs (see Figure 3.23); protein expression during 2.5 h at 13°C was used as a basis for normalization. Error bars indicate standard error. (After Tomanek and Somero 1999.)

occurred (T_{max}) and the temperature at which HSP synthesis ceased (T_{off}) (**Figure 3.24**) (Tomanek and Somero 1999). Acclimation led to some changes in T_{on} and T_{max}, but T_{off} showed no acclimation-induced change.

One important question that can be addressed by these data concerns the likelihood that HSP synthesis is induced in the species' different habitats. *C. brunnea* and *C. montereyi* would rarely, if ever, see temperatures above ~18°C in their habitats (see Figure 3.22B). Thus, it is unlikely that the HSR would be triggered by thermal stress in these species, at least as long as they did not stray into higher reaches of the intertidal zone. In contrast, *C. funebralis*, which has body temperatures that rise to at least 32°C, is likely to induce the HSR with some frequency. For Hsp70, the most strongly expressed HSP, induction of synthesis occurs above 24°C for specimens acclimated to 13°C and 23°C (see Figures 3.23 and 3.24). This interspecific difference in the frequency with which the HSR is likely to be induced in nature suggests that energy costs for proteostasis are higher in *C. funebralis* than in the other two temperate congeners.

The time-course of HSP synthesis also differed among congeners. When gill samples from *C. brunnea* and *C. funebralis* were exposed to 30°C for 2.5 h, *C. funebralis* induced the HSR almost immediately, whereas *C. brunnea* showed a slower response and took several hours to accumulate HSPs to significant levels (**Figure 3.25**) (Tomanek and Somero 2000). In addition, HSP synthesis in *C. funebralis* returned to basal levels more quickly following recovery than in *C. brunnea*, for example for Hsp70 and Hsp77. For *C. funebralis*, the HSR would be completed—and thus proteostasis would be restored—before the following daytime low tide occurred; this would not be the case for *C. brunnea*, were it

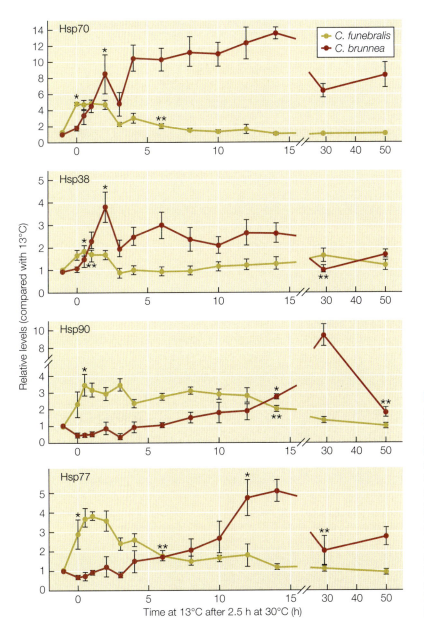

Figure 3.25 Patterns of HSP expression in response to acute heat stress vary between *C. brunnea* and *C. funebralis*, as measured by ^{35}S-labeled protein synthesis in excised gill tissue followed by gel separation via SDS-PAGE. Excised gill tissues were submerged and exposed to 2.5 h of heat stress at 30°C followed by recovery at 13°C for the time indicated on the x axis. * = time when synthesis has increased significantly above basal levels (13°C specimens); ** = time at which synthesis returns to basal levels. Error bars indicate standard error. (After Tomanek and Somero 2000.)

to face rare exposures to extremely high temperatures. Thus, the more rapid kinetics of the HSR in the more warm-adapted species can be viewed as an adaptively important element in the species' ability to cope with heat stress associated with the tidal cycle: Restoration of proteostasis is rapid enough to prepare the animal for coping with a second consecutive day of high temperatures.

Taken together, these studies indicate that evolved differences in whole-organism thermal tolerances of these species appear to be supported by differences in the regulation of the HSR. Moreover, the different intrinsic structural stabilities of orthologous proteins of the congeners (see Figure 3.5) suggest that the differences in the HSR are, at least in part, the consequence of proteomes whose thermal stabilities differ. For example, the higher T_{on} of *C. funebralis* may be due to thermal stability differences across the proteome, as indicated by the higher thermal stability of cMDH of *C. funebralis* compared with orthologs of *C. brunnea* and *C. montereyi*. In the context of biogeography, these molecular-level differences are likely to influence the distinct vertical distribution ranges of these species in the intertidal and the latitudinal distribution patterns as well, as shown by the extreme heat tolerance of the congener from the Gulf of California, *C. rugosa*.

Studies of limpets in the genus *Lottia* have revealed further aspects of inter-congener variation in the HSR that likely are instrumental in influencing upper thermal limits and vertical distribution patterns. *Lottia* limpets are gastropods common in the central California intertidal zone and, like *Chlorostoma*, comprise multiple species that inhabit distinct vertical distributions and microhabitats (**Figure 3.26**). *Lottia scabra* and *L. austrodigitalis* are both found in the mid to high intertidal, but they inhabit distinct microenvironments. *L. scabra* typically occupies horizontal rock surfaces that are fully exposed to the sun, whereas *L. austrodigitalis* occupies vertical or overhanging surfaces that are often protected by shade. Thus, we would expect *L. scabra* to experience a higher frequency of heat stress than *L. austrodigitalis*, despite the fact that they inhabit the same vertical distribution ranges. Two other congeners, *L. pelta* and *L. scutum*, have lower vertical distribution ranges in the mid intertidal, and the distribution of *L. scutum* extends lower in the intertidal than that of *L. pelta*.

The distinct thermal environments of these four congeners are associated with temperature adaptive differences in the HSR. Acute heat stress induced Hsp70 expression

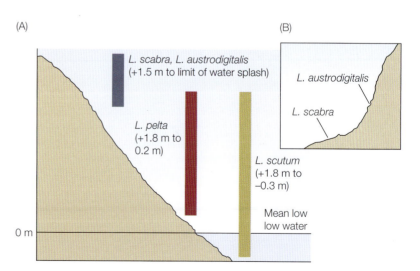

Figure 3.26 Vertical and microhabitat distributions of *Lottia* species differ. (A) Vertical distributions of *Lottia scabra*, *L. austrodigitalis*, *L. pelta*, and *L. scutum* (see text for details). (B) *L. scabra* and *L. austrodigitalis* differ in exposure to the sun. (After Dong et al. 2008.)

(A)

L. scabra, *L. austrodigitalis*
(+1.5 m to limit of water splash)

L. pelta
(+1.8 m to
0.2 m)

L. scutum
(+1.8 m to
−0.3 m)

Mean low
low water

0 m

(B)

L. austrodigitalis

L. scabra

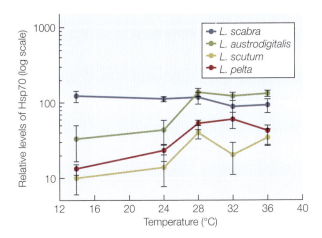

Figure 3.27 Hsp70 expression in response to acute heat stress differs among *Lottia* species that inhabit distinct thermal environments. Limpets were exposed to an aerial acute heat ramp of +8°C h⁻¹ up to the temperatures indicated on the x axis, followed by a recovery ramp of –8°C h⁻¹ down to 14°C. Hsp70 levels are expressed relative to a purified standard and were assayed via SDS-PAGE followed by Western blotting with an anti-Hsp70 antibody. Note the log scale of the y axis. Error bars indicate standard error. (After Dong et al. 2008.)

in three of the four species (**Figure 3.27**); however, the highest induction response was exhibited by the high-intertidal *L. austrodigitalis*, with *L. pelta* and *L. scutum* showing more muted induction responses (Dong et al. 2008). These muted responses suggest that the more cool-adapted *Lottia* congeners may exhibit a lag in the HSR similar to that observed in the cool-adapted *C. brunnea* (see Figure 3.25) (Tomanek and Somero 2000). In contrast, the species that lives in the hottest microenvironment, *L. scabra*, showed a unique pattern: Acute heat stress did not lead to changes in Hsp70 abundance in this species. Rather, *L. scabra* maintained extremely high levels of Hsp70 across all exposure temperatures. This molecular physiological pattern indicates a **preparative defense strategy** that *L. scabra* has evolved to cope with the extreme thermal environment of the exposed high intertidal in which it lives. Indeed, such an environment would lead to frequent bouts of heat stress that cause recurring disruptions to proteostasis. By having a high constitutive level of expression of Hsp70—as seen in some of the laboratory evolution studies of *E. coli* discussed earlier—*L. scabra* is prepared to cope with thermal stress as soon as it occurs, rather than face a delayed response as the HSR is activated. Even though the HSR is one of the fastest known responses to environmental stress, having the components of the HSR "on hand" for coping with heat stress appears to be of selective advantage. Last, as a caveat, it is worth noting that when a species fails to increase levels of HSPs in response to heat stress, the experimenter should not, in fact, view this as a "failure." Organisms that employ the preparative defense strategy are, indeed, well prepared for heat stress, even though they do not exhibit elevated production of HSPs beyond already highly expressed levels.

Adaptations beyond the HSR: Graded responses to increasing heat stress

Up to now, we have focused our analysis of cellular responses to *acute* heat stress on the HSR, which as described above is a fast response to disruptions in proteostasis and thus constitutes one of the cell's "first lines of defense" for coping with heat shock. We now consider cellular responses that go beyond the HSR, which allow cells to cope with extreme and/or prolonged thermal stress events. In the context of the induction of the HSR, the immediate threat that denatured proteins pose to the cell is reduced by the action of molecular chaperones. But the kinetics of transcriptomic responses to heat stress suggests that, if the duration or severity of the thermal stress persists, then the levels of macromolecular damage may outstrip the ability of molecular chaperones to ameliorate the problem. We briefly examine some of the additional aspects of the cellular stress response that have been characterized in ectotherms exposed to a range of heat stress conditions. Highly eurythermal species are especially valuable experimental systems for this type of study and serve as the focus for much of our analysis.

The longjaw mudsucker (*Gillichthys mirabilis*), a native of estuaries of California and Baja California, was briefly introduced in Chapter 1. There we emphasized the "eury" nature of this animal's physiology, namely its ability to cope with wide ranges of temperature, salinity, and dissolved oxygen. The mudsucker can be acclimated to wide ranges of temperature, and it tolerates acute changes in temperature over a range of nearly 40 degrees Celsius (Buckley et al. 2006; Logan and Somero 2011). The effects of acclimation of the mudsucker to different temperatures, followed by exposure to acute increases in temperature, included several changes in gene expression that go beyond the HSR (Figure 3.28) (Logan and Somero 2011). Gene expression induced by acute heat stress followed a tiered response in which particular functional classes of genes were induced at different points along a heat ramp ($+4°C\ h^{-1}$). Regardless of what temperature these fish were acclimated to (9°C, 19°C, or 28°C), genes encoding molecular chaperones were among the first to be induced, as expected from the fast nature of the HSR (Ritossa 1962; Ashburner 1970). As seen for congeners of *Chlorostoma*, the temperature of induction (T_{on}) of the HSR increased with increasing acclimation temperature. Further heating beyond T_{on} for the HSR led to the induction of sets of genes that carry out a variety of stress-related functions other than chaperoning. The genes with these higher T_{on} values encoded proteins involved in ubiquitin-dependent proteolysis and cell cycle arrest. Again, the T_{on} for these additional sets of stress-related genes varied with acclimation temperature. Understanding why acute heat stress causes organisms to express genes involved in pathways of proteolysis and cell cycle arrest can provide important insights into the mechanisms—and costs—of the cellular stress response.

The **ubiquitin-proteasome system** is the mechanism by which much of the cellular protein pool is degraded and recycled under nonstressful conditions (Ciechanover 2005). The sudden induction of genes involved in this pathway under conditions of acute heat stress indicates that the cell has incurred a level of protein denaturation that necessitates further upregulation of proteolytic pathways—a process that would address the problem of protein unfolding by degrading thermally denatured proteins. Thus, the increased expression of the ubiquitin-proteasome system is an indicator of severe macromolecular damage under conditions of extreme heat stress in *G. mirabilis*. The fact that these genes are not induced at lower temperatures suggests that molecular chaperones are able to successfully mitigate the effects of heat stress at lower temperatures to help the cell regain proteostasis. This interpretation is supported by comparative transcriptomics studies of differently thermally adapted species. For example, in marine mussels of the genus *Mytilus*, an acute heat stress of 28°C ($+6$ degrees Celsius h^{-1} heat ramp) led to the induction of genes encoding subunits of the proteasome in the more cool-adapted *Mytilus trossulus* but not in the warm-adapted *M. galloprovincialis* (Lockwood et al. 2010).

The expression of genes involved in **cell cycle arrest** suggests a more broadscale response to heat stress that extends beyond the realm of individual cells. Halting the cell cycle could signify an attempt to prevent the proliferation of macromolecular damage; if heat-damaged cells undergo mitosis, they may doom the survival of their daughter cells, and mitosis itself may go awry (Vidair et al. 1993). Furthermore, as we discuss later in this chapter (see Section 3.5), heat-induced double-strand breaks in DNA could be mutagenic and cause significant downstream damage to the organism (Yao and Somero 2012). Other studies have found evidence for heat-induced apoptosis in fishes (Buckley

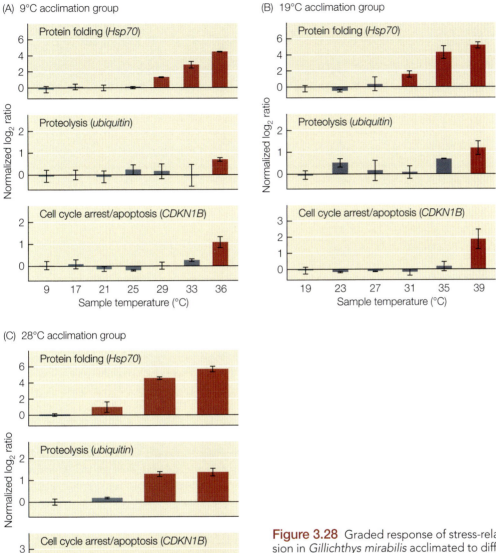

(A) 9°C acclimation group

(B) 19°C acclimation group

(C) 28°C acclimation group

Figure 3.28 Graded response of stress-related gene expression in *Gillichthys mirabilis* acclimated to different temperatures: 9°C (A), 19°C (B), and 28°C (C). Data were obtained using cDNA microarrays. Normalized $\log_2$-ratio expression profiles are shown with means and standard deviations. Red bars indicate expression levels that differ significantly from the control (one-way ANOVA, false discovery rate ≤ 0.05, Tukey's honestly significant difference). (After Logan and Somero 2011.)

et al. 2006; Buckley and Somero 2009; Logan and Buckley 2015) and mussels (Lockwood et al. 2010). Similar to cell cycle arrest, apoptosis may be a response that allows the organism to sequester heat-induced cellular damage and prevent cellular necrosis that would have negative consequences at the tissue- and organ-system levels.

In summary, to understand the mechanism and consequences of thermal stress at the cellular level, one must maintain a broad focus that looks at the sequential changes in the cellular stress response components as the intensity of thermal stress increases.

Of particular importance in these studies of graded stress responses is the potential for developing at least a relative sense of the costs of thermal stress. This question, addressed in Box 3.9, is a major emerging issue in ecological physiology.

BIOINDICATORS OF PHYSIOLOGICAL STATUS   Moving up through levels of biological organization, we begin to appreciate the complex effects of heat stress on physiological systems that extend beyond protein folding and the induction of the HSR. Indeed, physiological responses to heat stress comprise a multitude of molecular pathways, and it is beyond the scope of this chapter to review them all. However, there is one study we will highlight here that connects patterns of the HSR to responses of organ systems and the regulation of metabolic pathways in an ecologically relevant context of acute heat stress in the intertidal zone.

Han and colleagues (2013) examined the effects of heat stress on the intertidal limpet *Cellana toreuma*, a widely distributed species of the shores of the western Pacific Ocean (Dong et al. 2012) that inhabits a dynamic thermal environment that regularly exceeds 45°C during aerial emersion. These researchers looked at coordinated responses of heart rate and expression of genes encoding HSPs and key regulators of metabolism during acute heat stress (from 20°C to 40°C at a rate of +6 degrees Celsius h^{-1}). Increases in temperature initially led to increases in heart rate, but temperatures beyond 36°C led to sharp reductions in cardiac activity (**Figure 3.29**). Animals sampled during the acute stress were also examined for changes in gene expression, using quantitative PCR (qPCR) methods to characterize T_{on} values for selected stress-related genes. At temperatures near 30°C (i.e., well below the critical temperature that caused decreases in heart rate), the limpets induced expression of *Hsp70* and *Hsp90* genes, as well as genes important in the regulation of metabolism: genes encoding AMP-activated protein kinase (AMPK) and the histone/protein deacetylase sirtuin 1 (SIRT1). The T_{on} values for all of these genes were approximately the same, whereas T_{max} and T_{off} differed among the genes.

The induction of transcription for genes related to energy metabolism suggests a coupling between heat-induced damage to the cell and elevation of energy metabolism, presumably to provide ATP for cellular repair or proteolysis, processes that both have high ATP demands. AMPK is a sensor of the AMP:ATP ratio (Hardie and Sakamoto 2006), and SIRT1 is an upstream regulator of metabolism in response to changes in energy

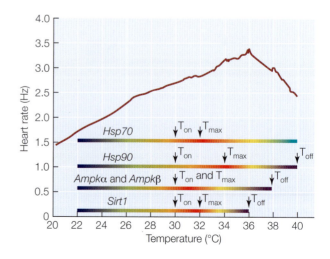

Figure 3.29 Heat stress causes simultaneous cardiac responses and induction of genes encoding HSPs (*Hsp70* and *Hsp90*) and regulators of metabolism (*Ampk* and *Sirt1*) in the intertidal limpet *Cellana toreuma*. Limpets were acclimated to a tidal cycle for 1 week at 20°C and then exposed to a heat ramp of 6°C h^{-1}. Heartbeats were detected by an infrared sensor affixed to the outside of the limpet's shell; the red line represents mean heart rate of three limpets. Gene expression was assayed via qPCR, and colored bars indicate mRNA levels. Red and blue represent highest and lowest levels of mRNAs, respectively. (After Han et al. 2013.)

stores (Cantó et al. 2009). These proteins interact to maintain the energy homeostasis of the cell (see Chapter 2). The fact that genes encoding molecular chaperones and genes involved in energy metabolism are induced in response to acute heat stress is a phenomenon that has been shown by other studies of intertidal invertebrates (Gracey et al. 2008; Connor and Gracey 2011), and indicates that the energetic costs of heat stress and the induction of the HSR are likely to be substantial.

The reproducible molecular responses to heat stress that exist across a vast array of organisms, and the connection of these molecular-level responses to higher levels of biological organization, as shown for heart rates in the intertidal limpet *C. toreuma* (Han et al. 2013) and for whole-organism heat tolerance in *Drosophila* (Welte et al. 1993; Feder et al. 1996), show that molecular responses of key genes and proteins can be used as proxies for physiological status, as **bioindicators**. Often, particularly in the context of field-acclimatized organisms, well-established bioindicators can be used as a more convenient physiological assay than the measurement of cardiac function or whole-organism heat tolerance. In the next section we will explore how the expression of Hsp70 correlates with thermal environment across broad geographic scales and how this bioindicator can be used as a tool for predicting suitable thermal habitats.

BROADSCALE PATTERNS IN BIOGEOGRAPHY CORRELATED WITH STRESS RESPONSES

Even though the induction of the HSR confers heat tolerance, the HSR is energetically expensive and frequent induction of the HSR may impose metabolic costs that impede growth and reproduction (Krebs and Loeschcke 1994). Thus, the HSR may serve as an informative bioindicator of physiological status in field-acclimatized organisms and provide a predictive framework for assessing the physiological factors that influence biogeography.

A recent study by Lima and colleagues (2016) illustrates how Hsp70 protein expression can be used as a bioindicator across broad geographic scales. *Patella vulgata* is a limpet species that inhabits a variety of thermal microenvironments in the mid- to high-intertidal zone of the eastern Atlantic Ocean. This species has a broad latitudinal distribution, from Norway to the Iberian Peninsula, but its abundance is greatest in the more northern stretches of its range (**Figure 3.30B,C**). The researchers sampled 17 locations across the species' range, from southwest Scotland to southern Portugal, where they found limpets in both shaded and sun-exposed locations. They measured local abundances, body temperatures of limpet mimics, and Hsp70 levels. The results of this study suggest that the physiological costs associated with heat stress and the induction of the HSR in the hottest (sun-exposed) environments pose thermal limits on limpets regardless of latitude. There was a striking correspondence of microhabitat (sun versus shade) to Hsp70 levels, where limpets sampled from sun-exposed habitats were found to have the highest Hsp70 levels (**Figure 3.30A**). This pattern was likely driven by the maximum temperature experienced by these animals in the 2 weeks prior to sampling (**Figure 3.30D**). Throughout the species' range, local abundances indicate that limpets found thermal refuge in shaded habitats, particularly at locations that experience more frequent extreme heat events. This conclusion is supported by the positive correlation between environmental temperature and Hsp70 levels (see Figure 3.30D) and the negative correlation between the frequency of hot temperatures and limpet abundance (**Figure 3.30E**).

It is important to note that the use of HSPs as bioindicators for the focal species should be validated in the laboratory in order to establish the environmental conditions that correspond to the induction of the HSR. For example, consider species that

Figure 3.30 Environmental temperature correlates with levels of Hsp70 protein expression ▶ and limpet abundances in field-sampled *Patella vulgata* in the eastern Atlantic Ocean. (A) Mean Hsp70 levels among 8 limpets sampled at each location (1–17) and microhabitat (shaded versus sun-exposed). Protein levels were measured via Western blotting and expressed as mg g^{-1} of loaded protein. Error bars indicate standard error. (B) Mean abundances of limpets at each location and microhabitat. Error bars indicate standard error. (C) Map of the sampling locations. (D) Correlation of temperature prior to sampling and mean Hsp70 levels. Solid and dashed lines represent a linear regression and 95% confidence limits, respectively. The vertical dashed line indicates the threshold temperature of 27.5°C used in part (E). (E) Correlation of the frequency of temperatures above 27.5°C and limpet abundance (mean density; number of individuals per square meter). Solid and dashed lines represent the 95th regression quantile and 95% confidence limits, respectively. (After Lima et al. 2016.)

have evolved to employ the preparative defense strategy—maintenance of high basal levels of HSPs—like *Lottia scabra* in central California (Dong et al. 2008). In that case, as emphasized earlier, Hsp70 expression would not be an informative bioindicator because the abundance of this protein does not change in response to heat stress. In this context, a more informative bioindicator might be ubiquitin conjugation or protein carbonylation, which are direct markers of protein damage. In the case of the aforementioned study on Hsp70 expression in field-acclimatized *P. vulgata* (Lima et al. 2016), there was a significant amount of variation in Hsp70 expression, and this variation correlated with thermal history. Thus, it appears that Hsp70 expression may be a good bioindicator in this system, but future studies should ground-truth this interpretation by conducting controlled experiments in the laboratory under common garden conditions. This may further reveal the presence of temperature-adaptive variation in the HSR among populations of *P. vulgata* that experience different thermal environments.

Finally, to appropriately use HSPs or other stress-related proteins as bioindicators, one must take the time-course of the stress response into account. As shown, for example, by the data in Figure 3.26, what one "sees" depends on when one samples. Thus, induction of HSPs exhibits a distinct temporal pattern, one that may differ among species. By sampling only at a single time point, the investigator may observe what could be a misleading "snapshot" of the cellular stress response. Sampling too early or too late in the response might lead one to conclude "nothing happened." Another complication concerns the molecular indicator that is chosen for use as a bioindicator. For example, the time-course of mRNA synthesis and degradation typically differs from the time-course of protein expression (see, e.g., Buckley et al. 2006). Thus, what one observes will depend on the molecules under consideration—for example, mRNAs versus proteins—as well as on the temporal pattern of sampling. Unfortunately, due to considerations of costs (financial and amount of human labor required), only a minority of studies of stress-induced bioindicators have examined time-course effects and considered whether the molecular system under consideration as a bioindicator is telling the complete story.

We now move beyond the realm of proteins to explore the effects of temperature on another class of macromolecules: nucleic acids. We will discover that many of the rules of thermodynamics and molecular physiology that govern protein thermal relationships also apply to nucleic acids. We will also explore the ways in which temperature change perturbs nucleic acid function and how temperature adaptation has dealt with these challenges (stress in the negative sense) and exploited these opportunities (stress in the positive sense) over evolutionary time.

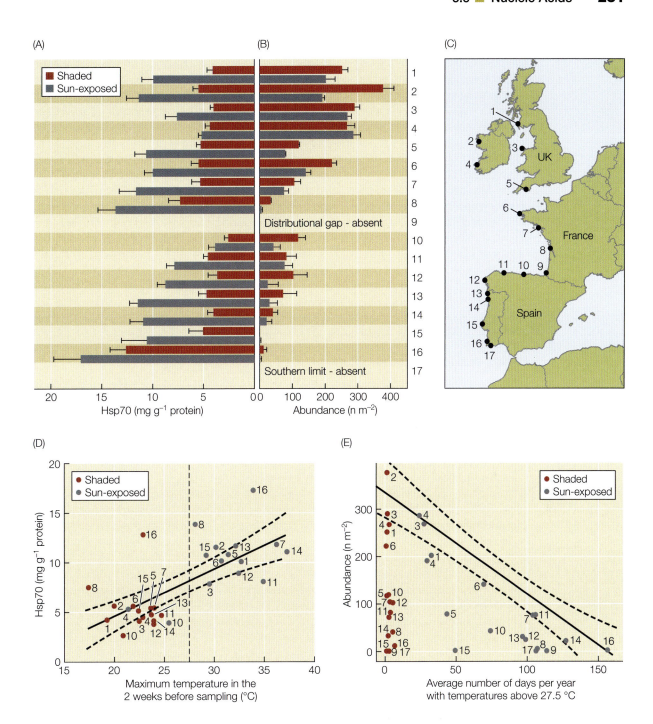

3.5 Nucleic Acids

As in the case of proteins, the covalently bonded primary structures of nucleic acids—DNA and the several classes of RNAs—are inherently quite resistant to thermal damage. Direct temperature-driven cleavage (hydrolysis) of the phosphodiester bonds that link nucleotide bases together is not the primary threat to nucleic acid structure posed by changes in temperature, although the ROS that sometimes result from thermal stress can lead to cleavage of these bonds (see Chapter 2). As is similar to what we have seen for proteins, the higher-order structures, notably the complex secondary and tertiary structures of nucleic acids, are readily altered by even minor changes in temperature. And as in the case of proteins, these direct effects of temperature on secondary (or tertiary) structure may either be damaging to cellular function or help modulate cellular functions in temperature-adaptive manners. Thus, much as leucine zipper proteins can function as thermally modulated switches for temperature-mediated control of transcription, thermal alteration of secondary structure of certain types of RNAs may be an important element in the regulation of transcription and translation. Regulatory systems that make use of the thermal sensitivity of nucleic acids have evolved in many contexts—ranging from activation of the HSR, as discussed earlier, to triggering expression of genes involved in bacterial pathological processes—and seem likely to be widespread in governing the responses of cells to alteration in temperature. To date, we likely have seen only the tip of a very large iceberg of such effects (Kortmann and Narberhaus 2012; Schumann 2012). Furthermore, as is typical in advances in molecular biology, initial progress has been made principally with bacterial model systems. Thus, we know little about the ways in which eukaryotic or archaeal species have come to exploit nucleic acids' thermal sensitivities in regulatory capacities. Here, then, is another field ripe for cultivation.

Basics of nucleic acid structure

Before we examine the challenges and opportunities that arise from the temperature sensitivities of nucleic acid structure, it is useful to review some of the basic properties of nucleic acids, especially the weak bond interactions that establish stabilities of higher-order structures. DNA and RNA are polymers of **nucleotides** (one of the nucleic acid bases [nucleobases] linked to a sugar and a phosphate group) whose primary structures are formed of alternating phosphate and sugar residues (**Figure 3.31**). Nucleotides are linked by **phosphodiester bonds** between phosphate groups and the third and fifth carbon atoms of adjacent sugar rings. The sugar in DNA is 2-deoxyribose; in RNA the sugar is ribose. In DNA, the bases are the **purines**, **adenine** (A) and **guanine** (G), and the **pyrimidines**, **cytosine** (C) and **thymine** (T) (see Figure 3.31A); in RNA, thymine is replaced by **uracil** (U). DNA forms a double-strand, coiled structure—a double helix—with a major groove and a minor groove (see Figure 3.31C).

The basis of the temperature sensitivity of nucleic acids stems in large measure from the central importance of base pairing in stabilizing DNA and RNA secondary structures. The DNA double helix is stabilized primarily by hydrogen bonds between nucleotides and base-stacking interactions among aromatic nucleobases (see Figure 3.31B,C). The hydrophobic effect is involved in the latter interactions. In the aqueous environment of the cell, the nucleotide bases align perpendicular to the axis of the DNA molecule, which minimizes their interaction with water. The complex folded structures of different forms of RNA are also stabilized by hydrogen bonding and hydrophobic interactions (see Figure 3.31D).

To a large degree then, the stabilization of secondary structures of DNA and RNA relies on the same types of bonding interactions discussed above for proteins. From this it follows that nucleic acids are likely to manifest many of the general types of thermal sensitivities we have discussed for proteins. As is the case for proteins, a balance between stability and flexibility—marginal stability—will be seen to be crucial for many aspects of nucleic acid function. And as in proteins, the different types of residues in the polymer contribute differently to thermal sensitivity. In particular, note that G:C pairs are

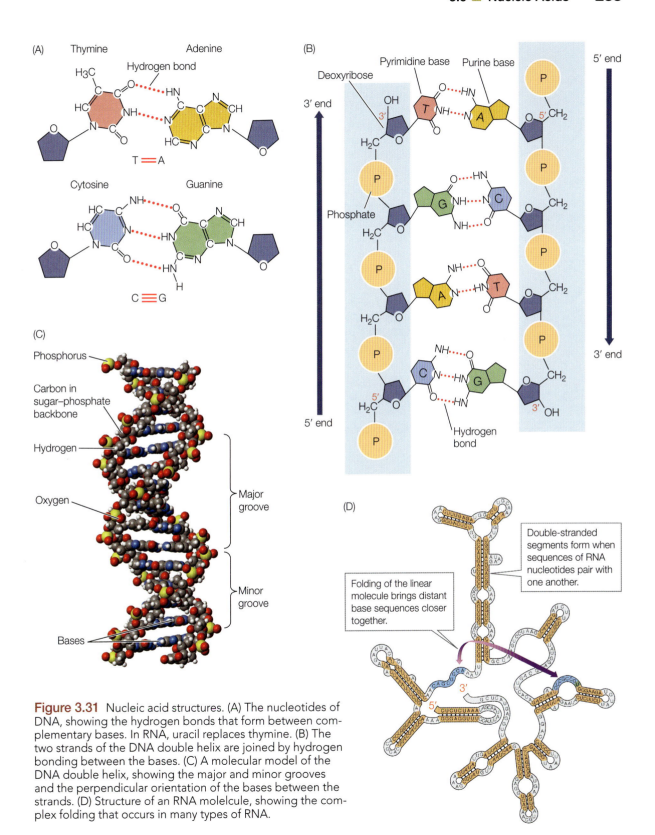

Figure 3.31 Nucleic acid structures. (A) The nucleotides of DNA, showing the hydrogen bonds that form between complementary bases. In RNA, uracil replaces thymine. (B) The two strands of the DNA double helix are joined by hydrogen bonding between the bases. (C) A molecular model of the DNA double helix, showing the major and minor grooves and the perpendicular orientation of the bases between the strands. (D) Structure of an RNA molelcule, showing the complex folding that occurs in many types of RNA.

stabilized by three hydrogen bonds whereas A:T pairs are stabilized by only two (see Figure 3.31A). The percentage of G+C in the genome varies across taxa. The implications of this variation in G+C content have been subjected to considerable analysis, notably in the context of species adapted to different temperatures. This topic serves as a good entry point into a broad discussion of how nucleic acids cope with—and exploit—temperature-dependent changes in secondary structure.

G+C content and adaptation to temperature: Are nucleic acids from warm-adapted organisms stabilized by more G:C base pairs?

Once Watson and Crick (1953) elucidated the double-helical structure of DNA and showed how base pairing stabilizes its higher-order structure, it occurred to molecular biologists interested in evolutionary questions that the percentage of G+C in a genome might exhibit a regular increase with adaptation temperature. Temperature-dependent variation in percentage of the more stable G:C base pairs could lead to conservation of a consistent stabilization state of DNA across the full spectrum of temperatures at which life exists. DNA, then, would show a pattern of conservation of a corresponding state of structure akin to what has been found for proteins.

GENOMIC DNA SEEMS "UNADAPTED" TO TEMPERATURE   Despite the a priori appeal of the above hypothesis, a relationship between adaptation temperature and G+C content appears not to pertain in the case of total genomic DNA (Galtier and Lobry 1997; Hurst and Merchant 2001). As is usually the case in broad surveys of genomic characteristics, the bulk of relevant data come from studies of bacteria and archaea. Galtier and Lobry (1997) examined DNA base compositions of 764 species of bacteria and archaea with different optimal growth temperatures (OGTs). Thermophiles were defined as species with OGT values from 47°C to 105°C; mesophiles had OGT values of 17°C to 41°C. (Note: These definitions differ somewhat from the canonical definitions now used; see Box 3.1.) Despite this wide range of thermal optima for growth, the overall percentage of G+C in the genomes showed no significant trend with OGT (Figure 3.32A). Hurst and Merchant (2001) reached a similar conclusion through a detailed and rigorous analysis of G+C of the total genome, of protein-coding regions, and of the third sites in codons where variation in G+C content is highest. They found "no hint" of any temperature-related variation in G+C in any of these comparisons. Thus there appears to be no selection for overall thermal stability of DNA, at least as mediated by variation in G+C percentages.

Studies of base compositions of genomes of eukaryotes are more limited in number, but again, no consistent trends in percentage of G+C with evolutionary thermal history are apparent. The percentage of G+C varies widely among animals, ranging from ~37% to 50% (Mooers and Holmes 2000). A complexity in evaluating the G+C percentages of eukaryotic genomes is that certain sections of genomic DNA comprise what are termed **isochores**. These are lengthy fragments of DNA (usually greater than 200 kilobases in length) that have highly uniform base compositions (Mooers and Holmes 2000). In some cases, isochores have G+C percentages that differ substantially from the overall G+C percentage of the entire genome. It has been proposed (Bernardi 1995) that G+C isochores in endothermic homeotherms reflect selection for nucleic acid stability at high body temperatures. However, there have been few comparative studies in which orthologous genes from organisms adapted to widely different temperatures have been compared in terms of G+C content. One study in which this type of analysis was performed examined the orthologous genes for lactate dehydrogenase A (*ldh-a*) and α-actin (*α-actin*) in 51 species of vertebrates adapted to temperatures spanning a range of almost 50 degrees Celsius—from –1.9°C in the case of Antarctic notothenioid fishes

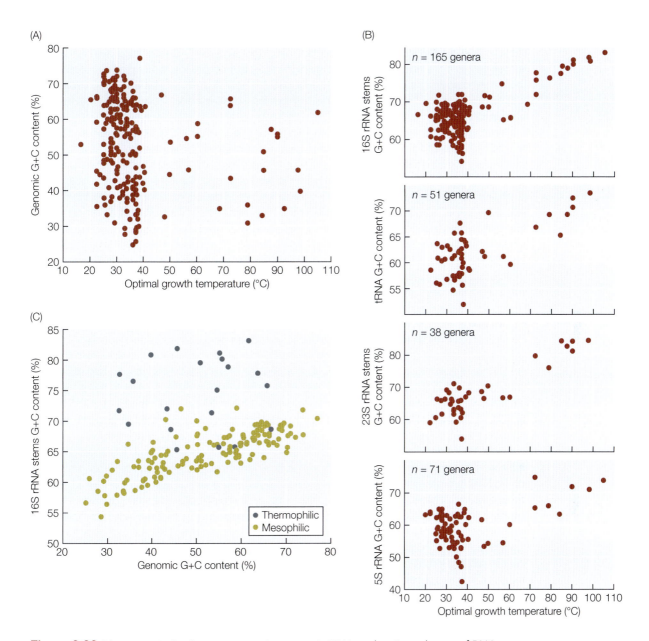

Figure 3.32 Variations in G+C percentages in genomic DNA and various classes of RNAs. (A) Genomic G+C content and optimal growth temperature (OGT). No significant relationship exists among the 224 genera of bacteria and archaea examined. (B) G+C content of different types of RNA structures of bacteria and archaea. From top to bottom: 16S rRNA stems; tRNA; 23S rRNA; and 5S RNA. In all cases, a significant increase in G+C content in these RNAs occurs with rising OGT. (C) Relationship between genomic G+C percentage and G+C content of stem portions (double-stranded regions) of 16S rRNA for 165 genera of bacteria and archaea. Blue symbols are for 21 thermophilic genera (OGT 47°C–105°C). Yellow symbols are for 144 mesophilic genera (OGT 17°C–41°C). Genomic G+C content does not significantly correlate with OGT (see part A). For mesophiles, G+C of 16S rRNA stem regions shows a positive relationship with genomic G+C; this is not seen for thermophiles. Thermophiles do have a significantly higher G+C content in 16S rRNA stem regions than mesophiles, which suggests selection for enhanced stability of double-stranded regions of 16S rRNA. (A and B after Galtier and Lobry 1997; C after Mooers and Holmes 2000, based on data from Galtier and Lobry 1997.)

to ~45°C for a thermophilic desert reptile (*Dipsosaurus dorsalis*) (Ream et al. 2003a). No correlation between adaptation temperature and G+C percentage was found for either gene. This finding, while limited to two sets of orthologous genes, certainly is inconsistent with the prediction of the isochores-based thermal stability hypothesis that genomes of endothermic homeotherms exhibit selection for thermal stability through elevated G+C percentages.

What does this lack of correlation imply for the temperature dependence of DNA function? One of the most significant requirements for corresponding states of DNA secondary structure lies in the context of transcriptional regulation, where the appropriate capacities of DNA structure to open up and engage in interactions with regulatory proteins must be maintained at all temperatures. The lack of a correlation between evolutionary adaptation temperature and G+C content shows that conservation of this corresponding state likely is not a consequence of DNA base composition. Thus, DNA of cold- and warm-adapted organisms appears not to differ in the lability of secondary structure. Instead, as we will see below, specific DNA-binding proteins may be regulated in a temperature-dependent manner to allow DNA to retain the appropriate poise for transcriptional regulation in the face of changing body temperature.

SOME RNAs REFLECT STRONG ADAPTATION IN STABILITY OF SECONDARY STRUCTURE
Even though the overall genomic G+C content does not reflect the evolutionary thermal histories of bacteria and archaea (or probably those of eukaryotes), significant positive trends have been found between G+C percentage and OGT of bacteria and archaea in the case of several types of RNAs: 16S ribosomal RNA (rRNA), transfer RNAs (tRNAs), 23S rRNA stems, and 5S rRNA (Galtier and Lobry 1997; Hurst and Merchant 2001; **Figure 3.32B,C**). These RNAs possess secondary structural elements that are critical for functions such as thermosensing (see below). An appropriate thermal sensitivity for such a function is suggested by the observed positive relationship between OGT and G+C content: The differences in intrinsic thermal stability of secondary structures of mesophilic and thermophilic RNAs might give them a corresponding state of thermal stability at normal growth temperatures. These corresponding states of stability might provide the regulatory regions of the different RNAs with just the right poise for responding to changes in temperature and other regulatory signals that exert their control by modifying RNA secondary structure and, thus, RNA function. Temperature sensing that involves this sort of mechanism represents an excellent example of how thermal lability of macromolecular structure can be exploited to assist cells in coping with rapid changes in temperature. RNA thermometers, discussed next, provide a fascinating case study of this type of molecular responsiveness.

Bacterial RNA thermometers: Using thermal perturbation of RNA structure to regulate transcription and translation

All organisms benefit from having abilities to detect even small changes in temperature and then to use this information to regulate a diverse array of processes. The temperature-regulated processes include transcription, translation, and modulation of protein activities. Some of the most thermally sensitive and rapidly responding temperature-sensing systems are based on thermal perturbation of macromolecular structure, which can lead to a nearly instantaneous response by the process undergoing regulation. Temperature-dependent changes in the secondary structures of a variety of forms of RNA provide excellent examples of these sorts of rapid influences of temperature changes on regulation. Here we consider **RNA thermometers** (**RNATs**), portions of mRNA molecules whose secondary structures are highly sensitive to changes in temperature, even to changes as small as 1 degree Celsius (Kortmann and Narberhaus 2012; Schumann 2012). Later in this section we discuss how these RNATs might adapt to fit the thermometer to the thermal range in which it must function. RNATs have been discovered in bacteria, but there is every reason to suppose that they will be found to play numerous roles throughout all domains of life. The "raw material" for fabricating RNATs is present in all RNA molecules, and the effectiveness of RNATs argues for their wide exploitation across all taxa.

RNATs of bacteria are essential components of regulatory mechanisms that control several processes whose roles change with increases or decreases in temperature. These processes include the heat-shock response, the cold-shock response, and pathogen virulence (the activation of pathogenesis-related genes of a bacterial pathogen when it invades a host with a high body temperature, for example, a bird or mammal). The complexity of RNATs differs greatly among different regulatory systems (Kortmann and Narberhaus 2012; Schumann 2012), but in all cases a change in temperature causes alterations in RNA secondary structure that lead to rapid changes in the processes being modulated, for example, the translation of a preexisting mRNA.

THE BASICS OF RNAT FUNCTION   RNATs of bacteria operate through two fundamental processes, both of which involve temperature-dependent change in secondary structure: the zipper mechanism and the switch mechanism (**Figure 3.33**). The **zipper mechanism** involves disruption (melting) by increasing temperature of base pairings that stabilize a hairpin structure within the RNA (see Figure 3.33A). This hairpin structure contains a regulatory region of the RNA that is blocked from function when the hairpin is intact. Melting of the hairpin exposes the regulatory region (see Figure 3.33B), which leads to

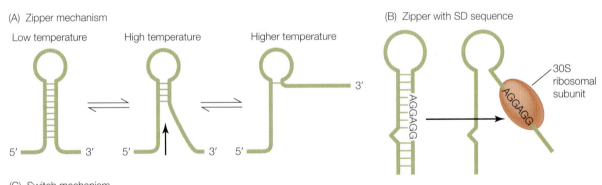

(A) Zipper mechanism

(B) Zipper with SD sequence

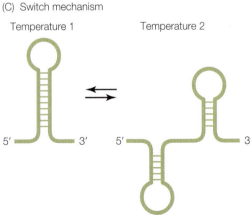

(C) Switch mechanism

Figure 3.33 Basic temperature-sensing mechanisms found in RNA thermometers (RNATs). (A) In the zipper mechanism, RNA hairpin structures melt at higher temperatures, exposing buried regulatory elements such as the Shine-Dalgarno (SD) sequence (AGGAGG; see text). Exposed SD elements allow mRNA to bind to the small subunit of the ribosome and initiate translation. Melting of the zipper region is graded and reversible, conferring on these switches an ability to grade the response in proportion to the level of heat stress and to return the RNAT to the hairpin (inactive) state when temperature is reduced. (B) Details of a zipper that contains the SD sequence. (C) In the switch mechanism, the RNAT's conformation alternates between two temperature-dependent structures. Switch mechanisms are especially important in RNATs that respond to reductions in temperature (see text). (After Kortmann and Narberhaus 2012.)

modulation of the process under temperature-dependent control. The **switch mechanism** (see Figure 3.33C) involves a shift in the equilibrium between two alternative secondary structures for the RNA. Both types of mechanisms afford a graded response to temperature stress, rather than a sharp on-off response at some particular temperature. Thus, as temperature rises or falls, the fraction of the RNAs capable of effecting regulation shifts proportionately to the amount of change in temperature. We now consider examples of each type of regulatory mechanism, to illustrate the speed and sensitivity of this type of temperature regulation of biochemical function.

ROSE ELEMENTS AND THE HEAT-SHOCK RESPONSE   Based on the discussion of the HSR given earlier, one might predict that the abilities of RNATs to exert such rapid regulatory responses to small shifts in temperature would suit them especially well to the control of expression of heat-shock proteins or other proteins upregulated under heat stress. Upregulation of heat-shock protein synthesis might be particularly rapid if the cell contained constitutively synthesized heat-shock mRNAs that could be activated through temperature-dependent changes in RNA secondary structure. In fact, RNATs are commonly found in the mRNAs that encode small heat-shock proteins in bacteria; these mRNAs are synthesized constitutively but remain in a nontranslatable form until temperature increases to the point that the HSR is needed.

These RNATs are called **ROSE elements** (repression of heat-shock gene expression), and their molecular structures and modes of action have been elucidated in considerable detail in several types of bacteria (Kortmann and Narberhaus 2012; **Figure 3.34**). ROSE elements are between 60 and more than 100 nucleotides in length and commonly contain two, three, or four individual hairpins. Not all of the hairpins are involved in temperature-dependent melting, however. The 5′ hairpin(s) remain stable under heat shock,

Figure 3.34 Structure of the ROSE1 zipper-type RNAT found in the mRNA for a small heat-shock protein of the bacterium *Bradyrhizobium japonicum*. The RNAT contains four hairpin structures (I–IV), but only hairpin IV is subject to melting with rising temperature. The region of hairpin IV contained in the box is the functional ROSE domain. Red lines indicate hydrogen-bonding interactions. The thermal lability of hairpin IV's ROSE domain is a consequence of several noncanonical base pairings that lower the melting temperature of secondary structure (see text). "O" indicates *syn-anti* base pairing. Melting temperatures of different regions of the hairpin structure are color-coded. Melting begins near 37°C and is complete near 42°C. The Shine-Dalgarno sequence (AGGAGG) is nucleotides 90–95. The start codon (AUG) comprises nucleotides 103–105. Details on chemical structures and base interactions in a key regulatory region (nucleotides 94–97) are shown at the bottom of the figure. (From Kortmann and Narberhaus 2012.)

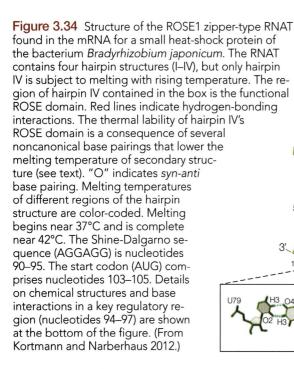
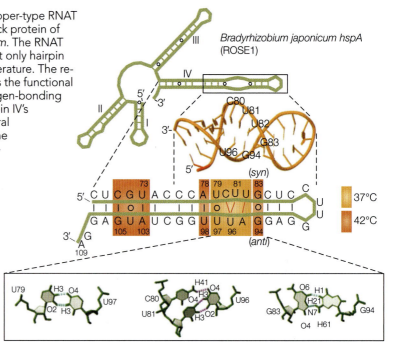

but the 3′-most hairpin is stable only at lower temperatures, that is, at temperatures below the threshold induction temperature for the HSR. The 3′-most hairpin contains a **Shine-Dalgarno (SD) sequence** (see Figure 3.33B). The SD sequence is positioned 6–10 nucleotides upstream of the start codon (AUG, nucleotides 103–105; see Figure 3.34) and functions as the binding site for 16S RNA of the 30S subunit of the bacterial ribosome. For a message to be translated, the SD sequence must be accessible for this binding. At low temperatures, the secondary structure of the ROSE element precludes such binding; the SD sequence is effectively buried within the folded structure of the ROSE element. Increasing temperature melts the ROSE element and exposes the SD sequence; rapid translation of the message then occurs.

The fine structure of the ROSE element in certain bacteria (the ROSE1 element of *Bradyrhizobium japonicum* is shown in Figure 3.34) shows how the temperature sensitivity of RNA secondary structure is adjusted so as to poise the RNAT for an appropriate response to changes in temperature that could damage the proteome. Note that, in addition to canonical Watson-Crick base pairing (G:C and A:U pairings in the case of RNA secondary structure) there are unusual (noncanonical) pairings in the ROSE element structure. For example, nucleotides G83 and G94, which lie opposite each other in the folded hairpin structure, cannot undergo normal Watson-Crick base-pairing interactions. What is termed a *syn-anti* base pairing occurs instead; this particular orientation of the bases leads to a less stable secondary structure. Other noncanonical base interactions include a three-base interaction of U96 with C80 and U81. This feature precludes regular base stacking and leads to a weak spot in the secondary structure; it is here that the zipperlike melting of the RNAT is initiated. The noncanonical base pairing of residues U79 and U97 likewise destabilizes the secondary structure; these residues, too, are important in the early phase of zipper melting. Last, there is weak base pairing between the AUG start codon and the nucleotide triplet C71-G72-U73. *All of these noncanonical pairings poise the zipper for melting at a relatively low temperature compared with what would be the case with complete canonical base pairing in the 3′ hairpin regions.*

The multiple ways in which noncanonical base pairing is used to modulate the melting temperature of the ROSE1 element suggest that evolution has great amounts of raw material at its disposal for adjusting the temperatures at which RNATs carry out their regulatory functions. For instance, adaptation at relatively low temperatures might favor addition of increasing amounts of noncanonical base pairing, in order to further destabilize the zipper's secondary structure. In contrast, evolution at high temperatures might reduce the amount of noncanonical base pairing in order to elevate the temperature at which the RNAT's zipper melts. Discerning how evolution has exploited these options might be possible by examining DNA sequence databases, albeit RNATs show little sequence conservation (Kortmann and Narberhaus 2012). Modeling of RNA secondary structures might be an even more effective way to shed light on this interesting facet of molecular evolution.

TEMPERATURE SENSING BY PATHOGENIC BACTERIA   Pathogenic bacteria that attack mammals typically encounter a large and rapid change in cell temperature when they enter their host. From an air, water, or soil temperature perhaps near 10°C–20°C, the bacterium is suddenly warmed to mammalian body temperatures (~37°C). This rapid increase in temperature not only can modify the rates of the bacterium's biochemical processes, but also can serve as a signal that leads to activation of virulence genes that confer the bacterium's pathogenic properties. Pathogenic bacteria generally do not express virulence genes unless lodged in a host; this strategy probably saves energy and helps the bacterium avoid the innate immune response of mammals. In light of what has been presented above about RNA thermometers, it should come as little surprise that temperature-driven changes in RNA secondary structure are a critical early component of the pathogenic response.

An example of RNAT-mediated regulation of virulence gene expression has been described in the bacterium *Listeria monocytogenes* (Johansson et al. 2002). This bacterium produces a **virulence regulator protein**, PrfA, in a temperature-dependent manner. Like the mRNAs for certain heat-shock proteins, the mRNA for PrfA is transcribed constitutively and is capable of rapid activation when a sufficient rise in temperature occurs. Like many other RNATs, the *prfA* mRNA has a thermometer region at its 5' end. The SD sequence is actually exposed in the *prfA* RNAT, but the structure of the mRNA is so rigid at lower temperatures that the SD sequence is prevented from binding to the small subunit of the ribosome; thus, translation of *prfA* mRNA is blocked at low temperatures. When temperature rises to 37°C, the RNAT region is destabilized, the *pfrA* mRNA binds to the ribosome, and translation commences. The PrfA protein upregulates 490 other proteins, including several virulence genes, some of which repress the host's innate immune response. Like the heat-shock response ROSE element system, the *pfrA* RNAT allows rapid translation of a protein that is critical for downstream regulation of a complex temperature-related process.

A variation on this type of regulatory response has been observed in the bacterium *Neisseria meningitidis*, which is a normal and generally benign commensal of the human nasopharynx region, but which can become pathogenic under some circumstances (Loh et al. 2013). When in its human host, the bacterium normally lives at a temperature near ~37°C and virulence genes are not expressed. At temperatures near 42°C, however, expression of genes that can lead to virulence begins to increase. The genes induced at 42°C include ones that encode proteins that help the bacterium evade the innate immune response of the host. Three distinct RNA thermometers are involved in this upregulation. Loh and colleagues proposed the following model to account for this temperature-related shift in gene expression. If the nasopharynx is invaded by a virus, the host is likely to develop a fever and increase production of immune system effectors. By activating genes that defend against the host's immune effectors, *N. meningitidis* may avoid succumbing to an immune response triggered by the virus. In addition, some of the fever-induced proteins, notably those involved in capsule formation, could increase the likelihood that the otherwise benign commensal bacterium would survive and develop its pathological functions if it entered the bloodstream. This conversion of a benign bacterium into a pathogen is thus another example of how RNAT-mediated mechanisms enable microorganisms to adapt to changes in circumstances. In this case, it is only under conditions of a fever triggered by a viral infection that the conversion occurs, challenging the animal host with not just one, but two pathogens at the same time.

TEMPERATURE-SENSITIVE RIBOSWITCHES AND CONSERVATION OF LIGAND BINDING   A recurring theme in this chapter is the importance of maintaining ligand-binding abilities and other regulatory functions in the face of changes in cell temperature. We dealt with this topic in some detail in discussing enzyme-temperature interactions. Here we examine how temperature effects on RNA conformation can lead not only to shifts in mRNA translation, as discussed above, but also to stability of regulation in the face of temperature change, namely to temperature compensation in the regulation of gene expression.

This type of temperature compensatory phenomenon was first discovered in *Vibrio vulnificus*, a marine bacterium that also can be a human pathogen. *V. vulnificus* must

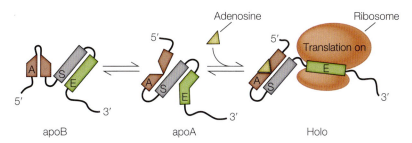

Figure 3.35 A temperature-dependent riboswitch. The mRNA that encodes adenosine de-aminase has a temperature-dependent riboswitch that exists in two conformational states: apoB (left) and apoA (middle). Only apoA can bind adenosine to form the holo (right) state of the mRNA, which can bind to the ribosome and initiate translation of the *add* gene. A, aptamer domain; S, switching sequence; E, effector sequence (see text). (After Serganov and Patel 2007.)

function across a wide range of temperatures, from ~10°C in its native marine habitat to ~37°C when it enters the human gut. The ability to regulate gene expression across this wide range of temperatures is maintained in part by temperature-dependent RNA structures (**Figure 3.35**; Reining et al. 2013). Expression of the *add* gene that encodes the essential enzyme adenosine deaminase is controlled by the binding of adenosine to a regulatory region of the *add* mRNA, a **riboswitch**. Riboswitches are regulatory elements contained in an mRNA molecule that bind specific ligands that in turn trigger changes in riboswitch secondary structure. The change in riboswitch structure alters the ability of the mRNA to be translated. Riboswitches contain two domains: an **aptamer** domain (A in Figure 3.35), which is an oligonucleotide capable of specific ligand binding; and an **expression platform** (this may comprise a switching sequence [S] and an effector sequence [E]), which changes conformation upon binding or release of the regulatory ligand to the aptamer domain (Serganov and Patel 2007). In the case of *add*, once adenosine binds to the mRNA, forming the holo complex (riboswitch + ligand), the mRNA can bind to the ribosome and initiate translation of the *add* message.

The conformation of the riboswitch that governs expression of the *add* gene in *V. vulnificus* exists in two conformational states, termed apoB and apoA. The regulatory ligand, adenosine, can bind only to apoA. Temperature compensation of adenosine binding, which maintains production of adenosine deaminase across the full range of temperatures the bacterium encounters, depends on the following mechanism. Increases in temperature favor the apoA conformation; decreases in temperature shift the equilibrium in favor of apoB. This temperature-dependent shift in the equilibrium between binding (apoA) and nonbinding (apoB) forms of the riboswitch offsets the direct effects of temperature on adenosine binding, which is favored at low temperatures (recall the increase in binding of substrates to enzymes at reduced temperature) and reduced at high temperature. Without temperature compensation, decreased binding of adenosine at high temperatures would lead to decreased expression of adenosine deaminase. By reducing the fraction of *add* mRNA in the apoA configuration at low temperatures and increasing it at high temperatures, the temperature sensitivity of the riboswitch stabilizes binding of adenosine (and subsequent translation of the *add* mRNA) across the range of temperatures the bacterium encounters.

It is not known how commonly this type of RNA-mediated temperature compensation of ligand-binding ability occurs. However, this mechanism would appear to have great evolutionary potential in regulatory contexts. Furthermore, as suggested in the case of RNATs, it would appear likely that the structure (base composition) of a given type of riboswitch may adapt to have different temperature sensitivities of conformational change depending on the temperatures at which the system evolves. The adaptation to temperature of riboswitches and RNA thermometers will be a fascinating area of molecular evolution to explore.

RNATs IN THE COLD: REGULATION OF EXPRESSION OF COLD-SHOCK PROTEIN A   RNATs are also important in sensing reductions in temperature. This mode of temperature sensing cannot of course be based on the heat-driven melting reactions found for zipper-type RNATs. Instead, for cold sensing, RNATs typically rely on the switch type mechanism illustrated in Figure 3.33C. As temperature changes, the RNAT undergoes a substantial reorganization of secondary structure that leads to alteration in its function. A well-understood example of this mechanism is provided by the mRNA encoding cold-shock protein A (CspA) (Kortmann and Narberhaus 2012).

CspA is one of the proteins induced most strongly in bacteria when a drop in cell temperature occurs. CspA is part of a larger suite of **cold-induced proteins (Cips)** that are encoded by the **cold-shock stimulon**. Cips are relatively small, acidic proteins that contain a **cold-shock domain** that has strong RNA-binding ability. The ability to reversibly bind to mRNAs during transcription and translation, which proceed in tandem in bacteria, enables CspA to function as an **RNA chaperone**. Its role in the cell at low temperatures is to prevent cold-induced formation of secondary structures in newly transcribed mRNAs that are being translated as they are transcribed (**Figure 3.36**).

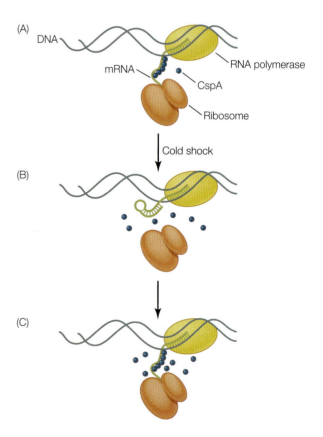

Figure 3.36 Cold-shock protein A (CspA) functions as an RNA chaperone during transcription and translation and assists in the effective coupling of these processes in bacteria. (A) Constitutively expressed CspA binds reversibly to nascent mRNA molecules and maintains the newly transcribed messages generated by RNA polymerase in a linear conformation that can bind to the small subunit of the ribosome and undergo translation. Under normal thermal conditions, sufficient CspA molecules are present to enable close coupling of mRNA transcription and its translation. (B) A sudden drop in temperature shifts two equilibria: that between linear and nonlinear conformations of the mRNA, and that of binding of CspA to the mRNA. Cold shock induces RNA secondary structures that preclude binding of the mRNA to the ribosome, and translation is blocked. (C) Low temperature rapidly induces synthesis of additional CspA molecules (see text for mechanism), leading to an increase in the amount of mRNA that is bound to CspAs. The increased binding of CspA molecules to nascent mRNA restores the linear conformation of the mRNA needed for binding to the ribosome, and translation recommences. (After Graumann and Marahiel 1998.)

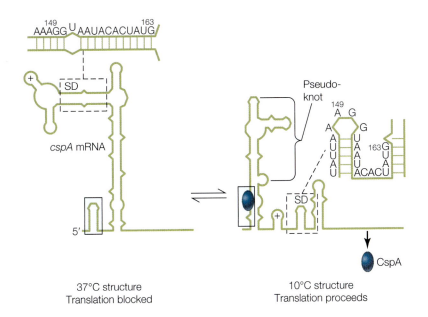

Figure 3.37 Temperature-dependent regulation of CspA synthesis by an RNAT switch mechanism. A decrease in cell temperature from 37°C to 10°C causes a major change in the secondary and tertiary structure of the 5′ untranslated region of the mRNA, the RNAT. The SD sequence becomes exposed, a pseudoknot that stabilizes the 10°C structure is formed, and the position of the nucleotides forming the cold box region (blue box) is altered. The 10°C structure binds readily to the 16S RNA of the 30S ribosomal subunit and translation of CspA commences. As the level of CspA rises, feedback inhibition of translation occurs. CspA binds to the cold box region, and this may decrease the probability of formation of the pseudoknot that is important in stabilizing the 10°C form of the RNAT. (After Kortmann and Narberhaus 2012.)

By preventing secondary structure—for example, hairpins—in the newly transcribed mRNA molecules, CspA fosters retention of translationally competent states of mRNAs and prevents termination of translation in the cold. The regulation of CspA synthesis is highly dependent on temperature, for two principal reasons. First, the mRNA for CspA is extremely unstable at high temperatures; at 37°C the half-life of the message is measured in seconds. A drop in temperature, for example, to 10°C, induces a shift in secondary structure (**Figure 3.37**) that dramatically increases the half-life of the *cspA* mRNA. Second, this cold-induced change in mRNA secondary structure also exposes the SD sequence needed for ribosome binding and initiation of translation. This mechanism of enhancing *cspA* mRNA translation is thus similar to that seen in ROSE elements, except that exposure of the SD sequence involves a major change in the RNAT's secondary structure, not a localized melting of a zipper region. One of the notable cold-induced changes in structure of the *cspA* mRNA is formation of what is termed a **pseudoknot**. A pseudoknot is a type of tertiary RNA structure that is involved in many types of processes. The pseudoknot found in the *cspA* mRNA is thought to play an important role in stabilizing the cold-temperature conformation. Further control of *cspA* mRNA translation is provided by feedback inhibition of translation as the level of CspA rises in the cell. One proposed mechanism of feedback inhibition involves binding of CspA to the **cold box** region of the low-temperature form of the mRNA. This binding interferes with formation of the pseudoknot that stabilizes the low-temperature form of the mRNA.

The complicated regulatory responses associated with temperature-dependent production of CspA illustrate the potential for changes in RNA structure to fine-tune the synthesis of proteins needed only under specific temperature conditions. Stabilization of the mRNA for CspA is seen in terms of the mRNA's secondary structure—which governs its capacity to be translated—and its vulnerability to degradation, which rises rapidly with increases in temperature. Thus, mRNA turnover and secondary structure help govern the bacterium's potential for producing an important protein, an RNA chaperone, in a temperature-specific manner.

TEMPERATURE-RNA INTERACTIONS: OTHER TYPES OF TEMPERATURE-SENSING EFFECTS
Most studies of RNATs have focused on bacteria, where this type of regulatory process was initially discovered. As pointed out earlier, however, the temperature sensitivities of RNA structure would seem to lend themselves to the evolution of diverse RNAT-like systems in all domains of life. These temperature-sensing systems are likely to involve a host of different types of structures and mechanisms in addition to the *cis*-acting mechanisms used by the RNATs discussed above that exploit temperature sensitivity of mRNA structure. (Note that *cis*-acting mechanisms involve structural changes at sites within a single molecule [intramolecular effects] whereas *trans*-acting mechanisms involve intermolecular interactions.)

One example of *trans*-acting regulation mediated by a temperature-dependent RNA structural change involves a small (~2kb, ~600 nucleotide) **noncoding RNA** (ncRNA) named heat-shock RNA-1 (HSR-1) that helps regulate the HSR in mammals (Shamovsky et al. 2006). HSR-1 is a highly conserved ncRNA that is constitutively expressed in mammalian cells. Its regulatory role is mediated by a temperature-dependent change in conformation from a closed state at low temperatures to an open state at temperatures where the HSR is needed. The high-temperature conformation of HSR-1 enables it to bind translation elongation factor eEF1A; this complex then interacts with HSF1 and leads to formation of the active trimeric form of this key transcriptional regulator of heat-shock genes. As discussed above, only the trimeric form of HSF1 can bind to the regulatory HSE of heat-shock genes and trigger expression of the genes. In the regulatory system involving HSR-1, temperature-driven changes in RNA geometry do not involve the types of zipper or switch mechanisms characteristic of *cis*-acting RNATs embedded in mRNAs. Rather, a small ncRNA with a temperature-dependent conformation serves as the regulator of gene expression. Such *trans*-acting mechanisms involving many different types of small ncRNAs may be of widespread importance in regulating temperature-dependent processes like the HSR (de la Fuente et al. 2012), as well as serving in their diverse and better understood roles in a range of processes involved in control of gene expression (Serganov and Patel 2007).

DNA: Temperature-sensitive structures and capacities for transcription

As in the case of RNA, the higher orders of structure of DNA are temperature sensitive because of the thermal labilities of the individual base-pairing interactions that stabilize these structures. These structures include the double helix and certain types of complex tertiary structures, for example, supercoiled regions, which DNA can form reversibly. DNA also exhibits temperature dependence in its interactions with the proteins that regulate gene expression and replication. These DNA-protein interactions, in turn, can be altered by temperature-dependent shifts in DNA configuration and in the conformations of the DNA-binding proteins. It is clear, then, that temperature's direct effects on DNA's structure and on its associations with gene-regulatory proteins will figure importantly in thermal biology. As in the case of RNA and proteins, thermal perturbation of DNA structure may bode ill in some cases, but offer important options for regulatory responses in other contexts. And as in the case of proteins, the ability of DNA to carry out its functions under different thermal conditions is attendant on maintenance of a particular marginally stable state of structure. Like proteins, DNA must be neither too rigid nor too unstructured if the reactions associated with transcription and replication are to be carried out with accuracy and efficiency.

Many of our most basic insights into temperature's effects on DNA structure have come from study of what might be termed "DNA at its simplest," the "raw" DNA molecules that occur as plasmids in bacteria. Here, DNA does not form a complex with proteins, so it is possible to obtain an especially clear picture of what a change in temperature can do to DNA structure per se. We will first examine temperature effects on plasmids from bacteria and archaea, to lay the groundwork for an analysis of temperature's influences on gene regulatory events that entail alterations in DNA structure and DNA-protein interactions. Despite the different levels of complexity we will examine in these disparate systems, common principles will emerge, notably the importance of maintaining DNA's ability to reversibly alter its structure in response to changes in temperature.

DNA SUPERCOILING: BASIC PROPERTIES   The prevailing conformation of DNA in the cell is termed **B-DNA**. B-DNA is a right-handed helix with a period of ~10.5 base pairs per turn (see Figure 3.31C). When the DNA strand is constrained such that free rotation of its ends is impossible—as in the cases of circular DNA in bacteria, mitochondria, and chloroplasts, and noncircular DNA in the nucleus that is anchored to the nuclear matrix— a more complex topology is generated. Twists in such constrained DNA favor what is termed **supercoiling** (Figure 3.38; Mirkin 2001). DNA supercoiling, which is found in all organisms, refers to the over- or underwinding of a strand of DNA; supercoiling is thus an indication of the amount of strain on the DNA strand, relative to the relaxed B-form of DNA. Extra helical twists generate what is termed **positive supercoiling**; positively supercoiled DNA is overwound compared with relaxed DNA. Reduced helicity is called **negative supercoiling**; here the DNA is underwound compared with relaxed DNA. In most organisms, DNA is negatively supercoiled, which facilitates conformational changes in the DNA needed for replication and gene expression (see below).

WHY SUPERCOILING IS IMPORTANT   Supercoiling is important for several reasons. One concerns the need to package DNA in the bacterial or archaeal cell, mitochondrion, chloroplast, or nucleus. Long DNA molecules must be compacted efficiently to accommodate them in the limited space available, and supercoiling greatly reduces the volume taken up

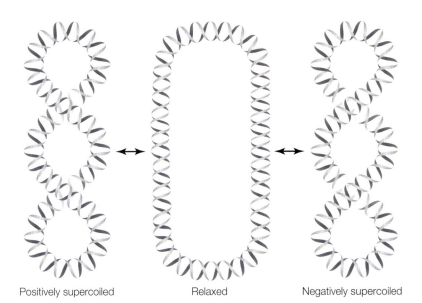

Positively supercoiled   Relaxed   Negatively supercoiled

Figure 3.38 Supercoiling of DNA. Positive and negative supercoiling compared with the structure of relaxed circular DNA.

by DNA. Another reason supercoiling is important is that the change in compactness that is caused by reversible supercoiling influences DNA's ability to be transcribed. Reversible supercoiling is catalyzed by topology-modifying enzymes such as **topoisomerases**. The localized changes in DNA topology that these enzymes cause can facilitate transcription of specific genes. Many topoisomerases can detect the degree of supercoiling and then either increase it or decrease it, as the need for change in DNA topology demands. Finally, supercoiling is also required for DNA (and RNA) synthesis. Because DNA must be unwound to allow RNA or DNA polymerases to function, reversible formation of supercoils accompanies these processes. The section of the nucleic acid sequence just ahead of the moving polymerase complex must be unwound to allow transcription or duplication. The stress resulting from this negative supercoiling is compensated by formation of positive supercoils ahead of the complex. Behind the polymerase complex, DNA is rewound; compensatory negative supercoils are generated. Topoisomerases such as DNA gyrase (a type II topoisomerase) are critical for determining the stress in nucleic acid structure during DNA and RNA synthesis.

TEMPERATURE AND DNA SUPERCOILING: ANOTHER TYPE OF REGULATORY SWITCH   One effect of a change in temperature on DNA structure is to induce a reversible change in supercoiling level (López-Garcia and Forterre 2000). These temperature effects on DNA topology may lead to significant changes in gene expression. Thus, as in the case of RNA thermometers, changes in DNA topology caused by temperature can function as sensors of temperature change and act as signal transduction mechanisms that influence gene expression. Because negatively supercoiled DNA is unstable and tends to unwind, increases in temperature can destabilize its topological state and thereby lead to changes in function. For example, there is evidence that the transient relaxation of DNA structure in negatively supercoiled DNA can lead to expression of heat-shock proteins (López-Garcia and Forterre 2000). The direct effects of temperature on the topology of negatively supercoiled DNA are enhanced by activities of topoisomerases, which can also be elevated by a rise in temperature. Furthermore, the heat-shock proteins that are rapidly synthesized may stabilize and activate certain topoisomerases, such as gyrases (Ogata et al. 1996). The activities of these topoisomerases may be needed to restore the degree of negative supercoiling that existed prior to thermal stress.

TEMPERATURE AND DNA SUPERCOILING: ARCHAEA AGAIN FOLLOW A UNIQUE PATH   One interesting example of how DNA supercoiling evolves in response to temperature is found in comparisons of hyperthermophilic archaea with mesophilic bacteria (Mirkin 2001). Whereas mesophilic species (including eukaryotes) have largely negatively supercoiled DNA, hyperthermophilic archaea have positively supercoiled DNA. Why should this difference exist? The likely answer has to do with maintaining a consistent marginal stability of DNA structure. Negatively supercoiled DNA predisposes DNA to open up its structure in response to regulatory proteins or increases in temperature. Therefore, negatively supercoiled DNA would be disadvantagous to a hyperthermophilic archaeon, because its strong tendency to have an open structure would lead to extreme unwinding in the heat. Unwinding of positively supercoiled DNA is energetically unfavorable, which creates an energy barrier to DNA strand separation. Of course, at the high temperatures at which hyperthermophilic archaea thrive (>80°C), there is abundant energy to overcome the topological rigidity barriers imposed by positively supercoiled DNA.

It will be interesting to see how the balance between negative and positive supercoiling varies among organisms evolved at different temperatures. For example, the propensities of DNA for forming these two types of supercoils could vary regularly with adaptation temperature. A similar—and complementary—variation in the activities of enzymes involved in regulating DNA topology might be anticipated as well.

DNA-binding proteins affect DNA structure and its flexibility

DNA's propensity for undergoing the conformational changes needed for transcriptional regulation is based not only on DNA topology; it also can be modified by proteins that interact with the DNA double-helical structure and govern its accessibility to gene regulatory proteins. To understand these regulatory interactions, it is helpful to briefly review the components of chromatin and how these molecules regulate gene expression. The basic chromatin unit of eukaryotes, the **nucleosome**, comprises about 150 base pairs of DNA and the four core histones (H2A, H2B, H3, and H4). Two other classes of proteins, **histone-1 (H1)** and **high mobility group (HMG) proteins**, bind transiently to the nucleosome and influence the conformation of the DNA. Through these interactions, the transcriptional poise of the DNA is established. H1 (termed a *linker* histone) has long been recognized as an inhibitor of transcription (Bianchi and Agresti 2005; Štros 2010). H1 tends to stabilize the condensed higher-order structures of chromatin and thereby reduce access of transcription factors to gene regulatory sites. HMG proteins, which are found in all eukaryotes, enhance transcription by inducing a more open conformation in chromatin, thus allowing transcription-regulating proteins access to gene regulatory sites. H1 and HMG proteins show competitive binding to nucleosomes, reflecting their opposing effects on transcriptional poise (Ueda and Toshida 2010).

HIGH MOBILITY GROUP PROTEINS: GLOBAL MODULATORS OF DNA STRUCTURE   The HMG proteins were named for their rapid movement in electrophoretic media, but one might view their name as appropriate for another reason: their enhancement of DNA's structural mobility (fluidity). All HMG proteins are relatively small (<30 kDa) and are commonly modified through posttranslational processes, notably lysine acetylation. Next to histones, HMG proteins are the most abundant proteins in the nucleus. The different classes of HMG proteins (HMGA, HMGB, and HMBN) are named based on the types of DNA-binding domains they contain (Štros 2010).

Here we focus on the **HMGB proteins**, which have a fluidizing effect on DNA structure and likely play an important role in compensating DNA structural flexibility to changes in temperature (Bianchi and Agresti 2005; Štros 2010; Ueda and Toshida 2010). HMGB proteins, the most abundant class of HMG protein, occur in at least four isoforms (HMGB1, -2, -3, and -4) and contain a diagnostic ~80-amino acid Box domain (hence the *B* in HMGB) that binds the minor groove of DNA with no (or only limited) DNA sequence specificity. We will chiefly examine HMGB1, which has received the most study to date. The principal function of all HMGB proteins is likely the bending of DNA, which leads to an openness of DNA structure and, thereby, an enhanced access of gene regulatory proteins to regulatory sites (Figure 3.39). Because HMGB proteins bind nonspecifically to different types of genes, they tend to have a *global* effect on the propensity of DNA to be transcribed. However, HMGB proteins also can bind transiently to specific transcription factors, leading to gene-specific activation of transcription. Last, HMGB1 proteins also regulate the activities of ATP-dependent chromatin remodeling proteins, which are important in regulation of transcription. HMGB proteins thus have two functions: They enhance the overall poise of chromatin for gene transcription; and they form complexes with specific transcription factors that determine which specific genes are transcriptionally activated. Note that in bacteria the protein **HU** may play a role similar to that of HMGB

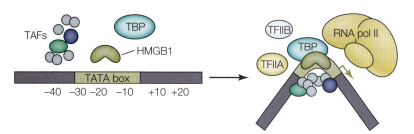

Figure 3.39 Interaction of high mobility group B1 protein (HMGB1) with DNA. HMGB1 binds to the TATA box in the promoter region of a gene. Binding creates a bend in the DNA, which enhances binding of regulatory proteins, including the TATA box binding protein (TBP) and TBP-associated factors (TAFs). As these regulatory proteins bind to the DNA, proteins required for transcription join the DNA-multiprotein complex, including transcription factors IIA and IIB (TFIIA and TFIIB) and the RNA polymerase complex (RNA pol II). In general, the DNA bending caused by HMGB1 is thought to be critical for HMGB1's ability to enhance and stabilize binding of transcription factors to DNA (Ueda and Yoshida 2010). The bending may also bring into juxtaposition formerly distant regions of DNA, further enhancing the ability of HMGB1 to modulate gene expression. (After Bianchi and Agresti 2005.)

proteins in eukaryotes, which implies that setting the transcriptional poise of DNA is important in all domains of life (Thomas and Travers 2001).

In view of the effect that temperature changes have on DNA structural stability, one might expect that conservation of the poise of DNA for transcription would be conserved if the concentrations of HMGB proteins in the nucleus were regulated in a temperature-specific manner: Decreases in temperature, which tend to tighten up DNA structure, would elicit increased expression of the genes encoding HMGB proteins; increasing temperature would have the opposite effect. Such temperature-dependent expression of HMGB proteins has been seen in a highly eurythermal killifish, *Austrofundulus limnaeus*, shown at left (Podrabsky and Somero 2004) (**Figure 3.40**). This South American tropical species has an unusual life cycle: It is an "annual" fish whose offspring endure the dry season as an encysted form in the

Austrofundulus limnaeus

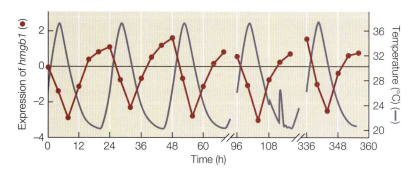

Figure 3.40 Temperature-correlated changes in the expression level of the *hmgb1* gene in the annual fish *Austrofundulus limnaeus*. Cycling of water temperature between 20°C and 37°C (continuous blue curves) was accompanied by a corresponding cycling in expression of the *hmgb1* gene (red dots with connecting red lines): Increasing temperature reduced expression, and falling temperature increased expression. Expression levels in steady-state acclimated fish (20°C or 37°C) resembled those found in fish sampled at the peaks or troughs, respectively, of expression in the cycling temperature regimen. (After Podrabsky and Somero 2004.)

dried mud that, in the previous wet season, had been the bottom of water-filled pools. In addition to having an ability to alternate between active, fully hydrated states as adults and metabolically quiescent desiccated embryos, *A. limnaeus* is able to tolerate a wide range of temperatures on a diurnal basis. In the field, water temperatures may vary between ~24°C and ~37°C over the course of several hours (Podrabsky et al. 1998). The fish's regular, circadian changes in body temperature are accompanied by changes in the expression of hundreds of genes, which lead to altered levels of mRNA for proteins such as molecular chaperones and enzymes involved in regulating membrane composition, cell proliferation, and organic osmolyte synthesis (Podrabsky and Somero 2004).

The role of HMGB1 proteins in the temperature-dependent regulation of circadian gene expression in this eurythermal species is suggested by the data in Figure 3.40. Expression of the gene encoding HMGB1 protein follows the exact pattern predicted above: Falling temperature strongly activates expression and rising temperature truncates expression of the *hmgb1* gene. The same inverse correlation between temperature and *hmgb1* gene expression was observed in fish during long-term acclimation to 20°C and 37°C. If temperature-related changes in *hmgb1* mRNA reflect changes in the levels of the HMGB1 protein, an assumption yet to be tested in this species, then the openness or fluidity of DNA seems to be modulated homeostatically; that is, changes in HMGB1 levels allow gene regulatory proteins to exert their effects across the almost 20-degree-Celsius range of body temperatures the fish encounters each day during the wet season. Upregulation of HMGB1 levels at low temperatures may help maintain homeostasis in gene regulatory activity for another reason: This type of temperature-dependent regulation might favor homeostasis—temperature compensation—in chromatin remodeling activity.

It is interesting that the promoter of the *hmgb1* gene has a binding site for HSF1 (Tang et al. 2005). This suggests a potential relationship between the HSR and expression of *hmgb1*. The fact that other studies of temperature-dependent gene expression in ectotherms have shown upregulation of *hmgb1* in the cold (Gracey et al. 2004) adds strength to the conjecture that maintaining the openness or fluidity of DNA in a stable fashion across all body temperatures—another example of the corresponding state phenomenon—is an important and perhaps ubiquitous feature of acclimation to change in temperature.

Temperature-dependent chromatin remodeling and regulation of gene expression

In the discussion of HMGB1 proteins, we pointed out that these molecules influence chromatin remodeling, which is of central importance in regulating gene expression. The opening up of DNA's structure by HMGB1 binding may facilitate several types of remodeling, including changes in the types of histones associated with the nucleosome and alterations in the extent of DNA methylation. The types of histone proteins associated with the nucleosome also help govern gene expression in various ways, including establishing the poise of DNA for transcriptional regulation, one of the effects already noted for HMGB proteins.

Temperature-dependent remodeling of nucleosomal histone composition is coming to be appreciated as an important mechanism for regulating gene transcription in response to alterations in ambient temperature. In some respects, the effect of temperature-mediated shifts in histone composition resembles what we have described for HMGB1 proteins, namely a change in the transcriptional poise for thousands of genes that are expressed in a temperature-specific manner. Here we focus on an example from a vascular plant (*Arabidopsis thaliana*), to show how changes in abundance of one type of nucleosomal protein, H2A.Z, enable gene expression to be up- and downregulated by changes in temperature.

We have discussed how replacement of the H1 linker (non-nucleosomal) histone by HMGB1 leads to a loosening-up of DNA structure and a concomitant rise in the regulatory responsiveness of genes. In contrast, when the H2A histone conventionally found in most nucleosomes is replaced by H2A.Z, the DNA in the nucleosome becomes more tightly wound and thereby less capable of binding transcription factors (Kumar and Wigge 2005). H2A.Z also disfavors transcription through a second mechanism: physical blocking. During normal transcription, RNA polymerase II (RNA pol II) relies on fluctuations in DNA structure to locally expose DNA and allow transcription to progress. Relatively rigid H2A.Z-containing nucleosomes prevent these fluctuations and thus can serve as a block to the progression of RNA pol II along the DNA. Nucleosomes containing H2A.Z tend to be most prevalent at the +1 position of genes; these H2A.Z-containing nucleosomes are thus just downstream of the transcriptional start site and are particularly well positioned to physically block transcription.

In view of these effects of H2A.Z-containing nucleosomes on the poise of the genome for transcription, it is pertinent to inquire how the prevalence of this type of nucleosome varies with temperature. In principle, temperature dependence in occupancy of DNA by H2A.Z-containing nucleosomes could lead to a global change in propensity for gene expression, somewhat akin to the effects noted for HMGB1.

Kumar and Wigge (2005) investigated temperature-dependent changes in expression of thousands of genes in the plant model organism *Arabidopsis thaliana*. Plants, like the bacteria discussed in the context of RNA thermometers, are able to sense changes in temperature of ~1 degree Celsius, albeit with differences in the precise gene regulatory switches involved. In *Arabidopsis* a large shift in growth temperature, from 12°C to 27°C, led to upregulation of ~2500 genes and downregulation of ~2900 genes. Central to the processes of both up- and downregulation was the involvement of H2A.Z-containing nucleosomes (Figure 3.41). The fraction of H2A.Z-containing nucleosomes bound to DNA decreased regularly with increase in temperature. This effect was noted across all genes examined. Tellingly, plants that were genetically engineered to lack H2A.Z-containing nucleosomes showed high-temperature transcriptional profiles at 17°C. For genes whose transcription was increased at high temperature, the dissociation of H2A.Z-containing nucleosomes from the DNA removed the physical blockage of transcription by RNA pol II and allowed specific transcription factors to access gene regulatory sites (see Figure 3.41A). For genes whose expression was reduced at high temperature, the loss of H2A.Z nucleosomes from the DNA led to increased exposure of gene regulatory sites to repressors, which replaced activators previously associated with the cold-expressed genes (see Figure 3.41B). In addition, downregulation could be enhanced by chromatin modifications, such as DNA methylation, which can lead to changes (decreases or increases, depending on the context) in the transcriptional ability of DNA.

In summary, chromatin remodeling represents an important way for organisms to alter the transcriptional poise of their genes for temperature-dependent expression. A variety of proteins that interact with DNA, including HMG proteins and nucleosomal and non-nucleosomal histones, work together in this complex regulatory mechanism.

DNA strand breakage: Influences of temperature

Throughout our discussion of thermal perturbation of macromolecular structure, we have emphasized the lability of the noncovalent (weak) chemical bonds that establish higher orders of structure; we have said relatively little about disruption of the covalent bonds that link individual structural units (amino acids or nucleotides) together. Whereas extremes of high temperature can lead to several types of damage to protein's covalent bonds, little

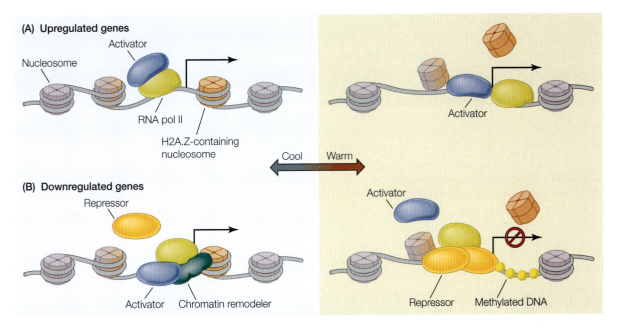

Figure 3.41 Temperature-dependent modulation of transcription by H2A.Z-containing nucleosomes. The nucleosomes contain different amounts of the conventional histone H2A and its variant, H2A.Z. Color shading indicates the relative amount of H2A.Z. Blue nucleosomes lack H2A.Z; red nucleosomes contain high amounts of H2A.Z. Mixed red-blue shading denotes fractional representation of the two histones. Other types of histones (H2B, H3, and H4) are present as well, but they appear not to play a role in temperature-dependent regulation of transcription. H2A.Z favors tighter compaction of the ~150 base pairs of DNA in a nucleosome. (A) For genes that are upregulated at high temperatures, increases in temperature favor release of H2A.Z nucleosomes from DNA, allowing progression of RNA pol II and binding of transcriptional activators to regulatory sites. (B) In contrast, for genes that are downregulated at high temperature, release of H2A.Z nucleosomes allows binding of repressors that block transcription. Further regulation may involve chromatin remodeling, for example, DNA methylation, which leads to reduction in transcriptional activity of heat-repressed genes. (After Kumar and Wigge 2005.)

is currently known about the relevance of these effects in real-world circumstances. Much the same is true for thermal damage to the covalent bonds that establish the backbone of the two helices of genomic DNA. However, recent work has shown that exposure to low as well as high temperatures can lead to accumulation of breaks in DNA strands, and these breaks can have prominent downstream effects on cellular function and integrity.

TYPES OF DNA STRAND BREAKAGE AND THEIR CHALLENGES TO CELLULAR INTEGRITY
Even in the absence of thermal stress, the bonds joining nucleobases frequently break. It is estimated that millions of **single-strand breaks** (**SSBs**) occur each day in a single human cell (Jackson and Loeb 2001). SSBs typically are repaired quickly and lead to no lasting damage to the genome. **Double-strand breaks** (**DSBs**) are far more threatening to the integrity of the genome. DSBs can interrupt transcription, DNA replication, and chromosome segregation. For example, DSBs can eliminate the types of supercoiling and topological complexity that are essential for DNA condensation in the nucleus and for regulation of gene expression. Both SSBs and DSBs, if they are extensive enough and are not repaired rapidly, can lead to downstream effects like apoptosis. Initiation of this process is marked by activation of caspase-3, the "executioner" caspase involved in apoptosis.

DNA STRAND BREAKAGE IN THE FIELD: EXAMPLES FROM MUSSELS   Evidence that SSBs and DSBs accumulate in a temperature-dependent pattern has come from studies of intertidal molluscs of the genus *Mytilus*, which experience large changes in body temperature as a consequence of tidal rhythms, latitude, and season (Yao and Somero 2012). These studies used hemocytes removed from mussels that had been subjected to increases or decreases in temperature for different periods of time. Hemocytes are free cells in the body fluids that participate in several processes, including immune defense and wound repair. The fact that the hemocytes were exposed to the temperature changes while in their normal hemolymph environment increases the likelihood that the effects observed in these studies are biologically meaningful. SSBs and DSBs were quantified using what is termed a *comet assay* (**Figure 3.42A**). In this procedure, the amount of DNA that occurs in the "tail" of the comet provides a quantitative measure of DNA strand breakage; the comet's tail contains fragments of DNA that have separated from the undamaged nuclear DNA that forms the comet's "head."

Decreases as well as increases in temperature led to increased amounts of SSBs and DSBs (**Figure 3.42B,C**). The accumulation of SSBs and DSBs showed a similar pattern. At

(A)

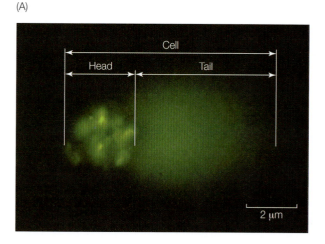

Figure 3.42 Influence of different thermal exposures (combinations of temperature × time) on single- and double-strand breakage of DNA. (A) The comet assay. SSBs and DSBs are quantified by the percentage of DNA found in the "tail" (% DNA-T) of the "comet" that appears when single hemocytes are subjected to electrophoresis in an agar medium and DNA is stained with SYBR green. Different cell treatments prior to electrophoresis are used to distinguish SSBs (neutral comet assay) and DSBs (alkaline comet assay). (B,C) The amounts of SSBs (B) and DSBs (C) in hemocytes of specimens of the mussel *Mytilus galloprovincialis*. * = significantly different from 13°C individuals ($P < 0.05$). (After Yao and Somero 2012.)

(B)

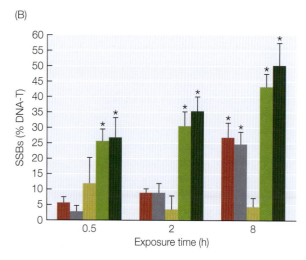

(C)

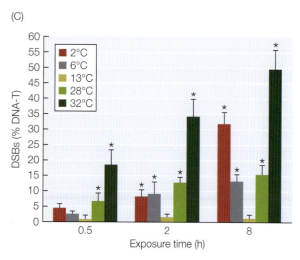

13°C, which is a typical water temperature for the mussel population sampled, the amounts of SSBs and DSBs were at a minimum, and SSBs were more prevalent than DSBs. For both types of strand breakage, the extent of DNA damage depended on the duration of exposure to either high or low temperature. There are undoubtedly multiple reasons that SSBs and DSBs increase over time at both high and low temperatures. Initially, increased strand breakage might be caused by changes in normal metabolic processes. For example, rising temperature would be expected to elevate rates of oxygen consumption and, therefore, production of ROS, which are known to disrupt DNA structure (see Chapter 2). The effectiveness of DNA repair systems also is likely to play a role in establishing the observed patterns. If, for example, damage caused by ROS production exceeds the capacity of repair systems to heal strand breakage, temperature- and time-dependent accumulation of damage would be expected. At low temperature, a deceleration in repair activities might be especially problematical for achieving DNA repair.

Correlated with damage to the hematocytes' DNA were reductions in cell viability (**Figure 3.43A**) and membrane integrity (**Figure 3.43B**), and an increase in caspase-3 activity (**Figure 3.43C**). Increases and decreases in temperature both led to cellular damage, but the effects of elevated temperatures were more severe in all cases. These effects may be causally related to DNA damage. Loss of membrane integrity, notably of the mitochondrial membrane, can lead to enhanced production of ROS (see Chapter 2). In turn, higher amounts of ROS may elevate the amount of damage to DNA and activate the enzymes involved in apoptosis, notably the executioner caspase, caspase-3. The hemocytes may have lost viability through a combination of direct temperature effects on membrane systems and downstream effects associated with apoptosis that is triggered by DNA damage.

The responses of hemocytes to the highest exposure temperature, 32°C, merit additional comment, for these data speak to the issue of the temperature range over which an ectotherm can mount an appropriate response to address thermal damage. Mussel body temperatures may exceed 35°C in the field. These high temperatures have profound effects on gene transcription and protein expression (Gracey et al. 2008; Lockwood et al. 2010; Tomanek and Zuzow 2010). In the context of apoptosis,

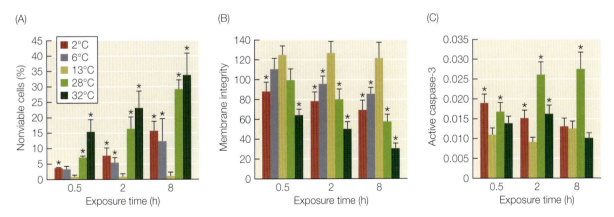

Figure 3.43 Effects of thermal exposure of *Mytilus galloprovincialis* on (A) hemocyte viability (indexed by the cell's ability to exclude the dye trypan blue), (B) lysosomal membrane integrity (indexed by the retention time of the dye neutral red in lysosomes), and (C) activation of the apoptosis executioner caspase, caspase-3. * = significantly different from 13°C individuals (*P* < 0.05). (After Yao and Somero 2012.)

caspase-3 activity rose to higher levels at 28°C than it did at 32°C (see Figure 3.43C). This observation suggests that an organism's ability to mount an appropriate response to damage from heat has a distinct upper thermal limit, beyond which the processes of transcription and translation are unable to achieve the level of activity needed to restore homeostasis or, if damage is too severe to be repaired, to trigger apoptosis. Thus, thermal death may result not only from direct perturbation of the structures and activities of biochemical systems, but also from a breakdown in the mechanisms that support a return to homeostasis.

Perhaps the key implication of the findings from the hemocyte studies is that, even at temperatures within the normal thermal range of a species, both increases and decreases in body temperature can lead to significant cellular damage that, in turn, would be expected to increase the costs of living. These increases in costs for cellular maintenance or replacement of cells that are killed by changes in temperature are clearly greater when temperatures rise than when they decrease. This observation speaks to a question we will return to later in this chapter in the context of compensation to temperature change by ectotherms: How does the cost of living vary with temperature? Perhaps a more significant challenge arising from accumulation of increased amounts of DNA strand breakage at extremes of temperature concerns the integrity of the genome itself. Does a rise in DNA strand breakage, DSBs in particular, lead to increased rates of mutation at temperature extremes? This is a question that definitely merits investigation as ectothermic organisms encounter a rapidly warming world.

3.6 Lipids

Lipid-temperature interactions provide an especially clear example of the need for establishing and defending a particular state (or range) of fluidity for molecular systems. As initially defined in Chapter 1 and illustrated by examples involving proteins and nucleic acids, fluidity is a broadly useful concept for describing the degree of fluctuation among structural states that is possible in a macromolecule or in a complex multimolecular structure such as a nucleosome. In this section we extend this analysis to lipid-containing systems, with a principal focus on cellular membranes. As with proteins and nucleic acids, lipid-rich systems will be seen to adapt or acclimatize to different temperatures by adjusting their composition to retain (or regain) an optimal level of fluidity—a corresponding state of structure. Studies of lipid-temperature relationships were, in fact, some of the earliest investigations of the importance of conserving fluidity in living systems (Hazel 1995). And as in the case of proteins and nucleic acids, we are now in a strong position to explain the thermal sensitivities of lipid-rich systems in precise molecular detail. In a sense, lipid-based systems have more raw materials to work with than proteins—whose constituents number but 20 amino acids—and nucleic acids—whose building blocks are even smaller in number. Lipids comprise a much larger number of structural forms (**Figure 3.44**), some of which are unique to a

Figure 3.44 Different types of lipids and their assemblage into components of cellular membranes. (A) Stearic acid and oleic acid, two 18-carbon fatty acids that are common constituents of membranes and depot (storage) lipids. A fatty acid is defined by the length of its carbon chain, total number of double bonds, and location of the first double bond relative to the terminal methyl group of the acyl chain (n). Thus oleic acid is designated as 18:1, n-9. (B) Glycerol, a triacylglyceride, and a phosphoglyceride (phosphatidylcholine). (C) Choline, ethanolamine, serine, and inositol are common head groups. (D) A tetraether lipid from a thermophilic archaeal species. (E) Cholesterol (eukaryotic) and a hopanoid (diploptene) found in bacterial membranes.

single domain of life. We will see below just how effectively this set of chemical raw materials has been exploited to permit critical membrane-based functions to occur over an almost 200-degree-Celsius range of cellular temperatures.

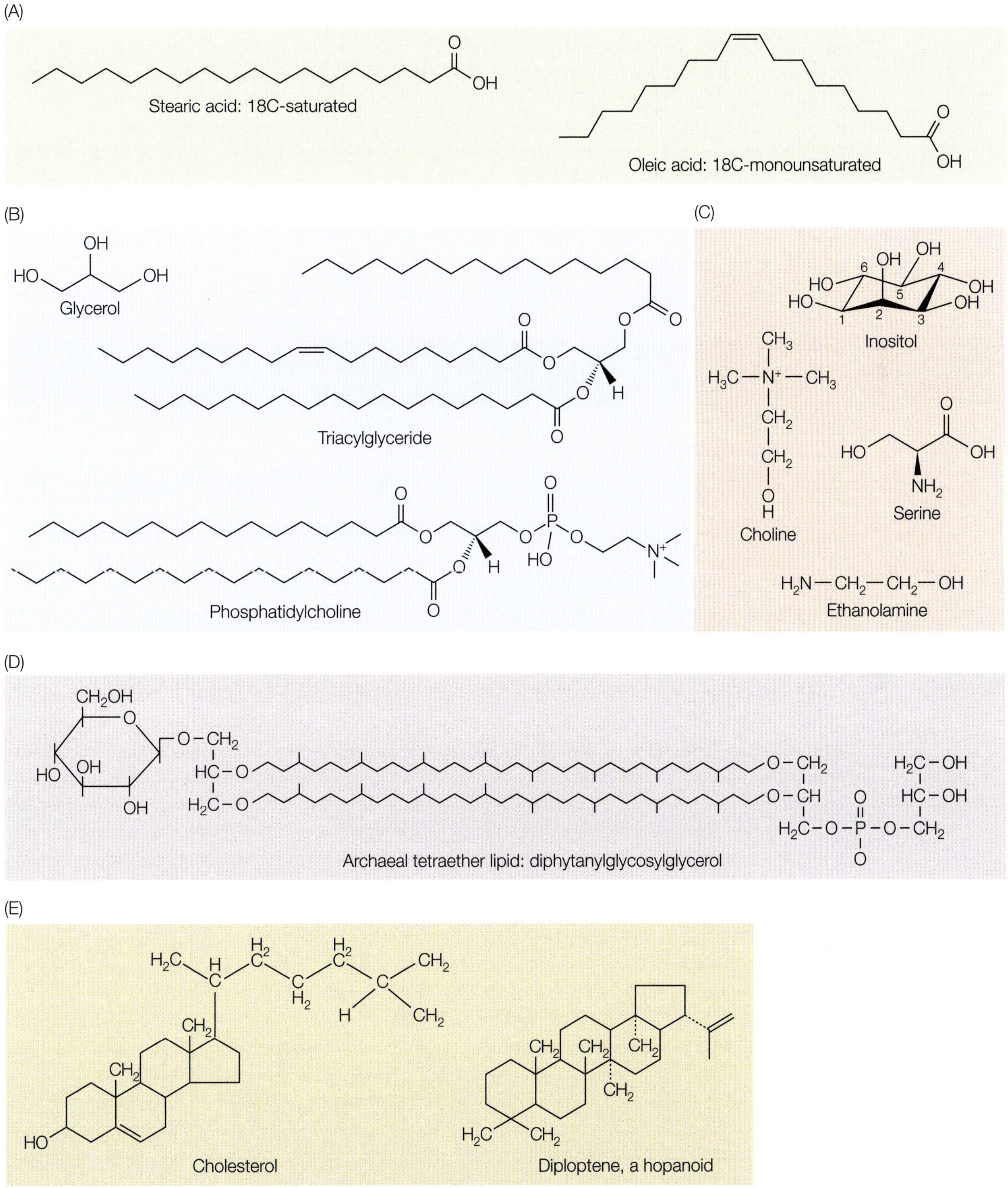

(A)

Stearic acid: 18C-saturated

Oleic acid: 18C-monounsaturated

(B)

Glycerol

Triacylglyceride

Phosphatidylcholine

(C)

Inositol

Choline

Serine

$H_2N-CH_2-CH_2-OH$
Ethanolamine

(D)

Archaeal tetraether lipid: diphytanylglycosylglycerol

(E)

Cholesterol

Diploptene, a hopanoid

Lipids: Their structures and multiple functions

To appreciate the temperature relationships and adaptational plasticity of lipids, it is useful to review the wide variety of chemical species that constitute lipid-containing structures like cellular membranes and storage (depot) lipids (see Figure 3.44). Lipids may be defined generically as relatively small organic molecules that are strongly hydrophobic or, in some cases, **amphiphilic** (one part of the molecule is polar or charged and readily interacts with water; the remainder of the molecule is hydrophobic and tends to separate from the aqueous phase). Lipids fulfill a variety of functions, including the fabrication of structures like membranes; the provision of fuel depots for exploitation when food is not readily available or when sustained physical activity precludes food intake; the supply of essential nutrients, including fat-soluble vitamins (A, D, E, and K); the defense against water loss, for example, through production of waxes that coat the cuticles of terrestrial arthropods; thermal insulation from cold; and the regulation of a great many physiological processes, including reproduction and inflammation, through actions of steroid hormones.

In the context of temperature-lipid interactions, the principal categories of lipids we will consider are fatty acids, whose acyl chains are incorporated into membrane **phosphoglyceride lipids** (phospholipids) and **depot glycerolipids**; sterols such as cholesterol (and their bacterial counterparts, the hopanoids), which have important effects on governing membrane order; and prenol lipids (derived from condensation of isoprene subunits), which occur in all domains of life but are particularly important in cell membranes of thermophilic archaea. These diverse types of lipids serve as the raw material for fabricating membranes with biophysical properties appropriate for the thermal conditions the organism faces.

With this brief review of lipid chemistry as a foundation, we now examine one of the most critical areas of the thermal biology of lipids: conservation of membrane structure and function. An analysis of temperature-membrane interactions will provide important insights into the fundamental underpinnings of thermal optima and thermal tolerance limits, as well as into the critical role played by phenotypic plasticity in enabling organisms to acclimatize to temperature change.

Membrane structure and function: The fundamental bases of membrane temperature sensitivity

To provide a foundation for understanding the broad importance of adaptation to temperature in membrane systems, it is useful to review the several basic functions of membranes and see how these functions are dependent on the composition and physical state of membrane lipids. Only by understanding the complex and dynamic characteristics of membrane structure is it possible to appreciate why alterations in membrane lipid composition can play such important and wide-ranging roles in maintaining membrane function.

THE MULTIPLE ROLES OF MEMBRANES   One function of membranes is to serve as a physical barrier between the cell and the surrounding medium (air, water, or extracellular fluids) or between an organelle and the cytosol in which it occurs. As we will discuss in Chapter 4, the fluids on either side of a membrane almost invariably differ in chemical composition, and the barrier function of membranes is crucial in sustaining these compositional differences. Membranes also contain transport systems that are essential for governing movement of solutes between body compartments or between the organism and the external solution. Such transport activities also generate transmembrane potentials needed for driving other types of transport and allowing conduction of nerve impulses. Signal transmission between nerves involves release of neurotransmitters contained within membrane-bound vesicles, which undergo exocytosis at the synapse in response to transmembrane ion flux. Certain membranes support bioenergetic processes such as mitochondrial ATP generation and photosynthetic electron transport. Membranes are also involved in diverse signal transduction systems. Signal transduction may involve an initial binding of a hormone to the outer surface of the membrane, followed by metabolic changes in the cell. In some cases, a membrane constituent, for instance, phosphatidylinositol, is chemically modified to form a signaling molecule. In many cases, movement of signal transduction proteins across the plane of the membrane is part of the signaling process, which is perhaps one of the best illustrations of the structurally dynamic nature of membranes.

MEMBRANE STRUCTURE IS DYNAMIC   A full understanding of the mechanisms by which membranes fulfill their diverse functions demands that we view membranes as highly dynamic structures. As in the case of drawings that illustrate the static structures of proteins and nucleic acids, the typical textbook illustrations of membrane structure (Figure 3.45) belie the dynamic features of the system. Both the protein and lipid components of membranes are in continuous motion. These motions involve fluctuations in the geometries of individual protein and lipid molecules—their structures "flicker" among a wide spectrum of microstates—and the lateral movement of proteins and lipids within the plane of the bilayer. Lipids can also "flip-flop" between the two hemilayers (leaflets) of the bilayer membrane. As discussed above in the case of soluble proteins, integral membrane proteins must possess the correct geometries to recognize and bind their ligands and to undergo the changes in conformation that are requisite for function. The abilities of membrane proteins to undergo conformational changes may be strongly influenced by the fluidity of the bilayer. Too rigid a bilayer may inhibit conformational changes; too fluid a bilayer may allow the protein to drift into conformational microstates that fail to support optimal function.

The interactions between membrane proteins, especially **integral** (**intrinsic**) **proteins**, and lipids comprise several phenomena that range from bulk lipid (membrane fluidity) effects on protein activity to specific effects arising from the lipids in the protein's immediate vicinity, so-called **annular lipids** (Qi et al. 2006). Integral membrane proteins exhibit different functional properties and structural stabilities when the annulus of lipids surrounding them changes (Lee 2004; Haviv et al. 2007; Phillips et al. 2009; Cornelius et al. 2015). Thermal disruption of functionally optimal aggregates of proteins and annular lipids thus may be one important consequence of changes in cellular temperature. Lipid-protein interactions may also be resistant to temperature change. Certain lipids that bind extremely tightly to proteins are termed **cofactor lipids** (the expressions *non-annular lipids* and *specifically bound lipids* are also used in the literature) and may not be susceptible to thermally induced dissociation from proteins (Lee 2004; Cornelius et al. 2015). Cofactor lipids may fit tightly into grooves between the α-helices of integral membrane proteins or be bound between subunits. Without cofactor lipids, the protein is likely to be nonfunctional, hence the appropriateness of the adjective *cofactor*.

A protein's movement within the plane of the membrane is also affected by the fluidity of the bilayer. Lateral movement of membrane proteins is often an essential event in membrane-localized processes like signal transduction. The fluidity of the bilayer must be adequate to allow lateral movement of proteins (and protein complexes) but not be too fluid such that proteins that interact can "float away" from their partners in a multiprotein complex. Thus the composition of the lipid bilayer, which typically contains a few hundred different types of phosphoglycerides as well as sterols like cholesterol, plays an important role in establishing the organization and functional properties of membrane proteins.

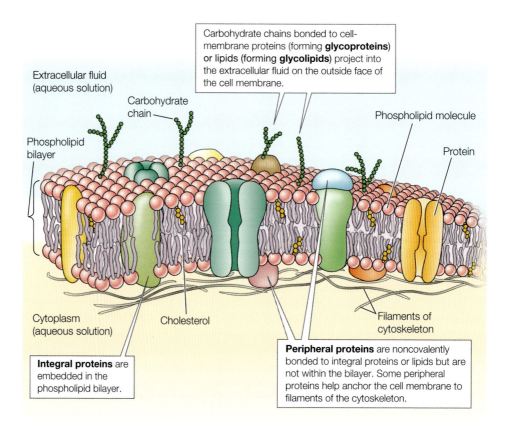

Extracellular fluid
(aqueous solution)

Carbohydrate
chain

Carbohydrate chains bonded to cell-
membrane proteins (forming **glycoproteins**)
or lipids (forming **glycolipids**) project into
the extracellular fluid on the outside face of
the cell membrane.

Phospholipid molecule

Phospholipid
bilayer

Protein

Cytoplasm
(aqueous solution)

Cholesterol

Filaments of
cytoskeleton

Integral proteins are
embedded in the
phospholipid bilayer.

Peripheral proteins are noncovalently
bonded to integral proteins or lipids but are
not within the bilayer. Some peripheral
proteins help anchor the cell membrane to
filaments of the cytoskeleton.

Figure 3.45 Structure of an animal cell membrane. The lipid content (percentage of mass) of membranes varies among different types of membranes. Cell membranes of eukaryotes are approximately 40% lipid and 60% protein by mass (Engelman 2005; Phillips et al. 2009). In animal cell membranes, cholesterol may constitute 30–50 mol percent of total lipids, the remainder being phosphoglycerides (phospholipids). The hydrophobic core region of the membrane, which corresponds to the region occupied by the acyl chains of phosphoglycerides, is ~30Å thick. The region containing the head groups on each side of the bilayer is ~12 Å thick. Phosphoglycerides of different lipid classes, as defined by the head group of the phosphoglycerides, are asymmetrically distributed between the two hemilayers (leaflets) of the bilayer. The outer leaflet of cell membranes consists primarily of phosphatidylcholine and sphingomyelin; phosphatidylethanolamine and phosphatidylserine are the major phosphoglycerides of the inner leaflet. Phosphatidylinositol, which generally is less abundant than other phospholipids, is localized in the inner leaflet of the cell membrane, where it plays important roles in signal transduction. Proteins are classified as integral (= intrinsic) or peripheral (= extrinsic) based on where they occur. Sugars and lipids may be attached to membrane head groups, yielding a further complexity in lipid composition. Within the membrane, proteins are generally closely juxtaposed. It is estimated that the mean center-to-center spacing among membrane proteins is ~10 nm; this distance is comparable to the estimated distances separating proteins in the cytoplasm (Phillips et al. 2009). As we will see in Chapter 4, close packing of proteins—termed *crowding*—has a strong effect on protein stability.

LIPID STRUCTURE AND MEMBRANE BIOPHYSICAL STATE   To understand how changes in lipid composition affect the thermal sensitivities of membranes, it is necessary to look in more detail at the differences among lipids that lead to their varied influences on membrane biophysical state. In the case of phosphoglycerides (see Figure 3.44B), among the characteristics important in establishing membrane properties, including temperature effects, is the **double-bond content** of acyl chains. Both the total number of double bonds and their positions along an acyl chain are important. If all carbons in an acyl chain contain the maximum number of hydrogens, there will be no double bonds between carbon atoms and the acyl chain is referred to as being **saturated**. Introduction of one or more double bonds creates an **unsaturated** acyl chain whose physical properties may differ substantially from a saturated acyl chain with the same number of carbons. The most important effect of the introduction of one or more double bonds is on the geometry of the acyl chain. A *cis* double bond (the two absent hydrogens are from the same side of the lipid chain; see Figure 3.44A) introduces a bend or kink in the acyl chain. Saturated acyl chains thus can pack together more tightly and engage in more extensive van der Waals interactions that favor a rigid membrane structure. This difference in packing density and acyl chain interactions is manifested in large differences in melting temperature between saturated and unsaturated acyl chains with the same number of carbons. For example, the 18-carbon fully saturated fatty acid stearic acid has a melting temperature of 69°C. The 18-carbon lipid oleic acid, which contains a single *cis* double bond, melts at 13°C. From this one example, it should be apparent how adjusting the double-bond contents of membrane lipids could achieve adaptive alterations in membrane fluidity to compensate for effects of temperature.

In general, introduction of the initial double bond has a stronger effect on melting temperature than the introduction of additional double bonds. However, the position of the double bonds along the carbon chain also can help define the temperature sensitivities and functional importance of a lipid. For example, long-chain polyunsaturated acyl chains such as docosahexaenoic acid (DHA), which has 22 carbons and 6 double bonds, with the first double bond in the *n*-3 position, will be seen to have extremely important influences on membrane function because one or more of the double bonds are located in the deeper region of the membrane bilayer. Through its effects on activities of enzymes involved in ATP generation, DHA may help serve as a "metabolic pacemaker" that influences the rate of mitochondrial oxygen consumption, as we discussed in the context of metabolic scaling and the evolution of high metabolic rates in birds and mammals (see Chapter 2).

Temperature effects on the geometry of the acyl chains of phosphoglycerides are also important in establishing the functional properties of membranes. The acyl chains of membrane lipids undergo continuous rotation about carbon-carbon single bonds, generating what are termed **gauche rotamers**. As temperature increases, the acyl chains tend to fan out more, thereby converting a lipid with an essentially cylindrical geometry at low temperatures into one with a conical geometry at increased temperatures (**Figure 3.46**). This effect of high temperature on lipid geometry enhances the probability of forming what are termed **inverted hexagonal phase II (H_{II}) structures**. These H_{II} structures serve important roles in certain processes that require membrane fusion, such as exocytosis and endocytosis. We will see below that conserving an optimal propensity for generating H_{II} phases is an important component of the homeostatic modification of membranes in the face of changing temperatures.

CHOLESTEROL AND ITS CRITICAL ROLES   Cholesterol is also important for establishing the functional properties of membranes. In cell membranes of eukaryotes, cholesterol (see Figures 3.44 and 3.52) may constitute 30–50 mol percent of the total lipids. Because cholesterol is a smaller molecule than a typical phospholipid, it contributes only about 20% to the total lipid mass. Cholesterol is an amphiphilic molecule, and its partitioning within the membrane reflects this fact. The rigid, nonpolar steroid ring system of

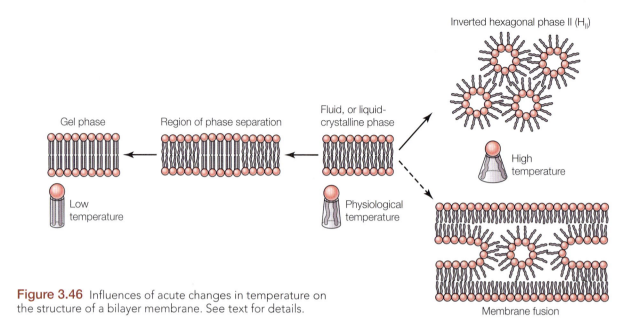

Figure 3.46 Influences of acute changes in temperature on the structure of a bilayer membrane. See text for details.

cholesterol intercalates into the bilayer, where it interacts with acyl chains and tends to favor their linear organization. The organizing effect of the steroid ring system is thought to be especially strong for saturated acyl chains, which lack the kink found in unsaturated acyl chains (see Figure 3.44A). The polar 3-β-hydroxyl group of cholesterol resides near the membrane surface and lies in close contact with ester linkages that join the acyl chains to glycerol. Localization of this hydroxyl group near the membrane surface helps keep the head groups of phosphoglycerides separated.

A critical role of cholesterol is in establishing functionally important degrees of order in the acyl chains of phosphoglycerides. Without cholesterol in their immediate vicinity, the fluid-phase acyl chains of the bilayer phosphoglycerides are in what is termed a **liquid-disordered (l_d) state**. When cholesterol is present, the acyl chains tend to form a more organized **liquid-ordered (l_o) state**, which is intermediate in order between the l_d state and the **gel phase** (see Figure 3.46). By stabilizing the l_o state, cholesterol disfavors the formation of the gel phase. Within a single membrane, both l_d and l_o states may coexist. Note that the expression *liquid-crystalline* state now is generally used to refer to the l_d state; however, earlier literature that predates a deep understanding of the effects of cholesterol tended to use this expression to contrast fluid states in general with gel-like states. In discussing examples from the older literature, we will follow the convention used by the authors, but with the realization that the fluid states they describe as "liquid-crystalline" likely represent combinations of l_d and l_0 phases.

Cholesterol's effects on the organization of acyl chains are thought to be critical for the proper functioning of **membrane rafts**, which are organized assemblages (microdomains) of cholesterol, **glycosphingolipids** (lipids in which acyl chains are linked to a long-chain amino alcohol, **sphingosine**; **Figure 3.47**), and proteins. Retention of the l_o state of raft lipids is regarded as essential for their functions, which include signal transduction, membrane protein trafficking, and control of neurotransmission.

Cholesterol also increases the thickness of the membrane bilayer and reduces its surface area. Cholesterol-mediated thickening of the bilayer provides a nonpolar lipid

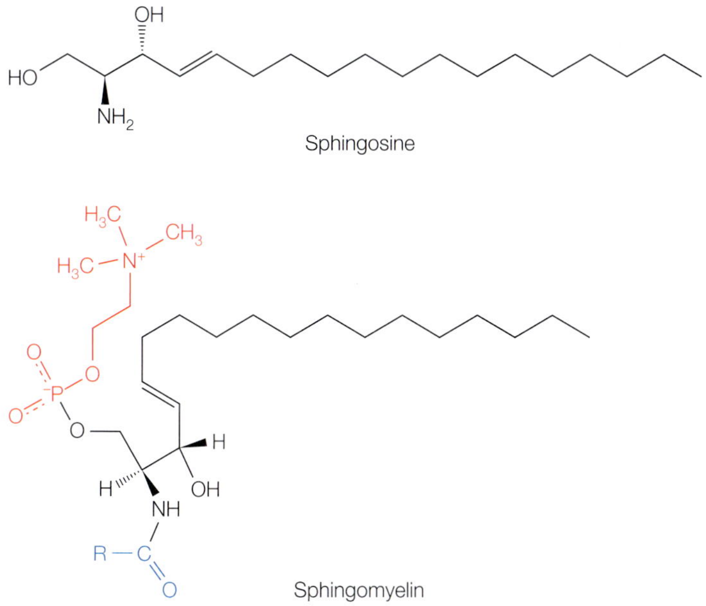

Figure 3.47 The amino alcohol sphingosine and a sphingolipid, sphingomyelin. Colors indicate the components of sphingomyelin: black, sphingosine; red, phosphocholine; blue, fatty acid.

environment along the full length of the transmembrane (TM) domains of membrane-spanning proteins. These TM domains are strongly hydrophobic, and membrane-spanning proteins are apt to function optimally when their TM domains are completely buried within the bilayer.

H_{II} VERSUS BILAYER STRUCTURES: WHAT CONTROLS THIS DECISION?   One final lipid-protein interaction that merits our attention is the influence that proteins have on the tendencies of lipids to form H_{II} structures versus bilayer structures. This balance is related in part to the lipid composition of a membrane. A mixture of two types of lipids, one with a propensity for forming bilayers and one with a preference for H_{II} structures, will form a bilayer if the mixture has more than ~20% bilayer-preferring lipid (Boni and Hui 1983). Addition of an integral membrane protein into the mixture leads to a stronger preference for bilayer formation. In other words, lipids that might normally enter an H_{II} structure in a pure lipid system adopt a bilayer structure in the presence of integral proteins. Thus, much as lipids can influence the functional properties of integral membrane proteins, these proteins can help determine whether or not lipids form bilayers or H_{II} structures. How changes in temperature affect this ability of proteins to influence phase behavior may depend on how changes in temperature alter the distributions of lipids and proteins in the membrane.

This brief account of the interdependence of lipids and proteins in establishing membrane function, and of the ways lipid-protein interactions might be disrupted by changes in temperature, indicates that adaptive changes in several membrane properties are likely to be essential for conserving membrane function in the face of temperature change during evolution and acclimatization. We now examine these conservative adaptations and the types of changes in lipid composition that bring them about. This analysis, which includes examples from all three domains of life, will illustrate a striking evolutionary convergence in function across taxa, which in some instances is supported by radically different chemistries in different evolutionary lineages. Thus, as we saw in our examination of protein evolution, the same result can be reached by different chemical routes.

Homeoviscous and homeophasic adaptation of membranes

The conservation of the appropriate biophysical characteristics of membranes in the face of changes in temperature, which is observed in all domains of life, is commonly discussed in the context of what are termed *homeoviscous adaptation* and *homeophasic adaptation*, which we will define shortly. This perspective on temperature adaptation of membranes has been developed through the efforts of a large number of researchers over several decades, but perhaps there remains no better general presentation of these ideas than that found in a review published by Jeffrey R. Hazel more than 20 years ago (Hazel 1995).

THERMAL CHALLENGES TO MEMBRANE HOMEOSTASIS   The gist of the challenges posed by changes in temperature to membrane structure and function is illustrated in Figure 3.46. The liquid-crystalline (liquid-ordered [l_o]) state that is optimal for membrane function is challenged by both decreases and increases in temperature. A fall in temperature can lead to phase separation, in which lipids with high melting temperature cluster together and form gel phase regions within the otherwise liquid-crystalline bilayer. At the junctions between gel phase and liquid-crystalline regions, the integrity of the membrane barrier may be compromised. Furthermore, the highly viscous gel phase lipids may hinder protein conformational changes and the lateral movement of proteins within the plane of the bilayer. Formation of gel phase regions, which are thicker than liquid-crystalline regions, will influence the hydrophobic interactions between the lipid bilayer and integral membrane proteins. When embedded in a bilayer, a membrane protein will, if possible, assume a conformation that matches the thickness of the TM domain to that of the membrane's nonpolar region (Lee 2004). Thus, membrane thickening caused by formation of gel phases would favor a change in the conformation of transmembrane proteins, which could alter their activity.

Increases in temperature disfavor gel phase structures, but may lead to maladaptive increases in the fluidity of bilayer. A highly fluid microenvironment for membrane proteins may lead to disaggregation of functional protein-protein and protein-lipid complexes. The proteins may also become less stable because of the increased freedom to undergo fluctuations in conformation. With a broader range of conformational microstates available to it, a protein may lose some of its binding capacity, as discussed earlier for soluble proteins. The effects of decreases and increases in temperature on membrane fluidity thus point to a need to regain membrane homeostasis in fluidity (or viscosity): **homeoviscous adaptation** (Sinensky 1974).

Temperature changes also alter the propensity of the membrane to form nonlamellar structures, notably inverted H_{II} structures of lipids, because of thermal effects on the geometries of acyl chains (cylindrical versus conical shapes; see Figure 3.46). Shifts in the poise for forming H_{II} structures can lead to maladaptive changes in the tendency for membranes to fuse. For example, high temperatures can enhance fusion of neurotransmitter-containing vesicles to the presynaptic membrane of neurons; the resulting increased release of neurotransmitters could lead to neural malfunction and death (Macdonald et al. 1988). Maintenance of the appropriate balance between lamellar and nonlamellar phases in the face of changing conditions is achieved through what is termed **homeophasic adaptation**.

LIPID MODIFICATIONS THAT RESTORE HOMEOSTASIS   How are these two general types of membrane homeostasis defended? Can a common set of lipid changes lead to both homeoviscous and homeophasic adaptation? Or do these two processes depend on

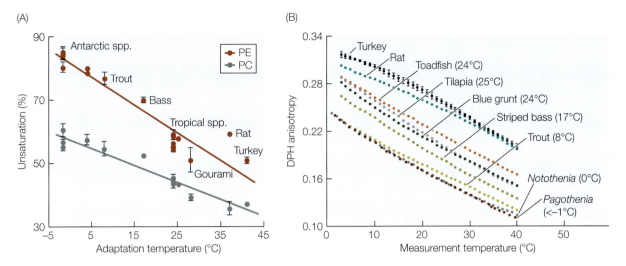

Figure 3.48 Adaptation temperature-related variation in acyl chain saturation and static order of brain synaptosomal membranes of vertebrates adapted to temperatures ranging from –1.9°C (Antarctic notothenioid fishes) to 39°C (turkey). (A) Variation in the percentage of unsaturated acyl chains in membrane phospholipids belonging to two classes, phosphatidylethanolamine (PE) and phosphatidylcholine (PC). Error bars represent standard deviation. (B) Relationship between membrane static order, as indexed by the fluorescence anisotropy of 1,6-diphenyl-hexatriene (DPH), and measurement temperature. Each data point represents a value determined at a single temperature during a temperature scan between 3°C and 40°C. Error bars for turkey represent standard deviation among 3 replicate preparations. High values of DPH anisotropy reflect high static order (low fluidity) of the lipid bilayer. (After Logue et al. 2000.)

unique alterations in lipid composition? **Figure 3.48** provides important insights into the types of changes in phospholipid acyl chains that foster homeostasis (Logue et al. 2000; Hayward et al. 2014). These experiments focused on the properties of neural synapses (**Figure 3.49**), and in particular of synaptic membranes. The studies used what are termed **synaptosomes**, which comprise synaptic vesicles and a small amount of associated cytoplasm, that were isolated from vertebrates whose body temperatures spanned an approximately 40-degree-Celsius range: from Antarctic notothenioid fishes to birds and mammals. A similar protocol was used to look at differently acclimatized ectotherms and yielded similar results (see Figure 3.50).

As shown in Figure 3.48A, the double-bond contents of phosphoglyceride acyl chains show a regular dependence on temperature: As adaptation temperature decreases, the double-bond content increases in both phosphatidylcholine- (PC-) and phosphatidyl-ethanolamine- (PE-) containing lipids. The proportion of unsaturated acyl chains rose from 35% to 60% for PC and from 55% to 85% for PE. In PC-containing lipids, the adaptation temperature-related rise in unsaturation level was due almost entirely to increased proportions of **polyunsaturated fatty acids (PUFAs)**, which increased from 7% to 40%. These changes involved replacement of monounsaturated fatty acids (MUFAs) with PUFAs at the *sn*-2 position (the second carbon of the glycerol component). In PE, a different temperature-related pattern was observed. The major change in saturation was due to replacement of saturated fatty acids with MUFAs at the *sn*-1 position (first glycerol carbon) as temperature decreased. Note that the slopes of the relationships between acyl chain unsaturation and temperature differ between head groups (PE and PC). We return to the role of changes in lipid class (head group composition) in temperature adaptation below.

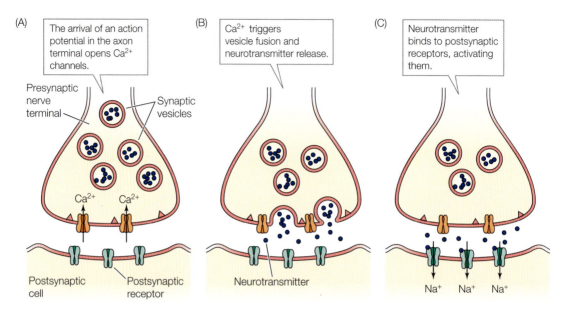

Figure 3.49 Structure and function of a neural synapse.

These changes in acyl chain composition are reflected in a close, though not complete, conservation in membrane fluidity (see Figure 3.48B). (Note that the terminology used to describe membrane physical state differs somewhat among authors and has evolved over time. *Static order* has become the preferred way to refer to *fluidity*. The two terms are reciprocally related: High static order equates to low fluidity.) The index of fluidity (or static order) used in this study is the freedom of movement of a probe molecule, **1,6-diphenyl-hexatriene** (**DPH**), that intercalates into the bilayer. DPH is a fluorescent molecule, so the polarization of its fluorescence emission can be used to quantify the ability of the probe to move within the bilayer. Thus, when a beam of polarized light is directed on the membrane preparation, the amount of light that is reemitted in the same plane as the light impinging on the probe reflects how much the probe has been able to move. A highly fluid membrane will permit a high level of probe rotation, leading to decrease in the polarization signal. The variable measured in such experiments is known as **fluorescence anisotropy**, where anisotropy (r) is related to the strength of polarization (P) by the equation $r = 2P/(3 - P)$. It is important to realize that fluorescence anisotropy is a bulk property of the membrane that reflects the *average* static order of the system. Local regions of the membrane—for example, lipids interacting with the TM domains of integral proteins—may have different fluidities from that shown by bulk anisotropy measurements. Another ambiguity associated with use of fluorescent probes concerns the fact that different probes give different estimates of fluidity and suggest different degrees of homeostasis in fluidity (Behan-Martin et al. 1993). These ambiguities notwithstanding, fluorescence anisotropy has proven useful for evaluating the effects of changes in membrane lipid composition on membrane static order, as shown by the data in Figure 3.48. The question we now must address concerns whether these alterations in lipid composition and physical properties manifest at the level of animal behavior.

LINKING CHANGES IN BRAIN LIPIDS TO COMPENSATION IN BEHAVIOR   In view of the role of synaptic membranes of brain neurons in neurotransmitter release and uptake, one would predict that alterations in brain synaptosome properties could lead to alterations in behavior. This type of linkage was demonstrated in a classic study by Cossins, Friedlander, and Prosser (1977), in which they investigated acclimation of goldfish to different water temperatures. This study included time-courses of changes in synaptosomal fluidity and three behavioral traits: hyperexcitability, equilibrium loss, and coma (**Figure 3.50**). The data reveal tight linkages between membrane biophysical properties, lipid composition, and behavior. Moreover, this study remains one of the best illustrations of the temporal aspects of the acclimation process and the different kinetics of acclimation to warm and cold temperatures. Transfer of goldfish from 5°C to 25°C led to a much more rapid acclimation in membrane fluidity and behavioral traits than the reciprocal transfer. In the cold-to-warm transfer, acclimation was largely complete by ~20 days. In the warm-to-cold

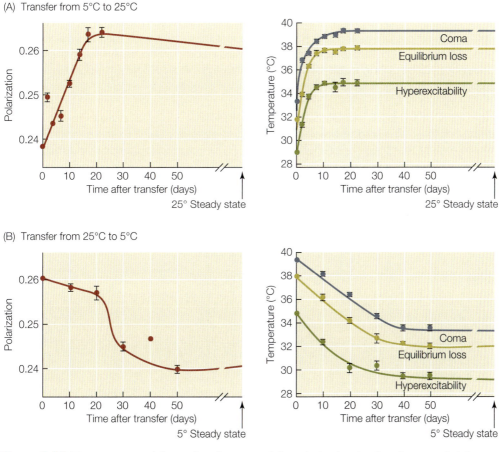

Figure 3.50 Time-courses of thermal acclimation of three behavioral traits—hyperexcitability, equilibrium loss, and coma—and fluidity of brain synaptosomal membranes. For behavioral assays, goldfish that had been at their acclimation temperature (5°C or 25°C) for a given time were removed from the acclimation tank, placed in 22°C water for 30 min, and then subjected to increasing temperatures from 22°C up to 40°C. The researchers recorded the lowest temperature that elicited the behavioral abnormality of interest. Membrane fluidity is indexed by fluorescence polarization, using the fluorescent probe 1,6-diphenyl-hexatriene (DPH). Error bars represent standard error. (A) Data for goldfish transferred from 5°C to 25°C. (B) Data for goldfish transferred from 25°C to 5°C. (After Cossins et al. 1977.)

transfer, acclimation took approximately twice as long. In both cases, however, there was a strong correlation between the amount of change in membrane fluidity (polarization signal) and the degree to which the three behavioral responses approached the final steady-state values for the two acclimation temperatures.

Although these results are fully consistent with a central role of membrane fluidity in establishing the thermal sensitivities of behavior, one might argue that the changes in membrane fluidity are only correlative and not causal. Evidence that disruption of brain synaptic function can cause these types of behavioral anomalies comes from work of MacDonald and colleagues (1988) with Antarctic notothenioid fishes. These extremely stenothermal species show extreme hyperactivity at temperatures above ~10°C and die within ~10 min at temperatures near 15°C (**Figure 3.51A**; Somero and DeVries 1967; Podrabsky and Somero 2006). Concomitant with the onset of hyperactivity is a rapid increase in the release of quanta of neurotransmitter (acetylcholine), as shown in **Figure 3.51B**. (A quantum of neurotransmitter is the amount contained within a single synaptic vesicle.) The release of neurotransmitter is not sustained at higher temperatures and collapses completely near 16°C. These data, in conjunction with the studies of Cossins and colleagues (1977), provide what we feel is a compelling case for linking thermal disruption of behavior to alterations in neurotransmitter release that, in turn, reflect underlying changes in membrane structure.

HEAD GROUP EFFECTS IN HOMEOPHASIC ADAPTATION   To what extent are these neural effects due to shifts in membrane phase properties as opposed to shifts in fluidity? One would expect that the propensity for forming H_{II} structures would rise with temperature, enhancing membrane fusion (see Figure 3.46) events such as those involved in exocytotic

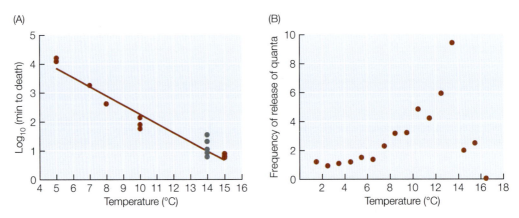

Figure 3.51 Thermal tolerance and synaptic activity in stenothermal Antarctic fishes. (A) Upper lethal temperatures of several species of field-acclimatized Antarctic fishes of McMurdo Sound (78° S). Red circles are data for *Trematomus bernacchii* exposed to a series of temperatures (Somero and DeVries 1967). Blue symbols are data for *T. bernacchii* and four other species, the zoarcid *Lycodichthys dearborni*, and the notothenioids *Pagothenia borchgrevinki*, *Trematomus hansoni*, and *Trematomus pennellii* exposed to 14°C (Podrabsky and Somero 2006). (B) Release of quanta of neurotransmitters from a trigeminal nerve in *P. borchgrevinki*. The Q_{10} for neurotransmitter release between 3°C and 13°C is ~9, a value much greater than expected on the basis of simple thermodynamic effects on rate. Rather, this high Q_{10} is suggestive of alterations in membrane properties due to heating. Above ~13°C the rate of transmitter release plummets quickly; death at these temperatures occurs in ~10 min (see part A). (A after Podrabsky and Somero 2006; B after MacDonald et al. 1988.)

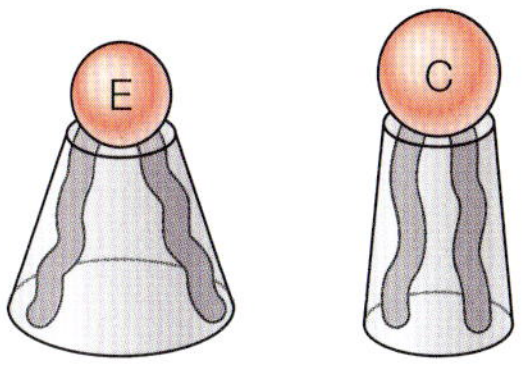

release of transmitters. We view this type of phase effect to be central to the phenomena shown in Figures 3.50 and 3.51. Thus, the fanning-out of lipid structures that occurs with rising temperature would lead to an increased propensity for H_{II} structures and membrane fusion. Because of the different geometries of phosphoglycerides with ethanol (E) and choline (C) head groups (see figure at right), it would seem likely that adaptation or acclimation to increased temperature would favor a shift from PE to PC phosphoglycerides, a shift that would tend to counterbalance the heat-induced increase in propensity for formation of H_{II} structures.

PE   PC

There have, in fact, been several studies that have demonstrated that a shift from PE to PC occurs during acclimation of ectotherms to higher temperatures (fishes, Hazel and Carpenter 1985; *Drosophila*, Cooper et al. 2012, 2014; **Box 3.10**). A reciprocal shift occurs during cold acclimation (Hazel and Carpenter 1985). These changes in head group composition (phosphoglyceride class), in conjunction with changes in acyl chain composition and cholesterol content (see below), may allow membranes to maintain a consistent propensity for forming nonlamellar structures at different temperatures. The increased levels of docosahexaenoic acid (DHA) in membranes of cold-acclimatized animals favor a propensity for forming H_{II} structures. Therefore DHA can be viewed as an enhancer of fluidity, through its effects on acyl chain order near the center of the bilayer, and as a contributor to retention of an ability to form H_{II} structures in the cold. Thus, to return to the question raised earlier about whether certain types of lipid changes can lead to both homeoviscous and homeophasic adaptation, there appears to be overlap in the effects of some types of lipid alterations—for example, DHA content—on both homeostatic processes.

BOX 3.10

Two lessons from *Drosophila*: The importance of the PE/PC ratio and selection for acclimatory plasticity

Here we briefly highlight recent field and laboratory studies of *Drosophila melanogaster* that address two important issues in membrane adaptation to temperature: (1) the relative importance of different types of membrane remodeling mechanisms (changes to head groups versus shifts in acyl chain composition), and (2) the influence of temperature variation during a population's evolutionary history on its phenotypic plasticity—that is, its ability to modify membrane composition to effect homeoviscous or homeophasic adaptation.

Cooper and colleagues (2012, 2014) have done especially informative analyses of both questions. Their field and laboratory studies of *D. melanogaster* demonstrated that changes in head group composition—the phosphatidylethanolamine (PE)/phosphatidylcholine (PC) ratio—were more sensitive to selection for acclimatory capacity than was the composition of the pool of membrane acyl chains.

They also showed that flies evolved in the laboratory under stable thermal conditions (16°C or 25°C) had reduced abilities to alter the PE/PC ratio during warm acclimation relative to flies that had evolved for dozens of generations under variable thermal conditions (shift between 25°C and 16°C every 4 weeks; Cooper et al. 2012). A complementary study of natural populations collected in Vermont, Indiana, and North Carolina gave similar results: Flies from Vermont, the most variable thermal environment, had the greatest ability to shift the PE/PC ratio during warm acclimation. The three populations also showed temperature-compensatory differences in acyl chain composition, but acclimatory plasticity in this trait did not differ among populations. These studies thus suggest that selection on acclimatory plasticity favors only certain lipid properties (i.e., the PE/PC ratio) that influence membrane biophysical state in *D. melanogaster*.

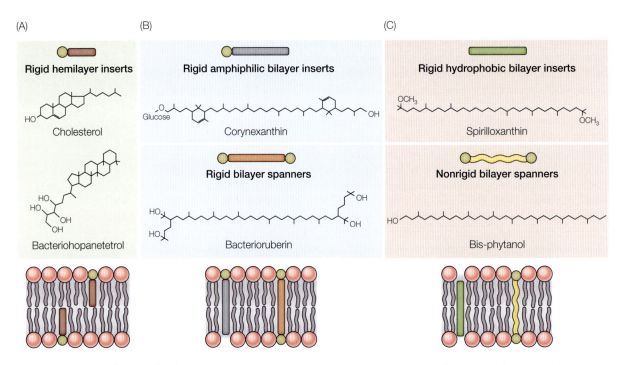

Figure 3.52 Structures of membrane stabilizers termed *inserts* found in eukaryotes, bacteria, and archaea. (A) Rigid amphiphilic inserts that insert into the membrane hemilayer (leaflet). In eukaryotes, cholesterol serves key roles in membrane stabilization (see text). In bacteria, similar roles are played by hopanoids, such as bacteriohopanetetrol. (B) Rigid amphiphilic bilayer inserts (e.g., corynexanthin) and rigid bilayer spanners (e.g., bacterioruberin). (C) Rigid hydrophobic bilayer inserts (e.g., spirilloxanthin) and nonrigid bilayer spanners (e.g., bis-phytanol). The bilayer-spanning lipids may function like the tetraether lipids of archaea, which generate an extremely strong membrane that is, to a considerable extent, a monolayer rather than a typical bilayer (see text). (After Robertson and Hazel 1997.)

CHOLESTEROL AND HOPANOIDS: ADDITIONAL CONTRIBUTORS TO MEMBRANE ADAPTATION   A variety of lipids other than phosphoglycerides also are important in adjusting the thermal sensitivities of membranes, across all domains of life (**Figure 3.52**). As mentioned above, in eukaryotes cholesterol plays an important role in stabilizing the liquid-ordered (l_o) state of membrane lipids in the face of changes in temperature. At low temperature, cholesterol impedes the formation of the gel phase, an action that prevents phase discontinuities that can lead to rupture of the membrane's barrier function. At high temperatures, cholesterol tends to stabilize the l_o state and counteract the tendency for lipids to enter the liquid-disordered (l_d) state. The cholesterol content of membranes of ectotherms has been observed to increase with rising acclimation temperature, consistent with its role in stabilizing the l_o state (Robertson and Hazel 1995). The importance of elevated cholesterol concentrations for defending the l_o state at high temperatures is also indicated by the significantly higher cell membrane cholesterol contents in birds and mammals than in ectotherms.

Cholesterol may be especially important in rapid adjustments of membrane biophysical characteristics in ectotherms. In a study of the eurythermal annual fish *Austrofundulus limnaeus*, long-term acclimation to constant temperatures of 20°C, 26°C, and 37°C was marked by changes in expression of genes associated with alterations in acyl chain composition, for example, in acyl saturation state (Podrabsky and Somero 2004). In contrast, when fish were subjected to environmentally realistic diurnal cycles of temperature between 20°C and 37°C, among all genes associated with lipid metabolism, the expression of genes related to cholesterol biosynthesis showed the strongest dependence on temperature. These findings suggest that there may be more than one mechanism for achieving homeostasis of membrane function in the face of changes in temperature, and that the mechanism chosen may be determined in part by the speed with which the alteration in membrane properties must be achieved.

Bacteria lack cholesterol, but they have another set of molecules, the **hopanoids**, which play similar roles. Hopanoids are relatively rigid, amphiphilic (amphipathic) hemilayer inserts that stabilize membrane properties in the face of changes in temperature (see Figure 3.52). As in the case of cholesterol, the contents of hopanoids may increase when bacteria are cultured at higher temperatures (Poralla et al. 1984). Why do bacteria use hopanoids rather than cholesterol to stabilize their membranes? Biosynthesis of cholesterol requires molecular oxygen, suggesting that cholesterol is a relatively recent biochemical invention that emerged only after adequate oxygen became available for use in biosynthetic processes (see Chapter 2). By that point, bacteria had been around for at least 2–3 billion years. Hopanoids do not require molecular oxygen for their biosynthesis. They are thought to have evolved prior to the advent of oxygenic photosynthesis, and are thus an extremely ancient type of membrane stabilizer.

A host of other types of membrane stabilizers are found in eukaryotes, bacteria, and archaea (see Figure 3.52). These range from rigid inserts that intercalate into one hemilayer (leaflet) of the membrane to membrane-spanning lipids that lead to a very high degree of membrane stability. Cholesterol and hopanoids fall into the first category. Membrane spanners include a variety of amphiphilic and nonpolar molecules. **Carotenoids** like corynexanthin and spirilloxanthin are long-chain molecules that insert into bilayers but do not span the full dimension of the bilayer. If a long-chain molecule has polar groups at both ends, it may function as a bilayer-spanning stabilizer that has a linear dimension equal to that of the bilayer itself (e.g., bacterioruberin and bis-phytanol). Further examples of the latter type of stabilizer are the membrane-spanning lipids of extremely heat-tolerant bacteria that are fabricated from a 30-carbon dicarboxylic acid, to which glycerol is esterified at both ends. The occurrence and concentrations of these membrane stabilizers reflect the adaptation temperature of the organisms: Higher temperatures of exposure are associated with elevated levels of these highly stabilizing molecules.

HEAT-RESISTANT MEMBRANES OF HYPERTHERMOPHILIC ARCHAEA   The extreme temperatures tolerated by hyperthermophilic archaea demand even stronger membrane stabilizing chemistry than found in the most heat-tolerant species of the other two domains of life. As discussed in the context of protein evolution, archaea are thought to have arisen in high-temperature environments where strong selection for heat-resistant membrane structures would have existed (Boussau et al. 2008). Whereas extremophilic archaea use some of the membrane inserts found in bacteria and eukaryotes, the nature of their major bilayer-forming lipids is unique. These lipids have three characteristics that confer a high level of thermal stability. First, the chemical bonds that attach the long nonpolar chains to glycerol are **ether linkages**, not the ester linkages found in most lipids of eukaryotes and bacteria. Although the latter two groups have some ether linkages, ether lipids represent only a small fraction of their total lipids (Lombard et al. 2012). Ether linkages are more chemically stable than ester linkages in the face of heat stress and thus have been selected for in the archaea. Second, the nonpolar chains of archaeal ether lipids are **fully saturated isoprenoid alcohols** rather than the acyl groups of fatty acids. Isoprenoid chains are found in all domains, so they are not strictly an archaeal characteristic. Furthermore, archaea have fatty acids, but isoprenoid chains are by far the greatest contributor to lipid chemistry (Lombard et al. 2012). The isoprenoid chains of archaea typically occur in two lengths: 40-carbon and 20-carbon. The 40-carbon chain is, in fact, a head-to-head condensation product of two 20-carbon chains (see figure below). Whereas the 20-carbon lipids are bilayer forming, the 40-carbon chains, to which glycerol is esterified at both ends, generate a lipid that is long enough to span the bilayer. The membrane-spanning nature of these monolayer-forming lipids is another mechanism for enhancing the thermal stability of the membrane. Third, when acclimation to temperature occurs in hyperthermophilic archaea, the saturation of lipids is not modified. Instead, the relative amounts of tetraether membrane-spanning lipids and diether bilayer-forming lipids are adjusted such that the fraction of membrane-spanning lipids increases with temperature.

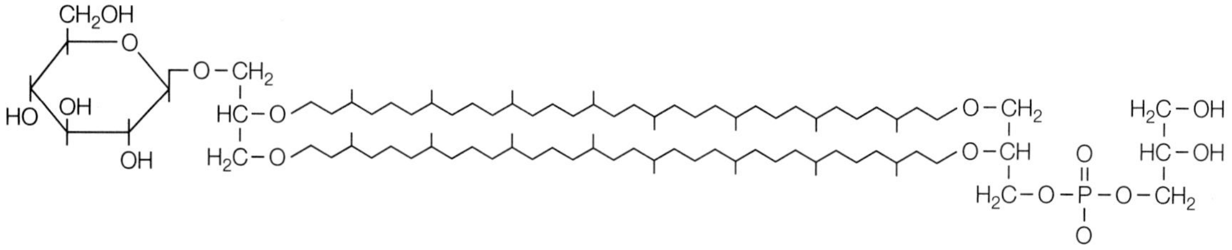

A final distinction between membrane lipids of archaea and those of the other two domains of life—a difference that, to our knowledge, is not related to temperature adaptation—is in the stereochemistry of the glycerol molecule to which the lipid chains are joined. Two stereoisomers of glycerol phosphate are used in lipid synthesis: Archaea use glycerol 1-phosphate, and bacteria and eukaryotes use glycerol 3-phosphate. Both molecules are produced from dihydroxyacetone-phosphate (DHAP), but the glycerophosphate dehydrogenase (GPDH) enzymes that catalyze the conversion of DHAP to glycerol phosphate, G1PDH in archaea and G3PDH in bacteria and eukaryotes, yield products with different stereochemistries. The evolutionary histories of these alternative pathways for glycerol phosphate synthesis are not known, but it has been conjectured that the earliest cell type—the **cenancestor** (*cen-* is a Greek prefix that can mean "common") of the current domains of life—may have produced lipids with both types of glycerol phosphate. Subsequent divergence of the archaeal lineage and the lineage leading to bacteria and eukaryotes led to differential reliance on the two GPDH enzymes, with G1PDH retained in the archaea and G3PDH dominating in the other two domains (for review, see Lombard et al. 2012).

In summary, archaeal membranes reflect a distinct evolutionary solution to the problems raised by thermal effects on membranes. Archaea fabricate membranes with high intrinsic resistance to disruption by high temperatures. The lipids are linked to glycerol with strong ether bonds, and 40-carbon, membrane-spanning lipids give the membranes of high-temperature-adapted archaea further resistance to disruption by heat. However, despite archaea on the one hand, and bacteria and eukaryotes on the other hand, taking different routes to generate heat-resistant membrane lipids, the three groups have adapted successfully, yielding membranes with physical states appropriate for their different thermal environments. Membranes thus provide an excellent illustration of convergent evolution in functional properties through different underlying mechanisms of biochemical adaptation.

Small heat-shock proteins and membrane stability

We have been emphasizing how changes in lipids help conserve the critical structural and functional states of membranes in the face of thermal stress. Molecular chaperones, however, also may be involved. **Small heat-shock proteins (sHSPs)** such as Hsp17 and α-crystallin have been shown to play diverse roles in stabilizing the physical states of membranes, including bilayer fluidity and the propensity for formation of nonlamellar structures (Tsvetkova et al. 2002; Horváth et al. 2008). Both Hsp17 and α-crystallin stabilize the liquid-crystalline state in membranes composed of phosphatidylglycerol and phosphatidylserine lipids (Tsvetkova et al. 2002). Both sHsps also inhibit the formation of hexagonal II structures and thereby stabilize the lipid bilayer under heat stress. Importantly, these membrane-stabilizing effects can lead to preservation of the optimal physical state of membranes at both low and high temperatures. The favorable effects of these sHsps are thus not restricted to their roles in preserving or restoring protein structure under heat-shock conditions. Furthermore, some sHsps mediate interactions between the cytoskeleton and cellular membrane that facilitate cell cleavage and changes in cell shape during development. Thus, sHsps are pivotal elements in maintenance of cytoskeletal architecture under nonstressful and stressful conditions (see Lockwood et al. 2010).

The mechanistic basis of these stabilizing effects on membrane structure and function appears to involve transitory binding interactions between lipid head groups and sHSPs. These binding interactions are highly specific and depend not only on head group composition but also on acyl chain unsaturation. The localization of sHSP-binding lipids in membrane rafts may be important for governing where membrane-sHSP interactions take place (Horváth et al. 2008).

The importance of membrane-sHSP interactions appears to go well beyond adaptive responses to thermal stress. For example, binding of heat-shock proteins, including larger HSPs such as Hsp70, to the outer surface of the cell membrane may be important in immunological phenomena and the etiology of cancer (Horváth et al. 2008).

In the lens of the vertebrate eye, interactions between the α-**crystallin** class of sHSPs and other lens proteins (generically known as crystallins) also figure in changes in vertebrate eye structure with aging (Makley et al. 2015). With aging, lens proteins can lose native structure and form insoluble amyloid-like aggregates that impede lens function. α-crystallins help maintain the native structures of other crystallins to ensure that they remain soluble and function effectively in diffraction of light. Certain sterol compounds related structurally and biosynthetically to cholesterol also can bind to sites on different types of crystallins and maintain or restore their native structures and functions. These sterols may also benefit lens proteins indirectly by influencing the extent of membrane binding by crystallins. The binding of crystallins, including α-crystallins, to the eye lens membrane increases with age. Sterol compounds appear to reverse this binding. In the case of α-crystallins, this sterol-mediated increase in solubility allows them to more effectively conduct their chaperone functions on other lens proteins. These sterol-lens protein interactions are, then, another illustration of the importance of interactions between lipids and proteins in sustaining physiological activities. There is a certain reciprocity of assistance, with sHsps stabilizing lipid structures, and selected types of lipids fostering the native structures of sHsps.

In summary, sHsps, and possibly other size classes of HSPs, likely play important roles in defending membrane structure and function in the face of thermal stress. This defensive role must be considered in a temporal framework. Thus, one proposed function for sHsp-membrane interactions is to be a rapid response system to thermal stress. Stabilizing effects of sHsps likely can occur very quickly relative to the time required for adaptive changes in head group or acyl chain composition. Similar to their roles as hold-ases in molecular chaperone function, sHsps could be an effective "first line of defense" against the effects of thermal stress on membranes.

The influence of diet on lipid composition: "We are what we eat," up to a point

To what extent is the lipid composition of a cell governed strictly by the cell's biosyn-thetic processes, rather than by the input of lipids obtained through the diet? Diets vary widely in the amounts and types of lipids they contain, so ingested lipids would seem to provide a rich source of materials for incorporation into lipid-containing structures. Might it be possible, then, that at least some lipid-containing structures of the cell—depot lipids, if not cellular membranes—would have compositions that track the lipid content of the diet?

In the context of conserving the appropriate biophysical properties of membranes, it would seem essential for the organism to disallow a strict reliance on dietary input in governing the composition of the bilayer lipids. In fact, studies of ectotherms and endothermic homeotherms present a complex picture of the influence of diet on mem-brane composition and membrane functional properties. We are what we eat, but only up to point.

Even though diet can affect membrane lipid composition, these changes may not lead to alterations in membrane function if there are balancing changes in lipid chem-istry that preserve a stable biophysical state. For example, Martin and colleagues (2013) examined the effect of different mixtures of dietary lipids on the membrane composition and function of mitochondria of red muscle from rainbow trout (*Oncorhynchus mykiss*). The lipid compositions of mitochondrial membranes reflected, to a certain extent, the types of lipids that were presented in the diet. However, there was clear evidence of close regulation of membrane composition. In some cases, the types of lipids presented in the diet at high levels were reduced in their contribution to the pool of bilayer lipids. Conversely, diets lacking certain types of lipids, for example certain long-chain PUFAs, led to compensatory increases in other classes of long-chain PUFAs that were synthesized by the cell. The most important finding of this study was that, despite some diet-driven effects on membrane composition, the functional characteristics of the mitochondria—for example, maximum rates of substrate oxidation, total oxygen consumption, and enzyme activities—were essentially unchanged. In other words, the adaptive responses associ-ated with homeoviscous and homeophasic adaptation are not thwarted by the types of lipids an animal ingests in its diet.

Studies of depot lipids, **triacylglycerides** (see Figure 3.44 and below), have shown that diet can influence not only the lipid composition of the adipocytes that house the depot lipids but also the behavior of the animal in the face of low temperatures. Studies of small mammalian hibernators have provided some fascinating stories in this context. The body temperatures of ground squirrels may approach −1.9°C during deep-winter hibernation (Barnes 1989), ~30 degrees Celsius lower than the melting tempera-tures of the depot lipids present in normothermic, active squirrels. To be acces-sible to metabolism, depot lipids must be in a liquid-crystalline state. Thus, if a squirrel is to metabolize its lipid stores during hibernation or during the initial stages of arousal (i.e., before normothermic body temperature is attained), it needs to accumulate lipids with low melting temperatures before hibernation. This is precisely what Frank (1991) found in studies of the ground squirrel *Sper-mophilus beldingi*, in which hibernation-competent individuals contained lipids with melting temperatures ~25 degrees Celsius lower than those of animals not

$$H_3C-(CH_2)_n-\overset{\overset{\displaystyle O}{\|}}{C}-O-CH_2$$
$$H_3C-(CH_2)_n-\overset{\overset{\displaystyle O}{\|}}{C}-O-CH$$
$$H_3C-(CH_2)_n-\overset{\overset{\displaystyle O}{\|}}{C}-O-CH_2$$

Triacylglyceride

entering hibernation. Frank conjectured that selective feeding during preparation for winter could alter depot lipid composition in an adaptive manner. Subsequent laboratory studies of a congener, *Spermophilus lateralis*, showed that animals given diets rich in PUFAs accumulated high levels of these lipids, had increased probability of entering a state of hibernation, and once in hibernation, had lower body temperature set points and improved ability to survive hibernation (Frank 1992).

In summary, dietary lipids are not the primary drivers of membrane lipid composition; organisms maintain control of the types of lipids incorporated into membranes by regulating use of ingested lipids and de novo synthesis within the cell. However, in the case of depot lipids, selective feeding may provide an effective avenue for acquiring lipids with melting temperatures appropriate to severe conditions like deep hibernation.

Lipids, membranes, and upper thermal limits for life

Before we examine the wide range of upper thermal limits for organisms and discuss some of the biochemical mechanisms that set these limits, an important caveat must be sounded. When evaluating thermal tolerance data, one must keep in mind that the observed tolerance of thermal extremes is governed not only by the basic biochemistry of a species, but also by the experimental protocols used to measure tolerance. Among the important elements of experimental design are duration of exposure to extreme temperatures and the rate at which temperature is changed. Thus, the experimental conditions employed must be carefully evaluated in order to allow meaningful comparisons across species (and laboratories) to be made. Where relevant, we touch on these issues in the sections that follow.

HOW HOT CAN EUKARYOTES GET (AND SURVIVE)?   Eukaryotes fall far short of bacteria and archaea in their tolerance of high temperatures. As pointed out near the beginning of this chapter (see Figure 3.1), archaea from deep-sea hot springs can grow at temperatures at least as high as 121°C and survive for short periods at ~130°C (Kashefi and Lovely 2003). The most thermophilic eukaryotes have upper lethal temperatures that are approximately 60 degrees Celsius lower than the highest temperatures tolerated by hyperthermophilic archaea and ~40 degrees Celsius lower than those of the most thermophilic bacteria (Stetter 1999). All indications are that eukaryotes are limited to temperatures no higher than 55°C–60°C. Among the most heat-tolerant eukaryotes are fungi from hot springs, which tolerate temperatures up to ~60°C on a continuous basis (Tansey and Brock 1972); desert ants of the genus *Cataglyphis*, which withstand temperatures near 55°C, albeit only for short periods (Gehring and Wehner 1995); some intertidal invertebrates like snails of the genus *Echinolittorina*, which have upper lethal temperatures near or slightly above 55°C, and may experience these temperatures for several hours (Li 2012); the unicellular red alga *Galdieria sulphuraria*, which is found in acidic hot springs and has an upper lethal temperature near 56°C (Schönknecht et al. 2013); and Alvinellid polychaete worms from deep-sea hydrothermal vents, which exhibit preferred temperatures up to ~50°C when placed in a thermal gradient (20°C–61°C) and withstand temperatures of 55°C for ~15 min (Girguis and Lee 2006).

LIPIDS AND MEMBRANES AS LIMITING FACTORS IN THERMAL TOLERANCE   The classes of biochemicals that set the upper thermal limits for life remain a topic of debate (Cowan 2004). The abilities of proteins to evolve highly heat-resistant structures may mean that proteins are not the limiting factor. Instead, some authors point to small molecules, rather than macromolecules, as the principal limiters of heat resistance. For example, key metabolites such as nicotinamide cofactors become unstable (at least in vitro) at temperatures near 100°C (Cowan 2004). Here we consider the limits that might be set by lipids, especially the lipids that occur in membranes. We begin by examining what might be the primary basis of the large difference in heat tolerance between eukaryotes and thermophilic archaea and bacteria.

A major distinction between these groups is the suites of lipids that are available to them for adjusting membrane stability. Archaea, in particular, are equipped with lipids that have extraordinary stabilities at high temperatures, as discussed above. Heat-resistant ether bonds and fully saturated lipid chains, which may be long enough to span the entire membrane, give archaea lipids that can work at temperatures close to 130°C. Bacteria lack these lipids, although some archaea-like features have been discovered in lipids of the most thermophilic bacteria. For example, a novel glycerol ether lipid has been found in the hyperthermophilic bacterium *Thermotoga maritima*, which has one of the highest growth temperatures known for bacteria (90°C–95°C; for review, see Stetter 1999). This ether bond may serve this bacterium's lipids much as ether bonds stabilize archaeal lipids. Thermophilic and hyperthermophilic bacteria also employ a variety of membrane inserts, including those that span the width of the bilayer, to add thermal stability to the cell membrane's structure. It may well be that each domain of life has its upper thermal limits established by the repertoire of membrane lipids it is able to produce. The absence in eukaryotes of thermally resistant membrane components of the types found in archaea and bacteria thus is proposed to limit eukaryotic life to temperatures between 50°C and 60°C. To the best of our knowledge, this conjecture was first presented by Tansey and Brock (1972), who placed special emphasis on the challenges of sustaining the structure and function of organellar membranes, such as those of the mitochondrion, at extremely high temperatures.

MITOCHONDRIAL THERMAL TOLERANCE: IS THIS THE ACHILLES' HEEL?   Studies of temperature effects on mitochondrial structure and function offer support for the conjecture of Tansey and Brock. Mitochondrial function, for example, respiration rate, shows a strong dependence on temperature and distinct, species-specific thermal limits (**Figure 3.53**). Over a span of temperatures that includes the normal physiological temperature range of a species, increases in temperature lead to a regular increase in mitochondrial oxygen consumption rate, following Q_{10} relationships. However, at a certain high temperature that varies among species and, at least in eurythermal ectotherms, with acclimation state, the slope of the rate versus temperature function changes sign: Further increases in temperature lead to a fall in respiration rate. This is shown especially clearly by Arrhenius plots (the natural log [ln] of rate versus the reciprocal of temperature in kelvin [K]) of the sort shown in Figure 3.53A. The temperature at which the slope of an Arrhenius plot changes is termed the **Arrhenius break temperature** (**ABT**).

Mitochondria exhibit considerable variation in ABT among species adapted to different temperatures and across different acclimation conditions. This is exemplified in Figure 3.53B by the differences among four congeners of abalone (genus *Haliotis*). (We examined these same species earlier in the context of protein [cMDH] adaptation to temperature; see Figure 3.5.) ABT of mitochondrial oxygen consumption varied among the *Haliotis* species and acclimation treatments by more than 10 degrees Celsius, a clear reflection of the plasticity of this organelle to adapt and acclimate to temperature. The biophysical properties of the mitochondrial membranes varied as well, reflecting

(A)

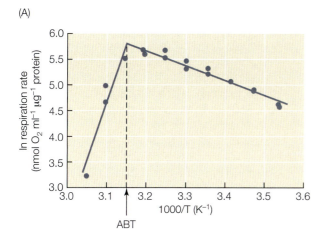

(B)

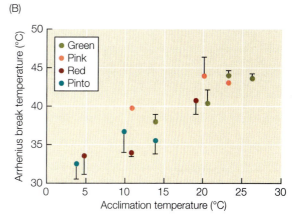

(C)

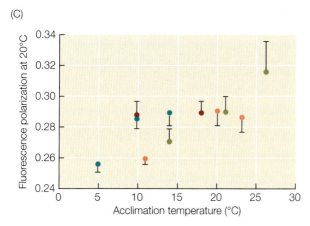

Figure 3.53 Effects of temperature on mitochondria of four congeners of abalone (genus *Haliotis*). Species names and temperatures where maximum abundance of each species occurs are as follows: *H. fulgens* (green abalone), 18°C–23°C; *H. corrugata* (pink abalone), 12°C–20°C; *H. rufescens* (red abalone), 10°C–16°C; and *H. kamtschatkana* (pinto abalone), 6°C–10°C. (A) An Arrhenius plot of mitochondrial respiration versus temperature for mitochondria of *H. fulgens*. Oxygen consumption rises with increasing temperature, following a strict Q_{10} relationship, up to the Arrhenius break temperature (ABT). (B) Variation in ABT among differently acclimated specimens of four species of abalone. (C) Fluorescence polarization of mitochondrial membranes determined using DPH. Error bars in (B) and (C) represent standard error. (After Dahlhoff and Somero 1993b.)

the differences in ABT (see Figure 3.53C). Using the fluorescence polarization signal from 1,6-diphenyl-hexatriene (DPH) as an index of membrane fluidity, it is seen that intrinsic fluidity of the mitochondrial membranes rises (polarization intensity decreases) as adaptation or acclimation temperature falls. These effects parallel those observed in homeoviscous adaptation in synaptosomal membranes (see Figure 3.48).

To gain a broader view of the relationship between ABT and adaptation temperature, ABT values for mitochondrial respiration of ectotherms adapted or acclimated to different temperatures were plotted against adaptation or acclimation temperature (**Figure 3.54**). This plot revealed a positive correlation between ABT and adaptation or acclimation temperature similar to that shown in Figure 3.53B. However, the slope of this regression line (which does not include the very low ABT for the Antarctic notothenioid fish *Trematomus bernacchii*) is shallower than the slope of the line of unity (ABT = adaptation temperature; slope = 1). This difference in slope indicates that adaptation temperature and ABT will eventually converge; above this adaptation temperature, mitochondrial function would seem to be precluded (ABT < adaptation temperature). This would be an impossible condition for eukaryotic life, at least aerobic life. It bears emphasizing that the two lines cross near a temperature of 55°C–60°C, which, as mentioned above, is near the upper thermal limit of eukaryotic life. Therefore, this analysis offers support for the hypothesis that membrane stability limits eukaryotic thermal tolerance.

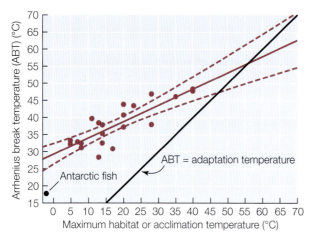

Figure 3.54 Arrhenius break temperatures (ABTs) of mitochondrial respiration from animals adapted to different temperatures. The regression line (shown with 95% confidence intervals) is based on all data except those for the Antarctic notothenioid fish *Trematomus bernacchii*, which has an extremely low ABT in reflection of its cold adaptation and stenothermy. A line of unity (ABT = adaptation temperature) is also shown. (After Weinstein and Somero 1998.)

CAN HEAT TOLERANCE BE GREATER IN THE DEEP SEA? Because elevated hydrostatic pressures tend to reduce membrane fluidity (Macdonald 1984), it is conceivable that high pressures might stabilize eukaryotic membrane function at temperatures higher than 55°C–60°C. Thus, given the combinations of high temperatures and high pressure experienced by certain hydrothermal vent eukaryotes, it seems possible that these species might withstand higher temperatures than do species adapted to 1 atmosphere of pressure. However, a study of Alvinellid polychaete worms by Girguis and Lee (2006) argues against this possibility. Their studies involved thermal preference experiments done in an aquarium with a thermal gradient and a pressure equal to the in situ pressure (~180 atm) for the hydrothermal vent site studied, Explorer Ridge. Worms exhibited preferred temperatures between 40°C and 50°C, but rapidly became dysfunctional at 55°C and died within minutes at 60°C. This finding contradicts an earlier report that Alvinellid worms can withstand exposure to waters with temperatures between 80°C and 90°C, at least temporarily (Cary et al. 1998). The latter observations were made from the submersible DSV *Alvin* (namesake of the worms) using thermal probes, and did not entail direct measurements of the animals' body temperatures. Thus, for eukaryotes, temperatures between 55°C and 60°C appear to represent the upper thermal limit of life, at whatever depth it occurs.

The extent to which high pressures might extend the thermal tolerance ranges of archaea beyond their current limit of ~125°C is not clear. Culture pressure did have an effect on thermal tolerance of one hyperthermophilic archaeon, the methanogen *Methanopyrus kandleri*. When culture pressure was increased from 0.4 MPa to 20 MPa, the maximum temperature for cell proliferation increased from 116°C to 122°C (Takai et al. 2008). This result is consistent with stabilization of cellular membranes by pressure, either through induction of acclimatory alterations in lipid composition or through direct offsetting of heat-induced changes in biophysical properties. Changes in pressure, as well as temperature, have been shown to alter membrane composition in deep-sea bacteria (DeLong and Yayanos 1986), but the importance of this type of acclimatization in hyperthermophilic archaea remains to be elucidated.

3.7 Endothermy and Homeothermy

The foregoing discussion of the diverse and often pronounced effects of temperature on biochemical systems should provide a good perspective for appreciating the benefits of maintaining a stable body temperature. If an organism is well adapted evolutionarily to a specific range of temperatures, one in which its macromolecular systems function optimally, then clearly the ability to maintain its temperature within this range is to its advantage.

Routes to a stable body temperature

Although the analysis of endothermy and homeothermy presented below focuses on biochemical mechanisms—most notably systems for heat production that have been exapted from preexisting biochemical systems that were originally selected for benefits unrelated to body temperature regulation—it is important to emphasize that there are several potential routes to control of body temperature. These routes are differentially accessible to different taxa.

BEHAVIORAL ADAPTATIONS   There are three principal means for attaining a stable body temperature, the **homeothermic** state. One is based on behavior. The ability to control body temperature through one or more behavioral mechanisms, such as altering exposure to the sun during the course of a day, can lead to a high degree of stability in core body temperature. In animals capable of long-distance travel, seasonal migrations across wide ranges of latitude can stabilize body temperatures and, for birds and mammals, reduce the requirements for generating heat to keep the body warm. Behavioral adaptations that involve long-distance migrations are common in birds and mammals but relatively rare in ectotherms, which typically cannot sustain long periods of locomotion powered by aerobic ATP generation. Notable exceptions are leatherback sea turtles and some highly aerobic fishes, such as tunas, which have crossed the line from being strict ectotherms and have gained substantial abilities for endothermic homeothermy in certain regions of their bodies.

ANATOMICAL ADAPTATIONS   A second general strategy for controlling body temperature involves a wide variety of anatomical adaptations. Common examples include the development of insulative surface structures (fur and feathers) and internal lipid layers (blubber). Changes of body color (reflectivity) can enhance or reduce absorption of solar radiation. Circulatory mechanisms that regulate the rate of heat loss to the environment are another important means for controlling body temperature. In mammals, for example, peripheral vasoconstriction may prevent loss of heat from the body during cold conditions. When heat must be dissipated, peripheral circulation is increased. A similar mechanism for stabilizing core body temperature has been observed in certain flying insects, for example, the sphinx moth *Manduca sexta* (Heinrich 1993), and in toucans, which use their enormous bill as a radiator (Tattersall et al. 2009).

Countercurrent heat exchangers are another means for regulating exchange of heat between the body and environment. These thermoregulatory structures, which are found in birds, mammals, leatherback sea turtles, and certain fishes, represent a fascinating example of convergent evolution at the anatomical level. Regardless of the organism in which they occur, countercurrent heat exchangers share the same fundamental anatomical organization. Warm blood flowing from regions of the body where substantial heat is produced gives up its heat to blood flowing in the opposite direction, blood that may be several degrees cooler than the metabolically heated blood (see figure at right). The exchange of heat occurs in an intricate web of small blood vessels termed a *rete mirabile* ("wonderful net"). The enormous exchange surface provided by these minute, intertwined blood vessels enables heat to move down the thermal gradient and warm the cooler blood flowing in the opposite direction. This type of circulatory adaptation enables rapidly swimming fishes like tunas and certain sharks to maintain the temperatures of portions of their locomotory musculature several degrees

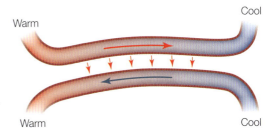

Warm   Cool

Warm   Cool

above ambient temperature (Dickson and Graham 2004). A relatively deep-living fish, the opah (*Lampris guttatus*), can keep most of its body at above-ambient temperatures by using countercurrent heat exchangers in its gills (Wegner et al. 2015). For almost all fishes, metabolic heat is quickly lost to the surrounding water as blood moves through the gills. The heat exchangers in the opah's gills enable it to retain a sizeable fraction of metabolically generated heat and gain a remarkable level of homeothermy. This capacity for homeothermy enables the opah to function over a wide range of temperatures as it moves between cooler deep waters and warmer shallow waters.

BIOCHEMICAL ADAPATATIONS   The third primary strategy for sustaining a stable and, typically, a relatively high temperature of all or part of the body comprises biochemical adaptations that produce high amounts of metabolic heat, either constitutively or on an as-needed basis. As we discuss these heat-generating biochemical adaptations, bear in mind that their success almost invariably depends on anatomical adaptations that facilitate retention or dissipation of metabolic heat. Organisms that are, in this sense, "self-heating" are called endotherms, to reflect the fact that the heat used to warm the body is generated internally by metabolic reactions. If the heat-generating abilities are paired with effective physiological and anatomical mechanisms for closely regulating exchange of heat with the environment, such that the body (or parts thereof) is maintained at a stable temperature, the organism is termed an endothermic homeotherm. When the entire body of an animal is held at a stable temperature through closely balancing heat production with heat loss, the state of systemic endothermic homeothermy is attained. Birds and mammals exemplify this condition, at least under normothermia, when environmental conditions allow a normal body temperature to be defended. Among the animal taxa of fishes, amphibians, reptiles, and invertebrates, the opah, mentioned above, is the only animal thought to be a systemic endothermic homeotherm. That being said, thermogenic systems that allow restricted regions of the body to maintain high and stable temperatures—regional endothermic homeothermy—are widely distributed across both the animal and the plant kingdoms. Regional homeothermic endothermy provides a wide variety of benefits to these species, even though the remainder of the organism has a temperature that varies with—and may essentially equal—that of the environment.

Systemic endothermic homeothermy provides several advantages in addition to the ability to maintain the body's temperature within a range where biochemical systems have evolved to functional optimally. A stable body temperature removes the need for acclimatory adjustments to restore an optimal state of biochemical systems—for example, adjusting the lipid compositions of membranes—and therby results in savings in metabolic energy. Energy savings accrue in other ways as well. The influences of temperature on rates of biochemical reactions, Q_{10} effects, enter into cost-benefit analyses of systemic endothermy in two important ways. First, the relatively high body temperatures of birds and mammals, compared with the body temperatures of most ectotherms, lead to higher specific activities (rate of function per molecule) of enzymes. Thus, a given amount of metabolic machinery can generate a higher level of output in systemic endothermic homeotherms. Second, the ability to sustain stable body temperatures in the face of changes in environmental temperature largely frees mammals and birds from short-term Q_{10} effects. Systemic endothermic homeotherms thus can sustain appropriate rates of physiological function over considerable ranges of temperature, and the organism is able to select habitats and times of activity based on criteria other than ambient temperature. The latter capacity may be especially important for small mammals like rodents, because it enables them to adopt a nocturnal activity pattern; they can carry out their feeding activities under cool nighttime conditions that, because of Q_{10} effects, hinder the activity of large ectothermic predators that can be active only under warmer daytime conditions. Because most early mammals were small and likely to have been easy prey items for much larger ectotherms, the ability to be night-active conferred by endothermic homeothermy was a major boon for mammalian evolution. A related capacity is one of the most important benefits of systemic endothermic homeothermy: the conservation of nervous system function across a range of temperatures. This is a fundamental ability in that it enables perception and locomotion, both key to finding food and avoiding predation, to remain largely independent of ambient temperature.

Regional endothermic homeothermy also confers a wide range of benefits, certain of which are shared with systemic endothermic homeothermy. As discussed below, heater organs that warm the brain and eye of certain large pelagic fishes give these organisms one of the key advantages just mentioned for birds and mammals—the ability to keep the sensory and information-processing systems functioning at stable rates over a range of temperatures. Regional endothermy also helps sustain high rates of swimming activity in tunas and certain large sharks because the deeper regions of the locomotory musculature are maintained at temperatures up to several degrees above ambient temperatures. Some flying insects, notably several species important for pollination, maintain their flight muscles at elevated and relatively stable temperatures. This enables the organisms to feed and gather pollen at relatively cool temperatures and to gain additional rates of function due to Q_{10} effects. In certain flowering plants, heat generated by the plant's metabolism is used to warm the regions where insects alight to attain pollen. These warm chambers help the pollinating insect keep its body temperature above ambient, thereby improving flight performance (and pollen transport) when it leaves the flower. The heat-producing sites in the plant's reproductive apparatus also may assist the synthesis and volatilization of insect-attracting chemicals. During peak reproductive season, some plants exhibit a high level of endothermic homeothermy in the pollination site. For example, the sacred lotus (*Nelumbo nucifera*) maintained the temperature of the pollination chamber between 30°C and 35°C as ambient temperature varied between 10°C and 30°C (Seymour and Schultze-Motel 1996).

COSTS OF HEAT GENERATION   Balancing the benefits of endothermic homeothermy, whether systemic or regional, is the cost of heat generation. Additional metabolic activity—often a very high rate of aerobic metabolism—is generally needed to provide the heat required for regulation of body temperature (or regions thereof). For example, mammals and birds have mass-specific rates of oxygen consumption that are several-fold greater than those of similar-sized ectotherms (see Chapter 2). This higher rate of metabolism means that an equivalently greater amount of food must be acquired and processed. Thus, birds and mammals generally expend much greater amounts of energy and time acquiring food than do ectotherms. Feeding activities are apt to expose endothermic homeotherms to risks of predation, another potential cost that offsets the benefits of maintaining a high and stable body temperature. Because the higher metabolic rates of mammals and birds are powered largely by mitochondrial ATP production, the risks of damage from ROS are likely to be much greater than in the case of ectotherms, as discussed in Chapter 2.

In addition to lower requirements for obtaining and processing food, ectothermy has other major advantages. Ectothermy permits the body to be of smaller size, so ectotherms can exploit niches unavailable to endothermic homeotherms. There is a lower size limit to endothermic homeotherms arising from surface-to-volume considerations: Below a certain mass, the animal's surface area would become too large for effective retention of metabolic heat. Another factor setting a lower size limit to endothermic homeotherms arises from space constraints associated with packaging enormous amounts of metabolic enzymes and organelles into a finite cellular volume. The high density of mitochondria in the cells of the smallest birds and mammals suggests that this packaging limit likely has been reached.

The large literature on the evolution of endothermic homeothermy focuses on both its ultimate causes—what factors led to selection for this ability?—and its proximate causes—what types of biochemical adaptations have allowed the high metabolic rates and relatively stable body temperatures of these organisms? In the discussion of endothermy below, we focus primarily on the second question, which concerns the mechanisms used to enhance and regulate heat production, in order to achieve regional or systemic endothermic homeothermy. This analysis will provide important insights into some of the core principles of biochemical adaptation, especially the role of exaptation in exploiting certain types of membrane-localized processes for orchestrating the appropriate levels of heat production needed to warm the organism.

Sources of heat for endothermic homeotherms: The central role of membranes in ATP turnover

In Chapter 2 we discussed the central role of membrane-localized processes in generating ATP and in its subsequent use as the cell's energy currency for performing work. It is helpful in the context of endothermic homeothermy to review certain facets of ATP turnover, in order to gain a clear sense of the types of raw material that are available to cells for developing the endothermic state. Seeing how cells use oxygen is an informative part of this analysis. Mitochondrial ATP production is estimated to account for about 70% of total oxygen consumption under resting metabolic conditions for mammals (Rolfe and Brown 1997). Mitochondrial proton leakage—inward (matrix-directed) movement of protons across the inner mitochondrial membrane that is not coupled to ATP synthesis—may account for another ~20% of oxygen consumption. The final ~10% of oxygen use is through non-mitochondrial processes such as oxygen-dependent biosynthetic reactions. Overall, then, approximately 90% of oxygen consumption involves membrane-based processes.

HOW IS ATP USED IN THE CELL?   How is ATP use distributed across physiological processes, and where are these processes primarily located? A large percentage of ATP turnover, estimated to be ~20%, is linked to membrane-localized processes of ion transport. The sodium-potassium ATPase (or Na^+-K^+-ATPase) pump may use ~15% of the ATP produced by the cell (Rolfe and Brown 1997). Calcium pumping by ATP-using pumps may account for an additional ~5% of ATP turnover. Protein synthesis is another process that demands high levels of ATP turnover, depending on the state of growth. Under low-level metabolic conditions where growth is not occurring, but where protein homeostasis is maintained (degraded proteins are replaced with new ones), protein synthesis may demand about 20% of ATP production. Whereas protein synthesis (translation) is not itself a membrane-localized process, the supply of building blocks (amino acids and their precursors) is dependent on transmembrane uptake that is powered largely by the sodium ion gradient maintained by the Na^+-K^+-ATPase pump. ATP-driven transmembrane exchanges may be involved in the movement of proteins between cellular compartments, for example, entry of proteins into the endoplasmic reticulum, where further processing and distribution of proteins occur. Membranes, then, are of pivotal importance in both the production and use of ATP.

What accounts for the high mass-specific oxygen consumption rates of endothermic homeotherms?

With this brief overview of oxygen use and ATP turnover as background, we are better prepared to address a question of central importance in the evolution of endothermic homeothermy: How do we account, mechanistically, for the several-fold higher mass-specific oxygen consumption rates of endothermic homeotherms relative to those of ectotherms of similar body mass? Comparisons of mammals and reptiles of similar body size and normal body temperature (e.g., see Hulbert and Else 1989) have shown that most, and probably all, of the oxygen-using systems listed earlier have higher activities in mammals and birds relative to ectotherms. It follows that capacities for ATP generation are similarly higher in endotherms as well (Else et al. 2004).

Going down the list of processes that account for use of oxygen, we find that proton leakage rates across the inner mitochondrial membrane are far greater per mass of tissue in endothermic homeotherms than in ectotherms (Brand et al. 1991). In part this is a consequence of greater mitochondrial abundance in endotherms. The fraction of the cellular volume occupied by mitochondria is significantly larger in mammals than in reptiles of similar body mass (Else and Hulbert 1985). The total inner mitochondrial membrane surface area is about fourfold larger in mammals than in non-avian reptiles. This enhanced mitochondrial presence cuts both ways: More ATP can be generated, but increased amounts of uncoupled proton leakage can occur as well.

Leakage of sodium and potassium ions across the cell membrane shows a pattern similar to that seen with mitochondrial proton leakage: A significantly higher flux rate is present in endothermic homeotherms (Else and Hulbert 1987). This increased passive flux of sodium and potassium ions must be counterbalanced by increased activity of the Na^+-K^+-ATPase. In fact, the activities of Na^+-K^+-ATPase pumps in the cell membrane are significantly higher in endothermic homeotherms compared with ectotherms (Else et al. 2004). This difference in Na^+-K^+-ATPase activity is due to higher k_{cat} values for the Na^+-K^+-ATPases of endothermic homeotherms, not to the presence in the membranes of greater numbers of sodium pumps. There is evidence that the higher specific activities of the endotherm Na^+-K^+-ATPases are caused by differences in lipid composition between endotherms and ectotherms. In particular, the amount of docosahexaenoic acid (DHA), the 22:6 PUFA found to increase in membranes of cold-acclimated ectotherms, is elevated in cell membranes of mammals relative to non-avian reptiles. The ability of DHA to fluidize the interior of the bilayer is conjectured to foster a higher specific activity of enzymes that span the bilayer, such as the Na^+-K^+-ATPase (Hulbert et al. 2007). DHA effects are at the core of the "metabolic pacemaker" hypothesis proposed by Hulbert and colleagues (see Chapter 2). Wu and colleagues (2004) tested this conjecture directly by exchanging the lipids in which Na^+-K^+-ATPases from a mammal (cow) and a reptile (crocodile) were embedded. Surrounding the mammalian Na^+-K^+-ATPase with crocodile-derived lipids generated an enzyme with reptilian levels of function. Replacing the lipids around the crocodile Na^+-K^+-ATPase with mammalian-derived lipids converted the enzyme to one with mammalian activity.

A final reason for the higher mass-specific oxygen consumption rates of endothermic homeotherms is that their tissues tend to have higher protein concentrations (Hulbert and Else 1989). Protein synthesis consumes large amounts of ATP in synthesizing peptide bonds and in supplying the translational apparatus with amino acids. Therefore, the higher protein concentrations in endothermic homeotherms mean an increased need for ATP.

In summary, all of the major metabolic processes that are dependent on oxygen use—including uncoupled flux of protons across the inner mitochondrial membrane as well as coupled proton flux that drives mitochondrial ATP production—are elevated in birds and mammals relative to ectothermic vertebrates. These differences thus go a long way toward explaining the proximate causes of the higher basal metabolic rates of endothermic homeotherms.

There is, however, one important aspect of endothermic homeothermy that is not directly explained by these differences in oxygen-dependent biochemical systems: the capacities of birds and mammals to greatly increase heat production when faced with low ambient temperatures. In other words, our brief review of oxygen-dependent metabolism offers insights into the higher *constitutive* levels of metabolism and body temperature found in birds and mammals, but it does not provide a full account of how the demands for additional *facultative* heat production are met during challenges from cold. Below we examine several mechanisms for supplementary heat production and find, again, that exaptation of systems fulfilling other functions has led to evolution of highly efficient thermoregulatory systems that occur not only in systemic homeotherms like mammals and birds, but also in diverse ectotherms that display regional endothermy.

Brown adipose tissue (BAT): A mammalian exaptation for thermogenesis

The ability to convert the free energy contained in the chemical bonds of foodstuffs into heat could, in principle, be achieved in two primary ways, both of which involve the electron transport chain of mitochondria. In the first mechanism of thermogenesis, mitochondrial ATP production would remain tightly coupled and large quantities of ATP would continue to be produced. The release of heat would occur only when this ATP is used to do work, such as transmembrane pumping of ions. Facultative thermogenesis involving high rates of ATP turnover is a common strategy in both endothermic homeotherms and in regionally endothermic ectotherms. This enhanced rate of ATP turnover involves what are known as **futile cycles**, which involve high rates of ATP turnover with no net work being done. We provide examples of such Sisyphus-like (see figure at left) biochemical cycles later in this section.

An alternative way to generate heat using the mitochondrial electron transport system skips the ATP production step entirely. In this case, electron transport still drives the formation of a proton gradient across the inner mitochondrial membrane, but the dissipation of the proton motive force (Δp) through proton movement back into the mitochondrial matrix is not coupled to synthesis of ATP by ATP synthase. Instead, protons cross the inner mitochondrial membrane through a different route, one that results in the "wasting" of all of the energy stored in the proton gradient. Clearly, such a thermogenic mechanism could not be exploited on a systemic basis because most tissues and organs need a continuous supply of ATP to accomplish their work, be it locomotion, membrane transport, or biosynthesis. However, in mammals (and as far as we know, only in mammals) specific types of lipid-rich (adipose) tissues have the potential for "wasting" essentially all of the energy released in aerobic metabolism. The primary tissue serving this thermogenic function is termed **brown adipose tissue** (**BAT**), a name that derives from its deep brownish-red color, which is due to cytochromes in the abundant mitochondria (**Figure 3.55**). Unlike white adipose tissue (WAT), which functions as an energy storage site—a triacylglyceride lipid depot—and plays roles in the regulation of several aspects of metabolism (Peirce et al. 2014), BAT functions principally as a thermogenic tissue, one whose abundance and heat-generating activity are tightly regulated in accordance with the body's need for

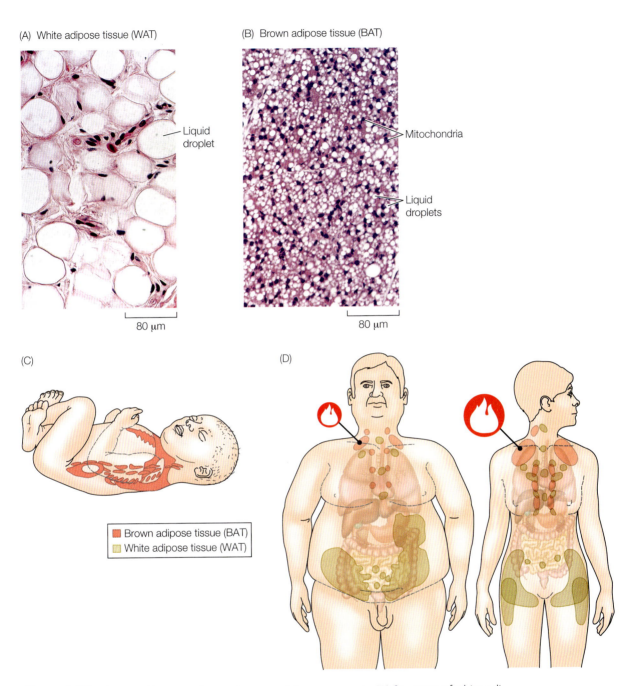

Figure 3.55 White and brown adipose tissues and thermogenesis. (A) Structure of white adipose tissue (WAT). WAT adipocytes are filled with large globules of triacylglycerides and are relatively depleted of mitochondria. (B) Structure of brown adipose tissue (BAT). BAT cells are rich in mitochondria and contain smaller lipid droplets. (C) Distribution of BAT in newborn humans. BAT deposits in newborn humans are large, relative to deposits in adults, and are positioned where warming of important organ systems, for example, the kidneys, is facilitated. (D) Occurrence of BAT in adult humans. In adult humans the amounts of BAT vary widely. Females tend to have larger BAT deposits than men. Lean individuals show a more pronounced development of BAT following exposure to cold than obese individuals. (A and B from Ross et al. 2009; C after Hull 1973; D after Farmer 2009.)

additional heat production (Cannon and Nedergaard 2004). BAT's other primary function, helping maintain body mass, will be considered later when we discuss the coupling that has been discovered between the metabolic functions of WAT and BAT.

CLASSES OF ADIPOCYTES (FAT CELLS) AND THEIR ONTOGENIES AND PLASTICITIES   The study of BAT has a long and interesting history, and the enormous efforts now being placed on health-related studies of lipid metabolism are revealing numerous unanticipated characteristics of the tissue (Peirce et al. 2014). The first reported observation of BAT was by the Swiss naturalist Konrad Gessner in 1551 (for this history, see Cannon and Nedergaard 2008). Gessner apparently was puzzled by the nature of this tissue, and declared that it was "neither fat, nor flesh" (*nec pinguitudo nec caro*). This observation may have been prescient, because recent studies of the ontogeny of BAT have revealed that BAT adipocytes derive from myoblastic-like Myf5-positive precursor cells that can differentiate into either skeletal muscle ("flesh") or thermogenic tissue (BAT; Seale et al. 2008; Trajkovski et al. 2012). If we assume that the myogenic (muscle-forming) developmental pathway is ancestral, then we can speak of exaptation at the level of cellular ontogeny; a cell type whose original fate was development into a skeletal muscle cell becomes exapted for a new function—thermogenesis.

The molecular details of the factors governing this binary choice of ontogenetic fates are being worked out, and are helping elucidate how the decision not to become "flesh" leads to a remarkable type of thermogenic cell. A key determinant of the fate of the common precursor cells is a zinc-finger transcription factor, Prdm-16 (Trajkovski et al. 2012), which can drive the ontogeny of the precursor cells toward BAT adipocytes. In turn, the activity of Prdm-16 is regulated by a specific microRNA, miR133, which negatively regulates Prdm-16 function. miR133 is strongly downregulated by cold exposure in both BAT and in subcutaneous lipids, where beige thermogenic adipocytes (see below) can form under the influence of Prdm-16 regulation. miR133's downregulation is mediated through cold-induced downregulation of its transcriptional regulator, Mef2 (Trajkovski et al. 2012). A role for RNA thermometers in this regulatory process would seem plausible, but to date the precise mechanism underlying temperature-mediated transcription of Mef2 has not been discovered.

BAT's ontogeny differs from that of white adipocytes, in keeping with these tissues' different roles. Until quite recently, it was thought that BAT and WAT were the only types of adipose tissues in mammals. We now know, however, that certain WAT cells can, under the appropriate circumstances, differentiate into thermogenic cells. This third type of adipocyte has been variously termed beige, breit (brown + white), or inducible adipose tissue (Peirce et al. 2014). Here we will use the term *beige*, in keeping with what seems to be the most common terminology. The formation of **beige adipose tissue** is triggered by cold stress, and the heat-generating mechanisms it exploits are similar to those of BAT, as we discuss below. When cold stress is relieved, beige adipose tissue can dedifferentiate and, again, become a lipid-storing adipocyte within a mass of WAT.

Before we examine the biochemistry of thermogenesis in BAT and beige adipose tissue, a review of the anatomical location of these tissues will help us develop a more comprehensive picture of the role they play in endothermic homeothermy. The locations of these two types of thermogenic tissues reflect several considerations, including the necessity of providing heat to organs like the heart, brain, and kidneys, whose activities typically must be maintained at high rates even if other regions of the body sustain cooling, and reducing thermal gradients between the body's core and its surface. BAT is typically positioned in the body core, to provide heat for critical organs. In some marine cetaceans, however, BAT may occur in the inner layer of the blubber and function as an "electric blanket" that covers much of the body. BAT with this distribution pattern has been reported in four species of dolphins and porpoises and may account for the abilities of these species to remain warm even under conditions of minimal locomotory activity (Hashimoto et al. 2015). Beige adipose tissue is similarly distributed as a thermogenic blanket near the body surface. Thus, beige adipose tissue may be found chiefly in subcutaneous lipid deposits. There, heat generation can reduce the thermal gradient between the surface of the organism and its internal organs (Ye et al. 2013).

The amount of thermogenic adipose tissue present in an organism varies with body size and physiological status. Small mammals, with their relatively high surface-to-volume ratios, contain a higher relative amount of BAT than larger mammals. Similarly, newborn mammals, which also have relatively high surface-to-volume ratios, enter the outside world with relatively large amounts of BAT compared with levels characteristic of adults (see Figure 3.55C). Up until a few years ago, physiologists believed that the abundant BAT of newborns was largely if not entirely lost by adulthood. This conclusion was shown to be false; adult humans typically do contain some BAT, with the amounts varying with physiological state (see Figure 3.55D; Cannon and Nedergaard 2008). Obese humans have lower amounts of BAT than lean humans, a finding that has helped precipitate a huge amount of health-related study of BAT (and beige adipose tissue), in hopes of finding ways to combat obesity. Females tend to retain more BAT than males.

PLASTICITY OF THERMOGENIC FAT CELLS   The amounts and the locations of thermogenic adipose tissues are quite plastic and subject to modification in response to ambient temperatures and diet. In healthy nonobese humans, a lipid-rich diet can engender increased amounts and activities of BAT, a response that assists in avoidance of obesity. Cold exposure can lead to increases in the mass of BAT and the activities of the biochemical systems in BAT that are responsible for thermogenesis. A variety of regulatory mechanisms control the thermogenic potential of BAT and beige adipocytes. An interesting recent discovery about the plasticity of thermogenic adipose tissue is that cold exposure leads to proliferation of beige adipose cells in subcutaneous fat deposits, which, as we discussed above, help retain heat in the body core. Two regulatory mechanisms (at least) are responsible for this increase in beige adipose tissue. One involves the canonical cascade of regulatory processes that control the mass and thermogenic activity of BAT cells (see Figure 3.56). Here, thermosensory cells in the hypothalamus lead to stimulation of the BAT cells by the sympathetic nervous system (SNS). Downstream events lead to enhanced thermogenesis. The second stimulatory mechanism is based on direct temperature sensing by the adipose cells themselves. The second stimulatory mechanism is based on direct, SNS-independent temperature sensing by certain classes of adipose cells. In a study by Ye and colleagues (2013), exposure of mouse-derived white, beige, and brown adipocyte cell lines to reduced temperatures (27°C–33°C) led to upregulation of thermogenic systems in subcutaneous white adipocytes and beige adipocytes. In contrast, the BAT-derived cell line did not exhibit enhanced thermogenesis. Thus, the thermogenic function of brown adipocytes appears to be regulated by different mechanisms than the thermogenic systems in certain types of white and beige fat cells. In whole animal experiments, where SNS regulation of thermogenic activity would be possible, exposure to cold led to upregulation of thermogenic function in BAT and in subcutaneous fat cells but not in visceral white adipose cells (Ye et al. 2013). This observation suggests that, much as brown adipocytes differ from beige and white adipocytes in regulatory control of thermogenesis, differences may also exist among different types of white adipocytes. These findings support the conjecture that visceral and subcutaneous adipocytes, which have been grouped together as one cell type may, in fact, differ in physiology (Peirce et al. 2014).

UNCOUPLING PROTEINS: A KEY MECHANISM FOR "WASTING" ENERGY   The thermogenic mechanisms found in BAT and beige adipocytes are extremely complex and involve neural, hormonal, immune system, bile acid, transcriptional, translational, and posttranslational regulation. However, the secret of success of these cells' thermogenic capacities is easily identified: a transmembrane protein of the inner mitochondrial membrane termed **uncoupling protein 1** (**UCP1**). This protein was the first of several uncoupling proteins (UCPs) to be discovered, and its discovery and putative thermogenic role in BAT led to its alternate name, **thermogenin**. All UCPs are members of a class of transmembrane metabolite transporter protein called **mitochondrial solute carriers** that appear to have arisen at the time of origin of the eukaryotic cell (Echtay 2007; see Chapter 2). Among the better-understood members of this class of protein is the **adenine nucleotide translocator** (**ANT**), which transports adenylates across the inner mitochondrial membrane. ANT is also capable of transporting fatty acids and can serve as an uncoupling factor, albeit its contribution to this process is less than that of UCP1.

UCP1 has been studied extensively to uncover its mechanism of uncoupling and the complex regulatory mechanisms that govern its activity in a temperature- (or diet-) dependent manner (Ricquier and Bouillaud 2000; Cannon and Nedergaard 2004; Echtay 2007). Several regulatory mechanisms for modulating the kinetics of UCP1 function are well established and accepted by all members of the large research community studying this protein. For example, UCP1's rate of activity is reduced by binding of purine nucleotides (ADP, ATP, GDP, and GTP), a kinetic control that is important in short-term regulation of UCP1 activity. Researchers also agree that UCP1's activity is stimulated by the presence of fatty acids, but how this stimulation fits into the uncoupling process is an ongoing matter of debate (Divakaruni and Brand 2011). One school of thought conjectures that fatty acids function within UCP's transport channel, where they reversibly bind protons and "guide" them though the channel. A second model (shown in Figure 3.56) builds on the observation that another mitochondrial solute carrier, ANT, can transport fatty acids through the inner mitochondrial membrane. If one assumes that UCP1 possesses a similar ability, and if one takes into account the pH gradient between the extramitochondrial space and the mitochondrial matrix, then a proton translocation mechanism involving fatty acids as proton carriers (protonophores) can be envisioned. The extramitochondrial space has a substantially lower pH than the matrix due to the proton transporting processes that generate the proton motive force. In the extramitochondrial space, the carboxylate groups of fatty acids have a higher likelihood of being bound to protons—and thus of being neutralized—than in the matrix. Neutral fatty acids can move through the inner mitochondrial membrane without assistance from a transporter. Once in the more alkaline matrix, the protons dissociate from the carboxylate groups of the fatty acid protonophore. The fatty acid, now with a negative charge, moves back to the extramitochondrial space through UCP1. A transporter is required for this movement because charged molecules do not move freely through the lipid bilayer. (A third model, which appears to be falling out of favor, involves fatty acid stimulation of UCP1 activity through allosteric effects on UCP1 [Divakaruni and Brand 2011]. We will not consider this third alternative further.)

WHAT WAS THE ORIGINAL ADVANTAGE OF UNCOUPLING?   However UCP1 transports protons—whether transport involves a "free" proton hopping from one channel carboxylate group to the next, or a fatty acid "revolving door" mechanism that uses a fatty acid as a protonophore—the use of this solute transporter in thermogenesis represents a fascinating case of exaptation. Mitochondrial solute transporters were present in cells long before endothermic homeothermy appeared. What, then, was the original function (the "adaptation" in the strictest sense of that term) of UCPs? The fact that some 20%–25%

of basal oxygen consumption in mammals (and a sizeable percentage in ectotherms as well) is not coupled to ATP generation suggests that uncoupling has a substantial role in cell physiology (Divakaruni and Brand 2011). What this role is likely to be was already suggested in Chapter 2, when we examined mitochondrial production of ROS. When flux of electrons through the electron transport chain is not closely matched by the supply of ADP for ATP synthesis by ATP synthase, a buildup of electrons can lead to univalent reduction of molecular oxygen, forming superoxide as the primary initial ROS. As we discussed in Chapter 2, superoxide can be converted to other, even more toxic types of ROS, including the hydroxyl radical. Prevention of excessive formation of ROS thus is viewed as a major—and perhaps the principal—role of uncoupling proteins. This conjecture is further supported by the fact that superoxide generated in the mitochondria stimulates activity of UCP1 (Echtay 2007). Superoxide thus functions as a signal for its own destruction.

The subsequent exploitation (exaptation) of UCPs as thermogenic factors in mammals is likely to have played a pivotal role in the evolution of these organisms. Many early mammals were small, nocturnal species that would have had difficulty maintaining a warm and stable core temperature, were it not for the ability of thermogenic adipocytes to generate large amounts of heat. Neonatal mammals likely were especially dependent on this mechanism for internal heat production. Of course it would be incorrect to state that one class of protein, UCPs, was responsible for the early evolutionary success of primitive mammals. Yet without the thermogenic potential provided by UCP1, it seems doubtful that the evolution of mammals would have enjoyed the success it has.

REGULATION OF THERMOGENIC ADIPOSE TISSUES   Research over the last two decades has revealed a wide variety of mechanisms for regulating the amount of thermogenic adipose tissue in an organism, its location in the body, and the heat-producing capacities of its cells. The canonical regulatory pathway involving the sympathetic nervous system has been studied in great detail in several mammals for many years. As shown in Figure 3.56, temperature-sensing cells in the hypothalamus trigger activity of sympathetic nerves that innervate BAT. These nerves release the catecholamine norepinephrine (NE), which binds to β_3 adrenoceptors on the cell membrane of the BAT adipocytes. Binding of norepinephrine initiates a cascade of regulatory reactions. The G protein-associated enzyme **adenylate cyclase** is activated, which leads to the production of cyclic AMP (cAMP). cAMP binds to the regulatory subunit of **protein kinase A** (**PKA**). The formation of this protein-ligand complex leads to dissociation of the regulatory and catalytic subunits of PKA. The free catalytic subunit then catalyzes the phosphorylation of three proteins involved in activating BAT metabolism. One is a protein termed **perilipin** ("around the lipid") that is thought to be involved in regulating access of the lipid stored in the BAT adipocyte by an enzyme termed **hormone-sensitive lipase** (**HSL**). This enzyme is another phosphorylation target of PKA. When activated by phosphorylation—and given access to triacylglyceride stores by phosphorylation of perilipin—HSL begins to generate free fatty acids (FFAs). FFAs serve two roles: They are substrates for oxidation and they are facilitators of transmembrane movement of protons, which dissipates the proton gradient. The effects of PKA-mediated phosphorylation of perilipin and HSL occur relatively quickly; they are evident within minutes. The third protein phosphorylated by PKA is the gene regulatory protein **cAMP response element-binding protein** (**CREB**). CREB activates transcription of the gene encoding UCP1, and thus is part of another, slower mechanism for regulating thermogenesis in the BAT cells, one that involves modifications of gene expression to increase the amount of heat-generating machinery in the tissue.

These regulatory events, which are initiated though SNS stimulation, have been a familiar part of the BAT regulatory landscape for more than a decade. Over the past several

Figure 3.56 Pathways involved in thermogenic activity in BAT and mechanisms used for their regulation. Regulation of heat production involves a suite of signal transduction processes whose time-courses differ (see text for details). (After Hochachka and Somero 2002.)

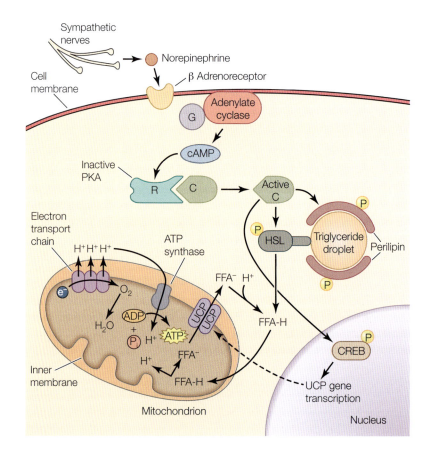

years, new regulatory mechanisms have been discovered as well, and these mechanisms reflect how intimately BAT metabolism is integrated into the physiology of the organism as a whole, including the dietary status of the individual. The effects of thyroid hormones offer an interesting illustration of how tightly BAT activity is integrated into nutritional status as well as into thermoregulatory needs. It has long been appreciated that thyroid hormone stimulates BAT function during cold stress, in part by facilitating upregulation of transcription of the UCP1-coding gene. Our understanding of the mechanisms of thyroid hormone action in BAT has recently taken a new turn, however. Thyroid hormone delivery by the blood is not the only way BAT obtains this hormone. BAT adipocytes themselves can convert a minimally active precursor of thyroid hormone, **thyroxin (T_4)**, into the more potent form, **triiodothyronine (T_3)**. This conversion is fostered by **bile acids** such as cholic acid, which are hormonal regulators of thyroid hormone synthesis. The conversion of T_4 to T_3 is catalyzed by **type 2 deiodinase (D2)**. Transcription of the gene encoding D2 is activated by bile acids in BAT but not in liver (Watanabe et al. 2006). The role of bile acids in fat metabolism has been investigated for decades, but not in the context of BAT. Bile acids are produced in the liver from cholesterol and stored in the gallbladder until needed. After ingestion of lipid-rich foods, bile acids are released into the intestine to promote the absorption of fat. Some bile acids also enter the general circulation and are carried to BAT. There they bind to a G protein-linked receptor (TGR5) on the brown adipocyte cell membrane. As in the case of binding of NE to the β_3 adrenoceptor, binding of bile acids to the TGR5 receptor leads to production of cAMP. This, in turn, leads to increased transcription of the gene encoding D2, and then to increases in the amount of D2 protein. The importance of this increase in D2 for thermogenesis is clear: Mice lacking the gene for this enzyme cannot sustain heat production under cold stress (Baxter and Webb 2006). The production of T_3 in the brown adipocyte represents an efficient mechanism for regulating UCP1 levels and, thereby, heat production. These bile acid effects reflect a tight integration between lipid ingestion and the burning of lipids by BAT. It is not known whether cold temperatures per se lead to increases in bile acid release from the gallbladder, independent of dietary lipid input. Such a regulatory mechanism would provide a further level of control of thermogenesis.

Another level of regulation of BAT involves an integration of BAT and WAT metabolisms that is effected through activities of immune system cells termed macrophages. These immune cells are highly mobile and are found throughout most tissues of the body. One of their primary roles is to destroy materials that might be damaging to the organism. When fighting pathogens, macrophages become transformed into proinflammatory cells that secrete catecholamines, which, along with cytokines, are important in regulating the intensity of the immune response at the infection site.

In the context of thermogenesis, macrophages have coordinated effects in both WAT and BAT (Whittle and Vidal-Puig 2011). When mice are subjected to cold stress, the macrophages found in WAT and BAT are activated to function in a noninflammatory mode (Nguyen et al. 2011). Because this alternate function is triggered by catecholamines released by the SNS, it perhaps should be regarded as an extension of the canonical BAT stimulation discussed above. The macrophages activated in BAT and WAT, in turn, release catecholamines that serve important functions in regulating lipid metabolism in both types of adipocytes. In WAT, catecholamines enhance the breakdown of depot triacylglycerides into FFAs, which are released into the bloodstream and taken up by BAT cells for use as fuel and protonophores. The catecholamines released by macrophages within BAT initiate the regulatory cascade shown in Figure 3.56, which leads to rapid increases in FFA supply and slower increases in UCP1 levels. The importance of macrophage-mediated thermogenesis was shown by genetic studies that removed the signaling mechanisms required for converting macrophages to an activated, noninflammatory state (Nguyen et al. 2011). Mice lacking the ability to develop these noninflammatory macrophages had impaired thermogenic abilities.

Several other mechanisms exist for rapid modulation of thermogenesis by BAT, mechanisms that are completely independent of increases in the amount of tissue. Primary among these mechanisms are responses to purine nucleotides (GDP, GTP, ADP, and ATP) and ROS. Binding of purine nucleotides to UCP1 can trigger a rapid (within seconds) inhibition of thermogenesis (Ricquier and Bouillaud 2000). Purine nucleotide binding, in turn, is subject to close regulation. Purine nucleotide binding is very sensitive to pH; falling pH stabilizes this binding and thereby facilitates rapid downregulation of BAT activity (Malan and Mioskowki 1988). This type of pH effect could be important in small mammalian hibernators, which commonly retain CO_2 and undergo respiratory acidosis during entry into deep hibernation, when body temperatures may be near 0°C (Snapp and Heller 1981). The shutdown of BAT during entry into hibernation is obviously essential, and this could be achieved quickly through pH "titration" of purine nucleotide binding to UCP1. During arousal from deep hibernation, a small hibernator hyperventilates and releases the excess CO_2 it held during hibernation. The resulting rise in pH could reduce binding of purine nucleotides to UCP1 and foster a rapid reactivation of BAT for restoring the normothermic body temperature.

Last, the activity of UCP1 is also modulated by ROS, and this effect is mediated, at least in part, through effects on purine nucleotide binding (Chouchani et al. 2016). In fact, the role of ROS in controlling BAT thermogenesis is an excellent illustration of the important and complex signaling roles played by ROS (see Chapter 2). During acute cold stress, ROS levels in BAT mitochondria rise rapidly, though the exact cause or origin of the increased ROS production is not known. The increased levels of ROS lead to oxidation of protein cysteine thiol groups that are of central importance in the functions of the target proteins. The tripeptide glutathione (GSH) is oxidized rapidly as well, and the pool of GSH is quickly depleted. This change in GSH levels would provide less protection for thiol groups, which can be oxidized to sulfenic and sulfinic acids. Furthermore, when GSH concentrations are low, reversible protein cysteine oxidation states (see Figure 2.19) that are key to signaling would tend to persist. A key cysteine oxidation event in early thermogenesis involves one of the seven cysteine residues (Cys253) of UCP1. Oxidation (sulfenylation) of Cys253 of UCP1 causes important downstream effects on thermogenesis. Cys253 is in the region of the protein where purine nucleotide binding is thought to occur (Chouchani et al. 2016). Sulfenylation of UCP1 thus is conjectured to disrupt binding of inhibitory purine nucleotides, which would rapidly activate thermogenesis.

In summary, regulation of BAT function is a multitiered process, one that includes relatively slow changes in the amounts of BAT found in the organism and rapid changes in the activity of BAT mitochondria in response to signaling from hormones, nucleotides, pH, and ROS. The fine-tuned regulation of BAT function is therefore a fascinating example of how exaptation of a given type of protein—an uncoupling protein—leads to profound changes in an animal's physiological capacities. However, emerging evidence suggests that an even broader type of regulation is involved, one that entails cooperation between the animal and its microbiome.

THE ROLE OF THE MICROBIOME IN THERMOGENESIS   The thermogenic potential of a mammal is not due entirely to physiological events stemming from activities of its own cells. Rather, the generation and the activity of thermogenic tissues, notably beige adipocytes, are also affected by the microbiome of the organism (Chevalier et al. 2015). Studies of the microbiome are revealing the scope and intimacy of animal-microbial interactions in a wide range of contexts (see McFall-Ngai 2015). Recent studies of the effects of low temperature on the composition of the microbiome of mammals are revealing yet another aspect of these interactions.

Mice maintained at different temperatures in the laboratory possess different microbiomes (**Figure 3.57**; Chevalier et al. 2015). The significance of these differences was demonstrated by transplantation of microbial populations obtained from mice held either in the cold or at room temperature into axenic (bacteria-free) recipients. Room temperature-acclimated mice that received the cold-acclimated microbiome showed increased generation of beige adipose cells from white adipose tissue as well as several other changes that led to a boost in metabolic rate. These changes included an increase in the absorptive potential of the small intestine. Total length of the intestine increased, as did the lengths of villi and microvilli. The enlarged absorptive area was paired with an increase in fermentative end products in the intestine. Sensitivity to insulin also increased, which fostered an elevation in cellular metabolic rates. These responses induced by the cold-acclimated microbiome in room temperature-acclimated mice largely mimicked the changes found in cold-acclimated mice. This is a fascinating example of shared control of the processes of cellular proliferation and differentiation between host and symbionts.

In summary, the rapidly growing understanding of thermogenic brown and beige adipose cells is providing insights into a host of important physiological processes that involve tight, intertissue regulation as well as interactions between the mammalian host and its temperature-dependent microbiome. The complexity of these control systems

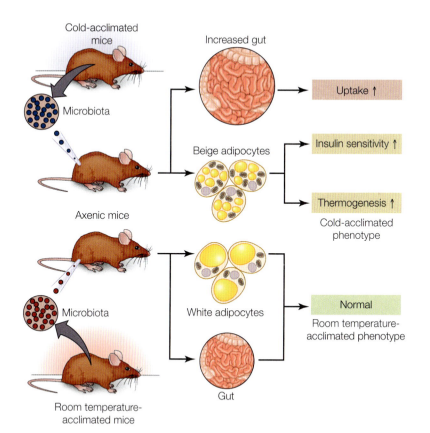

Figure 3.57 The intestinal microbiome changes during thermal acclimation and influences the development of thermogenic potential: differentiation of beige adipose cells and upregulation of intestinal surface area and insulin-dependent metabolic processes. Transplantation of the intestinal microbiome from cold-acclimated mice into room temperature-acclimated axenic mice triggered changes in adipose cell differentiation and intestinal morphology that resembled changes seen in cold acclimation. Transfer of the microbiota of room temperature-acclimated mice did not induce these changes. (After Chevalier et al. 2015.)

likely reflects both the benefits of adipose cell-localized thermogenesis and the dangers of allowing thermogenic activities to get out of control. Here, the analogy to a nuclear reactor seems accurate. The "invention" of UCP1 through exaptation of a mitochondrial solute carrier protein can be viewed as but an initial step in the development of mammalian non-shivering thermogenesis. To effectively exploit the heat-generating ability of membranes containing UCP1, a host of additional adaptations is necessary. These include provision of FFAs from non-thermogenic adipose tissue, WAT, and diverse regulatory mechanisms that can alter the amount of heat-generating tissue and the rate at which a given cell can produce heat. Regulation of heat-producing adipose tissue also can entail the conversion of non-thermogenic cells into thermogenic cells: beige adipose tissue. It seems fair to say that, whereas we now possess a deep understanding of thermogenesis by adipocytes, the large amount of effort currently being given to their study, largely in the context of concerns over human obesity, will lead to even further discoveries concerning how the thermogenic abilities of mammals have evolved—with the cooperation of their microbiomes.

Thermogenesis through making and then "wasting" ATP: Ion pump futile cycles

Heat-generating futile cycles that involve the formation and subsequent consumption of ATP, with no net work being accomplished in the overall process, are well known in a variety of taxa, including endothermic homeotherms and a variety of ectotherms. Typically these processes involve aerobic production of large amounts of ATP and its subsequent splitting in transmembrane ion-pumping systems, ion-dependent ATPases. All of these futile cycles use energy to establish ion concentration gradients and then facilitate a rapid dissipation of these gradients to enable maximum "wastage" of energy.

MUSCLE-BASED NONSHIVERING THERMOGENESIS IN BIRDS AND MAMMALS   We are all familiar with the process of shivering, in which splitting of ATP by the muscle contractile proteins leads to heat production but to no coordinated locomotory movements. Both birds and mammals possess this mechanism for on-demand heat generation. In addition, both groups of endothermic homeotherms have in their locomotory musculature a mechanism for nonshivering thermogenesis (NST) that relies on a futile cycle of calcium ion movement. Although this mechanism involves coupled ATP synthesis, it shares many similarities with other heat-generating mechanisms, notably the control systems that characterize BAT (**Figure 3.58**).

Both thermogenic systems are regulated by activity of the SNS. In each case, a SNS neuron releases NE at the surface of the cell (BAT or skeletal muscle), and this leads to the activation of a lipase whose activity increases FFA concentrations in the cell. In BAT, as we have seen, FFAs serve both as substrates for aerobic metabolism in the mitochondrion and as protonophores. In skeletal muscle, FFAs again serve as metabolic substrates, but have the additional function of activating calcium release channels in the **sarcoplasmic reticulum (SR)**. The calcium so released from the SR contributes to thermogenesis in two ways. First, it activates mitochondrial metabolism and thereby directly increases heat production; recall that only about 40% of the energy released in foodstuff oxidation is trapped in the bonds of ATP. Second, calcium plays a less direct role that involves splitting of the ATP whose synthesis it stimulates. This is accomplished by the SR membrane Ca^{2+}-ATPase pump, which transports cytosolic Ca^{2+} back into the SR. Overall, no work is done; calcium ion merely moves down its concentration gradient and then is pushed back up the gradient. This, then, is another instance of a Sisyphus-like molecular system that, like BAT, has a high capacity for heat generation plus a set of regulatory controls to keep the heat-generating activity in check.

BILLFISH CRANIAL HEATERS: EXAPTING EYE MUSCLE FOR THERMOGENIC FUNCTION   Our final example of a thermogenic tissue in animals takes NST a step further and represents an interesting case of exaptation of a muscle for heat generation, but at the cost

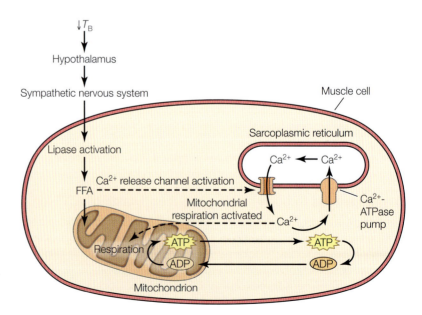

Figure 3.58 Nonshivering thermogenesis in skeletal muscle of birds and mammals. See text for details.

of losing all contractile function. We have emphasized the importance of stabilizing the temperatures of neurosensory systems, in order to allow these critical systems to maintain a high level of function over a range of temperatures. However, an elevated temperature for these systems can also be beneficial; Q_{10} effects can stimulate metabolism, including processes supporting conduction and transmitter activity. An example of regional endothermic homeothermy in billfishes (swordfish and marlins) illustrates a novel use of muscle tissue to warm the eye and brain (Block 1991, 1994).

Swordfishes and marlins have a cranial heater system, the heater organ, which is derived from sections of the extraocular muscles attached to the eye (**Figure 3.59**). As the muscle enters the orbit of the eye, it changes radically in appearance and function.

(A)

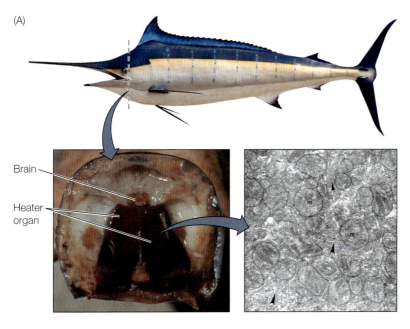

Brain

Heater organ

(B)

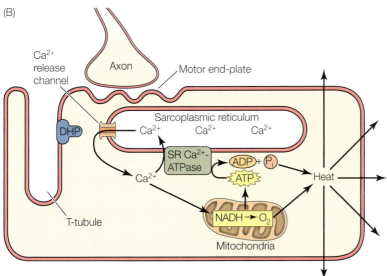

Figure 3.59 Cranial heater system of billfishes. (A) Portions of an ocular (eye) muscle are specialized for heat generation. The brain sits just above the heater organ and is surrounded by insulating fat deposits. Heat is retained in the cranial region by a countercurrent heat exchanger. The electron micrograph shows that most of the contractile machinery of the heat-generating region of the eye muscle is lost and is replaced by abundant mitochondria (m) and SR membranes (arrow heads). (B) As in the case of NST in birds and mammals, the heat generated by the heater organ comes from futile cycling of calcium ions (Ca^{2+}). In the cranial heater, a neuron of the SNS activates the t-tubule membrane receptor (DHP). This stimulates Ca^{2+} release from the SR via the Ca^{2+} release channel. Ca^{2+} stimulates substrate catabolism by mitochondria, which leads to heat production, as in NST. Pumping of Ca^{2+} back into the SR consumes ATP and completes the futile cycle. (A, swordfish tissue photo from Block 2011, transmission electron micrograph courtesy of Dr. Barbara Block; B after Hochachka and Somero 2002.)

Almost all of the contractile elements (actin and myosin fibers) are lost, and the density of mitochondria increases enormously. The mitochondria are closely juxtaposed to—they appear to be wrapped by—membranes derived from the SR. These membranes are rich in ATP-driven calcium pumps, which fulfill the same function in the cranial heater as they serve in NST. Missing from the cytoplasm of the cranial heaters are two calcium-binding proteins, parvalbumin and troponin C, which are generally present in high concentrations in skeletal muscle. Their absence in the heater organ means that the calcium released from the SR is more likely to remain free (unbound) in the cytoplasm and ready for reuptake into the SR by the Ca^{2+}-ATPase pump.

Regulation of heat production and its retention in the cranial region are due to a variety of biochemical and anatomical adaptations. As in the case of BAT and NST, the SNS is thought to provide an important stimulation for calcium release. SNS activation of the t-tubule membrane receptor leads to opening of calcium channels and release of Ca^{2+} into the cytoplasm. Electrical activity in surrounding regions of the ocular muscle that have not differentiated into heat-generating tissue may also stimulate calcium release. As in the case of NST in birds and mammals, Ca^{2+} released from the SR activates mitochondrial metabolism, generating ATP and heat. Return of calcium ion to the SR, driven by the Ca^{2+}-ATPase pump, completes the futile cycle and leads to more heat production. A countercurrent heat exchanger traps much of the heat generated by the heater organ in the cranial region of the fish.

The cranial heaters found in billfishes strongly resemble those found in a distantly related fish, the butterfly mackerel (Block 1994). The billfishes and butterfly mackerel heater organs fulfill the same function and appear to be highly similar biochemically. Yet they evolved independently and, in terms of function, through convergence. One distinction between the two independently evolved heater organs is the muscle from which they derived. Billfish heaters are derived from the *superior rectus* muscle; the butterfly mackerel heater is from the *lateral rectus* muscle. Exaptation processes thus have led to a convergence in function through exploitation of different starting materials.

Regional endothermy in plants

In our introduction to the topic of endothermy we mentioned that several plants have gained impressive abilities for regional endothermy, notably, but not exclusively, in their reproductive structures. Elevated temperatures benefit reproduction by driving the biosynthesis and volatilization of insect-attracting chemicals and by helping pollinating insects maintain the elevated temperatures needed for flight. Higher temperatures may also facilitate success of pollination and more rapid growth of pollen tubes,

further boons for reproduction (Seymour et al. 2009). In some plants, notably skunk cabbages of the genus *Symplocarpus*, elevated temperatures may help melt snow and ice around the plant, allowing earlier emergence—and hence reproduction—in the spring (Seymour 2004). Regionally endothermic plants may exhibit a high degree of homeothermy as well. The endothermic skunk cabbage *Symplocarpus renifolius* (shown at left) may keep its spadix temperatures within a 3.5-degree-Cesius temperature range (22.7°C–26.2°C) in the face of air temperatures that vary over a 37.4-degree-Cesius range (–10.3°C–27.1°C) (Seymour 2004, Seymour et al. 2009). Thermal regulation thus enables the reproductive systems of the plant to function near their evolved thermal optima near 23°C.

Intermembrane space

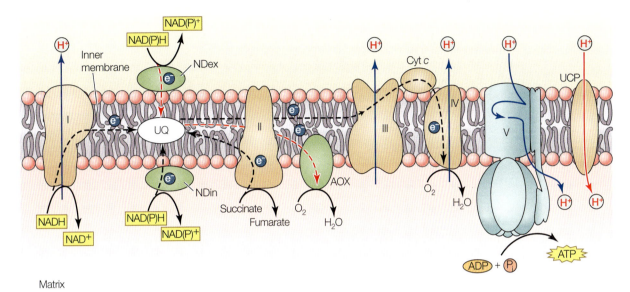

Matrix

Figure 3.60 The plant mitochondrial electron transport system and two energy-dissipating ("wasting") mechanisms. Red arrows indicate pathways of energy "wastage." Plant mitochondria have two mechanisms for dissipating energy without synthesizing ATP. One is a UCP-based system that dissipates the energy gradient established through pumping of protons across the inner mitochondrial membrane into the intermembrane space. The second is an alternative oxidase (AOX) system, which does not directly dissipate the transmembrane proton gradient. Instead, the AOX system accepts electrons from ubiquinone (UQ), which can be reduced through flows of electrons from complex I, succinate dehydrogenase (complex II), and NAD(P) dehydrogenases localized on the inner surface (matrix-facing; NDin) and outer surface (intermembrane space-facing; NDex) of the mitochondrial membrane. Electron flow through AOX bypasses complexes III and IV and therefore indirectly decreases the proton gradient by reducing the numbers of protons transported to the intermembrane space. Cyt c, cytochrome c; I, III, and IV, electron transport complexes I, III, and IV; V, ATP synthase. (After Zhu et al. 2011.)

THE ALTERNATIVE OXIDASE SYSTEM: ANOTHER HEAT-GENERATING EXAPTATION In plants, the principal generator of metabolic heat under cold stress involves the **alternative oxidase** (**AOX**) **system** (Figure 3.60; Zhu et al. 2011). The AOX system has been found in all plants so far examined, including heat-generating and non-heat-generating species. AOX proteins are encoded by two subfamilies of nuclear genes that exhibit complex regulatory responses to a variety of stresses, including decreased temperature (Feng et al. 2013). The ubiquitous distribution of AOX in plants strongly suggests that its original function was not thermogenesis. Thus, as in the case of UCPs, the thermogenic function of AOX is an exaptation.

What, then, was the original function of AOX? AOX is embedded in the matrix face of the inner mitochondrial membrane, where it accepts electrons from reduced ubiquinone (reduced coenzyme Q) and uses the electrons to reduce molecular oxygen to water (see Figure 3.60). Electrons that flow through the AOX system thus bypass complexes III and IV of the electron transport system. The AOX system also may include NAD(P) dehydrogenases that bypass complex I of the electron transport chain. Overall, electrons can flow from reduced NAD(P) to oxygen without driving any proton movement across the inner mitochondrial membrane. Although AOX may be able to reduce the production of ROS by channeling electron flow away from complexes III and IV, it is thought that this function is primarily the responsibility of UCPs, as discussed above. AOX is thought to have evolved as a mechanism for allowing the tricarboxylic acid (TCA) cycle to operate at required levels even if the reducing equivalents it contributes to electron transport cannot be fully used in the immediate production of ATP. Thus, as in the case of uncoupling proteins, dissipation of energy without production of ATP has a benefit that transcends attaining maximum levels of ATP production.

What are these benefits? Recall from Chapter 2 that the TCA cycle's role in metabolism is not just the provision of energy for ATP synthesis. In addition, intermediates of the TCA cycle can be drawn off and used in biosynthetic processes. For example, α-keto acids can be converted to amino acids by transaminases. The cell's requirements for TCA cycle intermediates must continue to be met, regardless of the needs for ATP production. If the need for these intermediates is high enough, the TCA cycle may generate reducing equivalents at a rate that exceeds the ability of the mitochondrial electron transport system to direct all of these electrons to ATP synthase (complex V). When this condition pertains, the AOX system allows excess electrons released through TCA cycle reactions to be shunted to oxygen through a pathway that does not involve generation of a proton gradient (see Figure 3.60).

AOX accepts electrons from reduced ubiquinone, a pivotal hub in mitochondrial electron transport. The redox state of ubiquinone is governed by the flow of electrons from several sources: complex I, NAD(P) dehydrogenases, and succinic dehydrogenase (complex II; see Figure 3.60). The AOX system can detect the redox state of the mitochondrion, as indexed by ubiquinone's redox state, and can alter its activity accordingly. Reduced ubiquinone may activate the AOX system posttranslationally by reducing a disulfide bond between the two subunits of the AOX dimer (Zhu et al. 2011). The oxygen level of the mitochondrion also helps govern AOX activity. The K_M of oxygen of the AOX system is 10- to 100-fold greater than that of the cytochrome c oxidase system (McDonald and Vanlerberghe 2004), indicating that only under conditions of high O_2 levels is AOX likely to function at maximum rate. High O_2 levels could result if rates of ATP synthesis are too low to allow rapid reoxidation of cytochrome c oxidase. High rates of photosynthesis could also lead to elevated levels of O_2 and, perhaps, to high rates of biosynthesis that require a large supply of TCA intermediates. Regulation of the genes that encode AOX is complex and is conjectured to involve ROS and specific signaling molecules, such as salicylic acid (SA) and nitric oxide (NO) (Feng et al. 2013). SA and NO may cause increases in ROS production by inhibiting cytochrome c oxidase or disrupting electron flow, so ROS signaling may be especially important in controlling levels of AOX protein in the mitochondrion.

The AOX system appears to be ubiquitous in plants, present in a variety of unicellular eukaryotes, but largely absent in animals. Genes encoding AOX-like proteins have been found in a few animal phyla (McDonald and Vanlerberghe 2004), but their roles remain to be clarified. The occurrence of these genes in certain intertidal invertebrates that experience large fluctuations in oxygen availability, and in some cases high concentrations of hydrogen sulfide, suggest metabolic roles different from those played by AOX systems in plants. Why might the AOX system be so crucial for all plants, yet not as essential for animals? A possible explanation relates to a major difference between autotrophs and heterotrophs. Animals are able to acquire many of the intermediates needed for processes like protein synthesis from their diets. Autotrophs must generate intermediates like amino acids "in-house," and the α-keto acids produced by TCA cycle activity can represent an important source of these biosynthetic materials. Thus, one can conjecture that animals would not need a system like the AOX system found in plants because animals have less need to decouple TCA cycle activity from cellular needs for ATP generation. And because animals generally lack the AOX system, their mechanisms for heat generation had to be based on exaptation of another type of system, the uncoupling proteins, which in plants likely play little (if any) role in thermogenesis.

The presence of two energy-dissipating systems in plant mitochondria raises the question of how their different roles are regulated as redox balance, metabolite concentrations, oxygen availability, and the proton gradient (Δp) change under different physiological states. In plants, it appears that the AOX and UCP systems do not attain maximum activity at the same time. Thus, whereas increases in levels of FFAs stimulate UCP activity, they may inhibit activity of AOX (Jarmuszkiewicz et al. 2010). AOX may "read" the metabolic status of the cell and adjust its activity accordingly. Thus, AOX activity is stimulated by pyruvate, the input substrate for the TCA cycle. Rising concentrations of pyruvate might be a signal to the AOX system that there is a blockage in TCA cycle activity—for example, feedback inhibition by excessive reduced NAD(P)—that needs to be removed. Reduced ubiquinone, the substrate for the AOX system, may stimulate both UCP and AOX activities. Both dissipation systems may function to reduce the formation of ROS, whereas AOX alone has assumed the role of thermogenesis (Zhu et al. 2011). Conversely, UCP activity, by dissipating the proton gradient and helping decouple electron flow originating in TCA cycle activity from ATP generation, could also assist in maintaining the biosynthetic functions of the TCA cycle, as done by the AOX system (Vercesi et al. 2006).

ENDOTHERMY: FINAL COMMENTS   In summary, the mechanisms used to generate large amounts of heat in systemic and regional endothermy reflect exploitation of several preexisting biochemical systems that are largely associated with membrane transport activities. These choices of raw material to fabricate thermogenic tissues reflect the large fraction of ATP turnover that is associated with transmembrane movement of ions—protons in the case of heat generation in BAT and beige adipocytes, and calcium ions in the case of heater organs found in fishes and of muscle-localized NST in birds and mammals. How the ontogeny of animal cells with ancestral functions of contraction has been reengineered to develop thermogenic tissues represents a fascinating and largely unexplored aspect of evolutionary biology at the cellular level. In plants too, many important discoveries related to evolution of regional endothermy await. Two areas that seem ripe for investigation are how the AOX system has acquired the needed regulatory capacities to become thermogenic, and how the AOX system functions in conjunction with UCP activity. Finally, the role of the AOX in organisms other than plants merits broad study to elucidate the full variety of roles played by this system.

3.8 Temperature Compensation of Metabolism

More than 99.99% of organisms are ectotherms that lack any capacity for regional endothermy. Short of being able to regulate body temperature through behavioral responses, these organisms would seem to have metabolic rates that are largely at the mercy of their habitat temperatures. That is to say, one would expect their metabolic rates to be governed by Q_{10} relationships rather than by production and dissipation of metabolic heat. However, in view of the likely benefits of escaping from strict adherence to the Q_{10} rule—as stated by Sir Joseph Barcroft in the opening quote of this chapter—it has been conjectured that metabolic compensation to temperature would be expected over both evolutionary and acclimatory time frames. Indeed, one of the longest-running discussions in thermal biology concerns how fully—and by what mechanisms—ectothermic species compensate for temperature (Q_{10}) effects on the rates of their physiological processes. Below we provide a brief review of this long-running debate and then explore some of the mechanisms, both known and conjectural, that could allow ectotherms to free their rates of metabolism from strict adherence to the Q_{10} rule.

Sources of the thermal challenge to metabolic rate in ectotherms

The sources of the challenges posed by changes in temperature to metabolic homeostasis—that is, to the maintenance of an adequate stability in rates of metabolic activity—are well understood. Because of the fundamental physics underlying the Q_{10} relationship (the Arrhenius relationship; Box 3.11), acute increases in cell or body temperature in the normal temperature range of a species are expected to lead to two- to threefold increases in rates of biochemical reactions with each 10-degree-Celsius rise in temperature. As a ballpark

BOX 3.11

The Arrhenius relationship

The **Arrhenius equation** expresses the effect of temperature on the rate constant (k) of a chemical reaction:

$$k = Ae^{-Ea/RT}$$

Here, T is temperature in kelvins, A is a preexponential factor, E_a is the activation energy, and R is the universal gas constant.

This equation was developed in 1889 by Svante Arrhenius (1856–1927), a Swedish physical chemist (and early Nobel laureate [1903]). Arrhenius based his analysis on studies done by a Dutch chemist, Jacobus Henricus van 't Hoff, who was interested in the effects of temperature on chemical equilibria. Of central importance in Arrhenius's treatment of temperature effects is the influence of a change in temperature on the fraction of the molecules in a population that have sufficient energy to be reactive. This minimal energy level is termed the **activation energy** and is usually symbolized as E_a. Units are kJ mol^{-1} or kcal mol^{-1}. Arrhenius showed that the fraction of molecules with energies $\geq E_a$ rises more rapidly with an increase in temperature than does the most probable energy level of the molecules, which serves as an index of temperature. As an example, consider the effect of a 10-degree-Celsius (10-degree-K) increase in temperature from 25°C (298 K) to 35°C (308 K). This ~3% increase in absolute temperature leads to a doubling or tripling (depending on the size of the

E_a value for the reaction) of the reaction rate.

A helpful way to visualize how the fraction of molecules in a population with energies $\geq E_a$ changes with temperature is to examine the **Maxwell-Boltzmann energy distribution curve** of the molecules (**Figure A**). With rising temperature, the most frequent energy level changes relatively little compared with the fraction of molecules with energies $\geq E_a$ (molecules under the shaded section of the two energy distribution curves).

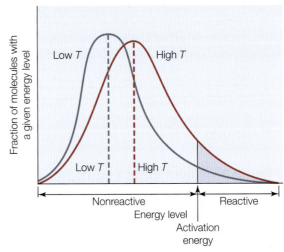

Figure A Effect of temperature on the Maxwell-Boltzman energy distribution curve.

BOX 3.11 *(continued)* ▮ ▮ ▬ ▬▬ ▬▬▬▬

The E_a for a reaction can be determined from what is known as an **Arrhenius plot**. This plot uses a natural logarithmic transformation of the Arrhenius equation:

$$\ln(k) = \ln(A) - E_a/R\,(1/T)$$

Thus, if the rate constant of a reaction obeys the Arrhenius equation, plot of $\ln(k)$ versus $1/T$ yields a straight line, whose slope and intercept can be used to determine E_a and A. The activation energy is defined as $(-R)$ times the slope of a plot of $\ln(k)$ versus $1/T$ at constant pressure (P):

$$E_a = -R\left[\frac{\partial \ln(k)}{\partial(1/T)}\right]_P$$

An illustration of an Arrhenius plot—$\ln(k)$ versus $1/T$—is given in **Figure B**.

The Arrhenius activation energy (E_a) is related to the activation enthalpy as follows:

$$E_a = \Delta H^{\ddagger} + RT$$

The most commonly used term for expressing temperature effects on rates of biological processes is the Q_{10} (sometimes referred to as the temperature coefficient). The general expression for Q_{10} is:

$$Q_{10} = \left(\frac{R_2}{R_1}\right)^{10/(T_2 - T_1)}$$

Where R is the rate, and T is the temperature in degrees Celsius or kelvins.

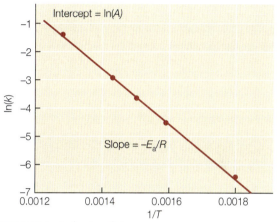

Figure B An Arrhenius plot.

estimate, each 1-degree-Celsius change in temperature shifts metabolism by about 8%; the exact percentage change depends on the Q_{10} value for the process. Keep in mind, however, that at temperatures outside the normal physiological range, determinants of rates other than the Arrhenius relationship may enter into play. Thus, rates may decrease rapidly due to factors like protein structural impairments in the heat or membrane viscosity changes in the cold. (Our discussion of thermal performance curves will examine these departures from strict adherence to the Arrhenius relationship in more detail; see p. 312.)

The central question in studies of temperature compensation of metabolism, then, concerns whether an ectotherm can overcome Q_{10} effects over its normal thermal operating range through compensatory adjustments in its biochemical systems. These adjustments may occur over evolutionary time-courses and involve changes to the genome, or over an organism's lifetime—for example, during seasonal acclimatization, when phenotypic plasticity allows compensatory adjustments in metabolic performance to offset Q_{10} effects. If there were no metabolic compensation to temperature, then metabolic rates of tropical ectotherms would be much higher (approximately eightfold higher, if a Q_{10} of 2.0 is used in the calculation) than rates of high-latitude species with similar activity levels and body mass. (**Box 3.12** discusses the role of latitudinal differences in metabolic rates in the context of global warming.) Similarly, at temperate latitudes where temperature varies greatly throughout the year, rates would speed up with the arrival of summer and slow down with the onset of winter. Without compensation, summer rates might be two to four times higher than winter rates. If one assumes that escape from the strict "tyranny" of the Arrhenius equation would benefit organisms—that is, if one assumes that a relatively high degree of stability in metabolic rate would benefit species from

BOX 3.12

Temperature compensation and effects of global warming

In the final chapter of this book—Adaptation in the Anthropocene—we will examine how greenhouse gas emissions are affecting organisms in diverse ways and at all levels of biological organization. As a "warm-up" to this analysis, we here examine how the effects of global warming will affect metabolic rates of ectotherms across latitude. An insightful analysis of this issue has been presented by Dillon, Wang, and Huey (2010). They show that climate warming and its effects on organisms differ strikingly at different latitudes. The greatest amount of warming has occurred in high-latitude environments (part 1 of the figure shows data for the Arctic; high rates of warming have also occurred in sections of Antarctica). However, in terms of the *absolute* amount of increase in metabolism due to rising temperatures

(part 2 of the figure), tropical ectotherms show larger increases than species from higher latitudes (temperate and Arctic regions). This latitudinal difference in increase in absolute metabolic rates stems from incomplete temperature compensation; as shown in Figure 3.61, tropical ectotherms have intrinsically higher metabolic rates than temperate or polar species at the different groups' normal environmental (body) temperatures. Thus the Q_{10} effects operating on tropical species will be increasing an already relatively high rate of metabolism, and the *absolute* change in metabolic rate will be greatest in the tropics (see part 2 of the figure), even though there may be relatively small differences across latitude in the percentage change in metabolic rate.

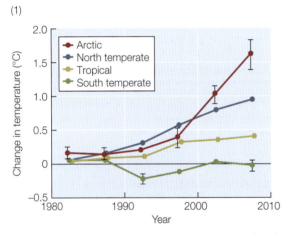

(1)

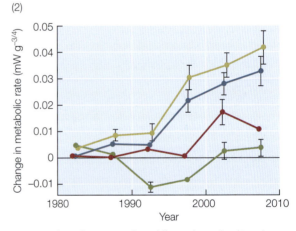

(2)

Changes in (1) mean temperature (5-year average) and (2) the predicted absolute change in the mass-normalized metabolic rates of ectotherms since 1980 (horizontal line gives 1980 estimates for reference) at four different latitudes. Error bars show S.E.M. (After Dillon et al. 2010.)

different latitudes and conspecifics encountering different seasonal temperatures—then one would expect to find metabolic compensation in nature.

In view of the relative ease with which rates of oxygen consumption can be measured, one might think that definitive tests of the metabolic compensation hypothesis would have been completed long ago, and that the issue would have been laid to rest. Surprisingly, this is not the case. Thus, before we examine the occurrence and extent of metabolic compensation, it is instructive to look first at the challenges that experimenters face in this line of comparative study—challenges that have led to considerable confusion about compensation's very existence.

Metabolic compensation: Pitfalls to definitive experimental analysis

Studies of temperature compensation of metabolism in ectotherms go back more than 100 years (Krogh 1916; Scholander et al. 1953; Clarke 1991, 2003). Nonetheless, the recent literature still reflects a lack of consensus as to whether metabolic compensation is real or artifactual, and if real, how complete compensation is. Are rates in cold-adapted (-acclimated) individuals equal to those found in warm-adapted (-acclimated) individuals, when measurements are made at the individuals' temperatures of adaptation or acclimation? The controversial nature of this subject stems from several sources (White et al. 2011).

A primary cause of confusion has been a failure to conduct "apples-to-apples" comparisons, that is, a failure to eliminate sources of variation in metabolic rate that are not related to evolutionary or acclimatory thermal history. A major source of confusion has resulted from comparisons of species that differ in locomotory habit and general levels of activity. Within a given habitat, species with different locomotory habits—for example, highly active pelagic fishes and sluggish demersal fishes—may differ by more than an order of magnitude in metabolic rate despite having the same body temperatures (Childress and Somero 1979; Childress 1995). Thus, to disentangle the effects of temperature per se from those caused by other variables that influence metabolic rate, careful choice of species must be made. Another variable with a strong effect on metabolic rate is body size. Metabolic scaling (see Chapter 2) accounts for large differences in oxygen consumption rates among organisms with different masses but identical body temperatures. Another possible bias in studies of metabolic compensation is the belief, spoken or otherwise, that ectotherms would, in the best of all possible worlds, be capable of complete ("perfect") compensation to temperature. This view fails to consider the possibility that costs of living may vary with temperature, such that life in the cold may require less energy than life at higher temperatures. Linked to this point is the broader ecological issue of how potentials for sustaining a given level of metabolism vary with latitude and, especially, with season. Thus, does food availability in winter conditions allow as robust a metabolic rate as summer conditions permit? If temperature compensation is absent or limited, does this reflect biochemical constraints or ecological circumstances?

Below we examine several studies that provide a compelling basis for accepting the reality of metabolic compensation to temperature. These studies also reveal some of the biochemical mechanisms that support compensation to temperature and raise some interesting questions that should drive additional investigation of this long-standing topic. One set of questions warranting additional analysis focuses on basic physical and chemical limits to compensation, limits that seem to preclude "perfect" compensation. If such roadblocks to complete compensation exist, then a fuller understanding of how metabolic needs might be adjusted for life at low temperatures is important.

To provide an integrated analysis, we address four questions that deal with important, but not fully resolved, issues within the broad context of temperature compensation. First, how fully have ectotherms overcome the "physics" of the situation? Can Q_{10} effects be overcome through acclimatization or evolutionary adaptation? Second, if compensation is documented convincingly, by what biochemical mechanisms is it achieved? Are these mechanisms the same over evolutionary time-courses and during phenotypic modifications on seasonal (or even daily) bases? Third, are there absolute barriers to achieving complete ("perfect") compensation to changes in temperature in ectotherms? And fourth, is complete compensation even necessary—or are the costs of living different at different temperatures?

A multilevel analysis of temperature compensation

We believe it can be helpful to examine these questions about metabolic compensation in a multilevel manner, one that comprises studies done at all levels of biological organization—from the molecule to the ecosystem. A multilevel analysis allows a reductionist approach to questions about metabolic compensation, yet also allows us to view the need for compensation in a broad context that includes factors like seasonal changes in food supply.

TEMPERATURE COMPENSATION: WHOLE-ORGANISM OXYGEN CONSUMPTION RATES   One way of setting the stage for this multifaceted analysis is to examine one of the foundational studies of metabolic rates of ectotherms native to different latitudes (Scholander et al. 1953). Per Scholander and colleagues conducted a broad survey of rates of oxygen consumption of Arctic and tropical ectotherms, including aquatic and terrestrial species. They measured rates of oxygen consumption over temperature ranges that characterized the species' environments. For almost all species, Q_{10} values were in the range of 2–3, indicating close adherence to the Arrhenius relationship. To evaluate the extent to which Arctic and tropical animals had compensated their metabolisms for temperature, the researchers then compared the rates of metabolism determined at typical Arctic and tropical temperatures, 0°C and 30°C, respectively.

The general patterns of metabolic rate that Scholander and colleagues observed in all ectothermic taxa (with the exception of some insects) are illustrated by their data on fishes and crustaceans (**Figure 3.61**). The data are presented as total rates of oxygen

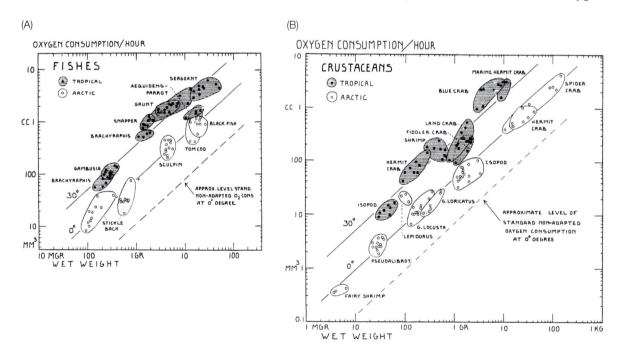

Figure 3.61 Temperature compensation of metabolism. Rates of whole-organism oxygen consumption by Arctic (data points in nonshaded areas) and tropical (data points in shaded areas) fishes (A) and crustaceans (B), measured at their normal habitat temperatures, ~0°C and ~30°C, respectively, and presented in relation to total body mass (wet weight). Whereas the metabolic rates of the tropical species at 30°C all lie above the rates for Arctic species of comparable body mass at 0°C, extrapolation of the rates for the tropical species to 0°C (dashed line) reveals considerable temperature compensation. (From Scholander et al. 1953.)

consumption as a function of body mass, for species of different masses. There is little overlap in the data from the tropical and Arctic species across the full mass range examined. In general, the metabolic rates of the tropical fishes are four to five times those of Arctic species of similar body mass at the organisms' normal habitat temperatures. For crustaceans a similar relationship is evident. From the vertical displacement of the Arctic and tropical data sets, it is obvious that "perfect" compensation to temperature has not occurred. However, the dashed lines in the figures show that considerable temperature compensation is present in both groups of animals. Thus, if the rates determined for the tropical species at 30°C are adjusted to a temperature of 0°C using the Q_{10} relationship, the rates fall well below the actual rates measured at 0°C for the Arctic species. This classic experiment provides strong evidence that at least partial compensation occurs in ectotherms adapted to different temperatures because of their latitudes of occurrence.

TEMPERATURE COMPENSATION IN TISSUE RESPIRATION RATES   Many of the relationships found by Scholander and colleagues in their study of whole-organism oxygen consumption rates are seen for oxygen consumption of isolated tissues (brain slices and gill filaments) from a cold-adapted stenothermal Antarctic teleost (*Trematomus bernacchii*) and differently acclimated (10°C and 30°C) goldfish (*Carassius auratus*) (Somero et al. 1968; **Figure 3.62**). Oxygen consumption by tissues of the Antarctic fish is substantially higher at its normal habitat temperature near −1.9°C than the rates predicted by extrapolation to this temperature for either acclimation group of goldfish. Thermal acclimation also influenced oxygen consumption rates of goldfish tissues, especially gill (Ekberg 1958). However, the 30°C-acclimated goldfish maintained a much higher rate of oxygen consumption at this higher acclimation temperature than found for either *T. bernacchii* at its normal habitat temperature or for 10°C-acclimated goldfish. Thus, as in the case of whole-organism oxygen consumption rates, compensation to temperature is present, but only partial. The finding that tissue respiration rates exhibit a pattern of compensation similar to that found in whole-organism studies indicates that interspecific differences in rates of whole-organism oxygen consumption likely have a strong signal of temperature compensation, and are not just driven by differences in locomotory activity, excitement,

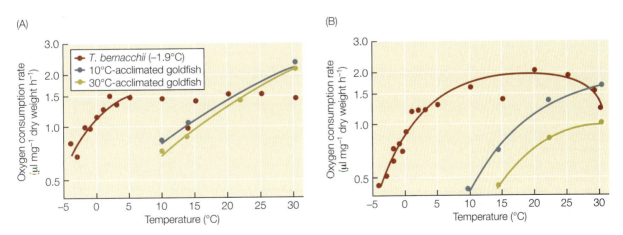

Figure 3.62 Metabolic compensation at the tissue level. In vitro oxygen consumption rates (O_2 consumed per milligram tissue dry weight per hour) of brain slices (A) and gill filaments (B) of the Antarctic notothenioid fish *Trematomus bernacchii* and differently acclimated goldfish (*Carassius auratus*). (After Somero et al. 1968; goldfish data from Ekberg 1958.)

or dietary history. The data on brain metabolism are especially revealing in this regard. Metabolic rate of brain tissue appears decoupled from factors like capacity for locomotion in fishes. Thus, activities of enzymes of ATP-generating pathways in brain tend to be highly similar among fishes, whereas the activities of these enzymes in white locomotory muscle vary by two to three orders of magnitude in species with different capacities for high-speed swimming (Childress and Somero 1979).

TEMPERATURE COMPENSATION IN CELLULAR LOCOMOTION   If we move down one level in biological organization, to the cellular level, we again find good evidence for temperature compensation. A useful cellular study system is provided by the rates of movement of **keratocytes**, cells that are common on the epithelial surface of fish. Keratocytes perform several functions, including wound healing, so an ability to sustain a temperature-compensated rate of movement would be advantageous to fish. As in the case of brain metabolism, differences in keratocyte movement rates would not be expected to derive from differences in swimming mode or other factors unrelated to temperature.

When the mobility (instantaneous speed in $\mu m\ min^{-1}$) of keratocytes isolated from species of fishes adapted to different temperatures is compared, a strong conservation of speed of movement is found at the respective adaptation temperatures of the species (**Figure 3.63**; Ream et al. 2003b). Rates of locomotion at common habitat temperatures for the three species were 11.4 $\mu m\ min^{-1}$ for *Gillichthys mirabilis* (15°C), 13.8 $\mu m\ min^{-1}$ for the desert pupfish (*Cyprinodon salinus*; 30°C), and 10.8 $\mu m\ min^{-1}$ for the tropical reef fish *Amphiprion percula* (30°C). The temperatures at which speed reached a maximum also differed among species according to their adaptation temperatures.

Acclimation of these species to different temperatures did not affect the mean speed of movement of keratocytes, but did influence directional aspects of mobility. The persistence of motion in one direction and turning angle varied with acclimation temperature, and were greatest at temperatures within the species' normal thermal ranges. The different responses of speed and direction and turning angle to acclimation temperature suggest that multiple temperature-sensitive mechanisms govern cell mobility (Ream et al. 2003b).

TEMPERATURE COMPENSATION OF METABOLISM: A MORE THOROUGH MULTILEVEL EXAMINATION   Although the early study of Scholander and colleagues (1953) provided strong support for the hypothesis that metabolic compensation to temperature was

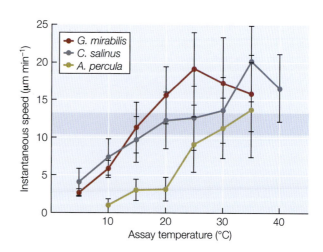

Figure 3.63 Temperature compensation in keratocyte mobility. Effect of measurement temperature on actin-driven movement (instantaneous speed in $\mu m\ min^{-1}$) of keratocytes (primary cultures) from three species of teleost fish: the longjaw mudsucker (*Gillichthys mirabilis*), the desert pupfish (*Cyprinodon salinus*), and a tropical clownfish, *Amphiprion percula*. The shaded area encompasses rates of movement at common habitat temperatures of the three species; strong conservation of movement speed at these temperatures is seen across species. Error bars represent standard deviation. (After Ream et al. 2003b.)

prevalent in differently adapted ectotherms, their work suffered from certain shortfalls in analysis. For example, it did not involve a phylogenetically corrected analytical perspective and did not delve into underlying mechanistic issues. A more mechanistic and analytically refined analysis of temperature compensation in fishes from different latitudes was made by White and colleagues (**Figure 3.64**; White et al. 2011). They conducted a meta-analysis using data in the literature for several dozen species of fishes (the exact number of species varied among different types of studies) with different latitudinal distribution patterns and locomotory habits. This especially thorough investigation of metabolic compensation to temperature took into account differences in metabolic rate that might be associated with locomotory habit, effects of phylogeny, and influences of body mass (metabolic scaling relationships). White and colleagues used the maximum latitude at which the species occurred as their index of adaptation temperature. Using experimentally determined Q_{10} values, all rates were adjusted to a common temperature to facilitate interspecific comparisons. By "leveling the playing field" in this way, the researchers could see whether there was a positive correlation between rate and latitude that would indicate temperature compensation. At the whole-organism level, they focused on standard metabolic rate (SMR)—the minimum rate of metabolism needed to sustain life at a specified temperature. SMR is measured using specimens that are resting and in a postabsorptive state (not digesting a recent meal). In the subsequent reductionist analysis, the comparisons of SMRs were complemented by analyses of mitochondrial respiration and activity of the TCA cycle enzyme citrate synthase (CS), which serves as a quantitative index of the activity of this pathway, and hence of capacity for aerobic generation of ATP (see Chapter 2).

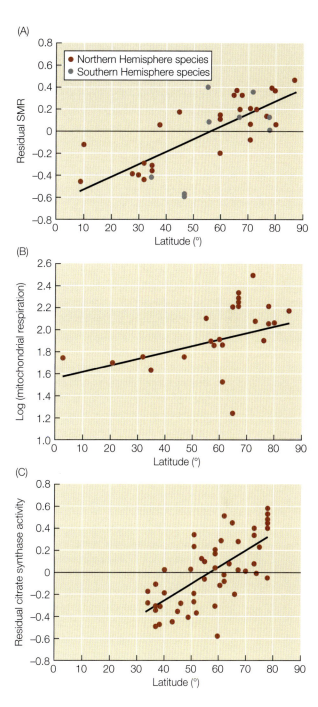

Figure 3.64 Relationships between latitude and three metabolic traits: (A) standard metabolic rate (SMR); (B) mitochondrial respiration; and (C) citrate synthase (CS) activity. SMR and mitochondrial respiration data are normalized to a common temperature of 20°C; CS data are normalized to 10°C. Because the Q_{10}-normalized values of SMR are positively related to body mass, and CS activity is related to minimum depth of occurrence (higher CS activity in shallower occurring species), residuals (deviation from the mean observed value) are presented. (After White et al. 2011.)

As shown in Figure 3.64, there were significant positive relationships between all three metabolic parameters and latitude. Thus there is consistent evidence for metabolic compensation to temperature across these three different levels of biological organization. The rigorous analysis of White and colleagues helps put metabolic compensation to temperature on a much firmer basis than many earlier studies were able to do, because it draws from studies of a wide number of species and attempts to remove influences on metabolic rate other than those of adaptation temperature. Nonetheless, there remains considerable unexplained variation around the regression lines. Some of the variation could reflect differences in experimental conditions among the studies examined—a common challenge in performing meta-analyses—although the authors did a careful job of selecting data gathered with highly similar protocols (pretreatment exposure of fish and methodologies for measuring oxygen consumption [SMR and mitochondria] and CS activity). Nonetheless, despite this inevitable limitation, the analysis of White and collaborators shows that whole-organism SMR is highly correlated with latitude (a proxy for temperature) and mechanistically determined, at least in large measure, by mitochondrial activity and levels of CS activity in mitochondria.

TEMPERATURE COMPENSATION IN CATALYTIC RATE CONSTANTS   Further insights into the aspects of protein evolution that favor metabolic compensation can be obtained by examining the intrinsic catalytic activities (k_{cat} values) of enzyme orthologs from differently thermally adapted species. Total measured enzymatic activity is a product of the k_{cat} of the enzyme times its concentration. For this reason, the total activity of an enzyme in a tissue homogenate, as in the case of the CS data shown in Figure 3.64, does not disclose whether any observed temperature-related variation among species or acclimatization groups is due to variation in the quantity of enzyme present or, instead, to variation in the intrinsic catalytic efficiency of the protein, that is, its k_{cat}. We have shown earlier (see Table 3.2) that the activation free energy ($\Delta G^{\pm}$) values of orthologous proteins show a regular variation with temperature, such that the energy barriers to catalysis (the magnitude of $\Delta G^{\pm}$) are reduced during evolutionary adaptation to cold. This indicates that k_{cat} values of orthologous proteins also should be negatively correlated with adaptation temperature. This relationship was shown for LDH-A$_4$ orthologs of congeneric damselfishes from temperate and tropical waters (see Figure 3.7; Johns and Somero 2006). A similar relationship is shown for LDH-A$_4$ orthologs of vertebrates adapted to temperatures between −1.9°C (Antarctic notothenioid fishes) and approximately 42°C (a thermophilic desert reptile) (**Figure 3.65A**). An even broader comparison is shown in **Figure 3.65B**, which illustrates how k_{cat} values differ between LDHs of an Antarctic fish (the icefish *Champsocephalus gunnari*), a mammal (pig), and a thermophilic bacterium (*Thermotoga maritima*), species whose preferred temperatures range from near 0°C to 80°C (Peng et al. 2015). (Note that the comparisons shown in Figure 3.65B do not necessarily involve orthologous forms of LDH. The pig enzyme is the heart-type [B-] ortholog, and the fish enzyme is likely an A-type ortholog. The bacterial enzyme's relationship to vertebrate LDH paralogs was not indicated in this study [Peng et al. 2015].)

The regular variation in k_{cat} values with adaptation temperature for LDH-A$_4$ orthologs indicates that the higher enzymatic activities measured in cold-adapted species may be due to intrinsically faster-working enzymes, rather than to higher enzyme concentrations. The trend seen for vertebrate LDH-A$_4$ orthologs reflects the general trends discussed earlier in this chapter, trends that are consistent with temperature compensation of metabolism across all domains of life.

(A)

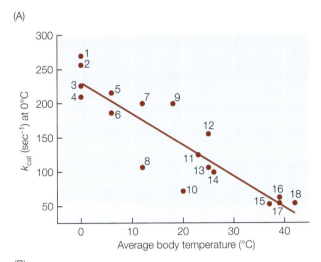

(B)

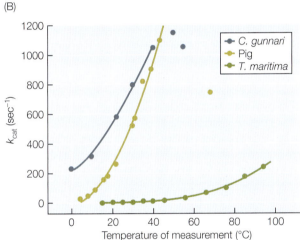

Figure 3.65 (A) Temperature compensation in k_{cat} values of orthologs of LDH-A$_4$ from vertebrates adapted to temperatures between –1.9°C and 42°C. Species studied included four Antarctic notothenioid fishes (*Parachaenichthys charcoti* [1], *Lepidonotothen nudifrons* [2], *Champsocephalus gunnari* [3], and *Harpagifer antarcticus* [4]); two South American temperate notothenioids (*Patagonotothen tessellata* [5] and *Eleginops maclovinus* [6]); a temperature rockfish, *Sebastes mystinus* (7); a halibut, *Hippoglossus stenolepis* (8); a temperate barracuda, *Sphyraena argentea* (9); a dogfish, *Squalus acanthias* (10); a subtropical barracuda, *Sphyraena lucasana* (11); a temperate goby fish, *Gillichthys mirabilis* (12); the bluefin tuna, *Thunnus thynnus* (13); a tropical barracuda, *Sphyraena ensis* (14); the cow, *Bos taurus* (15); the chicken, *Gallus gallus* (16); a turkey, *Meleagris gallopava* (17); and the desert iguana, *Dipsosaurus dorsalis* (18). (B) k_{cat} values as a function of measurement temperature for LDHs of an Antarctic fish (*Champsocephalus gunnari*), a pig, and a thermophilic bacterium (*Thermotoga maritima*). (A after Fields and Somero 1998; B after Peng et al. 2015.)

Why isn't metabolic compensation complete ("perfect")?

Taken together, these multilevel (SMR to k_{cat}) investigations suggest that metabolic compensation to temperature is unlikely to be an artifact, as some investigators have claimed (see the historical review of the field given in White et al. 2011). Nonetheless, a puzzle remains: Given that ectotherms can achieve a considerable degree of metabolic compensation to temperature across different levels of biological organization, why is compensation only partial rather than complete? Is this an indication that there are fundamental—and insurmountable—biochemical and biophysical barriers to full compensation? Or instead, is incomplete compensation an indication that complete (perfect) compensation is not necessary because metabolic demands differ with temperature? Might life be less metabolically expensive in the cold, allowing a lower metabolic rate in cold-adapted and cold-acclimatized individuals? Finally, do broader ecological conditions, notably the food supply available at different seasons or depths in the water column, provide an explanation for the seeming "failure" of metabolic rates to fully compensate to temperature? Is competition fiercer in low-latitude, high-temperature habitats, where ectotherms remain highly active all year?

A suitable starting point for addressing the possible bases of incomplete metabolic compensation to temperature is to calculate how fully the observed differences in k_{cat} can lead to compensation in enzymatic rates. The data in Figure 3.65A, which are normalized to a measurement temperature of 0°C, provide a basis for this analysis. The k_{cat} values for LDH-A$_4$ of Antarctic notothenioids, the most cold-adapted species studied, are approximately 250 sec^{-1} at 0°C; k_{cat} values of temperate-zone (~20°C) fishes lie near 150 sec^{-1} at 0°C. When one applies Q_{10} corrections to adjust the temperate fishes' k_{cat} values to those expected at 20°C, the incomplete nature of k_{cat} compensation is revealed. At 20°C, k_{cat} for the temperate fishes' LDH-A$_4$s would be near 600 sec^{-1}, that is, more than twofold higher than those of the Antarctic notothenioid fishes at their near-freezing habitat temperatures. Thus, although adaptive variation in k_{cat} values achieves substantial compensation to temperature, large differences in activity remain when comparisons are done at different species' normal cell/body temperatures.

ARE THERE LIMITS TO PROTEIN INSTABILITY?   The "failure" of enzymes to fully compensate for temperature effects during evolution at different temperatures suggests that there may be limits to how much $\Delta G^{\ddagger}$ can be reduced during adaptation to low temperatures. The relationship between $\Delta G^{\ddagger}$, k_{cat}, and enzyme flexibility (at least of the catalytically active regions of an enzyme) has been discussed earlier (see the section "Structural bases of adaptation in kinetic properties" in Section 3.4, p. 164). This relationship might lead one to conclude that adaptation to colder temperatures could be achieved by further increasing protein flexibility, thereby further reducing $\Delta G^{\ddagger}$ and increasing k_{cat}. This issue was analyzed by Feller and Gerday (2003) in studies of proteins of psychrophilic bacteria and archaea. They proposed that the most cold-adapted proteins may already be pushing the limits of protein instability. Because cold-adapted proteins are more cold-labile as well as more heat-labile, due in large measure to weakening of hydrophobic stabilization at low temperatures (hydrophobic stabilization is maximal near 20°C; see Box 3.6), further reductions in stability to gain lower $\Delta G^{\ddagger}$ values (higher k_{cat} values) might be impossible. There may, then, be a lower limit to intrinsic protein stability that serves as an insurmountable barrier to greater and greater increases in k_{cat} values during adaptation to low temperature. In fact, at any temperature of adaptation, there may be constraints on increases in k_{cat} arising from limitations in structural stability. The need for an enzyme to maintain the correct binding site geometry may affect how fully k_{cat} values can be elevated at any temperature of adaptation. As discussed earlier, if a protein has too flexible a structure, the distribution of conformational microstates may be shifted too far in the direction of nonbinding-competent configurations (see Figure 3.3). If the geometry of the binding site is too thermally labile, then appropriate binding ability, as indexed by K_M, may be compromised. Thus, regardless of adaptation temperature, the need to maintain the conformation of the binding site may preclude increases in conformational flexibility that favor higher k_{cat} values. If this entails having a more stable enzyme structure that leads to incomplete temperature compensation in k_{cat}, that may simply be the "compromise" that selection requires.

In summary, a lower limit to protein instability may be important in determining how fully k_{cat} values can compensate for temperature. There may be a lower limit to ΔG_{stab}, below which a protein cannot retain the correct conformational microstates to function adequately. One way to compensate for the barrier to further reductions in $\Delta G^{\ddagger}$ (= increases in k_{cat}) might be to increase enzyme concentrations, keeping in mind that enzyme activity is the product of k_{cat} times enzyme concentration. Is this a feasible strategy of adaptation to cold?

ARE THERE PACKAGING LIMITATIONS TO COMPENSATION?   If compensation for reduced temperature were to involve increasing concentrations of all enzymes, or at least the many enzymes responsible for generating ATP, there might be severe limits due to "packaging problems." If we assume that control of pathway flux involves multiple enzymes in a pathway (see Chapter 2 for a discussion of this point), then a significant packaging problem might arise during acclimation to cold. In an extreme (but unlikely) case where all enzymes of a pathway contribute to flux control, we can ask if it would be possible, for example, to overcome the effects of a 10-degree-Celsius decrease in temperature by doubling or tripling the concentrations of all ATP-generating enzymes. This seems unlikely. Although increases in mitochondrial volume and cristae area have been observed during cold acclimation in fishes (Tyler and Sidell 1984; Guderley 2004), these changes may reflect a shift toward more aerobic function in the cold, where oxygen concentrations are higher (but see the next section for an expanded view of this issue) and animals' activities may be reduced, than an attempt to overcome the "tyranny of the Arrhenius equation." There have been several studies in which cold acclimation has led to increases in activities of some enzymes, but in many cases such compensatory shifts are absent or very limited (Hazel and Prosser 1974). Furthermore, large-scale increases in protein concentration in the cold might be challenging to achieve due to costs of protein synthesis and protein turnover. Thus, packing in more enzymes to overcome decelerating effects of reduced temperatures seems unlikely to be a good general strategy for achieving metabolic compensation.

VISCOSITY AND TEMPERATURE: ANOTHER RATE-GOVERNING INFLUENCE   A further constraint on temperature compensation of enzymatic rates arises from temperature's influence on solution viscosity (Sidell 1998). The viscosities of fluids are strongly dependent on temperature, and increases in viscosity can lead to reduced rates of biochemical activity, for example, by increasing the energy costs of changes in protein conformation that are a necessary part of protein function (see Chapter 4). We already have seen that the viscosity (static order) of membrane bilayers can have strong effects on the activity of membrane-associated processes through what are termed viscotropic effects (see Chapter 2). Adjustments in membrane lipid composition help compensate for temperature's effects on bilayer viscosity, a homeostatic process observed across species from different thermal habitats and among conspecifics acclimatized to different temperatures.

   Similarly, in the aqueous phase, changes in viscosity can have strong effects on rates of biochemical processes. The temperature dependence of viscosity of cellular fluids has been measured in mammalian cells (Mastro and Keith 1984). At 20°C, viscosity was estimated to be 2.5 cP (centipoise units; 1 cP = 1 millipascal-second [mPa sec]); at 0°C, viscosity doubled to ~5 cP. What does this increased viscosity in the cold portend for rates of enzymatic activity? This question was addressed by Demchenko and colleagues (1989), using lactate dehydrogenase as the study system. The reaction rate measured in a sucrose solution with a viscosity near 6 cP was only about one-sixth the rate measured in a low-viscosity buffer. When temperature-dependent viscosity changes in the cytosol are added to the effects of temperature seen when enzymes are assayed in dilute buffers, the results of Demchenko and colleagues indicate that an even greater difference in LDH activity is likely to occur between cold-adapted and warm-adapted species at their normal body temperatures. In this context, it is appropriate to point out that orthologous LDH-A_4s of cold- and warm-adapted species exhibit similar responses to increases in viscosity at their normal functioning temperatures (Fields et al. 2001). Thus, the intrinsically more active (higher k_{cat}) but less stable orthologs of cold-adapted species are no less sensitive to temperature-viscosity relationships than orthologs of warm-adapted species, when measurements are conducted at normal body temperatures.

In Chapter 4 we will examine viscosity-related effects on protein stability and thereby gain further insights into the mechanistic basis for viscosity effects on catalytic rates. For our purposes here, the key point is that the strong retardation in LDH activity (and presumably in all enzyme activities; see Chapter 4) by increased viscosity at low temperatures may be one contributor to the incomplete compensation of biochemical rates to temperature. Here, then, we see another instance of a theme that runs throughout this volume (and is the principal focus of Chapter 4): that in vivo function of a biochemical system (here, an enzyme) is determined by the intrinsic properties of the biochemical component (k_{cat} in this case) and the extrinsic conditions established by the medium in which function occurs (the viscosity of the intracellular solution).

INTRACELLULAR OXYGEN DIFFUSION: ROLES OF THE MEMBRANE BILAYER AND DEPOT LIPIDS   During acclimatization and evolutionary adaptation to cold, fishes may accumulate higher levels of depot lipids and display mitochondrial proliferation (Sidell 1998). This observation speaks to another aspect of metabolic compensation to low temperatures: changes that facilitate the diffusion of oxygen to its site of use, the respiratory complex of the mitochondrion. In an extremely insightful analysis of this issue, Bruce Sidell considered the several different factors that influence diffusion of oxygen through the cellular fluids, including both the aqueous and lipid-containing systems of the cell. Reductions in temperature present a two-edged sword to the oxygen delivery systems of cells. Because the solubility of oxygen rises with a fall in temperature (a 1-degree-Celsius decrease in temperature increases oxygen solubility by about 1.4%), low temperature might be predicted to ease the oxygen supply problem and facilitate aerobic metabolism. However, the diffusion of oxygen within an aqueous phase decreases by about 3% for each 1-degree-Celsius drop in temperature. Thus, the diffusion constant for oxygen (K_{O_2}), which is a product of the solubility coefficient (α_{O_2}) and the diffusion coefficient for oxygen (D_{O_2}), decreases by approximately 1.6% per degree Celsius. For an Antarctic notothenioid fish with a body temperature near $-1.9°C$, the diffusion constant for oxygen would be ~42% lower than that for a temperate fish near 25°C (Sidell 1998).

The reduction in the diffusion coefficient for oxygen at decreased temperature is due to two factors: the reduction in the kinetic energy of the system, and an increase in the viscosity of the cytoplasm. As mentioned above, the viscosity of aqueous solutions is highly sensitive to temperature. For pure water, viscosity increases by more than 70% when temperature falls from 25°C to 5°C; for fish muscle cytoplasm, an increase in viscosity of approximately 82% was observed over this range of temperature (Sidell and Hazel 1987). The increase in water's viscosity at low temperature is the larger of the two determinants of temperature-dependent changes in D_{O_2}. The Q_{10} for viscosity is 1.35, a value roughly half that seen for thermochemical (metabolic) reactions, yet large enough to significantly affect the movement of oxygen through the cellular water. It is also noteworthy that the diffusion of myoglobin (Mb) in the cell is strongly dependent on viscosity (Sidell 1998). Therefore, whether oxygen is free in solution or bound to Mb, reductions in temperature will impede the movement of oxygen through the cell.

The proliferation of both mitochondria and depot lipids observed in skeletal muscle tissue of cold-acclimatized and cold-adapted fishes is likely to reduce the diffusion limitations to oxygen (and oxy-Mb) in the cold. Oxygen is much more soluble in lipids than in water, a fact that may come as a surprise to most readers. Studies by Battino and colleagues (1968) showed that oxygen is 4.4 times more soluble in olive oil (~85% triacylglycerides containing oleic acid) than in water at 25°C. Moreover, the solubility of oxygen in olive oil is relatively insensitive to change in temperature; solubility varied by less than 1% over a 30-degree-Celsius range of measurement temperatures. Because of these oxygen solubility properties of lipids, the nonpolar hydrocarbon phase of the cell (membrane bilayer and depot lipids) should contain three to five times the amount of oxygen that is present in an equivalent volume of aqueous cytoplasm. The lipid droplets present in depot lipid thus can be viewed as oxygen reservoirs and as low-resistance channels for the diffusion of oxygen through the cell. A similar role is played by membrane lipids. Movement through the nonpolar bilayer is faster than through the cytoplasm and is dependent in part of the acyl chain composition of the membrane phosphoglycerides; the more-fluid lipids accumulated at low temperatures provide a more effective medium for oxygen diffusion than the more rigid (viscous) lipids found in warm-acclimatized (-adapted) ectotherms (Sidell 1998). Therefore, the adaptive benefits of more fluid lipids extend beyond the effects commonly attributed to homeoviscous and homeophasic adaptation to another aspect of temperature compensation, the supply of oxygen to the mitochondria. The proliferation of mitochondria seen in cold-acclimatized and cold-adapted fishes (O'Brien 2011) thus has the potential for making a twofold contribution to metabolic compensation to cold: an increase in the aerobic ATP-generating capacity of the cell due to more enzymatic machinery and, thanks to an increase in total membrane area, an enhancement of the supply of oxygen to the systems of ATP generation in the mitochondrion.

DOES THE COST-OF-LIVING CHANGE WITH TEMPERATURE?   The potential limitations to complete compensation to temperature also need to be examined in the context of cost-of-living versus temperature. At reduced temperatures, several types of metabolic costs may possibly be reduced. The flux of water and ions across membranes could be reduced due to temperature effects on rates of diffusion. If so, then ion-pumping costs might be lower in the cold. Thermal damage to macromolecules might be reduced at low temperature. Heat-induced unfolding of proteins and damage to DNA (single- and double-strand breaks) suggest that life at high temperatures likely requires higher repair costs. Ecological considerations also must be brought into the picture to obtain a realistic analysis of temperature versus cost-of-living issues. For instance, in winter primary production may be drastically reduced—or even cease entirely in polar regions. With a reduced food supply, there would be an incentive to "hunker down" and significantly reduce metabolic rates, waiting for the return of spring conditions that drive primary production. This type of response has been characterized in an Antarctic notothenioid fish that enters a state of greatly reduced metabolism in the dark season, even though temperatures are at most only slightly lower than in summer (Campbell et al. 2008). Thus, complex interplay among temperature, activity levels (e.g., pelagic versus demersal habit), seasonal dependence of energy availability, and other physiological and ecological variables likely determines the optimal metabolic rate for an ectotherm. Full compensation of metabolism to changes in temperature, even if it were physically possible, might not be advantageous and favored by selection. Thus, the apparent "failure" of ectotherms to fully compensate for effects of temperature must instead be viewed as an outcome of the complex suite of physical, chemical, physiological, and ecological issues that ultimately govern the metabolic requirements of an organism.

Thermal performance curves: What determines thermal optima and tolerance limits?

In our treatment of temperature-metabolism interactions, we have been focusing on mechanistic analyses of individual classes of biochemical structures and processes. We will now step back and present a higher-level perspective that is centered on what are termed **thermal performance curves** (**TPCs**) (or **thermal reaction norms** [**TRNs**], as they are sometimes known) (**Figure 3.66**). A comprehensive analysis of TPCs for a variety of processes and species is given by Angilletta (2009); his volume is probably the best source available for a deep and critical understanding of TPCs and their overall use (and shortcomings) in characterizing the thermal relationships of organisms. Schulte (2015) has provided an especially well developed critical analysis of the use of TPCs in the context of aerobic metabolism.

TPCs are commonly used to represent an organism's responses to changes in temperature over its entire thermal tolerance range. They are used to identify temperatures at which performance attains its highest value and temperatures at which failure in performance begins to occur. The performance characteristic plotted on the ordinate of a TPC can be a major life history characteristic, such as fecundity, or a physiological function, such as growth rate, respiration, or locomotory activity. Some analyses of thermal responses may seek to determine the effect of temperature on Darwinian fitness (Deutsch et al. 2008), as indexed, for example, by survivorship at different temperatures.

The following treatment of TPCs (or TRNs) takes a critical look at their uses and misuses in characterizing the thermal relationships of ectotherms. This analysis is intended to provide a better understanding of TPC-based analyses and prepare the reader more fully for the discussion of global change presented in Chapter 5. One issue that merits focus is the general use of TPCs across species and processes. In view of the wide use of TPCs, one can question whether TPCs truly are a "one curve fits all" type of analytical framework that applies to all temperature-sensitive traits. The curve illustrated in Figure 3.66 has the shape generally thought to describe TPCs. However, one can ask whether this "Platonic form" is as broadly descriptive of temperature effects as many seem to think—or is reality a bit messier and harder to characterize than suggested by TPC

Figure 3.66 The "Platonic" shape of a thermal performance curve (TPC): the effect of an acute change in temperature on the rate of performance of an organism (or isolated physiological or biochemical system). Initially, the rate of performance rises with increasing temperature. Then the curve enters a plateau phase, whose width ("thermal breadth") may differ between stenothermal and eurythermal species and among different processes. The temperatures at which rate of performance is highest is termed the optimal temperature (T_{opt}). The T_{opt} determined in such an acute experiment may not, in fact, reflect what is optimal for the organism over a longer time period (see text). Above the T_{opt} the curve trends downward, and eventually the rate of performance falls to zero. If exposure times are sufficiently long, death may ensue when the rate falls to zero.

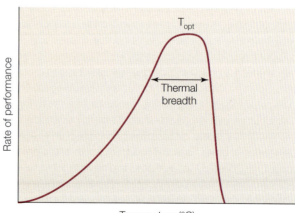

analyses? The information in **Box 3.13** shows that Platonism is ill-founded in the context of the shapes of TPCs. Thermal responses can be far more complex than conventional hypotheses about TPC relationships suggest.

We must also ask what the values for thermal optima and thermal limits presented by these types of curves tell us about the intrinsic properties of the organisms being studied, as opposed to effects contingent on the particular experimental treatments to which the organisms have been subjected. Thus, because of the extensive use of TPCs in thermal biology, it is important to deconstruct these analyses to uncover the factors that shape these curves.

"OPTIMAL" TEMPERATURES AND THERMAL LIMITS: WHAT CAN TPCS TELL US?   When employed to examine the effects of *acute* temperature changes on performance, TPCs

BOX 3.13

The thermophilic intertidal snail *Echinolittorina malaccana*: Thermal responses that defy the "Platonic form" of conventional TPCs

The intertidal snail *Echinolittorina malaccana*, which is indigenous to the tropical Pacific Ocean, is one of the most heat-tolerant animals known. Its upper lethal temperature is near 55°C, a temperature that we saw was near the predicted maximum temperature for mitochondrial function and, therefore, aerobic eukaryotic life (see Figure 3.54). Marshall and colleagues (2011) examined several physiological and biochemical characteristics of *E. malaccana* over a wide range of temperatures: 30°C to 58°C. Measurements of oxygen consumption over this wide range of body temperatures generated a thermal performance curve (TPC; see the figure) with a shape radically different from the shape we've been referring to as the TPC's Platonic form (see Figure 3.66). Rather than there being a regular, monotonic rise in oxygen consumption rate with increasing temperature and a single plateau region, the snail's oxygen consumption rate had a very complex response to rising temperature. Between 30°C and 35°C, the curve resembles a conventional sort of TPC. However, between ~35°C and 46°C, increases in temperature were not accompanied by a further rise in oxygen consumption rate. The Q_{10} value is close to unity, a puzzling finding in view of the dictates of the Arrhenius relationship. Then, between ~46°C and ~55°C, the snail's TPC has a more typical shape. Above 55°C, metabolic rate begins to decrease sharply as temperature rises.

Marshall and colleagues interpret the plateau region between 35°C and 46°C as a response that reflects energy limitations at elevated temperatures. Foraging activity, which is essential for energy acquisition, was highest at temperatures between ~32°C and 37°C. A rapid decrease in foraging activity occurred at high temperatures. Thus, the authors conjecture that the plateau region is not a consequence of an inability to acquire oxygen or a sign that

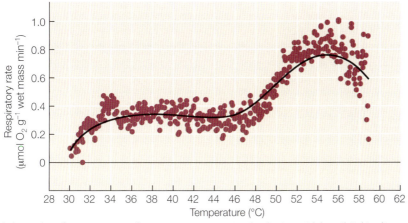

A thermal performance curve for oxygen consumption by the intertidal snail *Echinolittorina malaccana*. (After Marshall et al. 2011.)

BOX 3.13 *(continued)*

heat-induced damage to the metabolic apparatus (protein denaturation, for example) has taken place. Rather, this plateau is a reflection of physiological and behavioral responses driven by limited access to energy. What underlying mechanisms might account for the rise in oxygen consumption at temperatures above ~46°C? Measurements of the production of Hsp70 showed that this molecular chaperone was only synthesized at temperatures above 46°C. Thus, at least some of the rise in oxygen consumption rate at temperatures ≥ 46°C likely reflected activation of the energetically costly heat-shock response (HSR) and cellular stress response (CSR). Thermal damage to proteins at these high temperatures necessitated a shift from a state of metabolic quiescence to one of active aerobic ATP generation, which was needed in part to support the HSR and CSR.

Heat-tolerant limpets (*Cellana toreuma*) that co-occur with *E. malaccana* also show temperature-dependent shifts in gene expression that reflect alterations in the total amounts of catabolic activity

and in the balance between catabolic and anabolic processes (Han et al. 2013). These discoveries were discussed earlier (see p. 228) in the context of the CSR. Results from studies of both of these exceptionally heat-tolerant molluscs show that the animals are not "at the mercy" of the Arrhenius equation, as theoretical treatments like the metabolic theory of ecology (MTE) might suggest, but rather fine-tune their metabolic activities when environmental temperatures change. Both the absolute and the relative activities of catabolic and anabolic processes are modulated in the face of changes in body temperature. Clearly, there is much more than physics (the Arrhenius relationship) involved in governing the responses of organisms to changes in body temperature. Therefore, the shape of a TPC derives from numerous factors, ranging from thermally driven changes in behavior to complex adjustments in metabolic pathways that lead to shifts in total ATP turnover and changes in energy allocation between catabolic and anabolic processes.

have a form that, at least over some portion of the temperature range being studied, reflects the underlying physics of temperature effects we have discussed earlier (see Box 3.11). (Note that when TPCs are used for evaluating thermal dependence of such higher-level characteristics as fecundity, they generally are not showing acute effects of the sort discussed in the following analysis.) Over the lower part of an organism's temperature range, acute increases in temperature accelerate the rate of the process in an exponential manner, consistent with the predicted effects of Q_{10} relationships on enzymatic processes. However, direct links between the thermal responses of processes like growth or metabolism and the activities of enzymes remain to be adequately demonstrated (Schulte 2015). At some higher temperature, further increases in temperature lead to no further increase in rate. This plateau region, whose width varies among species, processes, and experimental treatments, is commonly spoken of as the range of optimal temperatures (T_{opt}). Whereas this use of the adjective *optimal* may be descriptive in the context of the shape of a TPC, one must not jump to the conclusion that the highest possible rates of performance for an organism are optimal in the sense of natural selection (with fecundity being a possible exception to this caveat). For example, merely because a physiological process like oxygen consumption rate reaches its highest value at a particular range of temperatures does not necessarily mean that a significant higher-level phenomenon like growth or fecundity will show the same thermal optimum. In fact, growth can fall off at temperatures well below the optimal temperature indicated by a TPC (Healy and Schulte 2012). The allocation of energy to different physiological processes may change with temperature, and at the high rates of metabolism found at "optimal" temperatures, a significant fraction of ATP turnover may be directed toward activities of the cellular stress response, rather than activities like cell division and growth. This conclusion is supported by the changes in gene transcriptional patterns noted earlier. For example, in the case of the eurythermal fish *G. mirabilis*, major batteries of genes associated with damage repair or cellular death are activated at the highest temperatures of acute heat stress (Logan and Somero 2011).

Another important caveat about the optimal temperatures found with TPCs concerns the dependence of T_{opt} on the experimental conditions used to generate a TPC. The T_{opt} that is observed may be influenced by the rate of heating or cooling and the duration of exposure to different temperatures, especially to high temperatures. (See Kingsolver and Woods [2016] for an excellent discussion of the time-dependent effects that can shape TPCs.) Recall that the stress resulting from exposure to a particular temperature regimen is a product of the temperature and duration of exposure to that temperature. Duration of exposure also comes into play in the context of the thermal ramp that is used to measure acute temperature effects. A slower ramping rate will lead to a longer total time at temperatures that create stress. Therefore, the TPC reflects a combination of the intrinsic thermal responses of the organism and the effects that experimental treatments impose on these responses. For other facets of thermal ramping experiments, see Terblanche et al. (2011).

Increases in temperature above T_{opt} cause a decrease in the rate of the process, and at sufficiently high temperatures, the process ceases entirely. If exposure to the highest temperatures is of long enough duration, death may occur. Again, the TPC one observes is strongly influenced by the conditions of thermal exposure used in the study—rate of heating, duration of exposure to new temperatures, and prior acclimatization state of the specimens. It is almost invariably challenging to tease apart the influences of such contingent factors from the role played by intrinsic properties of the organism in governing the shape of the TPC (Schulte 2015).

HOW TPC ANALYSES ARE USED   As mentioned above, TPCs have been used to describe processes that take place at many levels of biological organization, ranging from the activities of individual enzymes to the metabolism of entire ecosystems. TPCs are being used increasingly to predict the consequences of global warming. The proximity of the thermal optima of diverse performance characteristics to current thermal maxima may provide an index of how threatened a species is likely to be by further increases in ambient temperature (Tewksbury et al. 2008). We examine this issue in more detail in Chapter 5. Such TPC-based analyses have given major attention to aquatic species that face significant challenges from rising temperatures' interacting effects on metabolic rates and dissolved oxygen (DO) levels. In these analyses of temperature-DO and Q_{10} interactions, the **oxygen- and capacity-limited thermal tolerance (OCLTT)** hypothesis (Pörtner and Farrell 2008) has often been at center stage. The OCLTT hypothesis places strong emphasis on the factors that are responsible for the downward-sloping region of the TPC. What are the limiting conditions that prevent an organism from functioning well at temperatures above the T_{opt}? An important focus of the OCLTT perspective is the reduction in **aerobic scope**—the factorial increase in aerobic metabolism above standard or routine metabolism that can be called on to support such energy-demanding behaviors as prey capture or predator avoidance—that results from the combination of falling DO and increasing rate of standard metabolism as temperature increases. As we will discuss in Chapter 5, these dual effects of rising temperature put a "squeeze" on active metabolism and other energy-demanding processes like growth, and serve as important determinants of habitat suitability (see Deutsch et al. 2015). TPCs can be useful in characterizing this aspect of thermal tolerance, although the biochemical systems that underlie the shape of the TPC are undoubtedly multifarious and remain highly conjectural (see below). The OCLTT hypothesis also is applied to the lower end of the thermal tolerance range where, again, limitations in ability to take up oxygen can impede performance.

At the ecological level, the temperature dependence of metabolic processes has been used to develop a wide-ranging theory of energy flow in ecosystems, the **metabolic theory of ecology** (**MTE**; Brown et al. 2004). The MTE focuses on the portion of the TPC where rates increase regularly and exponentially with rising temperature, in accord with the Arrhenius relationship. MTE-based analyses attempt to link the fundamental temperature dependence of enzymatic reactions to the flow of energy through entire ecosystems. As Schulte (2015) points out in her analysis of the OCLTT hypothesis and the MTE, explaining higher-level effects—whether these are rates of aerobic respiration or energy flow through entire ecosystems—on the basis of thermal responses (E_a values) of individual enzymes is fraught with difficulties and ambiguities. She points out that E_a values usually are based on the response of an enzyme's maximum rate (V_{max}) to changes in temperature; that is, measurements are made at (or calculated for) saturating concentration of substrate. In the cell, substrate concentrations are almost invariably well below values that support V_{max}; K_M effects thus are critical in temperature relationships, as we have emphasized earlier, and E_a values based on V_{max} measurements are poor predictors of higher-level thermal responses (Schulte 2015). The temperature dependence of oxygen consumption may also vary with the organism's state of activity, further complicating efforts to rely strictly on thermodynamic approaches to predict metabolic rates at levels of biological organization ranging from the organismal to the ecosystem. Thus, oxygen consumption per se may not provide a satisfactory measure of metabolic energy turnover, further complicating models that rely on aerobic respiration as the key indicator of thermal effects on metabolic rate.

A further point that needs consideration in developing general concepts like the OCLTT hypothesis is that aquatic and terrestrial organisms may face very different challenges from rising temperatures. Terrestrial species experience much higher levels of ambient oxygen, and oxygen availability in the environment is minimally affected by change of temperature. In contrast, aquatic species face decreasing oxygen availability as temperature rises. Thus, the success of the OCLTT hypothesis in accounting for limitations in aerobic scope in aquatic species is not matched by its ability to account for effects in terrestrial organisms, notably insects (Schulte 2015).

Suffice it to say, in view of the broad use of TPCs in biology, it is very important to gain an understanding of the factors that govern the forms of these relationships. Below we briefly discuss some of the underlying physiological and biochemical mechanisms that might contribute to the shape of a TPC, notably its definition of "optima" and "limits," and its position on the temperature axis.

WHAT GOVERNS THE SHAPE OF A TPC? What factors cause the acute thermal responses exhibited in a TPC? To answer this question, one must deconstruct the different portions of a TPC, namely (1) the rise in performance up to a certain temperature, (2) the plateau that defines the "optimal" thermal range, and (3) the falling-off of performance above a certain temperature. The thermal range over which the rate of performance rises with temperature has commonly been attributed to the Arrhenius relationship (see Box 3.11). Thus, to a first (but only a first) approximation, the increasing rate of performance is simply a reflection of the basic thermodynamics of the system. A more challenging aspect of a TPC appears when we attempt to explain the plateau phase of the curve. What prevents a further increase in rate with rising temperature? Some have argued that this temperature insensitivity in rate reflects a balance between thermal acceleration of metabolic rates and denaturation of the enzymes that are responsible for metabolism. However, as we have seen in our earlier discussion of protein stability, one would not expect large-scale denaturation to commence at the relatively low temperatures where the TPC begins its plateau phase. What may be occurring, however, is a temperature-driven redistribution of protein conformational microstates, such that a decreasing fraction of the overall population of a given type of protein has the three-dimensional conformation to allow substrate binding and, hence, initiation of the enzymatic reaction. In other words, at the enzyme level, the plateau region of a TPC is more likely to be a reflection of effects on binding ability (K_M relationships) than of loss of protein molecules through denaturation. Theories like the MTE that employ only V_{max}-based temperature dependencies for enzymatic reactions thus fail to consider how enzymes function in vivo—that is, at the subsaturating substrate concentrations found in cells—where K_M effects play an important role in governing rate. Moreover, any analysis of TPC form that is based strictly on enzyme-level effects neglects other factors that may begin to be negatively affected as temperature rises. These other negative effects include disruption of membrane function, which may lead to direct effects on neural activity, especially on synaptic function, and downstream effects on cardiac function and locomotory activity.

A further challenge in interpreting TPCs arises from the fact that metabolism may change qualitatively with temperature, as well as in overall rate. Above we mentioned that ATP production at high temperatures may be increasingly directed toward repair processes of the cellular stress response, at the expense of energy provision for growth. Other metabolic costs may also shift with temperature. Thermal acceleration of ion movement across membranes may lead to increased ion regulatory costs. Among the temperature-dependent ion fluxes of importance here is proton leak across the inner mitochondrial membrane. Increases in temperature cause elevated rates of proton leakage, so a smaller fraction of the oxygen that is consumed leads to ATP synthesis at higher temperatures (Iftikar and Hickey 2013; Iftikar et al. 2014). A further oxygen-related challenge from rising temperatures stems from an increased production of ROS (Abele et al. 2002). The integrated signal provided by the rate of oxygen consumption thus will not allow an adequate appreciation of how energy is being allocated among processes across the organism's full thermal tolerance range. Therefore it is inappropriate to view the temperature at which an organism exhibits its maximum rate of metabolism as being physiologically optimal. Rather, at the temperature at which metabolic rate is greatest, there may be a suboptimal allocation of energy to processes like growth and reproduction that are of pivotal importance to what matters most—fecundity.

3.9 Water in the Solid State: Avoidance and Tolerance of Freezing

In this concluding section of our analysis of biochemical adaptation to temperature, we focus on the challenges organisms face when low temperatures threaten to cause ice formation in extracellular and intracellular fluids. This analysis of one critical aspect of water relationships will help build a foundation for Chapter 4, in which we will see that water plays a "center stage" role in a diverse range of biochemical functions. Indeed, much as temperature affects every biochemical system in an organism, so does water. As shown in Figure 3.1, a liquid state of the intracellular solution is essential if an organism is to remain metabolically active and able conduct all of its normal functions. Thus, for organisms that face a threat of freezing, biochemical adaptations that enable avoidance of ice formation in the cellular water play pivotal roles in allowing an organism to remain metabolically active at extremely low temperatures and avoid the damage that ice crystals cause to cellular structures (Costanzo and Lee 2013).

Overview: Basic strategies for controlling the physical state of water

There are two basic strategies for coping with the threats of ice formation in the extracellular and intracellular fluids. One strategy is termed **freeze-avoidance**. Here, ice formation is prevented and the organism remains in a liquid state even at potentially freezing temperatures. When freeze-avoidance is not possible—or when avoidance of ice formation is a suboptimal strategy for coping with long periods of cold temperatures—many multicellular organisms undergo partial freezing—**freeze-tolerance**—but generally only under closely regulated conditions that limit ice formation to extracellular fluids. In freeze-tolerance, ice formation in the extracellular fluids may be induced by **ice nucleators**, which, as we will show, reduce the risks of uncontrolled, spontaneous intracellular ice formation. The sizes of the ice crystals that do form in the extracellular fluids are kept small by the activities of **ice-binding proteins** (**IBPs**) and a "peripheral defense" system at the cell membrane that blocks propagation of ice from the extracellular compartment into the cell's internal water (Duman 2015). Researchers have reported tolerance of intracellular freezing by certain types of cells, for example, insect fat body cells, but this phenomenon remains poorly understood (see Sinclair and Renault 2010) and appears to be restricted to relatively few taxa.

Freeze-tolerant species include many plants and animals from high-latitude and high-elevation habitats that enter states of greatly reduced metabolism during winter dormancy (metabolic quiescence). By protecting the intracellular biochemical machinery from damage that would result from freezing of the cellular water, these organisms are prepared to rapidly regain metabolic function when temperatures rise above the freezing point. The cell's biochemical machinery is further protected by accumulation during acclimatization to cold of protein- and membrane-stabilizing molecules that range in size from small 3-carbon solutes like glycerol to stress-induced proteins. The establishment of a hospitable cellular microenvironment that stabilizes macromolecular and membrane structures is of key importance in the context of adaptation to a variety of environmental stressors, including adaptation to deep cold (see Chapter 4).

SOLID-STATE WATER: ICE VERSUS VITREOUS WATER   There is more than one form of solid-state water, and these different forms of "solid" water feature importantly in strategies of freeze-avoidance. To state that the formation of solid-state water may be critical for freeze-avoidance may sound contradictory, but this is in fact the case. To see why, we must compare the physical properties of different types of solid-state water (**Figure 3.67**). The most extreme cases of freeze-avoidance involve complete conversion of extracellular and intracellular water into a form of solid-state water termed **vitreous water** (see Figure 3.67D). An important distinction between this glasslike water and ordinary ice is that vitrification, unlike formation of normal ice, does not lead to an expansion of the solution's volume (see Figure 3.67A). Ordinary freezing, in contrast, can physically disrupt cells as spicules of ice grow and damage membranes and other structures (see Figure 3.67B). Ice formation also leads to increased concentrations of solutes in any remaining unfrozen water, because the ice crystal lattice essentially excludes all solutes. When extracellular freezing occurs, high solute concentrations in the extracellular space can establish steep osmotic gradients and lead to movement of water out of the cell. In some cases, this rise in osmotic concentration of the intracellular fluids can be advantageous because it reduces the freezing point of the remaining water in the cell. However, high

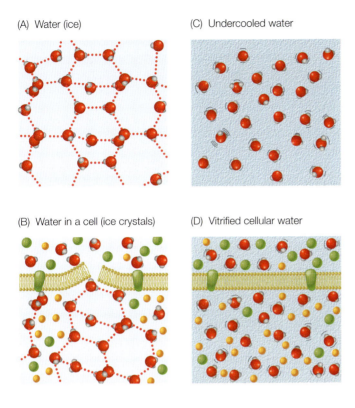

(A) Water (ice)

(C) Undercooled water

(B) Water in a cell (ice crystals)

(D) Vitrified cellular water

Figure 3.67 Low-temperature states of water. (A) Frozen water-ice. (B) Ice formation in cellular water. The cell membrane is disrupted by the increase in solution volume associated with the liquid-to-ice transition. Small solutes are excluded from the ice lattice and concentrated in small pockets of nonfrozen water. These high solute concentrations may damage proteins and membranes that are sensitive to osmolality or to high concentrations of specific solutes. (C) Undercooled water. (D) Vitrification of cellular water. The configuration of water molecules in undercooled water (C) poises water for vitrification. Vitrification does not lead to expansion of water volume and consequent damage to cellular membranes. During vitrification, small solutes are incorporated into the glasslike water lattice, so small volumes of water with high solute concentrations are not produced.

concentrations of solutes in the cellular water can damage the cellular machinery, for example, by perturbing protein structure and function (see Chapter 4). In contrast to what happens during ice formation, during vitrification small solutes "remain in place" and are incorporated into the vitrified water (see Figure 3.67C). Thus, vitrification protects cells in two important ways: It prevents the formation of damaging ice crystals, and it stabilizes the solute concentrations of intracellular and extracellular fluids.

WHAT DETERMINES THE OPTIMAL STRATEGY TO USE?   The optimal strategy for coping with a threat of freezing depends on the organism's particular set of circumstances, ranging from the broader ecological context in which low temperatures occur (e.g., the availability of food in cold seasons) to the biochemical tools available to the organism (Costanzo and Lee 2013). For example, does the organism have the ability to produce IBPs, and if so, how effective are they at controlling ice formation, ice crystal growth, and ice propagation? If an organism must remain active at all seasons, then avoidance of freezing is the strategy of choice. Marine fishes provide a good example of this type of situation. Their survival depends on maintaining cardiac and respiratory activities, even at the lowest temperatures they encounter. Although some Antarctic fishes enter a state of greatly reduced metabolism and locomotory activity ("hibernation") during the coldest times of the year (Campbell et al. 2008), they do not enter the type of deep dormancy and metabolic quiescence found for many terrestrial invertebrates and plants. Thus, in fish, any formation of internal ice that compromises these essential physiological activities is almost certain to be lethal. That being said, many high-latitude fishes do tolerate formation of small amounts of extracellular ice, and the mechanisms used to govern the further growth of this ice and to sequester it in a nonthreatening manner are some of the most fascinating stories in adaptation to cold (DeVries and Cheng 2005). Suffice it to say, even though an organism may be viewed as avoiding freezing, a more nuanced analysis of its biology may show that this categorization is something of a misnomer or only a partial truth. Freeze-avoiding organisms may not fully prevent formation of internal ice, but they do succeed in preventing this ice from doing damage—the organism does not freeze in its entirety.

If conditions such as limited food availability favor an organism's entry into dormancy during cold periods, and if the organism is capable of entering a state of metabolic quiescence in such circumstances, then freezing of some of the extracellular water may be advantageous. Many terrestrial animals, including vertebrates (e.g., some frogs) and invertebrates, adopt this strategy (Storey and Storey 1988; Costanzo and Lee 2013), as do many plants (Gusta and Wisniewski 2012; Gupta and Deswal 2014). In many cases, organisms that follow a freeze-avoidance strategy early in the cold season allow controlled formation of extracellular ice as conditions become more extreme. Commonly, these species accumulate small organic molecules, such as glycerol, to lower the freezing point by colligative mechanisms (see below) and, perhaps more important, to facilitate **undercooling** (also termed **supercooling**), a process whereby a solution remains in liquid state below its thermodynamic freezing point (see Figure 3.67C). Compounds like glycerol also appear to be critical for vitrification (see below), so accumulation of these solutes may help prepare the organism for transition into the vitreous state as winter conditions become even more extreme.

Freezing relationships are of interest not just to environmentally oriented biologists who are curious about adaptation to extreme conditions. Another reason this is such a vibrant field of research is its potential for advancing biomedicine, notably in the context of long-term cold-preservation of living materials. As basic biologists discover more and more examples of freeze-avoidance and freeze-tolerance in animals, plants, fungi, and bacteria, biotechnologists attempt to exploit these discoveries in applied contexts. Thanks to what has been discovered in such organisms as polar fishes and insects—species that are usually of no biomedical interest—great progress has been made in developing technologies for long-term, low-temperature storage of cells, organs, and even small organisms in vitrified form. Plant biologists interested in generating cold-resistant plants are also drawn to this field (Gusta and Wisniewski 2012). If genes that encode freeze-resistant traits were introduced into the genomes of freeze-sensitive plants, it might allow these plants to be cultivated under colder conditions, assuming other facets of their biochemistries permitted low-temperature function.

Colligative relationships and avoidance of ice formation

The simplest biochemical strategy for avoiding freezing is to exploit one of water's colligative properties: lower water's freezing point through increases in solute concentration. The **colligative properties** of solutions depend entirely on the number of osmotically active particles in solution. Freezing point depression is one of these colligative properties, as are lowering of vapor pressure, boiling point elevation, and osmotic pressure. The addition of 1 mole of particles (1 osmol or 1000 mosmol) to a kilogram of water lowers the equilibrium freezing point by 1.858 degrees Celsius. Oceanic seawater has a freezing point of −1.9°C, indicating the presence of approximately 1,030 mosmol of particles in a kilogram of seawater. As discussed in Chapter 4, marine invertebrates, bacteria, archaea, fungi, and algae almost invariably are close to being isosmotic to seawater. Thus, as long as seawater remains liquid, these organisms generally face no danger of freezing. Marine bony fishes, in contrast, are hyposmotic to seawater;

commonly, their extracellular and intracellular fluids contain only about 380–475 mosmol of particles per kilogram of water (Holmes and Donaldson 1969). Thus, most marine bony fishes have equilibrium freezing points in the range of –0.6 to –0.9°C; polar species may be more osmotically concentrated (~550–650 mosmol kg^{-1}; Dobbs and DeVries 1975; Raymond 1992) and have equilibrium freezing points near –1.0 to –1.1°C (DeVries and Cheng 2005).

COLLIGATIVE DEFENSES IN FISH: A COSTLY APPROACH TO FREEZE-RESISTANCE   There are a small number of exceptions to the general pattern of hyposmotic regulation in marine fishes. One example is the Arctic rainbow smelt (*Osmerus mordax dentex*; shown at right), which has the highest osmolality of any teleost studied to date, 1009 mosmol kg^{-1} (Raymond 1992). In common with the overwintering

strategy found in many terrestrial invertebrates and some amphibians, rainbow smelt seasonally accumulate glycerol, such that in midwinter glycerol concentrations reach ~0.4 mmol kg^{-1}, making the fish essentially isosmotic with seawater. Other Arctic teleosts also show elevated concentrations of glycerol in winter, but none are known to be as high as in rainbow smelt.

In view of the success of these Arctic fishes in using a colligative mechanism to avoid freezing, one might ask why this strategy is so rare in bony fishes. Glycerol is a metabolite common to all organisms, and its widespread use in cold-tolerant terrestrial invertebrates would seem to make glycerol accumulation a "smart" strategy in fishes as well. Glycerol readily moves through cell membranes and spreads among all body fluids. Thus, whatever its site of synthesis, glycerol can provide protection to the entire organism. Why, then, is glycerol not used more commonly by freeze-threatened bony fishes? The primary reason is probably the energy cost of this strategy. Unlike terrestrial invertebrates that accumulate multimolar concentrations of glycerol in winter and do not face loss of glycerol to the environment, the glycerol that fishes make readily leaks from the body into the environment through the gills. Loss of glycerol through the gut (defecation) and kidneys may also be extensive. The glycerol accumulated in smelt and certain other Arctic fishes therefore must be continuously replaced. This loss of glycerol is an important consideration in the context of energy budgets (as a reduced carbon molecule) and of the need for glycerol in a variety of metabolic processes such as lipid biosynthesis. As far as is known, Arctic smelt and a few other species are unique among bony fishes in using a colligative-effect freeze-avoidance strategy. This strategy clearly "works," but it carries a high cost in terms of energy metabolism.

COLLIGATIVE DEFENSES IN TERRESTRIAL ECTOTHERMS   Terrestrial ectotherms certainly exploit colligative relationships to defend against ice formation, but depression of the thermodynamic (equilibrium) freezing point is not likely to be the key factor in these adaptive strategies. To understand why, consider the amount of solute needed to reduce the thermodynamic freezing point of extra- and intracellular fluids to values below typical winter temperatures, which may dip below –40°C in cold temperate and polar regions. For each osmol kg^{-1} of particles added to their fluids, organisms gain an additional 1.9 degrees Celsius of protection. Thus, to gain protection against ambient temperatures near –40°C, a species would need to accumulate osmotically active solutes to a concentration near 20 osmol kg^{-1}. No organism is known to have this high an osmotic concentration (see Chapter 4). Nevertheless, many terrestrial species facing the threat of freezing in

BOX 3.14

Water and solute movement during formation of extracellular ice

How do terrestrial invertebrates regulate the concentrations of glycerol and other small molecules important in freezing relationships? On the production side of the equation, glycerol can readily be produced through core reactions of carbohydrate metabolism (see Chapter 2). Glycogen breakdown yields glucose, which can enter glycolysis to yield 3-carbon intermediates than can generate glycerol. Biosynthesis of glycerol commonly is upregulated early in the cold acclimatization process (Duman 2015), conferring on the cells a depressed freezing point and enhanced capacity for undercooling. In terrestrial invertebrates, concentrations of glycerol rise in proportion to the decrease in ambient (body) temperature (Bennett et al. 2005). Because glycerol may need to be transported to a site remote from its site of synthesis, mechanisms for regulating uptake of glycerol into cells are likely to be important in the acquisition of cold tolerance as well. One of the proteins involved in the uptake of glycerol and the release of water during extracellular freezing is **aquaporin** (Izumi et al. 2006). Aquaporins are membrane-localized water channel proteins that facilitate movement of small molecules like glycerol as well as water itself. When extracellular freezing removes liquid water from the extracellular fluids (e.g., hemolymph), this creates an osmotic gradient that favors movement of water from the intracellular fluids to the extracellular space. Most of this water likely exits the cell via aquaporin channels. Glycerol may enter through the same channels to increase intracellular osmolality and enhance tolerance of low temperature. In the rice stem borer (*Chilo suppressalis*), rates of water and solute transport through aquaporin channels were higher in larvae that were in overwintering condition and in a state of diapause than in larvae in the non-diapause state (Izumi et al. 2006). Only the overwintering/diapause larvae were freeze tolerant; they readily tolerated formation of extracellular ice. However, this freeze-tolerance was prevented when aquaporin channels were inhibited. These results indicate that the ability to synthesize cryoprotectant substances like glycerol is not sufficient to confer freeze-tolerance on an organism. Rather, mechanisms that facilitate uptake of glycerol and transport of water to the extracellular space must be viewed as part of the overall mechanism of freeze-tolerance.

---

deep winter do accumulate high levels of small organic molecules like glycerol (**Box 3.14**). The primary adaptive significance of these high solute concentrations in the context of freezing may lie in their ability to facilitate undercooling and, in some cases, prepare the cellular water for entry into a vitrified state. The undercooling that results from colligative effects—for example, through accumulation of glycerol—may be approximately twice the decrease noted in the thermodynamic freezing point (Duman 2015). The multiple roles of glycerol are illustrated particularly clearly by cold-tolerant arthropods.

A variety of terrestrial arthropods, including mites, spiders, and insects, have been studied to elucidate the types of adaptive mechanisms used to prevent ice formation in the cells and extracellular fluids (hemolymph) and to control the further growth of any ice that should form in the hemolymph. Among the taxa that have been studied to date, the red flat bark beetle (*Cucujus clavipes*; shown at left) is arguably the most remarkable in terms of tolerating extremes of cold temperature through reliance on colligative mechanisms. This species has a wide biogeographic distribution and, consequently, an extremely broad thermal tolerance range. It occurs from 35° N in North Carolina to 68° N in Alaska; in the latter habitat, temperatures in winter may fall to –60°C (Sformo et al. 2010, 2011; Duman 2015). As might be expected for a species with this broad a latitudinal distribution range, there is evidence of genetically distinct populations (subspecies) that differ physiologically. The West Coast subspecies (*Cucujus clavipes puniceus*), which occurs from

Cucujus clavipes

Alaska southward along the Pacific Coast, may have a different overwintering strategy than the eastern subspecies (*C. c. clavipes*), which occurs from the Great Plains to the Atlantic seaboard (Sformo et al. 2010). However, more interpopulation comparisons will be needed to link genetic and physiological variations. Overwintering larvae from the upper Midwest (Indiana) had undercooling temperatures between –20°C and –28°C, did not enter metabolic quiescence, and did not become dehydrated. In contrast, northern Alaska beetles under normal winter conditions had undercooling temperatures averaging about –40°C, entered metabolic quiescence, and underwent extreme dehydration (Bennett et al. 2005). *C. clavipes* larvae of both subspecies increase their ability to undercool as temperatures plummet, and this acclimatization involves elevation of glycerol concentrations. Under the conditions of extreme cold found at the northern extent of the western subspecies' biogeographic range, the increase in glycerol concentrations occurs concurrently with a large decrease in cellular water content. The western subspecies may lose approximately 80% of its water during acclimatization to extreme cold: Summer larvae had water contents of 2.0 mg H_2O mg^{-1} dry mass, and winter larvae had ~0.4 mg H_2O mg^{-1} dry mass (Sformo et al. 2010). Water is lost from the larvae as a consequence of the difference in vapor pressure between the unfrozen organism and the frozen external environment (Sformo et al. 2011). Larvae may be encased in ice during winter, which establishes favorable conditions for transfer of unfrozen water in the larva to the environment (Bennett et al. 2005). In this extremely dehydrated state the concentration of glycerol may reach concentrations of at least 4–6 mol l^{-1} and perhaps even higher (Sformo et al. 2010). These are among the most concentrated solutions known for any type of cell (see Chapter 4).

The cold, concentrated, and highly viscous solutions in the overwintering larvae not only facilitate undercooling, but also appear to poise the remaining cellular and extracellular water for vitrification. Highly viscous solutions with low water content are characterized by water with a greatly reduced mobility, which facilitates vitrification (Tarjus and Kivelson 2000). Thus, as temperatures fall below approximately –58°C, the remaining water in the larvae vitrifies (Sformo et al. 2010, 2011). This organism-wide vitrification prevents osmotic gradients from being established between the cellular and extracellular fluids, which eliminates one of the dangers associated with extracellular freezing. And as pointed out above (see Figure 3.67), vitrification does not damage cellular membranes. Vitrified larvae of *C. clavipes* survive at impressively low temperatures: Some larvae have been observed to survive cooling to –100°C, but the mean lower lethal temperature is near –70°C (Duman 2015). How widely distributed this type of vitrification strategy is among terrestrial ectotherms remains to be established. Whatever its breadth of occurrence in nature, the strategy has been of central interest in biotechnology laboratories, where treatment of biological materials with solutes that favor vitrification has become a common technique.

We leave the story of *Cucujus clavipes* at this point but will return to it later, after introducing the multiple roles of IBPs in strategies of freeze-avoidance and freeze-tolerance. Once IBP functions have been added to the *C. clavipes* story, we will see the true complexity of the strategies used to either prevent ice formation or to control the properties and location of the ice that may form in the body.

Noncolligative mechanisms for controlling ice formation: Ice-binding proteins (IBPs) and glycolipids

For several decades, physiologists have been fascinated by the fact that polar fishes, such as the notothenioids of the Southern Ocean, survive with equilibrium freezing points that are several tenths of a degree Celsius higher than the temperatures of the water in which they swim. As early as the mid-1950s, Scholander and colleagues (1957) sought to elucidate how polar teleosts remain nonfrozen under these conditions. Their studies of northern fishes living under the threat of freezing revealed one adaptive mechanism and hinted at another. For fishes living in deep fjords where ice was not present, survival during the winter was simply a matter of remaining in an undercooled state: living in deeper waters where no ice was present to seed freezing in the body fluids. However, when these fishes were brought to the surface and placed in aquaria that contained ice crystals, they rapidly froze. Undercooling is only effective when no ice is present to trigger freezing of the body fluids. Touching any point of the undercooled fish's body surface led to seeding of ice formation throughout the body.

Scholander and colleagues also examined shallow-occurring fishes that were normally exposed to ice in winter. These species, typical of bony fishes, had freezing point depressions that seemed insufficient to explain their continuing existence in a nonfrozen state. Attempts to identify the factors that might underlie this resistance to ice formation, which clearly seemed to defy explanation in terms of colligative relationships, did not succeed. It was only about a decade later that work by Arthur DeVries, carried out at the McMurdo Station in Antarctica with notothenioid fishes, led to discovery of what have come to be called "antifreeze" proteins. In fact, the first antifreezes discovered by DeVries were glycoproteins (**Figure 3.68**; DeVries and Wohlschlag 1969; DeVries and Cheng 2005) and are now appreciated as being but one member of a broad class of

(A)

Characteristic	AFGP	Type I AFP	Type II AFP	Type III AFP	Type IV AFP
Mass (kDa)	2.6 – 33	3.3 – 4.5	11 – 24	6.5	12
Key properties	AAT repeat; disaccharide	Alanine-rich α-helix	Disulfide bonded	β-sandwich	Alanine rich; helical bundle
Representative structure					
Natural source	Antarctic notothenioids, northern cods	Right-eyed flounders, sculpins	Sea raven, smelt, herring	Ocean pout, wolfish, eel pout	Longhorn sculpin

(B)

(i)

(ii) (iii) (iv)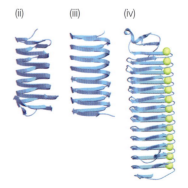

Figure 3.68 Structures of several classes of ice-binding proteins (IBPs). (A) Five classes of IBPs found in polar fishes. AFGP, antifreeze glycoprotein; AFP, antifreeze protein. (B) IBPs in other species and an ice-nucleating protein (INP): (i) model of an INP, (ii) insect (spruce budworm) IBP, (iii) plant (*Lolium perenne*) IBP, and (iv) bacterial (*Marinomonas primoryensis*) IBP. (A after Capicciotti et al. 2013; B, i from Garnham et al. 2011, ii–iv after Davies 2014.)

(A) Antifreeze

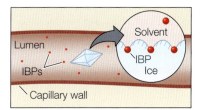

(B) Ice recrystallization inhibition

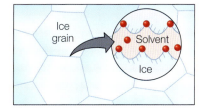

(C) Ice structuring

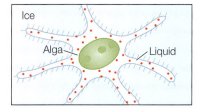

(D) Ice adhesion

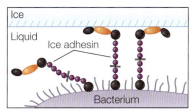

Figure 3.69 Roles of ice-binding proteins. (A) Antifreeze function and freeze-avoidance: IBPs bind to small ice crystals and prevent their growth to life-threatening size—that is, to a size that can clog fine capillaries. Binding of IBPs leads to a change in the geometry (radius of curvature) of the small ice crystal and a concomitant reduction in freezing point (the Kelvin effect; see inset circle and text). (B) Inhibition of ice recrystallization: Ice-binding by IBPs hinders loss of water from small crystals and gain of water by larger crystals. Ice crystal size is kept too small to endanger the organism. (C) Secreted IBPs—habitat liquefaction. IBPs secreted by microorganisms (bacteria, algae, fungi) maintain a zone of liquid water around the cell, helping ensure access to oxygen and nutrients. (D) Ice adhesion IBPs: IBPs (e.g., ice adhesin) found on the external surface (cell wall or cell membrane) of a bacterium "anchor" the organism to ice and allow it to remain in a microenvironment with adequate nutrients, oxygen, and light. (After Davies 2014.)

proteins that modulate ice formation: ice-binding proteins (Davies 2014). Moreover, ice-binding "antifreeze" molecules are now seen to include ice-binding glycolipids (IBGLs) as well as proteins and glycoproteins (Walters et al. 2009).

THERMAL HYSTERESIS   Before describing the diverse biological contexts in which IBPs and IBGLs contribute to freeze-avoidance or, when ice does form in freeze-tolerant species, to the management of ice crystal properties, it is useful to review the variety of proteins and glycolipids that have been shown to bind to ice crystals. This review will help establish the structure-function relationships common to these diverse molecules. IBPs and IBGLs have a diversity of chemical structures (see Figure 3.68) yet share the common property of being able to attach to ice crystals and influence any further growth of an ice crystal that might occur (**Figure 3.69**). One consequence of this interaction with ice by all types of IBPs and IBGLs is a lowering of the solution's observed freezing point below the true thermodynamic (equilibrium) freezing point that is predicted on the basis of colligative relationships. This disparity is termed the **thermal hysteresis (TH)** of the freezing point (**Figure 3.70**). Freezing point depression by IBPs may be two or three orders of magnitude greater than would be predicted strictly on the basis of colligative relationships (Gupta and Deswal 2014). A slight elevation in melting temperature of IBP-coated ice also occurs and is referred to as *thermal hysteresis of the melting temperature* (see Figure 3.70).

(A)

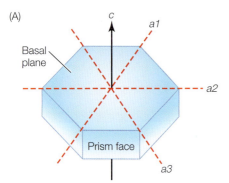

(B)

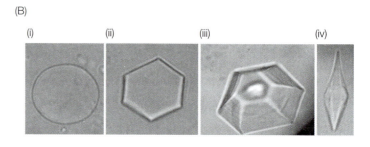

Figure 3.70 Ice crystal structures and thermal hysteresis caused by IBPs. (A) Addition of water to an ice crystal can occur along the three *a* axes (prism faces) and the *c* axis (basal plane). (B) Geometry of ice. (i) In the absence of IBPs, ice grows as a flat, round crystal. IBPs influence the sites of water addition to the growing crystal and thus its shape. (ii) In dilute solutions of most IBPs, IBP binding occurs primarily to the prism faces, leading to hexagonal ice crystals. (iii) Addition of IBPs to the prism face inhibits further binding of water to this face. Any additional water molecules added to the ice thus tend to bind to the basal plane, leading to growth along the *c* axis. (iv) At high concentrations of IBPs, IBP binding to all planes of the ice crystal occurs. Any further growth of the ice crystals leads to bipyramids that are hexagonal in cross section. (C) All classes of IBPs bind to ice crystals and lower the freezing point (FP) and increase the melting point (MP) of the ice. This change in FP and MP is termed the *thermal hysteresis gap*. Under conditions where only colligative effects on FP and MP are present, FP = MP. Low-TH IBPs and high-TH IBPs differ in their modes of interaction with ice, in terms of the preferred plane and axis of binding (see text). (A and B from Griffith and Yaish 2004; C after Davies 2014.)

(C)

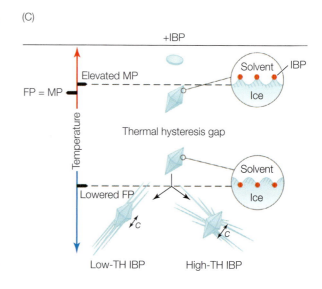

As shown in **Table 3.5**, IBPs differ in their **thermal hysteresis activities (THAs)**. High-TH IBPs are of greatest significance in freeze-avoiding species, where they reduce the likelihood of ice formation in the extracellular fluids. Low-TH IBPs are characteristic of freeze-tolerant species in which extracellular ice formation is part of the strategy of adaptation. In these species, the low-TH IBPs play an important role by limiting ice **recrystallization**—the growth of large ice crystals from water originating in smaller crystals (see Figure 3.69). Recrystallization is a common occurrence when ice forms from undercooled water. During the initial phase of the freezing process, many small crystals form. Because these tiny crystals have a relatively large radius of curvature and high surface free energy, water tends to be lost from the surface of these crystals and gained by larger crystals, to which addition of water is more energetically favorable. Whereas the smallest crystals may not represent a threat to the organism, larger crystals can cause a variety of damage, for example, blockage of capillaries, disruption of the extracellular matrix, or puncturing of membranes. In freeze-tolerant organisms, therefore, the size of the resulting ice crystals must be carefully regulated to avoid potentially lethal effects of extracellular ice. Low-TH IBPs have at least one other important function, that of preventing ice that does form in the extracellular fluids from seeding ice formation in cellular fluids, which is almost always

TABLE 3.5 ▮ ▬▬ ▬▬▬ ▬▬▬▬▬▬▬▬▬▬▬▬▬▬▬▬▬▬▬▬▬▬▬▬▬

Ice-binding proteins and glycolipids and their functions in preventing ice-induced damage to cells and death from freezing

High-thermal hysteresis IBPs (High-TH IBPs): True antifreeze proteins (AFPs) and antifreeze glycoproteins (AFGPs)
Proteins and glycoproteins with large thermal hysteresis activities (THAs) that function to prevent freezing in freeze-avoiding species. Fish antifreezes bind only to the basal plane of ice; arthropod antifreezes bind to both prism and basal faces of ice crystals and generally produce greater TH than fish antifreezes (see text). High-TH IBPs are generally present at relatively high (mmol l^{-1}) concentrations compared with low-TH IBPs.

Low-thermal hysteresis IBPs (Low-TH IBPs)
IBPs that have relatively low THAs and function in freeze-tolerant species to inhibit recrystallization (see Figure 3.69B) and, in some cases, prevent ice propagation from extra- to intracellular fluids. Although they have low THAs, low-TH IBPs are effective at relatively low (μmole l^{-1}) concentrations.

Ice-nucleating proteins (INPs)
Large proteins that induce ice formation at relatively high subfreezing temperatures in many freeze-tolerant species. INPs prevent deep undercooling, which can lead to catastrophic formation of internal ice.

Antifreeze potentiating proteins (AFPPs)
Small proteins that amplify the efficacy (raise the THA) of high-TH IBPs. AFPPs bind to the prism face of the ice crystal.

Antifreeze glycolipids (AFGLs)
Protein-free molecules that function primarily as high-TH antifreezes (Walters et al. 2009; Duman 2015).

Secreted IBPs of microorganisms
IBPs secreted by some algae, fungi, and bacteria into the surrounding water to ensure maintenance of a liquid state in their microenvironments (e.g., Raymond 2011; see Figure 3.69C). Secreted IBPs may function as antifreezes and inhibitors of recrystallization.

Ice adhesion proteins
Proteins with ice-binding ability that are attached to the external surfaces (cell walls) of some microbes. These ice adhesion proteins appear to be used to anchor the microbes to ice in regions where oxygen and nutrients are most readily available (see Figure 3.69D; Davies 2014).

Additional functions of IBPs
Inhibition of ice nucleators, protection of membranes under low-temperature perturbation of structure and function, reduction in ice propagation from environment to organism (see Figure 3.69), and in the case of plant IBPs, antipathogenic (antifungal) activities (Griffith and Yaish 2004). IBPs may also impede the crystallization of sugars such as trehalose at extremely low temperatures (Wen et al. 2016).

lethal for a cell. Some low-TH IBPs and IBGLs are attached to the surfaces of cells and act as a "peripheral defense" against transmembrane propagation of ice (Duman 2015). Interestingly, even though the same ice-binding sites are involved in THA and inhibition of recrystallization, the abilities to decrease FP (THA activity) and inhibit recrystallization are not strongly correlated; some IBPs that have a high THA are relatively ineffective as inhibitors of recrystallization (Gupta and Deswal 2014).

ANTIFREEZES HAVE PARTNERS: POTENTIATING PROTEINS   The THA observed with high-TH antifreezes like the antifreeze glycoproteins (AFGPs) found in Antarctic notothenioid fishes is increased in the presence of another class of proteins termed **antifreeze-potentiating proteins (AFPPs)** (DeVries 2004). These are relatively small (~15 kDa) proteins that lack carbohydrate. On their own, AFPPs have low THA, but when both AFPPs and AFGPs are present in the solution, THA attains very high values, approximately twice those observed with AFGPs alone. The basis of this enhancement of TH lies in the different ice-binding properties of the AFPPs and AFGPs. The former protein binds to the prism face of ice; the latter binds to the basal plane (see Figure 3.70A; Duman 2015). In effect, by having both IBPs present, ice growth over the entire surface of the ice crystal may be impeded. The higher TH observed with insect IBPs (THA of up to 13°C) relative to fish AFP or AFGPs (where THA is usually not more than ~1.5°C) reflects the abilities of insect IBPs to bind to both basal and prism faces of ice crystals (Pertaya et al. 2008).

CAUSING ICE TO FORM: ICE-NUCLEATING PROTEINS   Another class of IBP, **ice-nucleating proteins (INPs)**, which are common in freeze-tolerant organisms, may seem like paradoxical molecules for use in avoiding damage from freezing. How can it be to an organism's advantage to accumulate INPs that trigger freezing at relatively high subfreezing temperatures? Is it not better to deeply undercool, as seen with *Cucujus clavipes*? One answer to this question concerns the threats that undercooling can create as body temperature continues to drop. As long as the undercooled state of water is stable, the organism will not be threatened by ice formation. However, when a deeply undercooled solution does freeze, formation of ice will be rapid and complete. It is, therefore, advantageous for freeze-tolerant organisms to initiate ice formation at relatively high temperatures and avoid deep undercooling (Zacchariassen and Hammel 1976).

Several types of IBPs with ice-nucleation activity have been discovered and have been shown to function in triggering extracellular ice formation at temperatures only slightly below the thermodynamic freezing point (Davies 2014). As this ice forms, solutes are excluded from the ice lattice, and the freezing point of the remaining water is lowered due to colligative effects. The osmotic gradient that is established may draw water out of the cell, concentrating the solutes in the remaining cellular water and reducing its freezing point. Further cooling of the organism may lead to additional freezing in the extracellular water; however, as with the initial ice formation, the rate of freezing is relatively slow and not the catastrophic total freezing that might occur in the absence of the INPs. In freeze-tolerant animals and plants, low-TH IBPs work in conjunction with INPs by coating small ice crystals and preventing recrystallization and growth of large and potentially damaging crystals.

OTHER ROLES OF IBPs   Whereas much of the focus given to IBPs has been in cold-tolerant animals and plants, where control of extracellular ice formation is critical for survival, IBPs also occur in microorganisms. Unicellular algae, bacteria, and unicellular and multicellular fungi may secrete IBPs into the surrounding water in order to maintain this water in a liquid state that allows access to nutrients and dissolved gasses (Janech et al. 2006; Raymond 2011; see Figure 3.69). Secreted IBPs have been shown to alter the structure of sea ice (Raymond 2011), which might have effects on the microenvironments available to microbes. Another role of extracellular IBPs is attachment to ice surfaces (see Figure 3.69). Bacteria may employ these proteins to anchor themselves to the under-ice surface, thereby gaining access to light (for photosynthetic species) and oxygen.

In addition to helping govern ice formation in organisms, some types of IBPs also may play a broad role in the hydrologic cycle (Christner et al. 2008). Christner and colleagues found that biological INPs, many likely of bacterial origin, were abundant in all snowfall samples collected at sites in many different regions of the world. Ice generation in tropospheric clouds is necessary for snowfall and rainfall, and particulate matter is viewed as essential for seeding ice crystal formation. Biological materials, including bacterial cells, are abundant in the atmosphere and likely play an important role in seeding ice formation. This role of INPs is a nice illustration of the interplay between biology and broad atmospheric phenomena, a topic we inroduced in Chapter 2.

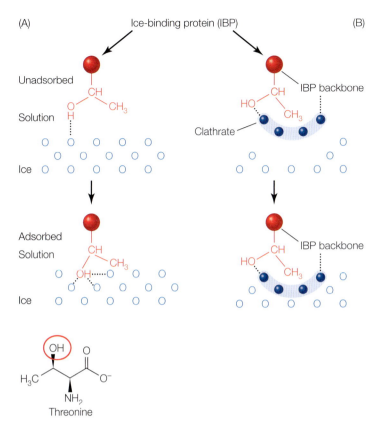

(A)

Ice-binding protein (IBP)

(B)

Unadsorbed

Solution

Ice

Adsorbed

Solution

Ice

IBP backbone

Clathrate

IBP backbone

OH

O

H₃C

O⁻

NH₂

Threonine

Figure 3.71 Two proposed mechanisms of binding between IBPs and water on or near the ice surface. The ice-binding site of the IBP is represented by a threonine side chain shown in red. Blue circles represent water molecules in the ice lattice or quasi-liquid water just above the ice surface. (A) Hydrogen bonding mechanism. A threonine hydroxyl group (circled in red in the threonine structure below) initially forms a hydrogen bond (dotted line) to the ice surface. The threonine hydroxyl subsequently integrates into the ice and occupies an ice lattice O atom site. This leads to formation of additional hydrogen bonds with the ice-associated waters. (B) Anchored clathrate mechanism. Water forms organized clathrate structures (indicated by dark blue dots) around an IBP's hydrophobic groups (e.g., methyl [–CH₃] groups). Water molecules in the clathrate hydrogen bond (dotted lines) to the protein backbone and side chain hydroxyls. This organized water merges with quasi-liquid water on or adjacent to the ice surface and becomes part of the ice crystal. A third hypothesized ice-binding mechanism (not shown) involves release of clathrate water molecules during IBP binding to ice (for details, see Davies 2014). (After Davies 2014.)

MECHANISMS OF ICE BINDING BY IBPs The interactions of IBPs with the different surfaces of ice (see Figure 3.70) may involve at least two types of mechanisms (**Figure 3.71**): (1) direct hydrogen bonding between the IBP and the ice lattice, as mediated, for example, by the hydroxyl (–OH) groups of threonine residues of AFGPs (see Figure 3.68); and (2) a water-fusion effect that involves merging of highly structured water (clathrates) around hydrophobic side chains and backbone regions of the IBP with quasi-liquid, ice-like water at the ice surface. The binding mechanism that pertains differs among IBPs and is determined by whether the **ice-binding surface** (**IBS**) is polar or hydrophobic. Both mechanisms may be important for classes of IBPs that possess both polar and non-polar surface features. An additional aspect of the energy changes associated with IBP adherence to ice may involve the hydrophobic effect (see the section "Thermodynamics of protein folding: Water and the hydrophobic effect," p. 149). Here, the shedding of organized waters from around hydrophobic groups leads to an increase in system entropy, which in the case of IBP-ice interactions would favor binding. Note that this hydrophobic interaction-based explanation of binding portrays the loss of organized water from around the IBP, rather than the incorporation of this clathrate of water into the growing ice surface. In general, the larger the IBP, the greater is the IBS available for attachment to ice. It follows that the efficacy of IBPs, their THA, is directly proportional to the mass of the IBP. This is seen especially clearly in the case of different-sized members of a single IBP family, for example, the notothenioid AFGPs, which have a similar unit structure but widely different numbers of ice-binding units in the IBP molecule (DeVries and Cheng 2005). The effective size of an IBP can also be influenced by IBP-IBP interactions that lead to aggregations of multiple IBPs on the ice surface (Duman 2015). As we will discuss in more detail later, the occurrence of high concentrations of small organic solutes that favor protein-protein interactions in IBP-containing organisms may be a reflection of this aggregation-enhancement effect. This pertains not only for glycerol but also (and perhaps especially) for the very strong facilitator of protein aggregation trimethylamine-N-oxide (TMAO), which occurs in unusually high concentrations in polar fishes (Raymond 1994; Raymond and DeVries 1998).

Whatever the exact mechanism of binding, attachment of the IBP to the ice surface reduces the likelihood that additional water molecules can attach to the ice crystal. This mechanism—the **adsorption-inhibition mechanism**—was initially described for the AFGPs of notothenioid fishes and attributed to hydrogen bonding of threonine residues to ice (Raymond and DeVries 1977). However, ice binding mediated by hydrophobic groups surrounded by water clathrates also leads to an adsorption-inhibition effect (see Figure 3.71B). As mentioned above, the organized water around hydrophobic (for example, methyl) groups and regions on the IBP backbone has a geometry that allows it to merge with the quasi-liquid water adjacent to the ice surface. This merged water then freezes onto the ice, anchoring the IBP.

The common outcome of IBP binding to ice, whether due to hydrogen bonding or hydrophobic/water clathrate interactions, is an alteration of the energy changes that occur during ice growth. The coating of the ice surface by IBPs, whether to the basal plane or the prism faces, makes it more difficult, thermodynamically, for water molecules to be added to the ice crystal (see Figure 3.70). This change in the energy of water binding is a consequence of the alteration in ice geometry that is caused by binding of IBPs. The spaces on the ice surface between the bound IBPs are the only sites where ice growth can occur, and the higher radius of curvature of the ice in these inter-IBP regions makes ice growth less energetically favorable (see Figure 3.70). This mechanism of inhibition of ice crystal growth is known as the **Kelvin effect** (Raymond and DeVries 1977; Davies 2014). It accounts for thermal hysteresis by explaining how the freezing point in the presence of an IBP differs from that observed when only colligative effects pertain.

Genomic origins of IBPs: Ancestral homologues and multitasking proteins

In light of the diversity of IBP structures (see Figure 3.68) that have been discovered in the wide range of taxa that produce these molecules, it is interesting to think about the genetic raw material that has been used by natural selection to generate these macro-molecules. What types of proteins make good candidates for developing ice-binding properties? Are IBPs simply proteins that can inhibit ice formation while conducting other functions as well? Or have genes for some proteins been modified to yield a protein that has gained an IBS?

Both sources of IBPs have been discovered. Because an essential requirement for IBP function is a region on the protein surface, the ice-binding surface, that can interact with ice—either directly or through interactions mediated by water organized around hydrophobic groups (see Figure 3.71)—it seems likely that a wide variety of proteins may serve as source material for generating IBPs. Thus, any protein that has—or could easily acquire through mutation—a surface that allows interaction with ice (or icelike organized water) could potentially serve as an IBP—or as a progenitor of an IBP. Proteins with a high fraction of β-sheet structure may be good sources for evolution of IBPs because the relatively flat surface provided by β-sheets might facilitate interactions with ice (Gupta and Deswal 2014). In fact, a substantial fraction of IBPs are rich in β-sheet secondary structures. Some IBPs, however, lack β-sheets (Wisniewski et al. 1999). As already mentioned, the types of molecular structures that can generate THA are not restricted to proteins and glycoproteins (see Figure 3.69). Molecules composed of carbohydrates and lipids—for example, the (lipo)xylomannan antifreeze found in an Arctic beetle (Walters et al. 2009) and the glycolipid IBPs found in certain plants and frogs (Walters et al. 2011)—may also contain IBS features that allow them to be highly effective IBPs.

SOURCES OF PLANT IBPs   Among the types of proteins that have been recruited to serve as IBPs (which in some cases has not led to loss of their original functions) are a suite of **pathogenesis-related** (**PR**) **proteins** in plants (Griffith and Yaish 2004). PR proteins are important in conferring resistance to fungal attack. This protection may be especially important during winter periods, when psychrophilic fungi prosper. Accumulating PR proteins prior to the onset of winter conditions thus may give plants a "preparative defense" against cold-adapted fungi. PR proteins include enzymes that degrade fungal cell walls and proteins that can block fungal enzymes released to attack plant tissues. The PR proteins whose synthesis is induced by exposure to low temperature and short day length have been discovered to have ice-binding ability as well as anti-fungal abilities. In fact, most of the IBPs identified in plants are homologous to PR proteins (Griffith and Yaish 2004). Plant IBPs appear to function as inhibitors of recrystallization; they block growth of ice crystals that form in intercellular spaces, xylem vessels, and tracheids. IBPs are found only in plants that are tolerant of freezing and occur in many different parts of the plant, including seeds, stems, bark, branches, leaves, flowers, roots, and tubers. In addition to conferring protection from freezing, these IBPs have retained their original PR protein activities as chitinases, glucanases, thaumatin-like proteins (thaumatins are stress-induced proteins that block fungal attack, among other functions), and inhibitors of fungal enzymes that degrade plant cells.

Another class of proteins found to exhibit cold-induced expression and thermal hysteresis activity in plants are **dehydrins** (Wisniewski et al. 1999). Dehydrins are intracellular, glycine-rich, hydrophilic proteins with high thermal stability. Dehydrins are widely distributed in plants and animals and are induced in response to a variety of stressors, including desiccation and extremes of temperature (Close 1997; Hughes et al. 2013). In the face of threat of freezing, dehydrins have been conjectured to play at least two roles. First, as their thermal hysteresis activities suggest, they may inhibit activities of ice nucleators found in the cell and depress the freezing point of the cytoplasm. Second, dehydrins protect other cellular proteins from denaturation due to loss of water (Wisniewski et al. 1999; see Chapter 4). Thus, when extracellular ice formation leads to loss of water from cells, dehydrins may facilitate conservation of the native structures of proteins that are sensitive to dehydration or to high concentrations of intracellular solutes. As in the case of PR proteins, therefore, dehydrins appear to have multiple functions in protecting cells from threats posed by low temperatures.

SOURCES OF FISH IBPs   Among marine fishes, four different types of protein IBPs and a glycoprotein IBP have been discovered to date (see Figure 3.68; Fletcher et al. 2001). The variety of IBPs found in fishes is a reflection of the repeated and independent evolution of freeze-avoidance in the face of falling water temperatures. Most orders and suborders of teleost fishes are thought to have arisen during a relatively warm period in Earth's history, roughly 255–250 million years ago (Figure 3.72). Thus, IBPs may not have been encoded by the genomes of the ancestors of contemporary fishes. Subsequent cooling in polar regions may have provided strong selection for acquisition of freeze-avoidance in the resident fish fauna. In the absence of an ancestral IBP progenitor, different routes

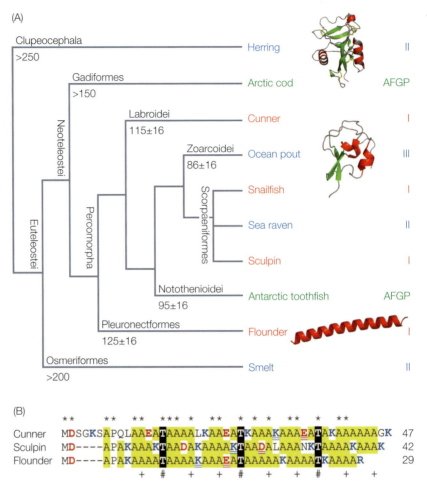

(A)

Clupeocephala >250 — Herring — II

Arctic cod — AFGP

Gadiformes >150

Labroidei 115±16 — Cunner — I

Zoarcoidei 86±16 — Ocean pout — III

Snailfish — I

Sea raven — II

Sculpin — I

Nototheniodei 95±16 — Antarctic toothfish — AFGP

Pleuronectformes 125±16 — Flounder — I

Osmeriformes >200 — Smelt — II

(Neoteleostei, Percomorpha, Scorpaeniformes, Euteleostei)

(B)

```
         **  **  **  *** *   **  * *  *  **   *   **
Cunner   MDSGKSAPQLAAEATAAAALKAAEATKAAAKAAAEATAKAAAAAAGK  47
Sculpin  MD----APAKAAAKTAADAKAAAAKTAADALAAANKTAAAAKAAAK    42
Flounder MD----APAKAAAATAAAAKAAAEATAAAAAKAAAATKAAAAR       29
               +   #   +    +   + #   +   +  #   +    +
```

Figure 3.72 Evolutionary relationships among AFP- and AFGP-containing fishes: evidence for convergent evolution. (A) Distribution of different classes of IBPs among lineages of bony fishes. Note the independent origins of AFPs of types I and II and AFGPs in two or more fish lineages. Dates (in millions of years before present) mark the origin of the lineages. (B) Partial alignment of sequences of the type I AFPs of cunner, sculpin, and flounder. Note the occurrence and spacing of the threonine (T) residues, which play critical roles in binding to ice. (After Graham et al. 2013.)

of IBP evolution have been taken. Nonetheless, **evolutionary convergence** clearly has occurred. When the different types of teleost IBPs are mapped onto a phylogeny of IBP-producing fishes (see Figure 3.72), it is immediately apparent that each type of IBP—with the exception of type III AFP, which is restricted to zoarcoid fish (ocean pouts, eelpouts, and wolffishes) (Deng et al. 2010)—has arisen more than once in different lineages (Graham et al. 2013).

When a particular type of IBP has evolved independently two or more times, it is pertinent to inquire if the same gene was exploited in all of the lineages that manifest this convergence in IBP structure. Or are some of the relatively simple fish IBP structures, for example, those of the alanine-rich type I AFPs and the notothenioid and cod AFGPs, readily generated de novo, again and again? Both evolutionary options have been used to produce the variation in IBPs we find in the contemporary fish fauna. In the case of the alanine-rich (> 50% of residues are alanine) type I AFPs, highly similar structures have arisen at least four times in fish as distantly related as flounders and sculpins (see Figure 3.72; Graham et al. 2013). All type I IBPs are α-helical and contain threonine and other hydrophilic residues at regular intervals along the IBP structure. These polar residues project from the side of the amphiphilic IBP structure. Recent sequencing work by Graham and colleagues (2013) has demonstrated that evolutionary convergence in the type I AFPs reflects multiple origins from nonhomologous genetic progenitors.

A different type of convergent evolution, **parallel evolutionary convergence**, is found in another type of fish IBP, the type II AFPs (Fletcher et al. 2001; Graham et al. 2008). These are globular proteins that are rich in cysteine residues. Type II AFPs, which are found in sea ravens, herrings, and smelt, have evolved from **C-type lectins**, proteins that bind to sugars in a Ca^{2+}-dependent manner. Retention of calcium dependence for THA is seen in herring and smelt type II AFPs. What is especially fascinating about this instance of convergence is that the spread of this type of AFP among the three groups of fishes possibly occurred by horizontal (lateral) gene transfer (HGT) (Graham et al. 2008). Two lines of evidence based on genetic relationships support this conjecture. First, the lectin-type genes that encode the type II antifreezes are not found in other fish lineages. Second, the close similarities in sequence and intron-exon structures of the type II AFPs in the three teleost lineages would not be expected if a progenitor gene in the last common ancestor of the three lineages led to the contemporary type II AFP structures. In other words, it is extremely difficult, for two reasons, to envision a scenario in which a deep ancestor of the three type II AFP-containing fish lineages had a gene for this type of IBP. First, it seems extremely improbable that subsequent evolution could have led to loss of this gene in all other lineages. Second, the remarkably high conservation of gene sequence and intron/exon structures in the lineages where the gene was retained makes HGT seem likely. Thus, the common reliance on a type II AFP in these diverse fishes appears not to be a result of repeated "invention" of a protein of this type, but rather an instance where a single origin event was followed by distribution of the IBP among the three lineages through horizontal exchange of DNA. The extent that HGT contributes to acquisition of IBP proteins in other lineages is not known. The discoveries made with type II AFPs may, in fact, be but the tip of the iceberg. Horizontal gene transfer of IBPs recently has been found in other evolutionary lineages. An IBP found in a sea ice crustacean may have originated in an algal species, although this observation requires further confirmation to unambiguously identify the source of the gene (Kiko 2010).

Yet another pattern of convergence is found in the AFGPs of Antarctic notothenioids and Arctic cods. The simple tripeptide repeat structure found in these molecules (AAT) arose from different progenitors in Antarctic and Arctic fishes. In notothenioids the AFGP gene evolved from a gene that encoded a trypsin-like protease (Chen et al. 1997; Cheng and Chen 1999). A fragment of the progenitor gene comprising the junction of intron 1 and part of exon 2 was duplicated many times, generating a gene that encoded the repetitive AAT sequence. Proteolytic sites were added to the gene, and multiple AFGPs are found within a single coding sequence; the gene thus encodes what is termed a **polyprotein**. Such polyprotein-coding genes allow high levels of expression of a given type of protein. An additional way of amplifying protein expression is to increase the number of copies of a gene through tandem gene duplication (see Chapter 1). As pointed out in Chapter 1, at least 118 protein-coding genes in notothenioid fishes have been increased in copy number by up to ~300 times, to facilitate adequate levels of expression of genes that play important roles in adaptation to near-freezing temperatures (Chen et al. 2008). Both notothenioid and cod IBPs have similar patterns of glycosylation, which involves addition of the disaccharide β-D-galactopyranosyl-(1→3)-2-acetamido-2-deoxy-α-D-galactopyranose to threonine residues (see Figure 3.68A). In notothenioids, AFGP synthesis occurs in specialized cells in the stomach and pancreas (Evans et al. 2011). The AFGPs pass through the pancreatic duct into the stomach and are subsequently absorbed into the blood. Arctic cods synthesize AFGPs in the liver and the pancreas (Duman 2015).

Fish type III AFPs, which have a predominantly β-sheet structure, also have their origin in an enzyme-coding gene. The enzyme sialic acid synthase (SAS) appears to be the progenitor of this category of antifreeze (Baardsnes and Davies 2001). Like the type II AFPs, type III AFPs originate from a protein that interacts with sugars and polysaccharides. Two isoforms of type III AFP have been identified in zoarcid fish (Nishimiya et al. 2005). One isoform lacks TH activity; the other possesses this function. Interestingly, addition of the TH-inactive antifreeze to a solution with the active isoform leads to enhancement of THA. Here, as in other instances of enhancement of IBP THA, complex formation between the two isoforms may lead to a larger effective IBP size and hence to increased THA.

Studies of type III AFPs have revealed another interesting facet of IBP evolution, which involves what is termed **escape from adaptive conflict (EAC)** (Deng et al. 2010). The ancestral SAS from which the first type III AFP evolved is thought to have had weak THA as well as its catalytic function in sialic acid synthesis. However, when only a single copy of the SAS-coding gene was present, there was an "adaptive conflict" based on different structural requirements for SAS and IBP activities that prevented optimization of both of the protein's functions. However, when gene duplication occurred, the two copies of the SAS-coding gene "escaped" from this adaptive conflict; it became possible for one gene to be optimized for SAS function and the other (which became *AFPIII*) for IBP function. Deng and colleagues (2010) have provided an elegant genetic analysis of this gene duplication process, which shows what types of changes were needed to permit the development of a highly effective IBP as well as further improvements in SAS enzymatic function. Nearly two-thirds of the *AFPIII* gene exhibits postduplication positive Darwinian selection, and the rate of accumulation of these adaptive mutations appears to have been rapid. The AFPIII protein also gained an extracellular secretory signal region, resulting in an IBP that could be transported from the site of intracellular synthesis to the body fluids where IBPs function. This, then, is an exceedingly good illustration of how development of a new function, **neofunctionalization**, occurs following gene duplication.

The origins and role(s) of type IV antifreezes identified in longhorn sculpin remain unclear (Duman 2015). This type of IBP is present in very low concentrations and consequently appears to lack biologically significant THA in vivo, although the expressed protein has THA.

In summary, the diversity of structures that enable organisms to control the formation and growth of ice reflects the innovative nature of biochemical adaptation and the pivotal role of exaptation—generating new functions for a variety of classes of proteins. Many different types of proteins are now known to have given rise to IBPs, and it seems certain that additional examples await discovery. As we come to understand what structural features are needed for a protein or glycoprotein to have ice-binding ability, we may more easily come to identify other members of this broad class of molecules.

Freezing relationships: A holistic analysis

A large fraction of the literature dealing with IBPs and other phenomena associated with control of ice formation in biological fluids comprises in vitro biochemical experimentation and molecular biological analysis of IBP structures and their evolutionary origins.

Whereas these studies have been very successful in achieving their aims, they may fall well short of accounting for the actual resistance to freezing that occurs in nature. For example, in vitro studies that monitor the growth of small ("seed") ice crystals to index the efficacy of IBPs may use crystals that are large relative to those that are found in organisms in nature. Because the potency of some IBPs to prevent ice growth is inversely proportional to seed crystal size (Duman 2015), in vitro experiments may give a misleadingly low value for THA. Furthermore, whereas experimentation in protein biochemistry commonly involves isolation of a single type of protein for detailed study, the solution in which the protein of interest occurs in vivo may contain solutes that have a strong influence on the protein's behavior (see Chapter 4). To obtain a biologically realistic estimate of the freeze-avoidance capacity of an organism, holistic studies must be designed that encompass the full set of behavioral, anatomical, and chemical influences on propensities for ice formation. Below we present selected examples of studies that illustrate the importance of this type of integrative experimentation.

ANTARCTIC NOTOTHENIOID FISHES   We have already discussed certain aspects of the freeze-avoidance strategies of Antarctic notothenioid fishes (DeVries and Cheng 2005). The discovery of AFGPs in these fishes by DeVries truly opened the door to the study of noncolligative antifreeze molecules and IBPs in general. Although the AFGPs of these fishes (and those of their Northern Hemisphere counterparts) are highly effective in conferring freeze-resistance on the organisms, these antifreezes do not operate in a vacuum; rather, other types of solutes can have a significant effect on their THA. The discovery that AFPPs can significantly increase the abilities of AFGPs to depress the temperature of ice crystal growth has already been discussed. In addition, small solutes may potentially assist in freeze-resistance. We have mentioned that blood of Antarctic notothenioids has a relatively high osmolality compared with that of most teleosts (Dobbs and DeVries 1975). Serum osmolalities range from approximately 550 to 650 mosmol l^{-1}, values that are approximately 200 mosmol l^{-1} higher than osmolalities of typical temperate-zone teleosts. In addition to having higher concentrations of inorganic ions like Na^+ and Cl^-, notothenioid blood contains high levels of trimethylamine-N-oxide (TMAO) (Raymond and DeVries 1998). Elevated TMAO has also been discovered in Arctic fishes (Raymond 1994, 1998). In these cold-adapted fishes, TMAO may play several roles, including its obvious colligative contribution to lowering of the freezing point. However, there is another intriguing possibility for TMAO function. TMAO is one of the strongest stabilizers of protein structure known; it also favors protein-protein interactions (see Chapter 4). It is thus conceivable that elevated levels of TMAO stabilize the native structures of IBPs and, perhaps more important for freeze-resistance, favor formation of complexes of multiple AFGPs. As noted earlier, the efficacy of IBPs in preventing growth of ice crystals is proportional to the size of the IBP molecule. The same would hold for aggregations of multiple IBPs (see discussion of insect IBPs below). Further insights into freeze-resistance in vivo are likely to be obtained from in vitro studies done with biologically realistic combinations of IBPs and the different macromolecular and low-molecular-mass solutes that influence their structures and functions.

To round out the notothenioid story, we must emphasize that despite the full spectrum of molecules that help prevent freezing, some ice forms in some parts of the fish. Thus, *freeze-avoidance* is an expression best used to describe whole-organism relationships only. Within the limited context of the blood (or hemolymph in invertebrates), freeze-tolerance applies: The presence of ice is tolerated because ice crystals do not attain lethally large sizes.

How do we know that an actively swimming and seemingly healthy polar fish contains ice? This was demonstrated by a study that compared the undercooling abilities of fish that were freshly collected with those of fish that had been held in the laboratory at higher temperatures that would be sufficient to melt any ice the fish had acquired in their habitats (DeVries and Cheng 2005). The presence of any ice crystals in the body fluids severely limits an organism's ability to undercool because the existing ice will rapidly induce further—and lethal—ice formation as temperatures fall. Fish brought in from the field and tested immediately had minimal undercooling ability; fish that had been held in the laboratory at higher temperatures (so that any ice in their fluids would have melted) were capable of deep undercooling. When these laboratory-acclimated fish were again placed in ambient-temperature seawater (–1.9°C) for a period of a few days and then tested for undercooling ability, they resembled field-collected specimens; they again had minimal ability to undercool. It is apparent, then, that Antarctic fishes normally contain ice crystals in their blood (and perhaps other fluids as well).

What becomes of these ice crystals? Are they a permanent, life-long feature of the organism's physiology, or is there a mechanism for melting these crystals? Studies have shown that ice crystals in the blood are coated with AFGP, and that the spleen likely removes the AFGP-coated ice crystals from the blood (Præbel et al. 2009); macrophages recognize the AFGP-coated ice as a foreign body and remove it from circulation (DeVries and Cheng 2005; Evans et al. 2011). It is not known, however, if processes in the spleen lead to actual melting of the ice that accumulates there. There is some reason to conjecture that this AFGP-coated ice remains throughout the fish's lifetime (perhaps of two decades). One of the challenges to removing this ice from the fish's system is the elevation of melting point that results from AFGP binding to ice (see Figure 3.71). Ice with this hysteretic melting point is referred to as **superheated ice** (Cziko et al. 2014). To melt the superheated AFGP-coated ice, temperatures have to be higher than the thermodynamic melting point; such temperatures rarely occur in waters as stably cold as those of McMurdo Sound (Cziko et al. 2014).

Last, in addition to extracellular antifreezes, some classes of IBPs occur intracellularly in certain fishes (Fletcher et al. 2001). These antifreeze proteins lack signal sequences and thus appear to be retained within the cells. IBPs inside skin cells may function as a "peripheral defense" in retarding or preventing ice propagation from the seawater to the internal fluids. Skin cell antifreezes in gill epithelia may be especially important in this regard because of the high surface area of gills and the likelihood that gills will contact small ice crystals in the surrounding water.

TERRESTRIAL ARTHROPODS   Terrestrial arthropods (insects, ticks, centipedes, spiders, collembola, and mites) found in cold climates provide some of the most striking examples of adaptations for coping with threats of freezing. Many of these organisms are freeze-tolerant and allow ice to form in their extracellular fluids. Again, we wish to emphasize that the terms *freeze-avoidance* and *freeze-tolerance* must be used in a nuanced way that recognizes the fact that even so-called freeze-avoiding species like polar fishes do tolerate small amounts of ice in their extracellular fluids. However, this ice does not prevent the fishes from remaining active at very cold temperatures and does not require them to enter a period of deep dormancy. In contrast, freeze-tolerant terrestrial species, both plants and animals, generally allow a substantial amount of ice formation to occur in their extracellular fluids, such that, in the case of animals, continued locomotory activity is likely to be precluded, and metabolic function is reduced to a very low level. Whereas this general picture is true for a wide variety of taxa, there is a diversity of responses to threats of freezing among species and, for a given species, during the course of the cold season.

Among insects we also find a rare example of tolerance of entry into the solid state. The beetle *Cucujus clavipes clavipes* avoids freezing, thanks to a combination of colligative mechanisms and TH from IBPs (THA in deep winter is approximately −13°C; Duman 2015), but allows its intra- and extracellular water to solidify in a vitreous state at low extremes of temperature. Thus *C. clavipes clavipes* is, strictly speaking, not freeze-tolerant; rather, if temperatures are extremely low (~−60°C or below) it is **vitrification-tolerant**.

Here we discuss strategies characteristic of both freeze-avoiding and freeze-tolerant insects, where a variety of behavioral, anatomical, and biochemical adaptations permit survival during extreme winter cold. Studies of the beetle *Dendroides canadensis* have provided important insights into the roles of behavior, anatomy, ice-binding proteins, ice nucleators, and small solutes like glycerol and trehalose in defining the freeze-avoidance and freeze-tolerance relationships that occur in terrestrial invertebrates. This freeze-avoiding species has a complex strategy for coping with threats of ice formation in its bodily fluids (Duman 2015). In autumn, habitats for overwintering are selected where the presence of ice crystals is less likely to occur. As temperatures begin to decrease, larvae (the overwintering life stage) begin to accumulate glycerol. They also cease feeding and clear their guts of bacteria that may trigger (nucleate) ice formation. Antifreeze proteins are produced as well on a seasonal basis, with synthesis upregulated in response to decreasing photoperiod and falling temperatures. In *D. canadensis* a complex suite of IBPs is produced. Approximately 30 IBPs are synthesized in a tissue-specific manner. These antifreezes function synergistically by enhancing each other's activities (Wang and Duman 2005). A thaumatin-like protein previously known only in plants also enhances THA, and this interaction is enhanced by the presence of glycerol (Wang and Duman 2006). The ice-binding glycolipids found in this species also enhance the THA of IBPs, as do citrate and some inorganic ions found in hemolymph (Li et al. 1998). The IBPs found in hemolymph also assist in undercooling through binding to INPs and inhibiting their activities (Duman 2001, 2002). The inhibition of INPs by IBPs is enhanced by the high concentrations of glycerol found in the hemolymph. As mentioned, glycerol enhances protein-protein interactions. IBPs also interact with the low-molecular-mass organic molecule trehalose, which may accumulate to levels near 0.09 mol l^{-1} in *D. canadensis* during winter (Wen et al. 2016). Wen and colleagues showed that these high levels of trehalose enhanced antifreeze activity of the insect's IBPs. In addition, the IBPs had significant effects on trehalose by inhibiting its precipitation at low temperatures. Trehalose tends to crystallize in the cold, which raises the threat of precipitation of trehalose crystals in the hemolymph of overwintering beetles. The IBPs of *D. canadensis* bind to the surfaces of small trehalose crystals and thereby inhibit their further growth. This is a fascinating example of a macromolecular-micromolecular interaction in which effects of the two components on each other afford multiple benefits to the organism.

Hemolymph IBPs also inhibit freezing caused by propagation of ice from the external environment into the organism, a phenomenon known as **inoculative freezing**. Pores in the insect cuticle are very tiny, so only small ice crystals are likely to propagate freezing of the organism. As mentioned earlier, the efficacy of IBPs is inversely proportional to ice crystal size, so hemolymph IBPs are very effective at preventing the tiny ice crystals that do make it through the cuticle from propagating further. This relationship fosters an effective "peripheral defense" at the cuticular surface. Gut IBPs inhibit ice-nucleating activities of any microbes that might still be present. Through these diverse types of adaptations, the undercooling points of *D. canadensis* larvae fall from approximately

–6°C in summer to –18°C to –26°C in winter, depending on the severity of the winter (Duman 2001). As seen for *C. clavipes*, the vitrification-tolerant beetle, populations of *D. canadensis* from different latitudes exhibit a wide range of capacities for undercooling and avoiding ice formation in winter.

In some terrestrial arthropods, the very act of going into diapause may contribute to freeze-avoidance. The large slow-down of metabolism that accompanies diapause leads to a corresponding decrease in cellular demand for substrates supplied by the circulation. This decreased need for catabolic substrates may facilitate resistance to freezing by reducing threats of ice nucleation that might arise from substances circulating in the hemolymph. Thus, in the stag beetle *Ceruchus piceus*, one facet of entry into the diapause state is the removal from hemolymph of lipid-transporting lipoproteins that have ice-nucleating activity and could hinder undercooling (Neven et al. 1986).

Freeze-tolerant terrestrial arthropods employ many of the mechanisms just described for *D. canadensis*, but nonetheless ice forms in their hemolymph. The IBPs that these arthropods produce have low THA and function principally as inhibitors of recrystallization (Duman 2015; see Figure 3.69). Some IBPs and ice-binding glycolipids occur on membranes and function to prevent propagation of intracellular freezing after ice formation in the hemolymph is initiated. Hemolymph freezing is closely regulated in freeze-tolerant species. Many species produce INPs that trigger ice formation at relatively high temperatures, to avoid the serious and perhaps lethal damage that ice formed from deeply undercooled water could cause.

In summary, the relationships seen among water, small solutes, and ice-binding proteins and glycoproteins in freeze-avoiding and freeze-tolerant species illustrate several important themes that are important for understanding biochemical adaptation—and for designing experimental procedures appropriate for gaining this understanding. First, evolutionary processes reflect striking instances of convergence; in many cases, different lineages "discover" a common means for solving a given type of environmental challenge. Second, acquisition of a biochemical trait important for survival under a particular set of environmental circumstances may entail "inventing" the trait oneself—here exaptation is common—or acquiring it from other, already adapted species through horizontal (lateral) gene transfer. Third, adaptation involves a collaborative effect among many different cellular constituents. This point is made very well by the enhancement of IBP activity by other proteins and by small solutes, which may exert their effects by enhancing protein-protein interactions. And last, the aqueous background in which these adaptations play out is an important part of the overall picture. The structures of water differ greatly in their physical properties and effects on cells and macromolecules. Freezing relationships are, of course, only one facet of water's pervasive roles in biology, and we now turn to a broader analysis of the many roles of water. We will show how the optimal functional and structural properties of macromolecules and large molecular assemblages depend strongly on the maintenance of a hospitable environment in the aqueous cellular solution. Water, which is too often neglected by biochemists, remains on center stage in the drama of adaptation, even if ice formation is not a threat.

CHAPTER 4

Water and Solutes: Evolution and Regulation of Biological Solutions

Liquid water is not a bit player in the theatre of life — it's the headline act.

Martin Chaplin

Biochemical research usually tends to focus most strongly on the least abundant constituents of the cell rather than on the types of molecules that are far and away the most numerous. Thus, for example, whereas there are only a relatively small number of molecules of DNA per cell (only one may be present in an archaeal or bacterial cell), the intensity of investigation and the level of research funding focused on these relatively rare macromolecules vastly exceed the effort and the support for research on the most abundant molecules, the small solutes and water that constitute the solution in which the macromolecules work. Whether one views this choice of focus as wise, naïve, or just ironic, biochemists tend to think "big" rather than "small" when they choose their targets for research. The same is true of molecular evolutionists interested in protein evolution, who tend to focus on amino acid substitutions as the driver of functional diversification at the protein level. As we show below, a broader and more holistic focus is needed if we are to understand the important interplay—during evolution and in short-term physiological regulation—between macromolecules like DNA, RNA, and proteins and the host of small molecules, including water, inorganic ions, and organic solutes, in the solutions that bathe them.

4.1 Establishing a "Fit" Solution for Life: Water and Micromolecules

Water is one of the most abundant molecules in the Universe (after H_2, protonated hydrogen [H_3^+], and CO; Scharf 2014) and is by far the most abundant molecule found in living systems. It occurs at ~55 moles per liter in active, nondesiccated cells, and its abundance in biological solutions is two orders of magnitude greater than the sum of all other molecules combined. However, water is commonly taken for granted as the "background" for biochemistry. As Gerstein and Levitt (1998) put it, "When scientists publish models of biological molecules in journals, they usually draw their models in bright colors and place them against a plain, black background. We now know that the background in which these molecules exist—water—is just as important as they are." Water is far from being a bit player in the drama of life. As Chaplin emphasizes in the quote above, when we fully grasp the key roles of water in contemporary organisms and come to more fully appreciate the critical influence that water has had during the evolutionary development of all biochemical systems, this small molecule must indeed be viewed as "the headline act."

This chapter shines a spotlight on water's broad evolutionary imprint on biological structures and processes. Rather than serving just as the background for the macro-molecules that carry out their functions of information transfer, transport, catalysis, and fabrication of biological structures, water has played a pivotal role in influencing the evolution of macromolecules and shaping the complex solute systems we find in con-temporary cells.

We begin by examining some of water's unique properties that suit it for serving as the medium of life (see the prescient text of Henderson [1913], *The Fitness of the Environment*, for what appears to be the first analysis of water's "fitness" for life). Then we examine the diverse small solutes—the **micromolecules**—that constitute the domi-nant inorganic and organic constituents of biological solutions. We will discover that, in conjunction with water, these small organic and inorganic solutes provide a suitable milieu for supporting the structure and function of macromolecules and large molecular assemblages such as membranes and nucleosomes. A key point we will discover when we examine the interactions among water, micromolecules, and macromolecules is that biological solutions are closely regulated to establish an optimal environment for macro-molecular structure and function. Furthermore, the critical influences of micromolecules on macromolecules often stem from the manners in which both classes of molecules interact with water. Suffice it to say, an adequate understanding of biochemistry and biochemical evolution demands a holistic focus that includes all constituents of the cellular fluids—water and the large and small molecules it contains—and the diverse interactions that take place among them.

4.2 What Properties of Water Make It So "Fit" for Life?

Water's varied and essential roles in biological systems stem from its unique physical and chemical properties and from the manners in which these properties are influenced by factors such as changes in temperature, hydrostatic pressure, and the solutes pres-ent in solution. Water assumes a variety of structures in the liquid state and under-goes transitions between liquid, solid, and gaseous phases in response to changes in temperature. Like liquid water, solid-state water exists in a variety of states that have different effects on biological systems. Water has an unusually low compressibility com-pared with many liquids, but even its slight compressibility may bear significance for extremely deep-dwelling organisms that experience hydrostatic pressures up to ~1100 atmospheres (1 atm = 0.101 megapascal [MPa]). Temperature and pressure also affect the integrity of the water molecule itself. Water dissociates into protons (H^+) and hydroxyl ions (OH^-), and this dissociation reaction is enhanced by increases in temperature and hydrostatic pressure. Thus, the pH of pure water at 25°C and 1 atm pressure, ~7.0, is by no means the "normal" pH value of aqueous solutions, as naïve textbook treatments of pH may sometimes imply. Rather, pH values of aqueous solutions fall regularly with increases in temperature or pressure. The pH values of biological fluids likewise vary with temperature (much less is known about pressure-pH relationships of cellular and extra-cellular fluids). These temperature-dependent values of pH, which might at first glance seem to represent challenges to biological function, in fact will be shown to occur in a manner that conserves many essential biochemical properties of the cell. Lastly, as we will see in considerable detail throughout this chapter, both polar and nonpolar solutes interact with water and affect its structure in complex and biologically important ways. As we will repeatedly emphasize, solutes' effects on water structure play important roles in evolutionary selection for the appropriate compositions of biological solutions.

An outstanding and remarkably comprehensive website that is maintained and regularly updated by Professor Martin Chaplin (see www.lsbu.ac.uk/water/index2.html) provides an abundance of additional information on the many facets of water; it is a highly recommended source for obtaining an in-depth understanding of this remarkable molecule's diverse properties and roles in biology. Below we discuss several of these critical properties to lay the foundation for the analyses that follow of the complementary evolutionary processes that shape the properties of macromolecules and those of the solutions that bathe them.

Water's polarity and hydrogen-bonding capacities

Many of water's most biologically important properties stem from its polarity. As illustrated in **Figure 4.1**, the negative poise of water's oxygen and positive poise of its two hydrogens lead to a strongly polar molecule that is very effective in establishing

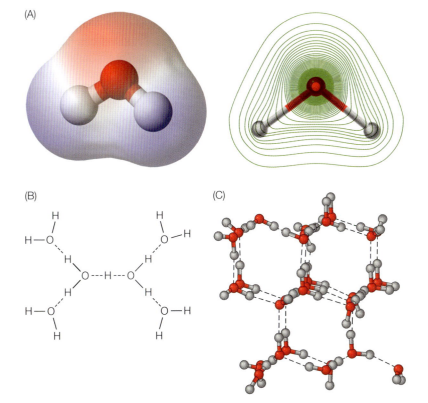

Figure 4.1 Structure of water. (A) The charge distributions around the oxygen and hydrogen atoms of a single water molecule. Left: Red indicates negative charge contributed by oxygen, white is neutral, and blue is positive charge due to hydrogen. Positive, negative, and neutral charges represent 37%, 33%, and 29%, respectively, of the total surface area of a water molecule (~30 Å^2, assuming a sphere of radius of ~1.5 Å) (Street et al. 2006). Right: The negative charge density is indicated by the green contour lines. (B) A hydrogen-bonding network among water molecules. The dashed lines represent hydrogen bonding; the solid lines represent the covalent bonds linking O and H atoms. (C) A cluster of water molecules (O is red; H is white) with hydrogen bonding among them.

a latticework of hydrogen-bonding interactions in solid and liquid states. In pure liquid water, each water molecule is, on average, hydrogen bonded to four other waters (Chandler 2005). Water's remarkable capacity as a solvent stems in great measure from its ability to form hydrogen bonds with charged and polar solutes. When solutes are added to water, reorganization of water's hydrogen-bonding networks occurs. Different solutes have dramatically different effects on water structure, effects that are critical for selection of an appropriate solute composition for biological solutions. How water is structured around solutes determines the effects of these solutes on protein stability, for example. When nonpolar solutes—molecules without capacities for hydrogen bonding to water—are introduced to an aqueous phase, the hydrogen-bonding networks among water molecules again are likely to change. A very small nonpolar (hydrophobic) solute may not reduce the number of water-water hydrogen bonds, but a larger hydrophobic molecule—or nonpolar group on an **amphipathic** molecule (one with both polar and nonpolar components)—may reduce the number of water-water hydrogen bonds to three or fewer per water molecule (Chandler 2005). Whether in pure water or in a complex biological solution, the hydrogen bonds linking water molecules are ephemeral. They form and break rapidly and continuously due to their low bond energies, which are of the same order of magnitude as the thermal energy of the system at most biological temperatures. At the highest temperatures at which organisms can survive when hydrated (somewhere near 130°C, which are temperatures at which culturing must be done at elevated hydrostatic pressure to ensure a liquid state of solutions), the ability of water molecules to form hydrogen-bonding networks is vastly reduced, with important consequences for solute-water interactions and macromolecular stability (see Chapter 3).

As essential as water is for life under almost all environmental conditions, survival under certain environmental extremes may require removal of virtually all cellular water, creating a state of **anhydrobiosis** ("life without water"). Thus, whereas active life depends absolutely on an abundance of water, quiescent life stages like spores, cysts, and seeds may survive only under conditions of a near absence of cellular water. In Section 4.7, we treat in detail the mechanisms of adaptation by which fully desiccated cells survive, often for prolonged periods (up to many centuries). The adaptations that allow anhydrobiosis commonly involve accumulation of "water substitutes" that themselves have strong hydrogen-bonding capabilities, which allow them to replace water in several contexts in which networks of hydrogen bonds are key to the stability of biological structures. Hydrogen-bonding networks, then, are essential elements in cellular structure and function, whether these weak bonds arise from water or from small organic molecules that can replace the "universal solvent" under conditions of environmental extremes.

Water has multiple structures in both solid and liquid states

It is customary to speak of three physical states for water: solid, liquid, and gaseous. Key differences among these three general states are the strengths and amounts of interactions among water molecules, which lead to differences in their physical spacing (**Figure 4.2**). Within the liquid and solid states of water lie many layers of fine-scale complexity. Below we examine aspects of this complexity that play important roles in a variety of biological processes.

SOLID-STATE WATER: ICE AND VITREOUS WATER   In the solid state, water may assume several different types of icelike structures (see Figure 4.2). Formation of ice invariably leads to an expansion in water volume. Solid-state water may also exist in a glasslike (vitreous) state. The rate of cooling is instrumental in determining whether water enters

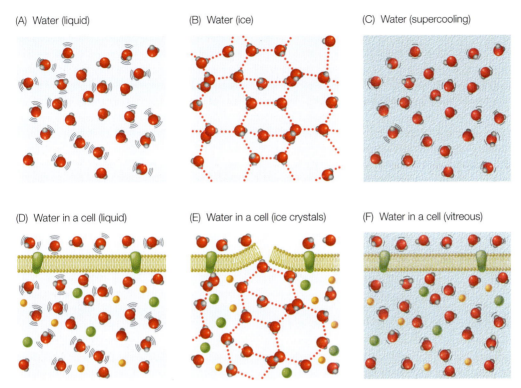

(A) Water (liquid)

(B) Water (ice)

(C) Water (supercooling)

(D) Water in a cell (liquid)

(E) Water in a cell (ice crystals)

(F) Water in a cell (vitreous)

Figure 4.2 Physical states of pure water (no solutes present; A–C) and of water in a cell (cytoplasmic solutes present; D–F). Small solutes in the cytoplasm are indicated by yellow and green dots. Note how the spacing among water molecules differs between the normal liquid state (A,D), ice (B,E), supercooled water (C), and vitreous water (F). The expansion of water during ice formation disrupts the cell membrane (E) and causes denaturation of proteins (not shown). Formation of vitreous water does not increase water volume. Many types of cells can exist in a vitrified state, which is often favored by the presence of certain types of organic micromolecules.

an icelike or vitreous state and whether water becomes supercooled (undercooled)—remaining liquid at a temperature below the thermodynamic freezing point (near 0°C for pure water at 1 atm). Some solutes tend to favor vitrification of solutions, as discussed in Section 4.7. Unlike what happens with the several different forms of ice, formation of **vitreous water** does not lead to an expansion in volume. The implications of this difference between ice and vitreous water are critical in the context of freeze-tolerance of organisms in nature (see Chapter 3), and in the biomedical realm where cryopreservation methods are of importance. Vitrification of aqueous solutions is also of great significance under conditions of desiccation, when cellular organization must be maintained under limiting conditions of solvent capacity.

WATER'S LIQUID STATES ARE INFLUENCED BY MICROMOLECULES AND MACROMOLECULES Liquid-state water also occurs in a range of distinct structural forms (see Figure 4.2). These varied types of water-water interactions often are a consequence of the types of micromolecules with which water interacts and the types of protein and membrane surfaces that are present in the immediate vicinity of a cluster of water molecules. Water near a polar surface may take on a structure that differs from the structure assumed by water near a nonpolar surface. Micromolecules differ strikingly in their effects on the hydrogen bonding taking place among water molecules. Some small solutes foster increased hydrogen bonding among the water molecules that immediately surround them. These solutes hydrogen bond strongly to water and thus exist as well-hydrated solutes. As discussed in more detail in Section 4.5, the structured water clustered around certain types of solutes may prevent them from penetrating the highly structured solvation layers around macromolecular surfaces (Street et al. 2006). This exclusion of solutes from a protein's surface, or to put it another way, the **preferential hydration** of macromolecules, is of central importance in protein stabilization (Bolen 2004). To a large degree, the extent to which a solute favors preferential hydration determines how effectively it stabilizes protein structure. Other micromolecules disrupt hydrogen bonding among water molecules. These solutes may interact strongly with the hydrated surfaces of proteins, fostering reductions in stability.

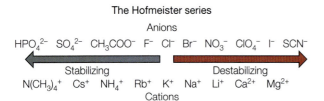

The Hofmeister series

The extent to which solutes stabilize or destabilize proteins is indicated by their positions in what is termed the **Hofmeister series**. This ranking of solute effects on protein solubility and stability was discovered in the late nineteenth century by the Czech-born chemist Franz Hofmeister. We will have much more to say about the Hofmeister series in conjunction with our discussion of mechanisms of protein stabilization and the effects of ions on the hydration states of proteins during their function. (For a masterful review of this topic, see von Hippel and Schleich 1969.)

Attaining the right balance between stabilizing and destabilizing effects of solutes on proteins is a primary outcome of the selective processes that have driven the evolution of biological solutions. One can view this aspect of micromolecular evolution as a process that is complementary to protein evolution per se. Thus, a protein's stability in vivo is due to its intrinsic stability, which is established by its amino acid sequence, and the effects of extrinsic stabilizers or destabilizers of protein structure. As discussed in Chapters 1 and 3, a state of marginal stability is necessary for proteins to function best. Their stabilities must be neither too great, so as to impede conformational changes, nor too low, so as to favor an unfolded and dysfunctional state of the protein. The marginally stable states that proteins gain during evolution through changes in amino acid sequence can be defended by adjusting the net stabilization potential created by the diverse set of micromolecules present in the solution. Because proteins have such low net stabilization free energies, this balance is a delicate one that also is influenced by the physical conditions the cell experiences, notably temperature and hydrostatic pressure. These two physical factors have strong effects on the intrinsic stabilities of proteins, and these effects may be offset through compensatory adjustments in the concentrations of micromolecular protein stabilizers. In conclusion, the numerous selective criteria that underlie the observed osmolyte compositions of biological solutions provide fascinating stories in biochemical evolution and form the chief focal point of this chapter. Evolution of the "small" things of the cell may often be neglected by evolutionary biologists, yet this is an extraordinarily important aspect of biochemical evolution. Indeed, macromolecular and micromolecular evolution are co-occurring and complementary processes.

Water's thermal physics: Heat capacity, density, and heats of vaporization and fusion

The occurrence of extensive hydrogen bonding among water molecules is manifested in several of water's physical properties, notably its thermal responses. Water has a high **heat capacity**; the increase in temperature that occurs as heat is added to an aqueous solution is small relative to many other liquids. This high heat capacity tends to buffer the temperatures of bodies of water and bodies of organisms, whether terrestrial or aquatic. Water's high **latent heat of vaporization** permits extensive cooling of terrestrial organisms when water transitions from the liquid to gaseous state, and in conjunction with a high heat capacity, water's latent heat of vaporization also tends to reduce risks of dehydration. Water exhibits a substantial **latent heat of fusion**. Thus, as water transitions from the liquid to the solid phase, considerable heat is released. The **density** of water is temperature-dependent in complex and biologically important ways. Pure water attains maximum density near 4°C at a pressure of 1 atm. Thus, as water cools below its temperature of maximum density and enters an icelike state, it expands, leading to a decrease in density (see Figure 4.2). As we all know, this density increase leads to the formation of ice on the surface of bodies of water, allowing aquatic organisms to persist in a liquid medium at depths below the ice layer. Surface ice has wide-ranging effects on the biosphere. The high reflectivity of ice causes more than 95% of incident solar radiation to be reflected back into space, helping maintain planetary temperatures. Global warming and loss of ice and snow cover are causing a higher fraction of solar radiation to be absorbed, leading to a positive feedback effect: even more rapid warming of water and land. The expansion that occurs as water is brought to its freezing point has another critically important effect on biology: Formation of ice in cells almost invariably leads to disruption of cellular structures (see Figure 4.2; also see Chapter 3). Notably, membrane integrity is lost and solute gradients are dissipated, leading almost inevitably to cellular death. Other thermal responses of water include effects on solvent capacity and rates of diffusion (including self-diffusion of water and solute diffusion). All of these thermal responses of water have wide-ranging impacts on organisms—impacts whose scope and severity are manifested by a wide range of adaptations.

Water's dissociation into protons and hydroxyl ions

At 25°C, pure water, which has a molarity of 55.56, contains H^+ (more precisely, the hydronium ion, H_3O^+) and OH^- each at a concentration of 10^{-7} molar. The miniscule fraction of water molecules that are dissociated into protons and hydroxyl ions at a given time belies the importance of this dissociation reaction in a vast range of biological processes. For example, removal or addition of protons occurs in numerous enzymatic transformations. In addition to serving as substrates for many reactions, protons also play important roles in regulating enzymatic activity. Protons are also critical in cellular energetics involving the electron transport system. Pumping of protons from the mitochondrial matrix into the intermembrane space generates a pH gradient of ~0.55 units; this electrochemical gradient is dissipated as protons reenter the matrix through the ATP synthase system, which drives production of ATP (see Chapter 2). Movement of protons within the aqueous phase also merits consideration. What may not be commonly appreciated in water-proton interactions is that movement of protons in a solution is rapid due to the ease with which H_2O and H_3O^+ can exchange a proton. The rapid movement of protons through an aqueous solution works to reduce proton gradients and prevent a buildup of protons at their sites of production.

Certain cellular compartments have pH values that are significantly lower than the average pH of the cytosol, and these low pH values have been selected to optimize the activity of the compartment in question. For example, lysosomes, which are sites of protein degradation, have low pH values that enhance protein degradation by favoring unfolding (denaturation) of proteins, making them better substrates for protein-digesting enzymes (proteases). In addition, proteases found in the lysosome have relatively low pH optima compared with most cytosolic proteins, which tends to optimize the function of these proteases in the lysosome environment.

Some of the most biologically important aspects of pH regulation stem from another fundamental property of water, the temperature dependence of its dissociation. As the temperature of an aqueous solution increases, more water molecules dissociate to form OH^- and H_3O^+. One way to characterize this relationship is to say that the neutral pH of water (pN)—the pH at which equal concentrations of OH^- and H_3O^+ are present—decreases with increasing temperature. The slope of the relationship between temperature and pN is not quite linear, but over the temperature range of ~0°C–100°C the slope is about –0.015 pH units per degree Celsius (see Section 4.6). Importantly, this is the same pH-versus-temperature relationship that characterizes biological fluids, at least in metabolically active organisms. We will see why this temperature dependence of extra- and intracellular pH is essential for biochemical function. Deviation from this relationship may severely perturb a wide spectrum of physiological and biochemical activities. However, under some conditions such as large-scale reduction in metabolic processes during dormant periods, regulating pH to be slightly acidic for the temperatures in question can provide the cell or organism with a "rheostat" for appropriately adjusting metabolic flux.

In conclusion, water's dissociation into protons and hydroxyl ions and the many ways in which proton concentrations are used to modulate biochemical activities provide another set of fascinating stories in biochemical adaptation and physiological regulation. In Chapter 5 we will broaden our analysis of these pH relationships by examining how anthropogenic addition of CO_2 to Earth's atmosphere and waters is reducing pH values in aqueous habitats. Ocean acidification is generating a diverse suite of influences, mostly negative, on biological structures and physiological processes.

Solvent capacity, solute reactivity, and cellular crowding: The cell's "packaging problems"

Water is commonly regarded as "the universal solvent," a designation that may suggest that water readily accommodates all of the different solutes needed for life, without any problem. Whereas water has the ability to dissolve an enormous range of chemical species, significant problems do emerge when we examine the challenges of accommodating, in a tiny volume of cellular water, the enormous diversity of solutes that are involved in intermediary metabolism, volume regulation, and defenses against stress from physical and chemical factors. Indeed, there are at least three fundamental sets of "packaging" problems that evolution needed to solve to enable a complex cellular solution and metabolic biochemistry to develop: (1) how to accommodate a wide range of small molecules, including all of the many substrates needed for intermediary metabolism and biosynthesis, in a small volume of water; (2) how to keep highly reactive metabolites at "safe" levels that minimize damaging chemical reactions not under close metabolic regulation (Atkinson 1969); and (3) how to reduce the probability of forming inappropriate aggregations of proteins in the "crowded" intracellular milieu, where interactions between both native and unfolded proteins are unavoidable (Fulton 1982; Levy et al. 2012).

SOLVENT CAPACITY   The first of these problems involves solvent capacity per se, namely ensuring that the full diversity of small organic solutes and inorganic ions needed by the cell under its particular set of physiological and environmental circumstances can be accommodated in the cellular water. Because of the need to package so many different types of small organic molecules and inorganic ions into a small volume of cellular water, the concentrations of individual solutes generally must be kept at low values, if the solvent capacity of the cellular water is not to be exceeded. It follows from this constraint on permissible solute concentrations that macromolecules like enzymes must be able to bind their ligands effectively at the very low concentrations, often 10^{-4} to 10^{-6} mol l^{-1} and below, at which almost all organic ligands are found in the cellular water. Thus, a tight pairing between ligand affinity (binding ability) and ligand concentrations has developed during protein evolution. Because binding ability is easily perturbed by physical and chemical stressors, sustaining this tight relationship between binding capacity and available ligand concentration is an important challenge in biochemical adaptation, as addressed in several contexts in this volume, notably in adaptation of enzymes to temperature (see Chapter 3).

A closely related solvent capacity issue concerns dissolved gasses. Solubilities of dissolved gasses, for example, oxygen, are influenced by the solute concentration and temperature of a solution: Increases in osmolality and temperature tend to drive gasses out of solution. Thus, to ensure that the mitochondria have access to sufficient amounts of oxygen to support required rates of aerobic ATP production, provisions have been made to enhance the oxygen-holding and diffusion properties of cells. Oxygen-binding proteins are present in many tissues of animals; myoglobins in muscle and neuroglobins in neural tissue help facilitate storage and transport of oxygen. Oxygen's relatively high solubility in lipids, which exceeds its solubility in water, is also reflected in cellular evolution: High lipid contents often assist in provision of oxygen to the mitochondria through elevating cellular oxygen concentration and increasing the rate of oxygen diffusion (Sidell 1998). As discussed in Chapter 3, the elevated tissue lipid levels found in some ectotherms adapted or acclimated to cold may represent a strategy for enhancing oxygen transport to mitochondria at low temperatures (Sidell 1998).

PACKAGING REACTIVE METABOLITES   A second packaging problem relates to the chemical reactivity of certain metabolites (Atkinson 1969). Unless suitably modified to allow them to occur in "safe" forms in the cell, certain organic compounds could pose threats to the integrity of the cell. An example is acetate ion. Acetic acid is a relatively weak acid, but the need for its anion, acetate, might require a concentration of this acid/anion that would threaten cellular integrity. In the cell, acetate ion is commonly present as acetyl-CoA. In this modified form, the free energy change that occurs during an acetylation reaction (the addition of an acetate group to another molecule) is negative enough to ensure that these reactions are thermodynamically feasible at low substrate (acetyl-CoA rather than acetate) concentrations. Free acetate would need to be maintained at much higher concentrations, based on the free energy change of the reaction. Acetyl-CoA thus provides a nice illustration of how substrate molecules evolve to allow metabolism to occur at exceptionally low intracellular substrate concentrations (see Atkinson 1969 for a lucid treatment of evolution of potentially "hazardous" chemicals).

MOLECULAR CROWDING   The third challenge that arises from packing so many different types of molecules into a small volume of water concerns avoidance of maladaptive interactions among proteins—the sticking together of proteins as a consequence of the macromolecular **crowding** that characterizes the cellular fluids (Fulton 1982). Whereas metabolic pathways that involve a series of enzyme-catalyzed reactions or multistep regulatory networks that, for example, transduce a membrane binding event to a change in gene expression commonly involve protein-protein interactions, these physiologically important interactions among proteins are clearly consequences of natural selection and involve highly specific, high-affinity protein-protein binding events. They are the opposite of random aggregations of proteins with unrelated functions. However, because of the occurrence of thousands of different proteins in the cell and a total protein concentration that may reach ~50% by mass in the mitochondrial matrix (Lin et al. 2002), there would seem to be a high probability that nonspecific ("promiscuous") binding among proteins would occur, leading to formation of protein aggregates that lack function and that could be damaging to the cell.

How is this promiscuity avoided? One mechanism for preventing inappropriate aggregations is to reduce the intrinsic "stickiness" of protein surfaces that are solvent-exposed but do not serve a role in specific protein-protein binding events. Levy and colleagues (Levy et al. 2012) investigated the amino acid compositions of solvent-exposed regions of proteins in bacterial (*Escherichia coli*), yeast (*Saccharomyces cerevisiae*), and human cells. They ranked the 20 amino acids in terms of stickiness and used this index to examine hundreds of different proteins for which three-dimensional structures were available. They discovered that the propensity for nonspecific protein-protein interactions involving solvent-exposed regions not involved in normal protein-protein interactions was inversely correlated with the abundance of the proteins. Thus, proteins that occur at high concentrations have a lower tendency for nonspecific aggregation with other proteins than do proteins that occur at low concentrations. Interestingly, in light of the high protein concentration of the mitochondrial matrix, mitochondrial proteins have a relatively low stickiness compared with proteins of other cellular compartments (Levy et al. 2012). Stickiness was also linked to protein size: Small proteins, which may face challenges in stability, have stickier interiors than large proteins. In addition, the regions of the protein surface that have evolved to possess low stickiness are also highly conserved in sequence. This is an interesting and perhaps surprising finding in view of the fact that these potentially sticky regions are not involved in specific biochemical functions like ligand binding or protein assembly. However, these highly conserved "nonfunctional" regions of proteins do emphasize the importance of minimizing promiscuous binding among proteins. This point was made earlier (see Chapter 3) in the context of molecular chaperones and their capacities for reducing the probability of inappropriate interactions between proteins with complementary geometries and a potentially high attractiveness to each other. In that discussion we emphasized the importance of minimizing nonspecific interactions between solvent-exposed hydrophobic groups that, in the native protein, are largely buried in the protein interior. Preventing these interactions between hydrophobic groups on newly synthesized—but not yet fully folded—proteins or between denatured proteins in which hydrophobic groups have moved from the oily interior of the protein to the protein's surface is a reflection of another of water's important roles, the driving of native macromolecular structures.

Water thermodynamically drives the formation of higher orders of macromolecular structure and assembly

Far from being an inert "background" medium that merely accommodates a diverse suite of macromolecules, water plays a fundamental role in driving the folding and assembly processes that characterize the functional three-dimensional configurations and assembly states of proteins and nucleic acids. The **hydrophobic effect**, defined as the tendency of nonpolar molecules (or parts thereof) to move from contact with water to a buried state where their contact with the aqueous phase is minimized or eliminated, accounts for much of the driving force involved in the folding of proteins into their native conformations, the assembly of multiprotein complexes, formation of the double-helical structure of DNA, and the generation of cellular membranes (Chandler 2005). The hydrophobic effect essentially reflects the rule that "oil and water do not mix." This is an oversimplified view, as we will see, but it does provide an appropriate starting point for examining how the "oil" and "water" of the cell attain their distributions and how these distributions are affected by other physical and chemical factors, including temperature and solute composition.

One of the many important aspects of the hydrophobic effect is its dependence on temperature. Whereas other types of noncovalent ("weak") bonds (see Chapter 3) are destabilized by increases in temperature (they form exothermically), burial of nonpolar groups requires input of heat—an endothermic process. An input of thermal energy is required because hydrophobic groups—for example, nonpolar side chains of the amino acids leucine, isoleucine, and valine—are surrounded by clusters of organized water molecules when in an aqueous phase (**Figure 4.3**). To accommodate the nonpolar side chains in an aqueous solution, even though these side chains are not truly dissolved in the same sense as, for example, a polar solute like glucose or an inorganic ion like K^+, water must alter its structure. This change in organization of water to accommodate nonpolar groups is exothermic. Thus, to reverse the process, to "melt" the array ("cage") of water molecules that surrounds a nonpolar group, heat must be added. Nonetheless, despite being characterized by a positive enthalpy change ($+\Delta H$), the burial of nonpolar side chains at biological temperatures is exergonic (negative free energy change [ΔG]) because of the substantial change in entropy (ΔS) that occurs when the organized water is dispersed into the bulk water (see Chapter 3). Due to the positive enthalpy change involved in forming hydrophobic interactions, protein stability may be compromised at low temperatures, much as it is challenged at high temperatures due to the exothermic nature of processes associated with forming hydrogen bonds and electrostatic interactions. Protein-water relationships thus affect protein stability across the full range of biological temperatures, with different classes of weak bonds playing temperature-specific roles in stabilization processes.

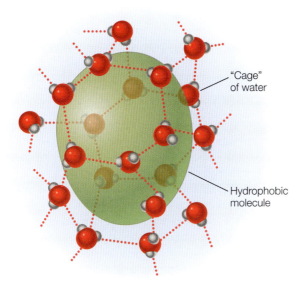

"Cage" of water

Hydrophobic molecule

Figure 4.3 Hydrophobic interactions between nonpolar solutes and water. An organized "cage" of water surrounds a hydrophobic molecule or nonpolar component of a molecule (e.g., a nonpolar amino acid side chain). The clusters of water molecules that surround a hydrophobic molecule remain hydrogen bonded to each other, but larger hydrophobic groups permit only a smaller number (< 4 per water molecule) of hydrogen bonds than found in pure water or in the presence of very small hydrophobic groups. As in the case of hydrogen bonding among water molecules in pure water, the hydrogen bonds stabilizing the "cage" continuously form and break; that is, the cage (sometimes termed a clathrate) is not stable over time. (Note the following issue in biochemical semantics: The so-called hydrophobic bond that may be referred to in the literature is not a bond per se but rather the avoidance by nonpolar entities of water. Nonpolar side chains may interact through van der Waals forces, however, and thereby stabilize macromolecular structure [see Chapter 3].)

Whereas the 20 common amino acids found in cells share much chemistry in common, they have widely different interactions with water due to their diverse side chains. Nonpolar (hydrophobic) side chains distribute away from contact with water, whereas charged and polar side chains tend to be hydrated. Thus, charged amino acid side chains like those of glutamate and aspartate (both negative) and lysine and arginine (both positive) are most commonly found in regions of the protein exposed to solvent. Polar but noncharged side chains likewise are commonly found in solvent-exposed parts of a protein. The network of hydrogen bonds that forms between charged or polar groups on the protein and the water adjoining the protein surface contributes importantly to protein stability (see Chapter 3). Loss of this surface water can destabilize the protein and lead to its denaturation. Thus, as discussed in Section 4.7, water substitutes are needed for protein stabilization under conditions of extreme cellular desiccation.

Similar water-based, thermodynamic driving of higher order of structure is seen in the case of DNA. Thus, the stacking of nonpolar bases in the interior of the double helix reflects the hydrophobic characteristics of these elements of the DNA structure. The exposure to solvent of charged and polar components of the DNA molecule—for example, the phosphate backbone groups that link nucleotides—reflects their propensities for hydrogen bonding with water. (As an interesting historical note, Linus Pauling initially failed to appreciate the role of exposure of phosphate groups to solvent in the formation of DNA's higher order of structures, an oversight that may have cost him a third Nobel Prize [Watson 1998].) The interactions between nucleic acids and the proteins that stabilize the genome's structure and regulate its activity during transcriptional and translational events reflect the same basic thermodynamic relationships that stabilize the individual components of these complex molecular systems.

In summary, water is a key factor in generating the native structures of proteins, nucleic acids, and membranes. Likewise, water is instrumental in governing how these large molecules interact and assemble within the cell. It is no exaggeration to say that the imprint of evolution in an aqueous medium is found throughout the structures of these large molecules and molecular assemblages. However, this is only part of a larger story. Thus, up to this point in our analysis of water-solute relationships we have considered only two components of biological solutions, water and large organic molecules. To develop the holistic and integrative analysis we alluded to earlier, it is essential to look at the micromolecules that are the numerically dominant solutes in all biological fluids and function as major determinants of the structures of large molecular systems and of water itself.

4.3 The Importance of Micromolecules

If one compares the qualitative and quantitative compositions of the intracellular solutions of diverse taxa—archaea, bacteria, plants, fungi, protists, and animals—one will find striking cases of evolutionary conservation among widely different species as well as fascinating instances of "exotic" chemistry that are unique to certain taxa (Yancey et al. 1982; Yancey 2001; Yancey 2005). Are there general principles that can be invoked to explain why certain features of the chemistry of intracellular solutions must be strongly conserved among taxa and, at the same time, account for highly unusual chemistries associated with particular types of organisms or particular environmental challenges? Answering this question is a central task of the remainder of this chapter.

One general principle we will examine in several contexts is the role that the inorganic ions and small organic solutes of biological solutions, especially those found at relatively high concentrations in intracellular fluids, play in providing a hospitable environment for macromolecular structure and function. We will see, again and again, how macromolecular and micromolecular evolution work in concert to provide cells with the appropriate types of macromolecule-based activities. As a corollary of this general principle about these complementary evolutionary processes, we will see how physiological responses to a diversity of macromolecule-perturbing physical and chemical stresses may entail modulating the types and concentrations of micromolecules in the cell. If, for example, by changing the concentration of one or a few kinds of small organic solutes, the structures and functions of thousands of different types of proteins can be conserved, then the organism is benefitting from what can fairly be called a simple or "global" solution to a complex problem. This type of global solution to threats posed by physical and chemical stressors arises in a broad range of contexts, including stress from extremes of osmotic concentration, temperature, hydrostatic pressure, and production of reactive oxygen species (ROS). Thus, the complex compositions of biological fluids are a reflection of the variety of tasks related to conservation of macromolecular structure and function that befall the micromolecules found in cellular and extracellular fluids. In some cases, the specific composition of micromolecules in a fluid reflects conservation of key properties of water structure. In other cases, notably under conditions of extreme cellular desiccation, micromolecules assume the roles of water in stabilizing cellular structures. We thus need to give micromolecules the recognition they deserve for allowing the "big" molecules of the cell to support life over a remarkably wide range of physical and chemical conditions.

In the following sections we examine in some detail the ultimate causes—the fundamental selective factors responsible for a trait (see Chapter 1)—of the composition of biological fluids. Insights from this analysis include an improved understanding of the principles of micromolecular evolution; a better sense of how macromolecules, micromolecules, and water function in close partnerships during evolution and in physiological regulation; and a deeper understanding of the chemical factors that set—and at times radically extend—the environmental ranges over which life, either active or dormant, can exist.

Perhaps the following—and somewhat strained—analogy will help put our treatment of water-solute relationships into an appropriate historical and philosophical context. The underappreciated "black background" of solution chemistry against which macromolecules are commonly drawn (Gerstein and Levitt 1998) is somewhat analogous to the dark energy and dark matter of physics. Both the physicist's and the biologist's "darknesses" represent most of what is present in the very different study systems they examine—the Universe as a whole and the chemistry of cells, respectively—and both types of phenomena have been slow to gain the level of understanding they merit in their respective fields. Furthermore, both "darknesses" can be notoriously hard to examine experimentally. Dark energy and dark matter cannot be "seen" directly; physicists must deduce their properties from putative effects of these forces on directly observable components of the Universe. Analogously in some ways, the importance of the solution "background" may only be discernable by examining changes in macromolecular structure and function that result from alterations of the aqueous milieu.

Inorganic ions

We begin our "enlightening" of the broad area of water-solute relationships by focusing on a question that seems rarely to have been asked by biochemists, and which may not have an easily proven single answer: Why did evolution select the particular suite of inorganic ions we find in virtually all contemporary cells?

CONSERVATION OF MONOVALENT CATION COMPOSITIONS OF INTRACELLULAR FLUIDS   The cytoplasm of most types of cells contains approximately the same mixture of inorganic ions (Kirschner 1991; Yancey 2005). Furthermore, the *absolute* concentrations of inorganic ions are often very similar across species that differ widely in total osmotic concentration (Box 4.1 defines the different units employed to express solute concentrations). The difference in total osmolality is due to the presence of organic micromolecules whose concentrations may exceed those of the summed concentrations of all inorganic ions. The *ratios* of concentrations of different inorganic ions in the cytoplasm are even more conserved across different taxa than are absolute concentrations. The dominant intracellular cation is potassium (K^+), whose concentration in the cytoplasm ($[K^+]_i$) ranges between ~100 and ~200 mosmol l^{-1} in most marine, freshwater, and terrestrial animals (Kirschner 1991). (Halophilic, or salt-loving, members of the domain Archaea are a striking exception to this pattern, as we discuss later.) Although concentrations of K^+ do not vary as widely as those of certain organic osmolytes, K^+ is an important inorganic osmolyte whose concentration often is adjusted during the early phases of cell volume regulation. K^+ may be a "first line of defense" in cell volume regulation and serve as a critical osmolyte until other volume-regulating processes, notably those involving organic osmolytes, can be activated. Sodium ion (Na^+) dominates in extracellular fluids in multicellular species and is present in much lower concentrations in cells, often at levels less than one-fifth or one-tenth those of $[K^+]_i$.

This relative stability in the inorganic ion compositions of intracellular fluids among diverse taxa is a striking example of the conservative nature of a great many key aspects

BOX 4.1

Units for expressing concentrations

Researchers use several different units for expressing concentrations of aqueous solutions, which you will encounter repeatedly in this chapter. We define the principal units here:

- **Molarity**: moles[*] per liter of solution (mol l^{-1}). H_2O (pure) is 55.56 molar.
- **Molality**: moles of solute per kilogram of solvent (mol kg^{-1}). Note: mol kg^{-1} solvent is usually close to mol l^{-1} if solutes have small effects on solution volume. However, in certain biological contexts, the difference is important if the solutes in question have appreciable influences on solution volume (see Box 4.2).
- **Osmolality**: moles of osmotically active particles per kilogram of water (osmoles kg^{-1}).

[*] One mole of a substance contains 6.02×10^{23} particles. This quantity is known as Avogadro's number. When Amedeo Avogadro (1776–1856) first proposed this concept, establishment chemists were hesitant to accept his conclusion and the calculations on which it was based (for this interesting history, see Jaffe 1976). History has proven Avogadro right, and his "number" is now part of the lore of chemistry and stored in the memory banks of millions of chemistry students.

- **Osmolarity**: moles of osmotically active particles per liter of solution (osmoles l^{-1}).

Osmolality is especially useful for discussing the osmotic relationships of physiological solutions. Osmolality is normally measured empirically, using one or more of the several colligative properties of solutions like vapor pressure or freezing point depression. Estimating osmolality by this means reflects two points about physiological solutions. First, ideality does not pertain in complex biological solutions, so even if the concentrations of every osmotically active solute could be determined, on a solute-by-solute basis, there is no guarantee that a true value of osmolality would result. Second, and following from the first point, measuring the concentrations of every solute in a physiological solution is tremendously demanding of time, energy, and financial resources. A physical means for quantifying osmolality is thus warranted on both theoretical and practical grounds. For a comprehensive and clear treatment of conversions between osmolality and osmolarity, see Šklubalová and Zatloukal (2009).

of biochemical evolution. This strongly conservative pattern emphasizes a central point that we make repeatedly in this chapter, namely that only a certain milieu creates an environment in which macromolecules attain their optimal structures and thereby are able to function most effectively. We will provide several examples of the validity of this principle and show what types of solution compositions can be used to reach this end-result—an optimally functioning metabolic system supported by macromolecules whose structures possess an appropriate balance between stability and rigidity.

The similarity in the inorganic ion compositions of cells across almost all taxa also reflects the varied effects of different inorganic ions on the structure of water. Thus, the choice of the inorganic ions that are accumulated at relatively high concentrations may in part be a reflection of the different demands these ions place on the solvent capacity of water—a point we touched on earlier in our initial treatment of the challenges cells face in packaging so many different types of solutes into a limited volume of water.

WHY POTASSIUM ION?   Can we elucidate the criteria that led to the selection of K^+ as the dominant intracellular monovalent cation and Na^+ as the major monovalent cation in extracellular solutions? (Note that we use the plural here—criteria—rather than criterion; there may be several advantages to the distributions of K^+ and Na^+ that we find in contemporary organisms.) One explanation we can propose is based on these two cations' different effects on the structure of water. Na^+ has a lower mass-to-charge ratio than K^+ (an atomic weight:charge ratio of 23:1 for Na^+ versus 39:1 for K^+). Thus, Na^+ is more water-structure-ordering than K^+. In terms of ion-water interactions, hydrogen bonding between K^+ and water is weaker than bonding of water molecules to each other, whereas Na^+ forms stronger hydrogen bonds with water than water molecules form with each other. These differences in ion-water interactions are reflected in the **partial molal volumes** ($\overline{V}^0$) of the ions (**Box 4.2**). In pure water at 25°C, the partial molal volumes of Na^+ and K^+ are -1.21 cm^3 mol^{-1} and $+9.02$ cm^3 mol^{-1}, respectively (Millero 1971). Thus,

BOX 4.2

Partial molal volumes

Partial molal volume ($\overline{V}^0$) (cm^3 mol^{-1}) is defined as the effect of a solute on the volume of an aqueous solution that is mediated through the solute's effects on water structure (see Millero 1971). An exceptionally clear presentation of the physical chemistry of partial molal volume effects is given by Withers et al. 1994. For example, at 5°C Na^+ reduces water volume by 2.93 cm^3 mol^{-1}. In contrast, at 5°C K^+ expands water volume by 7.59 cm^3 mol^{-1}. Other ions of biological importance may have very large effects on water organization and hence on solution volume. For example, at 5°C trimethylammonium ion $(CH_3)_3$–NH^+, has a $\overline{V}^0$ of +71.51 cm^3 mol^{-1}; ammonium ion (NH_4^+) has a $\overline{V}^0$ of +17.49 cm^3 mol^{-1}; and Cl^- has a $\overline{V}^0$ of +16.65 cm^{3}mol^{-1}. Partial molal volumes of divalent ions (at 5°C) may be large: Mg^{2+} (-21.66 cm^3 mol^{-1}), Ca^{2+} (-19.26 cm^3 mol^{-1}), and SO_4^{2-} (+11.70 cm^3 mol^{-1}). $\overline{V}^0$ varies somewhat between pure water and

seawater (Millero 1971). Ions with negative partial molal volumes exhibit less water-contraction ability in seawater; ions with positive partial molal volumes lead to slightly larger expansion of water in seawater.

Because temperature has such large effects on the hydrogen-bonded structures of liquid water, $\overline{V}^0$ may be strongly temperature-dependent. For example, $\overline{V}^0$ of Na^+ varies regularly with increasing temperature: $\overline{V}^0$ is -3.51 cm^3 mol^{-1} at 0°C, -2.90 cm^3 mol^{-1} at 5°C, -1.21 cm^3 mol^{-1} at 25°C, -0.30 cm^3 mol^{-1} at 50°C, and +0.8 cm^3 mol^{-1} at 100°C. Thus, at high temperatures the increased kinetic energy of water molecules precludes even a water structure-enhancing ion like Na^+ from having a strong effect on water structure and density. For NH_4^+, $\overline{V}^0$ is less temperature-sensitive than in the case of Na^+: At 0°C the value is 17.47 cm^3 mol^{-1}, at 5°C it is 17.86 cm^3 mol^{-1}, and at 25°C it is 19.20 cm^3 mol^{-1} (Millero 1971).

addition of Na^+ shrinks water, whereas addition of K^+ expands water (other changes to solute composition being equal). How might these differences in ion-water interactions influence natural selection of the best dominant monovalent cation to use in intracellular solutions? One factor may be the extent to which different ions tie up water. Thus, K^+ may be the preferable of these two monovalent cations because it ties up fewer water molecules in its hydration sphere than does Na^+, thus facilitating a greater solvent capacity of cellular water and helping cells solve the "packaging problem" discussed earlier.

However, as is the case with many attempts to look into the deep evolutionary past and deduce why one particular path was taken rather than another—our search for ultimate causes—we can pose alternate explanations for the choices of K^+ as the dominant intracellular monovalent cation and Na^+ as the primary extracellular monovalent cation. One conjecture involves the importance of **Na^+ gradients** for driving active transport in present-day cells. The energy gradient established by the much higher extracellular concentration of sodium ion is used to transport a variety of organic molecules, for example, free amino acids and sugars, into the intracellular space. It seems reasonable to propose that, at the dawn of cellular evolution, when organic molecules in the surrounding water may have been in relatively low concentrations, a process that could capture and bring into the cell molecules like amino acids or sugars would have been highly favorable for survival. Thus, a low intracellular concentration of sodium ion might have been a key advantage for early cells evolving in seawater, with its relatively high concentration of Na^+. This advantage continues in present-day organisms, which almost invariably rely on sodium ion gradients to drive energetically uphill accumulation of organic molecules across epithelial surfaces, whether in the guts of animals or at the external surfaces of soft-bodied marine invertebrates.

Another possible reason why K^+ was selected over Na^+ as the dominant intracellular monovalent cation is its more favorable effect on enzymatic activity. Recall the rankings of these two cations in the Hofmeister series: K^+ is a weak protein stabilizer and Na^+ is a weak protein destabilizer. These different effects on protein stability may underlie the ions' different effects on enzymatic activity. For example, the glycolytic enzyme pyruvate kinase (PK) is activated by K^+ (**Figure 4.4A**) and strongly inhibited by Na^+ (**Figure 4.4B**). The favorable response to increasing $[K^+]$ up to ~100 mmol l^{-1} is similar among the diverse species studied, which include a freshwater clam and crayfish, four marine crustaceans and an echinoderm, and a mammal. The total osmolalities of these animals span a wide range, from a low of ~75 mosmol l^{-1} for the clam (*Anodonta* sp.) to slightly more than 1000 mosmol l^{-1} for the marine invertebrates (Kirschner 1991). KCl effects on PK thus serve as a good illustration of how a given type of solute has similar effects on proteins across species, regardless of their total osmolalities. Similarly, different proteins typically show qualitatively similar responses to solutes, due to the fundamental relationships among solutes, water, and macromolecules that we discuss later in this chapter.

A caveat regarding the selective criteria just given for a preference for intracellular K^+ seems relevant at this point. One can ask whether the more favorable effects of K^+ relative to Na^+ on proteins reflect intrinsic differences between the two ions' effects on protein structure and function or, instead, reflect protein evolution that adapted proteins' responses to inorganic cations *after* natural selection had chosen the K^+-to-Na^+ concentration ratios we find in present-day cells and extracellular fluids. That is, did proteins adapt to the presence of high internal K^+ only after evolution had "decided," for reasons based on factors other than protein function and stability, that K^+ was better than Na^+ for use at high intracellular concentrations? Points raised at several junctures later in this chapter may help resolve this chicken-egg conundrum.

In evaluating the effects of a particular inorganic ion on biochemical systems, it is critical to pay attention to the *counterion* that is present in the experimental assay solution.

For instance, the use of a salt like KCl in studies of K^+ effects on biochemical systems neglects to take into account the fact that Cl^- concentrations in the cell are much lower than K^+ concentrations. Intracellular chloride ion concentrations vary among species, and range from about 15 to 30 mmol l^{-1} in vertebrates to values somewhat higher in osmotically concentrated marine invertebrates (Prosser 1973; Kirschner 1991). Thus, studying a biochemical system using KCl, that is, using equal concentrations of K^+ and Cl^-, is definitely not reflective of cellular chemistry.

The dominant intracellular anions comprise a diverse set of organic molecules (nucleic acids and many metabolites of intermediary metabolism, notably the adenylates ATP, ADP, and AMP) or larger (relative to Cl^-) inorganic ions like phosphate and bicarbonate. **Figure 4.4C** illustrates the types of artifacts that can result from uncritical use of simple inorganic salts like KCl in studies of solute effects (Weber et al. 1977). The activity of a mouse-derived in vitro protein synthetic system showed maximum activity at physiological K^+ concentrations (shaded area on figure) when the salt added to the reaction was potassium acetate. Acetate (when present in a well-buffered solution) is a stabilizing member of the Hofmeister series. When Cl^- was the counterion, however, protein

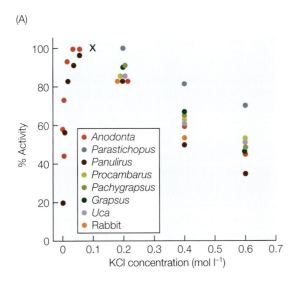

(A)

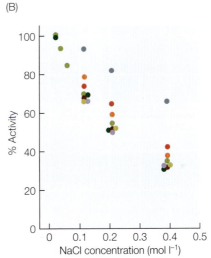

(B)

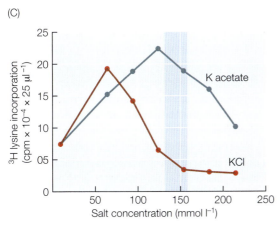

(C)

Figure 4.4 Salt effects on biochemical processes. (A) The effects of KCl on the activity of the glycolytic enzyme pyruvate kinase (PK) from a diverse array of invertebrates: freshwater clam (*Anodonta* sp.); freshwater crayfish (*Procambarus clarkii*); sea cucumber (holothurian; *Parastichopus parvimensis*); lobster (*Panulirus interruptus*); shore crabs (*Pachygrapsus crassipes, Uca crenulata,* and *Grapsus grapsus*); and a mammal (rabbit). The "x" represents superimposed data for *Parastichopus, Procambarus, Pachygrapsus, Grapsus, Uca,* and rabbit. (B) The effects of NaCl on PK from the same set of species. The assay mixture also contained 0.1 mol l^{-1} KCl. (C) The effects of potassium chloride and potassium acetate on cell-free protein synthesis (measured by incorporation of radiolabeled [^{3}H] lysine into cellular protein) by a mouse-derived preparation. The shaded area indicates the physiological range of intracellular K^+ concentrations in mice. (A and B after Bowlus and Somero 1979; C after Weber et al. 1977.)

synthesis was strongly inhibited at physiological concentrations of K^+. These findings not only provide another important illustration of ion-specific effects, but also issue a strong caveat about the need to use biologically realistic assay media for biochemical experimentation.

We will return to the central issue of what makes a "fit" solute after we examine a group of organisms that, at first glance, would seem to violate the rules given above on solute composition of the cytosol, notably conservation of inorganic ion concentrations. In fact, the "violation" we observe can be interpreted as a strong argument in support of the basic solute selection strategies we began to develop above and will return to in subsequent sections of this chapter.

Halophilic archaea: An exception to the rule about conservation of intracellular ion composition

When we stated that intracellular ion concentrations are strongly conserved among diverse taxa, we mentioned that there is a striking exception to this general rule, the extremely halophilic members of the domain Archaea (see upper photo at left). Halophilic archaea occur in environments like salt ponds (see lower photo at left) in which saturating concentrations of salt, principally NaCl, can occur. The reddish color of these ponds is largely due to rhodopsin pigments in the halophiles. These pigments absorb light to drive proton-pumping systems that establish a pH gradient across the cellular membrane, which is subsequently dissipated to drive ATP production. These extreme halophiles are unique in having intracellular concentrations of inorganic ions, K^+ in particular, that are at least an order of magnitude higher than the canonical ion concentrations found in most other types of organisms. Cells of these extreme halophiles are approximately isosmotic with the extremely saline medium in which they occur and thrive.

As one would predict on the basis of their high intracellular ion concentrations, the proteins of halophilic archaea are exceptionally salt tolerant (**Figure 4.5**). In fact, they *require* high concentrations of inorganic ions to function best; the activity of a halophile's protein may be maximal at salt concentrations that are fully inhibitory to the homologous protein of a nonhalophile. This difference can be seen in the effect of salt concentration on malate dehydrogenases (MDHs) from a halophilic archaeon and nonhalophilic marine bacterium: At 0.8 mol l^{-1} KCl, the bacterial MDH is ~90% inhibited, whereas the MDH of the halophile shows maximum activity (see Figure 4.5A).

Methanosarcina mazei, a halophilic archaean.

Salt ponds turned red by the resident halophiles.

The properties of proteins of halophiles—their abilities to function at extremely high salt concentrations and their requirements for high osmolality to assume their native folded structures—reflect their unusual amino acid compositions. Halophilic proteins have unusually high percentages of aspartate and glutamate residues, giving the proteins a net negative charge (see Figure 4.5B). Proteins of extreme halophiles also have relatively few strongly hydrophobic side chains. Due to the effects of charge repulsion among anionic side chains and a weak propensity for stabilization through the hydrophobic effect, the proteins of extreme halophiles are intrinsically unstable. How can this inherent instability of structure be rationalized? The basis of selection for intrinsically low structural stability is

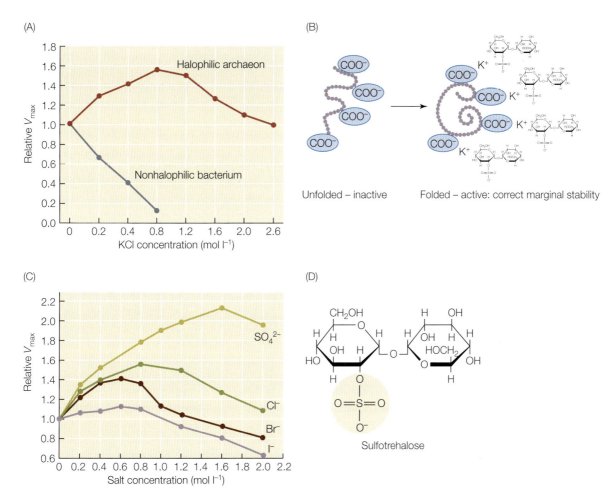

Figure 4.5 Salt effects on proteins of halophilic archaea. (A) Effects of increasing KCl concentration on the activity of malate dehydrogenase (MDH) of a halophilic archaeon and a nonhalophilic marine bacterium. (B) Folding of an intrinsically unstable protein of an extreme halophile. Potassium ions (K^+) titrate the charged carboxylate groups of aspartyl and glutamyl residues and facilitate their transfer into the protein's interior. Sulfotrehalose provides additional stabilization, leading to generation of an active, folded protein with the correct marginal stability for optimal function. (C) Hofmeister series effects of different potassium salts on the activity of MDH of a halophilic archaeon. (D) Sulfotrehalose, a protein-stabilizing solute that also serves as a counterion to K^+ (shown in B). (A and C, unpublished data of L. Borowitzka and G. Somero.)

the stabilizing influence that the concentrated intracellular milieu has on these proteins. The high concentrations of K^+, which is a weakly protein-stabilizing ion in the Hofmeister series, are nonetheless great enough to strongly increase protein stability. As emphasized at several junctures in this volume, proteins attain their optimal functional states when their structures are neither too rigid nor too labile. Were proteins of extreme halophiles to have a higher intrinsic stability, they might become too rigid to function effectively under the high osmolalities present in the cytoplasm. High osmolality, then, is needed to achieve the requisite level of stabilization of inherently labile halophilic proteins.

What types of solutes, in addition to high levels of K^+, are found in these extreme halophiles? In particular, is there a dominant anion to balance the high concentration of K^+? We approach this question by first examining how different potassium salts affect the activity of a halophilic MDH (see Figure 4.5C). As seen with the protein synthetic system presented in Figure 4.4C, different potassium salts have different effects on MDH function. The ranking of stimulation is $SO_4^{2-} > Cl^- > Br^- > I^-$, a ranking of ions that reflects their positions in the Hofmeister series. It is important to observe that the activation of MDH found with K_2SO_4 is maximal at ~1.6 mol l^{-1}, an extraordinarily high salt concentration for fostering optimal protein function. This activation reflects neutralization of anionic side chains on the protein by K^+ and a salt-driven burial of hydrophobic groups, which is due largely to the sulfate ion, a highly effective protein stabilizer.

Although sulfate ion is shown to be highly effective in activating the halophile's MDH in vitro, the cytoplasm of a halophile is not high in sulfate. Achieving the requisite stabilization falls to another anion, a sulfate derivative of the common disaccharide trehalose, **sulfotrehalose** (see Figure 4.5D). The concentration of sulfotrehalose in the cytoplasm of an extreme halophile varies with external osmolality. Under high osmolality, sulfotrehalose concentrations can reach ~1 mol l^{-1} (Martins et al. 1997). The fitness of sulfotrehalose in the context of stabilizing the intrinsically unstable proteins of extreme halophiles is based on at least two attributes of the molecule. First, as mentioned, sulfate is a strong protein stabilizer. Because sulfotrehalose is negatively charged, it will tend not to bind to the protein because of the high percentages of carboxylate side chains present in these glutamate- and aspartate-rich proteins. As discussed in Section 4.5, strong stabilizers of proteins, unlike denaturants like urea, do not bind to proteins, but rather are excluded from the protein surface. Second, the disaccharide trehalose is also an effective stabilizer of proteins (and of membranes), as discussed later in the chapter. When assembled into one molecule, these two protein stabilizers can exert a potent stabilizing effect on proteins (Martin et al. 1999). Thus, sulfotrehalose plays a major role in conferring on the intrinsically labile proteins of halophiles the right balance between stability and instability—the state of marginal stability.

Inorganic ions versus organic osmolytes: How best to respond to osmotic stress?

We have just discussed some of the biochemical adaptations used by halophilic archaea to provide an intracellular microenvironment conducive to optimal protein stability and function under varying but typically high external osmolalities. These adaptations share both similarities and differences with the volume regulatory responses of nonhalophilic species. Below we examine the roles of inorganic ions and small organic molecules in osmotic adaptations in diverse taxa to elucidate fundamental strategies of adaptation in organisms whose proteins are *not* adapted to function optimally at high concentrations of inorganic ions. As in the case of extreme halophiles, however, organic solutes play dominant roles in establishing a favorable environment for macromolecular activity. That such reliance on organic solutes should be necessary is suggested by the observed inhibitory effects of K^+ at concentrations higher than physiological values. Thus, for osmoconforming species—those species that allow the osmotic concentration of the cellular water to vary with environmental osmolality (Figure 4.6)—the need to increase cellular osmolality in the face of rising external salinity might not best be met by increasing the concentration of an inorganic ion like K^+ or, worse, Na^+. In fact, varying concentrations of inorganic ions during osmoregulation by euryhaline osmoconformers is seldom the long-term solution to problems of volume regulation. Osmoconformers characteristically exhibit a two-tiered response (Kültz 2012). Short-term adjustments to hyper- or hypo-osmolality of the environment may entail temporary increases or decreases, respectively, in inorganic ion

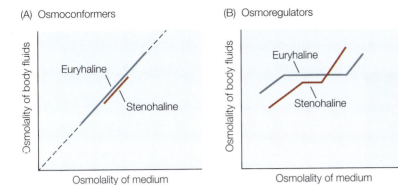

(A) Osmoconformers

(B) Osmoregulators

Figure 4.6 Osmoconforming and osmoregulating organisms: two different strategies of volume regulation. Species differ widely in their responses to changes in ambient osmolality. (A) Some species are osmoconformers: The osmolality of the intracellular and extracellular fluids (in multicellular species) closely tracks the osmolality of the medium. Euryhaline species (blue) can withstand wide ranges of internal osmolality compared with stenohaline species (red). At low extremes of osmolality, however, even osmoconformers regulate somewhat, to ensure that internal solute concentrations do not fall below tolerable limits. (B) Other species, notably vertebrates, are osmoregulators: Osmolality of the body fluids is held largely independent of external osmolality over some range of external osmolalities. Species differ in the range of medium osmolality over which they can survive and/or regulate the osmolality of their body fluids. Stenohaline species (red) are restricted to a relatively narrow range of external osmolalities compared with euryhaline species (blue). Deviations from a narrow range of internal osmolalities may be lethal (indicated by sharp increases or decreases in osmolality of body fluids). Note that this Figure is quite stylized; species may not strictly adhere to either strategy, especially at extreme ends of their osmotic tolerance ranges.

concentrations. Rapid changes in concentrations of inorganic ions during critical early stages of volume regulation might be viewed as a "holding down the fort" strategy that prevents dangerous shrinking or swelling of the cells when they first encounter osmotic stress. With cellular integrity ensured, solution conditions optimal for macromolecular activity and structure are reestablished. This second phase of volume regulation may require additional time, during which adjustments in concentrations of organic solutes supplant the changes in inorganic ions. This two-phase process is seen across a broad range of taxa, and a variety of organic osmolytes are used. For example, in the euryhaline teleost (bony) fish *Oreochromis mossambicus*, *myo*-inositol is accumulated during the second phase of acclimation to hyperosmotic challenge (**Box 4.3**; Gardell et al. 2013). Marine invertebrates commonly use amino acids or their breakdown products—for example, taurine—in this second phase. Why these particular types of organic micromolecules have been selected for use in volume regulation is the next focus of our analysis.

BOX 4.3

Activation of synthesis of organic osmolytes by spikes in inorganic ion concentrations and increases in intracellular pH (pH$_i$)

As mentioned in the text, volume regulation in the face of increases in external osmolality, a process known as **regulatory volume increase** (**RVI**) (Kültz 2012), commonly occurs in two phases: An initial, rapid increase in concentrations of inorganic ions, chiefly K^+ and Na^+, is followed by a fall in these ions, as organic osmolytes increase in concentration and assume the primary role in RVI. Studies of

BOX 4.3 (continued)

Mozambique tilapia (*Oreochromis mossambicus*) (**Figure A**) have revealed an important link between

Figure A Mozambique tilapia (*Oreochromis mossambicus*).

these two temporally distinct components of volume regulation and have shown an important effect of intracellular pH (pH$_i$) as well (Villarreal and Kültz 2015). The organic osmolyte of interest was *myo*-inositol (**Figure B**), which shows a rise in concentration within minutes of exposing tilapia to hyperosmotic conditions.

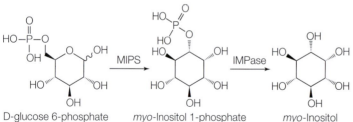

D-glucose 6-phosphate   *myo*-Inositol 1-phosphate   *myo*-Inositol

Figure B The MIPS and IMPase reactions.

The rapid rise in *myo*-inositol concentration contrasts with much slower increases in the biosynthesis (transcription and translation) of two enzymes that are instrumental in governing *myo*-inositol production, **myo-inositol phosphate synthase** (**MIPS**) and **inositol monophosphatase** (**IMPase**). Increased transcription of the mRNAs for these two enzymes and elevations in their protein levels require hours to days, whereas *myo*-inositol levels in gill cells may rise within minutes of exposure to increased external osmolality. This observation suggested that regulation of the specific activities of existing stocks of these two enzymes might be an important rapid-response mechanism for increasing *myo*-inositol concentrations. Indeed, this is what was observed. For one of the enzymes, IMPase (with inositol 3-phosphate

[Ins-3P] as substrate), rising concentrations of Na$^+$ and K$^+$ (Cl$^-$ was the counterion) led to sharp increases (up to ~250%) in enzyme activity due to effects on specific activity (k_{cat}) (**Figure C**). This is an unusual enzymatic response to an increase in Na$^+$ concentration, which usually leads to inhibition of catalysis (Yancey et al. 1982). MIPS was not stimu-

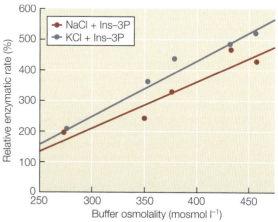

Figure C Activation of IMPase by monovalent cations. (After Villarreal and Kültz 2015.)

lated by increases in monovalent cations, and in fact its activity was slightly inhibited by these additions, in keeping with the general pattern found for enzymes.

It is noteworthy that changes in pH$_i$ also are likely to be involved in rapid upregulation of *myo*-inositol concentrations (Villarreal and Kültz 2015). During RVI mediated by uptake of Na$^+$ into the cell, a sodium-H$^+$ exchange (NHE) system plays an important role: Uptake of Na$^+$ leads to efflux of H$^+$, to maintain charge neutrality. As a result of the activity of the NHE system, the pH of the cell (gill, in the present case) rises during hyperosmotic stress, and this rapid and transient rise in pH may further stimulate *myo*-inositol production. Studies of MIPS (variant MIPS-160 was used) and IMPase have shown that both enzymes have alkaline pH optima (near pH 8.8 for both proteins). Activities of the enzymes rise sharply between pH values of ~7.25 and 8.8, which approximates the range of pH$_i$ values reported for fish gills (Villarreal and Kültz 2015). This activation by rise in pH is seen for both substrates of IMPase, inositol-3P and inositol-1P

BOX 4.3 *(continued)* ▮ ▰ ▰ ▰ ▰

(**Figure D**). Thus, the removal of protons from the cell, as well as the accumulation of Na^+ in the cell, leads to stimulation of these *myo*-inositol-generating enzymes. As *myo*-inositol concentrations rise, Na^+ is removed from the cell in exchange with H^+. Thus, there is an effective feedback relationship between changes in pH_i and $[Na^+]$ and concentrations of *myo*-inositol. Later in this chapter we discuss several other instances in which changes in pH_i "titrate" the activities of biochemical and physiological processes. In common with what has been observed for *myo*-inositol synthesis in gills, these pH_i-regulated processes exhibit responses to changes in proton activity that may represent one of the most rapid and sensitive forms of biochemical regulation.

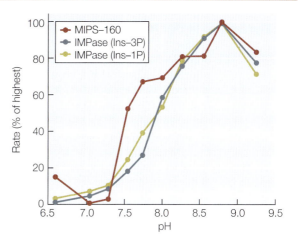

Figure D pH dependence of MIPS and IMPase activity. (After Villarreal and Kültz 2015.)

4.4 Organic Micromolecules

One of the central themes of this chapter is the complementary nature of the processes of macromolecular evolution and micromolecular evolution. Natural selection has "worked" on both classes of molecules, and the types of changes we find in cells' micromolecular constituents reflect how widely important these small molecules are in defending the properties of large molecules like proteins in the face of diverse types of environmental stressors. To understand the diverse roles of micromolecules, we first must address some issues in terminology, which can be complex and at times confusing.

Semantic issues: The multiple roles of organic micromolecules lead to complex terminology

Low-molecular-mass organic solutes—**organic micromolecules**—can be categorized in terms of their chemical structures (sugars, amino acids, methylammonium compounds, urea, nucleotides, etc.) and also named according to the functional roles they may play in diverse types of adaptive processes, including osmotic regulation, resistance to desiccation in anhydrobionts, temperature adaptation, resistance to hydrostatic pressure, and scavenging of ROS. The former, chemical structure-based classifications are conventional and straightforward; the functional categorizations, while quite logical, confront us with what can at times be a bewildering array of names.

In general, the functional category conferred on a particular small organic molecule reflects the focus of the experiment in which the molecule is studied, for example, osmotic regulation or temperature adaptation. However, despite the restricted functional context of any single study, it is important to keep in mind that *a given type of small organic solute can play multiple roles in adapting to environmental stress*. Thus, for example, when the process of cell volume regulation is being examined in an osmoconforming marine invertebrate, an organic molecule whose concentration varies with external osmolality is likely to be termed an **organic osmolyte**. Strictly speaking, the term "osmolyte" should be restricted to those solutes that *vary* their concentrations when the cell is confronted with a change in ambient osmolality. However, the term "osmolyte" is generally used to denote any osmotically active solute in a biological fluid. If an organic osmolyte is taken up from the external medium, a common process in bacteria facing osmotic stress, it may be termed an **osmoprotectant**. If, as we demonstrate below, an organic osmolyte does not disrupt macromolecular structure and function, it is commonly termed a **compatible solute**, to indicate that increases or decreases in its concentration do not perturb cellular function, at least over the normal range of concentrations found in cells. If two or more organic solutes with opposite effects on proteins are adjusted in concentration in response to stress, they are commonly termed **counteracting solutes** or **counteracting osmolytes**.

Other terms come into play in the context of other stressors. Under conditions of oxidative stress, when ROS are being produced at rates that threaten cellular integrity, a small organic molecule that plays a role in osmoregulation may be upregulated to serve an **antioxidant** function, as a **scavenger** of ROS. Under heat stress, an organic solute may be upregulated to protect protein structure; here the molecule may be referred to as a **chemical chaperone** or a **thermoprotectant**. When threats of freezing arise, certain small organic molecules, notably glycerol, exhibit huge increases in concentration to function as **antifreeze** molecules, **cryoprotectants**, or **supercooling** (**undercooling**) **agents** (see Chapter 3). Under conditions of extreme desiccation, when cellular water may be mostly lost, organic molecules may accumulate as **desiccation protectants**. Some organic solutes may be regulated in a depth-dependent manner to offset effects of hydrostatic pressure on proteins of marine organisms. These solutes are termed **piezolytes** (Yancey et al. 2014; Yancey and Siebenaller 2015). Lastly, it is important to keep in mind that we almost certainly don't fully understand the full suite of functions of organic micromolecules in adaptation to diverse physical and chemical stressors. The relatively recent discovery of piezolytes by Paul Yancey and colleagues is a good illustration of how unanticipated roles for organic micromolecules are still being discovered. There remain exciting frontiers to explore to gain a broader and deeper understanding of these diverse organic solutes.

Notwithstanding (or perhaps because of) this somewhat complicated lexicon based on function, many scientists tend to use the term "organic osmolyte" (or just "osmolyte") in a generic sense, whatever the functional role of the molecule being studied happens to be. This may not be the most logical and enlightening terminology (for a critical discussion of this issue, see Pais et al. 2009), but it has, for better or worse, become the convention. We will follow this tradition in some cases, but in many other contexts we will use a more function-denoting term to indicate the role of the solute(s) in adapting to a particular type of stressor.

In summary, organic micromolecules play diverse and, commonly, multiple roles in cellular responses to chemical and physical stressors. A key aspect of these micromolecular adaptations for sustaining macromolecular structure and function is that this type of adaptation may provide a *global* solution to a stressor-induced problem. Thus, through offsetting stressor-induced perturbation of macromolecular structure and function, the need for macromolecule-based adaptations such as changes in protein amino acid sequence may be reduced or even eliminated entirely. Macromolecular and micromolecular evolution thus work hand in hand to allow life to occur over a wide range of physical and chemical conditions. Below we examine these diverse roles of organic micromolecules in environmental adaptation and discuss, when applicable, how these adaptive processes incorporate the properties of the so-called background material of the system—water.

Organic osmolytes and osmotic regulation: An overview

The intracellular osmolalities of organisms differ greatly, ranging from values in excess of 6–7 osmoles l^{-1} in extremely halophilic archaea and glycerol-accumulating algae to ~50–100 mosmol l^{-1} in some freshwater invertebrates (Prosser 1973; Kirschner 1991). Among marine animals (**Figure 4.7**), most teleost fishes have osmolalities between ~300 and ~400 mosmol l^{-1} and are therefore hyposmotic to seawater. (Deep-sea teleosts are a striking exception to this generalization, as discussed later in the context of piezolytes and adaptation to hydrostatic pressure.) Almost all marine teleosts thus face a continuous loss of water to the environment and must combat this tendency by drinking and distilling seawater, a process with considerable energy costs due to ion-pumping activities. In contrast, marine chondrichthyan fishes and coelacanths (not shown) may be slightly hyperosmotic to seawater; there thus may be a slight tendency for water to enter their bodily fluids. (Note: *chondr-* is Greek for "cartilage," so "chondrichthyan" denotes fishes with cartilaginous skeletons [Smith 1936]. Two groups of chondrichthyan fishes exist: sharks, rays, and skates, which comprise the Elasmobranchii, and chimaeras, the Holocephali.)

Marine invertebrates, algae, and microbes (eukaryotic, archaeal, and bacterial) follow a different strategy: They are all osmoconformers. Whereas water may be lost or gained during the initial phases of volume regulation in response to change in external osmolality, once the isosmotic relationship with external water is reestablished, there is neither a net gain nor loss of water. In these isosmotic species (and in marine chondrichthyan fishes and coelacanths [not shown]), a variety of organic osmolytes make the dominant contributions (60%–70%) to intracellular osmolality (see Figure 4.7; Somero and Yancey 1997; Yancey 2005). And as emphasized earlier, despite the differences in total osmotic concentration between teleost fishes and isosmotic species, intracellular concentrations of inorganic ions, notably K^+, are highly conserved across these diverse taxa.

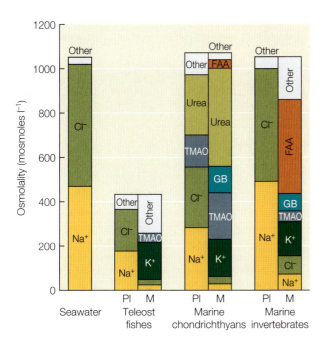

Figure 4.7 Contributions of inorganic ions and different categories of organic osmolytes to osmolalities of teleost fishes, marine chondrichthyans (sharks, skates, rays, and chimaeras), and marine invertebrates. Note the strong conservation of intracellular K^+ concentrations across all taxa and the dominant role of organic osmolytes in establishing osmotic balance in species isosmotic to seawater. Pl refers to plasma concentrations. M refers to intracellular concentrations in muscle. GB, glycine betaine; FAA, free amino acids; TMAO, trimethylamine-N-oxide. (After Yancey 2005.)

The classes of organic osmolytes found most commonly in osmotically concentrated plants, animals, and bacteria are illustrated in **Figure 4.8**. Additional types of organic micromolecules that, although contributing importantly to cell volume regulation, appear to have been selected primarily for thermoprotectant function will be discussed below when we examine bacteria and archaea that are extreme thermophiles. Organic osmolytes in mesophilic species can be grouped into only four general categories, based on their chemical structures: **carbohydrates**, notably polyhydroxy alcohols (polyols) like glycerol and disaccharides like trehalose; **free amino acids** and their derivatives (for example, taurine); **methylammonium** and **methylsulfonium** compounds like trimethylamine-N-oxide (TMAO), glycine betaine (GB), and β-dimethylsulfoniopropionate (β-DMSP); and **urea**.

Why have these particular classes of organic molecules been selected during evolution for use as the major organic osmolytes? We will see below that this question actually has two different components. First, what intrinsic properties make certain organic micromolecules fit for use at variable, and often extremely high, concentrations? Second, what accounts for the observation that certain combinations of osmolytes—often at distinct

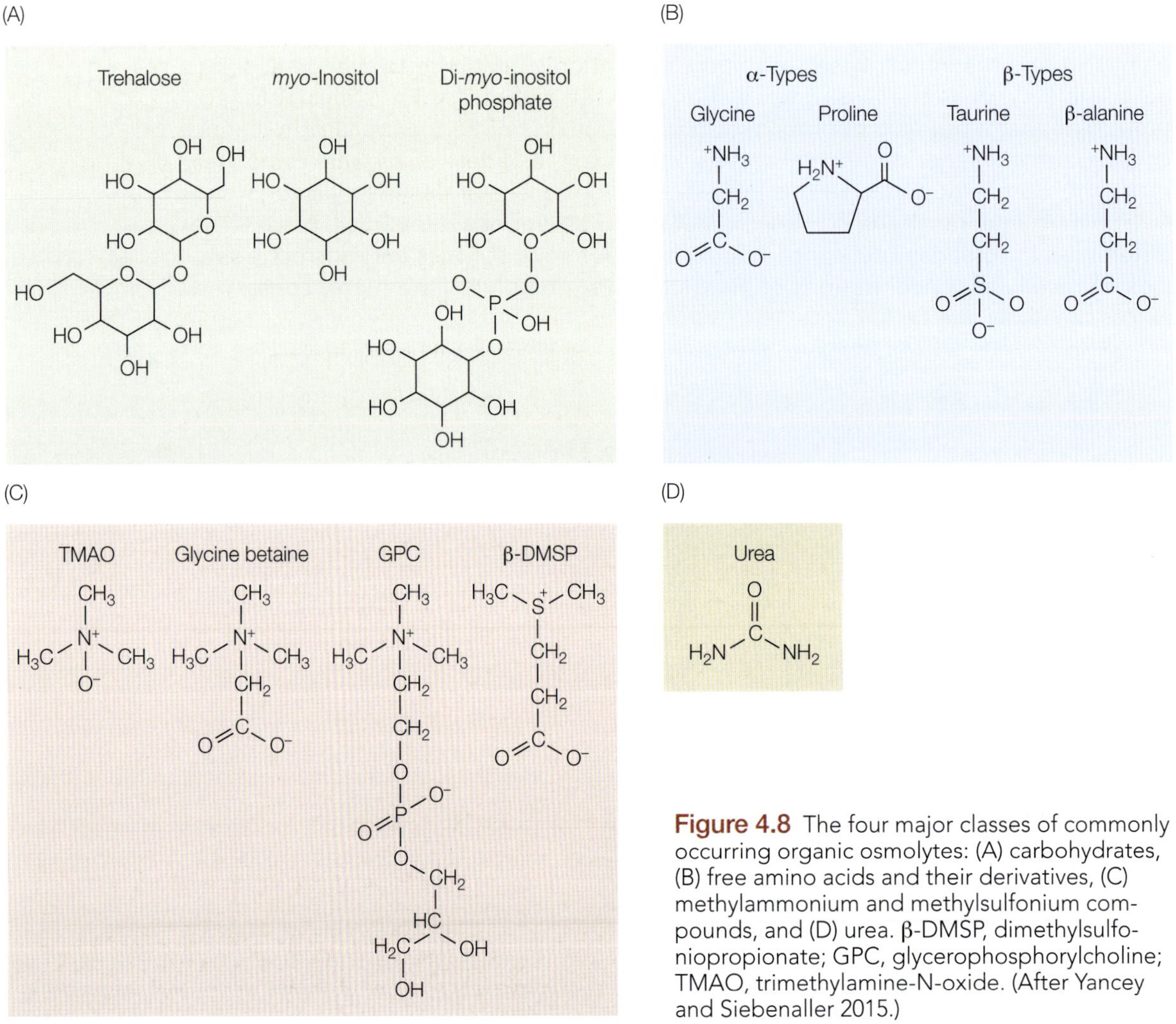

(A)

(B)

(C)

(D)

Figure 4.8 The four major classes of commonly occurring organic osmolytes: (A) carbohydrates, (B) free amino acids and their derivatives, (C) methylammonium and methylsulfonium compounds, and (D) urea. β-DMSP, dimethylsulfoniopropionate; GPC, glycerophosphorylcholine; TMAO, trimethylamine-N-oxide. (After Yancey and Siebenaller 2015.)

ratios of concentration—are commonly found in certain species? Answering these questions will not only shed important light on the interplay among macromolecules, organic micromolecules, and water, but also reveal some universal principles of evolution of biological solutions across all taxa.

Compatible organic osmolytes

Many years ago, Brown and Simpson (1972) coined a term that has stuck in the literature of organic osmolytes: *compatible solute* (see Brown 1976). As the adjective "compatible" denotes, compatible organic osmolytes are molecules whose concentrations can be varied over wide ranges without perturbing cellular physiology. Compatible osmolytes generally have only minimal effects on protein structure and function across the entire range of physiological concentrations observed for these solutes. Moreover, particular solutes or solute combinations—whether the solutes in question are stabilizing or perturbing in their effects—usually (but not invariably) affect different proteins in a similar way. Thus, the use of compatible osmolytes rather than perturbing solutes such as inorganic ions allows the entire proteome of the cell to avoid perturbation under conditions of changing osmolality. DNA structure may also be protected if compatible organic osmolytes are used to adjust cellular osmolality. Increases in inorganic ion concentrations in the cell can lead to double-strand breaks in DNA. This covalent damage to DNA may not be repairable (Dmitrieva et al. 2006) and can lead to cellular death through apoptosis (Burg et al. 2007). The use of compatible solutes allows total osmolyte concentration to be varied during osmotic regulation without perturbing metabolic processes as would occur if a noncompatible, protein-perturbing solute were used instead. Furthermore, osmoconformers that vary concentrations of compatible organic solutes rather than concentrations of inorganic ions like K^+ or Na^+ avoid disrupting membrane potentials established by the Na^+ and K^+ gradients.

Figure 4.9 illustrates the varied effects (or non-effects) of different solutes on the ability of the enzyme pyruvate kinase (PK) of a marine crab (*Pachygrapsus crassipes*) to

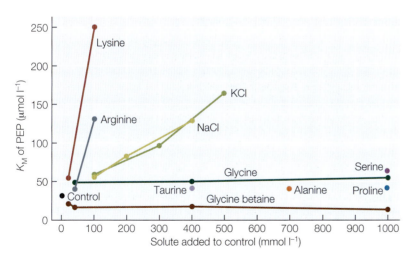

Figure 4.9 Effects of compatible and noncompatible solutes on substrate binding ability (indexed by the Michaelis-Menten constant, K_M, of phosphoenolpyruvate [PEP]) by the pyruvate kinase of a marine crab, *Pachygrapsus crassipes*. Control assays (black circle) contained 0.1 mol l^{-1} KCl. (After Yancey et al. 1982.)

bind one of its two substrates, phosphoenolpyruvate (PEP). Strong perturbation of PEP binding, as indexed by a sharp rise in K_M of PEP with increasing solute concentration, is noted for two inorganic salts, KCl and NaCl, as well as for two positively charged amino acids, lysine and arginine. The perturbing effects of KCl and NaCl show that inorganic salts are not good choices for use in balancing intracellular and extracellular osmolality by euryhaline osmoconformers. (Note: The perturbation observed with KCl and NaCl may also derive from the chloride ion; see below and Figure 4.4.)

In contrast to the effects of inorganic ions and positively charged amino acids, the other solutes examined had minor effects on the K_M of PEP. At concentrations up to 1 mol l^{-1}, glycine, alanine, serine, proline, and taurine failed to affect PEP binding. These solutes thus are excellent choices for varying intracellular osmolality in osmoconforming species. Glycine betaine (N,N,N-trimethylglycine; see Figure 4.8) slightly lowered the K_M of PEP, thus facilitating binding. However, this seemingly favorable effect on the enzyme may be disadvantageous, as we discuss in the next section. Recall that, in discussing temperature-K_M relationships (see Chapter 3), K_M values that are too high or too low can compromise metabolic function.

A variety of organic micromolecules, primarily carbohydrates and amino acids, have been found to be compatible in a broad range of protein systems (Yancey 2005). The similarities in the effects of these micromolecules on diverse proteins from a wide range of taxa suggest that compatibility is a basic feature of protein-solute interactions, independent of any adaptive variation in proteins themselves to facilitate compatibility. Unlike the halophilic archaea, most organisms accumulate compatible osmolytes not to drive protein folding, but rather to ensure that the intrinsic properties of the proteins—notably, the appropriate level of marginal stability—are conserved over a range of osmolalities.

Counteracting solute systems and protein stabilization: Algebraic additivity at the chemical level

Organic micromolecules create a spectrum of effects on protein stability, ranging from extreme stabilization to strong denaturation. In the previous section we focused chiefly on organic solutes that appear to have negligible effects on protein stability over the range of osmolyte concentrations found in cells. Here we turn to organic solutes that have either strongly stabilizing or strongly denaturing effects on proteins. Employment of either class of osmolyte might seem detrimental to cellular function, at least if the osmolyte is used as the principal or sole osmoregulatory solute. In appropriate combinations and concentrations, however, these solutes can be part of solute *systems* that are compatible with protein function and cellular physiology, thanks to effective **counteraction** of the different solutes' individual effects. Put another way, these counteracting solute systems exhibit algebraic additivity in solute effects: Stabilizers and destabilizers represent adding a plus and a minus to get a net effect of zero.

Some historical perspective seems appropriate for introducing the topic of counteracting solute systems because of an interesting mystery that took several decades to resolve. In the 1930s, physiologists began to characterize the unusual biological fluids of chondrichthyan fishes. The blood of chondrichthyans was found to have what appeared to be a pathologically high level of urea, typically in the range of 300–400 mmoles l^{-1} (see Figure 4.7). Later studies showed that the intracellular fluids of these fishes also maintained similarly high levels of this relatively toxic solute. As a point of reference, normal (healthy) human blood contains urea at concentrations of ~7–20 mmol l^{-1}; higher systemic levels of urea lead to several types of pathologies and, at extreme concentrations, death.

How do chondrichthyans and coelacanths (the West Indian Ocean coelacanth *Latimeria chalumnae* is shown in the photo at right) manage to survive with normal urea concentrations that are 20–50 times those found in healthy mammalian blood and seemingly high enough to kill them? The answer to this question came from studies of the effects of the second major organic osmolyte found in chondrichthyan and coelacanth body fluids, trimethylamine-N-oxide (TMAO; see Figure 4.8)—work that was done almost a half-century after the urea mystery first arose (Yancey and Somero 1980). And as we show later in this chapter, the solution of this urea mystery also led to insights into a second urea mystery: how some regions of the mammalian kidney manage to cope with urea concentrations that are several times those found in marine chondrichthyans and coelacanths.

The West Indian Ocean coelacanth, *Latimeria chalumnae.*

TMAO is one of the strongest naturally occurring protein stabilizers known (Street et al. 2006). It tends to be present in cells at elevated concentrations only under conditions of chemical or physical stress that would be expected to significantly perturb protein structure and function. Because of its exceptionally high ability to stabilize proteins, TMAO is not a compatible solute on its own, at least not at concentrations exceeding a few tens of mmol l^{-1}. The high degree of structural rigidity in proteins induced by TMAO is likely to impede protein function, which, as we have stressed repeatedly in this volume, commonly depends on sufficient structural flexibility to allow conformational changes required for function to occur with acceptably low energy barriers.

Figure 4.10 shows the urea-counteracting effects of TMAO on several protein systems. In all cases, urea "pushes" the system in one direction and TMAO "pulls" it in the opposite direction. The net effect on protein structure or function reflects this algebraic additivity of the solute effects. Figure 4.10A shows how urea and TMAO, separately and at a 2:1 ratio of urea:TMAO, affect the K_M of ADP for pyruvate kinase of a marine elasmobranch. At physiological concentrations found in shallow-living chondrichthyans (~400 mmol l^{-1}), urea significantly reduced ADP binding. TMAO at 200 mol l^{-1} strongly enhanced binding. Together, at the common 2:1 ratio found in shallow-living chondrichthyans, urea and TMAO fully counteract each other; the K_M of ADP is not significantly different from the control (neither urea nor TMAO; shaded area in Figure 4.10A) value.

Figure 4.10B shows the effects of urea, TMAO, sarcosine (N-methylglycine), and glycine betaine (GB) on thermal stability (denaturation temperature) of ribonuclease (RNase), a common "lab rat" protein used in studies of solute effects on protein stability. TMAO is a stronger stabilizer than sarcosine or GB; the stabilizing potency of the solutes reflects the amount of methylation of their nitrogen atoms. Once again, urea and TMAO exhibit counteraction, although the restoration of structural stability of RNase by TMAO is not complete at the canonical 2:1 physiological ratio of urea:TMAO. The GB commonly found in chondrichthyan, including marine elasmobranch, tissues (see Figure 4.7) may provide enough additional stabilization to effect full counteraction in vivo.

Figure 4.10C illustrates how urea and TMAO influence protein "breathing"—the rapid and continual opening and closing of the protein's conformation that is a normal protein structural phenomenon (see Chapter 3). The enzyme studied was glutamate dehydrogenase, which contains several cysteine residues that normally are buried in the protein's interior; that is, they are shielded from the surrounding solvent unless the protein opens up through transient "breathing" episodes. The reagent nbf-chloride reacts with the sulfhydryl (–SH) group of cysteine, yielding a color change (strong absorbance at a

(A)

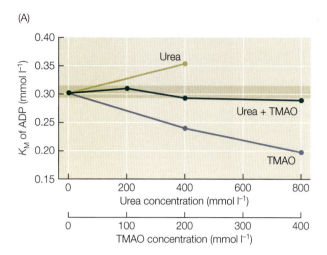

(B)

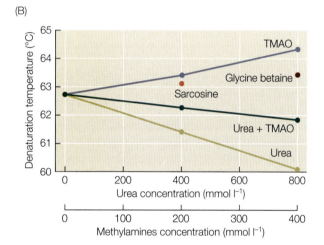

(C)

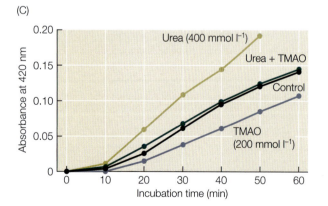

Figure 4.10 Counteracting effects of TMAO and urea on several protein systems. (A) Effects of urea, TMAO, and combinations of these two osmolytes on the K_M of ADP for pyruvate kinase of a marine elasmobranch. The shaded area shows the 95% confidence interval for the control value (no added TMAO or urea). (B) Effects of urea and methylamine osmolytes (TMAO, glycine betaine, and sarcosine) on the denaturation ("melting") temperature of ribonuclease. (C) Effects of urea and TMAO, singly and in combination, on the rate of labeling of sulfhydryl (–SH) groups of glutamate dehydrogenase that are transiently exposed to solvent during protein "breathing" events. (After Yancey et al. 1982.)

wavelength of 420 nanometers [nm]). Thus, by monitoring increase in absorbance at 420 nm, one can track the rate of labeling of –SH groups, a proxy for the extent of "breathing" the protein undergoes. Urea at 400 mmol l^{-1} enhances the rate of labeling of cysteine residues; TMAO reduces the labeling rate. At a ratio of 400 mmol l^{-1} urea to 200 mmol l^{-1} TMAO (2:1), the amount of breathing is not different from that of the control. Thus, the counteracting solute mechanism succeeds in maintaining the conformational stability of the protein, despite the increase in osmolality caused by the addition of 600 mosmol l^{-1} of organic osmolytes.

The mammalian kidney: Protein stabilization by GPC

In the inner medulla of the mammalian kidney, another methylammonium solute, **glycerophosphorylcholine** (**GPC**; see Figure 4.8), has been found to play a role analogous to the function of TMAO in chondrichthyans and coelacanths. For many years, a mystery existed about the mechanisms used by mammalian kidneys to cope with what should be lethally high urea concentrations, especially in the cells of the inner medullary region where urea concentrations reach their highest values. TMAO is not found at appreciable levels in mammals, so another type of counteracting solute system was conjectured to be present. Whereas kidneys upregulate the concentrations of certain polyols (sorbitol and *myo*-inositol), taurine, and glycine betaine when hyperosmotic conditions arise (Burg et al. 2007), GPC is the principal organic osmolyte that varies in parallel with urea (Yancey

2005). As in chondrichthyan fishes and coelacanths, the mammalian kidney maintains an approximately 2:1 ratio between urea and the protein-stabilizing solutes GPC and glycine betaine.

GPC is a potent counteractant of urea's effects on proteins in vitro (Burg et al. 1996). In fact, GPC's chemical structure suggests that it should be a very effective protein stabilizer because it comprises three different protein-stabilizing units: a methylammonium tip, as present in TMAO; a linking phosphate group, whose presence adds a stabilizing ion from the Hofmeister series; and a stabilizing polyol, glycerol, at the other tip of the molecule (see Figure 4.8). Thus, as in the case of sulfotrehalose in extreme halophiles, a single organic osmolyte with multiple stabilizing components is used in the kidney to offset urea-induced protein destabilization. As we discuss later in this chapter, the chemistry of GPC bears a striking resemblance to that of other types of organic micromolecules that play important roles in adaptation to both salinity and temperature in extremely heat-tolerant bacteria and archaea.

TMAO: A piezolyte that counteracts the effects of elevated hydrostatic pressure

The roles of powerful protein stabilizers like TMAO and GPC in establishing an optimal milieu for protein function extend well beyond counteraction during osmotic regulation. One of the most striking illustrations of how a stabilizing organic solute can be used to offset protein perturbation—here, by a physical force—is the regular increase in TMAO concentration found in marine fishes with increasing depth in the water column. In this section we examine this phenomenon, which has important implications in setting the depth limits of fishes.

HYDROSTATIC PRESSURE IS A SIGNIFICANT PERTURBANT OF BIOCHEMICAL SYSTEMS

Hydrostatic pressure increases linearly with depth. For each 10 m increase in depth, pressure rises by 1 atm (1 atm = 0.101 MPa). The average depth of the ocean is ~3800 m; the maximum depth is 10,911 m, in the Challenger Deep of the Mariana Trench. Because biochemical processes such as enzymatic reactions occur with changes in system volume, many of which arise from alterations in water structure (density), pressure can strongly perturb biochemical function. Indeed, pressure's perturbing influences on protein structure and function may become significant at pressures as low as 75–100 atm (Siebenaller and Somero 1978; Somero 1992; Yancey and Siebenaller 2015). It would seem necessary, then, for organisms living in most regions of the marine water column to evolve mechanisms for offsetting the perturbations caused by pressure.

INTRINSIC ADAPTATIONS TO PRESSURE

Two basic types of biochemical adaptation to elevated pressures have been discovered in studies of protein-pressure interactions: intrinsic and extrinsic. First, proteins of high-pressure-adapted species may be intrinsically more resistant to perturbation by elevated pressures than the orthologous proteins of shallow-living species (Somero 1992; Yancey and Siebenaller 2015). Pressure-resistance has been discovered in deep-sea fish orthologs of several proteins, including enzymes and contractile proteins. However, in most cases these proteins still show considerable sensitivity to pressure. Is this remaining sensitivity to pressure an unavoidable consequence of the basic physical relationships associated with volume changes? Certainly, the changes in system (protein + ligands + osmolytes + water) volume that accompany most biochemical processes and make them pressure sensitive may be difficult to reduce to zero in most cases. For example, changes in water organization (volume) during the binding of a ligand to an active site of an enzyme may be a necessary part of a biochemical process, regardless of the depth to which a species is adapted. We still do not adequately understand the sources—and the malleability—of volume changes that occur during function to know how fully they can be eliminated from biochemical processes.

Disruption of protein structure by elevated pressure is another problem faced by deep-living organisms. Proteins of deep-sea species might be able to greatly reduce these pressure sensitivities by developing exceptionally rigid structures, much like proteins of thermophilic species (see Chapter 3). There is evidence that orthologous proteins of deep-sea fishes possess more stable structures than their counterparts in shallower-living species (Siebenaller 1991). Enzyme orthologs of deep-sea animals also may have lower k_{cat} values, which is consistent with a more rigid structure (Somero and Siebenaller 1979; see Chapter 3). High structural rigidity may enable these pressure-adapted proteins to function well at the highest pressure the species encounters. However, at the species' shallowest depth of occurrence these rigid proteins might lack the flexibility to function optimally—or at least to function as well as the orthologous proteins of the shallower-occurring species they interact with while in the upper range of their vertical migrations. Natural selection might favor retention of adequate flexibility to allow a high level of function over a species' full depth range at the expense of less-than-complete adaptation to the highest pressures the species encounters. This accounting for the less-than-complete insensitivity to pressure is conjectural, but it is consistent with what is known about the structure-function relationships of protein orthologs from species adapted to high temperatures.

The snailfish *Notoliparis kermadecensis*.

EXTRINSIC ADAPTATIONS TO PRESSURE   The incomplete intrinsic adaptation of proteins of deep-living species suggests that one or more additional mechanisms for resisting the effects of high pressure may be required for life at great depths. In fact, complementing these macromolecular adaptations are extrinsic adjustments in the micromolecular environment of the cell that offset the perturbations induced by high pressure. Specifically, TMAO concentrations in teleost fishes, chondrichthyan fishes, and invertebrates increase regularly with depth of capture (**Figure 4.11A**). In bony fishes there is a proportionate rise in total osmolality with depth that is due largely to TMAO concentrations (**Figure 4.11B**); the deepest-living fish ever collected, the snailfish *Notoliparis kermadecensis* (shown at left), has the highest osmolality ever reported for a bony fish (~1000 mosmol kg^{-1}) and the highest concentration of TMAO as well (see Figure 4.11A) (Yancey et al. 2014).

In vitro studies have confirmed that TMAO can offset perturbation of proteins by high pressure (Yancey et al. 2001, 2004; Yancey 2005; Yancey and Siebenaller 2015). **Figure 4.12** shows TMAO's pressure-counteracting influences on protein stability and function of two glycolytic enzymes, lactate dehydrogenase (LDH) and pyruvate kinase (PK). For LDH-A$_4$ from a deep-sea rattail fish (*Coryphaenoides leptolepis*) (shown at left) that occurs to depths of ~5 km, an increase in pressure from 1 atm to 250 atm caused a significant increase in K_M of NADH under control (no added organic osmolyte) conditions and in the presence of glycine betaine, inositol, and glycine (see Figure 4.12A). In contrast, TMAO was able to fully offset the increase in K_M induced by high pressure. TMAO can also reduce the effects of high pressure on denaturation of protein structure (see Figure 4.12B). Incubation of LDH-A$_4$ under high pressures (500 and 1000 atm) led to a significant loss of activity in the absence of TMAO. With TMAO (at 250 mmol l^{-1}), 8 h of incubation at 1000 atm—a pressure twice the maximum pressure the species would ever encounter in its habitat—led to some

Coryphaenoides leptolepis, a deep-sea rattail fish.

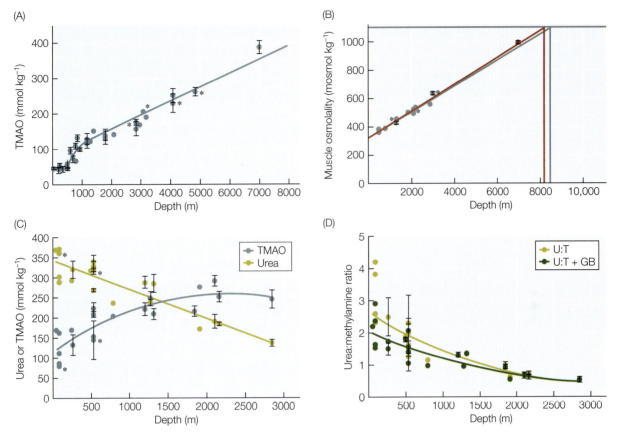

Figure 4.11 (A) Depth-dependent changes in TMAO content of muscle in 19 species of marine teleost fishes collected at different depths. The deepest-living species is the snailfish *Notoliparis kermadecensis*. Asterisks (*) indicate multiple values for a rattail species (*Coryphaenoides armatus*) collected at different depths. TMAO values in invertebrates (not shown) mirror the pattern found for bony fishes. (B) Muscle osmolality in teleost fishes as a function of depth. As in (A), the snailfish is the deepest-living species shown (red circle with standard deviation bars). Two regression lines are shown. The blue line uses earlier data, not including the snailfish data, and extrapolates to an isosmotic state at 8450 m (vertical blue line); the red line is a new regression line that incorporates the snailfish data and extrapolates to an isosmotic state at ~8200 m (vertical red line). The asterisks (*) denote specimens of *Antimora microlepis* that were collected at three depths. (C) Concentrations of urea and TMAO in muscle of chondrichthyans captured at different depths. The asterisks (*) are for specimens of a ray (*Raja rhina*) collected at different depths. (D) The ratio of urea to methylamine (TMAO or TMAO + glycine betaine) concentrations in specimens from (C). GB, glycine betaine; T, TMAO; U, urea. Throughout, error bars represent standard deviations. (A after Samerotte et al. 2007, Yancey et al. 2014, Yancey and Siebenaller 2015; B after Yancey et al. 2014; C and D after Laxson et al. 2011.)

loss of activity, but much less than was seen in the absence of TMAO. For PK of another deep-living fish, *Antimora rostrata*, the ability to bind ADP, as indexed by the K_M of ADP, was retained much better under high pressure when TMAO (250 mmol l⁻¹) was added to the solution bathing the enzyme (see Figure 4.12C).

The changes in TMAO concentration with depth in marine animals belonging to diverse taxa (see Figure 4.11) and the ability of TMAO to offset perturbation of proteins by high pressure (see Figure 4.12) suggest that a pressure-counteracting strategy based on TMAO's stabilizing effects on proteins enjoys widespread use among deep-living species and has evolved independently in many different evolutionary lineages. In other words, use of TMAO to offset pressure effects exhibits convergent evolution. In the context of high-pressure stress, TMAO is sometimes referred to as a piezolyte (*piezo-* from the Greek word *piezein*, meaning to squeeze or compress) to denote its ability to offset the "squeeze" that proteins face at rising pressure (Martin et al. 2002). The fact that organic micromolecules like TMAO and other common osmolytes have similar effects on most types of proteins suggests that elevating TMAO concentrations with increasing depth could be a global solution to perturbation from elevated hydrostatic pressure.

These depth-dependent changes in TMAO concentration are likely to be especially important for vertically migrating animals that may traverse several hundred meters of the marine water column during ontogeny or during vertical migrations as adults. Eurybaric species—those that tolerate a wide range of pressure—would seem to require a capacity for a graded response to change in pressure, such that the degree of pressure counteraction present in cells is in direct proportion to the severity of pressure stress. Thus, regulation of concentrations of a pressure-counteracting solute would be of significant benefit to the organism—a flexible means by which an organism could regulate the stability of the entire proteome.

A particularly clear illustration of the importance of sustaining an optimal balance between stabilizing and destabilizing effects is found in the case of the variation in TMAO and urea concentrations in chondrichthyan fishes from different depths in the marine water column (**Figure 4.11C,D**). Although the canonical ratio of [urea]:[TMAO] is approximately 2:1, this ratio actually pertains only to shallow-living

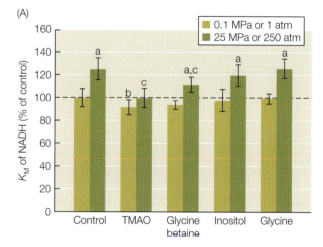

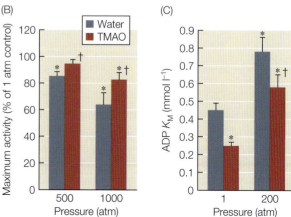

Figure 4.12 Stabilization by TMAO of protein structure and function under pressure. (A) Effect of increased pressure on the K_M of NADH of LDH-A$_4$ of the deep-living fish *Coryphaenoides leptolepis*. TMAO is the most stabilizing of the organic osmolytes tested. All osmolytes were present at 250 mmol l^{-1}. Significant differences are indicated with lowercase letters above the bars: a, significantly increased from 1-atm control; b, significantly reduced relative to 1-atm control; c, significant counteraction of perturbation by high pressure. (B) Percent remaining activity of lactate dehydrogenase (LDH) from the deep-sea fish *Coryphaenoides leptolepis* after 8 h incubation (5°C) at elevated pressures with and without ("Water") addition of TMAO. Values are normalized to incubation at 1 atm. * significantly different from 1 atm control; † significantly different from control at high pressure. (C) Effect of pressure and TMAO on K_M of ADP of pyruvate kinase (muscle) from the deep-sea fish *Antimora rostrata*. * significantly different from 1-atm value with no added TMAO (control); † significantly different from 200-atm control (no TMAO). (A after Yancey and Siebenaller 2015; B and C after Yancey et al. 2001.)

chondrichthyans. At greater depths, where hydrostatic pressure begins to exert a significant destabilizing effect of protein structure, the joint effects of this pressure-induced perturbation and that arising from urea favor a shift in the [urea]:[TMAO] ratio such that the concentration of TMAO eventually exceeds that of urea in the deepest-living species. Here, then, the concentration of the stabilizing solute, TMAO, is adjusted to reflect the need to offset structural perturbation arising from both chemical (urea) and physical (pressure) stresses. The conservation of an appropriate marginal stability for proteins is the end-result (**Box 4.4**).

BOX 4.4

Marginal stability: Optimal or "just good enough"?

Throughout this book we emphasize the importance of marginal stability for the preservation of biochemical functions. We discussed this topic in detail in Chapter 3, and address it again here in order to further demonstrate the importance of this concept. Perhaps not surprisingly, biochemists have long observed the phenomenon of marginal stability among proteins (Jaenicke 1991), and the pattern is particularly striking when we consider the energy required to stabilize the folded states of broad classes of proteins from a diverse array of species. The figure shows the distribution of Gibbs free energies (ΔG) for the folded state of more than 700 proteins from approximately 300 species from all domains of life (ProTherm database; Kumar et al. 2006). As these data demonstrate, the vast majority of proteins are stabilized by ΔG values that are less than 10 kcal mol^{-1}—equivalent to the energy of one or two hydrogen bonds (5% and 95% percentiles are indicated by the shaded area).

A priori we would expect natural selection to favor marginal stability as an optimal state. Some have argued, however, that marginal stability may be an artifact of mutation-selection balance rather than an adaptive trait per se (Arnold et al. 2001; Zeldovich et al. 2007; Harms and Thornton 2013). The major assumption of this argument is that mutations that lead to excess stability are functionally neutral, such that natural selection does not favor marginally stable proteins over hyperstable proteins. Under that assumption, through evolutionary time, mutation and genetic drift will push the distribution of energies of stabilization (ΔG) to the right (higher values of ΔG, as shown on the figure) and natural selection will push the distribution back to the left to the point where proteins are just stable enough to keep from unfolding (positive ΔG). The end-result is the evolution of proteins with a narrow distribution of slightly negative ΔG values, which fully explains the phenomenon of marginal stability.

Despite the elegant simplicity of this model, many issues arise when we consider the complexity of the adaptive landscape of protein evolution and the cellular environment in which proteins evolve. In order to evaluate the validity of this model, we first have to consider the premise that hyperstability is selectively neutral. This is a difficult hypothesis to test, primarily because hyperstable proteins are rarely observed in nature—as shown in the figure. Directed evolution studies have shown that it is possible to evolve hyperstable proteins without

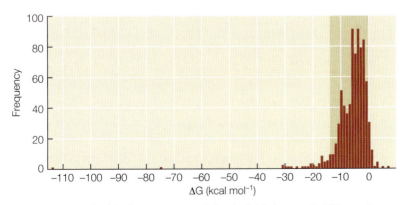

The distribution of Gibbs free energies (ΔG) for the folded state of >700 proteins from ~300 species from all domains of life. (After Kumar et al. 2006.)

BOX 4.4 *(continued)*

sacrificing protein function; however, these studies have also found that most mutations that increase stability actually lead to loss of function (Arnold et al. 2001). Thus, it is questionable whether hyperstable proteins are selectively neutral. Even if hyperstability is selectively neutral, we must also consider the other side of the stability distribution. A second major assumption of this model is that protein *instability* is deleterious and that natural selection acts to remove allelic variants that are prone to unfolding. This may not be a valid assumption because, although protein unfolding may lead to loss of function, in the context of the cellular milieu, protein instability may not lead to protein unfolding. Thus, protein instability may also be selectively neutral to a certain degree, due to the presence of molecular and chemical chaperones and stabilizing solutes. If this is the case, what maintains the narrow distribution of slightly negative ΔG values? How could proteins evolve to be marginally stable if both sides of the stability distribution are selectively neutral?

Perhaps the best answer to this question comes from data on the effects of counteracting osmolyte systems in chondrichthyan fishes, as presented in this chapter. Figure 4.10 shows how physiological ratios of the denaturant urea to the stabilizing solute

TMAO (2:1, urea:TMAO) preserve the functions and marginal stabilities of various proteins in species of shallow-living chondrichthyans. A striking pattern emerges when we examine how this ratio changes in the muscles of chondrichthyans captured at depth (see Figure 4.11C). The ratio of [urea]:[TMAO] is inversely proportional to the depth of capture, such that the destabilizing effects of pressure are offset by the tight regulation of the ratio of these two organic osmolytes. If protein hyperstability were selectively neutral, we would predict the ratio of [urea]:[TMAO] to be constant across depths. But that is not the case; thus, the data indicate that hyperstability is not selectively neutral. Otherwise, chondrichthyans would not have evolved to actively regulate the relative proportions of destabilizing versus stabilizing organic osmolytes in response to changes in pressure. The evolution of this flexible osmolyte system has allowed these species to exploit a wider environmental niche. Ultimately, this adaptation likely evolved due to the selective pressure to maintain marginal protein stability. Thus, at least for chondrichthyans (and by extension for all living organisms), it seems that marginal stability may in fact be optimal, as biochemists have long postulated, and not a product of neutral evolutionary processes.

WHY DON'T FISHES OCCUR IN THE OCEAN'S GREATEST DEPTHS? In view of the pressure-adaptive, depth-related changes in TMAO and urea shown in Figure 4.11, and the fact that bacteria, archaea, and invertebrates occur in the deepest regions of the ocean, one might expect to observe fishes in the deepest regions of the sea as well. However, neither bony fishes nor chondrichthyans have been captured or photographed in the deepest regions of the ocean, despite concerted efforts to do so. Bony fishes have not been observed at depths greater than ~8400 m, suggesting that they are absent from the deepest ~25% of the ocean (Jamieson et al. 2010; Yancey et al. 2014). The occurrence of chondrichthyan fishes decreases rapidly below ~3000 m, and no species of chondrichthyan has been reported below a depth of 4156 m (Priede et al. 2006). What causes these depth limitations? One proposal is that a putative limited food supply at great depth imposes depth limits on fishes. However, if food limitations were the cause of depth restrictions, why should bony fishes succeed at depths twice those at which chondrichthyans are found? And why should food limitations affect fishes so much more strongly than invertebrates? The work of Paul Yancey and colleagues points to another explanation, one involving the reliance of these two groups of fishes on piezolytes like TMAO.

Looking first at bony fishes, it would seem, in principle, that their abilities to elevate concentrations of TMAO to counteract effects of pressure would allow them to reach depths greater than ~8400 m. However, the analysis of Yancey and coauthors (2014), which incorporates considerations of osmotic regulation as well as pressure adaptation, suggests that there is an upper limit to TMAO concentrations in bony fishes. Thus, extrapolating the depth versus osmolality relationship for bony fishes developed using measured TMAO concentrations (see Figure 4.11A) as well as the concentrations of other osmotically active solutes in cellular and extracellular fluids, a bony fish would become isosmotic with seawater at a depth of 8000–8500 m (see Figure 4.11B). To penetrate to greater depths, a bony fish would need to become *hyperosmotic* to seawater, a condition that has never been observed in marine bony fishes. Bony fishes may have limited osmoregulatory abilities; for example, if they became hyperosmotic, they might be unable to extrude water entering down its concentration gradient from the sea to their more concentrated body fluids. Thus, they may be restricted to depths in the marine water column where they are, at most, isosmotic, but never hyperosmotic, to seawater.

In the case of chondrichthyans, Laxson and coauthors (2011) suggest several possible determinants of the relatively shallow maximum depth limits of this group. All of the proposed mechanisms involve, in one way or another, osmolyte or piezolyte effects. For example, the urea concentration of muscle tissue falls from a maximum near 370 mmol kg^{-1} in the shallowest species to 170 mmol kg^{-1} in the deepest-collected specimens. Might there, then, be a lower limit to the concentration of urea that an ancestrally urea-rich species can tolerate? The data in Figure 4.11C do not, in fact, suggest this; urea concentration shows a strong linear correlation with depth and no sign of leveling off at great depth. In contrast, TMAO concentration does seem to plateau at depths below ~2000–2500 m, albeit there are few data for fishes collected below this depth. What might restrict further increases in TMAO concentration at greater depth? Laxson and coauthors (2011) consider dietary restrictions (failure to obtain enough TMAO from ingested food) coupled with insufficient TMAO biosynthesis from choline as one possible, though untested, set of limitations. Two other sources of limitation to TMAO concentration are based on the biochemistry of TMAO itself. One limitation concerns TMAO synthesis from trimethylamine (TMA) by trimethylamine oxidase (TMAoxi). TMA is a toxic chemical and must be kept at low concentrations. If an organism with exceptionally high TMAO levels obtains its TMAO from "in-house" biosynthesis, high levels of TMA might also need to be present, which could have negative physiological impacts. To our knowledge it is not clear how bony fishes and chondrichthyans differ in their reliance on in-house biosynthesis of TMAO. If one group relies more heavily on de novo synthesis that requires high concentrations of TMA, then a lower level of TMAO might be dictated. A second limitation is that exceptionally high levels of TMAO may be "too much of a good thing." Thus, whereas TMAO can be viewed as a "good" molecule because of its strong abilities to stabilize proteins, nonphysiologically high concentrations of TMAO have been correlated with disease states, including atherosclerosis (Wang et al. 2011) and Alzheimer's-like plaque formation (Gazit 2002). Some and perhaps all of these pathological states could be consequences of overstabilization of proteins, including stabilization of disease-related aggregations of proteins that are thought to contribute to pathologies like Alzheimer's disease. The upper limits of TMAO concentration thus may be set by pathological changes induced by too high a level of this otherwise beneficial osmolyte. The highest measured levels of TMAO recorded in cells are near 390 mmol kg^{-1} in the deep-living snailfish (see Figure 4.11A) (Yancey et al. 2014). This TMAO concentration is ~100 mmol kg^{-1} greater than the highest value found in chondrichthyans, so the "too much of a good thing" hypothesis would not seem to apply to chondrichthyans. However, in light of the measurements made in bony fishes, might a concentration of TMAO near ~390 mmol kg^{-1} be the highest concentration that is physiologically tolerable (i.e., nonpathological)?

As Laxson and coauthors (2011) emphasize, the sources (dietary or de novo biosynthesis) of TMAO and its regulation in chondrichthyans, bony fishes, and invertebrates remain to be elucidated in detail. The finding that conspecifics of fishes and invertebrates collected at different depths have TMAO levels in their tissues that reflect depth of capture shows that marine animals have a capacity to regulate biosynthesis or dietary acquisition and retention of TMAO with a high degree of precision. How this regulation is achieved remains unknown. A minimal set of requirements for regulation must, of course, include a pressure-sensing mechanism that can elicit downstream effects on synthesis/acquisition or retention. Another open question is the rate at which piezolyte concentrations can be adaptively modified, for example, as a fish or invertebrate rises or sinks in the water column. The fact that diurnal vertical migration of the deep scattering layer is so prevalent in the world's oceans suggests that this type of regulatory response may be achieved in the time frame of minutes to hours. Studies of the mechanisms and velocities of regulation are clearly warranted.

Acanthogammarus victorii, a Lake Baikal amphipod.

TMAO IS A PIEZOLYTE IN DEEP-LIVING FRESHWATER ANIMALS TOO The discovery that TMAO concentrations in tissues of marine fishes and invertebrates begin to increase significantly at pressures slightly lower than 100 atm (depth of ~1000 m; see Figure 4.11) suggests that piezolytes may be important in deep freshwater lakes as well as in the sea. Lake Baikal, the world's deepest body of freshwater, has a maximum depth of 1642 m and animal life occurs in its deepest regions. Studies of the compositions of hemolymph and muscle of amphipod crustaceans (photo at left) collected in this lake at depths of up ~1200 m revealed that osmolality increased with depth of capture (Zerbst-Boroffka et al. 2005). Hemolymph osmolality rose due to increased levels of NaCl, and the osmolality of muscle cytoplasm increased due to rising concentrations of TMAO. These trends were evident in comparisons among different species with different depths of occurrence and among conspecifics collected at different depths. Thus, adjusting concentrations of piezolytes to offset perturbation by pressure appears to be important in deep freshwater lakes as well as in marine habitats.

DO OTHER ORGANIC MICROMOLECULES FUNCTION AS PIEZOLYTES? The abundance of bacteria and archaea throughout the ocean (there are in excess of 10^{33} individuals, more than the sum total of galaxies in the known Universe) and their enormous species diversity suggest that piezolytes not yet discovered in animals may be found in these two microbial domains. This expectation is also buttressed by the finding of novel thermoprotectant solutes in thermophilic archaea and bacteria, as discussed below. Unfortunately, there have been few studies of the organic solute compositions of deep-sea bacteria or archaea. However, Martin and colleagues (2002) cultured the marine bacterium *Photobacterium profundum* at 1 atm and 280 atm and found significant (about 2.5-fold) differences in total solute concentrations and qualitative shifts in which organic solutes were most abundant. Amino acids accounted for a substantial fraction of the organic solute pool, but their concentrations varied little between 1 atm and 280 atm. The organic solutes showing the largest changes were β-hydroxybutyrate (β-HB; shown at left) and its oligomers, which increased with growth pressure.

It is not known if β-HB, though termed a piezolyte by these authors, is capable of offsetting pressure perturbation of proteins. If the proteins of *P. profundum*

β-Hydroxybutyrate (β-HB)

are intrinsically stable enough under high pressure to enable optimal cell function—the optimal pressure for growth of *P. profundum* is 200–300 atm—then the upregulation of β-HB at these pressures may reflect other functions for this molecule. One alternative function of β-HB, known from studies of eukaryotic species, is in suppression of oxidative stress from ROS production (see Chapter 2). Whether ROS generation is higher at elevated pressures than at low pressures remains to be investigated, as is the antioxidant role of β-HB in *P. profundum*.

CAN REDUCTIONS IN PRESSURE BE STRESSFUL FOR DEEP-SEA ORGANISMS?   Up to this point in our discussion, we have taken a rather one-atmosphere-centric view of the issue by emphasizing only the damaging effects of *increases* in pressure on protein structure and function. We can turn the issue around and inquire whether proteins that have evolved adequate adaptations to allow optimal function at high pressure might be perturbed by *reductions* in pressure. Although few data are available on this question, studies of the synthesis and degradation of TMAO in a marine bacterium offer indirect evidence that shifts to either lower *or* higher pressures may be stressful for proteins.

Although the suites of piezolytes present in pressure-adapted bacteria and archaea remain to be characterized, involvement of TMAO is suggested by some studies. In *P. profundum*, an increase in culture pressure from 1 atm to 280 atm led to upregulation of the gene encoding TMAO reductase, an enzyme that reduces TMAO to trimethylamine (TMA) (Vezzi et al. 2005). This observation lends itself to multiple interpretations. One is that TMAO reductase is induced because TMAO is available in the cells grown under high pressure and that TMAO's concentration is regulated in a pressure-dependent manner. However, because TMAO reductase converts TMAO to TMA, induction of this enzyme at elevated pressure might, in fact, lead to reduced TMAO concentrations, assuming that other responses to pressure have not elevated TMAO levels, a possibility that remains to be studied. The importance of TMAO reductase in the context of adaptation to elevated pressure thus remains ambiguous. However, additional pressure-induced changes in gene expression in *P. profundum* suggest another interpretation of the upregulation of TMAO reductase. Specifically, culture of *P. profundum* at 1 atm pressure is marked by increased expression of several stress-response genes, including chaperones critical for protein folding, suggesting that, in this high-pressure-adapted bacterium, protection of proteins and the processes of translation and folding are more challenging at 1 atm than at the pressure optimum of the species, ~280 atm. This situation would arise if the proteins, and perhaps the protein synthetic apparatus itself, were optimized structurally for function at high pressure. If this is the case, then it might be advantageous for the cells to upregulate levels of TMAO at 1 atm, for example, by downregulation of TMAO reductase, as seen in this study. Therefore, for piezophiles with biochemistries adapted for optimal structure and function at high pressures, stress from low pressure might elicit the types of adaptive changes observed in less pressure-adapted species subjected to elevated pressures.

Additional support for this conjecture comes from a study of the deep-sea, pressure-adapted bacterium *Thermococcus barophilus*. In this bacterium, the concentration of mannosylglycerate (MG; see Figure 4.13), a strong protein stabilizer, decreased with increasing pressure of growth (Cario et al. 2010; see discussion in Lamosa et al. 2013), which is consistent with perturbation of protein structure by reductions in pressure. There is some evidence, then, that adaptation of proteins to pressure may share a common pattern with adaptation to temperature (see Chapter 3): Structure-function relationships are optimized, through evolutionary change in amino acid sequence, for a certain range of either physical factor, and exposure of proteins to values of temperature or pressure that are either above *or* below the normal physiological range are detrimental to activity. "Rescue" of protein function at either high or low extremes of these two physical stressors may depend on adjustments in concentrations of stabilizing organic micromolecules.

Buoyancy: Water-solute interactions and maintenance of depth

We turn now to a phenomenon that plays a major role in enabling pelagic organisms to maintain their position in the water column: the regulation of the densities of body fluids. We will see that the choice of organic osmolytes involves yet another criterion: the effects of the osmolyte on the structure and density of water.

BENEFITS ACCRUE FROM BEING ABLE TO ADJUST BUOYANCY   Pelagic organisms commonly benefit from an ability to maintain a particular depth in the water column, whether in a freshwater pond, a lake, or the sea. For example, phytoplankton benefit from being near the surface, which enables them to capture adequate sunlight to drive photosynthesis. However, nutrient levels may be higher in somewhat deeper water, so an ability to "dive down" to encounter better nutrient conditions and, when nutrient-replete, return to the surface is advantageous. Pelagic marine animals frequently make vertical excursions associated with feeding and ontogeny. Animals of the deep scattering layer seek depth during the day to escape predation and move upward in the dark to feed in food-rich shallow waters. Horizontal movement also benefits from a capacity to regulate buoyancy. For instance, costs of horizontal swimming by fishes are reduced significantly by maintaining neutral buoyancy. During horizontal swimming, the fraction of energy expenditure that is devoted to position (depth) maintenance varies with swimming speed (Alexander 1972). At 1 body length sec^{-1}, ~60% of power output is used for depth maintenance; at 3–4 body lengths sec^{-1}, ~20% of power is used for this function. These costs can be greatly reduced if buoyancy can be regulated by factors other than locomotory activity. Thus, whether for gaining food, nutrients, safety, or metabolic energy efficiency, regulation of body density—buoyancy—has critical advantages for aquatic organisms.

To achieve neutral buoyancy, a variety of mechanisms are used by different taxa. Buoyancy-regulating adaptations range from adjusting the volumes of gas-filled spaces (for example, teleost swim bladders or vacuoles in algal cells) to modulating the density of a portion of the body fluids, as seen in many invertebrates and phytoplankton (Denton 1963; Hochachka and Somero 1973; Sanders and Childress 1988; Boyd and Gradmann 2002). Here we examine solution-density mechanisms in which selection of appropriate inorganic ions and organic micromolecules enables diverse types of marine organisms to precisely adjust their densities and thereby remain neutrally buoyant at the depths where conditions are optimal. One of these mechanisms, the replacement of heavy ions like sulfate (SO_4^{2-}) with lighter ions like NH_4^+ without any change in solution osmolality relies in large measure on mass-adjustments of the solution (Denton 1963). However, even when the heavy-to-light ion replacement strategy is in play, there is a second contributor to solution density: the amount of water contained in a given volume of solution.

PARTIAL MOLAL VOLUME EFFECTS AND BUOYANCY REGULATION   Based on what has been presented earlier about the effects of different solutes on water structure, it should be apparent that solutes will differ in terms of whether they lead to an expansion or a contraction of water as they go into solution. Solutes that tightly organize water in their immediate vicinity may shrink the volume of water; conversely, solutes that disrupt water structure may increase water's volume. These differing effects can be quantified by deter-mining the partial molal volume ($\overline{V}^0$) of the solute (see Box 4.2). A water-structuring ion like Na^+, with its relatively low mass-to-charge ratio, will increase water density; it has a negative $\overline{V}^0$. Univalent cations of larger mass, for example, trimethylamine (TMA) or NH_4^+, reduce water density; they have a positive $\overline{V}^0$. Thus, for a given *volume* of solu-tion, a Na^+-containing solution will have more water molecules than a solution with TMA or another solute with a positive $\overline{V}^0$. Natural selection has exploited these differences in solute effects in the development of buoyancy mechanisms. This strategy is used by eukaryotic phytoplankton (Boyd and Gradmann 2002), cephalopods (Voight et al. 1994), and marine crustaceans (Sanders and Childress 1988).

The buoyancy regulating strategy used by the pelagic deep-sea shrimp *Notostomus gibbosus* (image at right) provides an excel-lent example of the exploitation of partial molal volume effects for modulating solution density (Sanders and Childress 1988). Much of the buoyancy regulation in this species is mediated through changes in composition of the large volume of fluid contained in a chamber formed by an enlargement of the dorsal carapace. Fluid in this chamber accounts for ~43% of the animal's total mass. Although this fluid is isosmotic with blood and seawater, its com-position yields a lift of 17.7 mg ml^{-1}, making the animal slightly positively buoyant in seawater. As is commonly seen in deep-sea organisms, heavy ions, chiefly divalent ions like sulfate and mag-nesium, are replaced with lighter ions, so some of the buoyancy

Pelagic deep-sea shrimp (*Notostomus gib-bosus*).

regulation is due to ion mass effects. However, the occurrence of TMA in the buoyancy chamber fluid at concentrations of ~130 mmol l^{-1} does not denote an ion-density adapta-tion. Thus, the molecular weight of TMA, 60, is almost three times that of the monovalent ion it primarily replaces, Na^+ (mass = 23). The basis for excluding Na^+ and accumulating TMA seems to be partial molal volume effects: TMA's large positive $\overline{V}^0$ (~ +70 cm^3 mol^{-1}, depending on temperature and other solutes present in the fluid) and the sodium ion's negative $\overline{V}^0$ (~ –3 cm^3 mol^{-1}) mean that this ion replacement provides a lift of 1.8 mg ml^{-1}. An even greater contribution to lift is provided by the high concentrations of NH_4^+ in the fluid in the buoyancy chamber. Ammonium ion concentrations in the fluid approach 300 mmol l^{-1}, generating a lift of 9.1 mg ml^{-1}. As in the case of TMA, lift generated by ammonium ion ($\overline{V}^0$ of ~ +18 cm^3 mol^{-1}) is due more to the ion's effects on reducing water density than to its slight mass difference from the sodium ion (18 versus 23). Reductions in concentrations of heavy divalent ions, notably magnesium ($\overline{V}^0$ of ~ –20 cm^3 mol^{-1}) and sulfate ($\overline{V}^0$ of ~ +14 cm^3 mol^{-1}), provide most of the remainder of the lift (5.4 mg ml^{-1}).

The reader may wonder at this juncture how TMAO, a close chemical relative of TMA, might fit into this picture of buoyancy regulation. TMAO, like other methylated nitrogen compounds, has a large positive $\overline{V}^0$, approximately +73 cm^3 mol^{-1} at 25°C (Withers et al. 1994). Withers and colleagues examined the contribution of TMAO to buoyancy regulation in elasmobranchs. They showed that TMAO and urea ($\overline{V}^0$ of ~ +60 cm^3 mol^{-1}) both contributed to buoyancy regulation, with the effect of the latter solute being approximately twice that of TMAO (3.7 g l^{-1} versus 1.8 g l^{-1}), in keeping with the [urea]:[TMAO] ratio found in the species.

In conclusion, buoyancy regulation mechanisms based on $\overline{V}^0$ differences among inor-ganic ions and organic micromolecules are widely employed by diverse marine taxa, and may be present as well in deep-living animals in bodies of freshwater like Lake Bai-kal, where [TMAO] has been found to increase with depth in amphipods. Convergent evolution thus is evident in multiple lineages, all of which have discovered the benefits of TMAO for maintaining position in the water column.

Before putting the TMAO-water structure story to rest, however, it is interesting to look more broadly at other possible influences of the volume-modifying effects of solutes. It seems possible that another process, one that is perhaps more fundamental than buoyancy adjustment, is also part of the story.

Conservation of organismal volume: Another role for TMAO and its kin?

The foregoing analysis of solute effects on solution volume and water density leads to an alternative perspective on the significance of the depth-related variation in [TMAO] observed among invertebrates, bony fishes, and chondrichthyans. The rise in [TMAO] with depth might suggest that buoyancy would rise regularly with depth across these taxa. However, one can ask, Why should a fish or an invertebrate necessarily need to increase its buoyancy when it increases its depth of occurrence? Perhaps the effects of depth-dependent variation in [TMAO] are strictly related to offsetting perturbation of proteins by rising pressure, and buoyancy effects are merely a by-product of this primary adaptive response. However, a different and arguably broader perspective on TMAO accumulation with depth can be developed by considering the consequences of compression of water by increases in hydrostatic pressure.

This alternative (or additional) rationalization of the [TMAO]-versus-depth relationship brings us back to the fundamental physiological challenge of maintaining cellular volume. In short, this new twist on the TMAO story proposes that the compression of an organism's fluids by pressure could be offset, at least in large measure, by the volume-expanding influences of TMAO and other solutes with large positive partial molal volumes.

To estimate the magnitude of the volume changes that would be caused by changing concentrations of water-expanding solutes, we can make the following calculation. Assume that [TMAO] in cellular fluids is ~400 mmol l^{-1} at a depth of 8000 m (equivalent to a pressure of 801 atms) (see Figure 4.11). Next, we use the $\overline{V}^0$ of TMAO, ~70 cm^3 mol^{-1}, to calculate how the addition of 400 mmol l^{-1} of TMAO to a biological solution would affect its volume. This addition would lead to an ~28-cm^3 expansion of solution volume per liter of solution. Glycine betaine (GB), because of its large $\overline{V}^0$ (+117 cm^3 mol^{-1}; Withers et al. 1994), could also contribute to the decompression effect. GB in chondrichthyans varies widely among species and shows no significant trend with depth. However, GB is found at concentrations up to ~25–35 mmol kg^{-1} in deep-living chondrichthyans, and this amount of GB could contribute an expansion effect near 3–5 cm^3 l^{-1}. Thus, the expansion due to methylammonium solutes (TMAO and GB) alone could be about 33 cm^3 l^{-1} of body fluids—obviously a "ballpark" estimate in the absence of values for volume-modulating effects of other solutes that may change with depth. Consider now the effect of pressure on the volume of water itself. The compressibility of water is ~46 parts per million atm^{-1}. Thus, 801 atm pressure would compress 1 liter of water by about 37 cm^3.

This close pairing of values leads us to ask: Do the largely offsetting effects of water expansion due to 400 mmol l^{-1} TMAO (plus a smaller amount of GB) and water compression by 801 atm pressure indicate that the linear rise in [TMAO] and its kin with depth represents a strategy for maintaining a near-constant volume for an organism, whatever its depth of occurrence? In view of the very tight conservation of cellular volume observed in the context of osmotic stress, might not the conservation of cellular volume in the face of compression stress from hydrostatic pressure be a critical element in the adaptive strategies of aquatic species? This is another issue that merits detailed study—for example, through measuring the net effects on solution volume of the complex mixture of organic and inorganic micromolecules in cellular fluids of deep- and shallow-living species.

It must be noted that selection for a strong depth-dependent rise in the concentration of TMAO, but not GB, speaks to the issue of whether protein stabilization or cell volume maintenance is the more important role of methylammonium solutes. Because GB has a much larger partial molal volume than TMAO, if volume effects were the more important contribution of methylammonium solutes to adaptation to pressure, then one would expect GB to be the solute that accumulates most regularly with depth. As shown, however, this is not the case. Thus, although we conjecture that cellular volume regulation and protein stabilization are both important contributions of methylammonium solutes in deep-living animals, the stronger protein-stabilizing effects of TMAO appear to make it the solute of choice for coping with increasing hydrostatic pressure, even though its partial molal volume is smaller than that of GB.

Hyperthermophilic (ultrathermophilic) microbes accumulate unique types of stabilizing organic micromolecules

Whereas upper thermal limits for Eukarya are near 55°C–60°C, heat-tolerant members of the domains Bacteria and Archaea can withstand much higher temperatures (see Chapter 3). These extreme heat tolerances are due in part to the types of macromolecular adaptations in proteins and membrane lipids discussed in Chapter 3. However, micromolecular adaptations also play important roles in extending the thermal tolerance ranges of bacteria and archaea. As a result of their combined adaptations in macromolecules and micromolecules, these microbes may be **thermophiles**, defined as species with optimal temperatures for growth between 60°C and 80°C, or **hyperthermophiles** (some authors instead refer to them as **ultrathermophiles**), with growth optima above 80°C. The most hyperthermophilic microbes belong to the domain Archaea and have the ability to grow at temperatures slightly above 120°C (Kashefi and Lovely 2003). Based on ribosomal RNA sequence analyses that have yielded a universal phylogenetic tree (Pace 1977), it appears that a hyperthermophile may have been the last universal common ancestor (LUCA) of contemporary life forms (Stetter 2006). However, this conjecture is still under debate and is far from having general acceptance among evolutionary biologists. In any event, thermophilic microbes definitely have a long and deep evolutionary history, and certain of the thermoprotectant molecules discussed below, notably a family of anionic organic solutes, may have appeared early in the history of life.

THE ORGANIC MICROMOLECULES THAT SUPPORT LIFE AT UPPER THERMAL EXTREMES   The micromolecules that confer high thermal tolerances in hyperthermophiles include certain of the uncharged (for example, trehalose) or zwitterionic (for example, amino acids) solutes that are also found in mesophilic species. However, in the most hyperthermophilic microbes, accumulation of unique negatively charged solutes (**Figure 4.13**) is often the dominant pattern observed (Martins et al. 1997; Empadinhas and da Costa 2006; Lamosa et al. 2013). Among these anionic solutes, di-*myo*-inositol 1, 3'-phosphate (DIP) appears to be especially critical in conferring thermal tolerance on cells of extremophiles (Martins et al. 1997; Faria et al. 2008). When the culture temperature of the hyperthermophilic archaeal species *Archaeoglobus fulgidus* was increased from 76°C to 87°C, the organic solute exhibiting the largest increase in concentration was DIP (**Figure 4.14A**). In the

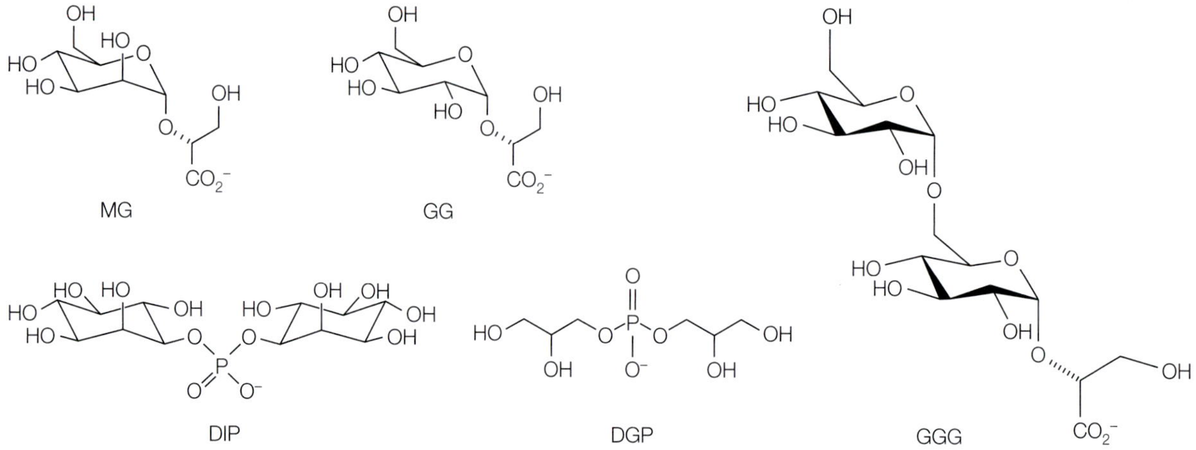

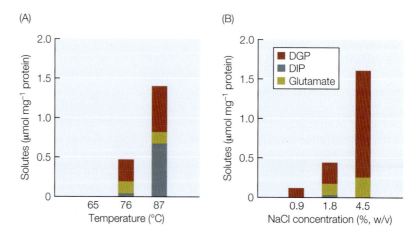

MG   GG

DIP   DGP   GGG

Figure 4.13 Chemical structures of anionic organic solutes accumulated in hyperthermophilic members of the domains Bacteria and Archaea. MG, mannosylglycerate; GG, glucosylglycerate; DIP, di-*myo*-inositol 1,3'-phosphate; DGP, diglycerol phosphate; GGG, α(1-6)glucosyl-α(1-2) glucosylglycerate.

most thermophilic species of archaea examined in the study of Martins and colleagues (1997), *Pyrodictium occultum*, which was cultured at 102°C, DIP was the only major organic solute accumulated. DIP and closely related solutes have in fact been detected only in hyperthermophilic archaea and bacteria (Martins et al. 1997). Another more recently discovered anionic solute in hyperthermophiles is α(1-6)glucosyl-α(1-2)glucosylglycerate (GGG), which is an even stronger stabilizer of some proteins than DIP (Lamosa et al. 2013). Discovery that hyperthermophilic archaea and bacteria use the same organic solutes in adaptation to extremely high temperatures suggests that there is something truly special about these particular molecules.

Because many hyperthermophilic microbes are tolerant of high salinities as well as extreme temperatures, the accumulation of an anionic organic solute like DIP may reflect adaptation to high osmolality instead of (or in addition to) adaptation to high temperature. To discern which types of solutes are involved in adaptation to these two forms of stress, culture experiments involving challenges from high salinity and high temperature

Figure 4.14 Effects of growth at different temperatures (A) and NaCl concentrations (B) on solute concentrations in the hyperthermophilic archaeon *Archaeoglobus fulgidus*. w/v, weight per volume. (After Martins et al. 1997.)

can be revealing. For example, culture of *A. fulgidus* at elevated [NaCl] led to an 11-fold increase in the intracellular concentration of diglycerol phosphate (DGP) and a slight rise in glutamate concentration (**Figure 4.14B**). However, no significant change occurred in the concentration of DIP, the solute whose concentration spiked most strongly during growth at high temperatures (see Figure 4.14A).

PROTEIN STABILIZATION BY DIP AND RELATED ORGANIC ANIONS   The intracellular accumulation of DIP and anionic solutes like DGP, mannosylglycerate (MG), glucosylglycerate (GG), and GGG during culture at high temperatures is, not surprisingly, a reflection of the extremely high stabilization of protein structures conferred by these organic anions. The sets of thermal stability measurements compiled in **Figure 4.15** illustrate several important points about the structure-stabilizing effects of a wide range of solutes, including the potent stabilizers DIP and GGG. Figure 4.15A shows how different solutes affect the melting temperature, T_m, of three proteins, malate dehydrogenase (MDH) from pig heart, egg-white lysozyme, and staphylococcal nuclease (SNase) from *Escherichia coli*. Figure 4.15B shows solute effects on MDH and SNase and includes data for GGG. The first point to note is the high degree of similarity in the effects on these three different proteins from widely different taxa (pig, bird, and bacterium). In almost all cases the ranking of stabilizing effects—the increase in T_m—is the same. This similarity in responses reflects the underlying thermodynamic mechanisms involving interactions among water, solutes, and the protein surface that are instrumental in establishing the stabilizing or

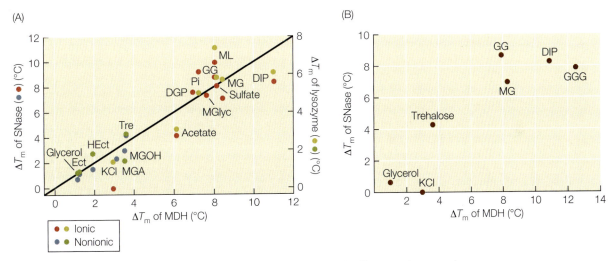

Figure 4.15 Solute effects on thermal stability. (A) The effects of different solutes on the thermal stabilities of three enzymes, staphylococcal nuclease (SNase) (blue and red symbols), lysozyme (light and dark green symbols), and malate dehydrogenase (MDH), against which data for the other two enzymes are plotted. The change induced in the melting temperature (ΔT_m) by 0.5 mol l^{-1} concentrations of different ionic (red and light green symbols) and nonionic (blue and dark green symbols) solutes provides an index of the stabilizing effects of the solutes. Note the similarities among the three enzymes in their responses to different solutes. DGP, diglycerol phosphate; DIP, di-*myo*-inositol 1,3'-phosphate; Ect, ectoine; GG, glucosylglycerate; GGG, α(1-6)glucosyl-α(1-2)glucosylglycerate; HEct, hydroxyectoine; MG, mannosylglycerate; MGA, mannosylglyceramide; MGlyc, mannosylglycolate; MGOH, mannosylglycerol; ML, mannosyl-lactate; Pi, phosphate; Tre, trehalose. (B) The effects of different solutes, including GGG, on stability of SNase and MDH. Solute concentrations were 0.5 mol l^{-1} except for GGG, which was present at 0.4 mol l^{-1} in the MDH assay. (A after Faria et al. 2008; B after Lamosa et al. 2013.)

destabilizing effects of solutes, as discussed below. A second major generalization can be based on these data: Ionic solutes, especially organic anions, are generally more effective stabilizers than uncharged or zwitterionic solutes, for example, glycerol and ectoine, respectively. The only ionic compound that has a relatively low stabilizing effect is KCl. Here, the chloride ion may be the "villain" responsible for the low level of stabilization found for this common salt. As discussed earlier in this chapter, increases in intracellular concentrations of Cl$^-$ during hyperosmotic stress are not common. Use of Cl$^-$ at high concentrations in in vitro studies does not, therefore, simulate biological responses to rising external osmolality and obscures a potentially stabilizing effect of its counterion, K$^+$.

The spectrum of stabilizing influences shown in Figure 4.15 suggests that the cell is able to effectively titrate the stability of proteins through modification of the types of solutes accumulated as well as their intracellular concentrations. As mentioned above, DIP and GGG are especially strong stabilizers—potent thermoprotectants—as one would predict from the selective accumulation of these solutes in the most hyperthermophilic species of archaea. The finding that DIP may not be accumulated in the cells of a hyperthermophile until culture temperatures are extremely high (see Figure 4.14A) suggests that there is a threshold level of stress involved in the choice of which organic micromolecule to use as a thermoprotectant or osmoprotectant. As stress from either physical or chemical perturbation increases, the cell may switch from less- to more-stabilizing solutes. The absence of DIP, DPG, and GGG in non-hyperthermophiles suggests that cell temperatures may need to exceed ~60°C–70°C before selection favors acquisition of the most powerful stabilizers. At lower temperatures, notably those within the range of eukaryotic species' thermal tolerance, highly stabilizing organic anions like DIP or DPG may confer excessive stabilization relative to what is needed for maintaining an appropriate marginal stability of protein structure. Thus, in eukaryotes the strongest known stabilizers occurring in this domain, TMAO and GPC, may be sufficient to provide the needed thermoprotection and osmoprotection. (Note: We are not aware of any side-by-side comparisons of the relative stabilizing potencies of methylammonium solutes like TMAO and GPC on the one hand, and DIP, DPG, and GGG on the other hand. However, the effects of trehalose on protein stability may provide a "calibration" function here. Trehalose is much less stabilizing than DIP, DPG, and GGG [see Figure 4.15B] but has similar stabilizing effects as glycine betaine when equal concentrations of the two solutes are compared [see Figure 4.19A]. Therefore, we conjecture that the methylammonium solutes prevalent in eukaryotes have much lower stabilizing potentials than the solutes—for example, DPG, DIP, and GGG—that hyperthermophilic archaea accumulate during growth at extremely high temperatures.)

There may be another reason why anionic stabilizers like DIP, DPG, and GGG are used in hyperthermophiles, however. This pertains to the possible loss of water-structure-based stabilization mechanisms at high extremes of temperature, a conjecture we return to in a later section of this chapter when we examine how loss of water structure at high temperatures may necessitate evolution of protein surface features and novel thermoprotectants that lead to protein stabilization by mechanisms somewhat different from those seen in mesophilic species (see the section titled "Protein surfaces, water structure, and thermoprotectants in hyperthermophiles: A case of coevolution?").

Missing from the picture of thermoprotectant micromolecular evolution is a clear understanding of the evolution of the pathways that lead to synthesis of the right sorts of thermoprotectant micromolecules for the thermal conditions experienced by a species. If the last universal common ancestor was a hyperthermophile, then the pathways for synthesis of highly stabilizing organic anions like DIP, DPG, and GGG may have arisen early in the evolution of life. Mesophilic and cryophilic species may have lost the ability to synthesize these compounds during subsequent adaptation to lower temperatures, much as many Antarctic ectotherms appear to have lost components of the heat-shock response during their long evolutionary history in stably ice-cold waters (see Chapter 5). It might be revealing to probe the genomes of mesophilic and cryophilic organisms, including bacteria, archaea, and eukaryotes, to see if there are remnants of the biosynthetic pathways in non-hyperthermophilic species.

Lastly in the context of synthetic pathways for protein-stabilizing micromolecules, laboratory syntheses of novel stabilizing molecules have shown significant success (Faria et al. 2008). Several of the stabilizers examined in Figure 4.15—including mannosyl lactate (ML), mannosylglycolate (MGlyc), and mannosylglycerol (MGOH)—are not known to occur naturally but were successfully synthesized in the laboratory, using the structure of a widely occurring thermoprotectant, mannosylglycerate (MG), as a guide. One of the synthetic stabilizers, ML, ranks highly among the natural stabilizers in its abilities to protect protein structure at elevated temperatures. Such studies done in controlled laboratory conditions would seem to have the potential for creating "super-stabilizers" and for shedding light on possible biosynthetic pathways operative in nature.

What are the roles of thermoprotectant organic micromolecules in eukaryotes?

Although several protein-stabilizing organic micromolecules occur in eukaryotes, it remains unclear as to how commonly the concentrations of these solutes are adjusted in the face of thermal stress. We have seen that there is a linear relationship between the concentration of the piezolyte TMAO and depth of occurrence of marine animals (see Figure 4.11), and that anionic solutes like DIP, DPG, and GGG are upregulated with increases in culture temperature for thermophilic and hyperthermophilic bacteria and archaea. Based on these findings, it would seem reasonable to expect that one or more stabilizing organic solutes would be upregulated in eukaryotic species exposed to rising temperatures. However, this possibility remains largely unexplored, especially in the case of animals that encounter large changes in cell temperature seasonally and/or diurnally.

Whereas there are few data on thermal responses of organic micromolecular systems in animals, studies of heat stress in yeast and vascular plants point to an important—and diverse—role for certain small organic solutes in rapid responses to elevated temperatures. Trehalose (shown at right), which we discuss at several junctures in this chapter as well as in Chapters 2 and 3, has many important functions under heat stress, including stabilizing membranes (Luo et al. 2010), protecting the native states of proteins (Singer and Lindquist 1998a,b), and reducing damage from ROS like superoxide and hydrogen peroxide (Luo et al. 2010). For example, in the yeast *Saccharomyces cerevisiae*, exposure to a relatively mild heat shock, 38°C, led to a rise in trehalose production and a concomitant scavenging by

Trehalose

trehalose of ROS (Benaroudj et al. 2001). Because increased production of ROS commonly occurs under heat stress, the ability of trehalose to scavenge ROS and prevent or at least reduce downstream damage to lipids, proteins, and nucleic acids is possibly as important a function as direct stabilization of proteins and membranes. As discussed in more detail later in this chapter, elevated trehalose concentrations may stabilize proteins in both their native and partially unfolded states and interact in important and complex manners with molecular chaperones—proteins, such as heat-shock proteins, that assist in the folding of nascent or denatured proteins (see Chapter 3) (Singer and Lindquist

1998a,b). The membrane-stabilizing effects of elevated trehalose concentrations under heat stress have been well characterized in the thylakoid membranes of plant (wheat) chloroplasts (Luo et al. 2010). As in yeast, increases in trehalose concentration achieve two important end-results: direct stabilization of membranes through interaction with the head groups of phospholipids, and reduction in damage from ROS. Trehalose's diverse roles in heat stress largely mirror those of a major organic osmolyte found in marine algae, β-dimethylsulfoniopropionate (β-DMSP), which exhibits heat-induced upregulation and protects cells from ROS damage (see Box 4.5; Raina et al. 2013). Lastly, in the context of trehalose's broad roles in helping cells cope with abiotic stress, upregulation of trehalose concentrations has been observed in response to both heat and cold stress. Thus, in the plant *Arabidopsis thaliana*, trehalose concentrations doubled after 4 h of heat stress at 40°C and increased eightfold after exposure to cold stress (4°C) for 4 days (Kaplan et al. 2004).

Other organic osmolytes with temperature-dependent accumulation patterns no doubt remain to be discovered, especially in taxa where trehalose is not regulated in environmentally responsive ways. Metabolomics technologies would seem well suited for this type of "molecular natural history" exploration. To date, however, metabolomic studies that characterize the composition of the pool of organic micromolecules in a cell have not given much attention to temperature effects. Where data are available, however, they do seem to point to a role for upregulation of thermoprotectant solutes under conditions of high temperature. A metabolomic study of two congeners of seagrasses from different latitudes showed upregulation of *myo*-inositol, sucrose, and fructose under heat stress, with greater increases observed in the northern species (Gu et al. 2012). None of these three solutes is an especially strong protein stabilizer, but the net effect of these increases in carbohydrate concentration could reflect what the authors term a reliance on "protective osmolytes" under heat stress. These authors also point out that chemical chaperones can influence the function of molecular chaperones. This fascinating phenomenon provides yet another example of why an inclusive, holistic perspective on cellular function that encompasses both macromolecules and micromolecules is critical for understanding how cells respond to stress.

Partnerships in protein folding and stabilization: Chemical chaperones as helpers of molecular chaperones

In Chapter 3 we discussed the varied and important roles of molecular chaperones in the context of thermal stress. Heat-shock proteins were seen to be critical elements in helping cells cope with temperature-induced unfolding of native protein structures. In the absence of stress from changes in temperature, constitutively expressed molecular chaperones play essential roles in helping guide the initial folding of nascent polypeptides and their translocation between cellular compartments. In view of the effects of organic micromolecules on protein stability, it seems likely that there could be an important interplay—perhaps a strong synergism—between the activities of molecular chaperones and chemical chaperones.

How, then, do these two distinct classes of chaperones interact to govern protein folding and stability in cells? The interactions discovered to date suggest that these interactions are ubiquitous and important for initial folding of newly synthesized proteins and for restoration of native structures of proteins denatured by stress. We will see once again that appropriate use of chemical chaperones can provide what may be a global solution to the problem at hand. For example, rapid upregulation of a protein-stabilizing solute—a chemical chaperone—may preclude the need to upregulate synthesis and activities of molecular chaperones, processes that require a great deal of energy and may occur more slowly than regulation of chemical chaperone concentrations. In the context of the time-course of cells' responses to heat stress, we will in fact see how sequential deployment of chemical and molecular chaperones plays out under conditions of cellular stress.

TMAO MODULATION OF HSP70 ACTIVITY IN ELASMOBRANCHS   In addition to the in vitro evidence for protein structural stabilization by TMAO (as in the studies shown in Figure 4.10), experimental manipulation of TMAO concentrations in living tissue has provided important insights into how this solute fits into the broader picture of protein stabilization. In particular, TMAO levels have been shown to influence induction of a common molecular chaperone, heat-shock protein 70 (Hsp70) (Villalobos and Renfro 2007). Using a tissue preparation from a marine elasmobranch, Villalobos and Renfro were able to study the effects of an acute heat stress on tissues that contained either normal TMAO levels or lacked TMAO. In the latter circumstance, heat stress induced Hsp70 to twice the level observed in TMAO-replete tissue. It thus appears that normal (constitutive) concentrations of TMAO provide considerable protection from heat stress.

A similar observation was made by Kolhatkar and colleagues (Kolhatkar et al. 2014), who studied the effects of TMAO on Hsp70 induction in red blood cells of the dogfish *Squalus acanthias*. The cells were subjected to heat stress in vitro in media that either lacked TMAO or contained physiological concentrations of the solute. Cells in media lacking TMAO exhibited induction of Hsp70, whereas cells in TMAO-replete medium did not. These authors concluded that TMAO has a thermoprotective function that can eliminate the need for mounting a heat-shock response.

This protection of the cell by the chemical chaperone TMAO could help save energy relative to the case where strong upregulation of Hsp70 occurs. As discussed in Chapter 3, the heat-shock response is energy-demanding; costs arise from upregulation of synthesis of mRNA and protein plus provision of the ATP needed to achieve refolding. Thus, to the extent that a protein-stabilizing micromolecule like TMAO can provide the needed protection of proteins, costs of dealing with thermal stress may be substantially lower than in situations where such chemical chaperone function is absent.

CHEMICAL CHAPERONES ESTABLISH OPTIMAL FOLDING ENVIRONMENTS FOR ACTIVITIES OF MOLECULAR CHAPERONES   The interactions between chemical chaperones and molecular chaperones have been examined in bacteria as well. Growth of *E. coli* in media with different osmolalities allowed establishment of widely different intracellular concentrations of organic osmolytes, notably glycine betaine (GB), a relatively strong protein stabilizer (Diamant et al. 2001). Cells grown in high-NaCl (0.5 mol l^{-1}) medium had four times the levels of GB found in cells grown in low-NaCl (0.17 mol l^{-1}) conditions. Concentrations of trehalose also increased, but only by about 1.2-fold. When the differently osmotically acclimated cells were subjected to heat stress at 47°C, the amounts of denatured and aggregated proteins differed substantially between acclimation conditions. Much less heat-induced aggregation was found in the cells grown at 0.5 mol l^{-1} NaCl, in which higher concentrations of protein stabilizers were present.

These in vivo studies were complemented with in vitro experiments that examined the effects of organic osmolytes (GB, trehalose, proline, and glycerol) on molecular chaperone function. The results of the in vitro studies suggest that chemical chaperones have important effects on the protein folding processes directed by molecular chaperones, and that these effects differ strongly among osmolytes. In a reconstituted chaperone system involving several molecular chaperones and co-chaperones, GB had the greatest stimulatory effect. Physiological concentrations of GB fostered higher rates of protein refolding than found with other osmolytes, some of which (notably, trehalose; see next section) actually interfered with refolding to the native state. At nonphysiologically high concentrations, all organic osmolytes tested reduced the ability of the molecular chaperone system to refold denatured proteins.

The explanation for these complex osmolyte-specific and concentration-dependent effects involves the need for proteins to have the correct marginal stability to allow optimal function, whether the function in question involves enzymatic activity or protein (re)folding. Nonphysiologically high concentrations of protein-stabilizing osmolytes could make the structures of the proteins comprising the chaperone refolding system too rigid to allow the system to work effectively. Molecular chaperoning entails binding of denatured proteins to the chaperone complex in a manner that allows the unfolded protein to try different conformational states until the native state is attained. Because multiple chaperones and co-chaperones are involved in refolding, it is also important for the refolding protein to be passed from one molecular chaperone to the next. If the concentrations of chemical chaperones are too high, the resulting high rigidity of the proteins in the refolding complex could impede—and perhaps block—the refolding process. Thus, a structural state for proteins that is neither too labile nor too rigid—a marginally stable state—may be important for protein refolding under conditions of heat stress.

Another factor that might be involved in the inhibition of refolding at nonphysiologically high osmolyte concentrations is solution viscosity. As the viscosity of a solution is increased, protein structural fluctuations ("breathing") are impeded (Beece et al. 1980). Conditions of high solution viscosity thus tend to stabilize protein structure, while at the same time inhibiting protein functions that rely on changes in conformation. Viscosity, then, is another factor that influences marginal stability. Whereas all stabilizing osmolytes increase solution viscosity, on a per-molecule basis trehalose increases viscosity much more than GB, glycerol, and proline (Diamant et al. 2001). Thus, in the reconstituted chaperone system we discussed above, the relatively strong inhibitory effects of elevated trehalose concentrations on protein refolding are conjectured to result from reduced conformational mobility. Interestingly, the protein-stabilizing effects of the four osmolytes examined by Diamant et al. (2001) are essentially the same when normalized to viscosity, rather than to solute concentration. We return to the potential importance of trehalose-induced viscosity effects in Section 4.7, when we examine anhydrobiosis—life without water.

TREHALOSE AND PROTEIN DENATURATION AND REFOLDING IN HEAT-STRESSED YEAST
In addition to stabilizing the conformations of native proteins under heat stress, trehalose may contribute to a minimization of heat-induced protein damage through its ability to stabilize partially unfolded proteins, preventing the occurrence of more extensive denaturation. These two roles of trehalose have been elucidated in studies of heat stress in yeast. In *Saccharomyces cerevisiae* (baker's yeast), trehalose concentrations quickly increase during heat stress, reaching concentrations near 0.5 mol l^{-1} (Hottinger et al. 1987). To examine the efficacy of these high concentrations of trehalose in protecting proteins under heat stress in vivo, Singer and Lindquist (1998a,b) subjected *S. cerevisiae* cells to heat stress and monitored protein refolding during recovery at a lower temperature. As expected, high concentrations of trehalose were protective of proteins. Trehalose led to stabilization of partially unfolded proteins, preventing large-scale denaturation and formation of aggregates of denatured proteins. After

cessation of heat stress, the trehalose-stabilized, partially denatured proteins could serve as substrates for heat-shock proteins, which restore native conformation. In essence, trehalose, as a chemical chaperone, reduced the energy needed for post-stress restoration of native structure through activities of molecular chaperones. However, for the heat-shock proteins to conduct their chaperoning activities, the high concentrations of trehalose induced by heat stress had to be removed from the system. When present at high concentrations, trehalose was so effective in stabilizing partially unfolded proteins that it interfered with their binding to heat-shock proteins and thereby hindered their refolding into the native conformation. The complex effects of trehalose in the context of heat shock are therefore appropriately viewed as "Yin and Yang effects" by Singer and Lindquist (1998a,b).

These studies with yeast emphasize the importance of examining heat stress in a holistic manner that incorporates a focus on both the small—chemical chaperones—and the large—molecular chaperones—protein-protecting constituents of cells. The efficacy of chemical chaperones and the dynamic temporal patterning of their synthesis and removal from the cell show an important facet of the response to heat stress that merits additional study. Thus, if their synthesis can be upregulated rapidly, faster than molecular chaperones can be produced, chemical chaperones may be an important "first line of defense" against cellular injury from heat. The contribution made by chemical chaperones to reductions in cellular damage from heat stress is almost certainly an underappreciated aspect of the thermal biology of organisms. In light of the earlier discussion of thermoprotectant micromolecules, we can view chemical chaperoning activity of chemicals like trehalose as one function of this broad set of organic micromolecules. The roles of these molecules in protecting other cellular structures, notably membranes, are also apt to be critical in establishing cells' abilities to cope with thermal stress. Further elucidation of how chemical chaperones and the broad class of micromolecules we have referred to as thermoprotectants contribute to cells' abilities to cope with challenges posed by temperature stress is likely to reveal other fascinating examples of how micromolecules contribute to retention of cellular integrity in the face of abiotic stress.

4.5 A Physical-Chemical and Thermodynamic Explanation for Solute Effects on Proteins

The discovery that organic micromolecules have such diverse effects on protein stability and function has led to an enormous amount of investigation into the underlying physical-chemical and thermodynamic bases of protein-solute interactions. These analyses have dealt with a wide variety of questions, but in all cases two points have been especially critical. One concerns the observed differences among organic micromolecules in their effects on proteins. What accounts for the range of stabilizing and destabilizing effects found among small organic molecules? The other involves the discovery that, for any given type of micromolecule, similar effects generally are seen among different types of proteins. What might the common element(s) be in these proteome-wide effects? Could water be the key player here? Are these common responses due to some ubiquitous element of protein structure, for example, certain types of amino acid side chains, or perhaps to the peptide backbone itself?

The analysis developed below provides at least partial answers to these and related questions about micromolecule-protein-water interactions. In addition to providing a physical-chemical and thermodynamic explanation for these interactions, this analysis will serve as a basis for interpreting the complementary evolutionary processes that have at once shaped protein structure and selected for the compositions of biological fluids. Much of the evolutionary activity in these two processes will be seen to involve the solvent-accessible surface areas of proteins, where interactions among water, organic micromolecules, and the peptide backbone establish optimal protein stability relationships. Through the detailed experiments and theoretical developments of Serge Timasheff, D. Wayne Bolen, and others, we are now able to describe in fine molecular detail the ways in which proteins, organic micromolecules, and water interact to establish the stabilizing or destabilizing effects of different biological solutions.

Preferential interaction/exclusion: Linking water to protein stabilization

The key unifying principle in these analyses is the concept of preferential interaction (see Bolen 2004 for a lucid presentation of this concept). In essence, this concept refers to the manners in which solvent (water) and a particular type of solute, for example, an organic micromolecule, distribute themselves around a protein in aqueous solution. There are two general distribution patterns that organic micromolecules can attain. When water is enriched near the protein surface relative to the surrounding bulk solution, there is a preferential interaction between protein and water. This condition is termed **preferential hydration**. From the standpoint of the distribution of the solute, this situation represents one of **preferential exclusion**. In the case of a solute that tends to be enriched near the protein surface relative to the solution as a whole, we speak of **preferential interaction** between protein and solute. We can also speak of solute binding in this case, keeping in mind that such binding is a weak and reversible interaction. From a large number of detailed analyses of preferential interaction/exclusion propensities for different organic solutes, an important general principle has emerged: *All of the naturally occurring stabilizing organic micromolecules used in adjusting the cell's osmolality and in fostering the appropriate structural stability of proteins, for example, in the face of stresses from temperature or hydrostatic pressure, are preferentially excluded from the protein surface.* That is, these stabilizing organic micromolecules favor preferential hydration of the protein. Conversely, denaturants like urea and guanidinium accumulate near and interact with the protein surface; they are preferentially interacting solutes. This simple dichotomy in solute distribution patterns has great explanatory power for accounting for differences in stabilizing/destabilizing effects among solutes, and therefore for explaining the selective bases for the solute accumulation patterns we observe in nature.

These differences in solute distribution tendencies are illustrated in **Figure 4.16**, which shows how the local osmolyte concentration near the protein surface (peptide backbone linkages in this case) differs among solutes with different stabilizing properties. Here, stability is indexed by a parameter termed the **transfer free energy**, Δg_{tr} (see below; Auton et al. 2011). Even though the total concentrations of the 10 osmolytes in solution are all 1 mol l^{-1}, the *local* concentrations near the protein's peptide backbone linkages span a range of ~0.2 mol l^{-1} to 1.8 mol l^{-1}. These data thus provide a striking illustration of the phenomena of preferential solute exclusion (local concentration < 1 mol l^{-1}) and preferential solute interaction (local concentration > 1 mol l^{-1}).

What mechanisms underlie preferential exclusion of stabilizing solutes?

The painstaking work done by Timasheff and others to demonstrate preferential exclusion of stabilizing solutes from the protein surface led almost at once to the next level of analysis: the search for the features of the protein-solute-water system that are responsible for this exclusion mechanism.

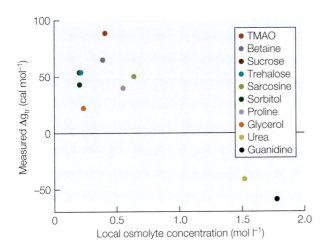

Figure 4.16 The concentrations of osmolytes near the protein surface (peptide backbone linkages) plotted against the transfer free energy (Δg_{tr}), which characterizes movement of a peptide backbone linkage from pure water to a 1 mol l^{-1} solution of the osmolyte. All protein-stabilizing solutes (e.g., TMAO and glycine betaine) are characterized by a positive Δg_{tr}, meaning that movement of the peptide backbone into a solution with these stabilizing solutes is thermodynamically unfavorable. Thus, stabilizing solutes are depleted—preferentially excluded—from solution adjacent to the backbone surface. In contrast, Δg_{tr} values for denaturing solutes (urea and guanidine) are negative; it is thermodynamically favorable for the backbone linkage to move into a solution containing urea or guanidine. These denaturants, then, are relatively enriched near the protein, as discussed in the text. Note the 2:1 ratio of Δg_{tr} for urea:TMAO (see text). (After Street et al. 2006.)

THE PEPTIDE BACKBONE IS KEY TO SOLUTE EFFECTS   The basis for the preferential exclusion of strongly stabilizing solutes from the protein surface can arise from several factors, including steric considerations—bulky solutes may have limited ability to directly interact with proteins—and water structure-mediated mechanisms. Thus, for example, because TMAO is a strong water-structurer, its tendency to be preferentially excluded from the water near the protein surface may be a reflection of the high degree of order found in the water that hydrates this solute. This organized water may not readily mix with the organized water associated with the surface of the protein.

How does the propensity for a solute to either separate itself from or interact with the surface of the protein determine its stabilizing or destabilizing effects, respectively, on the protein? To address this question, a thermodynamic analysis is helpful in order to build a foundation for understanding the energy changes that accompany the folding or unfolding of protein structure. The central thermodynamic variable here is the transfer free energy (Δg_{tr}) that accompanies movement of a protein group, either a specific amino acid side chain or a peptide backbone linkage, between a solution of pure water and one containing the solute of interest. If the addition of the solute leads to a negative Δg_{tr}, this means that the protein group is more soluble in the presence of the added solute than it is in pure water. Conversely, a positive Δg_{tr} means that the transfer of the protein group from pure water into the solution containing the solute is energetically unfavorable.

Through the use of models of protein structure based on amino acid sequence plus three-dimensional conformation, it is possible to make accurate predictions of what the total (net) transfer free energy would be during protein unfolding into either a pure aqueous phase or a solution with one or a combination of organic solutes. The work of Bolen and colleagues (see Bolen 2004, Auton et al. 2011) has shown convincingly that the transfer of the peptide backbone linkage, not amino acid side chains, is of dominant thermodynamic importance. As shown in **Figure 4.17**, the Δg_{tr} values of almost all amino acid side chains, whether polar or nonpolar, are negative in the presence of the structure-stabilizing solute TMAO. Thus, when we restrict analysis to only side chains, TMAO appears to be a denaturant (like urea). It is only the transfer of peptide backbone linkages from water to a solution of TMAO that is highly unfavorable energetically. Thus, the *net* effect of an organic solute like TMAO on protein stability represents the algebraic

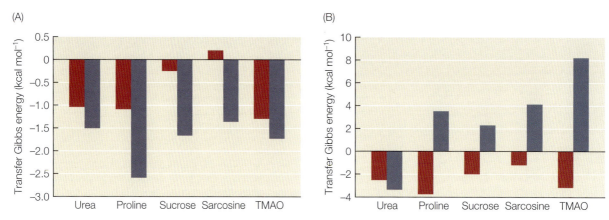

Figure 4.17 Transfer Gibbs energy accompanying transfer of amino acid side chains (A) and peptide backbone linkages (B) from pure water to 1 mol l^{-1} solutions of different organic solutes (urea, proline, sucrose, sarcosine, and TMAO). The protein studied was a modified ribonuclease. (A) Red bars denote hydrophobic side chains (W, F, Y, L, I, P, V, M, and A); blue bars denote polar/charged side chains (T, S, Q, N, K, R, H, E, D, and G). With the exception of transfer of hydrophobic side chains to 1 mol l^{-1} sarcosine, all water-to-solute transfers are thermodynamically favorable (Δg_{tr} is negative), illustrating that the stabilizing or destabilizing (urea) effects of these small organic molecules are not principally determined by the contributions of side chains to the overall Δg_{tr}. (B) The blue bars show the total contribution of the peptide backbone linkages to protein stabilization; the red bars show the total contribution to stabilization by side chains. Here, distinct differences are seen between the four stabilizing solutes and the denaturant urea, showing that the peptide backbone linkages are the more critical determinant of solute effects. For stabilizing solutes, Δg_{tr} is consistently positive, in accord with their stabilizing effects on proteins, and greater in absolute value than the Δg_{tr} due to side chain effects. Thus, a net stabilization (net positive Δg_{tr}) results when stabilizing solutes are present. Not illustrated in this figure is the algebraic additivity of effects between urea and stabilizing solutes observed on net Δg_{tr} (see Wang and Bolen 1997). (After Qu et al. 1998.)

sum of (1) the destabilizing effects of TMAO arising from amino acid side chains, which become more soluble in the presence of TMAO; and (2) the stabilizing effects arising from peptide backbone groups, whose exposure to a solution containing TMAO is energetically unfavorable.

In the broader context of the net effect on protein stability arising from all of the solutes present in the solution, it is important to reemphasize that stabilizing organic micromolecules differ in how strongly they influence Δg_{tr}. Many studies (see Bolen 2004; Street et al. 2006; Auton et al. 2011) have shown the following rank order of stabilizing effects: TMAO > sarcosine > sucrose > proline. As Bolen (2004) points out, there is truly a *continuum* of solute effects. Perhaps one reason for the complex mixtures of organic micromolecules found in cells is the need to balance the diverse influences of these solutes on stability and solubility. The balancing of urea and TMAO discussed earlier may be but one example of a more general evolutionary pattern.

SURFACE POLARITY OF SOLUTES GOVERNS PREFERENTIAL INTERACTION AND EXCLUSION   The next level of analysis asks, What structural properties of organic osmolytes govern their tendencies to either interact with or be excluded from the peptide backbone linkages of a protein? Studies of Street and colleagues (2006) have provided strong evidence that the dominant structural characteristic determining this tendency for interaction is the fraction of the organic solute's surface that is polar (**Figure 4.18**). The

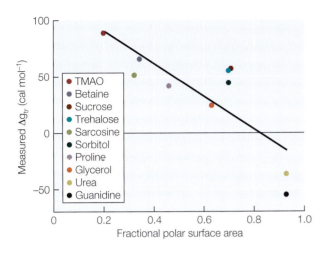

Figure 4.18 Relationship between the fraction of osmolyte surface that is polar and the stabilizing or destabilizing effect of the osmolyte on protein structure. The relationship shown in this figure mirrors the trend shown in Figure 4.16. Note the 2:1 ratio seen between Δg_{tr} values for urea and TMAO (see text). (After Street et al. 2006.)

relationship between the stabilizing or destabilizing influence of an organic solute at 1 mol l^{-1} concentration (indexed as Δg_{tr}) and its fractional polar surface area is highly significant (r^2 = 0.81). The larger the fraction of the micromolecule's surface that is polar, the greater is its propensity to interact with the peptide backbone linkages of the protein and thereby foster denaturation.

In the context of the differing effects of organic solutes on protein stability, the difference in Δg_{tr} between urea (–41 kcal mol^{-1}) and TMAO (+89 kcal mol^{-1}) mirrors the concentration ratio of approximately 2:1 at which these two solutes are essentially fully counteracting in their effects on protein structure and stability (at least under conditions of 1 atm pressure) (see Figures 4.16 and 4.18). Extending this point a bit further, if one were able to obtain a complete "recipe" for the full suite of organic solutes in a cellular solution, it should be possible, using Δg_{tr} information, to calculate the summed stabilizing influence of the entire pool of organic micromolecules.

VISCOSITY EFFECTS OF SOLUTES ALSO GOVERN THEIR STABILIZING OR DESTABILIZING EFFECTS   Although the phenomena of preferential exclusion and preferential interaction go a long way toward accounting for differences among organic solutes in their protein stabilizing or destabilizing abilities, other factors contribute to these abilities as well. As we touched on earlier, one factor is the influence of a solute on solution viscosity, which varies in response to the types and concentrations of solutes added to the solution. As shown by early studies of Beece and colleagues (1980), increases in solution viscosity tend to inhibit the breathing of proteins, and this effect can stabilize native structure. The data in **Figure 4.19** extend this type of observation by showing how four different stabilizing micromolecules—trehalose, proline, glycine betaine, and glycerol—stabilize proteins under conditions of heat stress and alter solution viscosity (Diamant et al. 2001). The per-molecule protein-stabilizing effects of these four micromolecules differ; trehalose and glycine betaine have greater stabilizing effects than glycerol and proline (see Figure 4.19A). Concomitantly, the four solutes differ in their effects on solution viscosity. Trehalose has by far the strongest effect on viscosity (see Figure 4.19B). When the protection of the protein from denaturation is plotted against viscosity (see Figure 4.19C), it is clear that the solutes are relatively similar in their stabilizing effects when normalized not to their concentrations but to their effects on viscosity. The effectiveness of trehalose in increasing viscosity, and thereby protein stability, may be important in a variety of biological contexts. We have already seen how trehalose can act as a chemical chaperone during heat stress. As we show later in the discussion of anhydrobiosis, accumulation of extremely high concentrations of trehalose may have such powerful

(A)

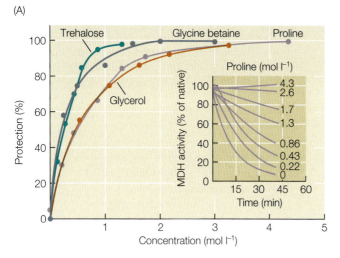

(B)

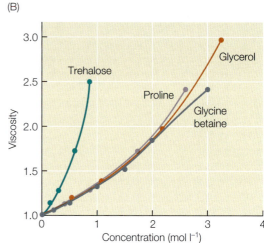

(C)

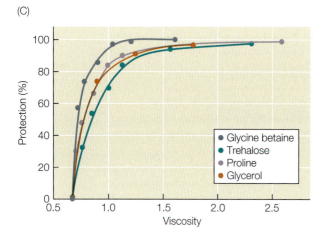

Figure 4.19 Effects of four organic micromolecules—glycine betaine, trehalose, proline, and glycerol—on the protection of mitochondrial malate dehydrogenase (mMDH) against heat denaturation. (A) Concentration dependence of thermoprotection by the four solutes. (B) Viscosity of the solution as a function of solute concentration. Viscosity is expressed as the time (minutes) needed for 1 ml of solution to flow through the capillary device used to measure viscosity. Viscosity units are thus proportional to time units. (C) Protection as a function of solution viscosity. All four solutes stabilize similarly when data are normalized to viscosity, but not to solute concentration (A). (After Diamant et al. 2001.)

effects on protein stability that (reversible) suppression of metabolic rates can occur. Protein stabilizers, then, can be exploited by cells as "on-off switches" of metabolism when it is advantageous for the organism to transition between periods of normal, active metabolism and greatly reduced metabolism (or even enter an ametabolic state). High concentrations of trehalose may establish a state of suspended animation in which the structures of macromolecules and membranes are highly stabilized such that, when trehalose concentrations are reduced during the arousal process, they are prepared to resume active function without delay.

How viscosity effects on protein stability play out in nature in response to different types of abiotic stresses is a question that merits additional study on several fronts. In the case of temperature effects, the increased viscosity of the cytoplasm at reduced temperatures might lead to increases in protein stability. Increased rigidity of structure might lead to increases in the energy barriers for conformational changes that occur during catalysis. Thus, viscosity-induced elevations in protein stability might increase the temperature dependence of biochemical processes (Q_{10} effects; see Chapter 3). The intrinsically less stable structures of orthologous proteins of cold-adapted species might be in part a reflection of the need to counteract viscosity-induced stabilization.

Such effects could be very great when viscosity-enhancing solutes are accumulated at low temperatures. For example, the accumulation of high concentrations of glycerol observed in many cold-tolerant ectotherms (see the discussion of freeze-resistance in Chapter 3) may have an important effect on protein stability and therefore on metabolic rates. Glycerol concentration may serve as a rheostat for down- or upregulating metabolism as environmental temperatures change.

PROTEIN SURFACES, WATER STRUCTURE, AND THERMOPROTECTANTS IN HYPERTHERMOPHILES: A CASE OF COEVOLUTION?   We end this discussion of the thermodynamic mechanisms of protein stabilization by raising a conjecture about the selective basis for the strong reliance of hyperthermophiles on negatively charged thermoprotectant solutes like DIP and GGG (see Figure 4.13). We hypothesize that the reliance on these strongly stabilizing anionic solutes may reflect a coevolutionary process in which the addition of charged groups to protein surfaces was linked with selection of appropriate thermoprotectants for interaction with these charged groups. This coevolution of protein surface features and thermoprotectants is driven by effects of extremely high temperatures on water structure.

The reasoning behind this conjecture is as follows. Because the relative stabilizing potentials of solutes are correlated with the degree to which the solutes are excluded from the protein surface (the preferential exclusion/interaction hypothesis), the prevalence of anionic thermoprotectants in hyperthermophiles suggests that there may be an unusually strong tendency for preferential exclusion of these anions. This exclusion could be driven by charge repulsion, rather than by effects mediated through water structure. A charge-repulsion mechanism for solute exclusion could prove to be very important at extremely high temperatures, where exclusion mechanisms based on water structure might be less feasible, in view of the reduction in hydrogen bonding and water structure in the presence of high levels of kinetic energy. At high temperatures, the distinction between being a water-structure breaker and a water-structure enhancer may be largely lost due to the effects of high temperature on the hydrogen bonding potentials of water. Thus, solute exclusion mechanisms, such as those discussed for TMAO, that are based on differences in water organization may not be very effective at the temperatures at which hyperthermophiles function optimally.

If charge repulsion accounts for preferential exclusion of solutes like DIP and GGG, one would predict that the surfaces of proteins of hyperthermophiles would be found to have more anionic (carboxylate) groups than proteins of less heat-tolerant species. Are there, in fact, differences in the surface properties of proteins from psychrophiles, mesophiles, thermophiles, and hyperthermophiles? Sælensminde and colleagues (2007) have examined the amino acid compositions of these four classes of microbes in the domains Bacteria and Archaea and have found that both the core and the surface of proteins show significant differences correlated with optimal growth temperature (OGT). The hydrophobicity of the protein core increases with OGT, as discussed in Chapter 3. The surface of the proteins examined also exhibits significant changes, notably in the number of charged residues. Across all species, glutamate shows the strongest positive correlation with OGT, consistent with the conjecture that increased numbers of anionic sites would facilitate preferential exclusion of anionic thermoprotectants like DIP and GGG from the protein surface. This, then, is evidence that loss of water structure at extremely high temperatures might favor evolution of protein surfaces and thermoprotectants that can enable native protein structures to be maintained at temperatures in excess of 120°C. In other words, the fundamental mechanisms by which proteins attain their native stabilities may differ between hyperthermophilic microbes and other, less heat-tolerant species.

Complicating the picture, however, is the observation that lysine content of the protein surface also shows a significant increase with OGT. Thus, both positively and negatively charged side chains increase in abundance during evolution at high temperatures. How does one explain this pattern? Parallel increases in anionic and cationic side chains would be expected if the primary function of these increases in charged residues was to establish additional salt bridges, a type of bonding that is highly stabilizing of protein structure (see Chapter 3). However, a subsequent analysis found that increasing the numbers of charged residues on the surface of thermophilic proteins is not accompanied by formation of additional salt bridges (Sælensminde et al. 2009). Thus, the charged amino acid side chains are principally involved in interactions with solvent, rather than binding to side chains of opposite charge.

Explanations for how the significant rise in surface charge, notably the rise in lysine residues, favors stabilization involves a novel mechanism of protein stabilization based on what has been termed **negative design** (Berezovsky et al. 2007). This concept focuses on changes in amino acid sequence that have the effect of making unfolded and partially unfolded states of proteins energetically unfavorable. If like-charged residues are well separated in the native state and if unfolding of the protein were to bring like-charged residues into closer juxtaposition, then protein stability would be enhanced by an increase in such charged side chains.

Another factor in governing stability of proteins with high surface charge involves a role for protein-thermoprotectant interactions, namely neutralization of lysine residues by DIP, GGG, and other anionic thermoprotectants. This interaction between anionic thermoprotectants and cationic side chains might contribute to protein stability by reducing charge repulsion among the lysine residues. Recall the importance of accumulating high concentrations of K^+ in cells of extreme halophiles to neutralize the negative charges of glutamyl and aspartyl side chains on the proteins and ensure proper folding.

More study of the surfaces of differently thermally adapted proteins is required to resolve how the demonstrated differences in surface charge lead to adaptive modification of stability. Are these surface changes driven by their effects on the intrinsic stability of proteins, through a mechanism like negative design, or by the need for a mechanism of preferential exclusion that can work effectively at high extremes of temperature? In view of the likely coevolution of protein surface properties and thermoprotection by organic micromolecules, we certainly need additional research that will enable us to create a more integrated view of protein-solute-water interactions in governing protein stability. As suggested by the discussion of anionic thermoprotectants in hyperthermophiles, the temperatures at which protein-solute-water interactions occur will also figure importantly in this evolutionary picture.

Temperature and selection of osmolytes: The example of β-DMSP

The temperature dependence of protein stabilization by organic micromolecules has received relatively little study. What often seems to be an unspoken assumption in this field of study is that, if an organic micromolecule stabilizes proteins at any one temperature, it is likely to be stabilizing at all temperatures. However, in at least some cases, it is known that an osmolyte that is stabilizing of protein structure at one end of the temperature spectrum may be destabilizing at other temperatures. This temperature dependence of stabilization relationship has been shown for β-**dimethylsulfoniopropionate** (β-DMSP), a multifunctional organic micromolecule that accumulates to high levels in marine algae, some seagrasses, and corals. β-DMSP plays important roles in ROS detoxification (Sunda et al. 2002; Raina et al. 2013) and grazer deterrence (Wolfe et al. 1997), as well as in osmotic regulation (**Box 4.5**).

β-DMSP concentrations in algae have been shown to be temperature-dependent, but in seemingly inconsistent ways. In some species of algae, β-DMSP concentrations rise in the heat; in other cases, concentrations are elevated in the cold. What might account

BOX 4.5

β-Dimethylsulfoniopropionate (β-DMSP)—a paradigm of multiple functions for a single micromolecule

For illustrating the multiple functions that a single type of micromolecule can perform, β-dimethylsulfoniopropionate (β-DMSP) provides a most appropriate example (**Figure A**). This molecule

Figure A β-Dimethylsulfonioproprionate (β-DMSP).

and its breakdown products play roles in osmotic and volume regulation, cryoprotection of proteins (Karsten et al. 1996), ROS detoxification (Sunda et al. 2002; Raina et al. 2013), grazer deterrence (Wolfe et al. 1997), signaling between corals and microbes, and provision of an overflow mechanism in situations where photosynthetic species face challenges from excess energy and levels of reduced compounds (for review, see Stefels 2000). β-DMSP is produced in some green plants, including seagrasses of the genus *Spartina* and sugarcane; marine macroalgae; free-living microalgae, including some coccolithophores; and symbiotic dinoflagellates of the genus *Symbiodinium*, which populate corals and provide their cnidarian host with the bulk of their requirements for reduced carbon compounds. Recent investigations have revealed a capacity for β-DMSP production in a limited set of marine invertebrates as well, notably certain reef-building corals (Raina et al. 2013). This finding was unanticipated because β-DMSP synthesis was thought to be restricted to photosynthetic species. As we will see, there may be a critical benefit to the host in the coral-dinoflagellate symbiosis in having this biosynthetic ability as well.

In the context of discerning adaptations from exaptations in biochemical evolution, the discovery that the synthetic pathways for β-DMSP may have evolved separately in at least three different lineages

of green plants and algae (Stefels 2000) raises the question of what was the original—that is, the true adaptive—function of β-DMSP in these different photosynthetic lineages. Was β-DMSP selected for the same primary function in each lineage, such that all other functions are derived exaptations? Or was β-DMSP's original (adaptive) function different in the different evolutionary lines?

In unicellular marine algae, β-DMSP is one of the most abundant organic molecules. It occurs intracellularly at concentrations up to ~400 mmol l^{-1} (Sunda et al. 2002), and it may account for up to 10% of the total carbon fixed by marine phytoplankton (Reisch et al. 2011). Because of β-DMSP's high concentration in marine algae, it has been logical to focus on its role as an intracellular osmolyte. However, the chemistry of β-DMSP and its breakdown products suggests another critical function, namely in the context of environmental stressors that lead to production of ROS. β-DMSP and some of its breakdown products—notably dimethyl sulfide (DMS), acrylate, dimethylsulfoxide (DMSO), and methane sulphinic acid (MSNA) (**Figure B**)—are potent antioxidants (see Chapter 2) that can scavenge hydroxyl radicals and other

Dimethylsulfoxide (DMSO)

Acrylic acid

Dimethyl sulfide (DMS)

MSNA

Figure B Some breakdown products of β-DMSP.

BOX 4.5 *(continued)*

types of ROS. The efficacy of β-DMSP, acrylate, DMS, DMSO, and MSNA is greater than that of the commonly occurring oxygen radical scavengers glutathione and ascorbate (Sunda et al. 2002). Regulation of β-DMSP production in algal cells reflects the amounts of ROS being produced due to various types of external abiotic stresses and intracellular chemical activities. Increased UV radiation, CO_2 limitations, increased water temperature, and iron limitation—four factors that promote ROS generation in algae—all lead to increased β-DMSP concentrations in algal cells (Sunda et al. 2002; Raina et al. 2013).

The studies by Raina and colleagues (2013) on symbiont-free cells of coral species of the genus *Acropora* provide deeper insights into the role of β-DMSP and its metabolites in protecting cells from damage from ROS. This study was the first to report that β-DMSP is synthesized in an animal and that the genes needed to achieve this biosynthesis were present in the cnidarian's genome. The importance of this host cell-localized capacity for synthesizing β-DMSP in the context of coral bleaching (expul-

sion of dinoflagellate symbionts) was demonstrated by the finding that heat (32°C)-stressed corals had severely damaged algal symbionts that were unable to synthesize high levels of β-DMSP. Heat stress of corals may lead to a form of symbiont swapping, whereby a more heat-tolerant strain of *Symbiodinium* is taken up by the host, to replace the more heat-sensitive strain that is lost during bleaching (see Chapter 5). To sustain the integrity of the host cell during bleaching and subsequent reinfection with new, better-adapted symbionts, production within the host's cells of β-DMSP may be critical.

In conclusion, the multiple evolutionary pathways leading to an ability to produce β-DMSP suggest the importance of this small molecule to a wide variety of photosynthetic organisms and even certain species of animals, namely reef-building corals. β-DMSP's many functions show, on the one hand, the multifunctionality that organic micromolecules can possess, but, on the other hand, make it challenging for students of biochemical evolution to distinguish adaptation from exaptation.

---

for these inconsistent trends? The general answer to this question is that β-DMSP serves different functions at different temperatures. The finding that β-DMSP concentrations rose with heat stress in a coral (a shift in culture temperature from 27°C to 32°C led to elevated β-DMSP concentrations in the cnidarian host and its algal symbionts; Raina et al. 2013) is consistent with β-DMSP serving as a protein stabilizer under heat stress. However, there may be another adaptive characteristic of β-DMSP that accounts for this shift, namely its ability to scavenge ROS. Elevated temperatures can cause large increases in ROS production, so this could be the primary advantage of increasing β-DMSP levels under heat stress (see Chapter 2).

The increase in β-DMSP at low temperatures may be more readily interpreted in the context of stabilization of native protein structures. In cold-adapted macroalgae and microalgae, low temperatures strongly upregulate β-DMSP, which protects these algae against damage from near-freezing conditions. For example, in a polar macroalga (*Acrosiphonia arcta*), β-DMSP levels were higher in cultures maintained at 0°C than at 10°C (Karsten et al. 1992, 1996; Stefels 2000). Likewise, in a unicellular alga (*Tetraselmis subcordiformis*) cultured at temperatures from 5°C to 23°C, β-DMSP levels increased eightfold as culture temperatures were reduced (Sheets and Rhodes 1996). The cold-protective function of β-DMSP is suggested by earlier findings made with a close relative of this molecule, dimethylsulfoxide (DMSO), which was found to protect proteins from cold denaturation (Arakawa et al. 1990). β-DMSP also has been shown to be an effective cryoprotectant of proteins that are susceptible to cold denaturation, for example, the glycolytic enzyme phosphofructokinase (Arakawa et al. 1990; Nishiguchi and Somero 1992) and malate dehydrogenase (Karsten et al. 1996). However, when examined in the context of stabilization against heat denaturation, β-DMSP failed to strongly stabilize proteins and could even be destabilizing in some instances. For one protein, lactate dehydrogenase

(LDH), 200 mmol l^{-1} β-DMSP did stabilize the protein, but a 300 mmol l^{-1} concentration gave no stabilization; for another enzyme, glutamate dehydrogenase (GDH), β-DMSP was destabilizing at all concentrations (Nishiguchi and Somero 1992).

The basis of this temperature dependence of protein stabilization was hypothesized to derive from the thermal dependence of the hydrophobic effect (Arakawa et al. 1990; see Chapter 3). Thus, because hydrophobic interactions are enhanced by elevated temperatures, interactions of the methyl groups on β-DMSP with nonpolar groups located on or near the protein surface would increase with rising temperature, perhaps favoring unfolding of protein structure. In fact, this mechanism has been proposed to explain the destabilizing effects of DMSO at high temperatures (Arakawa et al. 1990). Thus, preferential exclusion of osmolytes with strong potentials for hydrophobic interactions may be reduced with rising temperatures.

In summary, temperature-dependent effects of the sort observed with β-DMSP and DMSO may play important roles in the selection of cellular osmolyte pools. Notably, organic micromolecules with a strong hydrophobic moment—a site on the molecule with strong nonpolar composition that is separated from a region with a charge—may be selected against in high-temperature species. At low temperatures, the reduced probability of hydrophobic interactions might make solutes with a strong hydrophobic moment appropriate osmolytes because of exclusion of the solutes from the protein surface where nonpolar side chains are prevalent. Thus, as in our discussion of the appropriate organic micromolecules for hyperthermophiles to use at extremely high temperatures, the structure of the protein surface must be brought into the analysis. Are protein surfaces in cold- and warm-adapted species different in hydrophobic content? We know that amino acid compositions differ between orthologous proteins of cold- and warm-adapted species (see Chapter 3), but until we have a precise map of side chains with respect to surface accessibility, a convincing answer to this question will elude us.

Solute effects on the parts of proteins that change exposure to water during function

Our discussion of protein-solute-water interactions has, up to this point, focused chiefly on the equilibrium between the native and denatured states of a protein. There is, however, a much more dynamic context in which to investigate solute effects on protein structure. It is axiomatic that the exposure of side chains and peptide backbone linkages to the solution can change during protein function, because of the essential roles of conformational changes in protein activity and the reversible assembly of proteins involved in multiprotein reactions. How do these rapid changes in exposure and hydration state of protein groups contribute to the energetics of protein conformational changes? And how are the energetics of these reversible hydration events influenced by the solute composition of the medium?

The Δg_{tr} function we discussed above in the context of protein stabilization must also be considered during the rapid and reversible changes in protein conformation that accompany function. And as in the case of analyses of structural stabilization, the solute composition of the solution bathing the protein may help determine the energy changes that accompany processes like enzymatic catalysis, for which the change in conformation during binding may be rate-limiting to the reaction (see Chapter 3). We would predict, then, that solutes that differ in their effects on Δg_{tr} of amino acid side chains and peptide backbone linkages would have different effects on catalytic rates as a consequence of how these solutes influence both the ground-state structure of the protein and the structure of the protein in the activated enzyme-substrate complex (see the discussion of catalytic function in Chapter 3) (Low and Somero 1975; Greaney and

Somero 1979). Here, the expression "structure of the protein" must be appreciated to include the protein itself and the solvent that is structured around the protein surface.

One approach to analyzing how protein conformational changes and accompanying shifts in surface hydration contribute to the energetics of catalysis has been to study effects of inorganic ions on catalytic rates and the change in system (protein + water) volume that occurs during the rate-limiting step in catalysis, the **activation volume** ($\Delta V^{\ddagger}$). We will see below why $\Delta V^{\ddagger}$ can serve as a window into the changes in protein hydration that are induced by solutes under both ground-state and activated-complex conditions. As discussed in Section 4.2, inorganic ions, like organic solutes, differ in their effects on protein stability and solubility, according to their position in the Hofmeister series (von Hippel and Schleich 1969). At one extreme in this ranking are thiocyanate (CNS^-) and iodide (I^-), potent denaturants. At the other extreme of the series are phosphate (PO_4^{3-}) and sulfate (SO_4^{2-}), both strong protein stabilizers. The protein-stabilizing effects of the latter two anions were already alluded to when we discussed the protein-stabilizing effects of glycerophosphorylcholine (GPC) and sulfotrehalose. Furthermore, as one would predict from the effects of organic micromolecules on proteins, the differences in Hofmeister series ion effects are found across all types of proteins. This similarity among proteins in how they are affected by inorganic ions again suggests a common mechanism of action through differential ion effects on the solubilities of peptide backbone linkages and amino acid side chains.

The relationships shown in **Figure 4.20A** between maximum velocity (V_{max}) of the alkaline phosphatase (AP) reaction, activation volume ($\Delta V^{\ddagger}$), and salt composition of the assay medium show that the rate-limiting step in this enzymatic reaction occurs with a change in system volume—a decrease (negative $\Delta V^{\ddagger}$)—that is sensitive to the type of salt that has been added to the system. The more negative the value of $\Delta V^{\ddagger}$, the more rapid is the rate of the AP reaction. Why should this be?

The interpretation given to the inverse correlation between size of $\Delta V^{\ddagger}$ and catalytic rate, which is seen for AP and several other enzymes (Low and Somero 1975), is that exergonic hydration events occur during catalytic conformational changes. These hydration events lead to shrinkage of water around the amino acid side chains and peptide backbone linkages that increases their exposure to the surrounding solution during protein function. The extent of these hydration changes during the rate-limiting step will be influenced by the hydration of the ground state of the enzyme. In the presence of a protein-destabilizing ion like Br^- or I^-, the protein structure will be opened up and more hydrated than under conditions when a stabilizing ion like F^- is present in solution. Recall that protein stabilizers, whether organic or inorganic, tend to compact protein structure. During the rate-limiting conformational change, the exposure to solvent of buried (or partially buried) groups that interact favorably with water (recall the negative Δg_{tr} values found for amino acid side chains; see Figure 4.17A) will contribute more to stabilization of the activated complex than in the case where these groups are already hydrated in the ground state of the enzyme. Thus, negative values of $\Delta V^{\ddagger}$, as seen in the AP reaction, are interpreted to be a reflection of the changes in water structure that occur as a result of catalytic conformational changes during the rate-limiting step of the reaction. The reduction in $\Delta V^{\ddagger}$ is accompanied by a reduction in $\Delta G^{\ddagger}$ (see Chapter 3) due to the exergonic nature (negative Δg_{tr}) of these transfer reactions. The observed differences in $\Delta V^{\ddagger}$ among solutes thus are interpreted to reflect differences in change in hydration of amino acid side chains during conformational changes. The greater the increase in exergonic exposure of side chains, the greater the reduction in $\Delta G^{\ddagger}$. Because transfer of peptide backbone linkages into the surrounding solvent would be predicted to be

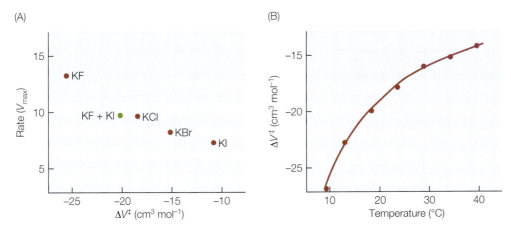

Figure 4.20 Hofmeister series effects on catalytic rate and activation volume of the alkaline phosphatase (AP) reaction. (A) The relationship between maximum velocity (V_{max}) of the AP reaction and activation volume ($\Delta V^{\ddagger}$, cm^3 mol^{-1}), as a function of solute composition. $\Delta V^{\ddagger}$ is determined from the effect of hydrostatic pressure on the velocity (rate-limiting step) of the reaction: $\Delta V^{\ddagger} = -RT\ d\ln k/dP$, where R is the universal gas constant (here, 82 cm^3 atm K^{-1} mol^{-1}), k is rate, and P is pressure (in atmospheres). The potassium salts of four anions of the Hofmeister series cause highly correlated changes in V_{max} and $\Delta V^{\ddagger}$ that reflect the anions' tendencies to favor the transfer of amino acid side chains and peptide backbone linkages from an organic milieu (as found in a protein) into water: $F^- < Cl^- < Br^- < I^-$. Salt concentrations were 0.4 mol l^{-1} for all salts except KF, which was present at 0.2 mol l^{-1}. The green symbol is for values obtained with 0.2 mol l^{-1} KF plus 0.2 mol l^{-1} KI; these data illustrate the commonly observed algebraic additivity of Hofmeister series ion effects, akin to the additivity of effects of urea and TMAO (see Figure 4.10). (B) The effect of measurement temperature on $\Delta V^{\ddagger}$. (After Greaney and Somero 1979.)

endergonic (Bohlen 2004), side chain effects may be dominant in the energy changes affecting catalytic rates, even though peptide backbone effects appear to be of greater importance in setting protein structural stability overall.

Additional, albeit indirect, evidence for hydration-based reductions in $\Delta V^{\ddagger}$ comes from measurements of the temperature dependence of $\Delta V^{\ddagger}$ (**Figure 4.20B**). The increased organization (= increase in density) of water molecules around protein groups during reversible protein hydration events is likely to be reduced as the thermal energy of the solution increases. Formation of densely organized water around water-structuring groups becomes less probable as the kinetic energy of the system rises. Thus, one might predict that increases in temperature would reduce the size of $\Delta V^{\ddagger}$. This was indeed what was seen in the case of AP. Higher temperatures might also favor a more open conformation of the ground-state enzyme, thereby reducing the amount of surface area that becomes exposed only during a conformational change.

The influences of micromolecules on the changes in hydration that accompany protein activity thus reveal another important aspect of protein-water-solute interactions. Much as solutes with different effects on protein stability can influence the balance between the native and denatured states, they can influence the energy changes that accompany rapid conformational changes that are part of protein function. More study of rapid hydration changes—and the manners in which these changes are influenced by solutes—could enhance our understanding of fundamental protein activities like catalysis. As the enzyme biochemist Gutfreund stated when discussing this point, "The overwhelming difficulty of interpreting thermodynamic parameters in terms of chemical change is due to the large effects of changes in solvation and solvent structure" (Gutfreund 1972). The changes he alludes to are not only a fascinating phenomenon in the study of enzyme mechanisms, but also an issue of deep importance in elucidation of the coevolution of proteins and the solutions that bathe them. It seems inevitable that the evolution of the internal milieu had as its charge, so to speak, not only the maintenance of optimal protein structure, but also the provision of a "working environment" where the energetics of rapid conformational changes could be optimal as well. In this latter context, reversible hydration events, whose energy changes are strongly influenced by the composition of the milieu, are likely to have played major roles in governing the evolution of the cellular solution. Whereas this hydration-based mechanism for explaining solute effects is quite conjectural, the energy changes associated with transient shifts in exposure to solvent of side chains and peptide backbone linkages almost certainly play a role in governing rates of protein function. It will be exciting to see if future efforts in elucidating these solute effects provide novel insights into the abilities of enzymes to reduce the energy barriers to their reactions, and how the micromolecular environment in which an enzyme works enters into the catalytic rate-enhancement abilities of enzymes.

4.6 pH Relationships

In our wide-ranging examination of the diverse constituents of biological solutions, we have left for last a tiny entity—the proton—that in fact is of utmost importance in evolution of biological systems and in the moment-by-moment regulation of cellular activities. We turn now to an examination of these pH relationships, beginning with a fundamental question about the selective processes that led to the evolution of the pH values we find in contemporary organisms.

The ultimate causes of biological pH values: What matters most in pH relationships?

All students of biochemistry and physiology are aware that the pH values of extra- and intracellular fluids are critical for sustaining and controlling a diverse suite of biological functions. However, relatively few readers may be aware of the ultimate causes of the particular pH values that have been favored by natural selection. In essence, we can ask, Why are some pH values optimal for life? An understanding of these ultimate causes can provide a unified context in which to appreciate the pH values found in *all* types of cells. Here again, as we have stressed throughout this volume, the types of conservative patterns found across diverse taxa reflect some of the most fundamental aspects of biochemical adaptation. In the case of pH values, the patterns one finds among species reflect an essential physiological need—the ability to use changes in proton activity to regulate, to "titrate" as it were, a host of physiological activities—and a strict constraint on the "raw material" available to evolution for achieving this regulatory capacity.

An appropriate starting point for an examination of pH relationships across all domains of life is to review what is known about the pH values of the intracellular fluids in unicellular and multicellular species. Once we understand what these pH values are—and how they vary among species with different cell temperatures—we will be in a good position to ask questions about the ultimate causes of these values and the linkages that exist between selection of pH values and establishment of solution conditions that are optimal for macromolecular functioning. Note that we will focus mainly on the cytosolic compartment and the mitochondrion, which together contain most of the metabolic machinery of the cell. Some compartments in the cell, for example, lysosomes, have pH values that differ greatly from the cytosolic pH, reflecting their unique functions (e.g., protein degradation). Thus, the low pH values of lysosomes (near 5.5) are maintained to destabilize proteins, making them better substrates for proteolytic enzymes, and to provide a pH close to the pH optimum of the proteolytic enzymes themselves. Lysosomal proteases may have pH optima between 5 and 6, values much lower than found for most other types of enzymes. These enzymes work effectively in the acidic lysosome, but would be largely inactive were they to "leak" into the cytosol. Thus, the pH optima of enzymes can be modified during evolution to fit the particular metabolic roles they serve and the cellular compartments in which they are lodged.

pH VARIES WITH TEMPERATURE IN A REGULAR MANNER   **Figure 4.21** shows how intracellular (cytosolic) pH (pH$_i$) varies among species in relationship to cellular temperature. Clearly, there is no such thing as "the" biological pH$_i$ value. Rather, over the greater part of the range of temperatures where organisms can remain active, pH$_i$ varies by about 2 pH units, from ~7.4 near 0°C to ~5.7 at 120°C. This corresponds to an approximately 100-fold difference in hydrogen ion activity between cryophilic species and hyperthermophiles. Because some species tolerate cell temperatures well below 0°C (Arctic insects may survive near −70°C; see Chapter 3), the range of pH$_i$ values may be even greater than shown in Figure 4.21. However, to our knowledge, there are no estimates of pH$_i$ for organisms held in deep, subzero cold. Likewise, we know of no pH$_i$ measurements for extreme hyperthermophiles, so the extrapolated relationship shown in this figure remains to be tested empirically.

   In view of the wide range of pH$_i$ values across species with different cell temperatures, it may not be immediately apparent how we can justify stating that a "conservative pattern" exists in pH relationships across all taxa. To understand what is strongly conserved

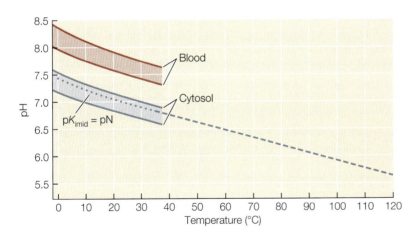

Figure 4.21 Temperature dependence of intracellular (cytosolic) pH (pH$_i$), extracellular pH (pH$_b$), pK$_{imid}$, and the neutral point of water (pN). The variation of pH$_i$ and pH$_b$ shown at any given temperature (between 0°C and 40°C) reflects the range of published values. Extracellular fluids of animals are typically about 0.4 pH units alkaline to pH$_i$, which facilitates removal of metabolically generated protons from the cell. Mitochondrial matrix pH (not shown) is several tenths of a pH unit alkaline to pH$_i$ and varies with temperature in parallel with the curves shown for other fluid compartments. The dashed line extending the pK$_{imid}$ versus temperature relationship out to the highest known temperatures of life, ~120°C, is a linear extrapolation that assumes a slope of −0.15 pH unit per degree Celsius. (After Hochachka and Somero 2002.)

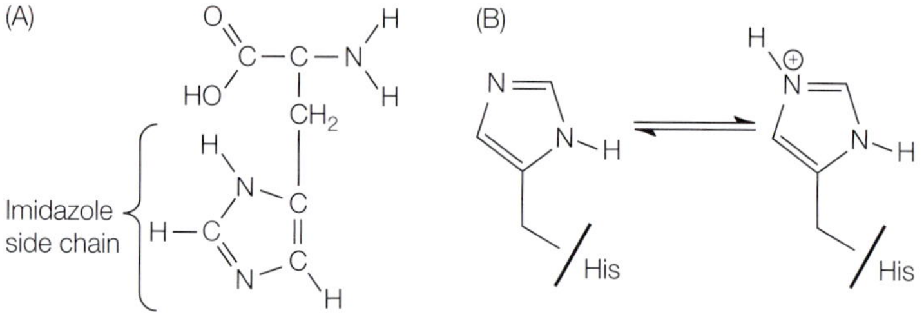

Figure 4.22 Histidine and its imidazole side chain. (A) The amino acid histidine. (B) Reversible protonation of the imidazole side chain.

and why the observed pH_i values are so critical for cellular function, we must examine not only the pH_i values themselves, but also the pK of an amino acid side chain, the **imidazole group of histidine (Figure 4.22)**, which we will show is the "tail that wags the dog" in biological pH relationships.

HISTIDINE'S IMIDAZOLE GROUP: A KEY TO BIOLOGICAL pH RELATIONSHIPS   The imidazole ring has a unique attribute that makes it of pivotal importance in pH relationships: With certain exceptions, it is the only amino acid side chain that can be titrated—can vary its charge state—at biological pH values. That is to say, the pK of imidazole (pK_{imid}) falls close to pH_i (see Figure 4.21), a property not found for any other amino acid side chain. In stating this key point, we are also stating that the pK of imidazole side chains varies with temperature in parallel with variation in pH_i. In fact, this parallel temperature dependence provided a critical clue to as to why biological pH values exhibit the temperature dependence we observe; that is, approximately −0.015 to −0.02 pH units per degree Celsius increase in temperature. Note that the temperature dependence of pK_{imid}, and therefore of pH_i and extracellular pH (pH_b), is not linear across the full range of biological temperatures. The relationship is slightly steeper at low temperatures: Near 0°C dpH/dT is ~0.021, whereas near 40°C dpH/dT is ~0.0147.

The extrapolation of the dpH/dT relationship to temperatures near the highest known temperatures at which life occurs, slightly above 120°C, was done using a value of dpH/dT of 0.015. This extrapolation predicts that a hyperthermophilic archaeon living near 120°C would have an intracellular pH near 5.7. To our knowledge, there have been no measurements of pH_i in hyperthermophiles, so this value remains to be determined. Whereas this value of pH_i might initially strike one as being a remarkably acidic pH for a cell to endure, in fact this low value of pHi would confer on the cells the benefits of the pH regulatory strategy—alphastat regulation—that we will now examine.

THE ALPHASTAT HYPOTHESIS   The relationship between pH_i and pK_{imid} led physiologists Robert Blake Reeves, Herman Rahn, and their colleagues to propose the **alphastat hypothesis** of pH regulation (Reeves 1977). They reasoned that the observed variation in pH of blood (pH_b) and pH_i with temperature reflected the stabilization of net protein charge. Because the sole side chain that could vary its charge state in this manner under biological pH conditions is the imidazole side chain of histidine, this one side chain took center stage in subsequent analysis of the evolution and regulation of pH values.

The overarching theoretical framework in which these analyses were developed employs a concept termed **alphastat regulation**. Here, alpha$_{imid}$ (α_{imid}) refers to the fractional dissociation state of imidazole side chains:

$$\alpha_{imid} = [imidazole]/([imidazole] + [imidazole^+])$$

What is being conserved by processes that regulate pH is the fractional dissociation state, α_{imid}, of the histidine side chain, not some single pH value that many earlier investigators of pH regulation believed to be the "correct" biological pH value under all circumstances. The recognition that α_{imid} was the focus of conservation in pH regulation, not some single pH value, revolutionized our understanding of biological pH values and their evolution.

It is worth digressing briefly into the history of this transformation in our thinking because of what we can learn about the ways in which science advances. This history is definitely consistent with Thomas Kuhn's ideas about the role played by dominant paradigms in shaping the course of experimentation and governing the rate of progress of science (Kuhn 1970). The prevailing paradigm may shape our beliefs concerning how experiments should be run, what experimental techniques are valid, and what data can be accepted as meaningful and reflective of reality. In short, adherence to the prevailing paradigm can be a block to the acceptance of new ("revolutionary") concepts. In the case of the development of our concepts of biological pH regulation, the choice of experimental system strongly influenced the core theory that was developed about "normal" pH values. Thus, because much of the earliest study of biological pH values and their regulation was done in a biomedical context, with mammals in normothermic conditions (body temperatures near 37°C–39°C), the pH of mammalian blood, which lies near 7.4, was viewed as "the" biological pH. It took decades of comparative study with ectothermic species to demonstrate that pH values observed at 37°C are not "normal" for organisms whose body (cell) temperatures differ from the mammalian and avian norm (see Reeves 1977 for details on this interesting scientific revolution). There was considerable skepticism about the "abnormal" pH values found, for example, in ectotherms measured at body temperatures well below 37°C. Furthermore, some of these "abnormal" pH measurements were made using newly available electrode technology, which was somewhat distrusted by investigators who had relied on older, chemical methods for measuring pH. Suffice it to say, overthrowing the "pH 7.4 paradigm" revolutionized our understanding of biological pH values and led to exploration of the advantages of a temperature-dependent pH_i and pH_b and alphastat pH regulation in general.

Alphastat regulation, imidazole titration, and the reversibility of biochemical processes

There are numerous instances in which the ability to vary (titrate) the charge state of an imidazole group is critical for physiological activities. Regulation of function through titration of imidazole side chains is common in enzymatic catalysis and in other protein-mediated activities, notably control of oxygen binding by hemoglobin and modulation of activities of entire metabolic pathways. Below we give examples of how a metabolic pathway of central importance in ATP generation, glycolysis, is regulated through changes in pH_i that influence imidazole charge state.

REVERSIBILITY OF GLYCOLYSIS: THE LACTATE DEHYDROGENASE REACTION   As every textbook on biochemistry emphasizes, enzymes do not change the equilibrium point of a chemical reaction, but only serve to catalyze rapid rates of the reaction in both directions. Many reactions in metabolic pathways must be reversible if the direction of the pathway in which they are embedded is to be controlled in concert with the status of the cell, for instance, its supply of oxygen and its needs for generation of ATP. Both of these aspects of cellular status play important roles in controlling flux through the glycolytic pathway. Here we focus initially on lactate dehydrogenase (LDH), whose direction of activity—either as a pyruvate reductase or a lactate dehydrogenase—will vary with the capacity of the cell to further process lactate anion.

As discussed in Chapter 2, glycolysis is an ancient, yet still central, pathway of intermediary energy metabolism that catabolizes glycogen and glucose for production of ATP. When oxygen availability is not limiting, the pyruvate that the pathway generates can enter the TCA cycle through the pyruvate dehydrogenase system to support aerobic ATP generation in the mitochondrion. Under limiting conditions of oxygen, however, pyruvate may be shunted to lactate, through the pyruvate reductase activity of LDH:

$$\text{Pyruvate} + \text{NADH} + \text{H}^+ \rightarrow \text{lactate} + \text{NAD}^+$$

This step regenerates the cofactor (NAD^+) needed to keep the glycolytic pathway active. The lactate produced may be retained in the tissue or released into the blood stream. Whether the lactate is directed toward gluconeogenesis, either in the tissue in which it is produced or in another organ, for instance, the liver, or toward ATP production, its further metabolism requires that it be oxidized to pyruvate, through the lactate dehydrogenase activity of LDH:

$$\text{Lactate} + \text{NAD}^+ \rightarrow \text{pyruvate} + \text{NADH} + \text{H}^+$$

As the presence of the proton in the two above equations indicates, the direction of the LDH reaction is strongly dependent on pH. If the rate of glycolysis exceeds the capacity of the cell for aerobic metabolism of pyruvate, pyruvate and protons will accumulate, leading to protonation of the LDH active site histidine. This protonation event favors binding of pyruvate and leads to regeneration of NAD^+ and production of lactate. When abundant supplies of oxygen again become available, the transient acidification of the cytoplasm that occurred during anaerobic glycolysis is reversed, the active site histidine loses its proton, and lactate binding is favored, leading to regeneration of pyruvate. The ability to titrate this imidazole side chain is clearly of central importance in regulating the LDH reaction and the overall flow of glycolytic carbon in the direction appropriate for the prevailing availability of oxygen (**Figure 4.23**).

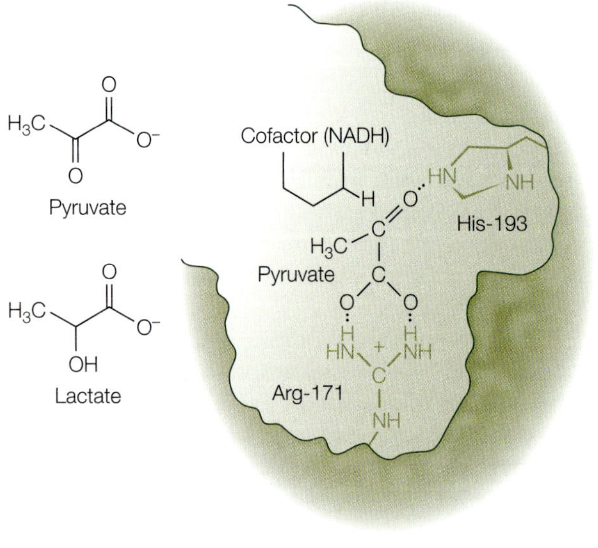

Figure 4.23 The binding of pyruvate to LDH. The imidazole side chain of the active site histidine (his) residue interacts with pyruvate (shown) and lactate (not shown). (Note: this histidine is numbered 193 in the figure; the actual residue number differs slightly among orthologs of LDH, which differ by a few amino acids in total sequence length.) Pyruvate binding requires a protonated imidazole side chain; lactate binding requires a deprotonated side chain. Changes in the protonation state of the active site histidine residue thus govern the direction of LDH function. Alphastat regulation maintains the regulatory ability of LDH across a species' range of body temperatures by stabilizing the fractional dissociation state of the imidazole ring as temperature changes (see text and Chapter 2). An active site arginine residue (Arg-171 in this figure) binds to the carboxyl group of both lactate and pyruvate.

The temperature dependence of pK_{imid} figures importantly into LDH-pyruvate interactions as well. Thanks to alphastat regulation, the effect of temperature on K_M of pyruvate (K_M^{pyr}) is greatly reduced from what it would be if **pHstat** conditions, temperature-independent pH, were present. Were pHstat conditions to pertain, rising temperatures would lead to reduced pyruvate binding by LDH due to the decreased likelihood of the active site histidine being protonated, as required for pyruvate binding (Yancey and Somero 1979). K_M^{pyr} stabilization by alphastat regulation helps conserve the regulatory ability of LDH discussed above over a species' full range of body temperatures (see Chapter 3).

REGULATION OF GLYCOLYSIS THROUGH CONTROLLING ASSEMBLY AND COMPARTMENTALIZATION OF PFK   Another mechanism through which changes in pH_i can modulate glycolytic activity involves a combination of effects on enzyme subunit assembly and enzyme localization (compartmentalization). The enzyme in question is phosphofructokinase (PFK), which plays an important role in governing flux through glycolysis (see Chapter 2). Each PFK subunit has several solvent-accessible histidines that can be titrated as pH_i changes. These reversible protonation events alter PFK's self-assembly and actin-binding equilibria. High pH_i favors assembly of PFK into a tetramer, the enzyme's active form, meaning that changes in pH_i help set the activity level of PFK. A slight drop in pH_i favors binding of PFK to the actin of thin filaments. Because all of the other glycolytic enzymes are already assembled on thin filaments, PFK binding completes assembly of the **glycolytic metabolon** and increases flux through glycolysis.

Here is how the pH-dependent modulation of PFK activity—and thus glycolysis—occurs in muscle. When cellular conditions first indicate a need for activating glycolysis—for example when ATP concentrations are decreasing and the concentration of citrate, a negative allosteric modulator of PFK, is falling—much of the PFK in the cytosol is free rather than bound to the glycolytic metabolon (**Figure 4.24A**). A slight drop in pH_i is sufficient, however, to shift the equilibrium between free and bound PFK in the direction of metabolon formation (**Figure 4.24B**), thus enhancing glycolytic activity. As glycolysis proceeds, there may be a need to "put the brakes" on the pathway if inadequate oxygen is present to allow pyruvate to be processed by the TCA cycle. In the face of low oxygen, anaerobic glycolysis is activated, lactate accumulates, and pH_i falls. This acidification of the cytoplasm shifts the equilibrium between tetrameric and lower assembly states of PFK in the direction of dimers and monomers, which are less active and inactive, respectively (**Figure 4.24C**). In turn, the reduced amount of tetrameric PFK reduces the amount of active enzyme available to bind to actin, leading to disassembly of the glycolytic metabolon and reduction in glycolytic flux. The pathway thus can be seen to have a type of autoregulation that relies on small initial decreases in pH_i to activate the pathway and subsequent larger falls in pH_i to reduce pathway flux.

REGULATION OF CARBOHYDRATE BREAKDOWN IN THE BRINE SHRIMP *ARTEMIA SALINA*
Another example of how carbohydrate breakdown is modulated by changes in pH_i comes from studies of gastrula-stage embryos (cysts) of the brine shrimp *Artemia salina*. Brine shrimp live in highly saline environments where oxygen levels can reach extremely low values. Under conditions of low oxygen availability, cysts of brine shrimp enter a state of metabolic dormancy (Busa and Crowe 1983), which is characterized by a cessation of carbohydrate breakdown. Exposure of cysts held under aerobic conditions to high levels of CO_2, which could lead to formation of carbonic acid and a rise in proton concentration, can also trigger cessation of carbohydrate breakdown, suggesting a central role for pH_i

Figure 4.24 The effects of changes in pH$_i$ on PFK subunit assembly, assembly of the glycolytic metabolon, and rate of glycolytic flux. (A) High pH$_i$. PFK exists primarily in the active, tetrameric state, but most PFK is not bound to the thin filament, where the other glycolytic enzymes are preassembled into the glycolytic metabolon. (B) Falling pH$_i$. PFK binding to thin filament is enhanced, allowing completion of the glycolytic metabolon and an increase in glycolytic flux from glucose-6-P to pyruvate or lactate. (C) Dissociation of active PFK tetramers as pH$_i$ continues to fall. Equilibrium of binding of PFK to thin filaments shifts toward free PFK (tetramers, dimers, and monomers). Glycolytic flux is reduced and further cellular acidification is minimized.

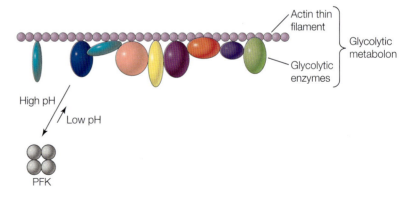

(A) High pH$_i$: PFK is mostly soluble. Glycolytic metabolon is in a low-activity state because it is missing PFK—a key "valve" in flux control.

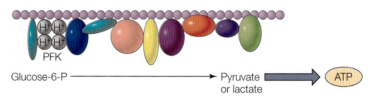

(B) Falling pH$_i$: pH$_i$ falls during activation of muscle contraction. PFK is protonated and binds more strongly to actin thin filament. Glycolytic metabolon is completed and glycolysis is activated.

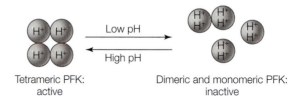

(C) Low pH$_i$: As pH$_i$ continues to fall during muscle function, PFK becomes more protonated, which promotes its dissociation. This reduces the rate of glycolysis and prevents excessive acidification of the muscle.

in regulation of energy metabolism. Carpenter and Hand (1986) identified the key pH-regulated enzymes of the carbohydrate catabolizing pathways: trehalase, which breaks down trehalose into two molecules of glucose, hexokinase, and phosphofructokinase. All three enzymes are inhibited by falling pH (Carpenter and Hand 1986).

Subsequent study of pH effects on trehalase showed that alterations in pH shifted the assembly state of the enzyme (Hand and Carpenter 1986). A fall in pH favored oligomerization of the enzyme into an aggregate with twice the mass of the enzyme present at higher pH values. The high-molecular-mass enzyme was strongly inhibited by acidic pH and trehalose. Thus, the shift in assembly equilibrium that was initiated when pH values fell below ~7.4 would appear to be an important mechanism for arresting trehalose-fueled metabolism during dormancy. Preemergence development of *Artemia* is fueled exclusively by trehalose, so a central role of a trehalose switch makes sense. The pH-driven changes in trehalase assembly state are fully reversible, such that restoration of normal pH values leads to reactivation of carbohydrate catabolism (Hand and Carpenter 1986). Thus, as in the case of PFK in muscle just discussed, pH_i titration of metabolic activity is a simple and rapid mechanism for attuning the metabolic state of the cell to the availability of oxygen.

pH effects on hemoglobin-oxygen interactions

Perhaps the most familiar pH-dependent process in physiology is the one involving pH-dependent binding of oxygen to vertebrate hemoglobin (Hb) (**Figure 4.25**). The effect of shifts in pH on $Hb\text{-}O_2$ interactions is commonly referred to as the **Bohr effect**, named after Christian Bohr, an early-twentieth-century Danish physiologist and father of the legendary physicist Niels Bohr. The **Root effect**, found particularly in Hb variants of bony fishes with gas gland-containing swim bladders, involves a larger-magnitude effect of pH on $Hb\text{-}O_2$ affinity and a drop in maximum saturation level (not shown). The discussion below focuses on the underpinnings of the Bohr effect and how histidine-mediated effects modulate oxygen binding.

How does a change in pH in the erythrocyte alter Hb's ability to bind oxygen? Hb's oxygen affinity is measured by the concentration of oxygen yielding half-saturation of the Hb (P_{50} value); thus P_{50} values are analogous to K_M values (see Figure 4.25A). An especially detailed analysis of $pH\text{-}P_{50}$ relationships has been done for mammalian hemoglobins, including a variety of mutant forms with altered oxygen-binding properties (Perutz et al. 1980). In adult humans, the dominant form of Hb in red blood cells is hemoglobin-A (Hb-A). It comprises four subunits, two alpha (α) and two beta (β). To understand how oxygen binding is regulated, the $\alpha_2\beta_2$ structure (see Figure 4.25B) should be viewed as a dimer of two $\alpha\beta$ dimers, $\alpha^1\beta^1$ and $\alpha^2\beta^2$ (Creighton 1984). Oxygen binding and responses

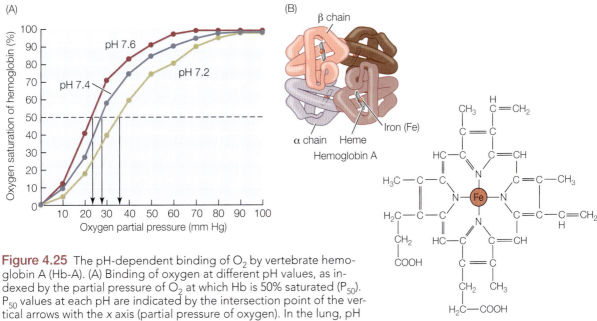

Figure 4.25 The pH-dependent binding of O_2 by vertebrate hemoglobin A (Hb-A). (A) Binding of oxygen at different pH values, as indexed by the partial pressure of O_2 at which Hb is 50% saturated (P_{50}). P_{50} values at each pH are indicated by the intersection point of the vertical arrows with the x axis (partial pressure of oxygen). In the lung, pH is relatively high and oxygen binding is facilitated. In a working tissue, such as skeletal muscle, blood pH is lower (pH 7.2 is shown here) and P_{50} increases, which facilitates release of O_2 at metabolically active sites where demands for O_2 are high. (B) Structure of Hb-A showing its two types of subunits and their associated heme groups. The structure of heme is shown to the right of the Hb-A model. Changes in orientation of subunits induced by alterations in pH lead to shifts in oxygen affinity (see text).

to modulators of oxygen affinity involve changes in the orientation of the two $\alpha\beta$ dimers. These changes in Hb-A's structure cause shifts in amino acid side chain interactions that allow transitions between two states of the protein, relaxed (R) and tight (T). The T-state structure has a lower oxygen affinity (higher P_{50}) than the R-state structure.

Changes in pH can shift the distribution between T and R states, and therefore modify oxygen affinity through mechanisms that include (yet are not restricted to) histidine imidazole-mediated effects. The imidazole side chain playing the largest role in the Bohr effect is on the C-terminal His-146 of the β chain (Perutz et al. 1980). In the T state of human Hb-A, the pK value of the imidazole side chain of His-146β is near 8.1; in the R state, the value is near 7.1. The higher pK_{imid} of His-146 in the T state is a consequence of the proximity of a carboxylate group on Asp-94β, which forms a salt bridge with His-146β in the T state. This anionic (COO^-) center stabilizes a positive charge on the imidazole ring, accounting for the rise in pK_{imid}. This salt bridge is absent in the R state, and consequently pK_{imid} has a lower value. The difference in pK_{imid} of His-146β in the R and T states leads to a difference in titration effects when pH_i of the erythrocyte changes. When Hb-A is in the R state, a decrease in pH is likely to substantially increase the protonation state of His-146 because its pK_{imid} value is close to pH_i of the red blood cell. Increased protonation of His-146 would favor a transition to the T state by increasing the likelihood that an ionic interaction could form with the carboxylate side chain that completes the salt bridge in the T-state molecule. The transition from the R to the T state mediated by protonation of His-146β is responsible for the largest individual contribution (~40%; Perutz et al. 1980) to the fall in affinity for O_2 (a rise in P_{50}) induced by decreasing pH. Other contributors to the Bohr effect include contributions from other histidine residues and valine residues.

The observed large increase (about one full pH unit) in pK of His-146 upon R-state-to-T-state transitions must not be regarded as some type of "violation" of alphastat regulation. Rather, these conformational change-associated shifts in pK_{imid} reveal even more powerfully how the pK of this one side chain can be used to fabricate diverse regulatory circuitries in biochemistry. The lower pK in the R state makes Hb-A highly sensitive to changes in pH, which can lead to oxygen release at sites where CO_2 production has caused a drop in pH. Transition to the T state keeps Hb-A in a low-affinity state that facilitates full unloading of the four oxygen molecules bound to the protein.

Production of metabolic CO_2 influences Hb-oxygen interactions through a mechanism separate from its effect on pH. CO_2 is able to bind to the α-amino groups of both α and β chains to form carbamate groups:

$$-NH_2 + CO_2 \leftrightarrow -NH-CO_2H$$

The ionized form of the carbamate group ($-NH-COO^-$) interacts with positively charged side chains of the T-state Hb-A. In the R state, these cationic groups are not in the same position in the protein, and the carbamate-side chain interaction is precluded. Rising levels of CO_2, like falling pH, thus favor a transition from the R state to the T state, leading to a reduction in oxygen binding ability and release of oxygen where it is needed in the body.

In summary, vertebrate Hb is arguably the best-studied protein from a structure-function perspective. The above discussion has presented only a short and very incomplete account of our vast understanding of how Hb's activity is modulated by pH, CO_2, and other micromolecules, including a suite of phosphorylated compounds like diphospho-glycerate and ATP. Furthermore, hemoglobin variants differ widely in their responsiveness to modulators. Some Hbs are largely pH-insensitive; others exhibit marked Bohr and

Root effects. The phosphorylated modulators of importance in governing oxygen binding differ among species. Hemoglobins thus illustrate how a given type of protein—even paralogs of a single species—have evolved to exploit changes in the concentrations of intracellular micromolecules to achieve precise regulation of oxygen binding in concert with changes in physiological state.

pHstat regulation: When and why can failure to follow alphastat regulation be beneficial?

When pH_i or pH_b is maintained at a constant or nearly constant value in the face of varying temperature—pHstat regulation—a number of consequences pertain for pH-sensitive physiological processes. Before we examine how pHstat regulation plays out in nature, however, it seems worth pointing out that, unfortunately, this type of pH regimen is used far too commonly—and inappropriately—in the biochemistry laboratory. Biochemists commonly fail to take into consideration the temperature dependence of the pK values of the buffers they use in in vitro studies of proteins. For example, phosphate buffers, which are commonly used in biochemical experiments with proteins, have pK values with only minimal dependence on temperature. In contrast, imidazole buffer, another common reagent, displays the type of temperature dependence noted in alphastat regulatory schemes. If a buffer with minimal temperature dependence of pK—for instance, phosphate buffer—is chosen, it is essential to prepare a unique buffer for each of the temperatures at which measurements will be made. Failure to do so is apt to lead to artifacts arising from nonphysiologically low pH values at low experimental temperatures (if the buffer is made to have a pH value typical of a normothermic mammal, for example) and unnaturally high pH values at high temperatures (if the buffer's pH is adjusted according to the pH expected for an ectotherm with a body temperature well below 37°C). For example, demonstrations of "cold denaturation" of proteins have often instead been instances where pHstat buffering leads to a pH value at cold temperatures that is much lower than the alphastat value and is sufficient to inactivate the protein in question. That is, the denaturation is due to a pH-mediated loss of structure, not cold-sensitivity of the protein itself. Studies of PFK have often yielded this type of artifact.

Whereas many biochemists may use pHstat regulation in their (naïve) experiments, organisms rarely do under environmental conditions that are permissive of normal activities. When external conditions preclude normal activities, however, then adjustment of pH values to downregulate metabolic flux becomes a useful regulatory strategy. The example given above of regulation of glycolysis is a good illustration of how effective this rapid titration of metabolism can be in the context of controlling metabolic activity under normal physiological conditions. pHstat regulation leading to a pH value below that expected for the prevailing cellular temperature is frequently one of the means by which the reversible ametabolic state is achieved. Small mammalian hibernators likewise may rely on pHstat regulation to downshift their metabolic rates. Some of these effects may be the result of inhibition of PFK activity by the mechanism illustrated in Figure 4.24 (Hand and Somero 1983).

Imidazole-based cellular buffers: Retaining buffering ability at different temperatures

The regulation of pH_i entails several active and passive processes. The active, ATP-consuming transport of protons (or proton equivalents) into or out of the cell and between the body and the external medium is one important component of this regulatory process (see Hill et al. 2016). However, intracellular buffers are also used to

Figure 4.26 Histidine-containing dipeptide buffers.

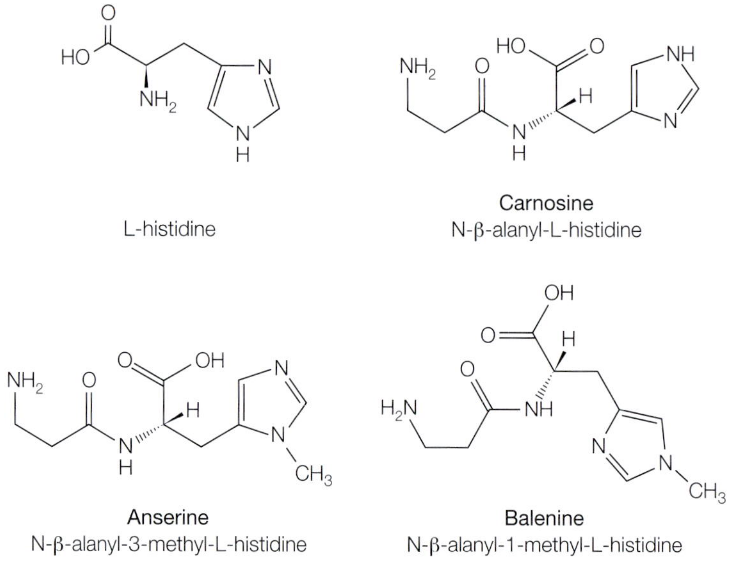

L-histidine

Carnosine
N-β-alanyl-L-histidine

Anserine
N-β-alanyl-3-methyl-L-histidine

Balenine
N-β-alanyl-1-methyl-L-histidine

stabilize pH_i in the face of changes in metabolic proton generation and to adjust pH_i to ensure the correct value at different cellular temperatures. Whereas bicarbonate/carbonate and phosphate-based buffers contribute to buffering to a greater or lesser degree in different cells, a unique role is played by **dipeptide buffers** that contain, as one of their two amino acid residues, the histidyl residue with its imidazole side chain (**Figure 4.26**). A primary advantage of these buffers is the fact that their pK value lies close to the pH of the cellular fluids. Furthermore, because the pK of the imidazole side chain plays the dominant role in governing pH_i variation with temperature, buffering capacity is conserved in the face of changes in body temperature. Thus, a eurythermal ectotherm that relies on imidazole buffering will retain a stable buffering ability across its range of body temperatures.

The commonly occurring dipeptide buffers have generally been named after the type of organism or the tissue in which they first were identified: anserine (ducks), balenine (whales), and carnosine (muscle). Most studies of these buffers have focused on vertebrates, but some dipeptide buffers occur in invertebrates as well (see Cameron 1989). The histidine moiety may be present in an unmodified form (carnosine) or as a methylated derivative (anserine and balenine). Histidine-containing buffers are commonly the dominant buffers in tissues that have an exceptionally high capacity for proton generation—notably locomotory muscle in powerful runners, swimmers, and flyers—and their concentrations are modulated relative to proton-generating capacity. Cellular buffering capacity is strongly correlated in vertebrate muscle with LDH activity, which is an index of the capacity for anaerobic glycolysis (Castellini and Somero 1982). Thermal acclimation of ectotherms also can influence the concentrations of dipeptide buffers. Cameron (1989) reported a ~60% rise in concentration of carnosine in leg muscle of blue crabs (*Callinectes sapidus*) when acclimation temperature was increased from 10°C to 30°C. This increase in carnosine levels could reflect an increasingly anaerobic poise to muscle glycolysis at elevated temperatures. As in many other instances of micromolecular adaptation, both the intrinsic properties of the micromolecule (here, the pK value and its temperature dependence) and its concentration reflect capacities for establishing a "fit" environment for protein structure and function.

The importance of being ionized: Another perspective on the ultimate causes of biological pH values

When we speak of the histidine imidazole side chain as being "the tail that wags the dog" in diverse protein-based processes involving pH-related activities, we are hardly exaggerating. The selection of histidine at an early stage in the evolution of life provided for cells a sensitive antenna for reading their acid-base status and a basis for effecting appropriate changes in protein-ligand interactions, protein subunit assembly, and protein compartmentalization, through titration of imidazole side chains. Despite the recognition we have given to histidine and its imidazole group, the astute reader might wish to raise a question concerning an even more basic ultimate cause of biological pH values. Granted that the pK of imidazole provides life with a capacity for various types of pH-dependent shifts in function and structure, was there something "magical" about setting pH_i to values that approximate the pK of imidazole? Why couldn't cells have evolved their biochemistry around pH values near the side chain pK values of the acidic amino acids glutamate and aspartate (pK near 4), or the basic amino acids lysine and arginine (pK near 11–12)?

An insightful analysis of this question was provided by the physiologist Bernard Davis (1958) in a paper titled "On the importance of being ionized." Here is Davis's interpretation of the evolutionary basis for selection of biological pH values. He proposed that, at the dawn of cellular evolution, the organic molecules that were becoming critical requirements of cells—whether for biosynthetic processes or substrates for catabolism—were in scarce supply in the surrounding water. When a cell was fortunate enough to obtain, say, a supply of amino acids, it would have been highly advantageous to prevent these molecules from diffusing out of the cell, back to the environment (probably seawater). As the reader will be aware, charged (ionized) molecules tend to have difficulty passing through membranes, unless specific pumps or channels are present for this function. Thus, the earliest cells presumably would have benefitted from having pH_i values at which metabolites like amino acids and carboxylic acids were charged. Thus, the title of Davis's classic 1958 paper.

Figure 4.27 illustrates how the charge states of a wide variety of organic molecules vary with pH. It is clear that at a pH_i between approximately 6 and 8, most cellular metabolites carry either a positive or a negative charge, which greatly reduces the likelihood that they will freely diffuse out of the cell. Davis thus proposes that the benefits of maintaining ionized metabolites led to selection for a pH_i that lies between pH 6 and pH 8. When we

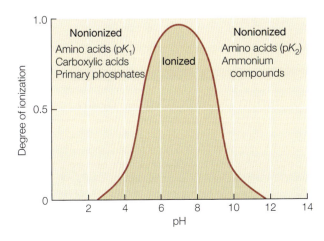

Figure 4.27 The importance of being ionized: how cellular pH influences the degree of ionization of organic metabolites. At cellular pH values (pH_i; see Figure 4.21), the vast majority of cellular metabolites are ionized, and therefore have reduced abilities to pass through the cell membrane and be lost to the environment. (After Davis 1958.)

TABLE 4.1

Processes that are regulated ("titrated") by changes in intracellular pH

Fertilization: Activation of zygote metabolism by postfertilization increase of the pH of cytoplasm. In sea urchins, for example, the pH_i of the newly fertilized egg rises by 0.3–0.5 pH unit as a result of proton extrusion from the cytoplasm (for review, see Epel 1997).

Hibernation in rodents (squirrels): Reduced pH leads to suppression of metabolism. Reduced ventilation during entry into hibernation leads to retention of increased amounts of CO_2, which leads to a fall in pH. During emergence from the metabolic depression associated with hibernation, hyperventilation leads to expulsion of CO_2 from the organism and a rise in pH to alphastat values (Snapp and Heller 1981). Muscle PFK of a ground squirrel shows the type of reversible, pH-dependent regulation illustrated in Figure 4.24 (Hand and Somero 1983).

Metabolic depression under anoxia in brine shrimp (Artemia *spp.*): Downregulation of metabolism under anoxia is facilitated by intracellular acidification; reactivation of metabolism is effected by an increase in pH (Busa and Crowe 1983).

Metabolism of locomotory muscle: During the initial activation of locomotory muscle, a transient fall in pH may activate glycolytic ATP generation. As pH continues to fall, glycolytic activity may be reduced to prevent an excessive fall in pH_i (see Figure 4.24).

Persistence of microbial spores in a hypometabolic state: Spores of bacteria have a low pH value, which likely is key in suppressing metabolism. Spore activation is associated with a rise in pH.

couple this requirement with the necessity of having an amino acid side chain that can be titrated at physiological pH values, we are left with a single candidate among the 20 commonly used amino acids: histidine.

ULTIMATE CAUSES AND BIOLOGICAL pH VALUES To briefly reiterate the causal chain we have presented here, ionization of metabolites was the driver of selection for biological pH values. Only a single amino acid, histidine, had a pK_a value in the pH range where ionization of metabolites was maximal. Because a large number of biochemical processes and structures benefit from having pH-dependent properties—functions that are capable of titration across the physiological pH range—histidine was beneficial for yet another reason. Perhaps there is no other story in the broad context of micromolecular evolution that has such wide-ranging consequences. Indeed, the number of processes that are regulated through changes in intracellular pH (**Table 4.1**), and the complex interactions between eukaryotes and certain of the pH-sensitive members of their microbiomes, speak to the enormously varied and important roles of the tiny proton in governing what organisms do—biochemically, physiologically, and as *Helicobacter pylori* has taught us, even behaviorally (**Box 4.6**).

BOX 4.6

pH, urea, and ulcers: Chemosensing and buffering by *Helicobacter pylori*

Approximately half of the human population worldwide harbors a stomach bacterium, *Helicobacter pylori* (see figure), that is responsible for gastric ulcers. Although this book does not have a strong biomedical focus, the fascinating story of how this bacterium prospers in the challenging environment of the stomach, where pH values range between 1.5 and 3.5, merits at least a brief presentation, in part because the story entails interactions between two micromolecules, urea and the proton, that we have already found to be important in a variety of other biological contexts.

BOX 4.6 *(continued)*

Perhaps surprisingly, *H. pylori* is not an acidophile; it definitely does not "love" an acidic environment, and in fact the bacterium exhibits strong negative chemotaxis in the face of low pH. Thus, the HCl of the stomach lumen is a chemorepellant to this bacterium. In addition to moving away from regions with low pH, *H. pylori* possesses a chemosensing mechanism that allows it to swim toward the site where it can thrive: the mucus-rich lining of the gut epithelium. The highly motile *H. pylori* cells move to this region and establish their colonies within ~25 µm of the stomach surface, either in the mucus that coats the stomach cells or attached to the cells themselves. There the bacteria have access to a rich supply of nutrients, but they can thrive only if the acidity of their microenvironment can be reduced.

Urea and the enzyme that degrades it, **urease**, the most highly expressed protein in this bacterium, play critical roles in both chemotaxis and pH regulation (Huang et al. 2015). A chemoreceptor (TlpB) in the cell membrane of the *H. pylori* cell detects urea that is released from stomach cells and binds the solute with extremely high affinity. TlpB allows urea to be detected at concentrations as low as ~50 nmoles l^{-1}. Using the guidance provided by TlpB, the bacterium swims up the urea gradient to reach the stomach lining. Urease on the bacterial cell wall degrades the urea in the bacterium's immediate vicinity, thus sharpening the urea gradient and increasing the accuracy of the chemotactic movement. Urease also plays an important role in establishing a pH microenvironment where the bacterial cells can

prosper. Once the *H. pylori* cell is near the stomach lining, continued high levels of urease activity work to buffer the pH of the microenvironment. The urease-catalyzed breakdown of urea yields ammonia and bicarbonate. These breakdown products thus increase the local pH and thereby establish a microenvironment where *H. pylori* cells can function well and, under certain conditions, generate stomach ulcers. As a historical note, the original proposal by Marshall and Warren (1984) that stomach ulcers were caused by a bacterium was greeted with some skepticism. However, this novel conjecture has been proven, and fittingly, its creators won the Nobel Prize in Physiology or Medicine in 2005.

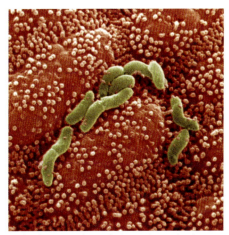

Helicobacter pylori cells (green) attached to a stomach lining.

4.7 Life at Low Water Activity: Desiccation and Anhydrobiosis

In view of the large number of critical roles of water in biochemical systems, it may seem paradoxical that any organism could survive in an essentially fully dehydrated state, by shoving water off "center stage," as it were. However, many types of organisms, including representatives of bacteria, fungi, mosses, vascular plants, and animals (notably insects, tardigrades, rotifers, nematodes, and crustaceans), are able to enter into reversible states of extreme desiccation, known as anhydrobiosis. Here, "extreme desiccation" is defined as a cellular water content of < 0.1 g H_2O g^{-1} dry weight (= less than 10% water content) (Alpert 2005). To provide a sense of what this level of dehydration means for cellular constituents normally bathed in water, it is estimated that this small amount of water is insufficient to provide a full hydration layer around the proteins and membranes of the cell (Billi and Potts 2002). It seems unlikely that such dehydrated proteins could function in enzymatic activity or even maintain their native structures, because of the importance of water in governing protein folding and assembly. Membranes, too, owe their structures to the interaction of their constituent molecules with water: The hydrophobic effect leads to burial of aliphatic lipid chains in the membrane interior, and polar interactions with water favor positioning of charged and polar head groups at the membrane-water interface. As Hoekstra and colleagues (2001) aptly remark, "If water completely dissipates from living matter, the driving force for cellular organization is lost."

Desiccation and suspension of life

Without cellular organization—native-state structures for proteins, lipids, and other large biomolecules—metabolic activity would seem to be precluded. In fact, it is a characteristic of completely desiccated cells that they are **ametabolic**. Metabolism is suspended—or at least reduced to such extremely low levels that the experimenter may not be able to detect oxygen consumption (indirect calorimetry) or heat production (direct calorimetry). As we show below, reducing metabolism to extremely low levels may be the "secret" to survival in the anhydrobiotic state, which often persists for periods of weeks to centuries, and at the extreme can persist for many millions of years. Among eukaryotes, the current record for survival in the anhydrobiotic state seems to be held by seeds of the lotus plant (*Nelumbo nucifera*). Seeds that were ~1100 years old were germinated successfully (Shen-Miller et al. 1995). Among bacteria, spores isolated from amber dated to 25–40 million years of age have been revived (Cano and Borucki 2012). Seeds, cysts, and spores are often the most desiccation-tolerant life stages of organisms, and can serve as a bridge

A tardigrade.

between good environmental times and bad. These life stages typically are highly resistant to other types of abiotic stresses as well, including extremes of temperature and ultraviolet (UV) radiation. For example, studies of completely desiccated tardigrades (photo at left) revealed remarkable resistance to UV radiation and tolerance of temperatures from about absolute zero (0 K) to over 100°C (Jönsson and Bertolani 2001). In the case of the midge *Polypedilum vanderplanki*, the most desiccation-tolerant insect known, completely dehydrated larvae survived temperatures between −270°C and 102°C (Hinton 1960).

The biochemical adaptations that permit survival under conditions of extreme desiccation, whether for short or very long time periods, commonly reflect convergent evolution among diverse taxa. Thus, the strategies of anhydrobiosis used by a fungal cell, a brine shrimp cyst, larvae of *P. vanderplanki*, the stress-resistant developmental (dauer) stage of the nematode *Caenorhabditis elegans,* and a seed of a vascular plant share many common features. These may include (1) the presence of high concentrations of protective organic micromolecules like trehalose, which stabilize proteins and membranes; (2) increased amounts of aquaporin proteins, which facilitate movement of water out of the cell; (3) elevated expression of molecular chaperones; (4) high titers of proteins like late embryogenesis abundant (LEA) proteins, which stabilize ("shield") the structures of other proteins; (5) increased levels of polyamines, whose exact function in anhydrobiosis remains to be demonstrated; (6) increased production of polyunsaturated fatty acids, perhaps to facilitate membrane stabilization; (7) high expression of enzymes that repair damaged amino acid residues in proteins, for example, protein repair methyltransferases; (8) increased levels of thioredoxins, which are involved in redox signaling and antioxidant function; and (9) elevated amounts of ROS scavenging enzymes like superoxide dismutase (SOD) (see Erkut et al. 2013 and Gusev et al. 2014 for comprehensive lists of these strategies). Although diverse species employ common strategies in anhydrobiosis, there nonetheless is a certain amount of variation in the mechanisms used; for example, in anhydrobiotic bdelloid rotifers, neither the commonly occurring disaccharide trehalose nor any other disaccharide could be detected in the dried state (Tunnacliffe et al. 2005).

Whatever the biochemical mechanisms are that permit extreme desiccation, these adaptations have two principal effects. First, they facilitate stabilization of the native structures of the metabolic machinery and membranes of the cell, thus keeping biochemical structures intact and poised for reactivation when the cell is rehydrated. Second, they reduce or eliminate the production and enhance the degradation of harmful metabolic by-products, notably ROS, that damage or destroy cellular structures (see Chapter 2). The ametabolic state can be viewed as one where scarce energy resources are conserved for more favorable times, and through this large reduction in metabolic flux, ROS generation is minimized and perhaps eliminated entirely.

Before we look at some specific examples of anhydrobiosis, it bears noting that there is a large amount of effort going into studies of the mechanisms that organisms use to survive extreme desiccation. The motivating interests include those of both the basic comparative biochemist, who might be especially interested in how these adaptations have evolved, and the biomedically focused researcher who seeks to discover improved methods for preservation of cells and organisms in a dry state. Recent studies have begun to reveal additional biochemical strategies for withstanding extreme desiccation, as well as genomic alterations that confer extreme tolerance of desiccation. Intriguingly, some of the genes that support desiccation tolerance may have been acquired by horizontal gene transfer (HGT). Thus, in the midge *P. vanderplanki*, a large set of genes that encode one class of protein that is key in desiccation tolerance—LEA proteins (discussed below)—appears to have been acquired from soil bacteria through HGT (Gusev et al. 2014). Subsequent amplification of the acquired genes has given this insect an unusually large number of LEA protein-encoding genes. Other genes important in anhydrobiosis also occur in large numbers of copies. In *P. vanderplanki* and certain other desiccation-tolerant species, the DNA fragmentation and disruption of cellular membranes that occur during cycles of entry into and exit from anhydrobiosis are conjectured to facilitate acquisition of foreign DNA and its integration into the genome (Gusev et al. 2014). Another major genetic characteristic of species with pronounced capacities for enduring anhydrobiosis is an exceptionally large upregulation of the expression of genes that encode the various proteins needed to defend the cell against the consequences of extreme desiccation (Gusev et al. 2014). Because the genes that encode desiccation-related proteins are often present in high copy numbers, transcriptional upregulation involves numerous paralogous genes, which can lead to extremely large increases in levels of the proteins needed for coping with desiccation stress.

Trehalose prevents cellular damage during desiccation by acting as a water substitute

Among the small organic molecules that undergo especially large changes in concentration during desiccation, the disaccharide trehalose stands out as a major player in several contexts, including structural stabilization and ROS scavenging. Trehalose is accumulated in a phylogenetically widespread suite of organisms to facilitate tolerance of desiccation. We have seen previously, in the discussions of in vitro protein stabilization (see Figure 4.19) and chemical chaperoning under heat stress in yeast (Singer and Lindquist 1998a,b), how effective trehalose can be in protecting proteins under conditions of full hydration.

Likewise, in the absence or near-absence of water, trehalose is again an effective stabilizer of native protein conformations and aggregation states.

Trehalose stabilizes proteins through two distinct mechanisms. One involves water replacement, where trehalose replaces the **vicinal water** (*vicinus* in Latin means "neighbor") that normally coats the protein surface under water-replete conditions. The other mechanism involves formation of a vitreous state of the cytoplasm (reviewed in Hoekstra et al. 2001; Fernandez et al. 2010; Hengherr et al. 2011). The mechanism that is operative depends on the extent of cellular dehydration. When cytoplasmic water content falls to ~0.3 g H_2O g^{-1} dry weight, there is essentially no bulk water left in the cell. One consequence of loss of bulk water is elimination of protein stabilization by preferential hydration. Under these water-limiting conditions, trehalose transitions from being an excluded solute to one that interacts directly with polar groups on the protein surface (**Figure 4.28**). This may seem paradoxical in light of the mechanistic analysis we have given of protein stabilization, in which stabilizing solutes are excluded from interacting with the protein surface and destabilizing solutes interact strongly and lead to unfolding. However, because of trehalose's capacity for hydrogen bonding to polar groups on the protein surface, it is able to act as a water substitute, forming a "hydration" layer around the protein and stabilizing its native structure. These trehalose-protein interactions through superficial hydrogen-bonding networks do not lead to unfolding of the native protein, unlike the effects of protein-binding denaturants like urea or guanidinium.

As water content continues to fall below 0.3 g H_2O g^{-1} dry weight, the mobility of molecules in the cytosol decreases by more than five orders of magnitude. Near 0.1 g H_2O g^{-1} dry weight, the cytoplasm vitrifies, entering a glasslike state in which rates of molecular diffusion and chemical reactions are drastically reduced. Trehalose is the only sugar known that is able to persist in a glasslike state under conditions of complete dehydration, making it the sugar of choice for stabilizing cellular organization through formation of the

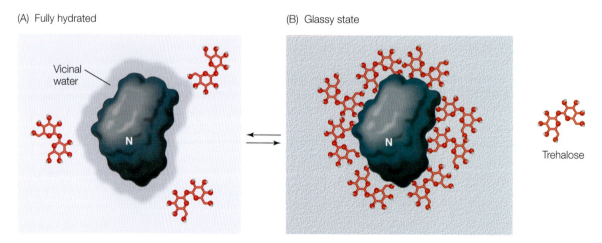

(A) Fully hydrated   (B) Glassy state

Vicinal water   N

Trehalose

Figure 4.28 Mechanisms of protein stabilization under desiccation and vitrification. In the fully hydrated state (A), trehalose is excluded from the surface of proteins, which are surrounded by a layer of vicinal water (dark blue). As water is withdrawn from the cell, trehalose replaces the vicinal water around proteins and the intracellular solution transitions into a vitreous (glassy) state (B). N (= native) indicates that the protein remains in its native structural state, despite the large change in water availability and the transition of the cellular water from a liquid to a vitreous state. (After Hoekstra 2005.)

vitreous state (Crowe et al. 1988; Fernandez et al. 2010). Vitrified cytoplasm is beneficial to desiccated cells in several ways. It prevents inappropriate aggregations of proteins, minimizes crystallization of solutes, inhibits fusion of membranes, and greatly reduces rates of metabolic activity. Thus, vitrification effectively stabilizes the organization of the cell and retards chemical activities that could be damaging during prolonged anhydrobiosis. Vitrification favors what is essentially an ametabolic state in which cellular organization is maintained, but in which biochemical function largely ceases. Thus, in the vitrified state, all of the biochemical structures and machinery of the cell are "locked in place" in a stable form and prepared to resume activity when an adequate hydration level is returned. As water returns to the cell, it displaces trehalose and brings the metabolic machinery back into a hydrated state that allows resumption of normal function. Removal of trehalose during rehydration is likely to be an important component of the return to normal biochemical function, similar to what was discussed in the context of trehalose's "holding action" influence in heat-stressed yeast (Singer and Lindquist 1998a,b).

Another important role of trehalose is stabilization of membranes and reduction in propensities for membrane leakiness during dehydration/rehydration processes. In our discussion of thermal effects on membranes (see Chapter 3), we emphasized how important the right structural state of the bilayer is for membrane function, including the barrier functions that separate different membrane-bound compartments in the cell or that separate the cellular fluids from the extracellular fluids. When the activity of cellular water decreases to very low values, trehalose becomes a water substitute for hydrating the surfaces of membranes (**Figure 4.29**). The spacing of the hydroxyl groups on trehalose is such that they allow a bridging of the polar head groups on the membrane's surface (Pereira and Hünenberger 2008). This bridging action keeps membrane phospholipids appropriately spaced in the bilayer, thus reducing the formation of aggregates of lipids and loss of barrier function.

Maintenance of membrane integrity during complete desiccation has important consequences for reducing or eliminating the production of ROS. Damage to mitochondrial membranes generally leads to increased production of ROS, which, as discussed

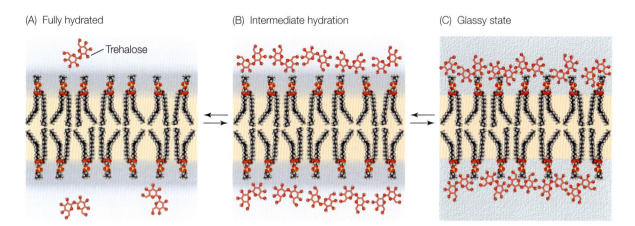

(A) Fully hydrated   (B) Intermediate hydration   (C) Glassy state

Trehalose

Figure 4.29 Trehalose-stabilized membrane structure during desiccation. As water is withdrawn from the cell and extracellular fluids (A,B), the vicinal water hydrating the polar ends of the membrane lipids (darker blue shading) is replaced by trehalose, which hydrogen bonds to the polar head groups of lipids (C). The spacing of the hydroxyl groups on trehalose is such that the head groups of lipids retain their spacing, and membrane integrity is preserved. (After Hoekstra 2005.)

in several contexts in this volume, can damage proteins, lipids, and nucleic acids, and at the extreme, prove lethal to cells. Through the complementary effects of greatly reduced metabolism and maintenance of membrane structure under conditions of extreme desiccation, ROS production may be reduced to the point that very long survival is possible for anhydrobionts. Any ROS that do form must be dealt with, however. Because proteins, including ROS-degrading enzymes, will not encounter adequate water to allow them to sustain their normal catalytic functions, low-molecular-weight organic molecules that can scavenge ROS become the primary line of defense (Hoekstra et al. 2001). We will return to this issue—how interactions among water content, metabolism, and ROS production play out in the context of longevity—after first examining another important set of biochemical players in the context of anhydrobiosis: desiccation-induced proteins.

LEA proteins and other desiccation-induced proteins

Protection of cells from damage during extreme desiccation is not achieved entirely by small solutes like trehalose. Several desiccation-induced proteins, including small heat-shock proteins and unique desiccation-related proteins, are also critical for protection of cells under conditions of low water activity (Clegg 2005). We begin our discussion with the LEA proteins.

LEA PROTEINS AND STABILIZATION OF CELLULAR STRUCTURES UNDER DESICCATION   In a wide range of taxa, including plants and animals (some members of the insects, tardigrades, crustaceans, nematodes, and rotifers; see Hand et al. 2011 for a review), LEA proteins accumulate during acquisition of desiccation tolerance. LEA proteins are named in recognition of the fact that they were first discovered in plants during late stages of embryogenesis. These proteins contribute importantly to stabilization of other proteins, membranes, and possibly DNA (Wise and Tunnacliffe 2004). LEA proteins may also help create, and then stabilize, the vitreous state of the cytoplasm (Shimizu et al. 2010).

Structurally, LEA proteins belong to a subset of proteins known as **intrinsically disordered proteins (IDPs)**. In solution these molecules tend to have low amounts of secondary and tertiary structure—that is, they are largely unfolded. It has been estimated that about one-third of all proteins in the proteome are, to a greater or lesser degree, intrinsically disordered (Hilser 2013; see Chapter 3). Some, like LEA proteins, only attain their mature, functional conformation when water is removed from the system. **Figure 4.30** shows how the three-dimensional structure of part of a LEA protein changes shape during drying, as water content is reduced from 83.5 weight percent down to 2.3 weight percent water. When folded into its native conformation at low water activities, the LEA protein exposes a surface rich in polar groups. LEA proteins thus are extremely hydrophilic and have been shown to bind water very tightly. It is conjectured that LEA proteins, in common with hydroxyl-rich solutes like trehalose, serve as water substitutes that coat (shield) proteins and membranes and thereby confer stabilization of structure. LEAs have been shown to associate with membranes and prevent the types of leakage that dissipate solute gradients and, in the case of mitochondria, lead to elevated rates of ROS production (Sales et al. 2000). Like small heat-shock proteins (see below) and trehalose, LEA proteins may bind to partially unfolded proteins and prevent further denaturation (Goyal et al. 2005). The full set of functions of LEA proteins and other desiccation-induced proteins remains an active topic of investigation in basic science and biotechnology, where long-term preservation of cells—perhaps immortality itself—remains something of a Holy Grail.

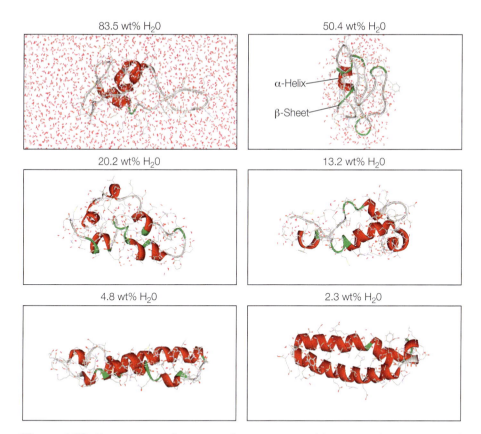

Figure 4.30 Changes in conformation of a 66-amino acid fragment of a LEA protein from the nematode *Aphelenchus avenae* as a function of water content. Water molecules are depicted in gray and red lines; the LEA protein fragment is shown using a solid ribbon (red, α-helix; green, β-sheet; gray, random coil). Note how the folded structure of the protein forms only when water content falls below 3%. (From Li and He 2009.)

BIOMEDICAL USE OF THE TOOL KITS OF ANHYDROBIONTS   Progress in stabilizing mammalian cells in extremely desiccated states has been made recently using genetically engineered human hepatoma cells that expressed two LEA paralogs, one targeted to the cytosol and nucleus and one targeted to the mitochondrion (Li et al. 2012). In some of the engineered cell lines, a trehalose transporter gene was introduced, leading to uptake of trehalose from the medium and its intracellular accumulation to levels near 20 mmol l^{-1}. The genes for the LEA paralogs were from the brine shrimp *Artemia franciscana*, which withstands extreme desiccation during the cyst phase of its life cycle. Hepatoma cells were dried to a very low water content, < 0.12 g H_2O per g dry weight of cells. Control cells lacking LEA protein-encoding genes or the ability to take up trehalose exhibited zero survival during a desiccation-rehydration cycle. However, the genetically engineered cells that accumulated trehalose and expressed the LEA proteins were 98% viable after rehydration. Perhaps the most remarkable finding was that the mitochondrially targeted LEA protein conferred 94% protection even when trehalose wasn't accumulated, a level of protection much greater than observed in trehalose-free cells expressing the cytosolic paralog. Mitochondria appear to play an especially important role in desiccation tolerance (Atkin and Macherel 2009), in part because of the importance of maintaining mitochondrial integrity to avoid excess production of ROS. The ability of mitochondria to rapidly restore respiratory activity during rehydration also may be crucial for cellular survival; energy demands may spike as cells "come back to life." Suffice it to say, the abilities of LEA proteins to protect membranes and to enhance vitrification of the intracellular fluid, especially when complemented by the same capacities of trehalose, suggest that long-term stabilization of mammalian cells in the dry state may be feasible. To survive, however, the cells must be maintained at a temperature below the glass transition temperature (T_g) (Li et al. 2012); survival of such engineered cells at room temperature remains problematic.

OTHER PROTEIN PARTNERS FOR LEA PROTEINS   Partnering with LEA proteins are a suite of other stress-induced proteins, including the small heat-shock protein Hsp26 (**artemin**), a desiccation-induced protein first identified in the brine shrimp *Artemia franciscana*, and heat-shock protein 70 (Hsp70), which is induced in response to a variety of stresses. These proteins further reduce damage to proteins during extreme desiccation (Clegg 2005). One of the functions of small heat-shock proteins like Hsp26 is to carry out a holding action, keeping partially denatured proteins from more fully unfolding and preventing aggregation of unfolded proteins. The high levels of trehalose commonly found in desiccation-tolerant cells likely assist here as well. The holding action function of trehalose was discussed above in the context of heat stress in yeast (Singer and Lindquist 1998a,b). Subsequently, when protein-damaging stress (whether high temperature or low water activity) is reduced, larger heat-shock proteins like Hsp70 can engage in the chaperoning activities that lead proteins along the productive folding pathways that generate their native three-dimensional structures.

In summary, whereas it is clear that there is no substitute for water as a medium for life (Henderson 1913), in the restricted context of low water activity in desiccation-tolerant species, there are a variety of water substitutes that can fulfill certain of the key functions of water as a driver and stabilizer of the native states of proteins and membranes.

Water activity, metabolic rate, ROS generation, and life span

What all does it take to enable anhydrobiotic organisms—or at least their most resistant life stages—to persist in anhydrobiotic states for years to millennia? An integrated perspective on this issue is beginning to emerge, one that focuses especially strongly on minimizing or eliminating ROS-induced damage to cellular structures during long periods of physiological quiescence. We have discussed above how the addition of small organic molecules like trehalose facilitates stabilization of proteins and membranes. By stabilizing membranes—perhaps with the help of macromolecules like LEA proteins—trehalose and other carbohydrates like sucrose (accumulated in many plants) reduce production of ROS. In addition, trehalose is an ROS scavenger. During desiccation, other ROS-destroying agents, including enzymes like superoxide dismutase and catalase and the tripeptide glutathione, accumulate to high levels (Kranner and Birtić 2005), leading to further protection of the cells from any ROS that might be generated during anhydrobiosis. ROS-degrading enzymes, of course, will begin to lose function as water content drops to values that preclude enzyme activity. Organic micromolecules will then assume the central role in ROS detoxification, as mentioned earlier. Whatever the relative roles of ROS-degrading enzymes and micromolecules during anhydrobiosis, the enormous reduction in metabolic rate (and thus oxygen usage), in and of itself, is likely to reduce ROS production to extremely low values (see Chapter 2).

LIPID ADAPTATIONS IN ANHYDROBIOSIS   An additional mechanism for reducing damage from ROS, and for further reducing the production of ROS, involves changes in the acyl chains of membrane phospholipids. In the discussion of adaptation of membranes to temperature (see Chapter 3), we emphasized the importance of sustaining the right state of membrane bilayer structure, a liquid-crystalline state, in the face of changes in cell temperature. During acclimatization to reduced temperature, the double-bond content of acyl chains of membrane phospholipids significantly increases in ectotherms (Hayward et al. 2014). In anhydrobionts, changes in membrane lipid saturation again emerge as an important adaptation to stress. However, in this context the main concern is not the maintenance of a physical state conducive to continued metabolic activity, but rather an avoidance of ROS-induced damage. As Hoekstra (2005) points out, production of a vitreous cytoplasm during extreme desiccation may stabilize structures and reduce biochemical activity in this fraction of the cell, but membranes are not stabilized in this way. The membranes under these conditions retain a high degree of molecular mobility and thus are susceptible to desiccation-induced damage, including attack from ROS. Membranes minimize the likelihood of ROS damage by reducing double-bond content of phospholipid acyl chains, thereby reducing the chances for ROS-induced peroxidation reactions to occur, as discussed in Chapter 2. The inverse correlation between double-bond content and longevity—for instance, the capacity for germination—is striking (**Figure 4.31**): The lower the double-bond content of membrane phospholipids, the greater is longevity, a relationship that peaks in the lotus plant with its 1100-year seed survival under complete desiccation. Here, then, is further evidence that protection of membranes from ROS-induced damage is a critical adaptation of anhydrobionts for attaining remarkably long life spans (Hoekstra 2005). The relationship between lipid composition (double-bond content), ROS damage, and longevity should recall the discussion of life span determination in birds and mammals in Chapter 2. Regardless of taxa, then, ROS-membrane interactions seem pivotal for determining length of life.

"TITRATING" METABOLISM WITH WATER CONTENT   Lastly, we examine two very different examples of the interactions between water content and either metabolism or longevity, to illustrate the degree of control that water's availability and physical state exert on these phenomena. The first example is from studies of an Antarctic insect (*Belgica antarctica*) that loses a large fraction of its body water during acquisition of tolerance to freezing

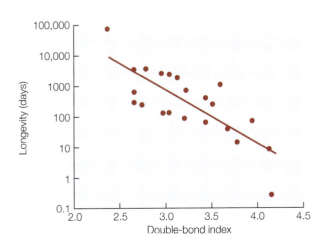

Figure 4.31 Relationship between double-bond content of membrane phospholipid acyl chains (double-bond index) and longevity for a broad range of plant propagules with different longevities. A significant negative correlation ($r = -0.78$) between double-bond content and longevity exists, with the extreme being seeds of the lotus plant (longevity near 100,000 days). (After Hoekstra 2005.)

Figure 4.32 Relationship between total body water content and rate of oxygen consumption by larvae of the Antarctic insect *Belgica antarctica*. Water content was manipulated by exposing larvae to solutions with different osmolalities. Red circles are for larvae acclimated to freshwater; blue circles are for larvae osmotically dehydrated in seawater. (After Elnitsky et al. 2009.)

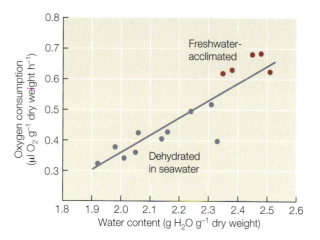

and desiccation (Elnitsky et al. 2009). This species is not a true anhydrobiont because its water content does not fall to the low value (<0.1 g H_2O g^{-1} dry weight) defining complete desiccation. However, it is instructive to examine the change in metabolism (oxygen consumption) that accompanies water loss in this species. The rate of oxygen consumption is strongly positively correlated with water content (**Figure 4.32**). Furthermore, as water content falls, trehalose levels rise almost twofold in this insect (Elnitksy et al. 2009), a trend similar to that observed in true anhydrobionts. The extent to which rising trehalose levels contribute to reducing metabolism—for example, through raising cellular viscosity—remains undetermined in this species. As discussed earlier and in Chapter 3, an increase in viscosity in the cold, especially for solutions with intrinsically high viscosities due to their micromolecular composition, is likely to be a potent retardant of metabolic activity. Regardless of specific mechanisms, however, the effectiveness with which changes in water content (or viscosity) can titrate rates of metabolism over a wide range of water content is evident.

If mass-specific metabolic rate is an important determinant of longevity, as discussed in Chapter 2 in the context of metabolic scaling relationships, then one might predict that reductions in metabolism in anhydrobionts would favor extreme longevity. The remarkable survival of desiccated lotus seeds (Shen-Miller et al. 1995) and bacterial spores (Cano and Borucki 1995) mentioned earlier in this chapter would seem to support this conjecture. However, a more mechanistic analysis, in which the hydration state and longevity of the same experimental organisms could be measured simultaneously, would provide a much more robust test of this proposition. This type of analysis has been performed on desiccation-tolerant plant seeds and pollen grains (**Figure 4.33**; Buitink et al. 2000). Buitink and colleagues examined an important consequence of changes in water activity—altered mobility of solutes in the cytoplasm. As stated above, as the hydration state of the cell decreases, molecular mobility is greatly reduced, falling by approximately five orders of magnitude when water content falls below 0.3 g H_2O g^{-1} dry weight (Hoekstra et al. 2001). When cellular water transitions to the vitreous state near a water content of 0.1 g H_2O g^{-1} dry weight, molecular motion is even more drastically reduced. Transition to the vitreous state significantly increases the Arrhenius activation energies of biochemical reactions (Buitink et al. 2000). Consequently, k_{cat} values of enzymes will be greatly reduced, leading to rapid, large-scale decreases in metabolism. These changes in molecular mobility (diffusion rate) and energy barriers to catalysis are primary driving factors behind the large reduction in biochemical activity in anhydrobionts. Thus, if

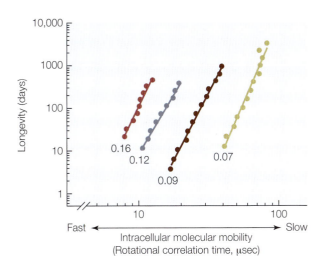

Figure 4.33 Relationship between life span (longevity in days) and intracellular molecular mobility, as indexed by the rotational correlation time (T_R, in μsec) of the probe CP (3-carboxy-2,2,5,5-tetramethylpyrrolidine-1-oxyl), measured in pea seeds with different water contents (values below each regression line represent g H_2O g^{-1} dry weight [g/g]). Points along each regression line represent measurements done at different temperatures: 0.16 g/g, 25°C–45°C; 0.12 g/g, 35°C–55°C; 0.09 g/g, 35°C–65°C; 0.07 g/g, 35°C–65°C. A low rotational correlation time corresponds to high solute mobility ("Fast ↔ Slow"). (After Buitink et al. 2000.)

metabolic rate is a primary determinant of life span, one would expect to find a strong negative correlation between molecular mobility and longevity.

This is precisely what Buitink and colleagues observed (see Figure 4.33). Their experimental procedure involved infusing plant pollen and seeds with an organic reporter molecule (3-carboxy-2,2,5,5-tetramethylpyrrolidine-1-oxyl [CP]), whose rate of rotation (rotational correlation time, T_R) could be monitored using a technique called saturation transfer electron paramagnetic resonance spectroscopy (Buitink et al. 2000). The researchers varied temperature and water content to create a wide range of T_R values. Their analyses showed that intracellular molecular mobility of CP was inversely proportional to life span. Much of this extension of life span probably resulted from the stabilization of cellular structures, especially membranes and proteins, under the conditions established by the glasslike state of water that existed at the lowest water activities. In addition, the large-scale reduction in metabolic rate, and thereby in ROS production, when the cytoplasm became increasingly glassy and rigid would have reduced and perhaps eliminated ROS damage of cellular structures. For seeds and pollen, then, slowing things down, literally, seems a sure way to facilitate a longer life.

4.8 Conclusions: The Need for Inclusivity in the Study of Evolution

Dobzhansky's famous remark "Nothing in biology makes sense except in the light of evolution" applies not only to the phenomena conventionally studied by evolutionary biologists—phylogeny, speciation, anatomy, morphology, genetics, behavior, and so forth—but also, and perhaps even more profoundly, to the pervasive and foundational influences of water on the fundamental constituents and processes of living systems. L. J. Henderson (1913), in his classic work *The Fitness of the Environment: An Inquiry into the Biological Significance of the Properties of Matter,* emphasized the unique "fitness" of the most abundant molecule in living systems, water. In the century that has elapsed since Henderson's book was published, we have learned an enormous amount about the many critical roles that water plays in two different—yet complementary—pathways of evolution, those involving macromolecules and micromolecules. The qualitative and quantitative compositions of cellular solutions—their fitness for living systems—can only be understood when viewed in the context of the critical interplay that exists among water, small solutes, and macromolecules (and large molecular assemblages like lipoprotein membranes). Likewise, the evolution of macromolecular structure can only be understood in the context of the stabilizing or destabilizing influences of the medium in which the macromolecules are bathed.

This chapter, then, has been written in large measure to bring water out of the shadows—to take it from the dark "background" role it is often assigned to its proper place as "the headline act" on the stage of molecular evolution. Water, of course, does not itself "evolve." However, it influences the evolution of biochemical systems of all types of organisms in an enormous variety of ways. Thus, we have attempted to show how profoundly the physical and chemical properties of water have influenced the evolution of macromolecules and, concurrently, have governed selection of the types and concentrations of micromolecules (inorganic ions, small organic solutes, and protons) found in biological solutions. The study of evolution must, then, be inclusive of both the big and small molecules of cellular fluids, if what we observe is truly to "make sense" in the most comprehensive and general manner possible.

CHAPTER 5

Adaptation in the Anthropocene

We have met the enemy and he is us.

Walt Kelly, cartoonist (1913–1973)

5.1 The Anthropocene: Unique Challenges for Biochemical Adaptation

We are living in an era that many scientists refer to as the **Anthropocene** (Crutzen and Stoermer 2000; Stager 2011; Ruddiman et al. 2015). During this period of Earth's ~4.5-billion-year history, human activities are affecting the planet on a scale that may exceed the effects of concurrent changes driven by natural geological and geochemical processes. Indeed, current *rates* of change in abiotic factors such as temperature and seawater pH are thought to exceed the rates that have occurred at most times in the past, when only natural processes—such as plate tectonics, weathering of rocks, volcanic activities, and the occasional strike from a large meteor, asteroid, or comet—were responsible for large-scale alterations of the environment. Even though we often cannot reconstruct past rates of change with great accuracy, it is unarguable that human activities are causing a rapid and widespread alteration in the biosphere that is atypical of Earth's history. As the great oceanographer Roger Revelle (1909–1991) once remarked, humans are conducting a massive-scale "natural experiment" in which release of heat-trapping gases like carbon dioxide is leading to unprecedented changes in the environment. The extent to which we can gain control of this ongoing "experiment" may well determine the future of the biosphere (for an insightful review of the scenarios that are possible, see Stager 2011).

There is an ongoing debate among scientists about just when the Anthropocene began (Stager 2011). Some argue that it began some 10,000 or so years ago when people first started clearing forests and planting crops, activities that would come to have large effects on the global carbon cycle. Others believe the Anthropocene should be dated from the beginnings of the Industrial Revolution in the late eighteenth and early nineteenth centuries, when combustion of fossil fuels like coal greatly accelerated release of CO_2 into the atmosphere. James Watt's development of the steam engine in the 1760s is sometimes regarded as the "smoking gun," as it were. Whatever starting date one prefers for the Anthropocene, what can no longer be a subject of dispute are the profound impacts of anthropogenic activities on the biosphere. The remarkably concurrent increases in atmospheric CO_2 and human population size provide an unequivocal basis for assigning responsibility for the rapid changes in greenhouse gas levels in the atmosphere over the past several centuries (**Figure 5.1**). These matching curves give weight to the quote from Walt Kelly that heads this chapter.

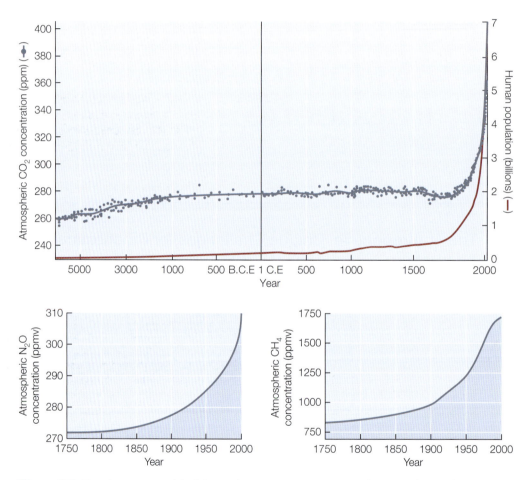

Figure 5.1 The changing world of the Anthropocene: Human population and greenhouse gas emissions. Changes in the size of the human population and atmospheric levels of greenhouse gases (CO_2, CH_4, and N_2O) over the past several thousand years. Note that the human population did not cross the 1 billion mark until well into the nineteenth century. The total number of humans ever to have lived has been estimated at ~108 billion. Preindustrial levels of CO_2 were near 280 parts per million (ppm); in 2015, levels of this gas surpassed 400 ppm. Microbial activities in rice paddies and in the digestive systems of ruminants account for a large share of the anthropogenic release of methane. On an equivalent mass basis, CH_4 has ~25 times and N_2O ~300 times the heat-trapping effect of carbon dioxide. (Data from the Environmental Protection Agency, https://www.epa.gov/ghgemissions/overview-greenhouse-gases.)

The anthropogenic release of CO_2 and other heat-trapping gases, such as methane (CH_4) and nitrous oxide (N_2O), is leading not only to a rise in the planet's temperature, but also to a suite of other changes that present significant challenges to life, especially in aquatic habitats. As we will see, many of these changes cause perturbation of the marine carbonate system (see Box 5.4), including decreases in pH (ocean acidification) and a decrease in dissolved oxygen concentrations. Unfortunately, the simultaneous effects of these multiple stressors on the biosphere remain poorly understood, and their analysis should be given a high priority in studies of global change.

"Zooming out": From the molecule to the biosphere at large

In the first four chapters of this volume, we used a "microscopic" lens for our analyses, one that was set to focus narrowly and deeply rather than broadly. We focused principally on the basic molecular-level phenomena that are of central importance in setting the environmental tolerance ranges and optima of organisms.

These relatively focused analyses have, we hope, prepared the reader to better understand why environmental change has the broad impacts that it does. This being said, we recognize that biochemical analysis cannot provide a complete picture of what global change portends (Box 5.1). In this final chapter, then, we "zoom out" our analytical lens and develop integrative analyses of environmental stress and adaptation that reach from the level of the molecule to the highest levels of biological organization—biogeographic

BOX 5.1 ■ ■ ■ ■

Beyond biochemistry: The need for an integrative analysis of global change

Even though the biochemical characteristics of organisms strongly influence how they will respond to a rapidly changing environment, there is a need to place biochemical traits with demonstrated sensitivities to abiotic factors into a much broader context, one that includes other traits of the individual species and, importantly, species interactions. For example, the ability of a species to move to a new habitat where its biochemical traits can retain or regain their optimal values may be strongly dependent on the species' mobility (dispersal capacity), reproductive rate, and ability to be an ecological "generalist," for example, to eat different foods if necessary (Angert et al. 2011). A species with insufficient dispersal ability might not be able to move to a new environment that allows it to function optimally at a biochemical level. However, even if a species can move to a new environment, it may not persist there if particular resources it needs are absent or if herbivores or predators prevent the species from surviving.

Interactions among species are likely to play a pivotal role in determining the ecological consequences of global change, and these interactions may be difficult, if not impossible, to predict from biochemical traits. As Gilman and colleagues (2010)

put it: "Precious little is known about the effect of species interactions on responses to climate change in open systems." The types of studies that will be needed to clarify these complex issues include comparisons of effects of global change, notably rise in temperature, on timing of natural events in a species' life cycle (the study of phenology). Decoupling of critical interspecific interactions is already occurring as a result of differential effects of temperature change on timing of important biological events, notably in the case of loss of synchrony between emergence of insect pollinators and flowering of plants that require these insects for reproductive success (Memmott et al. 2007). Another critical effect of global change on interspecific interactions stems from alterations in energy needs in a changing environment. For example, rising temperatures will lead to elevated metabolic rates in ectotherms, and these effects will differ with latitude (see Chapter 3; Dillon et al. 2010) and among species. Herbivores will need to consume larger quantities of their food plants, and carnivores may need to increase their levels of predation. Variations among species in responses to rising temperatures are almost certain to create imbalances between energy needs and energy sup-

BOX 5.1 *(continued)* ■ ■ ▬ ▬ ▬ ▬

ply. In the ocean, rising surface water temperatures may elevate metabolic rates and create the need for higher primary productivity in order to provide nutrition for organisms at higher levels of the food chain. The fall in oxygen content with rising temperature may limit metabolic rates and capacities for predation (see main text; Deutsch et al. 2015). Stratification of the water column caused by the development of a warm, less-dense surface layer may reduce upwelling of the nutrients needed to support enhanced primary productivity.

A meta-analysis by Sunday and colleagues (2012) has provided additional evidence that the effects of rising temperature on the distribution patterns of animals will be determined by more than intrinsic thermal sensitivities. These authors asked if species from terrestrial and marine habitats fully occupy the habitats where thermal conditions are tolerable for physiological function of the species. Overall, marine animals seemed to do so: Their distributions generally matched their potential thermal niches. In contrast, terrestrial ectotherms tended not to fill the lower-latitude edges of their potential thermal niches. The basis of this discrepancy is not clear, but terrestrial ecosystems differ from marine ecosystems in three relevant ways. First, moisture availability may be a critical abiotic factor that affects terrestrial distributions. Second, spikes of extreme heat may occur in terrestrial environments, and these relatively infrequent events may not be adequately taken into account in thermal niche determinations. Third, the relative species richness in low-latitude terrestrial ecosystems may increase the importance of biotic interactions in governing distribution ranges, weak-ening the correlation between thermal tolerance and latitude.

A more recent study by Stuart-Smith and colleagues (2015) found that marine fishes and invertebrates did not always occupy their full "fundamental" thermal niches, as determined in laboratory studies, but rather had more restricted "realized" thermal niches, as determined by field distribution data. This discrepancy further emphasizes the point that physiological tolerance limits are but one of many factors that will govern biogeographic shifts resulting from global change. It is also important to realize that the thermal tolerance of a species may differ across life stages; larval, juvenile, and adult forms may have different thermal niches as well as widely different capacities to move from a stressful environment to a more favorable one.

These are only very limited examples of the complex types of effects that will determine how global change influences individual species and, more broadly, the ecosystems in which they occur. It is outside the scope of this volume to comprehensively review these many supra-biochemical facets of global change. Excellent reviews can be found in papers by Gilman et al. (2010), Angert et al. (2011), Sunday et al. (2012), and Stuart-Smith et al. (2015) and the references contained within. Another excellent perspective on climate change analysis is given by Helmuth et al. (2014), who emphasize that interactions of multiple stressors should be examined on biologically relevant spatial and temporal scales. They caution that broadscale trends such as long-term averages may obscure the important relationships occurring at local, biologically relevant spatiotemporal scales.

---

patterning, ecosystem structure, and the general "health" of the biosphere. This mode of analysis also will be used to predict what lies ahead for the biosphere as the processes of global change continue and, indeed, rapidly accelerate.

Stressor effects and the genetic tools available for adaptive response

Our treatment of adaptation in the Anthropocene will focus on two primary questions. One concerns **stressor effects** in a broad biological context, and the other analyzes the relationship between the **genetic resources** available to a species and the species' likelihood of being able to cope successfully with rapid environmental change.

EFFECTS OF STRESSORS: HOW ABIOTIC STRESS AT THE BIOCHEMICAL LEVEL AFFECTS HIGHER LEVELS OF BIOLOGICAL ORGANIZATION   The first question asks, How do abiotic stressors affect the physiological performance of individual species and, thereby, their biogeographic distributions and biotic interactions? In Chapters 2–4, our analyses of stressor effects tended to focus on a single environmental variable at a time. Here, whenever available information allows us to do so, we will expand our scope to the more complex situation typically found in nature, where two or more variables change at the same time. We will examine in some detail several case studies, including ones that provide what we feel are especially clear illustrations of the types of complexities that arise when multiple stressors change simultaneously.

To a large extent, our multistressor analyses will emphasize work done with marine organisms. We believe that it is in the sea where the influences of multiple types of environmental changes may turn out to be most severe and where they can be most readily studied. One reason for this belief is that past episodes of major global change that led to mass extinctions appear in some instances to have had greater effects on marine ecosystems than on terrestrial ecosystems. A second reason, related to the first, is that it is in the marine realm where changes in multiple stressors can create the greatest problems for life. Many regions of the seas are undergoing changes in temperature, carbonate chemistry (pH and carbonate saturation state), surface salinity (due to enhancement of the hydrologic cycle), and dissolved oxygen (DO). From what we have seen in the discussions in Chapters 2–4, exposure to simultaneous changes in this set of abiotic factors is almost certain to present serious threats to a variety of physiological functions in all marine organisms. Indeed, the examples we have selected from the marine biological literature make it abundantly clear that the effects of global change on the marine portion of the biosphere are *already evident* at all levels of biological organization.

Global change is of course affecting the terrestrial realm as well, where large shifts in biogeographic patterning and in the timing of natural events (phenological changes) have been reported (e.g., Parmesan and Yohe 2003). In fact, we close this chapter with an example of how multistressor effects arising from increasing temperature and rising humidity are already presenting serious challenges to human populations in certain low-latitude regions of the planet. It is thus appropriate to ask, Can the species that is causing the environmental challenges of the Anthropocene continue to survive over its current biogeographic range?

HOW DOES THE GENETIC RAW MATERIAL AVAILABLE FOR FABRICATING ADAPTATIONS DETERMINE LIKELIHOOD OF SURVIVAL AND SUCCESS IN A CHANGING WORLD?   The second primary question we will examine in this chapter concerns the genetic raw material that organisms possess (or can access) for biochemical adaptation to global change. This analysis will place a strong emphasis on potential rates of adaptation and how these differ among organisms. Most of the adaptations we have been discussing in this book reflect long-term evolutionary processes that have played out over a great many generations. These adaptive processes were able to "take their time" because of the slow rate (on average) at which most abiotic factors were changing prior to the onset of the Anthropocene. Now the environment is changing vastly more quickly; the time frame for adaptation is measured not in millions of years or in millennia, but in decades and centuries. Will evolution be able to keep up?

If biochemical adaptation can occur rapidly enough, organisms may be able to persist in their current habitats and perturbation to ecosystem structure may be small. Adaptations that allow an organism to "stay put" would be the sorts of conservative adaptations we have examined in many contexts, namely adaptations that restore the optimal ranges of values for biochemical traits. However, if adaptation cannot occur quickly enough, then a species may need to seek a more appropriate habitat, one that has abiotic conditions similar to those the species encountered during its evolutionary history. In other words, behavioral responses—moving away from the problem, as it were—may be the only way a species can stay within the set of abiotic conditions to which its biochemistry and physiology are adapted.

For a species or population that cannot move to a new habitat, survival will depend on the capacity for acclimatization and on the ability to acquire new genetic information that facilitates rapid evolutionary adaptation. Our analysis thus will center on two primary issues. One is the species' genetically determined capacity for adaptive **phenotypic plasticity**, the altered use of an individual's existing genetic information; the second is the acquisition of new genetic tools over many generations and by a variety of processes. An organism's first line of biochemical defense, when challenged with a changing environment, is phenotypic acclimatization (phenotypic plasticity). This process can occur rapidly, perhaps over periods of hours to days, if the organism has the genetic tools needed to modulate the composition of its proteome and to regulate the proteome's activities in adaptive manners. However, the greater share of our attention will be given to the processes that change the composition of the genome, allowing appropriate shifts in the environmental optima and tolerance ranges of the cell's biochemical machinery. We will see how different types of organisms vary in the rates at which their genomes are likely to be able to facilitate adaptive responses to global change.

To conclude, we'd like to point out that these two potential mechanisms of adaptation to global change—phenotypic plasticity and acquisition of adaptive genetic material—should not be treated as being completely isolated from one another. There is the potential for plasticity to facilitate evolutionary adaptation, and vice versa. Ultimately, our integrated analysis of these population genetic parameters will help us address questions about "winners" and "losers" in the race to cope with global change.

To set the stage for our analysis of what the future may hold, we first take a look into the deep past and absorb a sobering lesson from our planet's prehuman history.

5.2 The Permian-Triassic Extinction Event: A "Distant Mirror" for the Anthropocene?

One of the primary goals of a student of history—whether the history in question entails human activities or paleontological events that predate the origin of humans—is to glean insights from study of the past so as to enable an understanding of the present and, ideally, support the formulation of well-grounded predictions about the future. The late Barbara Tuchman (1978) coined a most appropriate metaphor for such types of analysis in her highly regarded book *A Distant Mirror: The Calamitous 14th Century*. There, she examined catastrophic events in the fourteenth century in a context that provided a "mirror" in which to view contemporary events and their causes and consequences. We've all heard expressions to the effect that "those who are ignorant of history are bound to repeat its mistakes." We will see how this point might play out in the context of anthropogenic global change.

Perhaps the best "distant mirror" for the Anthropocene is the Permian-Triassic (PT) extinction event, which occurred ~252 million years ago (mya) and was marked by the greatest wave of extinctions ever to have occurred on our planet (Gastaldo et al. 2015). The massive extinctions that occurred in both aquatic and terrestrial habitats have earned for the PT extinction event the name *The Great Dying*. (For those interested in looking into other "distant mirrors," Stillman and Paganini [2015] provide an excellent and comprehensive review of other periods in Earth's history when changes more or less akin to those of the PT extinction took place.) Why is the PT extinction event especially relevant to the changes now occurring in the Anthropocene? The answer to this question is straightforward—and frightening. During the PT extinction, the largest volcanic event over the past half-billion years occurred in what is now called Siberia (**Box 5.2**). The Siberian Traps (*trappa* is Swedish for "stairs" or "steps") were formed by this massive ejection of basaltic material, which occurred on the shores of what was then the Tethys Sea. As is characteristic of volcanic activity in general, this eruption of basalt was accompanied by a massive release of CO_2 into the atmosphere; CO_2 levels may have reached ~2000 ppm, approximately five times current levels. The ejection of this greenhouse gas caused a major rise in global temperature (perhaps by up to 8 degrees Celsius) and an eventual decrease of perhaps 0.6–0.7 pH units in the ocean's pH (Clarkson et al. 2015). Oxygen levels in the atmosphere also plunged, after reaching new highs in the late Carboniferous (Pennsylvanian) and early Permian periods. Dissolved oxygen (DO) in seawater thus also decreased, and this decrease was amplified by the effects of rising water temperature; solubility of oxygen falls appreciably as temperature rises (see Box 5.3). For aquatic organisms, then, stress from hypoxia was added to the challenges posed by rising temperatures and falling pH. With these multiple stressor effects confronting aquatic life, it may not seem surprising that more than 90% of marine species disappeared. Moreover, the extraordinarily strong effects of the changes occurring at the PT transition on marine species and ecosystems may be an indication that the marine realm could be more affected by anthropogenic global change than terrestrial ecosystems due, in large measure, to the combined and interacting effects of multiple stressors: temperature, dissolved oxygen, pH, and salinity.

The similarities between the environmental changes that led to the PT extinction event and those now taking place in the Anthropocene should be obvious. The challenges

BOX 5.2

The Permian-Triassic extinction event

The release of CO_2 through volcanic activity at the Permian-Triassic boundary some 250 mya was, relative to current patterns of CO_2 release, both slow and huge. An initial period of volcanism, estimated to have lasted ~50,000 years, led to release of CO_2 at a rate that seems not to have caused a sharp fall in oceanic pH, although there would have been a global rise in air and water temperatures. Geochemical processes, such as the weathering of limestone, can buffer the effects of elevated CO_2 on oceanic pH if the rise in CO_2 occurs slowly enough. It was during a subsequent 10,000-year period that acidification spiked. During this second period of volcanism, it is estimated that ~24,000 gigatons (GT) of CO_2 were released into the atmosphere, a large fraction of which entered the ocean (Clarkson et al. 2015). How does the rate of CO_2 release some 250 mya compare with what is happening today? Here, a caveat is needed, of course: Whereas average rates seem well established over many millennia, there could have been periods of one or two centuries when rates were markedly higher than average. While recognizing this uncertainty, we develop our argument using the longer-term average values for CO_2 release. Whereas Earth's worst extinction event was associated with release of ~2.4 GT of CO_2 per year, current CO_2 release to the atmosphere from all sources is ~10 GT per year—a fourfold higher rate. The processes of geochemical buffering and biological adaptation thus are facing an unprecedented challenge. In view of the fact that CO_2 release during the PT extinction led to a fall in ocean pH of between 0.6 and 0.7 pH units, the current fourfold higher rate of CO_2 release will likely cause ocean acidification at a rate that will exceed the abilities of many species to adapt (see Stillman and Paganini 2015). And as we stress repeatedly in this chapter, marine species will have to cope with falling pH while simultaneously dealing with the challenges posed by rising temperatures and falling dissolved oxygen.

BOX 5.2 *(continued)*

What can we learn from paleontology that will help us predict how contemporary acidification, warming, and deoxygenation of the ocean will affect marine ecosystems? Among the groups showing the greatest loss during the PT extinction event were calcifiers. Several major groups disappeared, including trilobites. Molluscan diversity suffered greatly, and Earth lost its coral reefs for ~9 million years. The **figure** below illustrates the changing face of the tropical reef marine biota during the PT event. Currently, we are already seeing structural damage and population decreases among some calcifiers, for example, pteropod molluscs and reef-building corals (see text), most likely as a result of ocean acidification per se or from combinations of stress from alterations in pH, carbonate saturation state, and rising temperature.

(A)

(B)

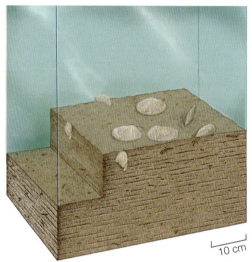

A conceptualization of the effects of the PT extinction on a tropical seafloor ecosystem, showing loss of reef-dwelling organisms. (A) Ecosystem composition before the PT extinction event. (B) Ecosystem structure after the PT extinction event. (From Benton and Twitchett 2003; artwork © J. Sibbick.)

organisms faced 252 mya are a reflection—or better, a predictor—of what life faces in the coming centuries. However, there is one particularly sharp distinction between the PT extinction event and the changes occurring at present: The rate of change in CO_2 levels, temperature, and oxygen during the PT extinction event is dwarfed by current rates of change in these biologically critical variables. These unprecedented rates of change in abiotic factors are especially noticeable in the oceans, where increases in temperature and decreases in pH and DO are already leading to marked changes in biogeographic distributions and ecosystem composition and function. What we see in the contemporary ocean is not only a situation that was portended by the PT extinction event, but also a "window" for viewing what the future is likely to hold for the biosphere at large.

5.3 The Changing Ocean: "Hot, Sour, and Breathless"

It has become customary for marine scientists to speak of the ocean as shifting to a state that is increasingly "hot, sour, and breathless." This is a good shorthand way—one that lay audiences might relate to—of stating that anthropogenic global change is causing increases in ocean temperature and proton activity (i.e., falling pH) and decreases in dissolved oxygen. The effects of these changes are extremely complex and influence not only the ocean's chemistry and biota, but also its physics (Somero et al. 2016). As an example, let's look at the effects of warming of the ocean surface. Warming has direct impacts on the biochemistry of marine organisms due to Q_{10} effects on rates of physiological processes and reductions in levels of dissolved oxygen, but also can have strong effects on large-scale oceanic phenomena like stratification and upwelling. Because warmer water is less dense, it tends to stratify at the surface. Stratification reduces vertical exchange of water through the upper regions of the water column, with important effects on water chemistry. Oxygen generated in shallow water by oxygenic photosynthesis may not be effectively transported to greater depths. Decreased oxygen entry into deeper waters can affect biological degradation of organic material sinking through the water column. Stratification and lack of adequate mixing can reduce the supply of nutrients upwelling from deeper water. Upwelling of cold, nutrient-rich water is a major factor in driving primary production in many coastal ecosystems, so reduced upwelling can greatly reduce surface productivity.

In sharp contrast to these productivity-diminishing effects of global change, other anthropogenic effects can stimulate primary productivity, for example, as a result of runoff of fertilizer-rich water from agricultural lands. When productivity in shallow waters is high, whether from natural or anthropogenic forcing, the settling of large quantities of dead organisms through the water column will trigger large amounts of microbial metabolism. This respiratory activity will further deplete oxygen levels and lead to the release of large amounts of CO_2. The growth of hypoxic or anoxic **dead zones** at the deltas where major rivers enter the oceans is a striking indication of the power of human activities to alter marine environments over large spatial scales (Box 5.3; for an excellent

BOX 5.3

Oxygen relationships

Commonly used terms and units

Oxygen partial pressure (P_{O_2}): The fraction of the total pressure exerted by all dissolved gases in seawater that is due to oxygen. Units: kilopascals (kPa). 101 kPa = 1 atm = 760 torr (mm Hg).

Oxygen concentration [O_2]: The number of moles of O_2 that are dissolved in seawater. Concentration is normally expressed as μmoles kg^{-1}. Oxygen solubility is influenced by temperature and salinity (Pilson 1998). At air saturation, 35 parts per thousand (ppt) salinity, and 10°C, seawater contains 275 μmoles kg^{-1} of dissolved oxygen (DO). At 20°C, [O_2] in 35 ppt seawater is 226 μmoles kg^{-1}. This concentration represents an 18% decrease from 10°C and, for an ectothermic organism, is apt to be paired with an approximate doubling of oxygen consumption rate (Q_{10} effect). [O_2] is also expressed in terms of mg l^{-1} and ml l^{-1}. One mole of gas occupies 22.4 l (at standard temperature and pressure), and the molecular mass of O_2 = 32. Thus, 1 mg l^{-1} = 0.70 ml l^{-1} = 31.3 μmoles kg^{-1}. A P_{O_2} of 24 torr = 1 ml l^{-1}. (See Verberk et al. [2011] for a comprehensive analysis of the roles of partial pressure and solubility in governing the oxygen relationships of aquatic ectotherms.)

Hypoxia: The level of DO at which physiological activities dependent on aerobic respiration become impaired. There is large variation among species in the level of DO at which impairment of function occurs. Thus, there is no single "threshold" oxygen concentration for hypoxia.

Oxygen minimum zone (OMZ): A region of the marine water column in which oxygen levels are extremely low and may in fact reach zero (these anoxic zones are sometimes referred to as oxygen deficient zones, or ODZs). OMZs and ODZs are caused by microbial decomposition of sinking organic matter. The relatively high primary productivity in certain coastal upwelling regions, where nutrient supply

BOX 5.3 *(continued)*

is high, is associated with strong OMZs (or ODZs) because of the large amount of organic material that sinks through the water column. Metabolism of these materials leads to major depletion of dissolved oxygen. Depletion of oxygen is coupled with release of metabolic CO_2, so OMZs typically have relatively low pH values. Depths of OMZs vary among oceanic regions. Commonly, OMZs begin near ~200 m and extend down to ~1000 m.

Critical P_{O_2} (P_{crit}): The oxygen partial pressure at which the rate of oxygen consumption starts to decline (O_2 consumption ceases to be independent of P_{O_2}).

Anoxia: Complete absence of O_2. Animals differ greatly in their tolerance of periods of anoxia.

Aerobiosis: The reliance on oxygen-dependent biochemical pathways, for example, the mitochondrial electron transport system, for energy metabolism (ATP generation) (see Chapter 2).

Anaerobiosis: The reliance on anaerobic metabolic pathways for energy metabolism. Note that anaerobic metabolism may be activated at or near the P_{crit} value, that is, well before anoxia is reached. Animals differ greatly in the oxygen level at which anaerobiosis is initiated and in the pathways of ATP generation that are used. Many invertebrates have anaerobic ATP-generating pathways with higher ATP yields than the glycolytic lactate-generating pathway found in vertebrates (see Chapter 2).

Anthropogenic changes in oxygen content of the ocean

The predicted decrease in total oxygen content ("oxygen inventory") varies among climate change models, as shown by the three different curves in **Figure A** (see Keeling et al. 2010). Whereas the predicted loss of total oxygen content is "only" 5%–8%, decreases in oxygen content are already causing biological effects, as discussed in the text. Predicted decreases in oxygen content differ among regions (see text).

OMZs also may be expanding due to anthropogenic effects such as release of CO_2 and increasing amounts of agriculture (which adds nutrients to waters flowing into the oceans). Vertical circulation through the water column may be reduced due to the occurrence of warmer, less dense surface waters, which leads to stratification of the water column. Increased runoff of nutrients from land where there is excessive use of fertilizers, coupled with higher water temperatures, leads to

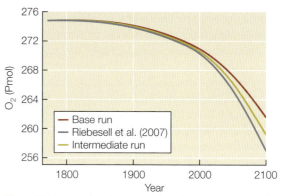

Figure A Historical and projected total oxygen content of the ocean over time. Projections based on three different models are shown. (After Keeling et al. 2010.)

increased surface primary production and thereby to a higher amount of sinking organic material and greater oxygen demands by microbes in the water column. Shoaling of OMZs could result from such combinations of increased stratification and greater surface productivity. Enhanced primary production due to runoff of nutrients also is expanding and intensifying coastal dead zones (Diaz and Rosenberg 2008; Stramma et al. 2008). Under conditions of extremely high surface productivity, anoxic conditions may develop in the bottom water of dead zones. In the absence of dissolved oxygen, animal life is absent and only bacterial and archaeal species adapted for life under anoxia are present. The dead zone at the Mississippi Delta is illustrated in **Figure B**.

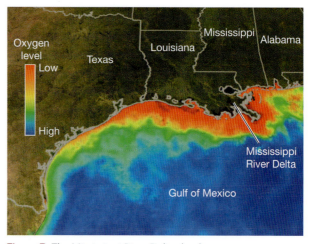

Figure B The Mississippi River Delta dead zone.

review, see Stramma et al. 2008). Enhanced shallow water productivity can also contribute to expansion and shoaling of oxygen minimum zones (OMZs; see Box 5.3).

Below we examine the influences of the warmer, sourer, less oxygenated ocean on selected marine organisms and ecosystems. We begin by providing some foundational information on oxygen and carbonate system relationships that will help bring the key biological issues into clearer focus and provide a basis for analyzing the effects of multiple, interacting stressors on marine life. The oxygen relationships summarized below reflect many of the principles developed in Chapter 2. Carbonate system effects were touched on in Chapter 4 when we discussed optimal pH values. Here we will expand the treatment of carbonate system parameters to include those that are central to processes of calcification.

The warming ocean and falls in dissolved oxygen

Ocean warming has complex effects on the quantity of dissolved oxygen that is accessible for the respiratory needs of marine ectotherms (see Box 5.3 for definitions related to oxygen relationships and for observed and predicted decreases in total oceanic DO over time). Both the oxygen availability in the surrounding water and an organism's capacity to take up and transport oxygen in the body fluids may be affected by changes in temperature. Looking first at the current full range of oceanic surface temperatures (approximately −1.9°C to 33°C), we find that oxygen's solubility in seawater decreases by ~50% (from 349 $\mu mol\ kg^{-1}$ to 190 $\mu mol\ kg^{-1}$) over this ~35-degree-Celsius range. This reduction in oxygen solubility is paired with an increase in oxygen requirements due to Q_{10} effects on metabolic rates.

It is important to examine the DO and metabolic (Q_{10}) effects of temperature changes of the magnitude predicted by Intergovernmental Panel on Climate Change (IPCC) (2014) models, in order to estimate the metabolic challenges that marine ectotherms might experience over the next few centuries (see Deutsch et al. 2015 and the discussion below). Although different models of greenhouse gas emission lead to significantly different predictions about future temperatures on land and in the sea, it is reasonable to predict a rise in sea surface temperature of 2–4 degrees Celsius over the next few centuries (Stager 2011; IPCC 2014). An increase in seawater temperature of 2 degrees Celsius would lead to decreases in DO of ~3%–5%, with the percentage decrease depending on the actual temperature of the seawater. Between −1°C and 1°C, DO falls by ~5%; between 14°C and 16°C, by ~4%; between 26°C and 28°C, by ~3.2%; and between 32°C and 34°C, by ~3%.

The smaller percentage fall in DO at higher temperatures might seem to imply lesser threats from oxygen limitations to ectotherms from warmer waters, but this is not the case. Recall from Chapter 3 that for a given increase in body temperature, the rise in oxygen consumption rates for warm-adapted ectotherms exceeds that for ectotherms with lower body temperatures because of the higher absolute metabolic rates of more warm-adapted species. Phrasing this differently, because metabolic compensation in ectotherms is not complete, global warming will have a stronger absolute influence on metabolic rates in warm-adapted species than in cold-adapted species (see Dillon et al. 2010). If, for example, global warming of 2 degrees Celsius leads to a Q_{10}-driven rise in oxygen consumption of ~15%, then a 25°C-adapted species with a metabolic rate of 100 $\mu mol\ kg^{-1}\ O_2\ h^{-1}$ might increase aerobic metabolism by 15 $\mu mol\ kg^{-1}\ O_2\ h^{-1}$. In contrast, a 15°C-adapted species with a metabolic rate of ~70 $\mu mol\ O_2\ kg^{-1}\ h^{-1}$ (reflecting partial temperature compensation) would increase its metabolism by ~10.5 $\mu mol\ O_2\ kg^{-1}\ h^{-1}$. The ~4% decrease in DO (~10 $\mu mol\ kg^{-1}$) encountered by the 15°C-adapted marine species contrasts with the 3.2% fall in DO (~7 $\mu mol\ kg^{-1}$) for the 25°C-adapted species. The difference in decrease in DO between the 15°C and 25°C waters, ~3 μmol kg^{-1}, contrasts with a 4.5 μmol O_2 kg^{-1} h^{-1} difference in the rise in demand for oxygen between these two species. Thus, the more warm-adapted ectotherm faces a larger challenge in elevating its oxygen consumption rate with rising temperature than does the more cold-adapted species (see Section 5.8; Deutsch et al. 2015).

From the lessons we learned in Chapters 2 and 3, this constellation of temperature-DO-metabolism effects seems likely to drive changes in physiological status in the short run and alterations in biogeographic distributions over longer timescales unless acclimatization or adaptation can be achieved (Deutsch et al. 2015; Stuart-Smith et al. 2015). Although the above analysis has emphasized changes in the surface waters, the effects of ocean warming on DO may be felt throughout the water column. Surface warming not only reduces oxygen's solubility—so that less oxygen enters the ocean in the first place—but also can cause stratification that prevents oxygenated surface waters from reaching deeper layers of the ocean. This reduction in oxygen transport can lead to shoaling and vertical expansion of oxygen minimum zones (Keeling et al. 2010).

When examining the DO relationships of aquatic species, it is important to realize that no single value of oxygen concentration can be used as a common "threshold" level for onset of physiological stress (see Box 5.3). This point should be apparent from the discussion in Chapter 2 of the different types of ATP-generating metabolic pathways and the circumstances under which they are recruited. Organisms differ widely in the degree to which they can rely on anaerobic pathways of ATP generation in order to continue functioning in the face of falling DO. The criteria used to define hypoxia thus need careful evaluation. One commonly used definition of hypoxia is that it begins when the DO concentration of seawater falls below 2 mg O_2 l^{-1} (62.6 μmol kg^{-1}; 22% saturation at 8°C; see Box 5.3). However, this value is not descriptive of the oxygen relationships of many and perhaps most marine fishes and invertebrates (Vaquer-Sunyer and Duarte 2008). Seibel (2011) appropriately emphasizes that critical O_2 thresholds for aquatic organisms must be based on criteria of organismal performance and survival, rather than on some single oxygen concentration. One commonly used performance criterion is based on the response of oxygen consumption rate to decreasing DO. The key parameter here is the **critical oxygen partial pressure (P_{crit})**, which is defined as the partial pressure of oxygen at which the rate of oxygen consumption initially begins to decrease as DO falls (see graph at left). Below P_{crit} the organism can no longer perform at its best, especially when demands arise for increased locomotory activity, as might be needed in predator-prey interactions. P_{crit} is established by several biochemical, physiological, and behavioral traits.

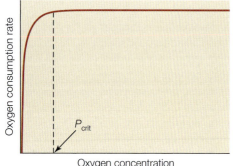

Values of P_{crit} differ widely among animals in relation to their activity levels. Highly active species with high demands for oxygen generally have higher P_{crit} values than sluggish species with relatively low metabolic rates. The levels of DO that characterize the habitat in which a species has evolved also are important determinants of P_{crit}. Organisms that evolved in habitats with low DO tend to have correspondingly low P_{crit} values, that is, they possess a relatively high ability to extract oxygen from the surrounding water (**Figure 5.2**; Seibel 2011). This ability may stem largely from oxygen-transport proteins such as hemoglobin and hemocyanin that have high oxygen affinities. Oxygen-binding affinities are customarily expressed as P_{50} values, the partial pressures of oxygen at which

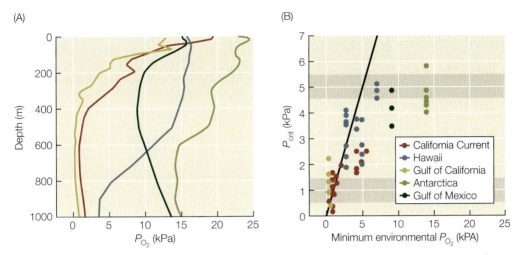

Figure 5.2 Dependence of respiration rate on DO, as indexed by P_{crit}, reflects ambient DO. (A) Oxygen profiles in water columns in five different regions. (B) P_{crit} values of pelagic fishes and invertebrates from the same five regions adapted to different minimum oxygen contents. In regions with relatively low oxygen concentrations, there is a close correlation between minimum oxygen level and P_{crit} (line of unity plotted in black). Species with P_{crit} values falling to the left of the line of unity experience oxygen levels that require partial reliance on anaerobic metabolism (or metabolic suppression at low oxygen levels). Species falling on the line of unity live near their evolved limits for oxygen extraction. The shaded areas portray two different types of oxygen thresholds. The plateau in P_{crit} noted above ~7 kPa indicates that further increases in oxygen concentration provide no additional benefit to the organism. The lower shaded area represents oxygen concentrations that appear not to have selected for lower P_{crit} values. (After Seibel 2011.)

the O_2-transport protein is half-saturated with O_2 (half of the binding sites are occupied by O_2). Thus, a low P_{50} value may be advantageous to a species living at low oxygen partial pressures. As shown in Figure 5.2, P_{crit} values for many pelagic marine animals are proportional to ambient DO levels typical of their habitats over a span of oxygen partial pressures between approximately 2 and 7 kilopascals (kPa). Also note that, at oxygen levels above ~7 kPa, further increases in oxygen content seem to provide no physiological advantage to organisms and the P_{crit} versus oxygen plot exhibits a plateau. Below an oxygen partial pressure of 2 kPa, further decreases in P_{crit} are not observed and may not be possible physiologically (see Seibel 2011). Treating O_2 levels independently of the effects of other stressors can, of course, lead to incorrect conclusions about metabolic relationships of ectotherms. As discussed below, ambient temperature and carbonate system variables such as pCO_2 can affect the ability of an animal to maintain an aerobic poise to its metabolism.

CO_2-driven changes in the marine carbonate system

Approximately 25% of the CO_2 added to the atmosphere enters the surface waters of oceans, where formation and subsequent dissociation of carbonic acid decrease the pH (Gattuso et al. 2015; **Box 5.4**). The decrease in pH that occurs in surface waters leads, over time, to decreasing pH of the entire water column (see Box 5.4, Figure B). This fall in pH is what is commonly referred to as **ocean acidification** (**OA**). The surface pH of the oceans, which now averages about 8.15, is projected to fall to ~7.8 by the end of the twenty-first century; the actual rate of change in pH will of course depend on the rate at which CO_2 is added to the atmosphere (see Box 5.4, Figure C).

BOX 5.4

Carbonate system variables and ocean acidification

When it enters the ocean's surface waters, CO_2 combines with a molecule of water to generate a molecule of carbonic acid (H_2CO_3). Carbonic acid quickly dissociates to form bicarbonate ion (HCO_3^-) and a proton; further dissociation of bicarbonate forms carbonate ion (CO_3^{2-}), which is critical for fabrication of calcified structures:

$$CO_2 + H_2O \leftrightarrow H_2CO_3 \leftrightarrow H^+ + HCO_3^- + CO_3^{2-} \quad (1)$$

HCO_3^- accounts for ~90% of the dissolved inorganic carbon (DIC) in the ocean; carbonate ion (CO_3^{2-}) is present at about one-tenth this level (Dickson 2010). Proton concentrations are expressed as pH values, but several pH scales are used in marine chemistry and biology (Dickson 2010). The *total* hydrogen ion concentration scale includes the reversible proton-binding effects of sulfate ion (SO_4^{2-}). The *free* hydrogen ion concentration scale does not include sulfate ion effects.

The protons arising from entry of atmospheric CO_2 into the surface ocean, along with protons entering seawater from other sources—for instance, from release of CO_2 generated by metabolic activities—may shift the equilibrium of biologically important reactions involving the carbonate ion. Among these reactions are those involved in formation of calcium carbonate ($CaCO_3$) structural minerals, which are important in many marine organisms:

$$H^+ + CaCO_3 \leftrightarrow HCO_3^- + Ca^{2+} \quad (2)$$

Dissolution of calcium carbonate structures is one of the most biologically important consequences of OA. Analogous reactions apply to magnesium carbonate ($MgCO_3$) structural minerals. The sensitivity of different forms of calcium and magnesium carbonates to dissolution by OA is indexed in part by the saturation state of seawater (Ω) for a particular calcium (magnesium) carbonate mineral (X):

$$\Omega\,(X) = [Ca^{2+}][CO_3^{2-}]/K_{sp}(X) \quad (3)$$

Where K_{sp} is the solubility product of the mineral in question:

$$K_{sp} = [Ca^{2+}]_{sat}[CO_3^{2-}]_{sat} \quad (4)$$

Equation 3 expresses the ratio between the observed ion product (numerator) and the value that would be expected if the solution were in equilibrium with the carbonate mineral in question. When the solution is in equilibrium with the mineral phase, $\Omega = 1$. If $\Omega > 1$, the solution is said to be supersaturated with respect to the mineral phase and mineral precipitation is favored. If $\Omega < 1$, the solution is undersaturated and dissolution of the calcium (magnesium) carbonate structure is favored.

Three carbonate-containing minerals with different sensitivities to pH are used by marine organisms to build structures: aragonite, calcite, and magnesium calcite. At 25°C, aragonite is ~1.5 times more soluble than calcite. In magnesium calcite, magnesium ions are randomly substituted for calcium ions, and solubilities vary in proportion to the number of such substitutions. Magnesium calcite is less resistant to pH-driven dissolution than calcite or aragonite. Solubilities of calcium and magnesium carbonates also are temperature sensitive. Reductions in temperature favor dissolution of carbonate structures. Dissolution is also favored by increases in hydrostatic pressure; stability of carbonate structures thus decreases with depth because of the combined influences of low temperature and elevated hydrostatic pressure. Below what is termed the *compensation depth*, carbonate structures tend to dissolve.

Ocean acidification refers to the changes in carbonate system chemistry that result from the entry of CO_2 into the ocean. Equation 1 summarizes these changes. Commonly, ocean acidification focuses on the decrease in pH that accompanies increased entry of CO_2 into the seas, but it must be remembered that effects on other components of the carbonate system, notably CO_3^{2-} concentrations, also have strong biological effects.

With rising CO_2 levels in the atmosphere (see Figure 5.1), roughly parallel increases in dissolved CO_2 (pCO_2) occur in the surface ocean, as shown in **Figure A** (Doney et al. 2009). The resulting effect of rising partial pressures of CO_2 on pH are shown as well. Projections of future values of seawater pH must take into account several factors, including depth-related effects of ocean acidification and the total amount of CO_2 that is generated by anthropogenic activities. **Figure B** provides projections for the time dependence of seawater pH at different depths (Caldeira and Wickett 2003). Over time, the effects of CO_2 entry into the ocean will be evident throughout the marine water column. The rate at which surface seawater pH will change over time depends strongly on two aspects of CO_2 emissions: the total amount of CO_2 released and the rate at which this release takes place. These combined effects are shown in **Figure C**.

BOX 5.4 *(continued)*

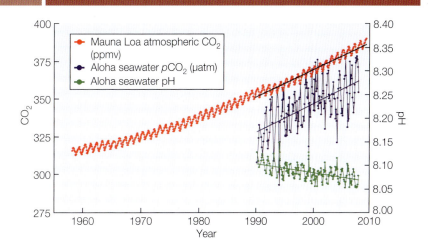

Figure A Example of coupled changes in atmospheric CO_2 levels and surface ocean pH. (After Doney et al. 2009.)

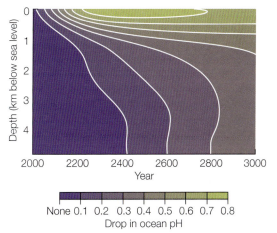

Figure B Projected changes in seawater pH as functions of time and depth. Note how the acidification of surface waters spreads, over time, to deeper waters. (After Caldeira and Wickett 2003.)

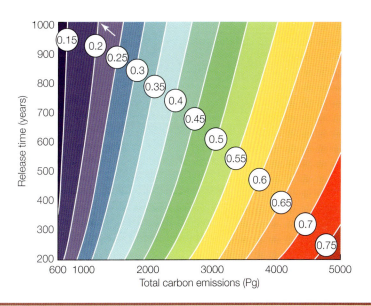

Figure C Influence of the rate of CO_2 emission on the decrease in ocean pH. The contour lines (white) indicate the predicted maximum decrease in pH in average ocean surface waters as a function of total anthropogenic CO_2 emissions (in petagrams [Pg] of carbon [C]; 1 Pg = 10^{15} g) and the time over which this CO_2 is released. The more rapidly a given amount of carbon emission takes place, the faster will be the decrease in surface ocean pH. If anthropogenic CO_2 release totals 1200 Pg C over a period of 1000 years, the surface ocean pH will decrease by ~0.2 units (see arrow). (After Zeebe et al. 2008.)

Although the expression "ocean acidification" is commonly used as shorthand for the effects of CO_2 entry into the seas, it is important to realize that changes in pH are only one component of the ongoing alterations in marine carbonate chemistry. The equilibria that characterize the marine carbonate system are shifting in ways that not only lead to increases in proton activity, but also to changes in the concentrations of bicarbonate (HCO_3^-) and carbonate (CO_3^{2-}) ions and in the carbonate saturation state (Ω) (see Box 5.4). The carbonate saturation state is instrumental in determining whether calcifying species (calcifiers), those organisms that build structures from calcium carbonate (or calcium carbonate + magnesium carbonate), can fabricate and maintain these structures.

Initial studies of OA focused strongly on calcifiers, but more recent work has included a variety of other species influenced by shifts in the marine carbonate system. These studies have demonstrated that different species and different life stages of an individual species may differ in their sensitivities to changes in carbonate chemistry (Kroeker et al. 2013; Nagelkerken and Connell 2015). Calcifiers, like many echinoderms and molluscs, may be most sensitive to changes in carbonate saturation state, especially during early life stages when delicate carbonate structures are being synthesized. Larval stages of calcifiers thus may be more sensitive to OA than adult stages (Kroeker et al. 2013). Other carbonate system variables may have direct effects on a host of other physiological processes, including metabolic rates and ion channel activities (Waldbusser et al. 2015). For example, for many animals, including calcifiers and non-calcifiers, increases in pCO_2 or decreases in pH reduce rates of respiration, and these effects are seen in both larval (Waldbusser et al. 2015) and adult (Rosa and Seibel 2008) life stages.

Interspecific differences in responses to changes in pCO_2 and pH may also be a consequence of the fact that species differ in their abilities to regulate extra- and intracellular pH. Strong pH regulators may be less influenced by OA than weak regulators. In some species, pH regulatory mechanisms change over the course of development, which can lead to differential effects of OA on pre-adult and adult life stages. For instance, adult fishes commonly have gill-localized systems that are highly effective at regulating pH, whereas early developmental stages (e.g., the yolk-sac stage) use different and less effective pH regulatory mechanisms. Because of this, fishes can be most sensitive to OA in early developmental stages (Tseng et al. 2013). Life stage-specific effects of OA thus are likely to occur for many species and for a variety of reasons. Understanding these temporal effects is important if one is to ascertain how changes in the marine carbonate system will affect a particular organism throughout its life history.

Because of the interspecific and ontogenetic variation in responses to changes in carbonate system variables, there is no single value of pH (or any other carbonate system component) that can be employed as a ubiquitous "threshold" for biologically significant effects. Thus, OA, like decreasing DO, must be evaluated on a case-by-case basis. Furthermore, the challenges posed by OA differ among oceanic regions that have different productivities and circulation patterns. OA effects may be particularly strong in productive coastal regions where large amounts of CO_2 are generated metabolically, as sinking organic materials are broken down in subsurface waters. Upwelling of these waters will expose shallow-living organisms to a pulse of water that is colder, of lower pH, and typically lower in DO than the waters they customarily experience (Booth et al. 2012). However, due to these same upwelling processes, coastal regions have routinely experienced lower pH and DO waters throughout evolutionary time, and so populations living in these regions may already be preadapted to life in "sour and breathless" conditions. Suffice it to say, organisms living in these dynamic coastal habitats are useful model systems for evaluating the kinds of multistressor effects that can arise from global change.

Metabolic challenges from life in a "hot, sour, and breathless" sea

Having now provided a brief overview of how greenhouse gases, and CO_2 in particular, are causing the oceans to grow "hot, sour, and breathless," we turn to an examination of how these multiple stressors are already affecting species and the marine ecosystems in which they occur. These studies provide particularly good examples of the linkages that have been identified between stresses at the biochemical level and shifts in biogeography and ecosystem structure.

THE HUMBOLDT SQUID   As individual stressors, oxygen, temperature, and pH are known to have a multitude of effects on organisms, as discussed in Chapters 2, 3, and 4, respectively. Predicting the *net* effects of simultaneous changes in these three stressors is difficult, however, and there has been only limited experimentation on these complex interactions. Therefore, we are still largely in the dark concerning how a warming, acidifying, and deoxygenating ocean will affect marine organisms. Obtaining this knowledge is crucial, however, if we are to make well-grounded predictions about the future state of marine ecosystems.

An especially clear illustration of the effects of exposure of a marine organism to simultaneous changes in temperature, pCO_2 (pH), and DO is provided by studies of the Humboldt squid (*Dosidicus gigas*) (see photo at right; Rosa and Seibel 2008; Seibel et al. 2014). The Humboldt squid is an ecologically important predator in the coastal waters of the northeastern Pacific Ocean. It is a vertically migrating species that returns to shallow waters at night after spending the day in the deeper waters (up to ~300 m depth) of the oxygen minimum zone (OMZ; see Box 5.3). It migrates into the OMZ in order to avoid predation by large fishes like tunas and billfishes. These predatory fishes have P_{crit} values that are much higher than the DO levels of the OMZ, and their energy metabo-

Dosidicus gigas

lism must remain aerobic. Thus, their vertical movements do not extend into the OMZ. Although it reduces dangers of predation during the daytime, the vertical movement of *D. gigas* exposes it to a substantial range of abiotic conditions that create their own set of challenges. Shallow waters are relatively warm and rich in DO, whereas the deeper, O_2-depleted waters of the OMZ are cooler and contain more CO_2. O_2 levels in the OMZ are lower than the squid's P_{crit}, but unlike tunas and billfishes, the squid is a facultative anaerobe (see Chapter 2), so P_{crit}:DO relationships do not limit dives into the OMZ. The squid switches to anaerobic pathways of ATP generation during the daylight hours spent in the OMZ (Seibel et al. 2014). During the squid's daily sojourn between oxygen-replete shallow waters and the OMZ, its behavior changes dramatically. At its daytime depths in the OMZ, the squid is sluggish because aerobic metabolism is critical for its high-speed locomotion. At night, when the squid leaves the OMZ to hunt in shallower waters, it is a vigorous swimmer and has one of the highest mass-specific O_2 consumption rates measured for a marine animal (Rosa and Seibel 2008; Seibel et al. 2014).

These behavioral, metabolic, and locomotory characteristics of the Humboldt squid, in concert with ongoing global change-driven alterations in the waters of the eastern Pacific Ocean, make it an excellent study system for examining multistressor effects arising from global change. The water column in which the Humboldt squid dwells is showing significant changes in temperature, DO, and pCO_2. Surface temperatures are increasing, the OMZ is expanding both vertically and horizontally, and pCO_2 is increasing as well (Booth et al. 2012). Thus, we might view the Humboldt squid as a "poster child" for the effects of multiple stressors on marine life.

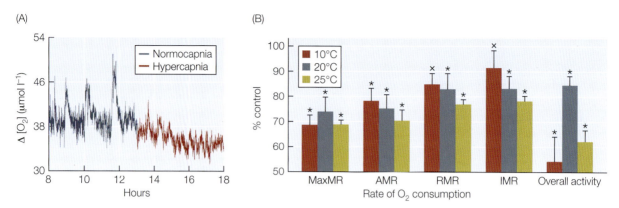

Figure 5.3 Effects of temperature and pCO_2 on rates of oxygen consumption and swimming activity of juvenile Humboldt squids (*Dosidicus gigas*). Rates were determined at a "normal" pCO_2 of ~330 ppm (normocapnia) and at an elevated pCO_2 of ~1000 ppm (hypercapnia). (A) Effects of temperature and pCO_2 on respiration rates and activity of *D. gigas*. The tracing indicates activity-dependent effects on oxygen consumption rate. Activity was measured by recording rates of mantle contraction, which powers swimming in squids. Spikes in oxygen consumption rate coincide with bouts of mantle contraction. Note the absence of large spikes in oxygen consumption under hypercapnia. (B) Temperature dependence of effects of elevated pCO_2 on oxygen consumption and activity levels. Data are expressed as the mean oxygen consumption rates (+ S.E.) under conditions of hypercapnia as a percentage of the normocapnic controls. Rates of oxygen consumption are plotted separately for inactive (IMR), routine (RMR), active (AMR), and maximum (MaxMR) activity levels. The overall effect of temperature and hypercapnia on activity level (number of active cycles per hour) is plotted on the far right. Asterisks (*) indicate significant reduction in rate compared with the controls, and "x" indicates no significant difference from control. Note that the effects of hypercapnia on metabolic rates depend on both temperature and activity level, with the greatest effects caused at the highest activity levels. (After Rosa and Seibel 2008.)

Rosa and Seibel (2008) studied the effects of changes in temperature, DO, and pCO_2 on several aspects of the Humboldt squid's biology: respiration rate, aerobic scope (capacity to elevate oxygen consumption rate during vigorous activity), and locomotory activity levels (**Figure 5.3**). Hypercapnia (elevated pCO_2) was associated with significant decreases in locomotory activity and aerobic scope. These effects were greatest at higher temperatures within the squid's normal temperature range, 20°C and 25°C, and at the highest activity levels (see Figure 5.3B). As in the case of most aquatic species, challenges to sustaining aerobic scope at high temperatures arise from the combined influences of Q_{10} effects on metabolic rates and reductions in DO with rising temperature. In the Humboldt squid, these effects likely are exacerbated by a reduction in the ability of the animal's oxygen-transport protein, hemocyanin, to bind O_2 under conditions of high pCO_2, and possibly under elevated temperature as well. Like enzymes, oxygen-transport proteins generally exhibit a reduced affinity for ligands when temperature increases.

To discern what these responses mean for the squid's performance under field conditions, the effects noted in these laboratory studies must be juxtaposed to the environmental changes that the squid experiences in its daily migratory patterns. Exposures to low DO led to a significant (~80%) decrease in oxygen consumption, a response that is adaptive for an animal that spends considerable time in the OMZ, where reduced demands for oxygen would be beneficial (Seibel et al. 2014). The transition to OMZ conditions is marked by an increased reliance on anaerobic glycolysis, accompanied by a substantial rise in octopine accumulation (Seibel et al. 2014). Total ATP turnover at depth is approximately 52% and 35% of that observed under normoxia in juvenile and adult squids, respectively. Thus, anaerobic generation of ATP is able to replace a considerable fraction of the aerobic ATP production that is lost. In transitioning between periods of high and low rates of oxygen consumption, however, the squid is likely to encounter challenges from reactive oxygen species (ROS) production. As discussed in Chapter 2, large changes in rates of oxygen consumption are typically accompanied by major alterations in rates of ROS production. In the laboratory, juvenile *D. gigas* had much higher activities of the ROS detoxifying enzymes superoxide dismutase (SOD) and catalase (CAT) at normal oxygen levels than they did under low oxygen (Trübenbach et al. 2013).

Another consequence of the squid's vertical migration into low-DO waters is the need to repay the oxygen debt that is incurred during time spent in the OMZ. The squid must reoxygenate its hemolymph oxygen-transport pigment (hemocyanin), and metabolic end products that have accumulated during periods of low oxygen availability must be converted to other compounds. For example, octopine is converted back to pyruvate and arginine. Pyruvate can enter the TCA cycle, and arginine can be phosphorylated to arginine-phosphate (see Chapter 2). These changes can restore the squid's ability to support the vigorous swimming required for prey capture and predator avoidance. The increases in temperature that are occurring in the upper region of the water column therefore represent a challenge to this aspect of the Humboldt squid's metabolic activities. Falling DO and thermal acceleration of metabolism overall may compromise the animal's capacity to restore aerobically poised metabolism and generate the high ATP turnover required for bursts of high-speed locomotion. Continuing decreases in pH in shallower waters may exacerbate this problem by impeding oxygen binding to hemocyanin.

The combined effects of changes in temperature, pCO_2 (pH), and DO may lead to alterations in the distribution range of the Humboldt squid. The vertical distribution range may be compressed by shoaling of low-pH OMZ waters combined with a reduction of DO and pH in warming surface waters, particularly at lower latitudes like those of the Gulf of California, where surface temperatures are already near the species' upper thermal tolerance limits (Rosa and Seibel 2008). Ongoing oceanographic changes may also alter the Humboldt squid's longitudinal and latitudinal ranges. Because of the role that access to OMZ waters plays in the species' normal diurnal activities, expansion of the OMZ across a broader region of coastal waters in the northeastern Pacific may permit the species' biogeographic range to expand (Stramma et al. 2008; Keeling et al. 2010). In fact, the latitudinal distribution of *D. gigas* recently has expanded northward to Canadian and Alaskan waters (Zeidberg and Robison 2007). This substantial range expansion could affect fish populations in the northeastern Pacific because the Humboldt squid is a voracious predator and, when environmental conditions are optimal, reproduces prolifically (Zeidberg and Robison 2007).

In conclusion, the complex interactions among changing temperature, pH, and DO on the physiology of the Humboldt squid translate into downstream changes in the species' vertical and horizontal distribution patterns. In turn, these changes in distribution of the species are likely to have wide-ranging impacts on coastal ecosystems and the economic activities—fishing in particular—that depend on them.

INTERACTING EFFECTS OF TEMPERATURE AND O_2 LEVELS ON ACTIVITIES AND DISTRIBUTION PATTERNS OF FISHES AND INVERTEBRATES   As discussed in Chapter 2, animals vary greatly in their reliance on aerobically powered locomotory activity, their abilities to extract oxygen from seawater and transport it to respiring tissues, and their capacities for sustaining metabolism using anaerobic pathways of ATP generation when hypoxia occurs, either from low DO (**environmental hypoxia**) or highly elevated metabolic rates (**physiological hypoxia**). Species that can use anaerobic metabolic pathways during bouts of hypoxia or anoxia (facultative anaerobes such as the Humboldt squid) may fare better during periods of low oxygen than species that are obligate aerobes (most fishes fall into this category, although they vary widely in hypoxia tolerance). Therefore, one would predict that different species, and different life stages of a single species, will vary in their responses to falling DO. Furthermore, the interacting effects of rising temperature and falling DO may differ greatly among species. For example, obligate aerobes with high respiration rates may be most sensitive to changes in these two stressors. Although temperature increases will lead to similar relative (fold) changes in metabolism among species, assuming that Q_{10} values are similar across species, the *absolute* change in metabolic rate will be greatest in the most active species. (We treated this point in Chapter 3, in the discussion of Q_{10} effects across latitude; see Box 3.10.) Thus, falling DO and rising temperatures are likely to affect species with high respiration rates more than sluggish species.

There is a large and growing literature on the importance of oxygen-temperature interactions on the physiological states and biogeographic patterning of marine fishes and invertebrates (Pörtner and Farrell 2008; Deutsch et al. 2015; Haigh et al. 2015). One especially critical focus of these analyses is on the ability of aquatic ectotherms to elevate their rates of oxygen consumption above resting levels during such critical activities as predation, avoidance of predators, and long-distance migration. The factorial increase in oxygen consumption above resting levels is commonly referred to as **aerobic scope** (or **metabolic scope**). We discussed this phenomenon in Chapter 3, in conjunction with the **oxygen- and capacity-limited thermal tolerance (OCLTT)** hypothesis, which emphasizes the importance of temperature-oxygen interactions in restricting rates of physiological activities (Pörtner and Farrell 2008). Above a certain temperature, aerobic scope begins to be lost because of limitations in oxygen availability.

More recently, the concept of **metabolic index** (Φ) has been developed to further explore the responses of aquatic ectotherms to changes in DO and/or temperature. The metabolic index expresses the ratio of O_2 supply to an organism's resting demand for O_2 (Deutsch et al. 2015). When Φ is 1.0, an organism has no aerobic scope; its normal resting metabolic needs demand all of the oxygen the organism is able to obtain from the surrounding water. When Φ is greater than 1.0, metabolic rate can increase by a factor of Φ above resting levels. For example, if Φ is 2.0, there is enough oxygen for the organism to double its aerobic metabolic rate above the resting level. This amounts to stating that the organism has an aerobic (metabolic) scope of 2. Each species has a critical value of metabolic index, Φ_{crit}, which is needed to allow normal function (**Figure 5.4**). Animals are handicapped in waters with Φ values lower than Φ_{crit} and will tend to avoid them if possible. In their broad analysis of marine fishes and invertebrates from many different latitudes and depths, Deutsch and colleagues discovered that values of Φ in the range of 2–5 were characteristic of most species. These two- to fivefold increases in oxygen consumption rates during active processes like predator-prey interactions agree with many earlier studies of aerobic scope in water-breathing ectotherms.

When these Φ-based relationships are examined in the context of global change, the combined effects of rising temperature on oxygen solubility and respiration rates (Q_{10} effects) suggest that major biogeographic shifts in distributions of marine ectotherms are on the near horizon. These biogeographic effects would be in addition to the changes in distribution associated with ectothermic species' attempts to find waters with temperature conditions amenable to maintaining the properties of their biochemical

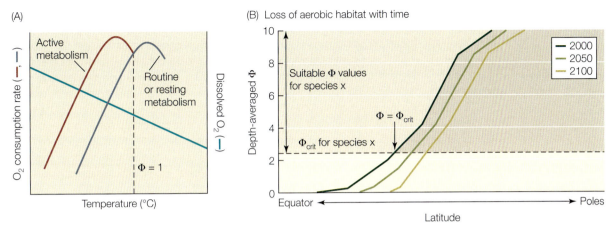

Figure 5.4 Effects of water temperature on dissolved oxygen, routine metabolic rate, active metabolic rate, metabolic index (Φ), and available aerobic habitat for an ectotherm with aerobically powered locomotory activity. "Aerobic habitats" are those that permit a species-specific fold rise in oxygen consumption when metabolism increases from routine to active levels, as occurs when locomotory activity rises. When Φ falls below a given value—the species' Φ_{crit}—the animal's performance is significantly handicapped. (A) Rising temperature leads to a reduction in dissolved oxygen (bright blue line) and an increase in oxygen consumption rate (red and dark blue lines). Both routine metabolism and active metabolism rise with temperature and reach maxima at some high temperature. The ability to increase active metabolism may be lost at a lower temperature than that at which routine metabolism becomes limited. Metabolic scope for activity is lost ($\Phi = 1$) above a certain combination of temperature and DO. (B) Loss of aerobic habitat due to global warming. Depth-averaged Φ is plotted against latitude for three different dates: 2000, 2050, and 2100. As shown here for a hypothetical species (x), rising temperatures decrease the available aerobic habitat—that is, habitat where Φ exceeds Φ_{crit} for the species. Species differ in Φ_{crit}, so this figure presents only a conceptual model of how depth-averaged Φ will influence the amount of aerobic habitat. (B after Kleypas 2015.)

systems—enzyme-ligand binding relationships and gene regulatory mechanisms, among others—at values optimal for function (see Chapter 3). Temperature-oxygen relationships thus must be seen as but one part of a web of temperature effects that organisms must cope with. However, even if we take a simplified view of global change's effects and focus strictly on metabolic index, major distributional changes are predicted. Ongoing ocean warming will reduce Φ throughout the upper part of the marine water column (0 to ~400 m), with model-dependent estimates ranging between 17% and 25% (average of ~21%; Deutsch et al. 2015). Most (~two-thirds) of the decrease in Φ will result from Q_{10} effects on respiration rates. Changes in DO and Φ will differ among oceanic regions. Decreases in Φ will be especially large in the Pacific Ocean, but the Φ problem is worldwide. Whatever the magnitudes of the Φ changes happen to be, these changes are predicted to drive distributions of many, albeit not all, species poleward or toward greater depths in the water column and to reduce the amount of suitable habitat in a time-dependent manner (see Figure 5.4B). In fact, several studies (e.g., Perry et al. 2005; Pinsky et al. 2013) have shown that many species of fish in Northern Hemisphere waters have already shifted their southern and northern distribution limits to higher latitudes, and some species have moved into deeper, cooler waters as well. Temperature effects on dissolved oxygen are only part of the story in terms of how anthropogenic change impedes respiratory performance in marine fishes: Falling seawater pH can also decrease aerobic capacity in fish

(Munday et al. 2009). Thus, ocean acidification, like temperature-driven decreases in DO, may influence fish metabolic performance and distribution patterns.

At any given location, variations in DO may determine the robustness of fish populations. For example, in an area along the coast of Oregon that experiences periods of hypoxia, ~70% of the observed variation in biomass for some species of fish was explained by variations in DO (Keller et al. 2010). In the Southern California Bight, where low-DO waters are increasingly reaching shallower depths, McClatchie et al. (2010) predict that shoaling of low-DO waters over the next two decades could lead to the loss of ~18% of available vertical water column habitat for some species of fish; shoaling of low-DO waters could also lead to hypoxic conditions in ~55% of these fishes' total habitat. These effects on the size and availability of suitable habitat do not include the potential influences of rising temperature. Increases in water temperature could shift biochemical systems out of their ranges of optimal function and reduce metabolic index; both effects could contribute to further loss of suitable habitat. The shoaling and horizontal expansion of coastal OMZs could increase the vertical distribution ranges of hypoxia-tolerant species of fish and invertebrates and increase their latitudinal biogeographic ranges as well

A lantern fish

(Bograd et al. 2008). Vertically migrating species like myctophid fishes (lantern fishes; see photo at left) and the Humboldt squid may, in this sense, benefit from global change because of increased habitat availability. Myctophids are the most abundant vertebrates in the biosphere. They may account for approximately two-thirds of deep-sea fish biomass, and they are a major dietary item for the squid. Thus, in this limited context of two ecologically important marine animals, the "balance of nature" may be retained if both types of organisms undergo similar changes in abundance and biogeographic distribution. In general, rising water temperatures, falling DO, and shoaling and widening of OMZs may lead to replacement of hypoxia-intolerant species by species that are adapted to low oxygen.

In conjunction with responses to falling DO, it is pertinent to inquire how much of a decrease in DO must occur before marine ectotherms seek refuge in waters with higher levels of oxygen. Many fishes and highly motile invertebrates move away from hypoxic areas, but this ability to escape low DO differs widely among species and life stages of an individual species. Demersal species (those living near the bottom) with sluggish locomotion may be less able to move away from areas of low DO than pelagic species (those inhabiting the water column in offshore areas; from the Greek *pélagos*, "open sea") with stronger swimming abilities. As we saw in the case of ocean acidification, the different life stages of aquatic ectotherms like fishes, and many invertebrates, may face different degrees of stress from low DO. Eggs and early developmental stages may have higher mass-specific oxygen consumption rates than adults and may generally be less able than adults to escape water with low DO. Thus, early life stages may be especially vulnerable to the decreases in oxygen concentrations and increases in metabolic rates being caused by global change.

Animals adapted to the extremely low DO of OMZs usually have low rates of oxygen consumption, low P_{crit} values, blood oxygen carriers (e.g., hemoglobins and hemocyanins) with affinities that are high enough to allow effective extraction of oxygen from low-DO waters (see Figure 5.2), and biochemical adaptations that facilitate shifts to anaerobic pathways of ATP production when DO levels are limiting for aerobic catabolic processes (Rosa and Seibel 2008; Seibel 2011; Torres et al. 2012). Some fishes have evolved anaerobic ATP-generating pathways that terminate in ethanol, an end product that does not perturb pH and can diffuse into the environment; these fishes may be especially able to tolerate periods of hypoxia and anoxia (see Chapter 2). Ethanol-producing glycolytic activity is common in vertically migrating myctophid fishes (Torres et al. 2012). As mentioned above, the ongoing vertical and horizontal expansion of OMZs may allow for an increase in the distribution ranges of myctophids and other hypoxia-tolerant species and a reduced abundance of species dependent on continuous access to high levels of oxygen. It remains to be seen what these changes will mean for marine ecosystems and for the abundance of the species on which fisheries most rely. From what we have already witnessed in terms of shifting biogeographic ranges, major ecological and socioeconomic effects are almost certain to occur (Gaylord et al. 2015).

Behavioral effects of OA: Behaving inappropriately while "on acid"

Escape from unfavorable conditions is commonly a first line of defense, especially if acclimatization cannot occur rapidly enough (or at all) to allow survival in an altered environment. One of the most surprising findings of recent studies of OA is that falling pH has widespread effects on a variety of behaviors. These behavioral effects derive from a suite of biochemical and physiological perturbations, including changes in blood ion composition that accompany regulation of internal pH, changes in certain calcareous structures that help mediate behavior, and shifts in ability to sense CO_2 levels in the surrounding water. As in the case of DO and temperature, OA's effects on behavior tell a story that extends from the level of micromolecules (the proton and small neurotransmitter compounds) to the level of ecological (predator-prey) interactions and ecosystem structure. Furthermore, some of the molecular-level effects of OA have a very broad taxonomic distribution and appear to poise numerous marine animals for significant behavioral disruptions from OA.

DISRUPTION OF BEHAVIOR THROUGH EFFECTS ON NEURAL ACTIVITY   OA has been observed to affect behavior in several marine fishes (Munday et al. 2012) and invertebrates (Briffa et al. 2012). In fishes, OA can influence sound detection, olfactory abilities, predator-prey interactions, locomotory function, and habitat choice (Munday et al. 2012). For example, tropical reef fishes that were exposed to CO_2 levels predicted for the end of this century had alterations in olfactory-mediated behaviors, notably homing and distinguishing kin from predators; hearing, such as ability to detect sounds that would be encountered in their native habitats; and learning. Effects of OA on behavior are not just a distant threat; behavioral effects can be induced by decreases in pH now found in certain oceanic waters, and the effects may be relatively long-lasting. Hamilton and colleagues (2014) exposed a northeastern Pacific rockfish (*Sebastes diploproa*) to pH values currently observed during episodic upwelling of low-pH waters. They found significant increases in anxiety levels in the rockfish, as indexed by swimming behavior. The heightened anxiety levels persisted for a week after exposure to low-pH water. Thus, even intermittent bouts of OA may have long-lasting effects on marine species.

These behavioral effects in fishes are caused, at least in part, by perturbation of ion gradients across nerve cell membranes (**Figure 5.5**; Nilsson et al. 2012). These changes in ion gradients, which arise from pH regulatory processes, alter the propensity of the nerves to fire. Under conditions of acidification, fishes increase the concentration of HCO_3^- in the blood; to maintain charge balance, HCO_3^- accumulation is coupled with reductions in plasma chloride ion concentrations (Hamilton et al. 2014). Under normal (non-OA) conditions, the concentration of Cl^- is slightly higher in extracellular fluids

(A) Normal conditions   (B) Ocean acidification

Figure 5.5 Effects of ocean acidification on ion flux through GABA-gated Cl⁻ channels in neural cell membranes. (A) Under normal seawater (non-OA) conditions, binding of GABA to the GABA_A receptor leads to entry of Cl⁻ into the neuron, causing hyperpolarization and inhibition of neural activity. (B) OA leads to pH-regulatory adjustments in plasma ion concentrations: Bicarbonate concentrations rise and chloride concentrations decrease. Under these conditions, Cl⁻ leaves the neuron when the GABA-gated channels open. The resulting depolarization leads to excitation of neural function. Behavioral abnormalities like those discussed in the text may result from this excitation. (After Hamilton et al. 2014.)

than in nerve cells. When ion channels in nerve cell membranes that transport Cl⁻ are opened as a result of binding of the neurotransmitter gamma-aminobutyric acid (GABA), Cl⁻ enters the cell, which leads to hyperpolarization and a reduction of neuronal activity. However, under OA stress, when plasma Cl⁻ concentrations are reduced, opening of the GABA-gated channels leads to an outflow of Cl⁻ from the neuron. This outward flux of Cl⁻ leads to depolarization of the nerve cells, an increased level of activity in the neural pathway, and disruption of normal behavioral patterns (Hamilton et al. 2014). The role of GABA-modulated ion channels in this OA response was confirmed using an antagonist of GABA, gabazine. Gabazine prevents binding of GABA to the regulatory site on the ion channel, preventing it from opening and allowing passage of Cl⁻ out of the nerve. When gabazine was employed to block the activity of the GABA-activated channel under conditions of OA, behavioral abnormalities were not observed (Nilsson et al. 2012; Hamilton et al. 2014). (Note: GABA-gated channels are the sites of action of benzodiazepine drugs that are used to treat anxiety in humans.) The widespread occurrence of GABA-modulated ion channels suggests that the types of effects seen in these studies of fishes could be of broad importance in establishing effects of acidification on animal behavior.

Plotosus japonicus

OA INFLUENCES ON A MARINE FISH'S SENSE OF SMELL   OA also can disrupt predator-prey interactions. The marine catfish *Plotosus japonicus* (see photo at left) can detect prey that inhabit marine sediments by "smelling" pH-associated increases in pCO_2 or H⁺ (Caprio et al. 2014). CO_2 generated by sediment-dwelling prey species like polychaete worms can provide a cue to the catfish that elicits predatory behavior. The catfish has a very acute pH-sensing ability: Changes in pH of less than 0.1 pH unit are enough to trigger predation behaviors. However, the pH range within which the pH-sensing system works is quite narrow. Sensitivity was maximal at the pH values of normal seawater, pH 8.1–8.2, but decreased rapidly below pH 8.0. In view of the prediction that surface seawater pH may decrease to ~7.8 by the end of the twenty-first century, OA may render these pH-sensing systems dysfunctional.

The cellular processes underlying the catfish's acute pH-sensing ability have not been identified. Furthermore, it is not yet known whether these pH sensors can acclimatize to different pH values, such that their sensitivities to prey-generated transients in pH could be maintained at the pH values expected with ongoing ocean acidification. If not, this pH-sensing mechanism could be a good example of an adaptive response developed over a great many generations under stable environmental conditions (pH remaining in the slightly alkaline range near 8.1–8.2) that is seriously threatened by rapid global change. In this sense, the long-term stability of ocean pH may have led to vulnerabilities analogous to those created by evolution in habitats like the Southern Ocean, where stability of water temperature has led to extreme stenothermy (see Section 5.7). It seems possible, then, that stenotolerance may extend to pH-sensing mechanisms in marine animals.

It is interesting to speculate about the potential evolution of a pH sensor that relies on titration of the imidazole side chain of histidine. As discussed in Chapter 4, the imidazole group is the "tail that wags the dog" in many biochemical pH relationships. For it to function with maximum sensitivity as a pH sensor, the imidazole side chain's pK value (pK_{imid}) should be close to the pH value of the water in which the sensing activity is performed. Because pK_{imid} can be modified through changes in nearby amino acids in the protein sequence, it should be possible to evolutionarily "tune" the pH sensitivity of an imidazole-based sensor to keep it in an optimal functional state in the face of OA. For example, if a negatively charged side chain of a glutamate or aspartate residue is close to an imidazole group, the likelihood of a proton binding to the imidazole chain is increased (pK_{imid} is higher). Conversely, a lysine or an arginine residue in the nearby sequence would disfavor proton binding to the imidazole, decreasing pK_{imid}. Amino acid substitutions that decrease the pK value of the imidazole-based sensor might allow the sensing system to maintain its sensitivity—to keep its pK value near the pH of seawater—as OA progresses. This type of amino acid substitution could provide one evolutionary route for adaptation to ocean acidification (Stillman and Paganini 2015).

pH EFFECTS ON BINDING OF SIGNALING MOLECULES (SEMIOCHEMICALS)   Molecules other than histidine's imidazole side chain may also be subject to pH titration and could affect an organism's vulnerability to ocean acidification. For example, binding of signaling molecules (**semiochemicals**) like pheromones to their receptors can be pH sensitive due to pH effects on the charge states of the semiochemicals themselves and on the semiochemicals' interactions with the ligand-binding sites on their receptors (Reisert and Restrepo 2009). Such pH effects have been discovered for a diverse set of small signaling molecules, including peptides, nucleosides, thiols, and organic acids (Hardege et al. 2011). The signaling function of these molecules is essential for a wide range of processes in marine organisms, including sperm-egg interactions (fertilization) and larval settling. Because these processes are so closely linked to successful reproduction, any perturbation of semiochemical binding by OA could have widespread impacts on marine ecosystems. pH-semiochemical interactions thus merit additional study, including investigations of how simultaneous changes in pH and temperature influence semiochemical-ligand interactions. The types of effects that arise from temperature's influences on pK values of imidazole groups, as discussed for enzymes in Chapter 4, may apply to other types of chemicals as well, if their pK values are significantly temperature sensitive. The role of imidazole side chains in semiochemical binding sites on proteins also merits exploration.

Acidification and energy budgets: Reallocation of energy under stress from OA

One of the most important effects of sublethal stress is a stress-induced change in energy allocation, such that an increased amount of ATP turnover is directed away from anabolic processes like growth and reproduction and used instead for redressing the problems resulting from stress. If sublethal stress is highly demanding of the cell's energy supplies, individuals may fail to reproduce and the population may eventually become extinct. Despite the importance of reallocation of energy in responding to sublethal stress, we know very little about the quantitative aspects of these responses.

One of the best-studied systems for addressing OA-induced changes in energy allocation is the purple sea urchin (*Strongylocentrotus purpuratus*). This species provides a highly tractable system for examining OA effects, from fertilization through development to the adult stage. Because this echinoderm is a calcifier, OA would be expected to have relatively strong effects on the organism's anatomy and physiology, especially during early development. However, studies of *S. purpuratus* larvae raised under conditions of elevated pCO_2 have yielded variable results, depending on the specific conditions of OA employed (pCO_2 and carbonate saturation state), the density of larvae in the cultures, and the amount of food provided to the growing animals (Pespeni et al. 2013a; Pan et al. 2015). This interstudy variability has made it difficult to draw general conclusions about OA's effects, even for this single species. However, particularly well designed studies are now beginning to clarify the energy costs of coping with OA and how these rising costs influence allocation of ATP.

Pan and colleagues (2015) have done what is perhaps the most quantitative analysis of the changes in energy allocation that accompany exposure of different life stages of *S. purpuratus* to elevated pCO_2. These investigators grew larvae under two pCO_2 regimens: 400 μatm (control) and 800 μatm. The latter value is not unrealistically high; the urchins can encounter pCO_2 values in this range during upwelling events in part of their normal distribution range (see Pespeni et al. 2013b and text below). Under the culture conditions employed in this study, OA led to no significant changes in size, metabolic rate (i.e., ATP turnover), biochemical composition (protein content), or gene expression.

Taken at face value, these results might be interpreted to mean that the species' development is insensitive to OA, at least up to values of pCO_2 near 800 μatm. This conclusion would be erroneous, however. When precise measurements were made of the fraction of ATP turnover directed to different physiological processes, striking effects of OA were noted. Even though protein content did not differ between control and OA-exposed individuals, the rate of protein synthesis increased ~1.6-fold and rate of protein turnover rose correspondingly in the specimens cultured under a pCO_2 of 800 μatm. These findings reflect a sharp decrease in protein depositional efficiency (i.e., the total amount of protein present in an organism normalized to the amount of protein synthesized over time) under OA. In control urchins, 34.3% of the proteins synthesized were retained; in OA-exposed individuals, depositional efficiency fell to 21.2%. In addition, costs of ion homeostasis rose ~1.4-fold under OA. Pumping of protons from the cell to maintain intracellular pH may require exchange of an inorganic cation like Na^+; hence OA can require more ATP for ion pumping. The total fraction of ATP turnover required for protein synthesis plus ion homeostasis rose from 55% to 84% in fed larvae; a slightly smaller increase was noted for unfed larvae. Because oxygen consumption rates (i.e., total ATP turnover) did not differ between control and OA-exposed organisms, the

OA-exposed individuals would almost certainly face limitations in ATP availability for other energy-demanding activities, including responses to other stressors. This is an important point in the context of multistressor analyses. Whereas the amount of ATP turnover an organism is capable of maintaining might be sufficient to cope with one type of stress, the large increase in energy directed to ion homeostasis and protein synthesis during exposure to OA conditions might severely limit the organism's ability to cope satisfactorily with another stress, for example, a rise in temperature. Suffice it to say, much more investigation of multistressor effects is needed to allow us to predict the consequences of global change in marine environments.

5.4 Calcification in a "Hot" and "Sour" Ocean

As mentioned earlier, studies of OA initially focused mainly on calcification processes. This focus reflects two key points, namely the wide range of biological structures—from coral reefs to mollusc shells to fish ear bones—that are built from carbonates, and the sensitivities of carbonates to the pH of the medium (see Box 5.4). Here we examine a limited number of case studies that illustrate not only the sensitivities of carbonate-based structures to OA, but also the complex interactions among environmental factors that determine how severe a challenge OA is likely to represent. This complexity raises a loud caveat about the design of OA experiments; failure to take into account variables other than carbonate chemistry can lead to erroneous conclusions about OA's effects.

Bivalve molluscs

Because of the importance of calcification processes in the life histories of bivalve molluscs, researchers have given considerable attention to the effects of changes in carbonate system variables on the growth of larval and adult life stages of several species. Closely controlled laboratory studies, in which all components of the carbonate system are tightly regulated, have begun to show how different components of the carbonate system influence construction of carbonate structures and other aspects of the organism's physiology. In oysters and mussels, carbonate saturation state, not pH per se, has been shown to be the most critical determinant of the success of calcification processes (Waldbusser et al. 2015). However, pH has been shown to have a significant effect on respiration rates in these bivalves (Waldbusser et al. 2015); thus reduced pH may limit provision of energy to drive calcification and other anabolic processes.

A potential role of energetics in responses to OA has also been observed in the interaction between food supply and response to elevated pCO_2 in bivalves. Melzner and colleagues (2011) showed that increased pCO_2 led to dissolution of the inner surface of the shell of adult *Mytilus edulis*, and this effect was exacerbated when less food was available. They conjectured that when mussels challenged by OA face limitations in food supply, energy is preferentially allocated to processes other than shell maintenance. Interactions between dietary status and responses to OA have also been seen in other species of bivalves and in different developmental stages. Larvae of the Olympia oyster (*Ostrea lurida*) exposed to different combinations of pCO_2 and food level exhibited negative effects on larval growth, metamorphic success, and total dry weight under elevated pCO_2 (Hettinger et al. 2013). At the highest feeding levels used, the effects of elevated pCO_2 were partially, but not fully, offset. Thus, OA may harm calcifiers even under conditions of enhanced primary productivity.

The impacts that changes in the carbonate system may have on an organism also may be influenced by exposure temperature (Kroeker et al. 2014). Temperature effects are complex because they combine Q_{10} effects on biochemical reaction rates and direct effects on the equilibria of the carbonate system (see Box 5.4). Carbonate structures are generally more stable at higher temperatures, so ocean warming, in and of itself, might favor calcification. However, this positive effect may be overridden by the decreases in pH and carbonate saturation state that characterize ocean acidification.

Oxygen availability is another variable that can be relevant in governing the effects of OA on bivalves and other species. Gobler and colleagues (2014) showed that acidification and hypoxia had negative synergistic effects on survival, growth, and metamorphosis in early developmental stages of a scallop (*Argopecten irradians*) and a clam (*Mercenaria mercenaria*).

The effects of OA are further complicated by diurnal and semidiurnal variations in the carbonate system and other abiotic factors. Bivalves and other intertidal species normally experience regular fluctuations in pH, CO_2, temperature, and DO due to the tidal cycle. In tide pools with limited connection to the ocean during low tide, photosynthesis may deplete CO_2 and lead to increases in pH during the day. At night, respiration may release substantial amounts of CO_2 into the water and cause pH to fall. Temperatures of isolated tide pools are also likely to vary diurnally, warming in the day and cooling at night. Dissolved oxygen may also show a diurnal pattern, with photosynthetically generated O_2 increasing DO during the daytime hours, and respiration causing a fall in DO at night. As a consequence of this diurnal variation, the ability to support aerobic metabolism may change during the course of a day.

Frieder and colleagues (2014) showed that natural semidiurnal fluctuations in pH, as might occur in an isolated tide pool, did not affect development in two congeners of *Mytilus*, *M. californianus* and *M. galloprovincialis*. However, exposure to a constant low pH (pH 7.6) had significant negative effects on development, delaying the transition from the trochophore to the veliger stage. In view of the likelihood that temperature, CO_2, and O_2 vary simultaneously with pH in isolated tide pools, expanded multistressor studies of bivalves and other calcifiers clearly are warranted if realistic assessments of effects of OA are to result from laboratory studies. In fact, it may only be through field studies that expose calcifiers to waters with different carbonate system variables along with variations in other abiotic factors that can influence calcification that we will obtain a "real world" picture of the effects of OA. Perhaps the most pertinent data of this sort currently available are from field studies of pteropods.

Pteropods

Shelled (thecosome) pteropods (sea butterflies) are small, pelagic snails (see photo at left) that provide an especially good study system for viewing consequences of OA on a single calcifying species and, more broadly, on the structure and status of the marine food webs in which the species plays a major role. Pteropod populations can be extremely large, especially in mid- to high-latitude oceans, and can contribute importantly to the food supply of larger species. For example, in the Southern Ocean pteropod biomass may exceed that of krill (Orr et al. 2005). Pteropods are a major food source for fishes and also serve as prey for higher-trophic-level species such as seabirds and whales.

Thecosome pteropods have a thin aragonite shell that can dissolve when exposed to conditions of high $p\mathrm{CO}_2$ and low carbonate saturation state (Bednaršek et al. 2014a). Low temperatures can

A sea butterfly

also decrease aragonite deposition by pteropods because of thermal effects on carbonate structural stability (see Box 5.4; Orr et al. 2005). In fact, pteropod populations are already experiencing the negative effects of OA. Field studies of the pteropod *Limacina helicini* in coastal waters along the Pacific coast from Washington to California have documented that this species is currently exposed to upwelled waters whose pH and aragonite saturation states are low enough to impair shell integrity and reduce population sizes (Bednaršek et al. 2014a). In the top ~100 m of the water column, the extent of shell dissolution was correlated with the percentage of water that was undersaturated with respect to aragonite (**Figure 5.6**). The percentage of specimens with severely damaged shells ranged from 24% in offshore specimens to 53% in nearshore specimens exposed to large upwelling events. In pelagic regions where more than 80% of the water column was undersaturated with respect to aragonite, up to 100% of pteropod shells exhibited dissolution. Since the beginning of the Industrial Revolution, the extent of the undersaturated nearshore waters along the northeastern Pacific coast is estimated to have increased by more than sixfold (Bednaršek et al. 2014a).

OA-induced dissolution of pteropod shells is likely to have a large ecological impact because of the pivotal role these organisms play in marine food webs, especially in high-latitude waters and in continental shelf regions where productivity is high. Pteropods are responsible for a substantial fraction of total carbonate production in many oceanic regions, and thus play a key role in the **marine carbon pump**—the processes that lead to sinking of carbon, fixed by photosynthesis or built into skeletal materials,

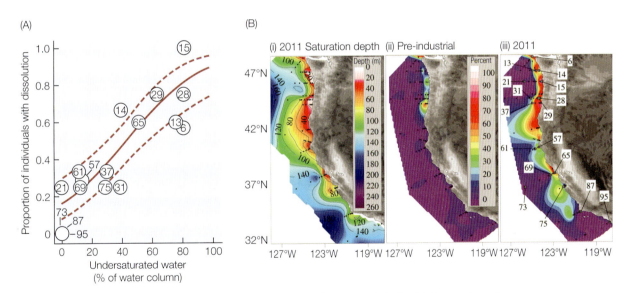

Figure 5.6 Effect of ocean acidification on the integrity of pteropod shells. (A) Proportion of pteropods (*Limacina helicini*) exhibiting severe shell dissolution as a function of the percentage of the water column (upper 100 m) that is undersaturated with respect to the carbonate aragonite. The numbers given within circles correspond to the sampling locations shown in squares in panel (iii) of part (B). The solid line is the logistic regression line fitted to the data; dashed lines are 95% confidence intervals. (B) Carbonate saturation patterns along the northeastern Pacific coast of the United States. (i) Depth of the aragonite saturation horizon in 2011. (ii,iii) Percentage of the upper 100 m of the water column in the California Current Ecosystem estimated to be undersaturated during (ii) preindustrial time and (iii) in August–September 2011. (From Bednaršek et al. 2014a.)

to greater depths. The marine carbon pump plays an important role in the transport of the carbon dioxide incorporated into biological materials to the deep ocean. In this context the effects of OA may be particularly catastrophic. Impaired shell production and falling pteropod population sizes will reduce the strength of the biological carbon pump because less ballast material (inorganic substances like calcium carbonates that have densities greater than seawater) will be produced in the upper regions of the water column (Bednaršek et al. 2014b). Models estimate that, with the decrease in aragonite saturation state in the upper 100 m of coastal northeastern Pacific waters projected by the year 2050, the resulting decrease in generation of carbonate materials may cut the amount of carbon that sinks into deeper waters by about one-half (Bednaršek et al. 2014b). A reduction in the carbon pump of this magnitude could potentially create a positive feedback loop that reduces the rate of biological removal of carbon from the upper ocean even as increasing amounts of CO_2 enter the ocean from the atmosphere.

Fishes

Above we discussed the effects of OA on various aspects of fish behavior. We interpreted these behavioral effects as arising from disruption of physiological systems involved in ion transport or detection of changes in water chemistry, not of calcification per se. Whereas one might not commonly think of fishes as calcifiers, they do have ear bones—**otoliths**—that are built from aragonite. Otoliths play roles in maintenance of balance, detection of acceleration, and sound reception. Thus, if OA caused anomalies in a fish's otolith structures, it might impair the fish's performance.

OA effects on otolith growth and structure have been studied in eggs and larvae of several marine fishes (Checkley et al. 2009; Maneja et al. 2013). In most cases, fishes exposed to elevated pCO_2 and decreased seawater pH had larger otoliths. For example, Checkley and colleagues (2009) showed that otolith masses of young white sea bass (*Atractoscion nobilis*) were 10%–14% and 24%–26% larger in fish cultured at 993 and 2558 μatm CO_2, respectively, compared with control fish maintained at 380 μatm CO_2. The enhancement of aragonite production could be a consequence of elevated blood bicarbonate ion concentrations, which, as discussed above, are part of the fish's pH regulatory processes. Higher bicarbonate levels could lead to higher carbonate levels as well (see Box 5.4), thus favoring deposition of aragonite. Whereas the effects of these changes in otolith mass on fish performance are not known, they might compound the behavioral effects of OA that are mediated through $GABA_A$ receptor function.

In summary, the effects of changes in the ocean's carbonate system on marine organisms are widespread among taxa, differ among life stages of a species, and involve perturbation of a broad range of biochemical, physiological, and behavioral processes. In many and perhaps most cases, the effects stemming from changes in the carbonate system are modified by alterations in other abiotic factors, including temperature and oxygen availability. It is clear that a great deal more attention needs to be given to these multistressor interactions if a realistic understanding of the effects of global change in the marine portion of the biosphere—which comprises more than 99% of the planet's "living space"—is to be obtained.

5.5 Invasive Species: Is There a Biochemical Explanation for Their Success?

We now turn to a different phenomenon—species invasion—that also begs for integrative analyses that take into account the influences of multiple abiotic stressors. A focus on invasive species is relevant for at least two reasons. First, human activities are causing a large increase in the numbers of invasive species found in both terrestrial and aquatic habitats. Thus, rises in numbers and impacts of invasive species are important features of the Anthropocene. Second, comparative study of closely related native and invasive species provides key insights into the role of abiotic factors in governing competitive success among species.

For well over a century there has been an expansion of the number of non-native species entering into new habitats, and there is virtually no habitat on the planet that hasn't been at least indirectly affected by this anthropogenic process (Mooney and Cleland 2001). In coastal marine and aquatic ecosystems, boats carry exotic species in their ballast water and adhered to their hulls (Geller et al. 2010), and can release these non-native species into new bodies of water. Species released in this way may become established in a novel ecosystem and thereby become **invasive species**. In some cases, invasive species become dominant members of the ecosystem, replacing native species and greatly altering the ecosystem's structure.

We begin with the question, What has taken place? and then move to a mechanistic analysis of possible causal relationships that govern the success of invaders. Physiological, biochemical, and genetic studies are helping us understand why invasive species are so good at becoming established, and in some cases replacing native species. These mechanistic analyses, when coupled with data on ongoing changes in abiotic factors such as water temperature and salinity, are also allowing us to make testable predictions about the future spread of invasives.

We will focus on two different invasions that have had major impacts on coastal marine ecosystems. The first involves closely related—but biochemically very distinct—congeneric species of blue mussels of the genus *Mytilus*. The other invasion, which we will discuss in Section 5.6, involves conspecifics of a remarkably successful invasive crustacean, the green crab (*Carcinus maenas*). This species has become something of a "poster child" of marine invasive species because of the success with which it has entered a wide range of habitats around the globe. Examination of these invasive animals will help frame our analysis of the types of genetic tools that foster successful responses to changing environments. In the case of blue mussels, we will look for genetically based interspecific differences in environmental optima and limits; in the case of the green crab, we will examine genetically distinct populations which have different environmental optima that enable the species to colonize environments having widely different thermal properties. Our discussion of the physiology and genetics of green crabs will provide a good foundation for our subsequent examination of the phenomenon of local adaptation and its role in sustaining a species in the face of anthropogenic global change.

Blue mussels (genus *Mytilus*)

Along the coastline of the northeastern Pacific Ocean there are three congeners of *Mytilus* (**Figure 5.7**). The ribbed or California mussel (*M. californianus*) has a broad latitudinal distribution from Alaska into subtropical waters off Mexico. *M. trossulus* is a native species of blue mussel that occurs at higher latitudes than a second blue mussel, the invasive *M. galloprovincialis* (see photo at right). The latter species entered coastal California waters at some point in the mid-twentieth century. Because the two blue mussel species are **cryptic**—they are extremely difficult to distinguish solely on the basis of morphological

Mytilus galloprovincialis

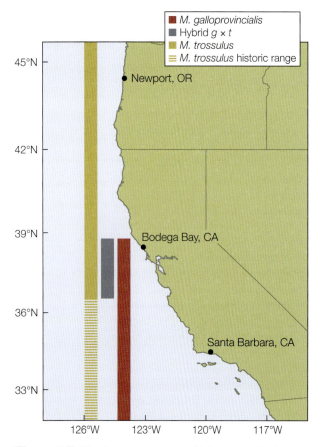

Figure 5.7 Distribution patterns of blue mussels along the coast of the northeastern Pacific Ocean. The historic range of the native species, *M. trossulus*, extended well south of the latitudes at which it now is found (striped bar). The invasive species from the Mediterranean, *M. galloprovincialis*, has displaced the native species in regions south of Monterey Bay (~37° N latitude). Where the two congeners co-occur, they frequently hybridize. After a recent episode of cooling of coastal waters, as part of a normal pattern in the Pacific Decadal Oscillation (PDO), the northern distribution of recruitment of *M. galloprovincialis* has receded to lower-latitude waters slightly to the north of Bodega Bay, California (see Hilbish et al. 2010). (After Lockwood and Somero 2011a.)

traits—the invasion went unnoticed for many years. In fact, most literature on blue mussels up until the late 1980s refers to "the" blue mussel on the West Coast of North America as *M. edulis*. This species is prevalent along the East Coast of the United States, but genetic markers have confirmed more recently that it is not found on the West Coast (McDonald and Koehn 1988). Instead, *M. trossulus* and *M. galloprovincialis* and their hybrids occupy the niches on the West Coast where blue mussels can thrive.

EVOLUTIONARY HISTORIES OF BLUE MUSSELS
The ancestral species that gave rise to the blue mussels originated in the northern Pacific Ocean. Approximately 3.5 mya, during a trans-Arctic movement of many marine species, members of this northeastern Pacific ancestral species moved into the northwestern Atlantic Ocean. Subsequent glaciation caused the Pacific and Atlantic populations to become geographically isolated, which led to the speciation of *M. trossulus* in the Pacific and *M. edulis* in the Atlantic. *M. edulis* subsequently expanded its range into the Mediterranean Sea. The Mediterranean population of *M. edulis* was isolated ~2 mya, giving rise to the third species, *M. galloprovincialis*. Even though these three species are closely related, they have evolutionarily diverged from one another in distinct environments. The waters of the Mediterranean Sea are warmer, more thermally stable, and have higher and more stable salinities than waters in the Atlantic and Pacific oceans (reviewed in Lockwood and Somero 2011a).

In the late 1980s, *M. trossulus* and *M. galloprovincialis* were identified, using genetic markers, as the two West Coast blue mussel species (McDonald and Koehn 1988). The further discovery that *M. galloprovincialis* was an invasive species that had replaced the native blue mussel over the southern portion of its former distribution range (Geller 1999) suggested that differences in thermal tolerance and thermal optima between the two cryptic species might have contributed to the observed biogeographic pattern (see Figure 5.7). Was *M. galloprovincialis* consistently more warm-adapted than *M. trossulus* across numerous biochemical and physiological systems? If so, these differences would go a long way toward explaining the species' latitudinal distributions in the twentieth century and provide a basis for predicting how these distributions might change as global temperatures continue to rise. In fact, as discussed below, in almost every case in which a comparison has been made between homologous systems of the two congeners, *M. galloprovincialis* displays the more warm-adapted phenotype. That being said, the full story on the linkage between global change and shifting biogeographic patterning of the two congeners involves more than temperature; a multistressor analysis is again needed to fully account for their shifts in distribution.

MULTILEVEL, ADAPTIVE DIFFERENCES BETWEEN CONGENERS   To start our examination of the interspecific differences in thermal responses between the two species, we will follow a reductionist approach, beginning with whole organism thermal tolerance and ending with amino acid differences between orthologous proteins. The three West Coast congeners of *Mytilus* showed significant differences in their ability to survive after exposure to 33°C, with the ribbed mussel (*M. californianus*) exhibiting the greatest survivorship (**Figure 5.8A**). Of the two blue mussels, *M. galloprovincialis* had a substantially higher survival rate than the native species *M. trossulus* (Dowd and Somero 2013). The differences in whole organism survival under heat stress were reflected by differences in cardiac performance (**Figure 5.8B**). The temperatures at which heart rate rapidly decreased during acute heating (termed the critical temperature [CT_{max}]) were 3 degrees Celsius and 4 degrees Celsius lower for *M. trossulus* compared with *M. galloprovincialis* in specimens acclimated to 14°C and 21°C, respectively (Braby and Somero 2006b). Intercongener differences in tolerance of low temperature were also found. Under acute cold stress at 0°C, heart function in *M. trossulus* continued while that of *M. galloprovincialis* ceased (Braby and Somero 2006b).

Studies using transcriptomic (Lockwood et al. 2010) and proteomic (Tomanek and Zuzow 2010) methodologies also revealed differences between the two species. In both species, acute heat stress led to more robust production of several mRNAs and proteins associated with stress responses, but intercongener differences were apparent as well. Genes encoding small heat-shock proteins, which are molecular chaperones that may be associated with sustaining homeostasis of the cytoskeleton, increased strongly in *M. galloprovincialis*

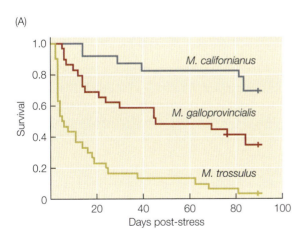
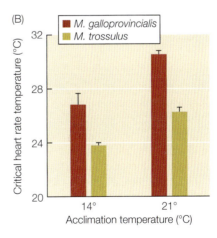

Figure 5.8 Differences in whole organism and cardiac heat tolerance between *Mytilus trossulus* and *M. galloprovincialis*. (A) Thermal tolerance of whole organisms. Survival after a 1-h exposure to 33°C in seawater. A third congener, the ribbed mussel (*M. californianus*), is also shown. (B) Thermal tolerance of heart function. The critical temperature at which a sharp decrease in heart rate occurs during acute warming is lower in *M. trossulus* than in *M. galloprovincialis*. Both species exhibited acclimation in cardiac heat tolerance. (A after Dowd and Somero 2013; B after Braby and Somero 2006b.)

(Lockwood et al. 2010). Damage to the cytoskeleton by abiotic stressors like temperature may be a critically important challenge to ectotherms (Lockwood et al. 2010), so the greater capacity of *M. galloprovincialis* to maintain integrity of the cytoskeletal apparatus may help account for the species' greater thermal tolerance. Whereas the stronger upregulation of small heat-shock protein genes could be interpreted as a greater level of perturbation in *M. galloprovincialis*, we conjecture that this response is instead a reflection of a higher capacity to mount a defensive response and not a sign of greater cellular damage from acute thermal stress. Figure 5.8A clearly shows that *M. galloprovincialis* is more heat toler-ant than *M. trossulus*, and as discussed in Chapter 3, heat-tolerant species often exhibit more rapid and robust gene and protein expression responses of molecular chaperones than their less heat-tolerant congeners, in what can be viewed as a defense strategy for dealing with heat stress. Consistent with this interpretation, levels of ubiquitinated pro-teins were higher in heat-stressed *M. trossulus* than in *M. galloprovincialis* (Hofmann and Somero 1996). This difference reflects higher levels of irreversible damage of proteins by heat in the native species of blue mussel. Transcriptomic and proteomic analyses also showed stronger upregulation of proteolytic proteins—for example, proteins found in the proteasome—in *M. trossulus* than in *M. galloprovincialis* under acute heat stress (Lockwood et al. 2010; Tomanek and Zuzow 2010). Therefore, the heat-induced changes in both the chaperone-mediated repair systems and the proteolytic degradation systems involving ubiquitin-tagging of irreversibly damaged proteins show interspecific differences consistent with the divergent evolutionary thermal histories of the two species.

The two congeners also show metabolic compensation to temperature (**Figure 5.9**). At a common temperature of measurement, physiological and biochemical functions occur at higher rates in *M. trossulus*. In comparisons of specimens acclimated to a common temperature, the intrinsic rate of heart function was found to be ~1.5-fold higher in the cold-adapted native species (see Figure 5.9A). Underlying this difference in organ-level performance were comparable interspecific differences in capacities for ATP generation: Tissue levels of activity of two enzymes involved in ATP generation, malate dehydroge-nase (MDH) and citrate synthase (CS), were 1.7 and 1.5 times greater, respectively, in the cold-adapted species (see Figure 5.9B,C). And as seen in comparisons of orthologous proteins from species adapted to different temperatures, the catalytic rate constant (k_{cat}) of the cytosolic malate dehydrogenase (cMDH) ortholog of *M. trossulus* was ~twofold higher than that of the cMDH of *M. galloprovincialis* (see Figure 5.9D; Fields et al. 2006). The differences in k_{cat} largely account for the observed differences in total MDH activity between the species. Finally, in accord with the patterns observed for several enzymes of differently thermally adapted species, the thermal stability of ligand binding was greater for the cMDH ortholog of the warm-adapted invasive species (see Figure 5.9E; Fields et al. 2006). The differences in thermal sensitivities of the cMDH orthologs were traced to a single amino acid substitution (Fields et al. 2006).

Overall, then, all data—ranging from whole organism heat tolerance to enzyme kinet-ics—indicate that the Mediterranean invasive is likely to outcompete the native blue mussel in a warming environment. Studies performed in the field have in fact shown that there is differential selection by temperature on the two species (Schneider and Helmuth 2007), data that are fully consistent with the northward expansion of the invasive species during the twentieth century. Furthermore, the recent southward retreat of *M. galloprovincialis* in conjunction with a cooling phase of the Pacific Decadal Oscillation (PDO) shows the dynamic nature of the invasion: Warm temperatures allow the invasive to expand its range to higher latitudes, whereas falling temperatures "push" it back toward lower latitudes (Hilbish et al. 2010).

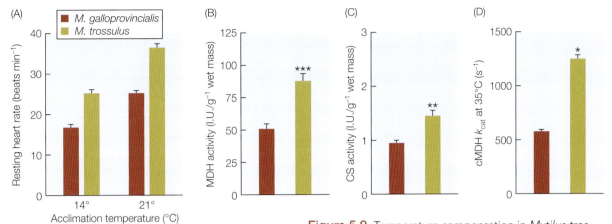

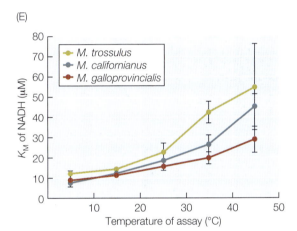

Figure 5.9 Temperature compensation in *Mytilus trossulus* and *M. galloprovincialis*. (A) Compensation of heart rates. *M. trossulus* has an intrinsically higher rate of heartbeat than *M. galloprovincialis*, reflecting adaptation to lower temperatures. Rates were measured at the temperatures of acclimation (14°C and 21°C). (B,C) Temperature compensation of enzyme activities. Activities of (B) malate dehydrogenase (MDH) and (C) citrate synthase (CS) are higher in *M. trossulus* than in *M. galloprovincialis* acclimated to common conditions. ** = $P < 0.01$; *** = $P < 0.001$. (D) Temperature compensation of enzyme activity by adaptation of catalytic rate constants (k_{cat}). k_{cat} for cMDHs of *M. trossulus* and *M. galloprovincialis*. * = $P < 0.05$. (E) Protein adaptation of K_M. The cMDH ortholog of *M. galloprovincialis* has a higher thermal stability of cofactor (NADH) binding, indexed by K_M^{NADH}, than the ortholog of *M. trossulus*. The thermal sensitivity of K_M^{NADH} is intermediate in the ribbed mussel (*M. californianus*). (A after Braby and Somero 2006b; B and C after Lockwood and Somero 2011a; D and E after Fields et al. 2006.)

Can we, then, be confident that global warming will lead to a continued expansion of the invasive species' range to higher latitudes, excluding what are likely to be temporary setbacks caused by natural fluctuations in ocean temperature due to the PDO? No; the picture is more complicated than this. A more inclusive, multistressor analysis suggests that temperature may not always be the dominant abiotic factor that governs the relative competitive abilities of these two species. This conclusion is based on an analysis of the relative abundances of the native and invasive species over a narrow range of latitude that contains habitats differing widely in both temperature and salinity (Braby and Somero 2006a). In the San Francisco and Monterey bays on the coast of California, there are numerous shallow habitats where freshwater enters via rainfall or release of water from water treatment plants. Entry of freshwater leads to episodes of relatively low salinity that have varying duration. These same habitats tend to have higher temperatures than other regions of the two bays, where oceanographic conditions driven by currents, as opposed to terrestrial runoff, are the dominant influences on water temperature and salinity. If temperature were invariably the major abiotic factor that determines the relative competitive abilities of *M. trossulus* and *M. galloprovincialis*, then one would predict that the prevalence of the invasive species would correlate positively with temperature,

(A)

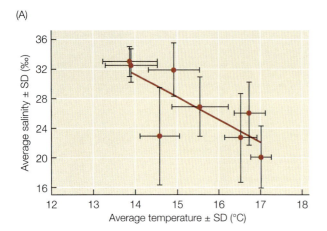

(B)

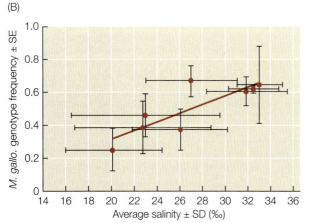

(C)

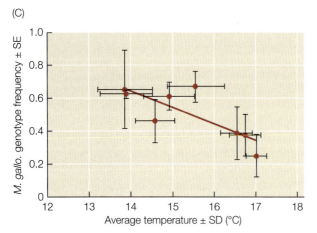

Figure 5.10 Relative abundances of *Mytilus trossulus* versus *M. galloprovincialis* in habitats in San Francisco and Monterey bays where large differences occur in average salinity and temperature. (A) Temperature and salinity covary strongly; the warmest habitats tend to have the lowest salinities. (B,C) The warm-adapted *M. galloprovincialis* shows greatest relative abundance in habitats with high salinity (B), not high temperature (C). (After Braby and Somero 2006a.)

independently of effects of salinity. This is not what was observed, however (**Figure 5.10**; Braby and Somero 2006a). The relative abundance of *M. galloprovincialis* was positively correlated with salinity but negatively correlated with water temperature. Therefore, it would appear that the evolutionary history of *M. galloprovincialis* as a Mediterranean species—one that evolved in waters that were relatively warm and with relatively high and stable salinity compared with the waters of the northeastern Pacific Ocean—has led to a complex response to variations in temperature and salinity. The stress caused by decreases in salinity appears to be sufficient to favor the native species in habitats where salinity is variable. Cardiac function is more impeded by reduced salinity in *M. galloprovincialis* than in *M. trossulus* (Braby and Somero 2006b). Studies of gene and protein expression in the face of acute changes in salinity revealed differences that are consistent with interspecific differences in sensitivity to hyposmotic stress (Lockwood and Somero 2011b; Tomanek et al. 2012). Specifically, expression of *ornithine decarboxylase* (a gene that is characteristic of animals' responses to osmotic stress) was significantly upregulated in *M. galloprovincialis*—a sign of stress—but not in *M. trossulus* (Lockwood and Somero 2011b).

In summary, comparisons of native and invasive congeners can be fruitful in teasing apart the relative roles of abiotic factors such as temperature and salinity in invasive species biology. The blue mussel study system has the advantage of allowing comparisons of two closely related species that have evolved under different conditions of temperature and salinity. Perhaps not surprisingly, stress from both of these factors can influence competitive interactions. Because models of global change predict rising temperatures and increasing precipitation for the northeastern Pacific Ocean, the latitudinal and local distributions of these blue mussel congeners are likely to be dynamic. Thus, the effects of global warming on the hydrological cycle, namely increases in rates of evaporation and rises in humidity in warmer air, must be incorporated into multistressor studies where relevant. Warming will favor a continued northward movement of the invasive species, albeit this movement to high latitudes will likely be affected by shifts in the PDO. Additionally, runoff from land during heavy rains could have strong effects on the distributions of the species as well, especially in shallow waters of bays and estuaries where decreases in salinity are greatest. Another factor that could possibly affect the distributions of the two species is the sensitivity of larval stages to abiotic factors. To our knowledge, there have been no direct comparisons of larvae of these two species, so we don't know how the sensitivities of different life stages to temperature and salinity variations may contribute to adult distributions.

Despite such gaps in our knowledge, one thing that is clear from the multilevel studies of blue mussels is that some of their biochemical and physiological systems may already be working under environmental conditions near the systems' thermal tolerance limits. This was seen for the CT_{max} of cardiac function (see Figure 5.8B). The question that heads the following section thus addresses a critical issue in the context of ongoing global change.

5.6 How "Close to the Edge" of Their Thermal Tolerance Limits Do Species Live?

When we examined thermal performance curves in Chapter 3, we showed that the expected Q_{10}-driven increases in reaction rate due to rising temperatures occurred only up to a certain temperature; beyond this temperature, which is often (and somewhat misleadingly) called the "optimal" temperature, rates begin to decrease, sometimes precipitously. We discussed some of the mechanisms that might underlie the complex shapes of thermal performance curves, but we postponed to this final chapter an important question in the context of global change: How "close to the edge" of heat tolerance limits do species currently live? To a first approximation, the threats posed by a warming climate may vary directly with the proximity of current maximum habitat temperatures to the upper thermal tolerance limits of a species. In fact, the same logic would seem to apply for other abiotic stressors. For example, if an aquatic species is currently experiencing dissolved oxygen levels that are close to its critical oxygen concentration (see Figure 5.2), then any decrease in oxygen availability would likely challenge the organism's capacity for sustaining aerobic scope and, at the extreme, its survival. We discussed this issue earlier in the context of the Humboldt squid and, more broadly, the role of metabolic index (Φ) in determining habitat suitability for marine fishes and invertebrates.

Below we focus strictly on temperature effects and examine data from studies of terrestrial and aquatic animals that were designed, in part, to address the question that heads this section. Analyses of terrestrial animals have provided strong evidence that tropical species are more likely to be threatened by global warming than are temperate-zone species (Tewksbury et al. 2008; Chan et al. 2016). For example, thermal performance curves for lizards show that tropical species live closer to their upper thermal limits than temperate-zone species (Figure 5.11). The greater heat tolerance of a tropical species does not make it less vulnerable to global warming—the proximity of its upper lethal temperature to current thermal maxima is what defines the danger of global warming (see Stuart-Smith et al. 2015). Furthermore, as we'll show in the case of certain marine animals, the ability to acclimatize to warmer temperatures may also be relatively limited in warm-adapted species. Tropical species may also have narrower thermal tolerance ranges than mid-latitude species, which may experience much higher levels of

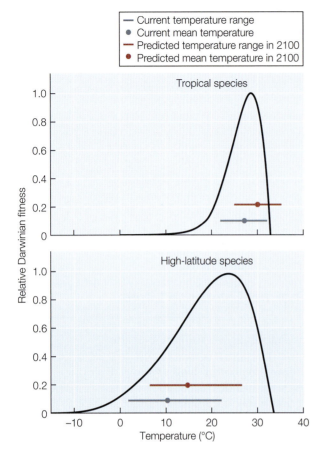

Figure 5.11 Relative vulnerabilities of tropical and higher-latitude terrestrial ectotherms to global warming. Darwinian fitness (the genetic contribution of an individual organism to the next generation's gene pool relative to the average for the population as a whole) is plotted as a function of equilibrium ("operative") body temperature for tropical and high-latitude species. Tropical species, despite being tolerant of relatively high temperatures, currently live closer to their upper thermal limits than do species found at higher latitudes, for example in thermally variable temperate-zone habitats. Tropical species thus may be more vulnerable to warming, despite their greater intrinsic resistance to heat. The thermal tolerance ranges of tropical and higher-latitude species differ as well. (After Tewksbury et al. 2008.)

temperature variation in their habitats due to strong seasonal effects on ambient temperature (Chan et al. 2016). Tropical species, then, share with the highly stenothermal species of the Southern Ocean (discussed in Section 5.7) a particularly strong challenge from rising temperatures: Both groups of species have upper thermal tolerance limits that are close to the maximum habitat temperatures they currently encounter, and their relatively limited abilities to acclimate to increases in temperature may further limit their abilities to cope with a warming world (Peck et al. 2014).

Can we identify a single physiological system that is shared by most animal species and that allows us to evaluate both the proximity of upper thermal tolerance limits to current habitat temperatures and capacities for thermal acclimatization? For several reasons, cardiac function is a good candidate for this type of mechanistic analysis. First, Pörtner and colleagues have demonstrated the broad importance of oxygen supply in determining thermal limits; in most species, the ability to sustain adequate cardiac activity is pivotal to sustaining aerobic metabolism (Pörtner 2010). Second, cardiac function typically exhibits a sharp decrease when temperature rises to a certain level (CT_{max}). If the CT_{max} for this organ's activity matches the upper thermal tolerance limit for a species, then a mechanistic explanation for heat death can be obtained, at least in cases where recovery of cardiac function after exposure to temperatures equal to or above the CT_{max} cannot occur. Third, measurement of heart rate as a function of temperature is an experimentally tractable procedure, one that allows a wide diversity of taxa to be examined. Below we discuss cardiac responses of an array of marine animals and show that species differ significantly in the proximity of the CT_{max} to upper habitat temperatures, in their abilities to recover after exposure to temperatures equal to or above the CT_{max}, and in their capacities for adjusting CT_{max} through acclimatization. These data allow us to begin to quantify the threat that rising temperatures pose to marine animals (and, by extension, to terrestrial species as well) and also provide further insights into the physiological characteristics that may help a species be a successful invader.

Congeners of turban snails (genus *Chlorostoma*)

We have already examined several aspects of the thermal biology of turban snails of the genus *Chlorostoma* (formerly *Tegula*) (see Chapter 3). We saw that the thermal stability of enzyme function and structure differed regularly among cMDH orthologs of congeners adapted to different temperatures. Likewise, the temperatures at which heat-shock protein synthesis was induced and the upper thermal limits of protein synthesis differed regularly among congeners, reflecting their adaptation temperatures. Similar interspecific differences have been observed in cardiac function (**Figure 5.12**; Stenseng et al. 2005). The acute effects of rising temperature on cardiac function result in a typical thermal performance curve (see Figure 5.12A). When snails were exposed to a heat ramp, the rate of cardiac function (beats min^{-1} [BPM]) increased, following the Q_{10} relationship. However, the rise in BPM with heating occurred only up to a particular temperature, the CT$_{max}$, after which BPM fell sharply. With further heating, BPM fell to zero at what is termed the flatline temperature (FLT).

The values of the CT$_{max}$ and FLT reflect both interspecific differences and the effects of acclimation. Looking first at interspecific differences among three mid-latitude turban snails with different vertical positioning in the intertidal zone (see Figure 5.12B), we see that the CT$_{max}$ and FLT values are highest for *C. funebralis*, the species found highest in the intertidal zone (see photo at right). The two congeners that occur in the lower intertidal and subtidal zone, *C. brunnea* and *C. montereyi*, are significantly less

Chlorostoma funebralis

(A)

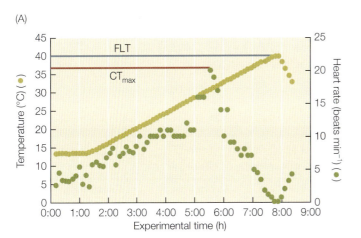

(B)

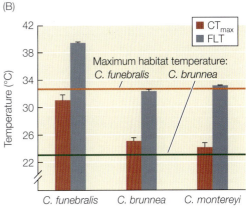

Figure 5.12 Temperature effects on cardiac function of intertidal snails of the genus *Chlorostoma* (formerly *Tegula*). (A) Effects of a heat ramp on the heart rate (BPM, beats min^{-1}) of a mid- to high-intertidal congener, *C. funebralis*. The temperature at which a sharp drop in BPM is observed during heating is termed the critical temperature (CT$_{max}$). The temperature at which further heating leads to a BPM of zero (heart failure) is termed the flatline temperature (FLT). (B) For the lower-occurring congeners, *C. montereyi* and *C. brunnea*, the maximum habitat temperature (indicated by the green horizontal line) is lower than the CT$_{max}$, whereas the maximum habitat temperature encountered by the highest-occurring congener, *C. funebralis*, (indicated by the red horizontal line) exceeds the CT$_{max}$ by several degrees Celsius and is close to the FLT. Thus, under field conditions, *C. funebralis* is apt to encounter temperatures close to the CT$_{max}$ on a regular basis (see text). Error bars indicate S.E. (After Stenseng et al. 2005.)

tolerant of high temperatures than *C. funebralis*; their CT_{max} and FLT values are both ~5–6 degrees Celsius lower than the values for *C. funebralis*. However, despite the greater heat tolerance of cardiac function in *C. funebralis*, this species is more likely to be threatened by heat perturbation of heart activity than the less warm-adapted congeners. This conclusion follows from the fact that body temperatures of *C. funebralis* in the field can reach at least 33°C (Tomanek and Somero 1999), a temperature that is ~2 degrees Celsius above the CT_{max}. For *C. brunnea* and *C. montereyi*, maximum body temperatures were never found to exceed 23°C. For these species, CT_{max} lies 1–3 degrees Celsius above maximum field body temperatures. Were individuals of these two more cold-adapted species ever to enter the highest zones where *C. funebralis* is abundant, maximum temperatures would exceed their CT_{max} values by several degrees and equal their FLTs. Even though heart function returns after temperatures are lowered below the FLT (see Figure 5.12A), this recovery is only temporary and mortality occurs over the next several hours. Thus, the FLT can be taken as a measure of upper lethal temperature. The differences in cardiac thermal sensitivity vis à vis normal body temperatures in the field reflect closely the patterns observed for the three congeners' heat-shock responses (see Figure 3.24). Thus, for the heat-shock response and cardiac function, the most warm-adapted species encounters thermal disruption of function (protein denaturation and fall in heart rate, respectively) at temperatures near the upper portion of its normal thermal range. The two lower-occurring species rarely, if ever, experience heat-induced perturbation sufficient to elicit the heat-shock response or a sharp fall in cardiac function.

The differences in thermal response of cardiac function among the turban snail congeners extend to their abilities to increase thermal tolerance during acclimation to high temperatures. As shown in **Table 5.1**, the two lower-occurring species, *C. brunnea* and *C. montereyi*, were able to elevate their CT_{max} values by 6.6 and 4 degrees Celsius, respectively, when acclimation temperature increased from 14°C to 22°C. In contrast, acclimation shifted the CT_{max} of *C. funebralis* by only 1.4 degrees Celsius. The increase in CT_{max} for *C. funebralis* did not raise CT_{max} above the maximum body temperature recorded in the field.

TABLE 5.1

Effects of thermal acclimation on the CT_{max} and FLT of cardiac function for three congeners of *Chlorostoma*.[a]

	C. funebralis	*C. brunnea*	*C. montereyi*
CT_{max}			
14°C acclimated	28.5°C ± 0.5°C	20.2°C ± 0.8°C	21.7°C ± 0.8°C
22°C acclimated	30.1°C ± 0.7°C	26.8°C ± 0.7°C	25.7°C ± 0.6°C
Field acclimatized	31.0°C ± 0.7°C	25.0°C ± 0.5°C	24.2°C ± 0.7°C
FLT			
14°C acclimated	39.8°C ± 0.2°C	31.7°C ± 0.4°C	33.6°C ± 0.3°C
22°C acclimated	40.6°C ± 0.2°C	32.8 C ± 0.3°C	34.1°C ± 0.2°C
Field acclimatized	39.4°C ± 0.2°C	32.4°C ± 0.2°C	33.1°C ± 0.1°C

[a] Data from Stenseng et al. 2005. Means ± SE are shown.

FLT varied among species with a pattern similar to that seen for CT_{max}: FLT was greatest in the most warm-adapted congener. However, FLT exhibited little change during acclimation. In particular, FLT of cardiac function of *C. funebralis* did not rise above 40.6°C, a temperature that southern populations of the species can encounter in the field (Gleason and Burton 2013). From these data on the CT_{max} and FLT, and from the results of experiments on the heat-shock response, it would appear that *C. funebralis* is unlikely to escape from thermal stress in a warming environment, despite its inherently greater heat tolerance compared with lower-occurring, more cold-adapted species. As in the case of certain terrestrial animals, then, adaptation to relatively high temperatures does not provide a protective preadaptation against the increases in temperature that will occur with global change. Indeed, it seems that *C. funebralis* will require further evolutionary adaptation in order to develop increased heat tolerance and persist in its current habitat in the face of future climate warming. The degree to which this is possible is uncertain, given the complexities of thermal tolerance physiology (see Chapter 3). Whether or not a population of a species can respond to thermal selection will also depend on several other factors, such as population size, generation time, mutation rate, and standing genetic variation (see below). One thing is certain: In the absence of such evolutionary adaptation, species like *C. funebralis* will likely need to shift their distributions to avoid local extinction.

Congeners of porcelain crabs (genus *Petrolisthes*)

Studies of marine crustaceans have revealed patterns similar to those observed in turban snails, but demonstrate an even stronger linkage between the CT_{max} of heart function and death from heating. An especially thoroughly studied group of crustaceans are the porcelain crabs (genus *Petrolisthes*). *Petrolisthes* comprises more than 100 species, which occur over a wide range of latitude (Stillman and Somero 2000). Congeners vary in their vertical distributions as well; some occur in the mid- to high intertidal, and others are subtidal. Studies of the effects of temperature on cardiac function have been performed on 21 eastern Pacific species from temperate and tropical habitats and from different vertical positions at these latitudinally different sites (**Figure 5.13**). As would be expected, the tropical species, which experience temperatures up to ~42°C, have greater heat tolerance (higher lethal temperature [LT_{50}] values) than the temperate species, regardless of the vertical position at which the species occur (see Figure 5.13A). Within each habitat, higher-occurring species exhibited greater heat tolerance than lower-occurring species.

When the LT_{50} values are plotted against maximum habitat temperature (see Figure 5.13B), a relationship similar to the one found with turban snails is observed: The most heat-tolerant congeners live closest to the upper limits of their thermal tolerance range. The line of unity shown in the figure is helpful in illustrating this relationship. If an LT_{50} value falls on or to the right of this line, then the animal may encounter lethally hot temperatures in its habitat. By this criterion, most of the 21 species are currently "safe"—their maximum habitat temperatures are considerably below the LT_{50} value. However, at least three of the tropical species, as well as two temperate-zone species found high in the intertidal zone, would appear to face challenges from temperatures near (or above) LT_{50} values. One of these temperate-zone species is *P. cinctipes* (yellow symbol in Figure 5.13B; see photo at right), for which the CT_{max}, LT_{50}, and maximum habitat temperature are

Petrolisthes cinctipes

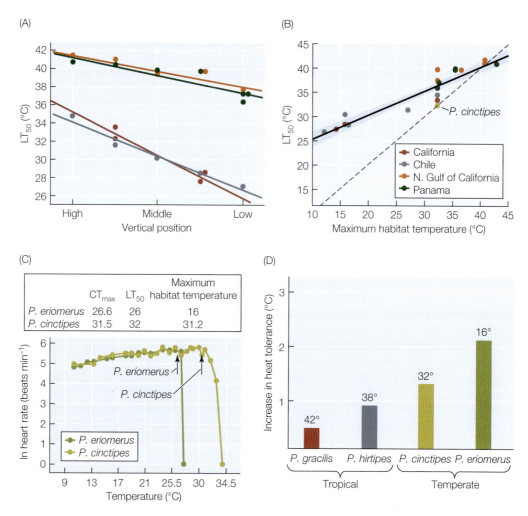

Figure 5.13 Thermal relationships of porcelain crabs (genus *Petrolisthes*). (A) Lethal temperatures (LT$_{50}$) of 21 species of *Petrolisthes* found at different vertical zones in tropical (northern Gulf of California and Panama) and temperate (Chile and California) intertidal and subtidal habitats. LT$_{50}$ values represent the temperature at which 50% of a population dies within the specified time period of the experiment. (B) Relationship between LT$_{50}$ and maximum habitat temperature for the different tropical and temperate zone collection sites. Solid line represents a linear regression of maximum habitat temperature versus LT$_{50}$ (R^2 = 0.88). Shading represents the 95% confidence band for the linear regression fit. The dashed line is a line of unity (LT$_{50}$ = habitat temperature). The yellow dot represents *P. cinctipes*; otherwise the symbols match those of part (A). (C) The effects of experimental temperature on heart rate of two temperate-zone porcelain crabs, *P. cinctipes* (mid-intertidal) and *P. eriomerus* (low intertidal/subtidal). The CT$_{max}$ values are indicated by arrows and were determined by regression analysis. The box above the graph presents temperature data that show how the CT$_{max}$, LT$_{50}$, and maximum habitat temperature compare. For *P. cinctipes*, the three values are essentially the same; for *P. eriomerus*, there is a ~10-degree-Celsius difference between maximum habitat temperature and temperatures of cardiac collapse and heat death, showing that the lower-occurring congeners have a large margin of safety in terms of thermal death. (D) Acclimation abilities differ among congeners of *Petrolisthes*. The higher the maximum habitat temperature (temperatures given on top of the bars), the lower is the increase in heat tolerance (indexed by the CT$_{max}$ of cardiac function) resulting from acclimation to high temperature. (A and B after Stillman and Somero 2000; C after Stillman and Somero 1996; D after Stillman 2003.)

essentially the same (see Figure 5.13C). In contrast, the lower-occurring congener *P. eriomerus* has an ~10-degree-Celsius margin of safety in terms of proximity of its LT_{50} to current maximum habitat temperatures. For *P. cinctipes*, the coincidence of the CT_{max}, LT_{50}, and maximum habitat temperature supports the conjecture that heart failure may be instrumental in establishing whole organism thermal tolerance. Cardiac function, then, provides a clear mechanistic window into the determinants of thermal tolerance in these crustaceans.

Another similarity between turban snails and porcelain crabs is seen in the capacities of differently thermally adapted congeners to increase CT_{max} during acclimation to higher temperatures (see Figure 5.13D; Stillman 2003). The ability to increase CT_{max} is inversely related to adaptation temperature. Tropical species were able to increase CT_{max} by less than 1 degree Celsius, whereas the most cold-adapted species, *P. eriomerus*, increased CT_{max} by more than 2 degrees Celsius.

An important difference exists between porcelain crabs, on the one hand, and turban snails and blue mussels, on the other hand. This difference lies in the abilities of the organisms to recover from exposure to temperatures equal to or greater than the CT_{max}; this difference may offer insights into how broad groups of species may differ in vulnerability to climate change. In crabs, unlike the two groups of molluscs, once the CT_{max} is exceeded and heart rate plummets, cardiac function does not recover after animals are returned to sublethal ($< CT_{max}$) temperatures. This difference in capacity to recover from heat stress may make crabs, and perhaps even arthropods in general, more vulnerable to rising temperatures than species like mussels and snails that can regain cardiac function if they experience cooler ($< CT_{max}$) temperatures after being subjected to temperatures equal to or greater than the CT_{max}. Here, then, is an example of how taxon-specific physiological characteristics may lead to sharply differential effects of global change on different taxa. The basis for these taxon-specific effects is currently not known, but might involve differences among taxa in abilities to rely on anaerobic pathways of ATP generation during stress. A sharp decrease in blood- or hemolymph-supplied oxygen when CT_{max} is surpassed and cardiac function falls might create fewer challenges for species capable of relying on anaerobic ATP production, like many molluscs, relative to obligate aerobes, such as arthropods. As discussed in Chapter 2, many molluscan lineages, including snails and mussels, are good facultative anaerobes. On the other hand, because crustaceans like *Petrolisthes* are more mobile than snails and mussels, they likely possess a greater capacity for behavioral thermal regulation by seeking out the coolest places in their habitats. In this manner, behavioral avoidance of stressful conditions may make some species less vulnerable to future climate change. However, this option may be limited in the intertidal zone where fierce interspecies competition for space may render suitable thermal microenvironments inaccessible.

Intraspecific (interpopulation) variation in green crabs

The highly invasive green crab (*Carcinus maenas*; see photo at right), like the invasive blue mussel *M. galloprovincialis*, evolved in European waters and then spread to a large number of habitats in other regions of the world's coastlines. What is viewed as the current native range of *C. maenas* spans waters from Iceland to North Africa (reviewed in Tepolt and Somero 2014). Throughout its biogeographic range, *C. maenas* experiences an ~25-degree-Celsius range of water temperatures (**Figure 5.14A**). Because *C. maenas* is a subtidal species that does not experience aerial exposure, this

Carcinus maenas

is its approximate range of body temperatures as well. Note one important difference between blue mussels and green crabs: Blue mussels are found over a range of temperatures similar to those experienced by green crabs, but in the case of the mussels, the ability to inhabit this wide range of temperatures is a consequence of differences in thermal optima and tolerance limits among a set of congeneric species. Although *C. maenas* is regarded as a single species, the differences in thermal responses observed among conspecifics are equivalent to, and likely exceed, the differences found among different congeners of blue mussels. Suffice it to say, *C. maenas* is extremely eurythermal and has a significant amount of genetic variation that underlies its ability to inhabit such a wide range of temperatures.

The thermal tolerance relationships of *C. maenas* are illustrated in Figure 5.14. In the native range, differences in heat and cold tolerance of heart function are found between northern (Norway) and southern (Portugal) populations (**Figure 5.14B**). Field-acclimatized crabs from Portugal had significantly higher CT_{max} values for cardiac function than crabs from Norway, and significant differences persisted after acclimation to different temperatures (5°C and 25°C; **Figure 5.14C**). Conversely, cold tolerance of cardiac function, indexed by heart rate at 0°C ($f_{H,0}$), was greater in the Norwegian population (**Figure 5.14D**). Cold tolerance, like heat tolerance, was influenced by acclimation temperature (**Figure 5.14E**). Differences were also observed among populations collected from sites of invasion. Green crabs from the East Coast of North America exhibited differences in CT_{max} and heart rate at 0°C that reflected their latitudinal distributions between New Jersey, Maine, and Newfoundland (see Figure 5.14B,D). The North American crabs likely represent at least two successive invasions, and the wide distribution of thermal tolerances could reflect genes entering from both southern and northern native populations of Europe. Alternatively, the differences among the Atlantic East Coast populations could reflect postinvasion divergence among the populations.

When the upper thermal tolerance limits of cardiac activity in green crabs are compared with their habitat (body) temperatures, it is clear that this species possesses a large margin of safety in terms of the difference between even the warmest habitat temperatures (near 25°C; the New Jersey population encounters the highest-temperature water) and CT_{max}. This margin of safety averages around 10–12 degrees Celsius in all populations at the maximum water temperatures in summer. This large difference between CT_{max} and maximum water temperature and the high absolute value for CT_{max} in green crabs contrast sharply with the thermal relationships found for many other species of crustaceans, including porcelain crabs (Tepolt and Somero 2014). Green crabs thus appear to be relatively well prepared for a warming ocean, and this capacity may ensure their continued success as an invasive species. Green crabs also exhibit substantial genetic

Figure 5.14 Thermal sensitivities of cardiac function in green crabs (*Carcinus maenas*) in their native range and at invasive sites in the western Atlantic Ocean. (A) Surface water temperatures over the course of a year at collection sites in Europe (NO, Norway; PT, Portugal) and along the East Coast of North America (NL, Newfoundland; ME, Maine; NJ, New Jersey). (B) CT_{max} values for field-acclimatized crabs from locations in the native range and the western Atlantic. CT_{max} correlates strongly with habitat temperature. Within each of the two regions, lines between points indicate significant differences between regions. (C) Effects of acclimation to 25°C and 5°C on CT_{max}. Lines connecting points for species from different populations indicate significant differences in CT_{max}. Acclimation effects within a population are indicated by symbols (*, **, ***) given at the top of each frame. (D) Cold tolerance ($f_{H,0}$) of heart function in field-acclimatized populations. Like CTmax, $f_{H,0}$ differs significantly between populations. (E) Effects of acclimation on $f_{H,0}$. Lines and statistical information are as in (C). In all graphs: * = $P < 0.05$; ** = $P < 0.01$; *** = $P < 0.001$, n.s. = not significant. (After Tepolt and Somero 2014.)

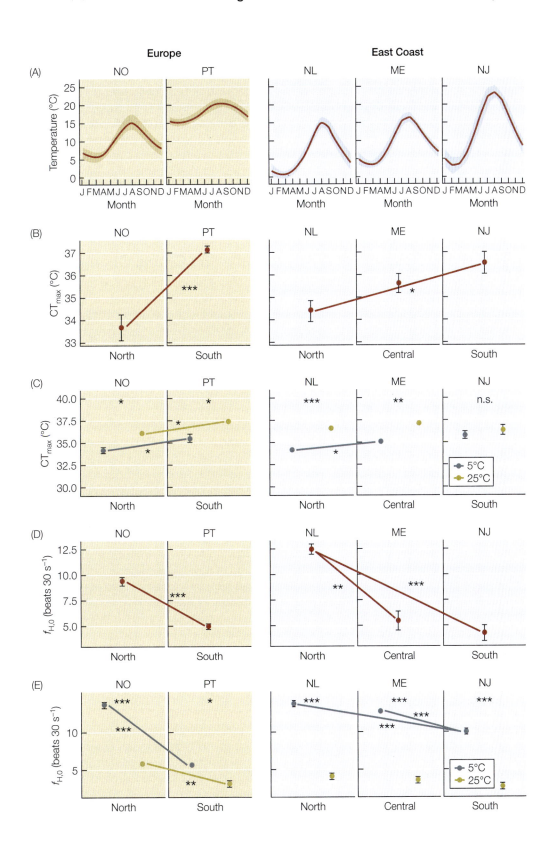

variation among populations, with the variation found in the native range showing strong signs of selection, perhaps reflecting the thermal gradient across this broad latitudinal range (Tepolt and Palumbi 2015). This high level of genetic variation may allow populations in the invasive range to more readily respond to natural selection for increased thermal tolerance on both the hot and cold ends of the thermal spectrum.

The differences found among populations of green crabs reflect what is termed **local adaptation**—the evolution of populations of a single species that differ genetically in environmental optima and tolerance relationships as a reflection of the particular constellation of environmental factors they face in their local environments. We now turn to a broader examination of this phenomenon as we look at the relationship between the capacity for adapting to global change and the ability to acquire the genetic tools needed for these responses.

5.7   Genetic Resources: A Key to Surviving Global Change

The rapidity and complexity of global change confront almost all types of organisms—microbes living deep in Earth's crust might be an exception—with a variety of challenges related to maintaining biochemical systems in optimal functional and structural states. How successfully a given species can respond adaptively to these challenges will depend largely on the information contained in its genome—its genetic tool kit—and the speed with which it can add to its repertoire of genetic tools in order to build a phenotype that functions well in a changing world. Whereas it is clear that behavioral responses to global change can sometimes offer a species a degree of escape from stressful conditions—allowing it to move to a new habitat with temperatures, salinities, pH, and oxygen levels characteristic of its evolutionary history—genetic changes that allow adaptation to altered abiotic factors are likely to be crucial for most species. These genetic changes could allow a species to "remain in place" in an ecosystem that might possess the right biotic interactions for optimal growth and propagation. As pointed out in Box 5.1, a new habitat with suitable abiotic characteristics might not necessarily have the biotic features, notably sources of food or absence of predators, that will allow the newly arriving species to succeed.

In Chapter 1 we listed several population genetic factors that play crucial roles in determining a species' chances for success in adapting to environmental change. Most of these genetic mechanisms involve the acquisition of new types of genetic information or, in many cases, recombining existing genetic variation into assemblages of alleles that confer adaptive advantages. The degree of adaptation, or the response to natural selection, that we would expect in any evolutionary scenario depends also on the strength of natural selection. And if past mass extinction events are any indication, the strength of natural selection in a changing climate is likely to be extraordinarily large. Thus, we must consider each species' actual and potential genetic resources in order to predict who will adapt versus who will go extinct. In responding to environmental changes that are occurring as rapidly as those of the Anthropocene, adaptation is more likely to depend on existing genetic variation than on new mutations (Lande and Shannon 1996). There may be exceptions to this general rule of thumb, particularly for species with enormous effective population sizes and a high degree of connectivity among geographically distant populations via migration—for example, larval transport in ocean currents among marine invertebrate populations. Recent work suggests that species such as these may not, in fact, be mutation-limited, because of their ability to access a large number of mutational events across the population (Karasov et al. 2010). However, the critical role of standing genetic variation in establishing adaptive capacity may apply to most multicellular species, not only because they tend to have smaller population sizes than unicellular species, but also because they have relatively limited access to new genetic information through horizontal gene transfer. Thus, in discussing multicellular life, we pay particular attention to the different means by which existing genetic variation can be tapped to facilitate adaptation.

Loss of genetic information, termed **DNA decay** (Harrison and Gerstein 2002), can also limit the ability of a group of organisms to cope with environmental change. Of particular interest is the DNA decay that can occur during long periods of evolution in highly stable environments. In the absence of environmental variation, genes essential for acclimatization may be selectively neutral, and if mutation disables them, they can be lost. In such cases, one of the species' first lines of defense against environmental stress—acclimatization—may disappear. DNA decay may also occur when a particular set of proteins is no longer needed under a specific set of environmental conditions. Much as cavefish that live in continuous darkness can withstand loss of their visual system (Rohner et al. 2015; see Chapter 1, p. 19), long-term evolution under other types of environmental extremes can allow loss of genetic information that could prove absolutely vital for survival if the environment shifts into another state. We will discuss the consequences of DNA decay in notothenioid fishes in the section "DNA decay in highly stable environments: Antarctic notothenioid fishes" (see p. 489).

The acquisition of new genetic information and DNA decay are both important in the context of global change. In particular, the rates at which new genetic information can be acquired—or existing genetic variation reshuffled into new, well-adapted combinations—may prove to be the ultimate determinants of a species' chances for survival in the face of rapid global change. Here we examine several different processes for altering the contents of genomes and try to identify species that appear relatively well positioned to adapt to global change and others that appear to face severe threats of extinction.

We again focus strongly on marine organisms, with special emphasis on rocky intertidal communities. Rocky intertidal communities are not only of particular experimental interest to the authors but, more important, are also uniquely suited to show how species living in a common setting can vary greatly in their sensitivities to environmental change and in their abilities to respond adaptively as a consequence of factors like behavioral capacities, generation time, standing genetic variation, and extent of gene flow among populations. By focusing on the rocky intertidal zone and on species with limited mobility, we will see how fine-scale spatial variation in the abiotic environment acts along with broader, latitudinally governed environmental change to influence species' chances for success in a changing world. This analysis will bring into clearer focus such concepts as *climate envelope models* and *mosaic models* of environmental tolerance relationships (see below). These are concepts that apply as well in terrestrial and freshwater ecosystems, so the analysis of marine communities will have important lessons to teach us about all environments.

Local adaptation: Can intraspecific variation "rescue" threatened species?

When species are distributed over environmental gradients, divergent patterns of **positive selection**—natural selection that favors a particular allele that increases fitness—can lead to genetic divergence among populations across the species' range. This is the phenomenon of local adaptation, and the end-result of this process is a population of individuals that have higher fitness, under their particular set of local conditions, than individuals (genotypes) of populations from distant sites in the species' range, which may be locally adapted to different conditions (Kawecki and Ebert 2004; Savolainen et al. 2007; Sanford and Kelly 2011). This mode of natural selection is also sometimes referred to as **spatially varying selection** because selection varies across spatial scales. By this mechanism, positive selection (i.e., spatially varying selection) that results in local adaptation increases standing genetic variation across the species, which may be beneficial in the context of global change and may facilitate species persistence (see following sections).

Genetic divergence among populations can also result from evolutionary processes that are distinct from local adaptation, such as balancing selection within populations. **Balancing selection** refers to selective processes that maintain multiple alleles within a single population. For example, spatially varying selection that varies across small spatial scales (e.g., selection driven by variation in temperature at different tidal heights in the intertidal zone) or temporally varying selection that varies across time (e.g., selection driven by variation in temperature or precipitation in different seasons) can result in a balanced polymorphism, such that distinct allelic variants are maintained in a population because they are each favored in different microenvironments or at different times of the year. Several examples of balanced polymorphisms have been documented in marine invertebrates (Sanford and Kelly 2011).

Local adaptation and balancing selection maintain allelic variation that would otherwise be lost by neutral evolutionary processes, such as gene flow or genetic drift. In the case of local adaptation, natural selection that causes divergence among populations is a force that works against the homogenizing effects of gene flow that make populations more similar. Whether or not local adaptation could serve as a mechanism of resilience to global change depends on the balance of evolutionary forces that create and maintain genetic diversity (e.g., mutation, recombination, and selection) versus forces that remove it (e.g., gene flow and genetic drift). If selection is strong, then local adaptation is possible. However, if gene flow among populations is high, then it may be difficult or impossible to establish locally adapted populations, even in the face of strong selection. Thus, dispersal is an important element in the equation that governs the likelihood of local adaptation. In marine ecosystems, it was long held that gene flow via long-distance dispersal of larvae would swamp out the effects of natural selection and prevent local adaptation. However, this view is changing now that examples of local adaptation are being discovered in species with widely dispersing pelagic larvae as well as in species with limited larval dispersal (Palumbi 2004; Sanford and Kelly 2011). Dispersal ability may also be key to enabling a species to reestablish itself at a site where it has gone locally extinct. This is a process by which, subsequent to local adaptation, local extinction of one population may lead to the replacement of the extinct population with conspecifics that are genetically better able to cope with the altered environment.

In the case studies presented below, we focus chiefly on the effects of temperature. This comparative analysis reveals wide differences among species in capacities for responding adaptively to heat stress. These differences mean that climate change is likely to have significantly different effects among coexisting species, with serious consequences for ecosystem structure. Some of the species we examine have been studied at multiple levels of biological organization, so it is becoming possible, in at least some cases, to "connect the dots" between molecular-level effects and broadscale shifts in distribution patterns. Some of the species also have been characterized in terms of their capacities for acclimatization—their phenotypic plasticity—as well as their interpopulation genetic differences. In such cases, one can compare the potential benefits of local adaptation and phenotypic plasticity in enhancing a species' chances for survival in the face of environmental change.

We would like to note that identifying true cases of local adaptation can be challenging, in large part because phenotypic plasticity can generate measurable phenotypic differences among populations that could be misinterpreted as local adaptation. A further complication is that phenotypic plasticity itself is a trait that can evolve to be locally adapted (Cooper et al. 2014). Distinguishing local adaptation from phenotypic plasticity is best done by comparing populations after culturing them under common garden conditions (all groups of organisms are raised under or acclimatized to the same conditions) for two or more generations, in order to eliminate effects of acclimatization and maternal effects.

As a final prefatory note, it is important to mention that a substantial fraction of predictions about effects of global change on species' distributions have been made within a particular theoretical framework, the **climate envelope model** (Pearson and Dawson 2003). Analyses within this framework typically assume that the environmental tolerance of a species is homogenous throughout the species' distribution range. In other words, local adaptation is commonly ignored in such analyses. The examples given below show that this assumption can be faulty; the effects of global change are more complicated than climate envelope analyses may suggest. More realistic analyses must take local adaptation into account and pay heed not only to effects of latitude but also to the **mosaic** nature of the environment, especially in intertidal habitats (Helmuth et al. 2002).

TURBAN SNAILS (*Chlorostoma funebralis*): LATITUDINAL VARIATION IN HEAT TOLERANCE AND GENE EXPRESSION   We turn first to an organism we have already discussed in several contexts in Chapters 3 and 5, the turban snail *Chlorostoma funebralis*. In Chapter 3 we focused on variation among *Chlorostoma* congeners in thermal responses of protein function and the heat-shock response. In this chapter we discussed differences among congeners in thermal resistance of cardiac function (see Figure 5.12). The wide latitudinal range of one of the congeners, *C. funebralis*, which occurs from Vancouver Island, Canada, to Baja California, Mexico, provides an opportunity to ask whether latitudinally separated populations exhibit differences in their thermal responses that reflect the types of interspecific differences that have been found in comparisons of congeners.

To determine whether local adaptation to temperature existed among turban snail conspecifics, Gleason and Burton (2013) examined the heat tolerance of six populations of *C. funebralis*, three from northern sites and three from southern sites under common garden conditions (Figure 5.15). They found significant differences in thermal tolerance among the populations: The northern conspecifics were significantly less heat tolerant than their kin from lower latitudes. These comparisons also revealed another example of the relationship that is frequently observed between intrinsic heat tolerance and proximity of upper lethal temperature to maximum habitat temperature. Southern populations, despite being more heat tolerant than northern populations, live closer to their upper thermal limit than the cold-adapted conspecifics found at higher latitudes. The most southern populations, those from La Jolla and Bird Rock in San Diego, California, encounter habitat temperatures of 40°C–41°C. Even though they survive longer at these thermal extremes than the northern populations, they clearly are "living on the edge" of their heat tolerance limits. In the absence of novel adaptations that increase heat tolerance, *C. funebralis* could face local extinction if global climate change increases

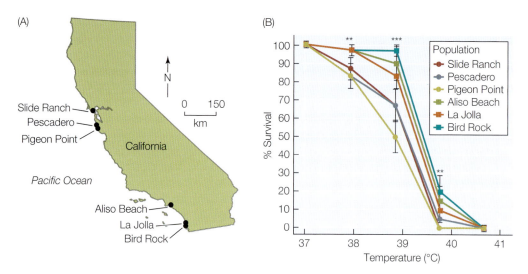

Figure 5.15 Intraspecific differences in heat tolerance (37°C to 41°C) between northern and southern populations of *C. funebralis*. (A) Collection sites for the northern and southern populations. (B) Survival following heating at a rate of 6 degrees Celsius h⁻¹. Circles denote northern populations; squares denote southern populations. Data are means ± S.E. Asterisks indicate a significant difference in survival between the three northern populations as a group compared with the grouped southern populations. ** = $P < 0.01$; *** = $P < 0.001$. (After Gleason and Burton 2013.)

maximum habitat temperatures by even 1 or 2 degrees Celsius. Recall that the flatline temperature (FLT), which corresponds to lethal temperature, for *C. funebralis* is near 40°C (for a population from Monterey, CA; see Figure 5.12 and Table 5.1) and shows little change during acclimation. Because the sampled southern California populations live near the southern end of the species' latitudinal range, it is possible that local extinction of these heat-adapted populations would not lead to replenishment of the species by an even more heat-tolerant population moving in from lower latitudes. There is, of course, some chance that populations of *C. funebralis* from Baja California might have thermal tolerances that exceed those of the San Diego populations, but the thermal tolerances of the Mexican populations have not yet been assessed. Thus, as in the case of other relatively heat-tolerant ectotherms, the persistence of *C. funebralis* in habitats at the low-latitude extreme of its distribution range may be jeopardized by a lack of yet more heat-adapted populations to replenish populations that go locally extinct. This is another general consideration in the context of latitudinal variation in effects of global change: There may be no populations to repopulate habitats in which the most heat-tolerant populations have gone locally extinct. Thus, ecosystems will not see a "replacing like by like" pattern of change, but rather a shift in community species composition.

Additional evidence suggesting that local adaptation to temperature is present in *C. funebralis* comes from studies of gene expression (transcriptomics). Differences in constitutive levels of gene expression and in heat-induced changes in expression were found between northern (Slide Ranch and Pigeon Point) and southern (Aliso Beach and La Jolla) populations under common garden conditions (Gleason and Burton 2015). The southern populations exhibited higher constitutive expression of several genes associated with the cellular stress response, including genes encoding heat-shock proteins, several components of the ubiquitin-mediated pathway of proteolysis, and enzymes essential for detoxification of ROS. This higher expression of stress-related genes in the southern populations was conjectured to represent a **preparative defense strategy**, a preadaptation that would prepare the populations for the extreme thermal stress that can occur in their low-latitude habitat. The southern conspecifics thus appeared to be "girding their loins" for exposure to damaging high temperatures. A similar response involving expression of heat-shock proteins has been observed in comparisons of *Chlorostoma* and *Lottia* congeners with different vertical distributions in the intertidal zone (*Chlorostoma*: Tomanek and Sanford 2003; *Lottia*: Dong et al. 2008). These authors, too, viewed the higher constitutive levels of heat-shock proteins, detected using Western blotting analysis, as an indication of a preparative defense strategy rather than an indication that these warm-adapted species had experienced an unusual level of heat stress.

One of the potential trade-offs of the preparative defense strategy may be a higher energy cost for living at high temperatures. This conjecture was raised in Chapter 3 when we considered some of the reasons why temperature compensation in ectotherms tends to be incomplete ("imperfect"); life at higher body temperatures simply might take more energy. The accumulation of higher levels of many types of stress-related proteins will obviously cost energy for their synthesis and turnover, as well as for their function, in that ATP hydrolysis is often used to drive the reactions they catalyze. Their higher constitutive presence may be a lifesaver under conditions of episodic extreme heat stress, but the day-in/day-out maintenance of high levels of these proteins may draw energy away from other processes, notably those of growth and reproduction, as discussed in the context of the heat-shock response in Chapter 3. In fact, lower-tidal-zone congeners of *Chlorostoma* do grow faster than *C. funebralis* (Watanabe 1982; Somero 2002). Whether this difference is due to factors such as feeding time and food availability or thermal stress is not clear. However, there can be little doubt that maintaining high concentrations of stress-related proteins could deplete cellular energy supplies and precursors for protein synthesis.

The southern and northern populations of *C. funebralis* also differed in the degree to which heat stress led to upregulation of different genes (Gleason and Burton 2015). Perhaps as a reflection of higher constitutive expression of stress-related genes, the southern population exhibited a smaller increase in expression of many stress-related genes during heat stress compared with the northern population. Large differences were found in upregulation of genes encoding heat-shock proteins and proteins associated with detoxification of ROS. The populations also differed in their expression of genes encoding proteins that govern the cell cycle. Heat-induced suppression of the cell cycle was predicted to be greater in northern populations, which may reflect a higher level of damage to DNA under heat stress in these conspecifics. The preparative or preadaptive gene expression pattern found in the southern populations may prevent the level of damage to DNA and other macromolecules that occurs in the more vulnerable northern population, where preparations for coping with heat stress are substantially lower.

TIDE-POOL COPEPODS: LATITUDE AND HEAT TOLERANCE   At the beginning of this section we stated that the "gold standard" for distinguishing acclimatization from genetic adaptation is to examine a particular trait such as thermal tolerance over multiple generations under common garden conditions. Excluding the possibility of inherited epigenetic modifications that alter environmental responses (see Chapter 1), this experimental approach allows one to definitively attribute population-level differences to genetic effects, thus ruling out the phenotypic effects of previous acclimatization to a particular set of environmental conditions. This type of multigeneration study is often challenging due to factors such as long generation times and difficulties in breeding and raising organisms in a controlled laboratory setting that allows clear distinctions to be drawn between "nature" and "nurture." Here we examine studies that help clarify a species' access to local adaptation and the role that intraspecific differences in thermal tolerance might play in ensuring that a species can maintain its biogeographic range.

Tigriopus californicus

Tigriopus californicus is an abundant intertidal harpacticoid copepod (see photo at left), which occurs over a wide (~3,000 km) latitudinal range, from Baja California (~27° N) to southeastern Alaska (57° N) (Kelly et al. 2012). The tiny copepod is found in tide pools that often are small, shallow, and located high up in the intertidal zone. In these pools, the species experiences wide ranges of temperature, salinity, pH, and oxygen concentrations. As one would predict, the species is extremely eurytolerant to these abiotic factors. Because of this species' short generation time and ease of laboratory culture, it has been possible to estimate the relative contribution of local adaptation versus plasticity to the phenotypic variance among its populations.

Across its wide biogeographic range, *T. californicus* comprises populations that differ in upper thermal tolerance limits (**Figure 5.16**; Willett 2010; Kelly et al. 2012). To demonstrate local adaptation in these populations—that is, to reduce the likelihood of acclimatization contributing to thermal tolerance—Kelly and colleagues reared specimens from eight North American populations (see map in Figure 5.16A) for multiple generations in the laboratory. The heat tolerance (LT_{50}) data shown in Figure 5.16A were obtained from second-generation individuals that were cultured in common garden at a constant 19°C. The latitudinal variation in LT_{50} is highly significant. Note, too, how little variation there is within a population, as indicated by small standard errors around the LT_{50} values, such that less than 1% of the total phenotypic variance for heat tolerance was attributed to within-population effects.

The different populations also manifested significant differences in potentials for laboratory acclimation to high temperature (see Figure 5.16B). In contrast to the pattern seen for cardiac function in turban snails and porcelain crabs, the acclimatory plasticity of LT_{50} in the *T. californicus* populations was negatively correlated with latitude; northern populations had a lesser capacity to acclimate than southern populations.

This study went a step further to characterize the potential of these copepod populations to adapt for survival at higher temperatures. The copepods were subjected to selection for thermal tolerance in the laboratory by 1-h exposure to near-lethal temperatures; such exposure over several generations led to some adaptive increases in LT_{50}, but this occurred in only six of the eight populations (see Figure 5.16C). Overall, the authors found very limited local adaptive potential for heat tolerance, as the responses to selection (blue bars in Figure 5.16C) were only a fraction of the total amount of variation in this trait in the species as a whole (indexed by the difference between the most versus least heat-tolerant populations; dotted lines in Figure 5.16C). At most, LT_{50} was increased by only 0.53 degree Celsius (population BR). The increase in heat tolerance in the northern populations was far from being enough to give them the pre-lab-selected level of tolerance found for the southern populations. Most of the increases in LT_{50} occurred over the first four or five generations; in only two populations (BD and SH) did a further increase in LT_{50} appear between generations six and nine. This pattern suggests that most of the available genetic variation for increased heat resistance had been exploited after one or a few generations.

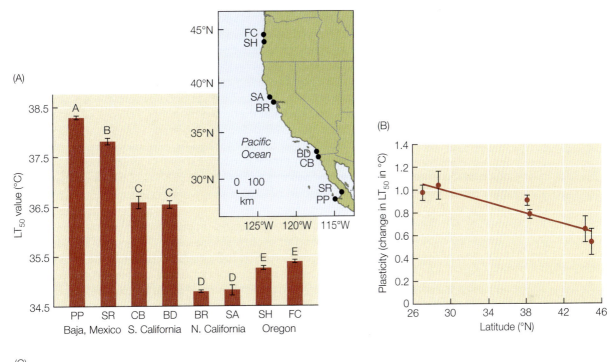

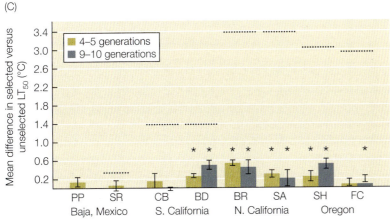

Figure 5.16 Local adaptation and acclimatory plasticity in eight populations of *Tigriopus californicus* from various latitudes. (A) Thermal tolerance (mean LT_{50} ± S.E.) of adult male specimens from the second generation of laboratory-raised populations that were cultured at 19°C. Sampling locations: Punta Prieta, MX (PP), Santa Rosalia, MX (SR), Cabrillo Point, CA (CB), Bird Rock, CA (BD), Bodega Marine Reserve, CA (BR), Salt Point, CA (SA), Strawberry Hill, OR (SH), and Fogarty Creek, OR (FC). Shared letters denote species for which LT_{50} values are not different. (B) Plasticity of thermal tolerance (change in LT_{50} with acclimation) as a function of latitude for six populations. Second-generation laboratory-reared copepods were held either at a constant cool temperature (19°C) or cycled between 19°C (18 h) and a high temperature, 28°C (6 h). (C) The responses of eight populations of *T. californicus* to selection on thermal tolerance, measured as the difference between selected versus unselected populations from a given site. Selection involved exposure to near-lethal temperatures for 1 h (see Kelly et al. 2012 for details). Asterisks indicate a significant difference from zero ($P < 0.05$). Dotted lines show the difference between the LT_{50} of the most tolerant population (PP) and the different unselected lines. PP and SR were subjected to only five generations of selection. (After Kelly et al. 2012.)

The lack of evolutionary potential for increased thermal tolerance in *T. californicus* populations may have been the result of low levels of genetic diversity present in each population—perhaps a consequence of population bottlenecks that these populations may undergo on a somewhat regular basis in isolated tide pools of the high intertidal zone. Nevertheless, this type of result has been corroborated by studies in *Drosophila* that have found upper thermal tolerance limits to be evolutionarily constrained, and oddly enough, more limited in adaptive potential than cold tolerance limits (Kellerman et al. 2012; Kristensen et al. 2015), despite the high degree of genetic variation that is present in many species of fruit flies (Karasov et al. 2010). Thus, by strictly relying on standing genetic variation it may be that upper thermal tolerance limits cannot quickly evolve in response to thermal selection. We conjecture that this may be due to the relatively rare occurrence of the correct combination of alleles that would confer adaptations in the diverse array of biochemical systems that underlie thermal tolerance, such as proteome stability, regulation of the heat-shock response, and maintenance of membrane-bound processes, such as the electron-transport chain (as discussed in Chapter 3). If indeed the evolution of thermal tolerance is constrained by low levels of standing genetic variation in populations of *T. californicus*, increased gene flow among *T. californicus* populations may lead to more potential for the evolution of thermal tolerance (see below).

What do these interpopulation differences in intrinsic heat tolerance and acclimatory capacity suggest about the influences of ongoing global warming on this species across its current biogeographic range? Measurements of daily temperature cycles in tide pools containing the copepod show that temperatures currently rise to within 1 degree Celsius of LT_{50} values for short periods at the six northern sites. The most southern sites likely attain temperatures equally close to LT_{50} values for the more heat-adapted populations. There is, then, a similar proximity of LT_{50} to maximum habitat temperature across a substantial fraction (Baja California to Oregon) of the species' latitudinal range. Therefore, global warming may pose similar threats to populations across this wide range of latitudes. This is different from the situations predicted for other species, for example, porcelain crabs and turban snails, where the warm-adapted species or populations seem most likely to be threatened by heat death. In short, the effects of global change are likely to differ among taxa because of variation among species in the proximity of LT_{50} to maximum habitat temperature.

Just as for the populations of *C. funebralis* (Gleason and Burton 2015) discussed above, latitudinally separated populations of *T. californicus* differed in their transcriptional responses to heat stress (Schoville et al. 2012). Populations from Santa Cruz, CA, (SC) and San Diego, CA, (SD) were compared in this study. In both populations, heat stress (35°C for 1 h) increased the expression of hundreds of genes: 867 in the SC population and 356 in the SD population. However, only 63 of these genes were upregulated in both populations. Among these were genes that encode proteins that are instrumental in protecting or restoring protein function (molecular chaperones) and in removing irreversibly damaged proteins from the cell (for example, proteins associated with the ubiquitin-mediated pathway of proteolysis). Although these 63 genes were upregulated in both populations, they were upregulated to different degrees: The SD population consistently showed a significantly higher level of upregulation.

This finding can be interpreted in two ways. First, it might be taken to mean that more heat-induced damage to the proteome occurred in the SD population. Thus, the degree to which these components of the cellular stress response are upregulated might be proportional to the amount of damage repair that is required. However, this is not a likely explanation because the SD population exhibits the higher heat tolerance at the phenotypic level, and organisms with lower heat tolerances also tend to incur greater heat-induced macromolecular damage. An alternative explanation, one favored by Schoville and colleagues (2012), is that the higher upregulation in the more warm-adapted and heat-tolerant population reflects a better ability to mount a cellular stress response than is found in the higher latitude population. The different responses of congeners of *Cholorostoma* to acute heat stress reflect the same pattern: The more warm-adapted *C. funebralis* mounts a larger and more rapid synthesis of heat-shock proteins than the more cold-adapted congener *C. brunnea* (see Figure 3.25). This pattern of more robust response to heat stress in more warm-adapted species and/or populations, then, should be interpreted as an adaptation to higher temperatures, rather than an indication of greater damage from heat. Thus, the underlying robust transcriptional response to heat stress is what may provide the SD population with the ability to survive at higher temperatures.

Studies of *T. californicus* have also provided insights into the potential benefits of hybridization among genetically distinct populations of a species. When formerly genetically isolated populations are interbred, the resulting hybrids can have environmental tolerances that exceed those of either of the parent populations (Pereira et al. 2014). This is an example of a process known as **transgressive segregation**, in which hybridization generates new allelic combinations that produce phenotypes that are extreme (here, in heat tolerance) relative to both parental populations. Pereira and colleagues examined populations of *T. californicus* from several sites along the California coast. When crosses were made between two genetically distinct populations collected near San Diego, the hybrid offspring displayed heat tolerances that were greater than those of the parental lineages. Moreover, this effect persisted, and even led to further increases in heat tolerance, after nine generations of subsequent inbreeding. Crosses between populations from more distant sites—that is, between populations with larger differences in thermal tolerance—did not show transgressive segregation. The increased LT_{50} values for the hybrids generated from the two heat-tolerant parental lines illustrate a potentially important role for hybridization in adapting to global change. If global change fosters contact between populations that have a potential for transgressive segregation, then it may be possible to generate individuals with enhanced tolerance of rising temperatures and other abiotic stresses associated with global change.

WHELKS: ADAPTATION IN A MOSAIC ENVIRONMENT   At the beginning of our discussion of local adaptation, we stressed that adaptive variation among populations might occur not only along latitudinal gradients of temperature but also across local regions where habitat temperatures are influenced by factors in addition to latitude. Even along latitudinal gradients there is by no means a uniform, linear variation in thermal stress. In intertidal habitats, for example, thermal stress is strongly influenced by latitudinal variation in the timing of the tidal cycle (Helmuth et al. 2002). Individuals at sites where midday low tides occur in summer months will face greater threats from heating and desiccation than individuals at sites where low tides occur during the night. The timing of tidal cycles thus will interact with latitudinal variation in temperature in determining abiotic stress, especially for species that are exposed during low tides.

In the intertidal zone, there is almost always a mosaic nature to the environment. Despite the broad influence of latitude on thermal conditions, there is fine-scale variation in temperature relationships due to sun exposure, amount of wave splash, height of occurrence along the subtidal-to-high intertidal axis, and shading by overlying organisms. Because of the fine-scale mosaic nature of intertidal habitats, variation in physical conditions, especially temperature and threats from desiccation, may be greater over a distance of a few meters than across tens of degrees of latitude (Dowd et al. 2015).

Nucella canaliculata

An illustration of how tidal cycle and latitude interact is given by studies of an intertidal whelk, *Nucella canaliculata* (see photo at left), that occurs over a wide range of latitudes. This species has direct development; it lacks a free-swimming, widely dispersing larval stage. In accord with this developmental strategy, there is a high level of genetic differentiation among populations (Sanford et al. 2003). Because of its distribution, life history, population genetic structure, and—importantly—its ability to be reared in the laboratory, *N. canaliculata* has proven to be a good species for studying local adaptation.

Kuo and Sanford (2009) examined seven populations of *N. canaliculata* from latitudinally separated sites along the coastlines of Oregon and California. They determined upper thermal tolerance limits for second-generation laboratory-bred individuals raised under common garden conditions to eliminate the effects of prior acclimatization. Unlike the pattern seen with *C. funebralis* (see Figure 5.15), there was no regular increase in LT_{50} with decreasing latitude (**Figure 5.17**). Rather, the most heat-tolerant populations of *N. canaliculata* were from the three sites along the Oregon coast (Fogarty Creek, Strawberry Hill, and Cape Arago) and the two northern California sites (Van Damme and Bodega Head).

What could account for this seemingly paradoxical relationship of LT_{50} and latitude? Kuo and Sanford (2009) conjecture that the higher heat tolerance of the northern populations reflects the timing of summer low tides. At these higher-latitude sites, midday

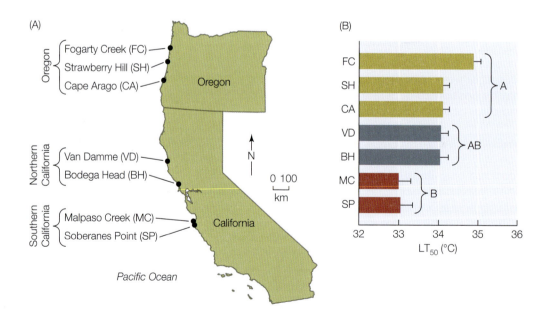

Figure 5.17 Local adaptation of upper thermal tolerance limits in latitudinally separated populations of the whelk *Nucella canaliculata*. (A) Sampling sites for the seven populations. (B) Upper thermal limits (LT_{50}; mean ± 1 SE) of second-generation whelks reared in the laboratory under common garden conditions. Shared letters to right of bars indicate LT_{50} values that do not differ between sites. (After Kuo and Sanford 2009.)

low tides are common, unlike the tidal pattern observed at the two southernmost sites, where summer low tides occur after dark. The midday low tides found at the northern sites would, then, be expected to lead to greater levels of heat stress than occurs at the southernmost sites (Malpaso Creek and Soberanes Point) in California. If this were the case, then one would predict that local adaptation could lead to the patterns of heat tolerance found in this study.

Standing genetic variation is a resource for adaption

We pointed out earlier that standing genetic variation could confer adaptive potential in the context of environmental change. Recall that levels of standing genetic variation depend on several factors, including population size, mutation rate, genetic recombination, gene flow, and natural selection. One mechanism of increasing standing genetic variation in species that span broad geographic distributions is the process of local adaptation, in which the homogenizing effects of gene flow are overcome by the force of positive selection for locally adapted genotypes. Mosaic environments, such as those present in the marine realm, vary across both space and time to create a complex selective regime, in which spatially and temporally varying selection can promote increased levels of genetic diversity in a local population (Sanford and Kelly 2011). As we will see below, local adaptation as well as spatially and temporally varying selection may promote levels of standing genetic variation that can then serve as a resource for evolutionary adaptive responses to environmental stress.

Strongylocentrotus purpuratus

SEA URCHIN LARVAE AND OCEAN ACIDIFICATION   An excellent example of how standing genetic variation can be a potential resource for resilience to environmental change is provided by the effects of ocean acidification on larvae of the purple sea urchin (*Strongylocentrotus purpuratus*; see figure at right). The purple sea urchin is one of the few marine invertebrates for which a sequenced and well-annotated genome is available (the genome of *Tigriopus* has recently been added to this roster: https://i5k.nal.usda.gov/Tigriopus_californicus), making it much more feasible to interpret genetic differences among populations or between populations exposed to different environments. The purple sea urchin has several other advantages as an experimental organism for study of ocean acidification. As a native of the California Current Ecosystem, *S. purpuratus* is exposed to wide spatial and temporal variation in carbonate chemistry that arises in large measure from episodic upwelling of waters with reduced pH and carbonate saturation. Exposure to these relatively acidic waters could impair the growth of the delicate carbonate-based skeletal elements of the larvae, and exposure to low-pH conditions over evolutionary time may have already led to adaptation in this species. The species also is highly fecund, has high genetic variability, and has a high dispersal potential.

Pespeni and colleagues (2013a) collected larvae from sea urchin populations at seven different sites with different upwelling characteristics across a distance of 1200 km. In a laboratory selection experiment, these larvae were exposed to contemporary surface ocean CO_2 partial pressure (pCO_2), 400 µatm, and to 900 µatm. The higher pCO_2 condition that these authors used in their experiment is, in fact, experienced by some populations during upwelling events, so it represents a high, yet biologically realistic, value. Because the upwelling conditions that lead to episodic exposure of urchin larvae to low pH have existed for millions of years, it seems probable that the species has acquired genetic tools for adaptation to episodic periods of OA. However, the high dispersal capacity of the species may reduce the probability of local adaptation of the types seen for *Chlorostoma* or *Tigriopus*. Rather, periods of strong upwelling might select for OA-resistant genotypes that may then disperse over much or all of the species' wide biogeographic range.

The results of the population genomics analysis (Pespeni et al. 2013a) and a follow-up study that compared the differential allelic changes among surviving larvae of these seven populations after experimental ocean acidification (Pespeni et al. 2013b) showed evidence for some degree of local adaptation, but all populations exhibited evolutionary potential for adaptation to low-pH conditions. Exposure of urchin larvae to low-pH water (900 µatm CO_2) caused differential mortality among genotypes that led to changes in the frequencies of numerous alleles over time. Among the 19,493 single nucleotide polymorphisms (SNPs) identified across the genomes of these larvae, significant changes in allele frequency were evident within the first day of exposure, and the effects of selection on different alleles increased through day 7 of the experiment. Control (400 µatm) populations did not show this type of time-dependent change in allelic variation.

The major functional categories of genes that showed time-dependent changes in allele frequencies in the OA-exposed cultures included genes known to play roles in biomineralization, lipid/membrane biochemistry, and ion homeostasis. These changes can be interpreted quite readily in the context of the challenges posed by reduced pH. Thus, biomineralization (production of calcium carbonate structures) is challenged by low pH, and pH regulation involves ion transport activities across membranes. Whereas these allelic changes do not directly indicate the molecular mechanisms involved in OA adaptation, they are consistent with selection for variants on biochemical themes that are critical in responding to OA effects on skeleton formation and pH regulation. A further argument for the proposed selective importance of the allelic changes is that 10.8% of the SNPs led to changes in amino acid sequence, and of these amino acid substitutions, 25.6% led to a change in protein charge. Such alterations of charge state would be expected to cause functional alteration of the proteins (see Chapter 3). In contrast to the OA-selected changes in allelic frequency, variations in pCO_2 exposure had little effect on gene expression levels. Neither was there any strong signal for allele-specific expression by individuals. These results argue for the importance of adaptive changes in the intrinsic properties of the proteins, not for their differential regulation between genotypes.

The ecological relevance of the observed allele frequency changes was addressed in a follow-up study by Pespeni and colleagues (2013b) that examined the link between environmental pH data and the degree of allele frequency changes among the seven populations of urchins that were sampled. Strikingly, there were differences among these seven populations in the degree of allele frequency change in the selection experiment, such that populations from locations that experience higher pH conditions in the wild showed greater shifts in allele frequency in response to experimental acidification than did populations that experience low-pH conditions (**Figure 5.18**). Thus, there appears to be significant local adaptation to low-pH conditions among *S. purpuratus* populations, despite the large amount of gene flow in this species. Low-pH populations already possess the adaptive alleles at high frequencies—evidence of natural selection in action. Importantly, because of the high degree of connectivity among populations of *S. purpuratus*, these adaptive alleles are shared among all populations, but they segregate at lower frequencies in populations that do not experience regular exposure to low-pH water.

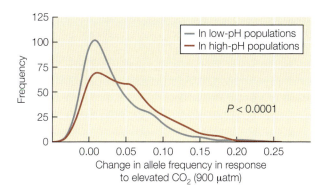

Figure 5.18 Allele frequency change in response to experimental ocean acidification among thousands of SNPs across the genome of the purple sea urchin (*Strongylocentrotus purpuratus*). The plot shows frequency histograms of allele frequency changes (the x axis indicates degree of shift in allele frequency, and the y axis indicates counts) after seven days of larval culture at a pCO_2 of 900 µatm in populations from low-pH locations (blue) and high-pH locations (red). The frequency change was greater in populations sampled from high-pH locations. (After Pespeni et al. 2013b.)

To summarize, these two studies by Pespeni and colleagues indicate that the purple sea urchin appears to have the capacity to adapt rapidly to ocean acidification because of the high amount of allelic variation it possesses across its genome, especially in genes encoding proteins that are involved in processes like pH regulation and biomineralization. This variation is likely the consequence of the mosaic nature of the environment—the wide, upwelling-caused variation in pH and other abiotic factors—in which this species has evolved over millions of years. The authors of the studies appropriately speak of the species' genetic variation as offering a "reservoir of resilience to climate change" (Pespeni et al. 2013a). How widespread this type of adaptive potential is among marine organisms remains to be established. Results from the limited number of published studies suggest that the potential to tap standing genetic variation to adapt to OA exists in other marine species, at least in populations with large population sizes, high fecundity, and high levels of existing allelic polymorphism (Sunday et al. 2011). Complementing the effects of selection on standing genetic variation are the effects of acclimatization. The studies on thermal biology of coral reefs we now consider reveal important roles for both of these processes in coping with another major challenge of global change.

CORAL REEFS AND HEAT STRESS   Coral reefs have received a great deal of attention in the context of global change, due to their multiple ecological and economic roles and their vulnerability to rising temperature, falling pH, and rising sea level (Hoegh-Guldberg et al. 2007; Nagelkerken and Connell 2015). Coral reefs, the largest biological structures in the world, are major centers of biodiversity and key sites of productivity in tropical waters. Humans enjoy many benefits from healthy coral reefs. They provide substantial amounts of food for fishing communities across the tropics, and the barrier function provided by reefs can ameliorate the effects of extreme storms. Because of the threats that rising temperatures and OA pose to coral reefs, many investigators have attempted to elucidate the magnitude of the challenges posed by these twin, interacting stressors and determine what potentials are present for adaptation and acclimatization.

Many corals have biogeographic distributions that span a wide enough range of latitude to allow comparisons of the thermal tolerances of populations locally adapted or acclimatized to differences in temperatures that are apt to be physiologically significant. Comparisons of conspecifics from different latitudes (thermal regimes) have shown that bleaching commonly occurs at temperatures only 1–2 degrees Celsius above local mean maximum temperatures in summer, at least when exposure to these high temperatures occurs for several weeks (Jokiel and Coles 1990). These observations suggest key roles for local adaptation and/or acclimatization in heat tolerance. If local adaptation exists and if larval dispersal occurs over wide enough distances, then "genetic rescue" of threatened reefs might be possible: Larvae from more heat-tolerant populations might be able to reach and then settle in the reefs that are facing significant heat stress.

One coral species with such a biogeographic range is *Acropora millepora*, illustrated in the photo below. Dixon and colleagues (2015) studied two populations of *A. millepora* from sites on the Australian Great Barrier Reef that are separated by ~5 degrees latitude. At the northern (warmer) site, Princess Charlotte Bay, mean monthly water temperatures vary between ~32°C and ~24°C. At the southern site, Orpheus Island, temperatures range between ~30°C and 21°C. In Dixon and colleagues' study, ten different intrapopulation crosses between parents obtained from the two sites were made, and the resulting larvae were tested for their tolerance of heat stress (odds of survival after exposure to 35.5°C for periods of 27 and 31 h). Survival rates varied significantly among crosses, and larvae produced by crossing adults from the Princess Charlotte Bay site had significantly higher heat tolerances than larvae generated by parents from Orpheus Island. The data did not provide strong indication that epigenetic effects were important in these differences. Rather, the data suggest that the majority of the differences in heat tolerance between the populations reflect past selection on standing genetic variation. In other words, the two populations may have locally adapted to the temperature conditions experienced during the separate reefs' evolutionary history. The biochemical systems responsible for the interpopulation differences remain to be elucidated in detail, but the classes of genes showing signs of strong selection included genes related to mitochondrial function. Dixon and colleagues emphasize that connectivity of coral populations may facilitate resilience and that human-assisted colonization efforts could enhance the effects of natural connectivity.

Acropora millepora

This study of thermal tolerance in latitudinally separated populations of *A. millepora* focused exclusively on genetic differences—adaptations—and did not examine the effects of acclimatization. The authors did point out, however, that in long-lived species like corals, acclimatization may potentially be more important than slower, genetically based adaptation. In fact, both adaptation and acclimatization have been shown to play important roles in coral thermal tolerance. In a study conducted in American Samoa, populations of another *Acropora* species, the tabletop coral *Acropora hyacinthus*, exhibited adaptive change mediated through both processes (Palumbi et al. 2014). Conspecifics from two reef sites were compared to tease apart fixed genetic differences from the influences of acclimatization. One site was a highly variable (HV) pool in which temperatures reached 35°C, which is ~5 degrees Celsius above the critical bleaching temperature for this species. The other site, the moderately variable (MV) site, rarely had temperatures exceeding 32°C. Corals in the HV site were more heat tolerant than conspecifics in the MV site, as evidenced by higher growth rates, higher survivorship, and higher symbiont photosynthetic efficiency after heat stress (Oliver and Palumbi 2011). Long-duration reciprocal transplantation experiments were performed using branches of *A. hyacinthus* from the HV and MV pools. Branches transplanted in this way produce new tissue that is genetically identical to the source coral, but grown under the conditions of the new site. Testing was done after 12, 19, and 27 months, so that the tissue sampled had not experienced the thermal regimen of the pool from which the transplanted coral branch had been removed. Significant changes in heat tolerance were observed during this long period of acclimatization. The heat resistance of MV branches transplanted to the HV pool increased, yet these branches failed to attain the heat tolerance of branches native to the HV pool. Heat tolerance of HV branches transplanted to the MV pool decreased to values of the native MV branches.

Studies of gene expression in the different native and transplanted populations of *A. hyacinthus* provided insights into the roles of acclimatization and fixed genetic differences in establishing heat tolerance (Palumbi et al. 2014). For several dozen genes, differences in expression levels between populations correlated with pool of origin, not with acclimatization at the transplant site. Other genes showed strong effects of acclimatization on expression levels. Overall, the studies of heat tolerance and gene expression suggested two important conclusions. First, acclimatization and fixed genetic effects contributed approximately equally to differences in heat tolerance. Second, during less than 2 years of acclimatization, heat tolerance increased by an amount that would be predicted from strong natural selection over many generations. Acclimatization, then, is an important "first line of defense" for reef-building (hermatypic) corals.

Interestingly, the effects of acclimatization appeared to be confined to the cnidarian host; the dinoflagellate symbionts (zooxanthellae) of *A. hyacinthus* did not alter their patterns of gene expression, and the types (clade memberships; see below) of symbionts did not change during acclimatization. Does this mean that the challenges of coping with rising temperatures must be borne entirely by the animal component of the holobiont (the host plus symbiont[s])—or are there cases where both the host and symbiont share this task?

Adaptation by symbiont shuffling: Having the right partners for the challenges at hand

Symbiotic partnerships between two or more organisms have played a central role in the evolution of biological novelty. Symbioses occur between members of the same domain (e.g., cnidarians and dinoflagellates) and in cross-domain associations (e.g., the enormous number of bacteria and archaea that constitute animal gut microbiomes). The importance of bacterial and archaeal symbionts to animal life is coming to be increasingly appreciated, as more and more roles for host-symbiont interactions are revealed (for outstanding reviews of this burgeoning field, see McFall-Ngai et al. 2013; McFall-Ngai 2015). Because the total number of cells in an animal's microbiome may be greater than the number of its "own" cells (for instance, the mass of our gut microbes exceeds the mass of our brain), it is logical to ask whether adaptation to environmental change might be achieved, at least in part, by altering the population of microbial symbionts that share metabolic duties with the organism's "own" cells. There are, of course, abundant cases demonstrating the importance of shifting microbial populations in the face of dietary changes (Gilbert and Epel 2015). However, in the context of adaptation to abiotic factors such as temperature or seawater pH, there are relatively few examples of changes in symbiont populations enabling the holobiont to regain an optimal state of function.

Reef-building corals have probably received the most attention in this regard, mostly because of the large genetic diversity and biogeographic patterning of the symbionts, all members of the dinoflagellate genus *Symbiodinium*. At least nine clades (A–I) of *Symbiodinium* exist (Pochon and Gates 2010), and the symbiont compositions of corals vary widely among species. In some cases, the type(s) of *Symbiodinium* present in a coral appear to reflect the particular environmental temperatures the coral faces. Unlike the results of the study of Palumbi and colleagues (2014) discussed above, other work suggests that a given cnidarian may sometimes alter the clade composition of its population of zooxanthellae, a process known as **symbiont shuffling**. When does this occur—and why?

Cunning and colleagues (2015a) have made one of the most thorough studies of the cause-and-effect relationships between thermal stress and symbiont composition. They investigated a Caribbean coral, *Orbicella faveolata*, that contains varying proportions of two genetically distinct clades of *Symbiodinium*, clade B1 and clade D1a (hereafter, clades B and D, respectively). Clade D's fractional contribution to the pool of symbionts ranged from 0 to 1. When corals were subjected to graded heat stress (low, medium, and severe) and then allowed to recover at different temperatures, large differences in bleaching levels were found among corals having different symbiont populations and symbiont densities. This high degree of variation may explain why studies done by different laboratory groups, using quite different thermal stress protocols, show such disagreement about symbiont-related effects and the occurrence of symbiont shuffling. The most severe level of heat stress given to *O. faveolata* favored corals having a high percentage composition of clade D, especially if recovery was at moderately high temperatures. This finding agrees with those of some earlier studies that clade D often is the most abundant clade of *Symbiodinium* under high-temperature conditions (Rowan 2004). For example, Oliver and Palumbi (2011) and Palumbi and colleagues (2014) found higher levels of clade D in their more heat-exposed (HV site) population of *A. hyacinthus* than in corals from the cooler habitat (MV site). However, Cunning and colleagues found that not all levels of heat stress of *O. faveolata* led to a preference for clade D. Under moderate heat stress, *O. faveolata* individuals tended not to enlarge the proportion of symbionts belonging to clade D, and under relatively low heat stress and low recovery temperatures, the proportion of clade B increased. Therefore, expansion of clade D seems to depend strongly on the severity of thermal stress; it's the right symbiont for the job under extreme heating, but it appears to be disadvantageous under less stressful temperatures. Why should this be the case?

The varied response of symbiont composition to graded heat stress reflects a trade-off between resistance to high temperature and physiological efficiency. Individuals with high percentages of clade D symbionts have lower photosynthetic performance and slower growth rates than individuals with symbionts belonging to other clades (B or C, for example). Photosynthetic performance is measured by the maximum quantum yield of photosystem II (F_v/F_m); a high value of F_v/F_m indicates high photosynthetic performance. Cunning and colleagues (2015a) found an inverse relationship between F_v/F_m and the proportion of clade D symbionts in *O. faveolata*. In the absence of thermal stress, therefore, it is likely to be to the coral's advantage to maintain a high percentage of B clade symbionts. Conversely, under severe heat stress, clade D seemed to protect individuals from bleaching; bleaching severity was inversely proportional to the fraction of clade D symbionts. A further complexity is that symbiont density in the host also affects propensity for bleaching, perhaps because higher symbiont densities can lead to greater damage from ROS when photosynthesis rates are high (Cunning et al. 2015a,b).

The apparent trade-off between resistance to high temperature (avoidance of bleaching) and physiological performance (maximizing F_v/F_m) that is associated with shuffling of clade B and clade D symbionts should remind the reader of the trade-offs between protein thermal stability and rates of protein function (see Chapter 3). Being "tougher" can mean working less efficiently; not all aspects of a physiological or biochemical system can be optimized simultaneously. The temperature responses of corals illustrate how this relationship plays out at a higher level of biological organization, that of an entire holobiont.

In conclusion, adaptation of corals to high temperatures involves changes in the cnidarian host and in the dinoflagellate symbiont. In the host, acclimatization and genetic adaptation are both important. Complementing these changes are alterations in the relative abundances of the different clades of zooxanthellae, which have different thermal tolerances, but also different photosynthetic efficiencies. Further experimental and modeling studies are needed to flesh out the ecological consequences of the different adaptive strategies of corals (see Cunning et al. 2015b). Of particular interest will be the net effect on reef growth caused by rising temperatures. Temperature-driven shifts in clade structure—for example, a shift from clade B to clade D under heat stress—may lead to slower growth. However, if Q_{10} effects are included in the models, then the net effect of rising temperatures on growth may be somewhat less marked than predicted strictly on the basis of symbiont shuffling. Also needed is a detailed analysis of how light availability will affect coral reefs under global warming (Muir et al. 2015). Sufficient levels of photosynthetically active radiation (PAR) are required throughout the year to support reef growth. In equatorial habitats, sufficient PAR is available year-round down to depths of 30–35 m (Muir et al. 2015). The depth at which sufficient PAR is present year-round becomes shallower with increasing latitude; with each 1-degree rise in latitude, this depth decreases by about 0.6 m. Because of these depth versus PAR relationships, the limited solar radiation in winter at higher latitudes could significantly constrain the latitudinal range extension of reef-building corals as ocean temperatures increase. At higher latitudes, hermatypic corals might be confined to depths that are too shallow to permit reef growth. Therefore, having the ability to acclimatize, genetically adapt, and shuffle symbionts in response to temperature change may provide some degree of resilience in tropical waters, but corals likely will be limited in their abilities to migrate to higher-latitude waters as sea surface temperatures rise.

DNA decay in highly stable environments: Antarctic notothenioid fishes

The notothenioid fishes of the Southern Ocean have taught—and continue to teach—important lessons about evolution in the face of environmental extremes. Earlier in this book (see Chapters 1 and 3) we discussed how additions to the genomes of these fishes, largely through extensive tandem gene duplications, facilitated their adaptation to extreme cold (Deng et al. 2010). Here we look at the flip side of the coin: the loss of genes during evolution in stable environments.

As we discussed earlier (see p. 473), genes that are not needed tend to be lost from the genome of a lineage—a process known as DNA decay (Harrison and Gerstein 2002). This loss can occur by several mechanisms, including point mutations that render the genes nonfunctional (even though they remain in the genome) and loss of some or all of the gene's sequence from the chromosome. Bacteria and archaea are notable for "streamlining" their genomes: When environmental conditions change, core genes are retained, but genes that are no longer needed are lost and new genes that satisfy specific environmental needs are acquired through horizontal gene transfer (see Box 3.7). The ease with which bacteria and archaea can remodel their genomes gives them a unique flexibility in adapting to environmental change (Fuhrman 2009). With the exception of unicellular species, eukaryotes generally lack this type of genetic flexibility. For eukaryotes, once genetic information is lost, it can be very challenging to regain should it again be needed when environmental conditions change.

The examples discussed below show how DNA decay may have poised notothenioid fishes to be extraordinarily vulnerable to global change. We introduced these fishes in Chapter 1, where we initially discussed stenotolerant and eurytolerant species. It was pointed out that the notothenioid fishes of the Southern Ocean are paradigmatic stenotherms based on the approximately 6- to 8-degree-Celsius range of temperatures they can tolerate (from –1.9°C, the freezing point of seawater, to 4°C–6°C, the highest temperatures to which McMurdo Sound species can be acclimated for indefinite periods of time)(Somero and DeVries 1967). During their evolutionary histories (**Figure 5.19A**), notothenioids have gained adaptations that make them fit for life at extremely cold temperatures (Beers and Jayasundara 2015; see Chapter 3). However, 10–15 million years of evolution in stably cold waters have led to critical types of DNA decay that could prove to be insurmountable barriers to survival in a warming ocean.

Perhaps the most striking—and certainly the most conspicuous—instance of DNA decay in notothenioids is the loss of DNA encoding hemoglobin (Hb) proteins in the family Channichthyidae (**Figure 5.19B,C**; Beers and Jayasundara 2015). These white-blooded icefishes have a strikingly anemic appearance because of the loss of Hb genes. In all 16 members of this family of notothenioids, the entire protein-coding region of the β-Hb gene as well as a portion of the α-Hb gene have been lost. Furthermore, another oxygen-binding protein, myoglobin (Mb), has been lost in six of these species (Borley and Sidell 2011). Whereas the loss of Hb seems to have been a single occurrence that took place

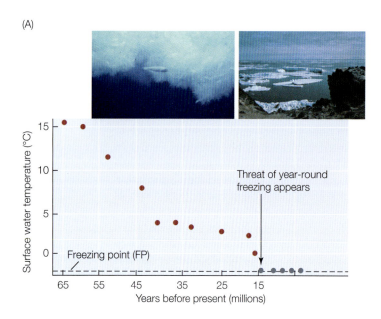

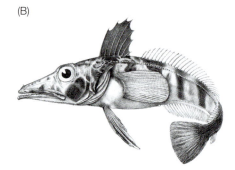

Figure 5.19 Cooling of the Southern Ocean and loss of hemoglobin in icefishes. (A) Cooling of the Southern Ocean's surface waters from 65 mya to the present. Between ~15 and 14 mya, surface water temperatures stabilized near the freezing point of seawater. (B) The channichthyid icefish *Chaenocephalus aceratus*. (C) Tubes of whole blood of hemoglobin-less icefish (*Chaenocephalus aceratus*; left) and a red-blooded notothenioid fish (right). (A, photos by G. Somero; B from Regen 1913; C, photo courtesy of Dr. Jody Beers.)

early in the evolution of channichthyids, the loss of Mb occurred on four different occasions during the subsequent radiation of the group. The environmental factor that may have been permissive of these DNA decay events is the high level of dissolved oxygen in Southern Ocean waters. In channichthyids, enough oxygen can be carried freely in solution to meet the respiratory needs of these relatively sluggish fishes. Nonetheless, this lineage has evolved what appear to be compensatory adaptations to a reliance on blood-dissolved oxygen: The circulatory system, including blood vessel diameter and heart size, is enlarged relative to that of red-blooded notothenioids. These anatomical changes, made subsequent to the loss of Hb, may be regarded as **disaptations**—characteristics of an organism whose contributions to fitness are less favorable than those of the phylogenetically antecedent traits that were lost and subsequently replaced by current traits (Montgomery and Clements 2000).

The loss of one or both oxygen-binding pigments in the channichthyids poses threats to the survival of this lineage as water temperatures rise. The increases in oxygen consumption rate and the falling DO that co-occur with increasing temperature could prove lethal to these species. Channichthyids have lower thermal tolerance than red-blooded notothenioids, so even among this highly stenothermal suborder of fishes, the icefishes stand out as being particularly vulnerable to rising water temperatures (Beers and Sidell 2011).

Other protein-coding genes have also been lost in some notothenioid lineages. For example, hearts of channichthyids lack the mitochondrial isoform of creatine kinase, an enzyme that is important in catalyzing the transfer of a phosphate group between creatine and ADP (O'Brien et al. 2014). As more genomic analyses of notothenioids are performed, it seems probable that other protein-coding genes will be found to either be lost from the genome or rendered dysfunctional because of mutations.

Compounding the challenges arising from loss of protein-coding sequences are losses in gene regulatory abilities, which may or may not derive from losses of DNA. The first of these losses to be discovered in the notothenioid fishes was the absence of a heat-shock response in the emerald notothen (*Trematomus bernacchii*), which we introduced in Chapter 1 in the context of stenotolerant versus eurytolerant species (Hofmann et al. 2000). Fish exposed to a series of elevated temperatures showed no increase in abundance of any size class of heat-shock protein (HSP), in sharp contrast to the response of the eurythermal goby *Gillichthys mirabilis* (**Figure 5.20**), and nearly every other species studied to date. The failure of HSPs to increase in abundance following acute heat stress was traced to the transcriptional level. No increase in mRNA for HSPs was observed following acute heat stress at 4°C (Buckley and Somero 2009). Loss of the heat-shock response, commonly viewed as a "universal trait" among organisms, would put notothenioids in a challenging position if rising temperatures lead to physiologically significant increases in protein denaturation. The factors that underlie failure of the transcriptional regulation system in these fishes remain to be identified.

Further insights into the possible fate of Antarctic ectotherms in a warming ocean have been provided by studies of a taxonomically diverse set of invertebrates and fishes (Coppes-Petricorena and Somero 2007; Peck et al. 2014; Beers and Jayasundara 2015). These studies suggest that the types of phenomena observed in notothenioids are likely to reflect the challenges that other ectotherms of the Southern Ocean, especially shallow-living species, will face as temperatures continue to rise. DNA decay and loss of gene regulatory capacities, notably for the heat-shock response, during millions of years of evolution under cold and stable temperatures may have predisposed these organisms for extinction as the Southern Ocean's temperatures continue to rise.

Figure 5.20 Effects of different thermal exposures of the stenothermal Antarctic notothenioid *Trematomus bernacchii* and the eurythermal longjaw mudsucker (*Gillichthys mirabilis*) on patterns of protein synthesis. Whole fish were injected, after thermal exposure, with ^{35}S-methionine and ^{35}S-cysteine, and protein synthesis was allowed to continue in fish returned to ambient temperature, –1.9°C and 16°C, respectively. Proteins were then isolated from the fish, separated by SDS-PAGE, and exposed to X-ray film to visualize the newly synthesized proteins. Note the strong induction of several size classes of heat-shock proteins (bands enclosed in boxes) in the mudsucker and the absence of any induction in *T. bernacchii*. Note also that *T. bernacchii* dies of heat death in less than 8h at 8°C and in less than 3h at 10°C (Somero and DeVries 1967). (After Hofmann et al. 2000.)

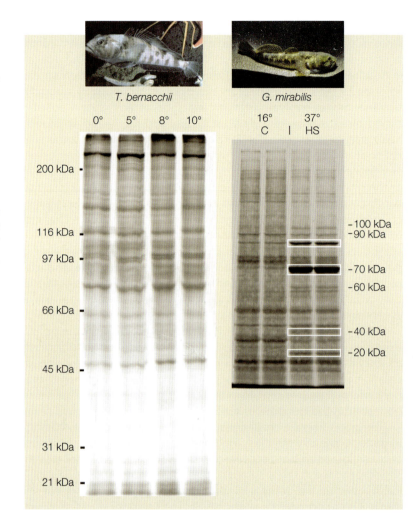

Horizontal gene transfer: The most efficient and quickest way to adapt?

After this rather pessimistic account of the future of genome-depleted cold-adapted stenotherms from the Southern Ocean, it may be reassuring to turn to the success with which a mechanism that in some ways is the opposite of DNA decay, namely **horizontal gene transfer** (**HGT**), can increase the amount of genetic information in a species' genome (for a concise review, see Zhaxybayeva and Doolittle 2011). In essence, HGT (referred to by some authors as lateral gene transfer, or LGT) involves exchange of genetic information across species lines, that is, across different vertical evolutionary lineages. As discussed briefly in Chapter 1 and again in some detail in Chapter 3, HGT is of vast importance in bacteria and archaea (López-García et al. 2015). In these two domains, adaptation is commonly achieved through HGT from other lineages, whereas in Eukarya, evolutionary innovations usually arise as a consequence of gene duplication and subsequent gain of new functions, which leads to expansion of existing families of genes (Innan and Kondrashov 2010). A major barrier to HGT in animals is the sequestration of the germline, which usually prevents any DNA acquired by somatic cells from being passed on to the next generation. However, this barrier is not insurmountable, and examples of HGT involving multicellular eukaryotes, including animals, are being discovered (see below).

The major contribution of HGT to eukaryotic evolution involved acquisition of symbionts that subsequently evolved into organelles (chloroplasts and mitochondria). These "wholesale" HGT events essentially involved acquiring another species' entire genome. Once the symbiosis was established, large amounts of **endosymbiotic gene transfer (EGT)** took place, leading to genetically depleted mitochondrial or plastid genomes (Keeling and Palmer 2008). In animal mitochondria, EGT was completed in the distant past; no evidence for any further EGT has been found over ~600 million years of animal evolution (Keeling and Palmer 2008).

Smaller-scale HGT of discrete pieces of genetic information also occurs in eukaryotes (reviewed in Keeling and Palmer 2008). Many examples come from protists, which are uniquely positioned among eukaryotes for HGT because DNA taken into the cell is in close contact with the nuclear genome. For example, dinoflagellate protists acquired genes for the light-harvesting protein proteorhodopsin from bacteria (Slamovits et al. 2011). HGT between eukaryotes, including animals, also can occur. Aphids have acquired from a fungus the genes needed for biosynthesis of carotenoids (Moran and Jarvik 2010). The potential for eukaryotes to adapt to abiotic factors via HGT is perhaps shown most dramatically by an extremophilic red alga that has gained genes from both bacterial and archaeal sources (see below; Schönknecht et al. 2013). These "adopted" genes provide this small eukaryote with remarkable tolerance of physical and chemical factors: It may be the "toughest" eukaryote we know of!

BUILDING A THREE-DOMAIN GENOME: A EUKARYOTIC EXTREMOPHILE RELIANT ON GENES SOURCED FROM BACTERIA AND ARCHAEA   The acquisition of genes from mesophilic bacteria by hyperthermophilic archaea (see Chapter 3) provides an excellent example of how an extremophilic group can evolve to function optimally in more mesic environments. Can the opposite process occur as well? Can a mesophilic species become an extremophile by accumulating genes from other species adapted to extremes of temperature, salinity, or acidity? We address this question by examining an extremophilic eukaryotic red alga, *Galdieria sulphuraria*, a member of the Cyanidiophyceae (see photos at right; Schönknecht et al. 2013). Members of the Cyanidiophyceae are among the most extremophilic eukaryotes known. They can grow at temperatures up to 56°C, which is near the upper temperature limit for eukaryotes (see Chapter 3), and at pH values of 0–4; they are highly tolerant of extreme salinities as well. *G. sulphuraria* occurs in hot volcanic springs where the abundant sulfides may be oxidized to sulfur (yellow color in lower photo). It is also found in sites where toxic metals are present in high concentrations due to either natural or anthropogenic activities. In environments that are rich in toxic metals like cadmium and mercury, *G. sulphuraria* can constitute up to 90% of total biomass and essentially all of the eukaryotic biomass. It is a eukaryote that can outcompete extremophilic bacteria and archaea, yet

G. sulphuraria

its success is based on past acquisition of a broad suite of genes from extremophilic members of those two domains. What, then, has *G. sulphuraria* acquired from HGT to make it such a remarkable extremophile?

Schönknecht and colleagues performed a highly detailed analysis of the genome of *G. sulphuraria* and discovered that at least 75 separate gene transfers had occurred from archaea and bacteria. These HGT events involved acquisition of genes from a wide variety of species, notably extremophilic bacteria. The genes that entered the *G. sulphuraria* genome have conferred tolerance of high temperatures, high salinities, and low pH values. The latter class of adaptations includes large alterations in the types of ion channels found in the cell membrane. Voltage-gated ion channels generally have a high conductance for protons, which could pose challenges for existence in environments with extremely low pH values. In the cell membrane of *G. sulphuraria*, proton-conducting ion channels are replaced by ion carriers, gained through HGT, that do not conduct protons. This assists the cell in maintaining a near-neutral intracellular pH—that is, the pH value expected on the basis of alphastat relationships (see Chapter 4)—in the alga's highly acidic environment. The alga's tolerance of extremely high salinities is enhanced by the acquisition of genes, probably from halophilic cyanobacteria, that encode proteins of the pathway that produces glycine betaine (see Chapter 4). This compatible osmolyte accumulates to high concentrations when *G. sulphuraria* grows at high salinities. Tolerance of high temperature is conjectured to be based in part on acquisition of archaeal genes encoding adenosine triphosphatases (ATPases). An ATPase-coding gene acquired from an archaeal source has undergone duplication and subsequent diversification into separate gene families. Here, then, HGT is followed by a canonical eukaryotic mechanism for generating biochemical diversity. The function of the archaeal ATPases is not known, but the number of gene copies for these enzymes in thermophilic and hyperthermophilic archaea correlates strongly with the optimal temperature for growth (**Figure 5.21**). Last, the capacities of *G. sulphuraria* to metabolize a wide range of organic substrates and to function in the presence of high concentrations of toxic metals also appear to be consequences of the acquisition of suites of genes from bacteria and archaea. It is appropriate, therefore, to refer to the genome of *G. sulphuraria* as a **pangenome**, an assemblage of genes from all three domains of life that allows a (mostly) eukaryotic species to thrive in environments where virtually all other eukaryotes would perish. This definition differs from the concept of pangenome proposed by Goldenfeld and Woese (2007), which focuses on the ease of exchange of genetic material among lineages (see Box 3.7). Nonetheless, both perspectives on the pangenome emphasize the possibilities for evolutionary novelty that can arise when a genome is expanded by information entering from HGT.

In summary, HGT is seen as an extraordinarily effective means for acquiring the genetic information needed to adapt to changing abiotic conditions. It is a mechanism for adaptation that is especially accessible to bacteria, archaea, and unicellular eukaryotes, but examples from the animal kingdom suggest a wider exploitation of this process. Differences among taxa in the ability

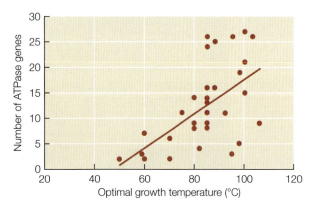

Figure 5.21 The copy number of archaeal ATPase genes found in the genomes of heat-tolerant archaea is correlated with heat tolerance (optimal temperature for growth). (After Schönknecht et al. 2013.)

to acquire new genetic information through HGT may be an important determinant of the relative abilities of different types of organisms to cope with the changes occurring in the Anthropocene.

5.8 Unique Challenges of Global Change to Endothermic Homeotherms, Including Humans

Mammals—and humans in particular—provide an interesting platform for discussing two critical features of adaptation to global change. First, in terms of genetic capacities for adaptation to a changing world, mammals—and large species in particular—may face serious limitations as a consequence of their life histories, especially their reproductive strategies. Mammals often have long generation times, notably in the case of larger species, and may take decades to reach reproductive maturity. Larger species also have relatively few offspring. Mammals generally exhibit what is termed K selection. In essence, relatively few offspring are produced and these often receive considerable parental assistance throughout their maturation. In contrast, r-selected species are prolific in reproduction. They may generate thousands or millions of offspring whose survival rate is extremely low. However, among this enormous number of offspring there may exist the genetic variation needed to adapt to changes in temperature, pH, oxygen availability, and so forth. In the examples given above, we focused on marine organisms that are r-selected species. In some cases—for example, sea urchins and tide-pool copepods—these genetically variable and highly fecund species seem relatively well poised to respond adaptively to environmental change. In contrast, K-selected large mammals seem less likely to be able to adapt biochemically or physiologically to global change. History seems to support this conjecture about the relative capabilities of r- and K-selected species to adapt to rapid climate change. During the Paleocene-Eocene thermal maximum (PETM), which we discuss later in this section, large K-selected foraminiferan species were replaced by small, rapidly reproducing r-selected species (Gingerich 2006).

Limits to evaporative cooling: The importance of wet bulb temperature

A second characteristic of mammals that makes them a very different study system from the ectothermic species we have been focusing on is their endothermic homeothermy. Whereas ectotherms may be forced to cope with direct effects of temperature changes on their biochemistry, rising global temperatures will affect mammals (and birds) in a different, less direct way. As discussed in Chapter 3, endothermic homeotherms have the ability to generate and release heat in a highly controlled manner, which enables them to conserve a stable body temperature over a wide range of ambient temperatures. One of the physical mechanisms on which endothermic homeothermy is founded is the **heat of vaporization** of water (see Chapter 3). A large amount of heat is required to change water from the liquid to the gaseous state; this is the heat that is exhausted to the environment when perspiration or water in moist nasal passages evaporates. A major challenge that mammals and birds face from rising temperatures concerns retention of the ability to cool evaporatively; this becomes more and more difficult as the

humidity of air increases along with temperature. One often hears the complaint, especially in warm climates in summer, "It isn't the heat, it's the humidity."

In fact, it's both. As we show below, limitations in evaporative cooling set in when temperatures exceed what is termed the **wet bulb temperature** (T_W). This temperature is the equilibrium temperature that is reached when a conventional glass thermometer bulb is wrapped in a moist piece of cloth and subjected to steady airflow. In other words, this is the temperature to which an object may be cooled by ventilation and evaporation. Net conductive and evaporative cooling is only possible if the organism's surface temperature is higher than the environmental T_W. The second law of thermodynamics prohibits loss of heat from an object to an environment whose T_W exceeds the temperature of the object. Were environmental T_W values to reach the temperature at which evaporative cooling is no longer possible, then global warming would be likely to prevent mammals—including humans—from inhabiting low-latitude regions where, currently, a substantial fraction of the human population is found. Birds too would face challenges, but their slightly higher body temperatures may give them a bit of additional leeway compared with mammals.

Most analyses of T_W relationships have focused on mammals, and the discussion below will similarly center on this one group of endothermic homeotherms. We will consider two primary questions in our analysis. Where do warm regions of the globe now stand in terms of proximity of maximum T_W values ($T_{W(max)}$) to the upper temperatures at which mammals can cool evaporatively? And what does the future portend for the biogeography of mammals, humans and otherwise, as both temperature and humidity levels increase?

The need for a long-term perspective

As a backdrop for this analysis, it is important to emphasize a critical point about global warming that has often been neglected in predictions about the future state of the biosphere, including human societies. Whereas the adverse effects of global change on humanity have often discussed such critical challenges as reductions in crop yields, shortfalls in supplies of freshwater, extremes in weather and their follow-on effects on agriculture and infrastructure, spread of infectious agents, rising sea levels and displacement of coastal populations, there has often been a failure to think beyond the end of the present century (Stager 2011). This focus on relatively near-term effects leads to a sense that, if emissions can be reduced strongly within the next few decades, we may be well on our way to stopping global warming. In fact, even if emissions of greenhouse gases are largely eliminated by the mid- to late twenty-first century, warming will continue for many centuries into the future due to greenhouse gases already in the atmosphere. It is likely that temperatures will continue to rise over at least the next few centuries, and increases in mean global temperature of 10–12 degrees Celsius are not out of the question. This is the range of increase if all fossil fuel deposits are used (McMichael and Dear 2010). As mean global temperatures rise—highest temperatures now are ~60°C—so will the maximum wet bulb temperature. As we show below, $T_{W(max)}$ will be instrumental in determining where humans and other mammals can survive. To project effects of global change on human populations, then, we need a multicentury perspective that encompasses both broad ecological issues (rising sea levels, falling crop yields, etc.) and physiological limits established by mammalian thermoregulatory mechanisms.

Wet bulb temperatures and human thermal limits: A status report

Limitations due to thermoregulatory requirements of mammals are well understood and provide a firm foundation for developing predictions about the consequences of rising heat and humidity. Under conditions of a 6-h exposure time, the highest T_W that humans can tolerate is ~35°C. Six hours is a realistic time of exposure to heat that might readily occur on an exceedingly hot day, especially for laborers working outdoors. This T_W barrier exists even under ideal conditions for cooling such as shade, good ventilation, and a minimal rate of heat production by the organism (Sherwood and Huber 2010). Human skin temperatures are closely regulated to values slightly below 35°C, to ensure that an adequate temperature gradient is present to favor conduction of heat from the warmer body core to the environment. When human core temperatures reach 42°C–43°C, death may occur unless cooling of the core can be done quickly. The average $T_{W(max)}$ near Earth's surface currently is 26°C–27°C and tends to be surprisingly similar across wide ranges of latitude due to the interacting effects of temperature and humidity. Currently, the hottest regions of the planet, low-latitude deserts, also are among the driest, which tends to keep $T_{W(max)}$ of these areas within mammalian tolerance limits. Thus, even though air and surface temperatures can exceed 40°C in desert regions, the low humidity associated with these temperatures keeps $T_{W(max)}$ to values of 31°C or less, and cooling from the 35°C skin surface is possible. Currently, then, humans in almost all regions (but see below) have, at worst, a ~4-degree-Celsius safety zone in terms of the rise in $T_{W(max)}$ that can be tolerated. Over a broader range of environments that are cooler and/or less humid, the safety margin is closer to 8–9 degrees Celsius.

As global temperatures continue to rise, however, many warm, low-latitude regions with conditions of high humidity will, on occasion at least, expose mammals to $T_{W(max)}$ values that surpass tolerance limits (Sherwood and Huber 2010). Increases in mean global temperature will be accompanied by increases in T_W at a ratio of approximately 4 degrees Celsius to 3 degrees Celsius (McMichael and Dear 2010). That is, if mean global air temperature rises by 4 degrees Celsius—a value well within the predicted range of warming for the next few centuries—T_W will go up by ~3 degrees Celsius. If global warming leads to a 10- to 12-degree-Celsius rise in temperature over the next several centuries—a range that is within the span of extreme predictions of some climate change models—then T_W may rise by nearly 9 degrees Celsius. What would such a spike in temperature and humidity mean for human biogeography?

Global change and human biogeography: Predictions for a hot and humid future

Sherwood and Huber (2010) have analyzed this issue in a detailed modeling exercise. They began by looking at the rise in temperature caused by doubling of CO_2 levels. Estimates for the upper limits of temperature increase vary between models and the assumptions used to create the models (e.g., the rate of fossil fuel combustion, fraction of total fossil fuel reserves that will be exploited, and ability of the oceans to absorb atmospheric CO_2). However, the rise in temperature associated with doubling of atmospheric CO_2 levels can reasonably be said to lie between 1.9 and 4.5 degrees Celsius per doubling. If all fossil fuels were burned, 2.75 doublings of atmospheric CO_2 would result. If the 4.5-degree-Celsius increase in temperature per doubling is used to predict future temperatures, then a 12-degree-Celsius rise could potentially occur as a result of fossil fuel combustion.

This estimate does not include warming that could arise from other greenhouse gas releases, such as degassing of methane from the tundra or the seafloor. The latter site contains the bulk of the planet's methane. Recall that on a per-molecule basis, methane is 24 times as potent a greenhouse gas as CO_2. Thus, Sherwood and Huber's estimate of potential warming is somewhat conservative. If average global temperature rises by 7 degrees Celsius—about one-half the maximum increase predicted to result in the case of complete exploitation of fossil fuel reserves—some low-latitude regions will experience $T_{W(max)}$ values that occasionally rise to 35°C or more. However, if warming should reach 10–12 degrees Celsius, then much of the area currently inhabited by humans would see $T_{W(max)}$ values reaching or exceeding the lethal level of 35°C. Sherwood and Huber put the biogeographic effects of this change into perspective by emphasizing that the area of land that would be rendered uninhabitable by rising $T_{W(max)}$ currently houses ~50% of the human population; the effect of $T_{W(max)}$ on human distribution would greatly exceed that caused by the loss of habitable land to rising sea level. Although climate change models differ in their projected rates of global warming, an increase in average global temperature of 10 degrees Celsius could occur by the year 2300 (McMichael and Dear 2010).

Is the future close upon us?

Although a long-term, multicentury perspective on effects of global change is critically important, it is also wise to keep an eye on the short term. Specifically, regional differences in temperature increases may identify areas where effects on human persistence may be felt in the near future, perhaps within the next few decades. For example, Pal and Eltahir (2016) used a high-resolution regional climate model to examine the effects of rising heat and humidity in southwest Asia. The Arabian Gulf (also called the Persian Gulf)/Arabian Peninsula (**Figure 5.22A**) is, literally, a hot spot of threat to human habitation. It may face the most extreme threats from rising temperatures because a rise in $T_{W(max)}$ to values above 35°C may occur there by the end of this century. Pal and Eltahir used two greenhouse gas accumulation scenarios to predict $T_{W(max)}$ out to the year 2100. The Intergovernmental Panel on Climate Change (IPCC) Representative Concentration Pathway (RCP) trajectories assumed either a "business as usual" increase in greenhouse gases (RCP8.5) or a scenario in which mitigation of greenhouse gas release is achieved (RCP4.5). A 6-h exposure of humans to a $T_{W(max)}$ of 35°C or higher was assumed to lead to death. Under the RCP8.5 trajectory of greenhouse gas accumulation, $T_{W(max)}$ exceeds the 35°C threshold during the hottest days of the year in several Arabian Gulf regions by 2100. Maximum temperatures in many regions also increase sharply, reaching values in excess of 60°C in summer. Typical summer (June, July, and August) temperatures exceeding 45°C may become common in low-lying cities near the Arabian Gulf. Two cities with especially high temperatures and $T_{W(max)}$ values are Dhahran, Saudi Arabia, and Abu Dhabi, UAE (**Figure 5.22B**). Pal and Eltahir emphasize that any outdoor activities are likely to be severely affected—and perhaps precluded—when $T_{W(max)}$ crosses the 35°C threshold.

The time frame for these changes in temperature and $T_{W(max)}$ represents a departure from the broader analysis of Sherwood and Huber (2010). The predictions for the Arabian Gulf and Peninsula region show lethal temperatures occurring by the end of this century. This estimate is likely to be realistic, in view of the extreme heat waves that have recently occurred in this region. For example, on 31 July 2015, T_W reached 34.6°C in one part of this region (Schär 2015). With ongoing global warming, it is not hard to imagine values of $T_{W(max)}$ greater than 35°C in the next few decades.

Once the $T_{W(max)}$ threshold of 35°C is crossed, the effects of heat waves change radically. When $T_{W(max)}$ values lie below 35°C, heat waves tend to cause death chiefly among old, ill, or ill-prepared individuals, for example, those who fail to move indoors to a cooler environment. Once $T_{W(max)}$ rises above 35°C, all individuals facing exposure outdoors or in uncooled buildings will be in danger of heat death, regardless of age or state of health. In addition to rising death rates from heat, there will be wide-ranging effects on economies and demographics. Mammals used for transport or food will be threatened by heat death. The economic costs arising from reduced outdoor labor, the effects on human migration to cooler regions, and the energy expenditures needed to keep indoor

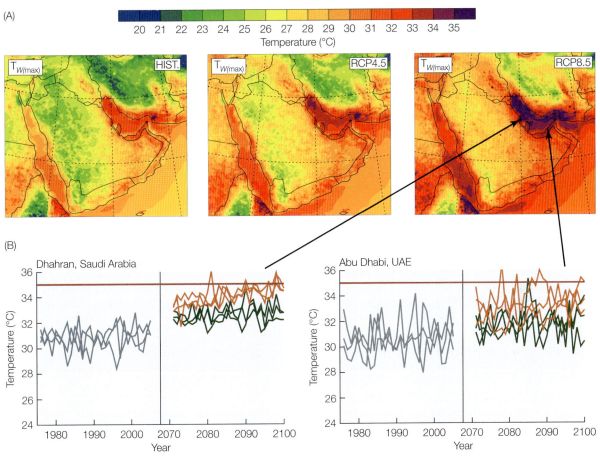

Figure 5.22 Increases in maximum wet bulb temperature ($T_{W(max)}$) in the region of the Arabian Peninsula and Arabian Gulf up to the year 2100. (A) Trends in $T_{W(max)}$: left panel, historic data (1975–2005); middle panel, projected $T_{W(max)}$ under RCP4.5, an IPCC scenario that involves mitigation; right panel, projected $T_{W(max)}$ under RCP8.5, the "business as usual" scenario for increases in greenhouse emissions. Note how the end-of-century (2071–2100) rise in $T_{W(max)}$ is ameliorated by the mitigation scenario. (B) Time series of the annual $T_{W(max)}$ for different greenhouse gas scenarios at two sites, Abu Dhabi, UAE, and Dhahran, Saudi Arabia (see arrows). Blue, green, and red tracings represent historical data (1975–2005), RCP4.5 (2071–2100), and RCP8.5 (2071–2100), respectively. $T_{W(max)}$ is the maximum daily value averaged over 6 h, an estimate of the time of lethal exposure of humans to T_W of 35°C or higher (horizontal red line). (After Pal and Eltahir 2016.)

spaces tolerable are all likely to be large. The implications of Sherwood and Huber's (2010) conclusion about the large fraction of low-latitude regions where human habitation is likely to be precluded are profound. The war-driven movements of humans seen in the Middle East during the second decade of this century may pale when compared to the flight from intolerable climate conditions that are likely to occur by century's end.

The specific effects of global warming will vary among other endothermic homeotherms, many of which face lower challenges than humans from $T_{W(max)}$ effects because of, for example, higher core body temperatures or surface area-to-mass ratios that are more favorable for removal of metabolic heat. It is clear, however, that mammals and birds, as groups, will exhibit large changes in biogeographic ranges because of rising heat and humidity unless they can adapt within evolutionarily minute periods of time.

The Paleocene-Eocene thermal maximum: Another "distant mirror"

What are the odds that rapid evolutionary change during the Anthropocene can alter the upper thermal tolerance limits of endothermic homeotherms through changes in their behavior, morphology, or biochemical systems? Is there any historical precedent for this type of adaptation? Although the Permian-Triassic extinction event is a good "distant mirror" for examining many aspects of global change, this 252-mya event cannot speak directly to the effects of global change on mammals and birds. A more relevant "mirror" is provided by the Paleocene-Eocene thermal maximum (PETM), which occurred ~55.5–50 mya and lasted ~120,000 years (Röhl et al. 2000; Gingerich 2006). During the Eocene there were several hyperthermal periods of comparable duration (lasting up to ~200,000 years), but the PETM appears to have had the greatest spike in temperature. Mean mid-latitude terrestrial temperatures are estimated to have increased by ~4–5 degrees Celsius over a period of ~20,000 years. High temperatures persisted for another 100,000 or so years. Polar ice caps disappeared—and will disappear again if mean global temperatures increase by 4–5 degrees Celsius. The cause of the rapid rise in temperature during the Eocene remains to be fully resolved, but it likely involved an initial release of massive amounts of CO_2 from structures known as kimberlite pipes, which form deep in the mantle and conduct large amounts of CO_2 to the surface. The CO_2 release caused warming of the shallow ocean, which led to heat-driven release of seabed methane (Gingerich 2006). This is the type of effect that can lead to a runaway greenhouse effect.

During this period of unusually high temperatures, it is estimated that average $T_{W(max)}$ values may have been as high as 32°C–33°C and spikes in $T_{W(max)}$ may have reached 36°C (Sherwood and Huber 2010). Radiation of mammalian groups was high during this period, and it seems likely that high $T_{W(max)}$ values per se did not have a broadly inhibitory effect on mammalian evolution. Nonetheless, high temperatures may have literally "shaped" the evolution of mammals during the PETM. For example, during this period the average body mass of mid-latitude mammals decreased ("transient dwarfing"; Gingerich 2006). When global temperatures subsequently fell, larger mammals appeared. Researchers have conjectured that this transient dwarfing was a response to changes in food supply, but the physics of heat transfer support an alternative hypothesis that dwarfing—and its attendant increase in surface area-to-mass ratio—could have been selected on thermoregulatory grounds. Regardless of the selective factors driving the shift in body size during the PETM, the change occurred slowly over an extremely long period of time—many tens of thousands of years—relative to the time frame relevant for adaptation to anthropogenic global change. If anthropogenic climate change continues at its current pace, values of $T_{W(max)}$ could reach levels that are lethal for large mammals within a few centuries, and possibly before the end of this century in some regions. Under these circumstances, it seems impossible that these species will have time to adapt—either by becoming smaller or by modifying all of the biochemical systems we discussed in Chapter 3—in order to increase their thermal optima and tolerance limits. The most likely species to continue to thrive in regions with high $T_{W(max)}$ will be smaller species that have the greatest potentials for cooling and, in many cases, natural histories that allow them to avoid exposure to the highest temperature-humidity combinations found in their environments. Fossorial species such as small burrowing mammals seem much more likely to weather the Anthropocene than large mammals, including the one species—ourselves—whose activities are driving global change.

5.9 Where Do We Go from Here?

History is a vast early warning system.

Norman Cousins (1915–1990)

The primary purpose of this volume has been to present a two-fold analysis of the interactions between biochemical systems and the diverse abiotic environments in which organisms function and evolve. First we examined how changes in different physical and chemical factors affect biochemical systems, commonly leading to stressful effects that force these systems into suboptimal states of structure and function. These sensitivities of biochemical systems to environmental stress were shown to be an inevitable consequence of a basic principle of biochemical structure-function relationships: They stem from the requirement that these systems remain in a state of marginal stability, in order to allow the changes in structure that are needed for carrying out their diverse functions.

Second, based on our analysis of the structural underpinnings of sensitivity to the environment, we asked how can biochemical systems "set things right" when experiencing stress that negatively affects their performances? We examined the types of molecular-level changes that are of central importance in adaptive responses, and whether the response occurs during the organism's lifetime (acclimatization) or over multiple generations (evolutionary adaptation). We reviewed the several ways in which genetic variation contributes to adaptation and showed how widely organisms differ in their capacities for adapting to environmental stress. This mechanistic analysis of environmental relationships laid the groundwork for discussing how environmental change leads to effects at higher levels of biological organization, including biogeographic patterning, ecosystem structure, and extinction events.

In the final chapter of this book we have attempted to use the mechanistic principles developed in earlier chapters to address critical issues that face organisms in the Anthropocene. The magnitude and rapidity of changes in the abiotic environment that are caused by human activities are presenting unprecedented challenges to life. In large measure, these challenges arise from the *speed* with which the world is changing as a result of the release of enormous amounts of greenhouse gases. The biological consequences of greenhouse gas releases are diverse and differ among species and ecosystems. Several factors are responsible for the differential effects seen among different species; species differ both in the threats posed to them by climate change and in their potential for adaptive change. The proximity of a species' environmental tolerance limits to current peak values of abiotic conditions, notably temperature, may determine the species' susceptibility to extinction. The amount of genetic information a species has that can support adaptation, and the facility with which this information can be exploited, assembled, exchanged, and increased, plays a key role in setting the species' capacity to adapt. Species are also challenged by the as yet poorly understood interactions that occur among different stressors. Differences among species in susceptibility to the effects of global change will inevitably alter ecosystem structure and function, but it remains difficult to predict what these changes will be—other than that they will be widespread and of great consequence for the biosphere.

To finish, we have brought our analysis "close to home" by examining the threats posed by global change to humans. The endothermic homeothermy that has enabled mammals like ourselves to be so successful in diverse environmental circumstances may also prove to be our undoing, at least in hot, humid, low-latitude regions of the planet—areas in which a high fraction of the human population lives. The near-term threats to human persistence in low-latitude habitats are, of course, only part of the ominous story of how global change is likely to affect our species. These threats are vitally important, however, because they illustrate very clearly how basic biochemical and biophysical principles represent insurmountable barriers to survival. We cannot thwart the basic constraints presented, for example, by the wet bulb temperature relationships that govern heat transfer. Foolhardy beliefs that "engineering will rescue us somehow" belie the fact that, as successful as engineering might be in many contexts, it is still bound by the basic laws of physics. Thus, our closing discussion of the effects of global change on human physiology, demography, and economics should be seen as a very sobering analysis. There is a key, literally life-threatening issue facing us, and it is an issue that seems minimally solvable by *biological* adaptation. Adaptation in the Anthropocene instead must involve adaptive changes in *philosophy* and *behavior*. We need to continue to engender in our society the sense of urgency to gain control of the "natural experiment" that Roger Revelle spoke of many years ago.

It is our hope that we are moving in the right direction on this front, because the majority of citizens in most countries do not agree with the naïve and misleading statements of global change deniers. Rather, there is a growing awareness that human-created changes in the environment pose great threats not only to the biosphere at large, but especially to the species—us—that is responsible for these changes. Indeed, global climate change will pose serious threats to human physiology and, more broadly, to the ecosystems we depend on for services like provision of food and of water. Past mass extinction events should serve as "early warning systems" about what happens when major changes occur in the abiotic environment. Past extinctions were, of course, due to "natural causes." The difference now, in the Anthropocene, is that we as a species may be able to control the outcome of ongoing, rapid global change, and perhaps reverse it, by changing our behavior. Is it not our responsibility to do so, in order to reduce the extent of further degradation of the biosphere and to help ensure the preservation of our own species?

Literature Cited

Chapter 1

Abele, D., J. P. Vázques-Medina, and T. Zenteno-Savin. 2012. *Oxidative Stress in Aquatic Ecosystems*. Wiley-Blackwell, Oxford, UK.

Baek, D., J. Villen, C. Shin, F. D. Camargo, S. P Gygi, and D. P. Bartel. 2008. The impact of microRNAs on protein output. *Nature* 433: 769–773.

Beers, J., and N. Jayasundara. 2015. Antarctic notothenioid fish: what are the future consequences of "losses" and "gains" acquired during long-term evolution at cold and stable temperatures? *J. Exp. Biol.* 218: 1834–1845.

Buckley, B. A., and G. N. Somero. 2009. cDNA microarray analysis reveals the capacity of the cold-adapted Antarctic fish *Trematomus bernacchii* to alter gene expression in response to heat stress. *Polar Biol.* 32: 403–415.

Burggren, W. W. 2014. Epigenetics as a source of variation in comparative animal physiology—or—Lamarck is lookin' pretty good these days. *J. Exp. Biol.* 217: 682–689.

Burggren, W. W. 2015. Dynamics of epigenetic phenomena: intergenerational and intragenerational phenotypic 'washout.' *J. Exp. Biol.* 218: 80–87.

Checkley, D. M., Jr., A. G. Dickson, M. Takahashi, J. A. Radich, N. Eisenkolb, and R. Asch. 2009. Elevated CO_2 enhances otolith growth in young fish. *Science* 324: 1683.

Chen, Z., C.-H. C. Cheng, J. Zhang, L. Cao, L. Chen, L. Zhou, Y. Jin, H. Ye, and 7 others. 2008. Transcriptomic and genomic evolution under constant cold in Antarctic fish. *Proc. Natl. Acad. Sci. USA* 105: 12944–12949.

Chevalier, C., O. Stojanović, D. J. Colin, N. Suarez-Zamorano, V. Tarallo, C. Veyrat-Durebex, D. Rigo, S. Fabbiano, and 7 others. 2015. Gut microbiota orchestrates energy homeostasis during cold. *Cell* 163: 1360–1374.

Dahlhoff, E. P., and N. E. Rank. 2000. Functional and physiological consequences of genetic variation at phosphoglucose isomerase: Heat shock protein expression is related to enzyme genotype in a montane beetle. *Proc. Natl. Acad. Sci. USA* 97: 10056–10061.

Dennett, D. C. 1995. *Darwin's Dangerous Idea: Evolution and the Meanings of Life*. Touchstone, New York.

Dowd, W. W., C. A. Felton, H. M. Heymann, L. E. Kost, and G. N. Somero. 2013. Food availability, more than body temperature, drives correlated shifts in ATP-generating and antioxidant enzyme capacities in a population of intertidal mussels (*Mytilus californianus*). *J. Exp. Mar. Biol. Ecol.* 449: 171–185.

Duman, J. G. 2015. Animal ice-binding (antifreeze) proteins and glycolipids: an overview with emphasis on physiological function. *J. Exp. Biol.* 218: 1846–1855.

Eanes, W. F. 1999. Analysis of selection on enzyme polymorphisms. *Annu. Rev. Ecol. Evol. System.* 30: 301–326.

Eastman, J. T. 1993. *Antarctic Fish Biology: Evolution in a Unique Environment*. Academic Press, San Diego.

Garrett, S. C., and J. J. C. Rosenthal. 2012. A role of A-to-I RNA editing in temperature adaptation. *Physiology* 27: 362–369.

Gleason, L. U., and R. S. Burton. 2015. RNA-seq reveals regional differences in transcriptome response to heat stress in the marine snail *Chlorostoma funebralis*. *Mol. Ecol.* 24: 610–627.

Goldenfeld, N., and C. Woese. 2007. Biology's next revolution. *Nature* 445: 369.

Gould, S. J., and R. C. Lewontin. 1979. The spandrels of San Marco and the Panglossian paradigm: a critique of the adaptationist programme. *Proc. R. Soc. B.* 205: 581–598.

Gracey, A., J. Troll, and G. N. Somero. 2001. Hypoxia-induced gene expression profiling in the euryoxic fish *Gillichthys mirabilis*. *Proc. Natl. Acad. Sci. USA* 98: 1993–1998.

Gusev, O., Y. Suetsugu, R. Cornette, T. Kawashima, M. D. Logacheva, A. S. Kondrashov, A. A. Penin, R. Hatanaka, and 20 others. 2014. Comparative genome sequencing reveals genomic signature of extreme desiccation tolerance in the anhydrobiotic midge. *Nat. Commun.* 5: 4784.

Helmuth, B., C. D. G. Harley, P. M. Halpin, M. O'Donnell, G. E. Hofmann, and C. Blanchette. 2002. Climate change and latitudinal patterns of intertidal thermal stress. *Science* 298: 1015–1017.

Ho, D. H. and W. W. Burggren. 2012. Parental hypoxic exposure confers offspring hypoxia resistance in zebrafish (*Danio rerio*). *J. Exp. Biol.* 215: 4208–4216.

Hochachka, P. W., and G. N. Somero. 2002. *Biochemical Adaptation: Mechanism and Process in Physiological Evolution*. Oxford University Press, New York.

Hofmann, G. E., B. A. Buckley, S. Airaksinen, J. E. Keen, and G. N. Somero. 2000. Heat-shock protein expression is absent in the Antarctic fish *Trematomus bernacchii* (Family Nototheniidae). *J. Exp. Biol.* 203: 2331–2339.

Isaksen, G. V., J. Aqvist, and B. O. Brandsdal. 2014. Protein surface softness is the origin of enzyme cold-adaptation in trypsin. *PLoS Comput. Biol.* 10: e1003813.

Jablonka, E., and G. Raz. 2009. Transgenerational epigenetic inheritance: prevalence, mechanisms, and implications for the study of heredity and evolution. *Q. Rev. Biol.* 84: 131–176.

Jayasundara, N., and G. N. Somero. 2013. Physiological plasticity of cardiorespiratory function in a eurythermal marine teleost, the longjaw mudsucker, *Gillichthys mirabilis. J. Exp. Biol.* 216: 2111–2121.

Kim, Y. E., M. S. Hipp, A. Bracher, M. Hayer-Hartl, and F. U. Hartl. 2013. Molecular chaperone functions in protein folding and proteostasis. *Annu. Rev. Biochem.* 82: 323–355.

Knight, J. C. 2004. Allele-specific gene expression uncovered. *Trends Genet.* 20: 113–116.

Kortmann J., and F. Narberhaus. 2012. Bacterial RNA thermometers: molecular zippers and switches. *Nat. Rev. Micro.* 10: 255–265.

Kültz, D. 2005. Molecular and evolutionary basis of the cellular stress response. *Annu. Rev. Physiol.* 67: 225–257.

Kültz, D., and G. N. Somero. 1995. Ion transport in gills of the euryhaline fish *Gillichthys mirabilis* is facilitated by a phosphocreatine circuit. *Am. J. Physiol. Regul. Integr. Comp. Physiol.* 37: R1003–R1012.

Lockwood, B. L., K. M. Connor, and A. Y. Gracey. 2015. The environmentally tuned transcriptomes of *Mytilus* mussels. *J. Exp. Biol.* 218: 1822–1833.

Logan, C. A., and B. A. Buckley. 2015. Transcriptomic responses to environmental temperature in eurythermal and stenothermal fishes. *J. Exp. Biol.* 218: 1915–1924.

Logan, C. A., and G. N. Somero. 2010. Transcriptional responses to thermal acclimation in the eurythermal fish *Gillichthys mirabilis* (Cooper 1864). *Am. J. Physiol. Regul. Integr. Comp. Physiol.* 299: R843–R852.

Logan, C. A., and G. N. Somero. 2011. Effects of thermal acclimation on transcriptional responses to acute heat stress in the eurythermal fish *Gillichthys mirabilis* (Cooper). *Am. J. Physiol. Regul. Integr. Comp. Physiol.* 300: R1373–R1383.

López-García, P., Y. Zivanovic, P. Deschamps, and D. Moreira. 2015. Bacterial gene import and mesophilic adaptation in archaea. *Nat. Rev. Microbiol.* 13: 447–456.

Marden, J. H., G. H. Fitzhugh, M. Girgenrath, M. R. Wolf, and S. Girgenrath. 2001. Alternative splicing, muscle contraction and intraspecific variation: associations between troponin T transcripts, Ca^{2+} sensitivity and the force and power output of dragonfly flight muscles during oscillatory contraction. *J. Exp. Biol.* 204: 3457–3470.

Mayr, E. 1982. *The Growth of Biological Thought: Diversity, Evolution, and Inheritance.* Harvard University Press, Cambridge, MA.

McFall-Ngai, M. J. 2015. Giving microbes their due—animal life in a microbially dominant world. *J. Exp. Biol.* 218: 1968–1973.

Müller, M., M. Mentel, J. J. van Hellemond, K. Henze, C. Woehle, S. B. Gould, R. Y. Yu, M. van der Giezen, and 2 others. 2012. Biochemistry and evolution of anaerobic energy metabolism in eukaryotes. *Microbiol. Mol. Biol. Rev.* 76: 444–495.

Nicholls, D. G., and S. J. Ferguson. 2013. *Bioenergetics 4.* Academic Press, San Diego.

Nilsson, G., D. L. Dixson, P. Domenici, M. I. McCormick, C. Sorensen, S.-A. Watson, and P. L. Munday. 2012. Near-future carbon dioxide levels alter fish behaviour by interfering with neurotransmitter function. *Nat. Clim. Change* 2: 201–204.

Ohno, S. 1970. *Evolution by Gene Duplication.* Springer-Verlag, New York.

Pastinen, T. 2010. Genome-wide allele-specific analysis: insights into regulatory variation. *Nat. Rev. Gen.* 11: 533–538.

Place, A. R., and D. A. Powers. 1984. Purification and characterization of the lactate dehydrogenase (LDH-B_4) allozymes of *Fundulus heteroclitus. J. Biol. Chem.* 259: 1309–1318.

Podrabsky, J. E., and S. C. Hand. 2015. Physiological strategies during animal diapause: lessons from brine shrimp and annual killifish. *J. Exp. Biol.* 218: 1897–1906.

Podrabsky, J. E., and G. N. Somero. 2006. Inducible heat tolerance in Antarctic notothenioid fishes. *Polar Biol.* 30: 39–43.

Richards, J. G., J. W. Semple, J. S. Bystriansky, and P. M. Schulte. 2003. Na^+/K^+-ATPase α-isoform switching in gills of rainbow trout (*Oncorhynchus mykiss*) during salinity transfer. *J. Exp. Biol.* 206: 4475–4486.

Rohner, N., D. F. Jarosz, J. E. Kowalko, M. Yoshizawa, W. R. Jeffery, R. L. Borowsky, S. Lindquist, and C. J. Tabin. 2013. Cryptic variation in morphological evolution: HSP90 as a capacitor for loss of eyes in cavefish. *Science* 342: 1372–1375.

Rose, M. R., and G. B. Lauder. 1996. *Adaptation.* Academic Press, San Diego.

Rosenthal, J. J. C. 2015. The emerging role of RNA editing in plasticity. *J. Exp. Biol.* 218: 1812–1821.

Rutherford, S., and S. Lindquist. 1998. Hsp90 as a capacitor for morphological evolution. *Nature* 396: 336–342.

Sanford, E., and M. W. Kelly. 2011. Local adaptation in marine invertebrates. *Annu. Rev. Mar. Sci.* 3: 509–535.

Schilder, R. J., S. R. Kimball, J. H. Marden, and L. S. Jefferson. 2011. Body weight-dependent troponin T alternative splicing is evolutionarily conserved from insects to mammals and is partially impaired in skeletal muscle of obese rats. *J. Exp. Biol.* 214: 1523–1532.

Schönknecht, G., W.-H. Chen, C. M. Ternes, G. G. Barbier, R. P. Shrestha, M. Stanke, A. Bräutigam, B. J. Baker, and 10 others. 2013. Gene transfer from bacteria and archaea facilitated evolution of an extremophilic eukaryote. *Science* 339: 1207–1210.

Schulte, P. M. 2004. Changes in gene expression as biochemical adaptations to environmental change: a tribute to Peter Hochachka. *Comp. Biochem. Physiol.* 139B: 519–529.

Somero, G. N. 2000. Unity in diversity: A perspective on the methods, contributions, and future of comparative physiology. *Annu. Rev. Physiol.* 62: 927–937.

Somero, G. N. 2010. The physiology of climate change: how potentials for acclimatization and genetic adaptation will determine "winners" and "losers." *J. Exp. Biol.* 213: 912–920.

Somero, G. N., J. M. Beers, F. Chan, T. M. Hill, T. Klinger, and S. Y. Litvin. 2016. What changes in the carbonate system, oxygen, and temperature portend for the Northeastern Pacific Ocean: A physiological perspective. *BioScience* 66: 14–26.

Somero, G. N., and A. L. DeVries. 1967. Temperature tolerance of some Antarctic fishes. *Science* 156: 257–258.

Spang, A., J. H. Saw, S. L. Jorgensen, K. Zaremba-Niedzwiedzka, J. Martijn, A. E. Lind, R. van Eijk, C. Schleper, and 2 others. 2015. Complex archaea that bridge the gap between prokaryotes and eukaryotes. *Nature* 521: 173–179.

Stenseng, E., C. Braby, and G. N. Somero 2005. Evolutionary and acclimation-induced variation in the thermal limits of heart function in congeneric marine snails (genus *Tegula*): Implications for vertical zonation. *Biol. Bull.* 208: 138–144.

Stillman, J. H. 2003. Acclimation capacity underlies susceptibility to climate change. *Science* 301: 65.

Storey, K. B. 2015. Regulation of hypometabolism: insights into epigenetic controls. *J. Exp. Biol.* 218: 150–159.

Suter, C. M., D. Boffelli, and D. I. K. Martin. 2013. A role for epigenetic inheritance in modern evolutionary theory? A comment in response to Dickins and Rahman. *Proc. R. Soc. B.* 280: 20130903.

Tewksbury, J. J., R. B. Huey, and C. A. Deutsch. 2008. Putting the heat on tropical animals. *Science* 320: 1296–1297.

Tomanek, L. 2010. Variation in the heat shock response and its implication for predicting the effect of global climate change on species' biogeographic distribution ranges and metabolic costs. *J. Exp. Biol.* 213: 971–979.

Tomanek, L. 2014. Proteomics to study adaptations in marine organisms to environmental stress. *J. Proteomics* 105: 92–106.

Tomanek, L. 2015. Proteomic responses to environmentally induced oxidative stress. *J. Exp. Biol.* 218: 1867–1879.

Tomanek, L., and G. N. Somero. 1999. Evolutionary and acclimation-induced variation in the heat shock responses of congeneric marine snails (genus *Tegula*) from different thermal habitats: Implications for limits of thermotolerance and biogeography. *J. Exp. Biol.* 202: 2925–2936.

Trajkovski, M., K. Ahmed, C. C. Esau, and M. Stoffel. 2012. MyomiR-133 regulates brown fat differentiation through Prdm16. *Nat. Cell Biol.* 14: 1330–1335.

Watabe, S. 2002. Temperature plasticity of contractile proteins in fish muscle. *J. Exp. Biol.* 205: 2231–2236.

Watt, W. B., and A. M. Dean. 2000. Molecular-functional studies of adaptive genetic variation in prokaryotes and eukaryotes. *Annu. Rev. Genet.* 34: 593–622.

Yancey, P. H., and J. F. Siebenaller. 2015. Co-evolution of proteins and solutions: protein adaptation versus cytoprotective micromolecules and their roles in marine organisms. *J. Exp. Biol.* 218: 1880–1896.

Zera, A. J. 2011. Microevolution of intermediary metabolism: evolutionary genetics meets metabolic biochemistry. *J. Exp. Biol.* 214: 179–190.

Chapter 2

Abe, H. 2000. Role of histidine-related compounds as intracellular proton buffering constituents of vertebrate muscle. *Biochemistry (Moscow)* 65: 757–765.

Abele, D., K. Heise, H. O. Pörtner, and S. Puntarulo. 2002. Temperature-dependence of mitochondrial function and production of reactive oxygen species in the intertidal mud clam *Mya arenaria*. *J. Exp. Biol.* 205: 1831–1841.

Abele, D., J. P. Vázques-Medina, and T. Zenteno-Savín. 2012. *Oxidative Stress in Aquatic Ecosystems*. Wiley-Blackwell, Oxford, UK.

Abnous, K., and K. B. Storey. 2007. Regulation of skeletal muscle creatine kinase from a hibernating mammal. *Arch. Biochem. Biophys.* 467: 10–19.

Agbor, T. A., A. Cheong, K. M. Comerford, C. C. Scholz, U. Bruning, A. Clarke, E. P. Cummins, G. Cagney, and C. T. Taylor. 2011. Small ubiquitin-related modifier (SUMO)-1 promotes glycolysis in hypoxia. *J. Biol. Chem.* 286: 4718–4726.

Al Hasawi, N., M. F. Alkandari, and Y. A. Luqmani. 2014. Phosphofructokinase: a mediator of glycolytic flux in cancer progression. *Crit. Rev. Oncol. Hematol.* 92: 312–321.

Ali, S. S., M. Hsiao, H. W. Zhao, L. L. Dugan, G. G. Haddad, and D. Zhou. 2012. Hypoxia-adaptation involves mitochondrial metabolic depression and decreased ROS leakage. *PLoS One* 7: e36801.

Aloia, R. C., and J. K. Raison. 1989. Membrane function in mammalian hibernation. *Biochim. Biophys. Acta* 988: 123–146.

Anbar, A. D., and A. H. Knoll. 2002. Proterozoic ocean chemistry and evolution: a bioinorganic bridge? *Science* 297: 1137–1142.

Anchordoguy, T. J., and S. C. Hand. 1994. Acute blockage of the ubiquitin-mediated proteolytic pathway during invertebrate quiescence. *Am. J. Physiol. Regul. Integr. Comp. Physiol.* 267: R895–R900.

Andrews, M. T. 2007. Advances in molecular biology of hibernation in mammals. *Bioessays* 29: 431–440.

Andreyev, A. Y., Y. E. Kushnareva, and A. A. Starkov. 2005. Mitochondrial metabolism of reactive oxygen species. *Biochemistry* 70: 200–214.

Appenzeller-Herzog, C. 2011. Glutathione- and non-glutathione-based oxidant control in the endoplasmic reticulum. *J. Cell Sci.* 124: 847–855.

Araki, K., and K. Nagata. 2012. Protein folding and quality control in the ER. In R. I. Morimoto, D. J. Selkoe, and J. W. Kelley (eds.), *Protein Homeostasis*, pp. 121–145. Cold Spring Harbor Press, New York.

Attwell, D., and S. B. Laughlin. 2001. An energy budget for signaling in the grey matter of the brain. *J. Cereb. Blood Flow Metab.* 21: 1133–1145.

Banavar, J. R., J. Damuth, A. Maritan, and A. Rinaldo. 2002. Supply-demand balance and metabolic scaling. *Proc. Natl. Acad. Sci. USA* 99: 10506–10509.

Barger, J. L., M. D. Brand, B. M. Barnes, and B. B. Boyer. 2003. Tissue-specific depression of mitochondrial proton leak and substrate oxidation in hibernating arctic ground squirrels. *Am. J. Physiol. Regul. Integr. Comp. Physiol.* 284: R1306–R1313.

Barnes, R. B. 1980. *Invertebrate Zoology*, 4th ed. Saunders, Philadelphia.

Barreto, F. S., and R. S. Burton. 2013. Elevated oxidative damage is correlated with reduced fitness in interpopulation hybrids of a marine copepod. *Proc. Biol. Sci.* 280: 20131521.

Bauer, V. W., T. L. Squire, M. E. Lowe, and M. T. Andrews. 2001. Expression of a chimeric retroviral-lipase mRNA confers enhanced lipolysis in a hibernating mammal. *Am. J. Physiol. Regul. Integr. Comp. Physiol.* 281: R1186–R1192.

Benedetti, M., M. Nigro, and F. Regoli. 2010. Characterization of antioxidant defences in three Antarctic Notothenoids species from Terra Nova Bay (Ross Sea). *Chem. Ecol.* 26: 305–314.

Bennett, K. A., B. J. McConnell, S. E. Moss, J. R. Speakman, P. P. Pomeroy, and M. A. Fedak. 2010. Effects of age and body mass on development of diving capabilities of gray seal pups: costs and benefits of the postweaning fast. *Physiol. Biochem. Zool.* 83: 911–923.

Berner, R. A. 2006. GEOCARBSULF: A combined model for Phanerozoic atmospheric O_2 and CO_2. *Geochim. Cosmochim. Acta* 70: 5653–5664.

Berner, R. A., J. M. Vandenbrooks, and P. D. Ward. 2007. Oxygen and evolution. *Science* 316: 557–558.

Bickler, P. E., P. H. Donohoe, and L. T. Buck. 2000. Hypoxia-induced silencing of NMDA receptors in turtle neurons. *J. Neurosci.* 20: 3522–3528.

Bienert, G. P., and F. Chaumont. 2014. Aquaporin-facilitated transmembrane diffusion of hydrogen peroxide. *Biochim. Biophys. Acta* 1840: 1596–1604.

Biggar, K. K., and K. B. Storey. 2012. Evidence for cell cycle suppression and microRNA regulation of cyclin D1 during anoxia exposure in turtles. *Cell Cycle* 11: 1705–1713.

Blackburn, E. H., E. S. Epel, and J. Lin. 2015. Human telomere biology: A contributory and interactive factor in aging, disease risks, and protection. *Science* 350: 1193–1198.

Blackstone, E., M. Morrison, and M. B. Roth. 2005. H_2S induces a suspended animation-like state in mice. *Science* 308: 518.

Bonekamp, N. A., A. Volkl, H. D. Fahimi, and M. Schrader. 2009. Reactive oxygen species and peroxisomes: struggling for balance. *BioFactors* 35: 346–355.

Braakman, I., and N. J. Bulleid. 2011. Protein folding and modification in the mammalian endoplasmic reticulum. *Annu. Rev. Biochem.* 80: 71–99.

Braakman, I., and D. N. Hebert. 2013. Protein folding in the endoplasmic reticulum. In S. Ferro-Novick, T. A. Rapoport, and R. Schekman (eds.), *The Endoplasmic Reticulum*, pp. 63–79. Cold Spring Harbor Press, Cold Spring Harbor.

Brand, M. D., L. F. Chien, and D. F. Rolfe. 1993. Control of oxidative phosphorylation in liver mitochondria and hepatocytes. *Biochem. Soc. Trans.* 21: 757–762.

Brand, M. D., P. Couture, P. L. Else, K. W. Withers, and A. J. Hulbert. 1991. Evolution of energy metabolism. Proton permeability of the inner membrane of liver mitochondria is greater in a mammal than in a reptile. *Biochem. J.* 275: 81–86.

Brand, M. D., N. Turner, A. Ocloo, P. L. Else, and A. J. Hulbert. 2003. Proton conductance and fatty acyl composition of liver mitochondria correlates with body mass in birds. *Biochem. J.* 376: 741–748.

Brocks, J. J., G. A. Logan, R. Buick, and R. E. Summons. 1999. Archean molecular fossils and the early rise of eukaryotes. *Science* 285: 1033–1036.

Buck, C. L., and B. M. Barnes. 2000. Effects of ambient temperature on metabolic rate, respiratory quotient, and torpor in an arctic hibernator. *Am. J. Physiol. Regul. Integr. Comp. Physiol.* 279: R255–R262.

Bulleid, N. J. 2013. Disulfide bond formation in the mammalian endoplasmic reticulum. In S. Ferro-Novick, T. A. Rapoport, and R. Schekman (eds.), *The Endoplasmic Reticulum.*, pp. 81–92. Cold Spring Harbor Press, Cold Spring Harbor.

Bulleid, N. J., and L. Ellgaard. 2011. Multiple ways to make disulfides. *Trends Biochem. Sci.* 36: 485–492.

Burggren, W. W. 2014. Epigenetics as a source of variation in comparative animal physiology—or—Lamarck is lookin' pretty good these days. *J. Exp. Biol.* 217: 682–689.

Burnett, L. E. 1997. The challenges of living in hypoxic and hypercapnic aquatic environments. *Am. Zool.* 37: 633–640.

Burns, J. M., K. Lestyk, D. Freistroffer, and M. O. Hammill. 2014. Preparing muscles for diving: Age-related changes in muscle metabolic profiles in harp (*Pagophilus groenlandicus*) and hooded (*Cystophora cristata*) seals. *Physiol. Biochem. Zool.* 88: 167–182.

Burton, R. S., C. K. Ellison, and J. S. Harrison. 2006. The sorry state of F2 hybrids: consequences of rapid mitochondrial DNA evolution in allopatric populations. *Am. Nat.* 168 Suppl 6: S14–24.

Butterfield, N. J. 2009. Oxygen, animals and oceanic ventilation: an alternative view. *Geobiol.* 7: 1–7.

Canfield, D. E. 2005. The early history of atmospheric oxygen: homage to Robert M. Garrels. *Annu. Rev. Earth Planet. Sci.* 33: 1–36.

Canfield, D. E. 2014. *Oxygen: A Four Billion Year History.* Princeton University Press, Princeton.

Canfield, D. E., S. W. Poulton, A. H. Knoll, G. M. Narbonne, G. Ross, T. Goldberg, and H. Strauss. 2008. Ferruginous conditions dominated later neoproterozoic deep-water chemistry. *Science* 321: 949–952.

Canfield, D. E., S. W. Poulton, and G. M. Narbonne. 2007. Late-Neoproterozoic deep-ocean oxygenation and the rise of animal life. *Science* 315: 92–95.

Carey, H. V., M. T. Andrews, and S. L. Martin. 2003. Mammalian hibernation: cellular and molecular responses to depressed metabolism and low temperature. *Physiol. Rev.* 83: 1153–1181.

Carey, H. V., C. L. Frank, and J. P. Seifert. 2000. Hibernation induces oxidative stress and activation of NK-kappaB in ground squirrel intestine. *J. Comp. Physiol. B.* 170: 551–559.

Castellini, M. A., and G. N. Somero. 1981. Buffering capacity of vertebrate muscle: correlations with potentials for anaerobic function. *J. Comp. Physiol. B.* 143: 191–198.

Champagne, C. D., D. E. Crocker, M. A. Fowler, and D. S. Houser. 2012a. Fasting physiology of the pinnipeds: The challenges of fasting while maintaining high energy expenditure and nutrient delivery for lactation. In M. D. McCue (ed.) *Comparative Physiology of Fasting, Starvation, and Food Limitation*, pp. 309–336. Springer-Verlag, Berlin

Champagne, C. D., D. S. Houser, and D. E. Crocker. 2005. Glucose production and substrate cycle activity in a fasting adapted animal, the northern elephant seal. *J. Exp. Biol.* 208: 859–868.

Champagne, C. D., D. S. Houser, M. A. Fowler, D. P. Costa, and D. E. Crocker. 2012b. Gluconeogenesis is associated with high rates of tricarboxylic acid and pyruvate cycling in fasting northern elephant seals. *Am. J. Physiol. Regul. Integr. Comp. Physiol.* 303: R340–R352.

Chandel, N. S. 2015. *Navigating Metabolism.* Cold Spring Harbor Press, Cold Spring Harbor.

Chapman, D. J., and J. W. Schopf. 1983. In J. W. Schopf (ed.), *Earth's Earliest Biosphere: Its Origin and Evolution,* pp. 302–320. Princeton University Press, Princeton, NJ.

Chicco, A. J., C. H. Le, A. Schlater, A. Nguyen, S. Kaye, J. W. Beals, R. L. Scalzo, C. Bell, and 4 others. 2014. High fatty acid oxidation capacity and phosphorylation control despite elevated leak and reduced respiratory capacity in northern elephant seal muscle mitochondria. *J. Exp. Biol.* 217: 2947–2955.

Childress, J. J., and G. Somero. 1990. Metabolic scaling: a new perspective based on scaling of glycolytic enzyme activities. *Am. Zool.* 30: 161–173.

Chouchani, E. T., V. R. Pell, E. Gaude, D. Aksentijevic, S. Y. Sundier E. L. Robb, A. Logan, S. M. Nadtochiy, and 21 others. 2014. Ischaemic accumulation of succinate controls reperfusion injury through mitochondrial ROS. *Nature* 515: 431–435.

Christensen, M., T. Hartmund, and H. Gesser. 1994. Creatine kinase, energy-rich phosphates and energy metabolism in heart muscle of different vertebrates. *J. Comp. Physiol. B.* 164: 118–123.

Clapham, M. E., and J. A. Karr. 2012. Environmental and biotic controls on the evolutionary history of insect body size. *Proc. Natl. Acad. Sci. USA* 109: 10927–10930.

Clegg, J. S. 2007. Protein stability in *Artemia* embryos during prolonged anoxia. *Biol. Bull.* 212: 74–81.

Clegg, J. S., L. E. Drinkwater, and P. Sorgeloos. 1996. The metabolic status of diapause embryos of *Artemia franciscana. Physiol. Zool.* 69: 49–66.

Cochran, R. E., and L. E. Burnett. 1996. Respiratory responses of the salt marsh animals, *Fundulus heteroclitus, Leiostomus xanthurus,* and *Palaemonetes pugio* to environmental hypoxia and hypercapnia and to the organophosphate pesticide, azinphosmethyl. *J. Exp. Mar. Biol. Ecol.* 195: 125–144.

Cotgreave, I. A., R. Gerdes, I. Schuppe-Koistinen, and C. Lind. 2002. S-glutathionylation of glyceraldehyde-3-phosphate dehydrogenase: role of thiol oxidation and catalysis by glutaredoxin. *Methods Enzymol.* 348: 175–182.

Couture, P., and A. J. Hulbert. 1995. Relationship between body mass, tissue metabolic rate, and sodium pump activity in mammalian liver and kidney. *Am. J. Physiol. Regul. Integr. Comp. Physiol.* 268: R641–R650.

Covi, J. A., W. D. Treleaven, and S. C. Hand. 2005. V-ATPase inhibition prevents recovery from anoxia in *Artemia franciscana* embryos: quiescence signaling through dissipation of proton gradients. *J. Exp. Biol.* 208: 2799–2808.

Cox, A. G., C. C. Winterbourn, and M. B. Hampton. 2010. Mitochondrial peroxiredoxin involvement in antioxidant defence and redox signaling. *Biochem. J.* 425: 313–325.

Crocker, D. E., J. D. Williams, D. P. Costa, and B. J. Le Boef. 2001a. Maternal traits and reproductive effort in northern elephant seals. *Ecology.* 82: 3541–3555.

Crowe, J. H., L. M. Crowe, J. F. Carpenter, S. Petrelski, F. A. Hoekstra P. De Araujo, and A. D. Panek. 1997. Anhydrobiosis: cellular adaptation to extreme dehydration.

In W. H. Dantzler (ed.), *Handbook of Physiology,* pp. 1445–1477. Oxford University Press, Oxford, UK.

Csala, M., E. Margittai, and G. Banhegyi. 2010. Redox control of endoplasmic reticulum function. *Antioxid. Redox Signal.* 13: 77–108.

Dalle-Donne, I., R. Rossi, G. Colombo, D. Giustarini, and A. Milzani. 2009. Protein S-glutathionylation: a regulatory device from bacteria to humans. *Trends Biochem. Sci.* 34: 85–96.

Darveau, C. A., P. W. Hochachka, D. W. Roubik, and R. K. Suarez. 2005a. Allometric scaling of flight energetics in orchid bees: evolution of flux capacities and flux rates. *J. Exp. Biol.* 208: 3593–3602.

Darveau, C. A., P. W. Hochachka, K. C. Welch, Jr., D. W. Roubik, and R. K. Suarez. 2005b. Allometric scaling of flight energetics in Panamanian orchid bees: a comparative phylogenetic approach. *J. Exp. Biol.* 208: 3581–3591.

Darveau, C. A., R. K. Suarez, R. D. Andrews, and P. W. Hochachka. 2002. Allometric cascade as a unifying principle of body mass effects on metabolism. *Nature* 417: 166–170.

Darveau, C. A., R. K. Suarez, R. D. Andrews, and P. W. Hochachka. 2003. Reply. *Nature* 421: 714.

Dean, J. B. 2010. Hypercapnia causes cellular oxidation and nitrosation in addition to acidosis: implications for CO_2 chemoreceptor function and dysfunction. *J. Appl. Physiol.* 108: 1786–1795.

Decker, H., and K. E. van Holde. 2011. *Oxygen and the Evolution of Life.* Springer-Verlag, Berlin.

Desagher, S., J. Glowinski, and J. Premont. 1997. Pyruvate protects neurons against hydrogen peroxide-induced toxicity. *J. Neurosci.* 17: 9060–9067.

Dilly, G. F., C. R. Young, W. S. Lane, J. Pangilinan, and P. R. Girguis. 2012. Exploring the limit of metazoan thermal tolerance via comparative proteomics: thermally induced changes in protein abundance by two hydrothermal vent polychaetes. *Proc. R. Soc. B.* 279: 3347–3356.

Divakaruni, A. S., and M. D. Brand. 2011. The regulation and physiology of mitochondrial proton leak. *Physiology* 26: 192–205.

Doeller, J. E., B. K. Gaschen, V. V. Parrino, and D. W. Kraus. 1999. Chemolithoheterotrophy in a metazoan tissue: sulfide supports cellular work in ciliated mussel gills. *J. Exp. Biol.* 202: 1953–1961.

Drew, K. L., C. L. Buck, B. M. Barnes, S. L. Christian, B. T. Rasley, and M. B. Harris. 2007. Central nervous system regulation of mammalian hibernation: implications for metabolic suppression and ischemia tolerance. *J. Neurochem.* 102: 1713–1726.

Drinkwater, L. E., and J. H. Crowe. 1987. Regulation of embryonic diapause in *Artemia*: environmental and physiological signals. *J. Exp. Zool.* 241: 297–307.

Duerr, J. M., and J. E. Podrabsky. 2010. Mitochondrial physiology of diapausing and developing embryos of the annual killifish *Austrofundulus limnaeus*: implications for extreme anoxia tolerance. *J. Comp. Physiol. B.* 180: 991–1003.

Dupont, C. L., A. Butcher, R. E. Valas, P. E. Bourne, and G. Caetano-Anollés. 2010. History of biological metal utilization inferred through phylogenomic analysis of protein structures. *Proc. Natl. Acad. Sci. USA* 107: 10567–10572.

Eanes, W. F., T. J. S Merritt, J. M. Flowers, S. Kumagai, E. Sezgin, and C.-T. Zhu. 2006. Flux control and excess capacity in the enzymes of glycolysis and their relationship to flight metabolism in *Drosophila melanogaster*. *Proc. Natl. Acad. Sci. USA* 103: 19413–19418.

Ebert, M. S., and P. A. Sharp. 2012. Roles for microRNAs in conferring robustness to biological processes. *Cell* 149: 515–524.

Echtay, K. S. 2007. Mitochondrial uncoupling proteins—what is their physiological role? *Free Radic. Biol. Med.* 43: 1351–1371.

Eddy, S. F., and K. B. Storey. 2003. Differential expression of Akt, PPARgamma, and PGC-1 during hibernation in bats. *Biochem. Cell Biol.* 81: 269–274.

Elia, M. 1992. Organ and tissue contribution to metabolic rate. In J. M. Kinney and H. N. Tucker (eds.), *Energy metabolism: tissue determinants and cellular corollaries*, pp. 61–79. Raven Press Ltd, New York, NY.

Ellington, W. R. 1989. Phosphocreatine represents a thermodynamic and functional improvement over other muscle phosphagens. *J. Exp. Biol.* 143: 177–194.

Ellington, W. R. 2001. Evolution and physiological roles of phosphagen systems. *Annu. Rev. Physiol.* 63: 289–325.

Ellison, C. K., and R. S. Burton. 2006. Disruption of mitochondrial function in interpopulation hybrids of *Tigriopus californicus*. *Evolution* 60: 1382–1391.

Else, P. L., M. D. Brand, N. Turner, and A. J. Hulbert. 2004. Respiration rate of hepatocytes varies with body mass in birds. *J. Exp. Biol.* 207: 2305–2311.

Erwin, D. H., M. Laflamme, S. M. Tweedt, E. A. Sperling, D. Pisani, and K. J. Peterson. 2011. The Cambrian conundrum: early divergence and later ecological success in the early history of animals. *Science* 334: 1091–1097.

Falkowski, P. G., M. E. Katz, A. J. Milligan, K. Fennel, B. S. Cramer, M. P. Aubry, R. A. Berner, M. J. Novacek, and W. M. Zapol. 2005. The rise of oxygen over the past 205 million years and the evolution of large placental mammals. *Science* 309: 2202–2204.

Farquhar, J., H. Bao, and M. Thiemens. 2000. Atmospheric influence of Earth's earliest sulfur cycle. *Science* 289: 756–759.

Farquhar, J., J. Savarino, S. Airieau, and M. H. Thiemens. 2001. Observation of the wavelength-sensitive mass-dependent sulfur isotope effect during SO_2 photolysis: Implications for the early atmosphere. *J. Geophys. Res.* 106: 32829–32839.

Fell, D. A. 1997. *Understanding the Control of Metabolism*. Portland Press, London.

Fell, D. A. 2000. Signal transduction and the control of expression of enzyme activity. *Adv. Enzyme Regul.* 40: 35–46.

Fields, P. A., C. Eurich, W. L. Gao, and B. Cela. 2014. Changes in protein expression in the salt marsh mussel *Geukensia demissa*: evidence for a shift from anaerobic to aerobic metabolism during prolonged aerial exposure. *J. Exp. Biol.* 217: 1601–1612.

Fields, P. A., M. J. Zuzow, and L. Tomanek. 2012. Comparative proteomics of blue mussel (*Mytilus*) congeners to temperature acclimation. *J. Exp. Biol.* 215: 1106–1116.

Finazzi, D., and P. Arosio. 2014. Biology of ferritin in mammals: an update on iron storage, oxidative damage and neurodegeneration. *Arch. Toxicol.* 88: 1787–1802.

Fowler, M. A., D. P. Costa, D. E. Crocker, W. J. Shen, and F. B. Kraemer. 2015. Adipose triglyceride lipase, not hormone-sensitive lipase, is the primary lipolytic enzyme in fasting elephant seals (*Mirounga angustirostris*). *Physiol. Biochem. Zool.* 88: 284–294.

Frank, C. L., S. P. J. Brooks, H. J. Harlow, and K. B. Storey. 1998. The influence of hibernation patterns on the criticial enzymes of lipogenesis and lipolysis in prairie dogs. *Exp. Biol. Online* 3: 1–8.

Fraser, K. P., D. F. Houlihan, P. L. Lutz, S. Leone-Kabler, L. Manuel, and J. G. Brechin. 2001. Complete suppression of protein synthesis during anoxia with no post-anoxia protein synthesis debt in the red-eared slider turtle *Trachemys scripta elegans*. *J. Exp. Biol.* 204: 4353–4360.

Fratelli, M., H. Demol, M. Puype, S. Casagrande, P. Villa, I. Eberini, J. Vandekerckhove, E. Gianazza, and P. Ghezzi. 2003. Identification of proteins undergoing glutathionylation in oxidatively stressed hepatocytes and hepatoma cells. *Proteomics* 3: 1154–1161.

Frerichs, K. U., C. B. Smith, M. Brenner, D. J. DeGracia, G. S. Krause, L. Marroe, T. E. Dever, and J. M. Hallenbeck. 1998. Suppression of protein synthesis in brain during hibernation involves inhibition of protein initiation and elongation. *Proc. Natl. Acad. Sci. USA* 95: 14511–14516.

Fridovich, I. 1998. Oxygen toxicity: a radical explanation. *J. Exp. Biol.* 201: 1203–1209.

Fuson, A. L., D. F. Cowan, S. B. Kanatous, L. K. Polasek, and R. W. Davis. 2003. Adaptations to diving hypoxia in the heart, kidneys and splanchnic organs of harbor seals (*Phoca vitulina*). *J. Exp. Biol.* 206: 4139–4154.

Garland, M. A., J. H. Stillman, and L. Tomanek. 2015. The proteomic response of cheliped myofibril tissue in the eurythermal porcelain crab, *Petrolisthes cinctipes*, to heat shock following acclimation to daily temperature fluctuations. *J. Exp. Biol.* 218: 388–403.

Goodell, M. A., and T. A. Rando. 2015. Stem cells and healthy aging. *Science* 350: 1199–1204.

Grabek, K. R., S. L. Martin, and A. G. Hindle. 2015. Proteomics approaches shed new light on hibernation physiology. *J. Comp. Physiol. B* 185: 607–627.

Grabowska, D., and A. Chelstowska. 2003. The ALD6 gene product is indispensable for providing NADPH in yeast cells lacking glucose-6-phosphate dehydrogenase activity. *J. Biol. Chem.* 278: 13984–13988.

Grieshaber, M. K., I. Hardewig, U. Kreutzer, and H. O. Pörtner. 1994. Physiological and metabolic responses to hypoxia in invertebrates. *Rev. Physiol. Biochem. Pharmacol.* 125: 43–147.

Griffiths, J. R. 1981. A fresh look at glycogenolysis in skeletal muscle. *Biosci. Rep.* 1: 595–610.

Grimm, D. 2015. Why we outlive our pets. *Science* 350: 1182–1185.

Grivennikova, V. G., and A. D. Vinogradov. 2013. Mitochondrial production of reactive oxygen species. *Biochemistry (Moscow)* 78: 1490–1511.

Guan, K. L., and Y. Xiong. 2011. Regulation of intermediary metabolism by protein acetylation. *Trends Biochem. Sci.* 36: 108–116.

Guderley, H., and H. O. Pörtner. 2010. Metabolic power budgeting and adaptive strategies in zoology: examples from scallops and fish. *Can. J. Zool.* 88: 753–763.

Guimarães-Ferreira, L. 2014. Role of the phosphocreatine system on energetic homeostasis in skeletal and cardiac muscles. *Einstein (Sao Paulo)* 12: 126–131.

Gupta, V., and R. N. Bamezai. 2010. Human pyruvate kinase M2: a multifunctional protein. *Protein Sci.* 19: 2031–2044.

Guzun, R., M. Gonzalez-Granillo, M. Karu-Varikmaa, A. Grichine, Y. Usson, T. Kaambre, K. Guerrero-Roesch, A. Kuznetsov, and 2 others. 2012. Regulation of respiration in muscle cells in vivo by VDAC through interaction with the cytoskeleton and MtCK within Mitochondrial Interactosome. *Biochim. Biophys. Acta* 1818: 1545–1554.

Halliwell, B., and J. M. C. Gutteridge. 2007. *Free Radicals in Biology and Medicine*. Oxford University Press, Oxford.

Halliwell, B., and J. M. C. Gutteridge. 2015. *Free radicals in Biology & Medicine*. Oxford University Press, Oxford.

Hamilton, N., and C. D. Ianuzzo. 1991. Contractile and calcium regulating capacities of myocardia of different sized mammals scale with resting heart rate. *Mol. Cell. Biochem.* 106: 133–141.

Hand, S. C., and M. A. Menze. 2008. Mitochondria in energy-limited states: mechanisms that blunt the signaling of cell death. *J. Exp. Biol.* 211: 1829–1840.

Hand, S. C., M. A. Menze, A. Borcar, Y. Patil, J. A. Covi, J. A. Reynolds, and M. Toner. 2011. Metabolic restructuring during energy-limited states: insights from *Artemia franciscana* embryos and other animals. *J. Insect Physiol.* 57: 584–594.

Harney, E., S. Artigaud, P. Le Souchu, P. Miner, C. Corporeau, H. Essid, V. Pichereau, and F. L. Nunes. 2016. Non-additive effects of ocean acidification in combination with warming on the larval proteome of the Pacific oyster, *Crassostrea gigas*. *J. Proteomics* 135: 151–161.

Harrison, J. F., A. Kaiser, and J. M. VandenBrooks. 2010. Atmospheric oxygen level and the evolution of insect body size. *Proc. Biol. Sci.* 277: 1937–1946.

Harrison, J. S., and R. S. Burton. 2006. Tracing hybrid incompatibilities to single amino acid substitutions. *Mol. Biol. Evol.* 23: 559–564.

Hatzivassiliou, G., F. Zhao, D. E. Bauer, C. Andreadis, A. N. Shaw, D. Dhanak, S. R. Hingorani, D. A. Tuveson, and C. B. Thompson. 2005. ATP citrate lyase inhibition can suppress tumor cell growth. *Cancer Cell*. 8: 311–321.

Hawkins, A. 1991. Protein turnover: a functional appraisal. *Funct. Ecol.* 5: 222–233.

Hawrysh, P. J., and L. T. Buck. 2013. Anoxia-mediated calcium release through the mitochondrial permeability transition pore silences NMDA receptor currents in turtle neurons. *J. Exp. Biol.* 216: 4375–4387.

Hayssen, V., and R. C. Lacy. 1985. Basal metabolic rates in mammals: Taxonomic differences in the allometry of BMR and body mass. *Comp. Biochem. Physiol. A* 81: 741–754.

Heise, K., M. S. Estevez, S. Puntarulo, M. Galleano, M. Nikinmaa, H. O. Pörtner, and D. Abele. 2007. Effects of seasonal and latitudinal cold on oxidative stress parameters and activation of hypoxia inducible factor (HIF-1) in zoarcid fish. *J. Comp. Physiol. B*. 177: 765–777.

Heise, K., S. Puntarulo, H. O. Pörtner, and D. Abele. 2003. Production of reactive oxygen species by isolated mitochondria of the Antarctic bivalve *Laternula elliptica* (King and Broderip) under heat stress. *Comp. Biochem. Physiol. C Toxicol. Pharmacol.* 134: 79–90.

Hemmingsen, A. M. 1960. Energy metabolism as related to body size and respiratory surfaces, and its evolution. *Reports of the Steno Memorial Hospital and Nordisk Insulinlaboratorium* 9: 1–110.

Henry, J. R., and J. F. Harrison. 2004. Plastic and evolved responses of larval tracheae and mass to varying atmospheric oxygen content in *Drosophila melanogaster*. *J. Exp. Biol.* 207: 3559–3567.

Hill, R. W., G. A. Wyse, and M. K. Anderson. 2016. *Animal Physiology*, 4th ed.. Sinauer Associates, Inc., Sunderland, MA.

Hindle, A. G., K. R. Grabek, L. E. Epperson, A. Karimpour-Fard, and S. L. Martin. 2014. Metabolic changes associated with the long winter fast dominate the liver proteome in 13-lined ground squirrels. *Physiol. Genomics* 46: 348–361.

Hittel, D., and K. B. Storey. 2002. The translation state of differentially expressed mRNAs in the hibernating 13-lined ground squirrel (*Spermophilus tridecemlineatus*). *Arch. Biochem. Biophys.* 401: 244–254.

Hochachka, P. W. 2003. Intracellular convection, homeostasis and metabolic regulation. *J. Exp. Biol.* 206: 2001–2009.

Hochachka, P. W., and G. N. Somero. 2002. *Biochemical Adaptation: Mechanism and Process in Physiological Evolution*. Oxford University Press, Oxford, UK.

Holland, H. D. 2006. The oxygenation of the atmosphere and oceans. *Philos. Trans. R. Soc. Lond. B Biol. Sci.* 361: 903–915.

Houser, D. S., and D. E. Crocker. 2004. Age, sex, and reproductive state influence free amino acid concentrations in the fasting elephant seal. *Physiol. Biochem. Zool.* 77: 838–846.

Hulbert, A. J., P. L. Else, S. C. Manolis, and M. D. Brand. 2002. Proton leak in hepatocytes and liver mitochondria from archosaurs (crocodiles) and allometric relationships for ectotherms. *J. Comp. Physiol. B*. 172: 387–397.

Hulbert, A. J., R. Pamplona, R. Buffenstein, and W. A. Buttemer. 2007. Life and death: metabolic rate, membrane composition, and life span of animals. *Physiol. Rev.* 87: 1175–1213.

Jinka, T. R., O. Toien, and K. L. Drew. 2011. Season primes the brain in an arctic hibernator to facilitate entrance into torpor mediated by adenosine A1 receptors. *J. Neurosci.* 31: 10752–10758.

Jo, S. H., M. K. Son, H. J. Koh, S. M. Lee, I. H. Song, Y. O. Kim, Y. S. Lee, K. S. Jeong, and 4 others. 2001. Control of mitochondrial redox balance and cellular defense against oxidative damage by mitochondrial NADP$^+$-dependent isocitrate dehydrogenase. *J. Biol. Chem.* 276: 16168–16176.

Johansson, B. W. 1996. The hibernator heart—nature's model of resistance to ventricular fibrillation. *Cardiovasc. Res.* 31: 826–832.

Johnston, D. T., F. Wolfe-Simon, A. Pearson, and A. H. Knoll. 2009. Anoxygenic photosynthesis modulated Proterozoic oxygen and sustained Earth's middle age. *Proc. Natl. Acad. Sci. USA* 106: 16925–16929.

Jungas, R. L., M. L. Halperin, and J. T. Brosnan. 1992. Quantitative analysis of amino acid oxidation and related gluconeogenesis in humans. *Physiol. Rev.* 72: 419–448.

Kaiser, A., C. J. Klok, J. J. Socha, W.-K. Lee, M. C. Quinlan, and J. F. Harrison. 2007. Increase in tracheal investment with beetle size supports hypothesis of oxygen limitation on insect gigantism. *Proc. Natl. Acad. Sci. USA* 104: 13198–13203.

Kamp, G., and H. P. Juretschke. 1987. An in vivo ^{31}P-NMR study of the possible regulation of glycogen phosphorylase a by phosphagen via phosphate in the abdominal muscle of the shrimp *Crangon crangon*. *Biochim. Biophys. Acta* 929: 121–127.

Kanatous, S. B., L. V. DiMichele, D. F. Cowan, and R. W. Davis. 1999. High aerobic capacities in the skeletal muscles of pinnipeds: adaptations to diving hypoxia. *J. Appl. Physiol.* 86: 1247–1256.

Kashiwaya, Y., K. Sato, N. Tsuchiya, S. Thomas, D. A. Fell, R. L. Veech, and J. V. Passonneau. 1994. Control of glucose utilization in working perfused rat heart. *J. Biol. Chem.* 269: 25502–25514.

Kay, L., K. Nicolay, B. Wieringa, V. Saks, and T. Wallimann. 2000. Direct evidence for the control of mitochondrial respiration by mitochondrial creatine kinase in oxidative muscle cells in situ. *J. Biol. Chem.* 275: 6937–6944.

Keller, M., A. M. Sommer, H. O. Pörtner, and D. Abele. 2004. Seasonality of energetic functioning and production of reactive oxygen species by lugworm (*Arenicola marina*) mitochondria exposed to acute temperature changes. *J. Exp. Biol.* 207: 2529–2538.

Kim, J. W., and C. V. Dang. 2005. Multifaceted roles of glycolytic enzymes. *Trends Biochem. Sci.* 30: 142–150.

Klok, C. J., A. J. Hubb, and J. F. Harrison. 2009. Single and multigenerational responses of body mass to atmospheric oxygen concentrations in *Drosophila melanogaster*: evidence for roles of plasticity and evolution. *J. Evol. Biol.* 22: 2496–2504.

Kooijman, S. A. L. M. 2010. *Dynamic Energy Budget Theory for Metabolic Organization*. Cambridge University Press, Cambridge.

Kooyman, G. L., and P. J. Ponganis. 1998. The physiological basis of diving to depth: birds and mammals. *Annu. Rev. Physiol.* 60: 19–32.

Krauss, S., C. Y. Zhang, and B. B. Lowell. 2005. The mitochondrial uncoupling-protein homologues. *Nat. Rev. Mol. Cell Biol.* 6: 248–261.

Krebs, H. A. 1950. Body size and tissue respiration. *Biochim. Biophys. Acta* 4: 249–269.

Krivoruchko, A., and K. B. Storey. 2010. Epigenetics in anoxia tolerance: a role for histone deacetylases. *Mol. Cell. Biochem.* 342: 151–161.

Kuzmiak, S., B. Glancy, K. L. Sweazea, and W. T. Willis. 2012. Mitochondrial function in sparrow pectoralis muscle. *J. Exp. Biol.* 215: 2039–2050.

Lalonde, S. V., and K. O. Konhauser. 2015. Benthic perspective on Earth's oldest evidence for oxygenic photosynthesis. *Proc. Natl. Acad. Sci. USA* 112: 995–1000.

Lane, N. 2002. *Oxygen: The Molecule that Made the World*. Oxford University Press, Oxford.

Le Boef, B. J., and R. M. Laws. 1994. *Elephant Seals: Population Ecology, Behavior, and Physiology*. University of Califonria Press, Berkeley.

Li, Q., R. Burgess, and Z. Zhang. 2012. All roads lead to chromatin: Multiple pathways for histone deposition. *Biochim. Biophys. Acta* 1819: 238–246.

Lopez-Bernardo, E., A. Anedda, P. Sanchez-Perez, B. Acosta-Iborra, and S. Cadenas. 2015. 4-Hydroxynonenal induces Nrf2-mediated UCP3 upregulation in mouse cardiomyocytes. *Free Radic. Biol. Med.* 88: 427–438.

Lunt, S. Y., and M. G. Vander Heiden. 2011. Aerobic glycolysis: meeting the metabolic requirements of cell proliferation. *Annu. Rev. Cell. Dev. Biol.* 27: 441–464.

Lushchak, V. I., H. M. Semchyshyn, and O. V. Lushchak. 2012. The classic methods to measure oxidative damage: lipid peroxides, thiobarbituric-acid reactive substances, and protein carbonyls. In D. Abele, J. P. Vásquez-Medina, and T. Zenteno-Savin (eds.), *Oxidative Stress in Aquatic Ecosystems*, pp. 420–431. Wiley-Blackwell, Chichester, UK.

Lutz, P. L., and S. L. Milton. 2004. Negotiating brain anoxia survival in the turtle. *J. Exp. Biol.* 207: 3141–3147.

Lutz, P. L., G. E. Nilsson, and H. Prentice. 2003. *The Brain without Oxygen: Causes of Failure of Molecular and Physiological Mechanisms for Survival*. Kluwers Press, Dordrecht, Netherlands.

Lyons, T. W., C. T. Reinhard, and N. J. Planavsky. 2014. The rise of oxygen in Earth's early ocean and atmosphere. *Nature* 506: 307–315.

Ma, Q. 2013. Role of nrf2 in oxidative stress and toxicity. *Annu. Rev. Pharmacol. Toxicol.* 53: 401–426.

Magarinos, A. M., B. S. McEwen, M. Saboureau, and P. Pevet. 2006. Rapid and reversible changes in intrahippocampal connectivity during the course of hibernation in European hamsters. *Proc. Natl. Acad. Sci. USA* 103: 18775–18780.

Mailloux, R. J., R. Beriault, J. Lemire, R. Singh, D. R. Chenier, R. D. Hamel, and V. D. Appanna. 2007. The tricarboxylic acid cycle, an ancient metabolic network with a novel twist. *PLoS One* 2: e690.

Mailloux, R. J., S. L. McBride, and M. E. Harper. 2013. Unearthing the secrets of mitochondrial ROS and glutathione in bioenergetics. *Trends Biochem. Sci.* 38: 592–602.

Mamady, H., and K. B. Storey. 2008. Coping with the stress: expression of ATF4, ATF6, and downstream targets in organs of hibernating ground squirrels. *Arch. Biochem. Biophys.* 477: 77–85.

Marden, J. H. 2013. Nature's inordinate fondness for metabolic enzymes: why metabolic enzyme loci are so frequently targets of selection. *Mol. Ecol.* 22: 5743–5764.

Marden, J. H., H. W. Fescemyer, R. J. Schilder, W. R. Doerfler, J. C. Vera, and C. W. Wheat. 2013. Genetic variation in HIF signaling underlies quantitative variation in physiological and life-history traits within lowland butterfly populations. *Evolution.* 67: 1105–1115.

Margittai, E., and G. Banhegyi. 2008. Isocitrate dehydrogenase: A NADPH-generating enzyme in the lumen of the endoplasmic reticulum. *Arch. Biochem. Biophys.* 471: 184–190.

Marín-Hernández, A., S. Rodríguez-Enríquez, P. A. Vital-González, F. L. Flores-Rodríguez, M. Macías-Silva, M. Sosa-Garrocho, and R. Moreno-Sánchez. 2006. Determining and understanding the control of glycolysis in fast-growth tumor cells. Flux control by an over-expressed but strongly product-inhibited hexokinase. *FEBS J.* 273: 1975–1988.

Martin, S. L., G. D. Maniero, C. Carey, and S. C. Hand. 1999. Reversible depression of oxygen consumption in isolated liver mitochondria during hibernation. *Physiol. Biochem. Zool.* 72: 255–264.

Mathieu, O., R. Krauer, H. Hoppeler, P. Gehr, S. L. Lindstedt, R. M. Alexander, C. R. Taylor, and E. R. Weibel. 1981. Design of the mammalian respiratory system. VII. Scaling mitochondrial volume in skeletal muscle to body mass. *Respir. Physiol.* 44: 113–128.

McKechnie, A. E., and B. O. Wolf. 2004. The allometry of avian basal metabolic rate: good predictions need good data. *Physiol. Biochem. Zool.* 77: 502–521.

Meir, A., S. Ginsburg, A. Butkevich, S. G. Kachalsky, I. Kaiserman, R. Ahdut, S. Demirgoren, and R. Rahamimoff. 1999. Ion channels in presynaptic nerve terminals and control of transmitter release. *Physiol. Rev.* 79: 1019–1088.

Metallo, C. M., and M. G. Vander Heiden. 2010. Metabolism strikes back: metabolic flux regulates cell signaling. *Genes Dev.* 24: 2717–2722.

Meyer, L. E., L. B. Machado, A. P. Santiago, W. S. da-Silva, F. G. De Felice, O. Holub, M. F. Oliveira, and A. Galina. 2006. Mitochondrial creatine kinase activity prevents reactive oxygen species generation: antioxidant role of mitochondrial kinase-dependent ADP re-cycling activity. *J. Biol. Chem.* 281: 37361–37371.

Michal, G., and D. Schomburg. 2012. *Biochemical Pathways: An Atlas of Biochemistry and Molecular Biology.* John Wiley & Sons, Inc., Hoboken, New Jersey.

Milton, S. L., J. W. Thompson, and P. L. Lutz. 2002. Mechanisms for maintaining extracellular glutamate levels in the anoxic turtle striatum. *Am. J. Physiol. Regul. Integr. Comp. Physiol.* 282: R1317–R1323.

Moore, M. J. 2005. From birth to death: the complex lives of eukaryotic mRNAs. *Science* 309: 1514–1518.

Morin, P., Jr., and K. B. Storey. 2005. Cloning and expression of hypoxia-inducible factor 1alpha from the hibernating ground squirrel, *Spermophilus tridecemlineatus. Biochim. Biophys. Acta* 1729: 32–40.

Müller, M., M. Mentel, J. J. van Hellemond, K. Henze, C. Woehle, S. B. Gould, R. Y. Yu, M. van der Giezen, and 2 others. 2012. Biochemistry and evolution of anaerobic energy metabolism in eukaryotes. *Microbiol. Mol. Biol. Rev.* 76: 444–495.

Murphy, M. P. 2009. How mitochondria produce reactive oxygen species. *Biochem. J.* 417: 1–13.

Murphy, M. P. 2012. Mitochondrial thiols in antioxidant protection and redox signaling: distinct roles for glutathionylation and other thiol modifications. *Antioxid. Redox Signal.* 16: 476–495.

Nelson, D. L., and M. M. Cox. 2008. *Lehninger Principles of Biochemistry.* W. H. Freeman and Company, New York.

Nguyen, V. D., M. J. Saaranen, A. R. Karala, A. K Lappi, L. Wang, I. B. Raykhel, H. I. Alanen, K. E. Salo, and 2 others. 2011. Two endoplasmic reticulum PDI peroxidases increase the efficiency of the use of peroxide during disulfide bond formation. *J. Mol. Biol.* 406: 503–515.

Nordgren, M., and M. Fransen. 2014. Peroxisomal metabolism and oxidative stress. *Biochimie* 98: 56–62.

O'Brien, K. M., I. A. Mueller, J. I. Orczewska, K. R. Dullen, and M. Ortego. 2014. Hearts of some Antarctic fishes lack mitochondrial creatine kinase. *Comp. Biochem. Physiol. A Mol. Integr. Physiol.* 178: 30–36.

O'Toole, P. W., and I. B. Jeffery. 2015. Gut microbiota and aging. *Science* 350: 1214–1215.

Olson, K. R., and K. D. Straub. 2016. The role of hydrogen sulfide in evolution and the evolution of hydrogen sulfide in metabolism and signaling. *Physiology* 31: 60–72.

Overgaard, J., H. Gesser, and T. Wang. 2007. Tribute to P. L. Lutz: cardiac performance and cardiovascular regulation during anoxia/hypoxia in freshwater turtles. *J. Exp. Biol.* 210: 1687–1699.

Overgaard, J., T. Wang, O. B. Nielsen, and H. Gesser. 2005. Extracellular determinants of cardiac contractility in the cold anoxic turtle. *Physiol. Biochem. Zool.* 78: 976–995.

Patil, Y. N., B. Marden, M. D. Brand, and S. C. Hand. 2013. Metabolic downregulation and inhibition of carbohydrate catabolism during diapause in embryos of *Artemia franciscana. Physiol. Biochem. Zool.* 86: 106–118.

Planavsky, N. J., P. McGoldrick, C. T. Scott, C. Li, C. T. Reinhard, A. E. Kelly, X. Chu, A. Bekker, and 2 others. 2011. Widespread iron-rich conditions in the mid-Proterozoic ocean. *Nature* 477: 448–451.

Planavsky, N. J., C. T. Reinhard, X. Wang, D. Thomson, P. McGoldrick, R. H. Rainbird, T. Johnson, W. W. Fischer, and T. W. Lyons. 2014. Earth history. Low mid-Proterozoic atmospheric oxygen levels and the delayed rise of animals. *Science* 346: 635–638.

Podrabsky, J. E., I. D. Garrett, and Z. F. Kohl. 2010. Alternative developmental pathways associated with diapause regulated by temperature and maternal influences in embryos of the annual killifish *Austrofundulus limnaeus. J. Exp. Biol.* 213: 3280–3288.

Podrabsky, J. E., and S. C. Hand. 1999. The bioenergetics of embryonic diapause in an annual killifish, *Austrofundulus limnaeus. J. Exp. Biol.* 202: 2567–2580.

Podrabsky, J. E., and S. C. Hand. 2015. Physiological strategies during animal diapause: lessons from brine shrimp and annual killifish. *J. Exp. Biol.* 218: 1897–1906.

Pollak, N., C. Dolle, and M. Ziegler. 2007. The power to reduce: pyridine nucleotides—small molecules with a multitude of functions. *Biochem. J.* 402: 205–218.

Porter, R. K. 2001. Allometry of mammalian cellular oxygen consumption. *Cell. Mol. Life Sci.* 58: 815–822.

Pörtner, H. O. 2010. Oxygen- and capacity-limitation of thermal tolerance: a matrix for integrating climate-related stressor effects in marine ecosystems. *J. Exp. Biol.* 213: 881–893.

Powell, M. A., and G. N. Somero. 1986. Hydrogen sulfide oxidation is coupled to oxidative phosphorylation in mitochondria of *Solemya reidi. Science* 233: 563–566.

Qureshi, I. A., and M. F. Mehler. 2012. Emerging roles of non-coding RNAs in brain evolution, development, plasticity and disease. *Nat. Rev. Neurosci.* 13: 528–541.

Ramnanan, C. J., and K. B. Storey. 2006. Glucose-6-phosphate dehydrogenase regulation during hypometabolism. *Biochem. Biophys. Res. Commun.* 339: 7–16.

Rasmussen, B., I. R. Fletcher, J. J. Brocks, and M. R. Kilburn. 2008. Reassessing the first appearance of eukaryotes and cyanobacteria. *Nature* 455: 1101–1104.

Rawson, P. D., and R. S. Burton. 2002. Functional coadaptation between cytochrome c and cytochrome c oxidase within allopatric populations of a marine copepod. *Proc. Natl. Acad. Sci. USA* 99: 12955–12958.

Rea, L. D., M. Berman-Kowalewski, D. A. Rosen, and A. W. Trites. 2009. Seasonal differences in biochemical adaptation to fasting in juvenile and subadult Steller sea lions (*Eumetopias jubatus*). *Physiol. Biochem. Zool.* 82: 236–247.

Ringwood, A. H., and C. J. Keppler. 2002. Water quality variation and clam growth: is pH really a non-issue in estuaries. *Estuaries.* 25: 901–907.

Robinson, C., D. K. Steinberg, T. R. Anderson, J. Arísteguid, C. A. Carlsone, J. R. Frostf, J.-F. Ghiglioneg, S. Hernández-Leónd, and 10 others. 2010. Mesopelagic zone ecology and biogeochemistry—a synthesis. *Deep Sea Res., Part II* 57: 1504–1518.

Rolfe, D. F., and G. C. Brown. 1997. Cellular energy utilization and molecular origin of standard metabolic rate in mammals. *Physiol. Rev.* 77: 731–758.

Rosen, D., and A. Hindle. 2016. Fasting. In M. A. Castellini and J.-A. Mellish (eds.), *Marine Mammal Physiology*, pp. 169–191. CRC Press Taylor & Francis Group, Boca Raton.

Ruf, T., and F. Geiser. 2015. Daily torpor and hibernation in birds and mammals. *Biol. Rev.* 90: 891–926.

Rydstrom, J. 2006. Mitochondrial NADPH, transhydrogenase and disease. *Biochim. Biophys. Acta* 1757: 721–726.

Rye, R., and H. D. Holland. 1998. Paleosols and the evolution of atmospheric oxygen: A crticial review. *Am. J. Sci.* 298: 621–672.

Ryslava, H., V. Doubnerova, D. Kavan, and O. Vanek. 2013. Effect of posttranslational modifications on enzyme function and assembly. *J. Proteomics* 92: 80–109.

Savage, V. M. 2004. Improved approximations to scaling relationships for species, populations, and ecosystems across latitudinal and elevational gradients. *J. Theor. Biol.* 227: 525–534.

Schmidt-Nielsen, K. 1984. *Scaling: Why is Animal Size so Important?* Cambridge University Press, Cambrdige.

Schönfeld, P., M. R. Wieckowski, M. Lebiedzinska, and L. Wojtczak. 2010. Mitochondrial fatty acid oxidation and oxidative stress: lack of reverse electron transfer-associated production of reactive oxygen species. *Biochim. Biophys. Acta* 1797: 929–938.

Schönfeld, P., and L. Wojtczak. 2008. Fatty acids as modulators of the cellular production of reactive oxygen species. *Free Radic. Biol. Med.* 45: 231–241.

Secor, S. M. 2009. Specific dynamic action: a review of the postprandial metabolic response. *J. Comp. Physiol. B.* 179: 1–56.

Selak, M. A., S. M. Armour, E. D. MacKenzie, H. Boulahbel, D. G. Watson, K. D. Mansfield, Y. Pan, M. C. Simon, and 2 others. 2005. Succinate links TCA cycle dysfunction to oncogenesis by inhibiting HIF-alpha prolyl hydroxylase. *Cancer Cell* 7: 77–85.

Semenza, G. L. 2010. Defining the role of hypoxia-inducible factor 1 in cancer biology and therapeutics. *Oncogene* 29: 625–634.

Shoubridge, E. A., and P. W. Hochachka. 1980. Ethanol: novel end product of vertebrate anaerobic metabolism. *Science* 209: 308–309.

Sokolova, I. M., M. Frederich, R. Bagwe, G. Lannig, and A. A. Sukhotin. 2012. Energy homeostasis as an integrative tool for assessing limits of environmental stress tolerance in aquatic invertebrates. *Mar. Environ. Res.* 79: 1–15.

Somero, G. N., and J. J. Childress. 1980. A violation of the metabolism-size scaling paradigm: Activities of glycolytic enzymes in muscle increase in larger-size fish. *Physiol. Zool.* 53: 322–337.

Somo, D. A., D. C. Ensminger, J. T. Sharick, S. B. Kanatous, and D. E. Crocker. 2015. Development of dive capacity in northern elephant seals (*Mirounga angustirostris*): Reduced body reserves at weaning are associated with elevated body oxygen stores during the postweaning fast. *Physiol. Biochem. Zool.* 88: 471–482.

Spang, A., J. H. Saw, S. L. Jorgensen, K. Zaremba-Niedzwiedzka, J. Martijn, A. E. Lind, R. van Eijk, C. Schleper, and 2 others. 2015. Complex archaea that bridge the gap between prokaryotes and eukaryotes. *Nature* 521: 173–179.

Sperling, E. A., C. A. Frieder, A. V. Raman, P. R. Girguis, L. A. Levin, and A. H. Knoll. 2013. Oxygen, ecology, and the Cambrian radiation of animals. *Proc. Natl. Acad. Sci. USA* 110: 13446–13451.

Sperling, E. A., A. H. Knoll, and P. R. Girguis. 2015. The ecological physiology of earth's second oxygen revolution. *Annu. Rev. Ecol., Evol. Syst.* 46: 215–235.

Starkov, A. A. 2013. An update on the role of mitochondrial alpha-ketoglutarate dehydrogenase in oxidative stress. *Mol. Cell. Neurosci.* 55: 13–16.

Stecyk, J. A., J. Overgaard, A. P. Farrell, and T. Wang. 2004. Alpha-adrenergic regulation of systemic peripheral resistance and blood flow distribution in the turtle *Trachemys scripta* during anoxic submergence at 5°C and 21°C. *J. Exp. Biol.* 207: 269–283.

Storey, K. B. 1997. Metabolic regulation in mammalian hibernation: enzyme and protein adaptations. *Comp. Biochem. Physiol. A Physiol.* 118: 1115–1124.

Storey, K. B. 2007. Anoxia tolerance in turtles: metabolic regulation and gene expression. *Comp. Biochem. Physiol. A Mol. Integr. Physiol.* 147: 263–276.

Storey, K. B. 2015. Regulation of hypometabolism: insights into epigenetic controls. *J. Exp. Biol.* 218: 150–159.

Storey, K. B., G. Heldmaier, and M. H. Rider. 2010. Mammalian hibernation: physiology, cell signaling and gene controls on metabolic rate depression. In E. Lubzens (ed.), *Dormancy and Resistance in Harsh Environments*, vol. 21, pp. 227–252. Springer-Verlag, Berlin.

Suarez, R. K. 1992. Hummingbird flight: sustaining the highest mass-specific metabolic rates among vertebrates. *Experientia* 48: 565–570.

Suarez, R. K., and C. A. Darveau. 2005. Multi-level regulation and metabolic scaling. *J. Exp. Biol.* 208: 1627–1634.

Suarez, R. K., C. A. Darveau, and J. J. Childress. 2004. Metabolic scaling: a many-splendoured thing. *Comp. Biochem. Physiol. B Biochem. Mol. Biol.* 139: 531–541.

Suswam, E. A., Y. Y. Li, H. Mahtani, and P. H. King. 2005. Novel DNA-binding properties of the RNA-binding protein TIAR. *Nucleic Acids Res.* 33: 4507–4518.

Szabo, C., C. Ransy, K. Modis, M. Andriamihaja, B. Murghes, C. Coletta, G. Olah, K. Yanagi, and F. Bouillaud. 2014. Regulation of mitochondrial bioenergetic function by hydrogen sulfide. Part I. Biochemical and physiological mechanisms. *Br. J. Pharmacol.* 171: 2099–2122.

Tahara, E. B., F. D. Navarete, and A. J. Kowaltowski. 2009. Tissue-, substrate-, and site-specific characteristics of mitochondrial reactive oxygen species generation. *Free Radic. Biol. Med.* 46: 1283–1297.

Talaei, F., H. R. Bouma, M. N. Hylkema, H. R. Bouma, A. S. Boerema, A. M. Strijkstra, R. H. Henning, and M. Schmidt. 2012. The role of endogenous H_2S formation in reversible remodeling of lung tissue during hibernation in the Syrian hamster. *J. Exp. Biol.* 215: 2912–2919.

Tavender, T. J., J. J. Springate, and N. J. Bulleid. 2010. Recycling of peroxiredoxin IV provides a novel pathway for disulphide formation in the endoplasmic reticulum. *EMBO J.* 29: 4185–4197.

Templeton, J. R. 1970. Reptiles. In G. C. Whittow (ed.), *Comparative Physiology of Thermoregulation*, vol. 1, pp. 167–221. Academic Press, New York.

Tessier, S. N., T. E. Audas, C. W. Wu, S. Lee, and K. Storey. 2014. The involvement of mRNA processing factors TIA-1, TIAR, and PABP-1 during mammalian hibernation. *Cell Stress Chaperones.* 19: 813–825.

Tew, K. D., and D. M. Townsend. 2012. Glutathione-s-transferases as determinants of cell survival and death. *Antioxid. Redox Signal.* 17: 1728–1737.

Tift, M. S., E. C. Ranalli, D. S. Houser, R. M. Ortiz, and D. E. Crocker. 2013. Development enhances hypometabolism in northern elephant seal pups (*Mirounga angustirostris*). *Funct. Ecol.* 27: 1155–1165.

Tomanek, L. 2015. Proteomic responses to environmentally induced oxidative stress. *J. Exp. Biol.* 218: 1867–1879.

Tomanek, L., and M. J. Zuzow. 2010. The proteomic response of the mussel congeners *Mytilus galloprovincialis* and *M. trossulus* to acute heat stress: implications for thermal tolerance and metabolic costs of thermal stress. *J. Exp. Biol.* 213: 3559–3574.

Tomanek, L., M. J. Zuzow, L. R. Hitt, L. Serafini, and J. J. Valenzuela. 2012. Proteomics of hyposaline stress in blue mussel congeners (genus *Mytilus*): implications for biogeographic range limits in response to climate change. *J. Exp. Biol.* 215: 3905–3916.

Tomanek, L., M. J. Zuzow, A. V. Ivanina, E. Beniash, and I. M. Sokolova. 2011. Proteomic response to elevated P_{CO2} level in eastern oyster, *Crassostrea virginica*: evidence for oxidative stress. *J. Exp. Biol.* 214: 1836–1844.

Torres, J. J., M. D. Grigsby, and M. E. Clarke. 2012. Aerobic and anaerobic metabolism in oxygen minimum layer fishes: the role of alcohol dehydrogenase. *J. Exp. Biol.* 215: 1905–1914.

Townsend, D. M., Y. Manevich, L. He, S. Hutchens, C. J. Pazoles, and K. D. Tew. 2009. Novel role for glutathione S-transferase pi. Regulator of protein S-glutathionylation following oxidative and nitrosative stress. *J. Biol. Chem.* 284: 436–445.

Tripodi, F., R. Nicastro, V. Reghellin, and P. Coccetti. 2015. Post-translational modifications on yeast carbon metabolism: Regulatory mechanisms beyond transcriptional control. *Biochim. Biophys. Acta* 1850: 620–627.

Trujillo, M., G. Ferrer-Sueta, and R. Radi. 2008. Peroxynitrite detoxification and its biologic implications. *Antioxid. Redox Signal.* 10: 1607–1620.

Ultsch, G. R. 2006. The ecology of overwintering among turtles: where turtles overwinter and its consequences. *Biol. Rev.* 81: 339–367.

van Breukelen, F., and S. L. Martin. 2001. Translational initiation is uncoupled from elongation at 18°C during mammalian hibernation. *Am. J. Physiol. Regul. Integr. Comp. Physiol.* 281: R1374–R1379.

van Breukelen, F., and S. L. Martin. 2002. Reversible depression of transcription during hibernation. *J. Comp. Physiol. B.* 172: 355–361.

van Breukelen, F., and S. L. Martin. 2015. The hibernation continuum: Physiological and molecular aspects of metabolic plasticity in mammals. *Physiology* 30: 273–281.

Verdin, E. 2015. $NAD(^+)$ in aging, metabolism, and neurodegeneration. *Science* 350: 1208–1213.

von der Ohe, C. G., C. Darian-Smith, C. C. Garner, and H. C. Heller. 2006. Ubiquitous and temperature-dependent neural plasticity in hibernators. *J. Neurosci.* 26: 10590–10598.

Wagner, J. T., and J. E. Podrabsky. 2015. Extreme tolerance and developmental buffering of UV-C induced DNA damage in embryos of the annual killifish *Austrofundulus limnaeus. J. Exp. Zool. A Ecol. Genet. Physiol.* 323: 10–30.

Wallace, D. C. 2005. A mitochondrial paradigm of metabolic and degenerative diseases, aging, and cancer: a dawn for evolutionary medicine. *Annu. Rev. Genet.* 39: 359–407.

Wallimann, T., M. Tokarska-Schlattner, D. Neumann, R. M. Epand, R. F. Epand, R. H. Andres, H. R. Widmer, T. Hornemann, and 3 others. 2007. The phosphocreatine circuit: molecular and cellular physiology of creatine kinases, sensitivity to free radicals, and enhancement by creatine supplementation. In V. Saks (ed.), *Molecular System Bioenergetics: Energy for Life*, pp. 195–264. WILEY-VCH Verlag GmbH & Co. KGaA, Weinheim.

Walsh, C. T. 2006. *Posttranslational Modification of Proteins: Expanding Nature's Inventory.* Roberts and Company Publishers, Englewood, Colorado.

Walsh, C. T., S. Garneau-Tsodikova, and G. J. Gatto, Jr. 2005. Protein posttranslational modifications: the chemistry of proteome diversifications. *Angew. Chem. Int. Ed. Engl.* 44: 7342–7372.

Wang, P., F. G. Bouwman, and E. C. Mariman. 2009. Generally detected proteins in comparative proteomics—a matter of cellular stress response? *Proteomics* 9: 2955–2966.

Wang, Y., and S. Hekimi. 2015. Mitochondrial dysfunction and longevity in animals: Untangling the knot. *Science* 350: 1204–1207.

Wang, Z., T. P. O'Connor, S. Heshka, and S. B. Heymsfield. 2001. The reconstruction of Kleiber's law at the organ-tissue level. *J. Nutr.* 131: 2967–2970.

Wasserman, D. H., L. Kang, J. E. Ayala, P. T. Fueger, and R. S. Lee-Young. 2011. The physiological regulation of glucose flux into muscle in vivo. *J. Exp. Biol.* 214: 254–262.

Weibel, E. R., L. D. Bacigalupe, B. Schmitt, and H. Hoppeler. 2004. Allometric scaling of maximal metabolic rate in mammals: muscle aerobic capacity as determinant factor. *Respir. Physiol. Neurobiol.* 140: 115–132.

Weltzin, M. M., H. W. Zhao, K. L. Drew, and D. J. Bucci. 2006. Arousal from hibernation alters contextual learning and memory. *Behav. Brain Res.* 167: 128–133.

West, G. B., J. H. Brown, and B. J. Enquist. 1997. A general model for the origin of allometric scaling laws in biology. *Science* 276: 122–126.

West, G. B., J. H. Brown, and B. J. Enquist. 1999. The fourth dimension of life: fractal geometry and allometric scaling of organisms. *Science* 284: 1677–1679.

West, G. B., W. H. Woodruff, and J. H. Brown. 2002. Allometric scaling of metabolic rate from molecules and mitochondria to cells and mammals. *Proc. Natl. Acad. Sci. USA* 99 Suppl 1: 2473–2478.

White, C. R., and R. S. Seymour. 2003. Mammalian basal metabolic rate is proportional to body mass$^{2/3}$. *Proc. Natl. Acad. Sci. USA* 100: 4046–4049.

White, P. C., D. Rogoff, D. R. McMillan, and G. G. Lavery. 2007. Hexose 6-phosphate dehydrogenase (H6PD) and corticosteroid metabolism. *Mol. Cell. Endocrinol.* 265–266: 89–92.

Whitford, W. G. 1973. The effects of temperature on respiration in the Amphibia. *Am. Zool.* 13: 505–512.

Widdel, F., S. Schnell, S. Heising, A. Ehrenreich, B. Assmus, and Bernhard Schink. 1993. Ferrous iron oxidation by anoxygenic phototrophic bacteria. *Nature* 362: 834–835.

Williams, C. T., A. V. Goropashnaya, C. L. Buck, V. B. Fedorov, F. Kohl, T. N. Lee, and B. M. Barnes. 2011. Hibernating above the permafrost: effects of ambient temperature and season on expression of metabolic genes in liver and brown adipose tissue of arctic ground squirrels. *J. Exp. Biol.* 214: 1300–1306.

Wong, K. K., A. C. Lane, P. T. Leung, and V. Thiyagarajan. 2011. Response of larval barnacle proteome to CO_2-driven seawater acidification. *Comp. Biochem. Physiol. Part D Genomics Proteomics* 6: 310–321.

Wourms, J. P. 1972. The developmental biology of annual fishes. 3. Pre-embryonic and embryonic diapause of variable duration in the eggs of annual fishes. *J. Exp. Zool.* 182: 389–414.

Wu, P., J. M. Peters, and R. A. Harris. 2001. Adaptive increase in pyruvate dehydrogenase kinase 4 during starvation is mediated by peroxisome proliferator-activated receptor alpha. *Biochem. Biophys. Res. Commun.* 287: 391–396.

Xiong, Y., Q. Y. Lei, S. Zhao, and K. L. Guan. 2011. Regulation of glycolysis and gluconeogenesis by acetylation of PKM and PEPCK. *Cold Spring Harb. Symp. Quant. Biol.* 76: 285–289.

Yamashita, K., and T. Yoshioka. 1991. Profiles of creatine kinase isoenzyme compositions in single muscle fibres of different types. *J. Muscle Res. Cell Motil.* 12: 37–44.

Yoshihara, T., T. Hamamoto, R. Munakata, R. Tajiri, M. Ohsumi, and S. Yokota. 2001. Localization of cytosolic NADP-dependent isocitrate dehydrogenase in the peroxisomes of rat liver cells: biochemical and immunocytochemical studies. *J. Histochem. Cytochem.* 49: 1123–1131.

Zhang, J., K. K. Biggar, and K. B. Storey. 2013. Regulation of p53 by reversible post-transcriptional and post-translational mechanisms in liver and skeletal muscle of an anoxia tolerant turtle, *Trachemys scripta elegans. Gene.* 513: 147–155.

Zhao, S., W. Xu, W. Jiang, W. Yu, Y. Lin, T. Zhang, J. Yao, L. Zhou, and 14 others. 2010. Regulation of cellular metabolism by protein lysine acetylation. *Science* 327: 1000–1004.

Zhegunov, G. F., Y. E. Mikulinsky, and E. V. Kudokotseva. 1988. Hyperactivation of protein synthesis in tissues of hibernating animals on arousal. *Cryo-Letters* 9: 236–245.

Chapter 3

Abele, D., K. Heise, H. O. Pörtner, and S. Puntarulo. 2002. Temperature dependence of mitochondrial function and production of reactive oxygen species in the intertidal mud clam *Mya arenaria. J. Exp. Biol.* 205: 1831–1841.

Ahern, T. J. and A. M. Klibanov. 1985. The mechanism of irreversible enzyme inactivation at 100°C. *Science* 228: 1280–1284.

Anfinsen, C. B. 1973. Principles that govern the folding of protein chains. *Science* 181: 223–230.

Angilletta, M. J. Jr. 2009. *Thermal Adaptation: A Theoretical and Empirical Synthesis.* Oxford: Oxford University Press.

Ashburner, M. 1970. Patterns of puffing activity in the salivary gland chromosomes of *Drosophila. Chromosoma* 31: 356–376.

Attrill, H., K. Falls, J. L. Goodman, G. H. Millburn, G. Antonazzo, A. J. Rey, and S. J. Marygold. 2016. FlyBase: establishing a Gene Group resource for *Drosophila melanogaster. Nucleic Acids Res.* 44: 786–792.

Baardsnes, J. and P. L. Davies. 2001. Sialic acid synthase: the origin of fish type III antifreeze protein? *Trends Biochem. Sci.* 26: 468–469.

Balasubramanian, S., S. Sureshkumar, J. Lempe, and D. Weigel. 2006. Potent induction of *Arabidopsis thaliana* flowering by elevated growth temperature. *PLoS Genetics* 2: e106.

Barcroft, J. 1934. *Features in the Architecture of Physiological Function.* Cambridge University Press, Cambridge, UK.

Bardwell, J. C., and E. A. Craig. 1984. Major heat shock gene of *Drosophila* and the *Escherichia coli* heat-inducible *dnaK* gene are homologous. *Proc. Natl. Acad. Sci. USA.* 81: 848–852.

Barnes, B. M. 1989. Freeze avoidance in a mammal: body temperatures below 0°C in an Arctic hibernator. *Science* 244: 1593–1595.

Battino, R., F. D. Evans, and W. F. Danforth. 1968. The solubilities of seven gases in olive oil with reference to theories of transport through the cell membrane. *J. Am. Oil Chem. Soc.* 45: 830–833.

Baxter, J. D., and P. Webb. 2006. Bile acids heat things up. *Nature* 439: 402–403.

Beeby, M., B. D. O'Connor, C. Ryttersgaard, D. R. Boutz, L. J. Perry, and T. O. Yeates. 2005. The genomics of disulfide bonding and protein stabilization in thermophiles. *PLoS Biol.* 3(9) e309.

Beece, D., L. Eisenstein, H. Frauenfelder, D. Good, M. C. Marden, L. Reinisch, A. H. Reynolds, L. B. Sorensen, and K. T. Yue. 1980. Solvent viscosity and protein dynamics. *Biochemistry* 19: 5147–5157.

Beers, J., and N. Jayasundara. 2015. Antarctic notothenioid fish: what are the future consequences of 'losses' and 'gains' acquired during long-term evolution at cold and stable temperature? *J. Exp. Biol.* 218: 1834–1845.

Behan-Martin, M., K. Bowler, G. Jones, and A. R. Cossins. 1993. A near perfect adaptation of bilayer order in vertebrate brain membranes. *Biochim. Biophys. Acta* 1151: 216–222.

Benkovic, S. J., and S. Hammes-Schiffer. 2006. Enzyme motions inside and out. *Science* 213: 208–209.

Bennett, V. A., T. Sformo, K. Walters, O. Toien, K. Jeannet, R. Hochstrasser, Q. Pan, A. S. Serianni, B. M. Barnes, and J. G. Duman. 2005. Comparative overwintering physiology of Alaska and Indiana populations of the beetle *Cucujus clavipes* (Fabricus): Roles of antifreeze proteins, polyols, dehydrations, and diapause. *J. Exp. Biol.* 208: 4467–4477.

Berezovsky, I. N., and E. I. Shakhnovich. 2005. Physics and evolution of thermophilic adaptation. *Proc. Natl. Acad. Sci. USA* 102: 12742–12747.

Bernardi, G. 1995. The human genome: organization and evolutionary history. *Annu. Rev. Genet.* 29: 445–476.

Bettencourt, B. R., M. E. Feder, S. Cavicchi, and F. Selmi. 1999. Experimental evolution of Hsp70 expression and thermotolerance in *Drosophila melanogaster*. *Evolution* 53: 484–492.

Bettencourt, B. R., Kim, I., Hoffmann, A. A., and M. E. Feder. 2002. Response to natural and laboratory selection at the *Drosophila Hsp70* genes. *Evolution* 56: 1796–1801.

Bettencourt, B. R., C. C. Hogan, M. Nimali, and B. W. Drohan. 2008. Inducible and constitutive heat shock gene expression responds to modification of *Hsp70* copy number in *Drosophila melanogaster* but does not compensate for loss of thermotolerance in *Hsp70* null flies. *BMC Biol.* 6:5.

Bianchi, M. E., and A. Agresti. 2005. HMG proteins: dynamic players in gene regulation and differentiation. *Curr. Opin. Genet. Dev.* 15: 1–11.

Block, B. A. 1991. Endothermy in fish: thermogenesis, ecology and evolution. In P.W. Hochachka and T. Mommsen (eds.), *Biochemistry and Molecular Biology of Fish*, Vol. 1, pp. 269–311. Elsevier, London.

Block, B. A. 1994. Thermogenesis in muscle. *Annu. Rev. Physiol.* 56: 535–577.

Blumberg, H., and P. A. Silver. 1991. A homologue of the bacterial heat-shock gene *DnaJ* that alters protein sorting in yeast. *Nature* 349: 627–630.

Boni, L. T., and S. W. Hui. 1983. Polymorphic phase behavior of dilinoleoylphosphatidylethanolamine and palmitoyloleoylphosphatidylcholine mixtures. Structural changes between hexagonal, cubic, and bilayer phases. *Biochim. Biophys. Acta* 731: 177–185.

Boussau, B., S. Blanquart, A. Necsulea, N. Lartillot, and M. Gouy. 2008. Parallel adaptations to high temperature in the Archaean eon. *Nature* 456: 942–946.

Brand, M. D, P. Couture, P. L. Else, K. W. Withers, and A. J. Hulbert. 1991. Evolution of energy metabolism. *Biochem. J.* 275: 81–86.

Brown, J. B., N. Boley, R. Eisman, G. E. May, M. H. Stoiber, M. O. Duff, B. W. Booth, J. Wen, and 33 others. 2014. Diversity and dynamics of the *Drosophila* transcriptome. *Nature* 512: 1–7.

Brown, J. H., J. F. Gillooly, A. P. Allen, V. M. Savage, and G. B. West. 2004. Toward a metabolic theory of ecology. *Ecology* 85: 1771–1789.

Buckley, B. A., and G. N. Somero. 2009. cDNA microarray analysis reveals the capacity of the cold-adapted Antarctic fish *Trematomus bernacchii* to alter gene expression in response to heat stress. *Polar Biol.* 32: 403–415.

Buckley, B. A., A. Y. Gracey, and G. N. Somero. 2006. The cellular response to heat stress in the goby *Gillichthys mirabilis*: A cDNA microarray and protein-level analysis. *J. Exp. Biol.* 209: 2660–2677.

Buckley, L. B., and R. B. Huey. 2016. How extreme temperatures impact organisms and the evolution of their thermal tolerance. *Integr. Comp. Biol.* 56: 98–109.

Campbell, H. A., K. P. P. Fraser, C. M. Bishop, L. S. Peck, and S. Eggington. 2008. Hibernation in an Antarctic fish: On ice for winter. *PLoS One* 3: e1743.

Cannon, B., and J. Nedergaard. 2004. Brown adipose tissue: Function and physiological significance. *Physiol. Rev.* 84: 277–359.

Cannon, B., and J. Nedergaard. 2008. Neither fat nor flesh. *Nature* 454: 947–948.

Cantó, C., Z. Gerhart-Hines, J. N. Feige, M. Lagouge, L. Noriega, J. C. Milne, P. J. Elliott, P. Puigserver, and J. Auwerx. 2009. AMPK regulates energy expenditure by modulating NAD$^+$ metabolism and SIRT1 activity. *Nature* 458: 1056–1060.

Capicciotti, C. J., M. Doshi, and R. N. Ben. 2013. Ice recrystallization inhibitors: From biological antifreezes to small molecules. In P. Wilson (ed.), *Recent Developments in the Study of Recrystallization*. InTech. DOI: 10.5772/54992.

Cary, S.C., T. Shank, and J. Stein. 1998. Worms bask in extreme temperatures. *Nature* 391: 545–546.

Cavicchi, S., Guerra, D., Natali, V., Pezzoli, C., and G. Giorgi. 1989. Temperature-related divergence in experimental populations of *Drosophila melanogaster* II. Correlation between fitness and body dimensions. *J. Evol. Biol.* 2: 235–252.

Chen, Z., C.-H. C. Cheng, J. Zhang, L. Cao, L. Chen, L. Zhou, Y. Jin, H. Ye, and 7 others. 2008. Transcriptomic and genomic evolution under constant cold in Antarctic fish. *Proc. Natl. Acad. Sci. USA* 105: 12944–12949.

Chen, L., A. L. DeVries, and C. C.-H. Cheng. 1997. Evolution of antifreeze glycoprotein gene from a trypsinogen gene in Antarctic notothenioid fish. *Proc. Natl. Acad. Sci. USA* 94: 3811–3816.

Cheng, C.-H.C., and L. Chen. 1999. Evolution of an antifreeze glycoprotein. *Nature* 401: 443–444.

Chevalier, C., O. Stojanović, D. J. Colin, N. Suarez-Zamorano, V. Tarallo, C. Veyrat-Durebex, D. Rigo, S. Fabbiano, and 7 others. 2015. Gut microbiota orchestrates energy homeostasis during cold. *Cell* 163: 1360–1374.

Childress, J. J. 1995. Are there physiological and biochemical adaptations of metabolism in deep-sea animals? *Trends Ecol. Evol.* 10: 30–36.

Childress, J. J., and G. N. Somero. 1979. Depth-related enzymic activities in muscle, heart and brain of deep-living pelagic marine teleosts. *Mar. Biol.* 52: 273–282.

Chouchani, E. T., L. Kazak, M. P. Jedrychowski, G. Z. Lu, B. K. Erickson, J. Szpyt, K. A. Pierce, D. Laznik-Bogoslavski, and 5 others. 2016. Mitochondrial ROS regulate thermogenic energy expenditure and sulfenylation of UCP1. *Nature* 532: 112–116.

Christner, B. C., C. E. Morris, C. M. Foreman, R. Cai, and D. C. Sands. 2008. Ubiquity of biological ice nucleators in snowfall. *Science* 319: 1214.

Ciechanover, A. 2005. Proteolysis: from the lysosome to ubiquitin and the proteasome. *Nat. Rev. Mol. Cell Biol.* 6: 79–87.

Clarke, A. 1991. What is cold adaptation and how should we measure it? *Am. Zool.* 31: 81–92.

Clarke, A. 2003. Costs and consequences of evolutionary temperature adaptation. *Trends Ecol. Evol.* 18: 573–579.

Close, T. J. 1997. Dehydrins: A commonality in the response of plants to dehydration and low temperature. *Physiol. Plant.* 100: 291–296.

Connor, K. M., and A. Y. Gracey. 2011. Circadian cycles are the dominant transcriptional rhythm in the intertidal mussel *Mytilus californianus*. *Proc. Natl. Acad. Sci. USA* 108: 16110–16115.

Cooper, B. S., L. A. Hammad, and K. L. Montooth. 2014. Thermal adaptation of cellular membranes in natural populations of *Drosophila melanogaster*. *Func. Ecol.* 28: 886–894.

Cooper, B. S., L. A. Hammad, N. P. Fisher, J. A. Karty, and K. L. Montooth. 2012. In a variable thermal environment selection favors greater plasticity of cell membranes in *Drosophila melanogaster*. *Evolution* 66: 1976–1984.

Coquelle, N., E. Fioravanti, M. Weik, F. Vellieux, and D. Madern. 2007. Activity, stability and structural studies of lactate dehydrogenases adapted to extreme thermal environments. *J. Mol. Biol.* 374: 547–562.

Cornelius, F., M. Habeck, R. Kanai, C. Toyoshima, and S. J. D. Karlish. 2015. General and specific lipid-protein interactions in Na,K-ATPase. *Biochim. Biophys. Acta* 1848: 1729–1743.

Cossins, A .R., M. J. Friedlander, and C. L. Prosser. 1977. Correlations between behavioural temperature adaptations by goldfish and the viscosity and fatty acid composition of their synaptic membranes. *J. Comp. Physiol.* 120: 109–121.

Costanza, J. P., and R. E. Lee, Jr. 2013. Avoidance and tolerance of freezing in ectothermic vertebrates. *J. Exp. Biol.* 216: 1961–1967.

Cowan, D. A. 2004. The upper temperature for life—where do we draw the line? *Trends Microbiol.* 12: 58–60.

Creighton, T. E. 1993. *Proteins: Structures and Molecular Properties*. W.H. Freeman, New York.

Cziko, P. A., A. L. DeVries, C. W. Evans, and C.-H. C. Cheng. 2014. Antifreeze protein-induced superheating of ice inside Antarctic fishes counters melting by seasonal warming. *Proc. Natl. Acad. Sci. USA* 111: 14583–14588.

D'Amico, S., C. Gerday, and G. Feller. 2001. Structural determinants of cold adaptation and stability in a large protein. *J. Biol. Chem.* 276: 25791–25796.

D'Amico, S., J.-C. Marx, C. Gerday, and G. Feller. 2003. Activity-stability relationships in extremophilic enzymes. *J. Biol. Chem.* 278: 7891–7896.

Dahlhoff, E. P., and G. N. Somero. 1993a. Kinetic and structural adaptations of cytoplasmic malate dehydrogenases of eastern Pacific abalone (genus *Haliotis*) from different thermal habitats: Biochemical correlates of biogeographical patterning. *J. Exp. Biol.* 185: 137–150.

Dahlhoff, E. P., and G. N. Somero. 1993b. Effects of temperature on mitochondria from abalone (genus *Haliotis*): adaptive plasticity and its limits. *J. Exp. Biol.* 185: 151–168.

Dahlhoff, E. P., and N. E. Rank. 2000. Functional and physiological consequences of genetic variation at phosphoglucose isomerase: heat shock protein expression is related to enzyme genotype in a montane beetle. *Proc. Natl. Acad. Sci. USA* 97: 10056–10061.

Davies, P. L. 2014. Ice-binding proteins: a remarkable diversity of structures for stopping and starting ice growth. *Trends Biochem. Sci.* 39: 548–555.

de la Fuente, M., V. Santos, and J. L. Martinez-Guitarte. 2012. ncRNAs and thermoregulation: A view in prokaryotes and eukaryotes. *FEBS Lett.* 586: 4061–4069.

de Nadal, E., G. Ammerer, and F. Posas. 2011. Controlling gene expression in response to stress. *Nat. Rev. Genet.* 12: 833–845.

Deatherage, B. L., and B. T. Cookson. 2012. Membrane vesicle release in bacteria, eukaryotes, and archaea: a conserved yet underappreciated aspect of microbial life. *Infect. Immun.* 80: 1948–1957.

DeLong, E. F., and A. A. Yayanos. 1986. Biochemical function and ecological significance of novel bacterial lipids in deep-sea procaryotes. *Appl. Environ. Microbiol.* 51: 730–737.

De Maio, A., M. G. Santoro, R. M. Tanguay, and L. E. Hightower. 2012. Ferruccio Ritossa's scientific legacy 50 years after his discovery of the heat shock response: a new view of biology, a new society, and a new journal. *Cell Stress Chaperones* 17: 139–143.

Demchenko, A. P., O. I. Ruskyn, and E. A. Saburova. 1989. Kinetics of the lactate dehydrogenase reaction in high-viscosity media. *Biochim. Biophys. Acta.* 998: 196–203.

Deng, C., C.-H. C. Cheng, H. Ye, X. He, and L. Chen. 2010. Evolution of an antifreeze protein by neofunctionalization under escape from adaptive conflict. *Proc. Natl. Acad. Sci. USA* 107: 21593–21598.

Denny, M. W. 2016. *Ecological Mechanics: Principles of Life's Physical Interactions*. Princeton University Press, Princeton, New Jersey.

Denny, M. W., and C. D. G. Harley. 2006. Hot limpets: predicting body temperature in a conductance-mediated thermal system. *J. Exp. Biol.* 209: 2409–2419.

Detrich, H. W. III, S. K. Parker, R. C. Williams, Jr., E. Nogales, and K. H. Downing. 2000. Cold adaptation of microtubule assembly and dynamics: Structural interpretation of primary sequence changes in the α- and β-tubulins of Antarctic fishes. *J. Biol. Chem.* 275: 37038–37047.

Deutsch C., A. Ferrel, B. Seibel B, H. O. Pörtner, and R. B. Huey. 2015. Climate change tightens a metabolic constraint on marine habitats. *Science* 348: 1132-1135.

Deutsch, C. A., J. J. Tewksbury, R. B. Huey, K. S, Sheldon, C. K. Ghalambor, D. C. Haak, and P. R. Martin. 2008. Impacts of climate warming on terrestrial ectotherms across latitude. *Proc. Natl. Acad. Sci. USA* 105: 6668–6672.

DeVries, A. L. 2004. Ice antifreeze proteins and antifreeze genes in polar fishes. In B. M. Barnes and H. V. Carey (eds.), *Life in the Cold*, pp. 275–282. Institute of Arctic Biology, University of Alaska, Fairbanks.

DeVries, A. L., and C.-H. C. Cheng. 2005. Antifreeze proteins and organismal freezing avoidance in polar fishes. In A. P. Farrell and J. F. Steffensen (eds.), *The Physiology of Polar Fishes. Fish Physiology*, vol. 22. Elsevier, Amsterdam.

DeVries, A. L., and D. E. Wohlschlag. 1969. Freezing resistance in some Antarctic fishes. *Science* 163: 1073–1075.

Dhaka, A., V. Viswanath, and A. Patapoutian. 2006. Trp ion channels and temperature sensation. *Annu. Rev. Neurosci.* 29: 135–161.

Dickson, K. A., and J. B. Graham. 2004. Evolution and consequences of endothermy in fishes. *Physiol. Biochem. Zool.* 77: 998–1018.

Dillon, M. E., G. Wang, and R. B. Huey 2010. Global metabolic impacts of recent climate warming. *Nature* 467: 704–707.

Dillon, M. E., G. Wang, P. A. Garrity, and R. B. Huey. 2009. Thermal preference in *Drosophila. J. Therm. Biol.* 34: 1009–119.

Divakaruni, A. S., and M. D. Brand. 2011. The regulation and physiology of mitochondrial protein leak. *Physiol.* 26: 192–205.

Dobbs, G. H., III, and A. L. DeVries. 1975. Renal function in Antarctic teleost fishes: Serum and urine composition. *Mar. Biol.* 29: 59–70.

Dong, Y.-W., L. P. Miller, J. G. Sanders, and G. N. Somero. 2008. Heat-shock protein 70 (Hsp70) expression in four limpets of the genus *Lottia*: interspecific variation in constitutive and inducible synthesis correlates with in situ exposure to heat stress. *Biol. Bull.* 215: 173–181.

Dong, Y.-W., and G. N. Somero. 2008. Temperature adaptation of cytosolic malate dehydrogenases of limpets (genus *Lottia*): differences in stability and function due to minor changes in sequence correlate with biogeographic and vertical distributions. *J. Exp. Biol.* 212: 169–177.

Dong, Y.-W., H. Wang, G. Han, C. Ke, X. Zhan, T. Nakano, and G. A. Williams. 2012. The impact of Yangtze river discharge, ocean currents and historical events on the biogeographic pattern of *Cellana toreuma* along the China coast. *PLoS One* 7: e36178.

Dorion, S., and J. Landry. 2002. Activation of the mitogen-activated protein kinase pathways by heat shock. *Cell Stress Chaperones* 7: 200–206.

Dowd, W. W., and G. N. Somero. 2013. Behavior and survival of *Mytilus* congeners following episodes of elevated body temperature in air and seawater. *J. Exp. Biol.* 216: 502–514.

Dowd, W. W., C. A. Felton, H. M. Heymann, L. E. Kost, and G. N. Somero. 2013. Food availability, more than body temperature, drives correlated shifts in ATP-generating and antioxidant enzyme capacities in a population of intertidal mussels (*Mytilus californianus*). *J. Exp. Mar. Biol. Ecol.* 449: 171–185.

Dowd, W. W., F. A. King, and M. W. Denny. 2015. Thermal variation, thermal extremes and the physiological performance of individuals. *J. Exp. Biol.* 218: 1956–1967.

Duman, J. G. 2001. Antifreeze and ice nucleator proteins in terrestrial arthropods. *Annu. Rev. Physiol.* 63: 327–357.

Duman, J. G. 2002. The inhibition of ice nucleators by insect antifreeze proteins is enhanced by glycerol and citrate. *J. Comp. Physiol. B.* 172: 163–168.

Duman, J. G. 2015. Animal ice binding (antifreeze) proteins and glycolipids: An overview with emphasis on physiological function. *J. Exp. Biol.* 218: 1846–1855.

Echtay, K. S. 2007. Mitochondrial uncoupling proteins—What is their physiological role? *Free Rad. Biol. Med.* 43: 1351–1371.

Ekberg, D. R. 1958. Respiration in tissues of goldfish adapted to high and low temperatures. *Biol. Bull.* 114: 308–316.

Ellis, R. J. 1993. The general concept of molecular chaperones. In R. J. Ellis, R. A. Laskey, and G. H. Lorimer (eds.), *Molecular Chaperones*, pp. 1–5. Springer, Netherlands.

Else, P. L., and A. J. Hulbert. 1985. An allometric comparison of the mitochondria of mammalian and reptilian tissues: The implications for the evolution of endothermy. *J. Comp. Physiol. B.* 156: 3–11.

Else, P. L., and A. J. Hulbert. 1987. Evolution of mammalian endothermic metabolism: "leaky" membranes as a source of heat. *Am. J. Physiol. Regul. Integr. Comp. Physiol.* 253: R1–R7.

Else, P. L., N. Turner, and A. J. Hulbert. 2004. The evolution of endothermy: Role of membranes and molecular activity. *Physiol. Biochem. Zool.* 77: 950–958.

Engelman, D. M. 2005. Membranes are more mosaic than fluid. *Nature* 438: 578–580.

Erickson, J. R., B. D. Sidell, and T. S. Moerland. 2005. Temperature sensitivity of calcium binding for parvalbumins from Antarctic and temperate zone teleost fishes. *Comp. Biochem. Physiol. A.* 140: 179–185.

Evans, C. W., L. Hellman, M. Middleditch, J. M. Wojnar, M. A. Brimble, and A. L. DeVries. 2012. Synthesis and recycling of antifreeze glycoproteins in polar fishes. *Antarctic Sci.* 24: 259–268.

Evans, C. W., V. Gubala, R. Nooney, D. E. Williams, M. A. Brimble, and A. L. DeVries. 2011. How do Antarctic notothenioid fishes cope with internal ice? A novel function of antifreeze glycoproteins. *Antarctic Sci.* 23: 57–64.

Fahey, R. C., J. S. Hunt, and G. C. Windham. 1977. On the cysteine and cystine content of proteins. Differences between intracellular and extracellular proteins. *J. Mol. Evol.* 10: 155–160.

Farias, S. T., and M. C. Bonato. 2003. Preferred amino acids and thermostability. *Genet. Mol. Res.* 2: 383–393.

Farmer, S. R. 2009. Be cool, lose weight. *Nature* 458: 839–840.

Feder, M. E., N. V. Cartano, L. Milos, R. A. Krebs, and S. L. Lindquist. 1996. Effect of engineering *Hsp70* copy number on Hsp70 expression and tolerance of ecologically relevant heat shock in larvae and pupae of *Drosophila melanogaster. J. Exp. Biol.* 199: 1837–1844.

Feller, G. 2008. Enzyme function at low temperatures in psychrophiles. In K. S. Siddiqui and T. Thomas (eds.), *Protein Adaptation in Extremophiles*, pp. 35–69. Nova Science Publ., New York.

Feller, G. 2010. Protein stability and enzyme activity at extreme biological temperatures. *J. Phys. Condens. Matter* 22: 323101.

Feller, G., and C. Gerday. 2003. Psychrophilic enzymes: hot topics in cold adaptation. *Nat. Rev. Microbiol.* 1: 200–208.

Feng, H., D. Guan, K. Sun, Y. Wang, R. Zhang, and R. Wang. 2013. Expression and signal regulation of the alternative oxidase genes under abiotic stress. *Acta Biochim. Biophys. Sin.* 45: 985–994.

Fields, P. A. 2001. Review: Protein function at thermal extremes: balancing stability and flexibility. *Comp. Biochem. Physiol. A.* 129: 417–431.

Fields, P. A., Y. Dong, X. Meng, and G. N. Somero. 2015. Adaptations of protein structure and function to temperature: there is more than one way to "skin the cat." *J. Exp. Biol.* 218: 1801–1811.

Fields, P. A., and D.E. Houseman. 2004. Decreases in activation energy and substrate affinity in cold-adapted A_4-lactate dehydrogenase: Evidence from the Antarctic notothenioid fish *Chaenocephalus aceratus. Mol. Biol. Evol.* 21: 2246–2255.

Fields, P. A., Y.-S. Kim, J. F. Carpenter, and G. N. Somero. 2002. Temperature adaptation in *Gillichthys* (Teleost: Gobiidae) A_4-lactate dehydrogenases: identical primary structures produce subtly different conformations. *J. Exp. Biol.* 205: 1293–1030.

Fields, P. A., E. Rudomin, and G. N. Somero. 2006. Temperature sensitivities of cytosolic malate dehydrogenases from native and invasive species of marine mussels (genus *Mytilus*): sequence-function linkages and correlations with biogeographic distribution. *J. Exp. Biol.* 209: 656–677.

Fields, P. A., and G. N. Somero. 1997. Amino acid sequence differences cannot fully explain interspecific variation in thermal sensitivities of gobiid fish A_4-lactate dehydrogenases (A_4-LDH). *J. Exp. Biol.* 200: 1839–1850.

Fields, P. A., and G. N. Somero. 1998. Hot spots in cold adaptation: Localized increases in conformational flexibility in lactate dehydrogenase A_4 orthologs of Antarctic notothenioid fishes. *Proc. Natl. Acad. Sci. USA* 95: 11476–11481.

Fields, P. A., B. D. Wahlstrand, and G. N. Somero. 2001. Intrinsic versus extrinsic stabilization of enzymes. The interaction of solutes and temperature on A_4-lactate dehydrogenase orthologs from warm-adapted and cold-adapted marine fishes. *Eur. J. Biochem.* 268: 4497–4505.

Fletcher, G. L., C. L. Hew, and P. L. Davies. 2001. Antifreeze proteins of teleost fishes. *Annu. Rev. Physiol.* 63: 359–390.

Frank, C. L. 1991. Adaptation for hibernation in the depot fats of a ground squirrel (*Spermophilus beldingi*). *Can. J. Zool.* 69: 22702–2711.

Frank, C. L. 1992. The influence of dietary fatty acids on hibernation by golden mantled ground squirrels (*Spermophilus lateralis*). *Physiol. Zool.* 65: 906–920.

Fuhrman, J. 2009. Microbial community structure and its functional implications. *Nature* 459: 193–199.

Galtier, N., and J. R. Lobry. 1997. Relationships between genomic G+C content, RNA secondary structures, and optimal growth temperature in prokaryotes. *J. Mol. Evol.* 44: 632–636.

Garnham, C. P., R. L. Campbell, V. K. Walker, and P. L. Davies. Novel dimeric β-helical model of an ice nucleation protein with bridged active sites. *BMC Struct. Biol.* 11: 36.

Garrett, S. C., and J. J. C. Rosenthal. 2012a. RNA editing underlies temperature adaptation in K^+ channels from polar octopuses. *Science* 335: 848–851.

Garrett, S. C., and J. J. C. Rosenthal. 2012b. A role for A-to-I RNA editing in temperature adaptation. *Physiology* 27: 362–369.

Garrity, P. A., M. B. Goodman, A. D. Samuel, and P. Sengupta. 2010. Running hot and cold: behavioral strategies, neural circuits, and the molecular machinery for thermotaxis in *C. elegans* and *Drosophila. Genes Develop.* 24: 2365–2382.

Gasch, A. P., P. T. Spellman, C. M. Kao, O. Carmel-Harel, M. B. Eisen, G. Storz, D. Botstein, and P. O. Brown. 2000. Genomic expression programs in the response of yeast cells to environmental changes. *Mol. Biol. Cell* 11: 4241–4257.

Gehring, W. J., and R. Wehner. 1995. Heat shock protein synthesis and thermotolerance in *Cataglyphis*, an ant from the Sahara Desert. *Proc. Natl. Acad. Sci. USA* 92: 2992–2998.

Georlette, D., B. Damien, V. Blaise, E. Depiereux, V. N. Uversky, C. Gerday, and G. Feller. 2003. Structural and functional adaptations to extreme temperatures in psychrophilic, mesophilic, and thermophilic DNA ligases. *J. Biol. Chem.* 278: 37015–37023.

Gillis, T. D., C. R. Marshall, and G. F. Tibbits. 2007. Functional and evolutionary relationships of troponin C. *Physiol. Genom.* 32: 16–27.

Girguis, P. R., and R. W. Lee. 2006. Thermal preference and tolerance of alvinellids. *Science* 312: 231.

Glauser, D. A., W. C. Chen, R. Agin, B. L. MacInnis, A. B. Hellman, P. A. Garrity, M.-W. Tan, and M. B. Goodman. 2011. Heat avoidance is regulated by transient receptor potential (TRP) channels and a neuropeptide signaling pathway in *Caenorhabditis elegans. Genetics* 188: 91–103.

Goldenfeld, N., and C. Woese. 2007. Biology's next revolution. *Nature* 445: 369.

Gong, W. J., and K. G. Golic. 2004. Genomic deletions of the *Drosophila melanogaster Hsp70* genes. *Genetics* 168: 1467–1476.

Gracey, A. Y., M. L.Chaney, J. P. Boomhower, W. R. Tyburczy, K. Connor, and G. N. Somero. 2008. Rhythms of gene expression in a fluctuating intertidal environment. *Curr. Biol.* 18: 1501–1507.

Gracey, A. Y., E. J. Fraser, W. Li, Y. Fang, R. R. Taylor, J. Rogers, A. Brass, and A .R. Cossins. 2004. Coping with cold: an integrative, multi-tissue analysis of the transcriptome of a poikilothermic vertebrate. *Proc. Natl. Acad. Sci. USA* 101: 16970–16975.

Graham, L. A., R. S. Hobbs, G. L. Fletcher, and P. L. Davies. 2013. Helical antifreeze proteins have independently evolved in fishes on four occasions. *PLoS One* 8: e81285.

Graham, L. A., S. C. Lougheed, K. V. Ewart, and P. L. Davies. 2008. Lateral transfer of a lectin-like antifreeze protein gene in fishes. *PLoS One* 3: e2616.

Graumann, P. L., and M. A. Marahiel. 1998. A superfamily of proteins that contain the cold-shock domain. *Trends Biochem. Sci.* 23: 286–290.

Graves, J. E., and G. N. Somero. 1982. Electrophoretic and functional enzymic evolution in four species of eastern Pacific barracudas from different thermal environments. *Evolution* 36: 97–106.

Griffith, M., and M. W. F. Yaish. 2004. Antifreeze proteins in overwintering plants: a tale of two activities. *Trends Plant Sci.* 9: 399–405.

Gu, J., and V. J. Hilser. 2008. Predicting the energetics of conformational fluctuations in proteins from sequence: a strategy for profiling the proteome. *Structure* 16: 1627–1637.

Gu, J., and V. J. Hilser. 2009. Sequence-based analysis of protein energy landscapes reveals nonuniform thermal adaptation within the proteome. *Mol. Biol. Evol.* 26: 2217–2227.

Guderley, H. 2004. Metabolic responses to low temperature in fish muscle. *Biol. Rev.* 79: 409–427.

Gupta, R., and R. Deswal. 2014. Antifreeze proteins enable plants to survive in freezing conditions. *J. Biosci.* 39: 931–944.

Gusta, L. V., and M. Wisniewski. 2012. Understanding plant cold hardiness: an opinion. *Physiol. Plant.* 147: 4–14

Hakoshima, T. 2014. Leucine zippers. *Encyclopedia of Life Sciences*. Wiley, New York. DOI: 10.1002/9780470015902.a0005049.pub2

Hamada, F. N., M. Rosenzweig, K. Kang, S. R. Pulver, A. Ghezzi, T. J. Jegla, and P. A. Garrity. 2008. An internal thermal sensor controlling temperature preference in *Drosophila*. *Nature* 454: 217–220.

Han, G.-D., S. Zhang, D. J. Marshall, C.-H. Ke, and Y.-W. Dong. 2013. Metabolic energy sensors (AMPK and SIRT 1), protein carbonylation and cardiac failure as biomarkers of thermal stress in an intertidal limpet: linking energetic allocation with environmental temperature during aerial emersion. *J. Exp. Biol.* 216: 3273–3282.

Haney, P. J., J. H. Badger, G. L. Buldak, C. I. Reich, C. R. Woese, and G. J. Olson. 1999. Thermal adaptation by comparison of protein sequences from mesophilic and extremely thermophilic *Methanococcus* species. *Proc. Natl. Acad. Sci. USA* 96: 3578–3583.

Hardie, D. G., and K. Sakamoto. 2006. AMPK: A key sensor of fuel and energy status in skeletal muscle. *Physiology* 21: 48–60.

Hashimoto, O., H. Ohtsuki, T. Kakizaki, K. Amou, R. Sato, S. Doi, S. Kobayashi, A. Matsuda, and 6 others. 2015. Brown adipose tissue in cetacean blubber. *PLoS One* 10: e0116734.

Haviv, H., E. Cohen, Y. Lifshitz, D. M. Tal, R. Goldshleger, and S. J. D. Karlish. 2007. Stabilization of Na$^+$, K$^+$-ATPase purified from *Pichia pastoris* membranes by specific interactions with lipids. *Biochem.* 46: 12855–12869.

Hayward, S. A. L., B. Manso, and A. R. Cossins. 2014. Molecular basis of chill resistance adaptations in poikilothermic animals. *J. Exp. Biol.* 217: 6–15.

Hazel, J. R. 1995. Thermal adaptation in biological membranes—is homeoviscous adaptation the explanation? *Annu. Rev. Physiol.* 57: 19–42.

Hazel, J. R., and C. L. Prosser. 1974. Molecular mechanisms of temperature compensation in poikilotherms. *Physiol. Rev.* 54: 620–677.

Hazel, J. R., and R. Carpenter. 1985. Rapid changes in the phospholipid composition of gill membranes during thermal acclimation of the rainbow trout (*Salmo gairdneri*). *J. Comp. Physiol. B.* 155: 597–602.

Healy, T. M., and P. M. Schulte. 2012. Thermal acclimation is not necessary to maintain a wide thermal breadth of aerobic scope in the common killifish (*Fundulus heteroclitus*). *Physiol. Biochem. Zool.* 85: 107–119.

Heinrich, B. 1993. *The Hot Bodied Insects. Strategies and Mechanisms of Thermal Regulation*. Harvard University Press, Cambridge.

Hilser, V. J. 2013. Signaling from disordered proteins. *Nature* 498: 308–310.

Hinton, H. E. 1960. A fly larva that tolerates dehydration and temperatures of –270°C to +102°C. *Nature* 188: 336–337.

Hochachka, P. W. 1965. Isoenzymes in metabolic adaptation in a poikilotherms: subunit relationships in lactic dehydrogenase of goldfish. *Arch. Biochem. Biophys.* 111: 96–103.

Hochachka, P. W., and G. N. Somero. 2002. *Biochemical Adaptation*. Oxford University Press, Oxford.

Hoekstra, L. A., and K. L. Montooth. 2013. Inducing extra copies of the *Hsp70* gene in *Drosophila melanogaster* increases energetic demand. *BMC Evol. Biol.* 13: 68.

Hoffmann, A. A. 2010. Physiological climatic limits in *Drosophila*: patterns and implications. *J. Exp. Biol.* 213: 870–880.

Holland, L. Z., M. McFall-Ngai, and G. N. Somero. 1997. Evolution of lactate dehydrogenase-A homologs of barracuda fishes (Genus *Sphyraena*) from different thermal environments: Differences in kinetic properties and thermal stability are due to amino acid substitutions outside the active site. *Biochemistry* 36: 3207–3215.

Holmes, W. N., and E. M. Donaldson. 1969. *Fish Physiology*, vol. 1, pp. 53–54. Academic Press, New York.

Horváth, I., G. Multhoff, A. Sonnleitner, and L. Vígh. 2008. Membrane-associated stress proteins: More than simply chaperones. *Biochim. Biophys. Acta*. 1778: 1653–1664.

Huang, J., X. Zhang, and P. A. McNaughton. 2006. Modulation of temperature-sensitive TRP channels. *Sem. Cell Devel. Biol.* 17: 638–645.

Hughes, S., L. Scharet, J. Malcolmson, K. A. Hogarth, D. M. Martynowic, E. Tralman-Baker, S. N. Patel, and S. P. Graether. 2013. The important of size and disorder in the cryoprotective effects of dehydrins. *Plant Physiol.* 163: 1376–1386.

Hulbert, A. J., and P. L. Else. 1989. Evolution of mammalian endothermic metabolism: mitochondrial activity and cell composition. *Am. J. Physiol.* 256 (*Regul. Integr. Comp. Physiol.* 25): R63–R69.

Hulbert, A. J., R. Pamplona, R. Buffenstein, and W. A. Buttemer. 2007. Life and death: metabolic rate, membrane composition, and life span of animals. *Physiol. Rev.* 87: 1175–1213.

Hull, D. 1973. Thermoregulation in young mammals. In G. C. Whittow (ed.), *Comparative Physiology of Thermoregulation*, vol. 3, pp. 167–200. Academic Press, New York.

Hurst, L. D., and R. Merchant. 2001. High guanine-cytosine content is not an adaptation to high temperatures: a comparative analysis among prokaryotes. *Proc. R. Soc. B.* 268: 493–497.

Iftikar, F., and A. J. R. Hickey. 2013. Do mitochondria limit hot fish hearts? Understanding the role of mitochondrial function with heat stress in *Notolabrus celidotus*. *PLoS One* 8: e64120.

Iftikar, F., J. R. MacDonald, D. W. Baker, G. M. C Renshaw, and A. J. R. Hickey. 2014. Could thermal sensitivity of mitochondria determine species distribution in a changing climate? *J. Exp. Biol.* 217: 2348–2357.

Isaksen, G. V., J. Aqvist, and B. O. Brandsdal. 2014. Protein surface softness is the origin of enzyme cold-adaptation in trypsin. *PLoS Comput. Biol.* 10: e1003813.

Izumi, Y., S. Sonoda, H. Yoshida, H. V. Danks, and H. Tsumuki. 2006. Role of membrane transport of water and glycerol in the freeze tolerance of the rice stem borer, *Chilo suppressalis* Walker (Lepidoptera: Pyralidae). *J. Insect Physiol.* 52: 215–220.

Jackson, A. L., and L. A. Loeb. 2001. The contribution of endogenous sources of DNA damage to the multiple mutations in cancer. *Mutat. Res.* 477: 7–21.

Jaenicke, R. 2000. Stability and stabilization of globular proteins in solution. *J. Biotechnol.* 79: 193–203.

Janech, M. G., A. Krell, T. Mock, J.-S. Kang, and J. A. Raymond. 2006. Ice-binding proteins from sea ice diatoms (Bacillariophyceae). *J. Phycol.* 42: 410–416.

Jarmuszkiewicz, W., A. Woyda-Ploszczyca, N. Antos-Krzeminska, and F. E. Sluse. 2010. Mitochondrial uncoupling proteins in unicellular eukaryotes. *Biochim. Biophys. Acta* 1797: 792–799.

Johansson, J., P. Mandin, A. Renzoni, C. Chiaruttini, M. Springer, and P. Cossart. 2002. An RNA thermosensor controls expression of virulence genes in *Listeria monocytogenes*. *Cell* 110: 551–561.

Johns, G. C., and G. N. Somero. 2004. Evolutionary convergence in adaptations to proteins to temperature: A_4-lactate dehydrogenases of Pacific damselfishes. *Mol. Biol. Evol.* 21: 314–320.

Jönsson, K. I., and R. Bertolani. 2001. Facts and fiction about long-term survival in tardigrades. *J. Zool.* 257: 181–187.

Kang, K., V. C. Panzano, E. C. Chang, L. Ni, A. M. Dainis, A. M. Jenkins, K. Regna, M. A. Muskavitch, and P. A. Garrity. 2012. Modulation of TRPA1 sensitivity enables sensory discrimination in *Drosophila*. *Nature* 481: 76–80.

Kashefi, K., and D. R. Lovely. 2003. Extending the upper temperature limit for life. *Science* 301: 934.

Kiko, R. 2010. Acquisition of freeze protection in a sea-ice crustacean through horizontal gene transfer. *Polar Biol.* 33: 543–556.

Kim, S.-Y., K. Y. Hwang, S.-H. Kim, H.-C. Sung, Y. S. Han, and Y. Cho. 1999. Structural basis for cold adaptation. Sequence, biochemical properties, and crystal structure of malate dehydrogenase from a psychrophile *Aquaspirillum arcticum*. *J. Biol. Chem.* 274: 11761–11767.

Kim, Y. E., M. S. Hipp, A. Bracher, M. Hayer-Hartl, and F. U. Hartl. 2013. Molecular chaperone functions in protein folding and proteostasis. *Annu. Rev. Biochem.* 82: 323–355.

Kingsolver, J. G., and H. A. Woods. 2016. Beyond thermal performance curves: modeling time-dependent effects of thermal stress and ectotherm growth rates. *Am. Nat.* 187: 283–294.

Kiss, A. J., A. Y Mirarefi, S. Ramakrishnan, C. F. Zukoski, A. L. DeVries, and C.-H. C. Cheng. 2004. Cold-stable eye lens crystallins of the Antarctic notothenioid toothfish *Dissostichus mawsoni* Norman. *J. Exp. Biol.* 207: 4633–4649.

Koninkx, J. F. J. G. 1976. Protein synthesis in salivary glands of *Drosophila hydei* after experimental gene induction. *Biochem. J.* 158: 623–628.

Kortmann, J., and F. Narberhaus. 2012. Bacterial RNA thermometers: molecular zippers and switches. *Nat. Rev. Microbiol.* 10: 255–265.

Koshland, D. 2004. Crazy, but correct. *Nature* 432: 447.

Krebs, R. A., and M. E. Feder. 1997. Natural variation in the expression of the heat-shock protein Hsp70 in a population of *Drosophila melanogaster* and its correlation with tolerance of ecologically relevant thermal stress. *Evolution* 51: 173–179.

Krebs, R. A., and V. Loeschcke. 1994. Costs and benefits of activation of the heat-shock response in *Drosophila melanogaster*. *Funct. Ecol.* 8: 730–737.

Krenek, S., M. Schlegel, and T. U. Berendonk. 2013. Convergent evolution of heat-inducibility during subfunctionalization of the *Hsp70* gene family. *BMC Evol. Biol.* 13: 49.

Krogh, A. 1916. *Respiratory Exchange of Animals and Man*. Longmans, Green and Co.: London.

Krylov, D., and C. R. Vinson. 2001. Leucine zipper. In *Encyclopedia of Life Sciences*. Wiley, New York. DOI: 10.1038/npg.els.0003001

Kültz, D. 2005. Molecular and evolutionary basis of the cellular stress response. *Annu. Rev. Physiol.* 67: 225-257.

Kumar, S. V., and P. A. Wigge. 2005. H2A.Z-containing nucleosomes mediate the thermosensory response in *Arabidopsis*. *Cell* 140: 136–147.

Lee, A. G. 2004. How lipids affect the activities of integral membrane proteins. *Biochim. Biophys. Acta* 1666: 62–87.

Lemaux, P. G., S. L. Herendeen, L. Bloch, and C. Neidhardt. 1978. Transient rates of synthesis of individual polypeptides in *E. coli* following temperature shifts. *Cell* 13: 427–434.

Lewis, M., P. J. Helmsing, and M. Ashburner. 1975. Parallel changes in puffing activity and patterns of protein synthesis in salivary glands of *Drosophila*. *Proc. Natl. Acad. Sci. USA* 72: 3604–3608.

Li, H.-T. 2012. Thermal tolerance of *Echinolittorina* species in Hong Kong. M.S. Thesis. The University of Hong Kong. URL: http://hdl.handle.net/10722/173874.

Li, N., C. A. Andofer, and J. G. Duman. 1998. Enhancement of insect antifreeze protein activity by solutes of low molecular mass. *J. Exp. Biol.* 210: 2243–2251.

Lima, F. P., F. Gomes, R. Seabra, D. S. Wethey, M. I. Seabra, T. Cruz, A. M. Santos, and T. J. Hilbish. 2016. Loss of thermal refugia near equatorial range limits. *Glob. Chang. Biol.* 22: 254–263.

Lindquist, S. 1986. The heat-shock response. *Annu. Rev. Biochem.* 55: 1151–1191.

Lockwood, B. L., J. G. Sanders, and G. N. Somero. 2010. Transcriptomic responses to heat-stress in invasive and native blue mussels (genus *Mytilus*): molecular correlates of invasive success. *J. Exp. Biol.* 213: 3548–3558.

Lockwood, B. L., and G. N. Somero. 2012. Functional determinants of temperature adaptation in enzymes of cold- vs. warm-adapted mussels (genus *Mytilus*). *Mol. Biol. Evol.* 29: 3061–3070.

Logan, C. A., and B. A. Buckley. 2015. Transcriptomic responses to environmental temperature in eurythermal and stenothermal fishes. *J. Exp. Biol.* 218: 1915–1924.

Logan, C. A., and G. N. Somero. 2011. Effects of thermal acclimation on transcriptional responses to acute heat stress in the eurythermal fish *Gillichthys mirabilis* (Cooper). *Am. J. Physiol. Regul. Integr. Comp. Physiol.* 300: R1373–R1383.

Logue, J. A., A. L. DeVries, E. Fodor, and A. R. Cossins. 2000. Lipid compositional correlates of temperature-adaptive interspecific differences in membrane physical structure. *J. Exp. Biol.* 203: 2105–2115.

Loh, E., E. Kugelberg, A. Tracey, Q. Zhang, B. Gollan, H. Ewles, R. Chalmers, V. Pelicic, and C. M. Tang (2013). Temperature triggers immune evasion by *Neisseria meningitidis*. *Nature* 502: 237–241.

Lombard, J., P. López-García, and D. Moreira. 2012. The early evolution of lipid membranes and the three domains of life. *Nat. Rev. Microbiol.* 10: 507–515.

López-Garcia, P., and P. Forterre. 2000. DNA topology and the thermal stress response, a tale from mesophiles and hyperthermophiles. *BioEssays* 22: 738–746.

López-García, P., Y. Zivanovic, P. Deschamps, and D. Moreira. 2015. Bacterial gene import and mesophilic adaptation in archaea. *Nat. Rev. Microbiol.* 13: 447–456.

Low, P. S., J. L. Bada, and G. N. Somero. 1973. Temperature adaptation of enzymes: roles of the free energy, the enthalpy, and the entropy of activation. *Proc. Natl. Acad. Sci. USA* 70: 430–432.

Low, P. S., and G. N. Somero. 1974. Temperature adaptation of enzymes: a proposed molecular basis for the different catalytic efficiencies of enzymes from ectotherms and endotherms. *Comp. Biochem. Physiol.* 49B: 307–312.

Ma, B.-G., A. Goncearenco, and I. G. Berezovsky. 2010. Thermophilic adaptation of protein complexes inferred from proteomic homology modeling. *Structure* 18: 819–828.

Macdonald, A. G. 1984. The effect of pressure on the molecular structure and physiological function of cell membranes. *Phil. Trans. Roy. Soc. Ser. B.* 304: 47–68.

Macdonald, J. A., J. C. Montgomery, and R. M. G. Wells. 1988. The physiology of McMurdo Sound fishes: current New Zealand research. *Comp. Biochem. Physiol.* 90B: 567–578.

Madden, P. W., M. J. Babcock, M. E. Vayda, and R. E. Cashon. 2004. Structural and kinetic characterization of myoglobins from eurythermal and stenothermal fish species. *Comp. Biochem. Physiol. B* 137: 341–350.

Makley, L. N., K. A. McMenimen, B. T. DeVree, J. W. Goldman, B. N. McGlasson, P. Rajagopal, B. M. Bunyak, T. J. McQuade, and 5 others. 2015. Pharmacological chaperone for α-crystallin partially restores transparency in cataract models. *Science* 350: 674–677.

Malan, A., and E. Mioskowski. 1988. pH-temperature interactions on protein function and hibernation: GDP binding to brown adipose tissue mitochondria. *J. Comp. Physiol. B.* 158: 487–493.

Marshall, D. J., Y.-W. Dong, C. D. McQuaid, and G. A. Williams. 2011. Thermal adaptation in the intertidal snail *Echinolittorina malaccana* contradicts current theory by revealing the crucial roles of resting metabolism. *J. Exp. Biol.* 214: 3649–3657.

Martin, N., D. P. Bureau, Y. Marty, E. Kraffe, and H. Guderley. 2013. Dietary lipid quality and mitochondrial membrane composition in trout: response of membrane enzymes and oxidative capacities. *J. Comp. Physiol. B* 183: 393–408.

Marx, J.-C., T. Collins, S. D.'Amico, G. Feller, and C. Gerday. 2007. Cold-adapted enzymes from marine Antarctic microorganisms. *Mar. Biotech.* 9: 293–304.

Mastro, A. M., and A. D. Keith. 1984. Diffusion in the aqueous compartment. *J. Cell. Biol.* 99: 180–187.

Matthews, B. W. 1987. Genetic and structural analysis of the protein stability problem. *Biochemistry* 26: 6885– 6888.

McDonald, A. E., and G. C. Vanlerberghe. 2004. Branched mitochondrial electron transport in the Animalia: Presence of alternative oxidase in several animal phyla. *Life* 56: 333–341.

McFall-Ngai, M. 2015. Giving microbes their due: animal life in a microbially dominant world. *J. Exp. Biol.* 218: 1968–1973.

McFall-Ngai, M. J., and J. Horwitz. 1990. A comparative study of the thermal stability of the vertebrate eye lens: Antarctic ice fish to the desert iguana. *Exp. Eye Res.* 50: 703–709.

McKemy, D. D. 2007. Temperature sensing across species. *Pflugers Arch. Eur. J. Physiol.* 454: 777–791.

Mirault, M., M. Goldschmidt-Clermont, L. Moran, A. P. Arrigo, and A. Tissières. 1978. The effect of heat shock on gene expression in *Drosophila melanogaster*. *Cold Spring Harb. Symp. Quant. Biol.* 42: 819–827.

Mirkin, S. M. 2001. DNA topology: fundamentals. In: eLS. John Wiley & Sons Ltd., Chichester. http://www.els.net [doi: 10.1038/npg.els.0001038

Mooers, A. Ø., and E. C. Holmes. 2000. The evolution of base composition and phylogenetic inference. *Trends Ecol. Evol.* 15: 365–369.

Morimoto, R. I., D. J. Selkoe, and J. W. Kelley (eds.). 2012. *Protein Homeostasis*, Cold Spring Harbor Perspectives in Biology. Cold Spring Harbor Laboratory Press, Cold Spring Harbor, New York.

Myers, B. R., Y. M Sigal, and D. Julius. 2009. Evolution of thermal response properties in a cold-activated TRP channel. *PLoS One* 4: e5741.

Nelson-Sathi, S., F. L. Sousa, M. Roettger, N. Lozada-Chávez, T. Thiergart, A. Janssen, D. Bryant, G. Landan, and 4 others. 2015. Origins of major archaeal clades correspond to gene acquisitions from bacteria. *Nature* 517: 77–80.

Neven, L. G., J. G. Duman, J. M. Beals, and F. J. Castellino. 1986. Overwintering adaptations of the stag beetle, *Ceruchus piceus*: Removal of ice nucleators in winter to promote supercooling. *J. Comp. Physiol.* 156: 707–716.

Nguyen, A. D., Gotelli, N. J. and S. H. Cahan. 2016. The evolution of heat shock protein sequences, cis-regulatory elements, and expression profiles in the eusocial Hymenoptera. *BMC Evol. Biol.* 16: 15.

Nguyen, K. D., Y. Qiu, X. Cui, Y. P. S. Goh, J. Mwangi, T. David, L. Mukundan, F. Brombacher, and 2 others. 2011. Alternatively activated macrophages produce catecholamines to sustain adaptive thermogenesis. *Nature* 480: 104–109.

Nishimiya, Y., R. Sato, M. Takamichi, A. Miura, and S. Tsuda. 2005. Cooperative effect of the isoforms of type III antifreeze protein expressed in notched-fin eelpout, *Zoarces elongates*. *FEBS J.* 272: 482–492.

O'Brien, K. M. 2011. Mitochondrial biogenesis in cold-bodied fishes. *J. Exp. Biol.* 214: 275–285.

Ogata, Y., T. Mizushima, K. Kataoka, K. Kita, T. Miki, and K. Sekimizu. 1996. DnaK heat shock protein of *Escherichia coli* maintains the negative supercoiling of DNA against thermal stress. *J. Biol. Chem.* 46: 29407–29414.

Ohno, S. 1970. *Evolution by Gene Duplication*. Springer-Verlag, New York.

Olufsen, M., A. O. Smalås, E. Moe, and B. Brandsdal. 2005. Increased flexibility as a strategy for cold adaptation. A comparative molecular dynamics study of cold- and warm-active uracil DNA glycosylase. *J. Biol. Chem.* 2880: 18042–18048.

Palladino, M. J., L. P. Keegan, M. A. O'Connell, and R. A. Reenan. 2000. dADAT, a *Drosophila* double-stranded RNA-specific adenosine deaminase is highly developmentally regulated and is itself a target for RNA editing. *RNA* 6: 1004–1018.

Panova, M., and K. Johannesson. 2004. Microscale variation in *Aat* (aspartate aminotransferase) is supported by activity differences between upper and lower shore allozymes of *Littorina saxatilis*. *Mar. Biol.* 144: 1157–1164.

Pastinen, T. 2010. Genome-wide allele-specific analysis: insights into regulatory variation. *Nat. Rev. Genet.* 11: 533–538.

Peck, L. S., S. A. Morley, J. Richard, and M. S. Clark. 2014. Acclimation and thermal tolerance in Antarctic marine ectotherms. *J. Exp. Biol.* 217: 16–22.

Peirce, V., S. Carobbioi, and A. Vidal-Puig. 2014. The different shades of fat. *Nature* 510: 76–83.

Peng, H.-L., T. Egawa, E. Change, H. Deng, and R. Callender. 2015. Mechanism of thermal adaptation in the lactate dehydrogenases. *J. Phys. Chem.* 119: 15256–15262.

Pertaya, N., C. B. Marshall, Y. Celik, P. L. Davies, and I. Braslavsky. 2008. Direct visualization of spruce budworm antifreeze protein interacting with ice crystals: basal plan affinity confers hyperactivity. *Biophys. J.* 95: 333–341.

Phillips, R., T. Ursell, P. Wiggins, and P. Sens. 2009. Emerging roles for lipids in shaping membrane-protein function. *Nature* 459: 379–385.

Place, A. R., and D. A. Powers. 1984. Purification and characterization of the lactate dehydrogenase (LDH-B$_4$) allozymes of *Fundulus heteroclitus*. *J. Biol. Chem.* 259: 1299–1308.

Podrabsky, J. E., T. Hrbek, and S. C. Hand. 1998. Physical and chemical characteristics of ephemeral pond habitats in the Maracaibo basin and Llanos region of Venezuela. *Hydrobiologia* 362: 67–78.

Podrabsky, J. E., and G. N. Somero. 2004. Changes in gene expression associated with acclimation to constant and fluctuating daily temperatures in an annual killifish *Austrofundulus limnaeus*. *J. Exp. Biol.* 207: 2237–2254.

Podrabsky, J. E., and G. N. Somero. 2006. Inducible heat tolerance in Antarctic notothenioid fishes. *Polar Biol.* 30: 39–43.

Poralla, K., T. Hartner, and E. Kannenberg. 1984. Effect of temperature and pH on the hopanoid content of *Bacillus acidocaldarius*. *FEMS Microbiol. Lett.* 113: 107–110.

Pörtner, H. O., and A. P. Farrell. 2008. Physiology and climate change. *Science* 322: 690–692.

Posé, D., L. Verhage, F. Ott, L. Yant, J. Mathieu, G. C. Angenent, R. G. H. Immink, and M. Schmid. 2013. Temperature-dependent regulation of flowering by antagonistic FLM variants. *Nature* 503: 414–417.

Powers, D. A., M. Smith, I. Gonzalez-Villasenor, L. DiMichelle, D. Crawford, G. Bernardi, and T. Lauerman. 1993. A multidisciplinary approach to the selectionist/neutralist controversy using the model teleost *Fundulus heteroclitus*. In D. Futuyma and J. Antonovics (eds.), *Oxford Surveys in Evolutionary Biology*, 9: 43–107. Oxford University Press, Oxford.

Præbel, K., B. Hunt, L. H. Hunt, and A. L. DeVries. 2009. The presence and quantification of splenic ice in the McMurdo Sound notothenioid fish *Pagothenia borchgrevinki* (Boulenger, 1902). *Comp. Biochem. Physiol. A* 154: 564–569.

Qi, S., M. Krogsgaard, M. M. Davis, and A. K. Chakraborty. 2006. Molecular flexibility can influence the stimulatory ability of receptor-ligand interactions at cell-cell junctions. *Proc. Natl. Acad. Sci. USA* 103: 4416–4421.

Raymond, J. A. 1992. Glycerol is a colligative antifreeze in some northern fishes. *J. Exp. Zool.* 262: 347–352.

Raymond, J. A. 1994. Seasonal variations of trimethylamine oxide and urea in the blood of a cold-adapted marine teleost, the rainbow smelt. *Fish Physiol. Biochem.* 13: 13–22.

Raymond, J. A. 1998. Trimethylamine oxide and urea synthesis in rainbow smelt and some other northern fishes. *Physiol. Zool.* 71: 515–523.

Raymond, J. A. 2011. Algal ice-binding proteins change the structure of sea ice. *Proc. Natl. Acad. Sci. USA* 108: E198.

Raymond, J. A., and A. L. DeVries. 1977. Adsorption inhibition as a mechanism of freezing resistance in polar fishes. *Proc. Natl. Acad. Sci. USA* 86: 881–885.

Raymond, J. A., and A. L. DeVries. 1998. Elevated concentrations and synthetic pathways of trimethylamine oxide and urea in some teleost fishes of McMurdo Sound, Antarctica. *Fish Physiol. Biochem.* 18: 387–398.

Ream, R. A., G. C. Johns, and G. N. Somero. 2003a. Base compositions of genes encoding α-actin and lactate dehydrogenase-A from differently adapted vertebrates show no temperature-adaptive variation in G + C content. *Mol. Biol. Evol.* 20: 105–110.

Ream, R. A., J. A. Theriot, and G. N. Somero. 2003b. Influences of thermal acclimation and acute temperature change on the motility of epithelial wound-healing cells (keratocytes) of tropical, temperate, and Antarctic fish. *J. Exp. Biol.* 206: 4539–4551.

Reining, A., S. Nozinovic, K. Schlepckow, F. Buhr, B. Fürtig, and H. Schwalbe. 2013. Three-state mechanism couples ligand and temperature sensing in riboswitches. *Nature* 499: 355–360.

Richter, K., M. Haslbeck, and J. Buchner. 2010. The heat shock response: Life on the verge of death. *Mol. Cell* 40: 253–266.

Ricquier, D., and F. Bouillaud. 2000. The uncoupling protein homologs: UCP1, UCP2, UCP3, StUCP and AtUCP. *Biochem. J.* 345: 161–179.

Riehle, M. M., A. F. Bennett, R. E. Lenski, and A. D. Long. 2003. Evolutionary changes in heat-inducible gene expression in lines of *Escherichia coli* adapted to high temperature. *Physiol. Genomics* 14: 47–58.

Ritossa, F. 1962. A new puffing pattern induced by temperature shock and DNP in *Drosophila*. *Experientia* 18: 571–573.

Ritossa, F. 1963. New puffs induced by temperature shock, DNP and salicilate in salivary gland of *Drosophila melanogaster*. *Drosoph. Inf. Serv.* 37: 122–123.

Ritossa, F. 1996. Discovery of the heat shock response. *Cell Stress Chaperones* 1: 97–98.

Robertson, J. C., and J. R. Hazel. 1995. Cholesterol content of trout plasma membranes varies with acclimation temperature. *Am. J. Physiol.* 269 (*Reg. Integr. Comp. Physiol. 38*): R1113–R1117.

Robertson, J. C., and J. R. Hazel. 1997. Membrane constraints to physiological function at different temperatures: Does cholesterol stabilize membranes at elevated temperatures? *Soc. Exp. Biol. Sem. Ser., Global Warming: Implications for Freshwater and Marine Fish.* 61: 25–49. (C. M. Wood and D. G. McDonald, eds.)

Rolfe, D. E. S., and G. C. Brown. 1997. Cellular energy utilization and molecular origin of standard metabolic rate in mammals. *Physiol. Rev.* 77: 731–758.

Rosenthal, J. J. C. 2015. The emerging role of RNA editing in plasticity. *J. Exp. Biol.* 218: 1812–1821.

Rosenzweig, M., K. Kang, and P. A. Garrity. 2008. Distinct TRP channels are required for warm and cool avoidance in *Drosophila melanogaster*. *Proc. Natl. Acad. Sci. USA* 105: 14668–14673.

Ross, M. H., W. Pawlina, and T. A. Barnash. 2009. *Atlas of Descriptive Histology.* Sinauer Associates, Sunderland, MA.

Rudolph, B., K. M. Gebendorfer, J. Buchner, and J. Winter. 2010. Evolution of *Escherichia coli* for growth at high temperatures. *J. Biol. Chem.* 285: 19029–19034.

Saito, H., and H. Uchida. 1977. Initiation of the DNA replication of bacteriophage lambda in *Escherichia coli* K12. *J. Mol. Biol.* 113: 1–25.

Scholander, P. F., L. van Dam, J. W. Kanwisher, H. T. Hammel, and M. S. Gordon. 1957. Supercooling and osmoregulation in northern fishes. *J. Cell. Comp. Physiol.* 49: 5–24.

Scholander, P. F., W. Flagg, V. Walters, and L. Irving. 1953. Climatic adaptation in Arctic and tropical poikilotherms. *Physiol. Zool.* 26: 67–92.

Schönknecht, G., W.-H. Chen, C. M. Ternes, G. G. Barbier, R. P. Shrestha, M. Stanke, A. Bräutigam, B. J. Baker, and 10 others. 2013. Gene transfer from bacteria and archaea facilitated evolution of an extremophilic eukaryote. *Science* 339: 1207–1210.

Schulte, P. M. 2015. The effects of temperature on aerobic metabolism: towards a mechanistic understanding of the responses of species to a changing environment. *J. Exp. Biol.* 218: 1856–1866.

Schumann, W. 2012. Thermosensory stems in eubacteria. In C. López-Larrea (ed.), *Sensing in Nature.* Landes Bioscience and Springer Science + Business Media, Austin TX.

Seale, P., B. Bjork, W. Yang, S. Kajimura, S. Chin, S. Kuang, A. Scime, S. Devarakonda, and 6 others. 2008. PRDM16 controls a brown fat/skeletal muscle switch. *Nature* 454: 961–967.

Seitz, P., and M. Blokesch. 2012. Cues and regulatory pathways involved in natural competence and transformation in pathogenic and environmental Gram-negative bacteria. *FEMS Microbiol. Rev.* 37: 336–363.

Serganov, A., and D. J. Patel. 2007. Ribozymes, riboswitches and beyond: regulation of gene expression without proteins. *Nat. Rev. Genet.* 8: 776–790.

Seymour, R. 2004. Dynamics and precision of thermoregulatory responses of eastern skunk cabbage *Symplocarpus foetidus*. *Plant Cell. Environ.* 27: 1014–1022.

Seymour, R., and P. Schultze-Motel. 1996. Thermoregulating lotus flowers. *Nature* 383: 305.

Seymour, R., Y. Ito, Y. Onda, and K. Ito. 2009. Effects of floral thermogenesis in Asian skunk cabbage *Symplocarpus renifolius*. *Biol. Lett.* 5: 568–570.

Sformo, T., J. McIntyre, K. R. Walters, B. M. Barnes, and J. G. Duman. 2011. Probability of freezing in the freeze avoiding beetle larvae *Cucujus clavipes puniceus* (Coleoptera, Cucujidae) from interior Alaska. *J. Insect Physiol.* 57: 1170–1177.

Sformo, T., K. Walters, K. Jeannet, B. Wowk, G. M. Fahy, B. M. Barnes, and J. G. Duman. 2010. Deep supercooling, vitrification and limited survival to −100°C in the Alaskan beetle *Cucujus clavipes puniceus* (Coleoptera, Cucujidae) larvae. *J. Exp. Biol.* 213: 502–509.

Shamovsky, I., M. Ivannikov, E. S. Kandel, D. Gershon, and E. Nudler. 2006. RNA-mediated response to heat shock in mammalian cells. *Nature* 440: 556–560.

Shetty, A., S. Chen, E. I. Tocheva, G. J. Jensen, and W. J. Hickey. 2011. Nanopods: a new bacterial structure and mechanism for deployment of outer membrane vesicles. *PLoS One* 6: e20725.

Siddiqui, K. S., and R. Cavicchioli. 2006. Cold-adapted enzymes. *Annu. Rev. Biochem.* 75: 403–433.

Sidell, B. D. 1998. Intracellular oxygen diffusion: the roles of myoglobin and lipid at cold body temperature. *J. Exp. Biol.* 201: 1118–1127.

Sidell, B. D., and J. R. Hazel. 1987. Temperature affects the diffusion of small molecules through cytosol of fish muscle. *J. Exp. Biol.* 129: 191–203.

Sinclair, B. J., and D. Renault. 2010. Intracellular ice formation in insects: Unresolved after 50 years? *Comp. Biochem. Physiol. A* 155: 14–18.

Sinensky, M. 1974. Homeoviscous adaptation—a homeostatic process that regulates the viscosity of membrane lipids in *Escherichia coli*. *Proc. Natl. Acad. Sci. USA* 71: 522–525.

Snapp, B. D., and H. C. Heller. 1981. Suppression of metabolism during hibernation in ground squirrels (*Citellus lateralis*). *Physiol. Zool.* 54: 297–307.

Somero, G. N. 1995. Proteins and temperature. *Annu. Rev. Physiol.* 57: 43-68.

Somero, G. N., and A. L. DeVries. 1967. Temperature tolerance of some Antarctic fishes. *Science* 156: 257–258.

Somero, G. N., A. C. Giese, and D. E. Wohlschlag. 1968. Cold adaptation of the Antarctic fish *Trematomus bernacchii*. *Comp. Biochem. Physiol.* 26: 223–233.

Somero, G. N., and P. W. Hochachka. 1968. The effect of temperature on catalytic and regulatory functions of pyruvate kinases of the rainbow trout and the Antarctic fish *Trematomus bernacchii*. *Biochem. J.* 110: 395–400.

Stefani, M., and C. M. Dobson. 2003. Protein aggregation and aggregate toxicity: New insights into protein folding, misfolding diseases and biological evolution. *J. Mol. Med.* 81: 678–699.

Stetter, K. O. 1999. Extremophiles and their adaptation to hot environments. *FEBS Letters.* 452: 22–25.

Storey, K. B., and J. M. Storey. 1988. Freeze tolerance in animals. *Physiol. Rev.* 68: 27–84.

Štros, M. 2010. HMGB proteins: interactions with DNA and chromatin. *Biochim. Biophys. Acta.* 1799: 101–113.

Takai, K., K. Nakamura, T. Toki, U. Tsunogai, M. Miyazaki, J. Miyazaki, H. Hirayama, S. Nakagawa, and 2 others. 2008. Cell proliferation at 122°C and isotopically heavy CH$_4$ production by a hyperthermophilic methanogen under high-pressure cultivation. *Proc. Natl. Acad. Sci. USA.* 105: 10949–10954.

Tang, D., Y. Shi, L. Jang, K. Wang, W. Xiao, and X. Xiao. 2005. Heat shock response inhibits release of high mobility group Box 1 protein induced by endotoxin in murine macrophages. *Shock* 23: 434–440.

Tansey, M. R., and T. D. Brock. 1972. The upper temperature limits for eukaryotic life. *Proc. Natl. Acad. Sci. USA.* 69: 2426–2428.

Tattersall, G. J., D. V. Andrade, and A. S. Abe. 2009. Heat exchange from the toucan bill reveals a controllable vascular thermal radiator. *Science* 325: 468–470.

Tehei, M., B. Franzetti, T. Madern, M. Ginzburg, B. Z. Ginzberg, M.-T. Giudici-Orticoni, M. Bruschi, and G. Zaccai. 2004. Adaptation to extreme environments: macromolecular dynamics in bacteria compared in vivo by neutron scattering. *EMBO Rep.* 5: 66–70.

Terblanche, J. S., A. A. Hoffmann, K. A. Mitchell, L. Rako, P. C. le Roux, and S. L. Chown. 2011. Ecologically relevant measures of tolerance of potentially lethal temperatures. *J. Exp. Biol.* 214: 3713–3725.

Tewksbury, J. J., R. B. Huey, and C. A. Deutsch. 2008. Putting the heat on tropical animals. *Science* 320: 1296–1297.

Thomas, J. O., and A. A. Travers. 2001. HMG1 and 2, and related 'architectural' DNA-binding proteins. *Trends Biochem. Sci.* 26: 167–174.

Tian, S., R. A. Haney, and M. E. Feder. 2010. Phylogeny disambiguates the evolution of heat-shock cis-regulatory elements in *Drosophila*. *PLoS One* 5: e10669.

Tokuriki, N., and D. S. Tawfik. 2009. Protein dynamism and evolvability. *Science* 324: 203–207.

Tomanek, L., and G. N. Somero. 1999. Evolutionary and acclimation-induced variation in the heat-shock responses of congeneric marine snails (genus *Tegula*) from different thermal habitats: implications for limits of thermotolerance and biogeography. *J. Exp. Biol.* 202: 2925–2936.

Tomanek, L., and G. N. Somero. 2000. Time course and magnitude of synthesis of heat-shock proteins in congeneric marine snails (genus *Tegula*) from different tidal heights. *Physiol. Biochem. Zool.* 73: 249–256.

Tomanek, L., and M. J. Zuzow. 2010. The proteomic response of the mussel congeners *Mytilus galloprovincialis* and *M. trossulus* to acute heat stress: implications for thermal tolerance and metabolic costs of thermal stress. *J. Exp. Biol.* 213: 3559–3574.

Tomazic, S. J., and A. M. Klibanov. 1988. Mechanisms of irreversible thermal inactivation of *Bacilllus* α-amylases. *J. Biol. Chem.* 263: 3086–3091.

Trajkovski, M., K. Ahmed, C. C. Esau, and M. Stoffel. 2012. MyomiR-133 regulates brown fat differentiation through Prdm16. *Nat. Cell Biol.* 14: 1330–1335.

Tsvetkova, N. M., I. Horváth, Z. Török, W. F. Wolkers, Z. Balogi, N. Shigapova, L.M. Crowe, F. Tablin, and 3 others. 2002. Small heat-shock proteins regulate membrane lipid polymorphism. *Proc. Natl. Acad. Sci. USA* 99: 13504–13509.

Tyedmers, J., A. Mogk, and B. Bukau. 2010. Cellular strategies for controlling protein aggregation. *Nat. Rev. Mol. Cell Biol.* 11: 777–788.

Tyler, S., and B. D. Sidell. 1984. Changes in mitochondrial distribution and diffusion distances in muscle of goldfish upon acclimation to warm and cold temperatures. *J. Exp. Zool.* 232: 1–9.

Ueda, T., and M. Yoshida. 2010. HMGB proteins and transcriptional regulation. *Biochim. Biophys. Acta* 1799: 114–118.

Van der Linden, M. G., and S. T. Farias. 2006. Correlation between codon usage and thermostability. *Extremophiles* 10: 479–481.

van Wolferen, M., M. Ajon, A. J. M. Driessen, and S.-V. Albers. 2013. How hyperthermophiles adapt to their lives: DNA exchange in extreme conditions. *Extremophiles* 17: 545–563.

Venkatachalam, K., and C. Montell. 2007. TRP channels. *Annu. Rev. Biochem.* 76: 387–417.

Vercesi, A., J. Borecký, I. de Godoy Maia, P. Arruda, I. M. Cuccovia, and H. Chaimovich. 2006. Plant uncoupling mitochondrial proteins. *Annu. Rev. Plant Biol.* 57: 383–404.

Vidair, C. A., S. J. Doxsey, and W. C. Dewey. 1993. Heat shock alters centrosome organization leading to mitotic dysfunction and cell death. *J. Cell. Physiol.* 154: 443–455.

Vieille, C., and G. J. Zeikus 2001. Hyperthermophilic enzymes: Sources, uses and molecular mechanisms for thermostability. *Microbiol. Mol. Biol. Rev.* 65: 1–43.

Walters, K. R. Jr., A. S. Serianni, T. Sformo, B. M. Barnes, and J. G. Duman. 2009. A non-protein thermal hysteresis-producing xylomannan antifreeze in the freeze-tolerant Alaskan beetle *Upis ceramboides*. *Proc. Natl. Acad. Sci. USA* 106: 20210–20215.

Walters, K. R. Jr., A. S. Serianni, Y. Voituron, T. Sformo, B. M. Barnes, and J. G. Duman. 2011. A thermal hysteresis-producing xylomannan glycolipid antifreeze associated with cold tolerance is found in diverse taxa. *J. Comp. Physiol. B* 181: 631–640.

Wang, L., and J. G. Duman. 2005. Antifreeze proteins of the beetle *Dendroides canadensis* enhance one another's activity. *Biochemistry* 44: 10305–10312.

Wang, L., and J. G. Duman. 2006. A thaumatin-like protein from larvae of the beetle *Dendroides canadensis* enhances the activity of antifreeze proteins. *Biochemistry* 45: 1278–1284.

Watabe, S. 2002. Temperature plasticity of contractile proteins in fish muscle. *J. Exp. Biol.* 205: 2231–2236.

Watanabe, M., S. M. Houten, C. Mataki, M. A. Christoffolete, B. W. Kim, H. Sato, N. Messaddeq, J. W. Harney, and 5 others. 2006. Bile acids induce energy expenditure by promoting intracellular thyroid hormone activation. *Nature* 439: 484–489.

Watson, J. D., and F. H. C. Crick. 1953. A structure for deoxyribose nucleic acid. *Nature.* 171: 737–738.

Watt, W. B. 1983. Adaptation at specific loci. II. Demographic and biochemical elements in the maintenance of the *Colias* PGI polymorphism. *Genetics* 103: 691–724.

Watt, W. B., and A. M. Dean. 2000. Molecular-functional studies of adaptive genetic variation in prokaryotes and eukaryotes. *Annu. Rev. Genet.* 34: 593–622.

Weake, V. M., and J. L. Workman. 2010. Inducible gene expression: diverse regulatory mechanisms. *Nat. Rev. Genet.* 11: 426–437.

Wegner, N. C., O. E. Snodgrass, H. Dewar, and J. R. Hyde. 2015. Whole-body endothermy in a mesopelagic fish, the opah, *Lampris guttatus. Science* 348: 786–789.

Weinstein, R. R., and G. N. Somero. 1998. Effects of temperature on mitochondrial function in the Antarctic fish *Trematomus bernacchii. J. Comp. Physiol. B.* 168: 190–196.

Welte, M. A., J. M. Tetrault, R. P. Dellavalle, and S. L. Lindquist. 1993. A new method for manipulating transgenes: engineering heat tolerance in a complex, multicellular organism. *Curr. Biol.* 3: 842–853.

Wen, X., S. Wang, J. G. Duman, J. F. Arifin, V. Juwita, W. A. Goddard III, A. Rios, F. Liu, and 4 others. 2016. Antifreeze proteins govern the precipitation of trehalose in a freezing-avoiding insect at low temperature. *Proc. Natl. Acad. Sci. USA* 113: 6683–6688.

Wheat, C. W., W. B. Watt, D. D. Pollock, and P. M. Schulte. 2005. From DNA to fitness differences; Sequences and structures of adaptive variants of *Colias* phosphoglucose isomerase (PGI). *Mol. Biol. Evol.* 23: 499–512.

White, C. R., L. A. Alton, and P. B. Frappell. 2011. Metabolic cold adaptation in fishes occurs at the level of whole animal, mitochondria and enzyme. *Proc. R. Soc. B.* 279: 1740–1747.

Whittle, A. J., and A. Vidal-Puig. 2011. Immune cells fuel the fire. *Nature* 480: 46–47.

Wintrode, P. L., and F. H. Arnold. 2001. Temperature adaptation of enzymes: lessons from laboratory evolution. *Adv. Protein Chem.* 55: 161–225.

Wisniewski, M., R. Webb, R. Balsamo, T. J. Close, X.-M. Yu, and M. Griffith. 1999. Purification, immunolocalization, cryoprotective, and antifreeze activity of PCA60: A dehydrins from peach (*Prunus persica*). *Physiol. Plant.* 105: 600–608.

Wrabl, J. O., J. Gu, T. Liu, T. P. Schrank, S. T. Whitten, and V. J. Hilser. 2011. The role of protein conformational fluctuations in allostery, function, and evolution. *Biophys. Chem.* 159: 129–141.

Wu, B. J., A. J. Hulbert, L. H. Storlien, and P. L. Else. 2004. Membrane lipids and sodium pumps of cattle and crocodiles: an experimental test of the membrane pacemaker theory of metabolism. *Am. J. Physiol. Reg. Integr. Comp. Physiol.* 287: R633–R641.

Xie, X. S. 2013. Enzyme kinetics, past and present. *Science* 342: 1457–1459.

Xue, Q., and E.S. Yeung. 1995. Differences in the chemical reactivity of individual molecules of an enzyme. *Nature* 373: 681–683.

Yamamori, T., K. Ito, Y. Nakamura, and T. Yura. 1978. Transient regulation of protein synthesis in *Escherichia coli* upon shift-up of growth temperature. *J. Bacteriol.* 134: 1133–1140.

Yang, S., L. Blachowicz, L. Makowski, and B. Roux. 2010. Multidomain assembled states of Hck tyrosine kinase in solution. *Proc. Natl. Acad. Sci. USA* 107: 15757–15762.

Yao, C.-L., and G. N. Somero. 2012. The impact of acute temperature stress on hemocytes of invasive and native mussels (*Mytilus galloprovincialis* and *Mytilus californianus*): DNA damage, membrane integrity, apoptosis and signaling pathways. *J. Exp. Biol.* 215: 4267–4277.

Ye, L., J. Wu, P. Cohen, L. Kazak, M. J. Khandekar, M. P. Jedrychowski, X. Zeng, S. P. Gygi, and B. M. Spiegelman. 2013. Fat cells directly sense temperature to activate thermogenesis. *Proc. Natl. Acad. Sci. USA.* 110: 12480–12485.

Yura, T., H. Nagai, and H. Mori. 1993. Regulation of the heat-shock response in bacteria. *Annu. Rev. Microbiol.* 47: 321–350.

Zacchariassen, K. E., and H. T. Hammel. 1976. Nucleating agents in the haemolymph of insects tolerant to freezing. *Nature* 262: 285–287

Závodszky, P., J. Kardos, Á. Svingor, and G. A. Petsko. 1998. Adjustment of conformational flexibility is a key event in the thermal adaptation of proteins. *Proc. Natl. Acad. Sci. USA* 95: 7406–7411.

Zhong, M., A. Orosz, and C. Wu. 1998. Direct sensing of heat and oxidation by *Drosophila* heat shock transcription factor. *Mol. Cell.* 2: 101–108.

Zhu, Y., J. Lu, J. Wang, F. Chen, F. Leng, and H. Li. 2011. Regulation of thermogenesis in plants: the interaction of alternative oxidase and plant uncoupling protein. *J. Int. Plant Biol.* 53: 7–13.

Chapter 4

Alexander, R. McN. 1972. The energetics of vertical migration by fishes. *Symp. Soc. Exp. Biol.* 26: 273–294.

Alpert, P. 2005. The limits and frontiers of desiccation-tolerant life. *Integr. Comp. Biol.* 45: 685–695.

Arakawa, T., J. F. Carpenter, Y. A. Kita, and J. H Crowe. 1990. The basis for toxicity of certain cryoprotectants: A hypothesis. *Cryobiology* 27: 401–415.

Arnold, F. H., P. L. Wintrode, K. Miyazaki, and A. Gershenson. 2001. How enzymes adapt: lessons from directed evolution. *Trends Biochem. Sci.* 26: 100–106.

Atkin, O. K. and D. Macherel. 2009. The crucial role of plant mitochondria in orchestrating drought tolerance. *Ann. Bot. (Lond.)* 103: 581–597.

Atkinson, D. E. 1969. Limitation of metabolite concentrations and the conservation of solvent capacity in the living cell. *Curr. Top. Cell. Regul.* 1: 29–43.

Auton, M., J. Rosgen, M. Sinev, L. M. F. Holthauzen, and D. W. Bolen. 2011. Osmolyte effects on protein stability and solubility: A balancing act between backbone and side-chains. *Biophys. Chem.* 159: 90–99.

Beece, D., L. Eisenstein, H. Frauenfelder, D. Good, M. C. Marden, L. Reinisch, A. H. Reynolds, L. B. Sorsensen, and K. T. Yue. 1980. Solvent viscosity and protein dynamics. *Biochemistry* 19: 5147–5157.

Benaroudj, N., D. H. Lee, and A. L. Goldberg. 2001. Trehalose accumulation during cellular stress protects cells and cellular proteins from damage by oxygen radicals. *J. Biol. Chem.* 276: 24261–24267.

Berezovsky, I.N., K. B. Zeldovich, and E. I. Shakhnovich. 2007. Positive and negative design in stability and thermal adaptation of natural proteins. *PLOS Comput. Biol.* 3: e52.

Billi, D., and M. Potts. 2002. Life and death of dried prokaryotes. *Res. Microbiol.* 153: 7–12.

Bolen, D. W. 2004. Effects of naturally occurring osmolytes on protein stability and solubility: issues important in protein crystallization. *Methods* 34: 312–322.

Bowlus, R. D., and G. N. Somero. 1979. Solute compatibility with enzyme function and structure: Rationales for the selection of osmotic agents and end-products of anaerobic metabolism in marine invertebrates. *J. Exp. Zool.* 208: 137–152.

Boyd, C. M., and D. Gradmann. 2002. Impact of osmolytes on buoyancy of marine phytoplankton. *Mar. Biol.* 141: 605–618.

Brown, A. D. 1976. Microbial water stress. *Bacteriol. Rev.* 40: 803–846.

Brown, A. D., and J. Simpson. 1972. Water relations of sugar-tolerant yeasts: The role of intracellular polyols. *J. Gen. Microbiol.* 72: 589–591.

Buitink, J., O. Leprince, M. A. Hemminga, and F. A. Hoekstra. 2000. Molecular mobility in the cytoplasm: an approach to describe and predict lifespan of dry germplasm. *Proc. Natl. Acad. Sci. USA* 97: 2385–2390.

Burg, M. B., J. D. Ferraris, and N. I. Dmitrieva. 2007. Cellular response to hyperosmotic stress. *Physiol. Rev.* 87: 1441–1474.

Burg, M. B., E. D. Kwon, and E. M. Peters. 1996. Glycerophosphorylcholine and betaine counteract the effect of urea on pyruvate kinase. *Kidney Int.* 50: Suppl. 57: S100–S104.

Busa, W. B., and J. H. Crowe. 1983. Intracellular pH regulates transitions between dormancy and development of brine shrimp (*Artemia salina*) embryos. *Science* 221: 366–368.

Cameron, J. N. 1989. Intracellular buffering by dipeptide at high and low temperature in the blue crab *Callinectes sapidus*. *J. Exp. Biol.* 143: 543–548.

Cano, R. J., and M. K. Borucki. 1995. Revival and identification of bacterial spores in 25- to 40-million-year-old Dominican amber. *Science* 268: 1060–1064.

Cario, A., N. Jebbar, N. Kervarec, and P. Oger. 2010. Influence of high hydrostatic pressure on the salt and heat stress response in the piezophilic archaeon *Thermococcus barophilus*. *Book of Abstracts of Extremophiles* P7: 108.

Carpenter, J. F., and S. C. Hand. 1986. Arrestment of carbohydrate metabolism during anaerobic dormancy and aerobic acidosis of *Artemia* embryos: determination of pH-sensitive control points. *J. Comp. Physiol.* 156: 451–459.

Castellini, M. A., and G. N. Somero. 1981. Buffering capacity of vertebrate muscle: correlations with potentials for anaerobic function. *J. Comp. Physiol.* 143: 191–198.

Chandler, D. 2005. Interfaces and the driving force of hydrophobic assembly. *Nature* 437: 640–647.

Clegg, J. S. 2005. Desiccation tolerance in encysted embryos of the animal extremophile, *Artemia*. *Integr. Comp. Biol.* 45: 715–724.

Creighton, T. E. 1984. *Proteins: Structures and Molecular Properties*. W.H. Freeman, New York.

Crowe, J. H., J. F. Carpenter, and L. M. Crowe. 1988. The role of vitrification in anhydrobiosis. *Annu. Rev. Physiol.* 60: 73–103.

Davis, B. 1958. On the importance of being ionized. *Arch. Biochem. Biophys.* 78: 497–509.

Denton, E. J. 1963. Buoyancy mechanisms in sea creatures. *Endeavor* 22: 3–8.

Diamant, S., N. Eliahu, D. Rosenthal, and P. Goloubinoff. 2001. Chemical chaperones regulate molecular chaperones *in vitro* and in cells under combined salt and heat stresses. *J. Biol. Chem.* 276: 39586–39591.

Dmitrieva, J. I., J. D. Ferraris, J. L. Norenburg, and M. B. Burg. 2006. The saltiness of the sea breaks DNA in marine invertebrates: possible implications for animal evolution. *Cell Cycle* 5: 1320–1323.

Elnitsky, M. A., J. B. Benoit, G. Lopez-Martinez, D. L. Denlinger, and R. E. Lee, Jr. 2009. Osmoregulation and salinity tolerance in the Antarctic midge, *Belgica antarctica*: seawater exposure confers enhanced tolerance to freezing and dehydration. *J. Exp. Biol.* 212: 2864–2871.

Empadinhas, N., and M. S. da Costa. 2006. Diversity and biosynthesis of compatible solutes in hyper/thermophiles. *Int. Microbiol.* 9: 199–206.

Epel, D. E. 1997. Activation of sperm and egg during fertilization. In J. F. Hoffman and J. D. Jamieson (eds.), *Handbook of Physiology: Cell Physiology*, pp. 859–884. Oxford University Press, Oxford.

Erkut, C., A. Vasilj, S. Boland, B. Habermann, A. Shevchenko, and T. V. Kurzchalia. 2013. Molecular strategies of the *Caenorhabditis elegans* dauer larva to survive extreme desiccation. *PLoS One* e82473.

Faria, T. Q., A. Mingote, F. Siopa, R. Ventura, C. Maycock, and H. Santos. 2008. Design of new enzyme stabilizers inspired by glycosides of hyperthermophilic microorganisms. *Carbohyd. Res.* 343: 3025–3033.

Fernandez, O., L. Béthencourt, A. Quero, R. S. Sangwan, and C. Clément. 2010. Trehalose and plant stress responses: friend or foe? *Trends Plant Sci.* 15: 409–417.

Fulton, A. B. 1982. How crowded is the cytoplasm? *Cell* 30: 345–347.

Gardell, A. M., J. Yang, R. Sacchi, N. A. Fangue, B. D. Hammock, and D. Kültz. 2013. Tilapia (*Oreochromis mossambicus*) brain cells respond to hyperosmotic challenge by inducing *myo*-inositol biosynthesis. *J. Exp. Biol.* 216: 4615–4625.

Gazit, E. 2002. The "Correctly Folded" state of proteins: Is it a metastable state? *Angew. Chem. Int. Ed.* 41: 257–259.

Gerstein, M., and M. Levitt. 1998. Simulating water and the molecules of life. *Sci. Am.* 279: 100–105.

Goyal, K., L. J. Walton, J. A. Browne, A. M. Burnell, and A. Tunnacliffe. 2005. Molecular anhydrobiology: Identifying molecules implicated in invertebrate anhydrobiosis. *Integr. Comp. Biol.* 45: 702–709.

Greaney, G. S. and G. N. Somero. 1979. Effects of anions on the activation thermodynamics and fluorescence emission spectrum of alkaline phosphatase: evidence for enzyme hydration changes during catalysis. *Biochemistry* 24: 5322–5332.

Gu, J., K. Weber, E. Klemp, G. Winters, S. U. Franssen, I. Wienpahl, A.-K. Huylmans, K. Zecher, and 3 others. 2012. Identifying core features of adaptive metabolic mechanisms for chronic heat stress attenuation contributing to systems robustness. *Integr. Biol.* 4: 480–493.

Gusev, O., Y. Suetsugu, R. Cornette, T. Kawashima, M. D. Logacheva, A. S. Kondrashov, A. A. Penin, R. Hatanaka, and 20 others. 2014. Comparative genome sequencing reveals genomic signature of extreme desiccation tolerance in the anhydrobiotic midge. *Nat. Commun.* 5: 4784.

Gutfreund, H. 1972. *Enzymes: Physical Principles*, pp. 160–165. Wiley, New York.

Hand, S. C., and J. F. Carpenter. 1986. pH-induced metabolic transitions in *Artemia* embryos mediated by a novel hysteretic trehalase. *Science* 232: 1535–1537

Hand, S. C., M. A. Menze, M. Toner, L. Boswell, and D. Moore. 2011. LEA proteins during water stress: Not just for plants anymore. *Annu. Rev. Physiol.* 73: 115–134.

Hand, S. C., and G. N. Somero. 1983. Phosphofructokinase of the hibernator *Citellus beecheyi*: Temperature and pH regulation of activity via influences on the tetramer-dimer equilibrium. *Physiol. Zool.* 56: 380–388.

Harms, M. J., and J. W. Thornton. 2013. Evolutionary biochemistry: revealing the historical and physical causes of protein properties. *Nat. Rev. Genet.* 14: 559–571.

Hayward, S. A. L., B. Manso, and A. R. Cossins. 2014. Molecular basis of chill resistance adaptations in poikilothermic animals. *J. Exp. Biol.* 217: 6–15.

Henderson, L. J. 1913. *The Fitness of the Environment: An Inquiry into the Biological Significance of the Properties of Water*. Harvard University Press, Cambridge, MA.

Hengherr, S., A. G. Heyer, K. Brümmer, and R. O. Schill. 2011. Trehalose and vitreous states: desiccation tolerance of dormant stages of the crustaceans *Triops* and *Daphnia*. *Physiol. Biochem. Zool.* 84: 147–153.

Hill, R. W., G. A. Wyse, and M. Anderson. 2016. *Animal Physiology*, 4th ed. Sinauer Associates, Sunderland, MA.

Hilser, V. J. 2013. Signaling from disordered proteins. *Nature* 498: 308–309

Hinton, H. E. 1960. A fly larva that tolerates dehydration and temperatures of –270°C to +102°C. *Nature* 188: 336–337.

Hochachka, P. W., and G. N. Somero. 1973. *Strategies of Biochemical Adaptation*. W. B. Saunders, Philadelphia.

Hochachka, P. W., and G. N. Somero. 2002. *Biochemical Adaptation: Mechanism and Process in Physiological Evolution*. Oxford University Press, Oxford, UK.

Hoekstra, F. 2005. Differential longevities in desiccated anhydrobiotic plant systems. *Integr. Comp. Biol.* 45: 725–733.

Hoekstra, F. A., E. A. Golovina, and J. Buitink. 2001. Mechanisms of plant desiccation tolerance. *Trends Plant Sci.* 6: 432–438.

Hottinger, T., T. Boller, and A. Wiemken. 1987. Rapid changes of heat and desiccation tolerance correlated with changes of trehalose content in *Saccharomyces cerevisiae* cells subjected to temperature shifts. *FEBS Lett.* 220: 113–115.

Huang, J. Y., E. G. Sweeney, M. Sigal, H. C. Zhang, S. J. Remington, M. A. Cantrell, C. J. Kuo, K. Guillemin, and M. R. Amieva. 2015. Chemodetection and destruction of host urea allows *Helicobacter pylori* to locate the epithelium. *Cell Host Microbe* 18: 147–156.

Jaenicke, R. 1991. Protein stability and molecular adaptation to extreme conditions. *Eur. J. Biochem.* 202: 715–728.

Jaffe, B. 1976. *Crucibles: The Story of Chemistry*, 4th ed. Dover Publications, New York.

Jamieson, A. J., T. Fujii, D. J. Mayor, M. Solan, and I. G. Priede. 2010. Hadal trenches: the ecology of the deepest places on earth. *Trends Ecol. Evol.* 25: 190–197.

Jönsson, K. I., and R. Bertolani. 2001. Facts and fiction about long-term survival in tardigrades. *J. Zool.* 257: 181–187.

Kaplan, F., J. Kopka, D. W. Haskell, W. Zhao, K. C. Schiller, N. Gatzke, D. Y. Sung, and C. L. Gray. 2004. Exploring the temperature-stress metabolome of *Arabidopsis*. *Plant Physiol.* 136: 4159–4168.

Kashefi, K., and D. R. Lovely. 2003. Extending the upper temperature limit for life. *Science* 301: 934.

Kasten, U., C. Wiencke, and G. O. Kirst. 1992. Dimethylsulphoniopropionate (DMSP) accumulation in green macroalgae from polar to temperature regions: interactive effects of light versus salinity and light versus temperature. *Polar Biol.* 12: 603–607.

Karsten, U., K. Küch, C. Vogt, and G. O. Kirst. 1996. Dimethylsulfoniopropionate production in phototrophic organisms and its physiological function as a cryoprotectant. In R. P. Kiene, P. T. Visscher, M. D. Keller, and G. O Kirst (eds.), *Biological and Environmental Chemistry of DMSP and Related Sulfonium Compounds*, pp. 109–119. Plenum Press, New York.

Kirschner, L. B. 1991. Water and ions. In C. L. Prosser (ed.), *Comparative Animal Physiology: Environmental and Metabolic Animal Physiology*, pp. 13–107. Wiley-Liss, New York.

Kolhatkar, A., C. E. Robertson, M. E. Thistle, A. K. Gamperl, and S. Currie. 2014. Coordination of chemical (trimethylamine oxide) and molecular (heat shock protein 70) chaperone responses to heat stress in elasmobranch red blood cells. *Physiol. Biochem. Zool.* 87: 652–662.

Kranner, I., and S. Birtić. 2005. A modulating role for antioxidants in desiccation tolerance. *Integr. Comp. Biol.* 45: 734–740.

Kuhn, T. S. 1970. *The Structure of Scientific Revolutions*, 2nd ed. University of Chicago Press, Chicago.

Kültz, D. 2012. The combinatorial nature of osmosensing in fishes. *Physiology* 27: 259–275.

Kumar, M. D. S., K. A. Bava, M. M. Gromiha, P. Prabakaran, K. Kitajima, H. Uedaira, and A. Sarai. 2006. ProTherm and ProNIT: thermodynamic databases for proteins and protein-nucleic acid interactions. *Nucleic Acids Res.* 34: D204–D206.

Lamosa, P., M. V. Rodrigues, L. G. Gonçalves, J. Carr, R. Ventura, C. Maycock, N. D. Raven, and H. Santos. 2013. Organic solutes in the deepest phylogenetic branches of the *Bacteria*: identification of α(1-6)glucosyl- α(1-2) glucosylglycerate in *Persephonella marina*. *Extremophiles* 17: 137–146.

Laxson, C. J., N. E. Condon, J. C. Drazen, and P. H. Yancey. 2011. Decreasing urea: trimethylamine N-oxide ratios with depth in Chondrichthyes: A physiological depth limit? *Physiol. Biochem. Zool.* 84: 494–505.

Levy, E. D., S. De, and S. A. Teichmann. 2012. Cellular crowding imposes global constraints on the chemistry and evolution of proteomes. *Proc. Natl. Acad. Sci. USA* 109: 20461–20466.

Li, D., and X. He. 2009. Desiccation induced structural alterations in a 66-amino acid fragment of an anhydrobiotic nematode late embryogenesis abundant (LEA) protein. *Biomacromolecules* 10: 1469–1477.

Li, S., N. Chakraborty, A. Borcar, M. A. Menze, M. Toner, and S. C. Hand. 2012. Late embryogenesis abundant proteins protect human hepatoma cells during acute desiccation. *Proc. Natl. Acad. Sci. USA* 109: 20859–20864.

Lin, J.-J., T.-H. Yang, B. D. Wahlstrand, P. A. Fields, and G. N. Somero. 2002. Phylogenetic relationships and biochemical properties of the duplicated cytosolic and mitochondrial isoforms of malate dehydrogenase from a teleost fish *Sphyraena idiastes*. *J. Mol. Evol.* 54: 107–117.

Low, P. S., and G. N. Somero. 1975. Protein hydration changes during catalysis: a new mechanism of enzymic rate enhancement and ion activation/inhibition of catalysis. *Proc. Natl. Acad. Sci. USA* 72: 3305–3309.

Luo, Y., F. Li, G. P. Wang, X. H. Yang, and W. Wang. 2010. Exogenously-supplied trehalose protects thylakoid membranes of winter wheat from heat-induced damage. *Biol. Plant.* 54: 495–501.

Marshall, B. J., and J. R. Warren. 1984. Unidentified curved bacilli in the stomach of patients with gastritis and peptic ulceration. *Lancet* 323: 1311–1315.

Martin, D. D., D. H. Bartlett, and M. E. Roberts. 2002. Solute accumulation in the deep-sea bacterium *Photobacterium profundum*. *Extremophiles* 6: 507–514.

Martin, D. D., R. A. Ciulla, and M. F. Roberts. 1999. Osmoadaptation in Archaea. *Appl. Environ. Microbiol.* 65: 1815–1823.

Martins, L. O., R. Huber, H. Huber, K. O. Stetter, M. S. Da Costa, and H. Santos. 1997. Organic solutes in hyperthermophilic *Archaea*. *Appl. Environ. Microbiol.* 63: 896–912.

Millero, F. J. 1971. The molal volumes of electrolytes. *Chem. Rev.* 71: 147–176.

Nishiguchi, M. K., and G. N. Somero. 1992. Temperature- and concentration-dependence of compatibility of the organic osmolyte β-dimethylsulfoniopropionate. *Cryobiology* 29: 118–124.

Pace, N. R. 1997. A molecular view of microbial diversity and the biosphere. *Science* 276: 75–83.

Pais, T. M., P. Lamosa, B. Garcia-Morena, D. L. Turner, and H. Santos. 2009. Relationship between protein stabilization and protein rigidification induced by mannosylglycerate. *J. Mol. Biol.* 394: 237–250.

Pereira, C. S., and P. H. Hünenberger. 2008. The effect of trehalose on a phospholipid membrane under mechanical stress. *Biophys. J.* 95: 3535–3534.

Perutz, M. F., J. V. Kilmartin, K. Nishikura, J. H. Fogg, and P. J. G. Butler. 1980. Identification of residues contributing to the Bohr Effect of human haemoglobin. *J. Mol. Biol.* 138: 649–670.

Priede, I. G., F. Rainer, D. M. Bailey, M. A. Bergstad, J. E. Dyb, C. Henriques, E. G. Jones, and N. King. 2006. The absence of sharks from abyssal regions of the world's oceans. *Proc. R. Soc. B* 273: 1435–1441.

Prosser, C. L. 1973. *Comparative Animal Physiology*. W. B. Saunders, Philadelphia.

Qu, Y., C. L. Bolen, and D. W. Bolen. 1998. Osmolyte-driven contraction of a random coil protein. *Proc. Natl. Acad. Sci. USA* 95: 9268–9273.

Raina, J.-B., D. M. Tapiolas, S. Forêt, A. Lutz, D. Abrego, J. Ceh, F. O. Seneca, P. L. Clode, and 3 others. 2013. DMSP biosynthesis by an animal and its role in coral thermal stress response. *Nature* 502: 677–680.

Reeves, R. B. 1977. The interaction of body temperature and acid-base balance in ectothermic vertebrates. *Annu. Rev. Physiol.* 39: 559–586.

Reisch, C. R., M. J. Stoudemayer, V. A. Varaljay, I. J. Amster, M. A. Moran, and W. B. Whitman. 2011. Novel pathway for assimilation of dimethylsulphoniopropinate widespread in marine bacteria. *Nature* 473: 208–211.

Sales, K., W. Brandt, E. Rumbak, and G. Lindsey. 2000. The LEA-like protein HSP 12 in *Saccharomyces cerevisiae* has a plasma membrane location and protects membranes against desiccation and ethanol-induced stress. *Biochim. Biophys. Acta* 1463: 267–278.

Samerotte, A. L., J. C. Drazen, G. L. Brand, B. A. Seibel, and P. H. Yancey. 2007. Correlation of trimethylamine oxide and habitat depth within and among species of teleost fish: An analysis of causation. *Physiol. Biochem. Zool.* 80: 197–208.

Sælensminde, G., Ø. Halskau Jr., R. Helland, N.-P. Willassen, and I. Jonassen. 2007. Structure-dependent relationships between growth temperature of prokaryotes and the amino acid frequencies in their proteins. *Extremophiles* 11: 585–596.

Sælensminde, G., Ø. Halskau Jr., and I. Jonassen. 2009. Amino acid contacts in proteins adapted to different temperatures: hydrophobic interactions and surface charges play a key role. *Extremophiles* 13: 11–20.

Sanders, N. K., and J. J. Childress. 1988. Ion replacement as a buoyancy mechanism in a pelagic deep-sea crustacean. *J. Exp. Biol.* 138: 333–343.

Scharf, C. 2014. *The Copernicus Complex*. Scientific American/Farrar, Straus and Giroux, New York.

Sheets, E. B, and D. Rhodes. 1996. Determination of DMSP and other onium compounds in *Tetraselmis subcordiformis* by plasma desorption mass spectrometry. In R. P. Kiene, P. T. Visscher, M. D. Keller, and G. O. Kirst (eds.), *Biological and Environmental Chemistry of DMSP and Related Sulfonium Compounds*, pp. 55–63. Plenum Press, New York.

Shen-Miller, J., M. B. Mudgett, J. W. Schopf, S. Clarke, and R. Berger. 1995. Exceptional seed longevity and robust growth: Ancient sacred lotus from China. *Am. J. Bot.* 82: 1367–1380.

Shimizu, T., Y. Kanamori, T. Furuki, T. Kikawada, T. Okuda, T. Takahashi, H. Mihara, and M. Sakurai. 2010. Desiccation-induced structuralization and glass formation of group 3 late embryogenesis abundant protein model peptides. *Biochemistry* 49: 1093–1104.

Sidell, B. D. 1998. Intracellular oxygen diffusion: the roles of myoglobin and lipid at cold body temperature. *J. Exp. Biol.* 201: 1118–1127.

Siebenaller, J. F. 1991. Pressure as an environmental variable: magnitude and mechanisms of perturbation. In P. W. Hochachka and T. P. Mommsen (eds.), *Biochemistry and*

Molecular Biology of Fishes, Vol. 1, pp. 323–343. Elsevier, Amsterdam.

Siebenaller, J. F., and G. N. Somero. 1978. Pressure-adaptive differences in lactate dehydrogenases of congeneric fishes living at different depths. *Science* 201: 255–257.

Singer, M. A., and S. Lindquist. 1998a. Thermotolerance in *Saccharomyces cerevisiae*: the Yin and Yang of trehalose. *Trends Biotech.* 16: 460–468.

Singer, M. A., and S. Lindquist. 1998b. Multiple effects of trehalose on protein folding in vitro and in vivo. *Mol. Cell.* 1: 639–648.

Šklubalová, Z., and Z. Zatloukal. 2009. Conversion between osmolality and osmolarity of infusion solutions. *Sci. Pharm.* 77: 817–826.

Smith, H. W. 1936. The retention and physiological role of urea in the Elasmobranchii. *Biol. Rev.* 11: 49–82.

Snapp, B. D., and H. C. Heller. 1981. Suppression of metabolism during hibernation in ground squirrels (*Citellus lateralis*). *Physiol. Zool.* 54: 297–307.

Somero, G. N. 1992. Adaptations to high hydrostatic pressure. *Annu. Rev. Physiol.* 54: 557–577.

Somero, G. N., and J. F. Siebenaller. 1979. Inefficient lactate dehydrogenases of deep-sea fishes. *Nature* 282: 100–102.

Somero, G. N., and P. H. Yancey. 1997. Osmolytes and cell volume regulation: physiological and evolutionary principles. In J. Hoffman and J. Jamieson (eds.), *Handbook of Physiology: Cell Physiology*, pp. 441–484. Oxford University Press, Oxford.

Stefels, J. 2000. Physiological aspects of the production and conversion of DMSP in marine algae and higher plants. *J. Sea. Res.* 43: 183–197.

Stetter, K. O. 2006. Hyperthermophiles in the history of life. *Philos. Trans. Roy. Soc. B. Biol. Sci.* 361: 1837–1842.

Street, T. O., D. W. Bolen, and G. D. Rose. 2006. A molecular mechanism for osmolyte-induced protein stability. *Proc. Natl. Acad. Sci. USA* 103: 13997–14002.

Sunda, W., D. J. Kieber, R. P. Kiene, and S. Huntsman. 2002. An antioxidant function for DMSP and DMS in marine algae. *Nature* 418: 317–320.

Tunnacliffe, A., J. Lapinski, and B. McGee. 2005. A putative LEA protein, but no trehalose, is present in anhydrobiotic bdelloid rotifers. *Hydrobiologia* 546: 315–321.

Vezzi, A., S. Campanaro, M. D'Angelo, F. Simonato, N. Vitulo, F. M. Lauro, A. Cestaro, G. Malacrida, and 5 others. 2005. Life at depth: *Photobacterium profundum* genome sequence and expression analysis. *Science* 307: 1459–1461.

Villalobos, A. R. A., and J. L. Renfro. 2007. Trimethylamine oxide suppresses stress-induced alteration of organic anion transport in choroid plexus. *J. Exp. Biol.* 210: 541–552.

Villarreal, F. D., and D. Kültz. 2015. Direct ionic regulation of the activity of *myo*-inositol biosynthesis enzymes in Mozambique tilapia. *PLoS One* 10: e0123212.

Voight, J. R., H.-O. Pörtner, and R. K. O'Dor. 1994. A review of ammonia-mediated buoyancy in squids (Cephalopoda: Teuthoidea). *Mar. Freshw. Behav. Physiol.* 25: 193–203.

von Hippel, P. H., and T. Schleich. 1969. The effects of neutral salts on the structure and conformational stability of macromolecules in solution. In S. Timasheff and G. D. Fasman (eds.), *Structure and Stability of Biological Macromolecules*, pp. 417–574. Marcel Dekker, New York.

Wang, A., and D. W. Bolen. 1997. A naturally occurring protective system in urea-rich cells: Mechanism of osmolyte protection of proteins against urea denaturation. *Biochemistry* 36: 9101–9108.

Wang, Z., E. Klipfell, B. J. Bennett, R. Koeth, B. S. Levison, B. DuGar, A. E. Feldstein, E. B. Britt, and 10 others. 2011. Gut flora metabolism of phosphatidylcholine promotes cardiovascular disease. *Nature* 472: 57–63.

Watson, J. D. 1998. *The Double Helix: A Personal Account of the Discovery of the Structure of DNA*. Scribner, New York.

Weber, L. A., E. D. Hickey, P. A. Maroney, and C. Baglioni. 1977. Inhibition of protein synthesis by Cl⁻. *J. Biol. Chem.* 252: 4007–4010.

Wise, M. J., and A. Tunnacliffe. 2004. POPP the question: What *do* LEA proteins do? *Trends Plant Sci.* 9: 13–17.

Withers, P. C., G. Morrison, G. T. Hefter, and T.-S. Pang. 1994. Role of urea and methylamines in buoyancy of elasmobranch fishes. *J. Exp. Biol.* 188: 175–189.

Wolfe, G. V., M. Steinke, and G. O. Kirst. 1997. Grazing-activated chemical defense in a unicellular marine alga. *Nature* 387: 894–897.

Yancey, P. H. 2001. Water stress, osmolytes and proteins. *Am. Zool.* 41: 699–709.

Yancey, P. H. 2005. Organic osmolytes as compatible, metabolic and counteracting cytoprotectants in high osmolarity and other stresses. *J. Exp. Biol.* 208: 2819–2830.

Yancey, P. H., M. E. Clark, S. C. Hand, R. D. Bowlus, and G. N. Somero. 1982. Living with water stress. *Science* 217: 1214–1222.

Yancey, P. H., A. L. Fyfe-Johnson, R. H. Kelly, V. P. Walker, and M. T. Auñón. 2001. Trimethylamine oxide counteracts effects of hydrostatic pressure on proteins of deep-sea teleosts. *J. Exp. Zool.* 289: 172–176.

Yancey, P. H., M. E. Gerringer, J. C. Drazen, A. A. Rowden, and A. J. Jamieson. 2014. Marine fish may be biochemically constrained from inhabiting deepest ocean depths. *Proc. Natl. Acad. Sci. USA* 111: 4461–4465.

Yancey, P. H., M. D. Rhea, K. M. Kemp, and D. M. Bailey. 2004. Trimethylamine oxide, betaine and other osmolytes in deep-sea animals: Depth trends and effects on enzymes under hydrostatic pressure. *Cell. Mol. Biol.* 50: 371–376.

Yancey, P. H., and J. F. Siebenaller. 2015. Co-evolution of proteins and solutions: protein adaptation versus cytoprotective micromolecules and their roles in marine organisms. *J. Exp. Biol.* 218: 1880–1896.

Yancey, P. H., and G. N. Somero. 1980. Methylamine osmoregulatory solutes of elasmobranch fishes counteract urea inhibition of enzymes. *J. Exp. Zool.* 212: 205–213.

Zeldovich, K. B., P. Chen, and E. I. Shakhnovich. 2007. Protein stability imposes limits on organism complexity and speed of molecular evolution. *Proc. Natl. Acad. Sci. USA* 104: 16152–16157.

Zerbst-Boroffka, I., R. M. Kamaltynow, S. Harjes, E. Kinne-Saffran, and J. Gross. 2005. TMAO and other organic osmolytes in the muscles of amphipods (Crustacea) from shallow and deep water of Lake Baikal. *Comp. Biochem. Physiol. A* 142: 58–64.

Chapter 5

Angert, A. L., L. G. Crozier, L. J. Rissler, S. E. Gilman, J. J. Tewksbury, and A. J. Chunco. 2011. Do species' traits predict recent shifts at expanding range edges? *Ecol. Lett.* 14: 677–698.

Bednaršek N., R. A. Feely , J. C. P. Reum, B. Peterson, J. Menkel, S. R. Alin, and B. Hales. 2014a. *Limacina helicina* shell dissolution as an indicator of declining habitat suitability owing to ocean acidification in the California Current Ecosystem. *Proc. R. Soc. B.* 281: 2014–2023.

Bednaršek, N., G. A. Tarling, D. C. E. Bakker, S. Fielding, and R. A. Feely. 2014b. Dissolution dominating calcification process in polar pteropods close to the point of aragonite undersaturation. *PLoS One* 9(10): e109183.

Beers, J. M., and N. Jayasundara. 2015. Antarctic notothenioid fish: what are the future consequences of 'losses' and 'gains' acquired during long-term evolution at cold and stable temperatures? *J. Exp. Biol.* 218: 1834–1845.

Beers, J. M., and B. D. Sidell. 2011. Thermal tolerance of Antarctic notothenioid fishes correlates with level of circulating hemoglobin. *Physiol. Biochem. Zool.* 84: 353–362.

Benton, M. J., and R. J. Twitchett. 2003. How to kill almost all life: The end-Permian extinction event. *Trends Ecol. Evol.* 18: 358–365.

Bograd, S. J., C. G. Castro, E. Di Lorenzo, D. M. Palacios, H. Bailey, W. Gilly, and F. P. Chavez. 2008. Oxygen declines and the shoaling of the hypoxic boundary in the California Current. *Geophys. Res. Lett.* 35: L12607.

Booth, J. A. T., E. E. McPhee-Shaw, P. Chua, E. Kingsley, W. W. Denny, R. Phillips, S. J. Bograd, L. D. Zeidberg, and W. F. Gilly. 2012. Natural intrusions of hypoxic, low pH water into nearshore marine environments on the California coast. *Cont. Shelf Res.* 45: 108–115.

Borley, K. A., and B. D. Sidell. 2011. Evolution of the myoglobin gene in Antarctic icefishes (Channichthyidae). *Polar Biol.* 34: 659–665.

Braby, C. E., and G. N. Somero. 2006a. Ecological gradients and relative abundance of native (*Mytilus trossulus*) and invasive (*Mytilus galloprovincialis*) blue mussels in the California hybrid zone. *Mar. Biol.* 148: 1249–1262.

Braby, C. E., and G. N. Somero. 2006b. Following the heart: temperature and salinity effects on heart rate in native and invasive species of blue mussels (genus *Mytilus*). *J. Exp. Biol.* 209: 2554–2566.

Briffa, M., K. de la Haye, and P. Munday. 2012. High CO_2 and marine animal behaviour: potential mechanisms and ecological consequences. *Mar. Pollut. Bull.* 64: 1519–1528.

Buckley, B. A., and G. N. Somero. 2009. cDNA microarray analysis reveals the capacity of the cold-adapted Antarctic fish *Trematomus bernacchii* to alter gene expression in response to heat stress. *Polar Biol.* 32: 403–415.

Caldeira, K. and M. E. Wickett. 2003. Anthropogenic carbon and ocean pH. *Nature* 425: 365.

Caprio, J., M. Shimohara, T. Marui, S. Harada, and S. Kiyohara S. 2014. Marine teleost locates live prey through pH sensing. *Science* 344: 1154–1156.

Chan, W.-P., I-C. Chen, R. K. Colwell, W.-C. Liu, C.-Y. Huang, and S.-F. Shen. 2016. Seasonal and daily climate variation have opposite effects on species elevational range size. *Science* 351: 1437–1439.

Checkley, D. M. Jr, A. G. Dickson, M. Takahashi, J. A. Radich, N. Eisenkolb, and R. Asch. 2009. Elevated CO_2 enhances otolith growth in young fish. *Science* 324: 1683.

Clarkson, M. O., S. A. Kasemann, R. A. Wood, T. M. Lenton, S. J. Daines, S. Richoz, F. Ohnemueller, A. Meixner, S. W. Poulton, and E. T. Tipper. 2015. Ocean acidification and the Permo-Triassic mass extinction. *Science* 348: 229–232.

Cooper, B. S., L. A. Hammad, and K. L. Montooth. 2014. Thermal adaptation of cellular membranes in natural populations of *Drosophila melanogaster*. *Funct. Ecol.* 28: 886–894.

Coppes-Petricorena, Z. L., and G. N. Somero. 2007. Biochemical adaptations of notothenioid fishes: Comparisons between cold temperate South American and New Zealand species and Antarctic species. *Comp. Biochem. Physiol. A.* 147: 799–807.

Crutzen, P. J., and E. F. Stoermer. 2000. The "Anthropocene". *IGBP Newsletter* 41: 17

Cunning, R., R. N. Silverstein, and A. C. Baker. 2015a. Investigating the causes and consequences of symbiont shuffling in a multi-partner reef coral symbiosis under environmental change. *Proc. R. Soc. B.* 282: 20141725.

Cunning, R., N. Vaughan, P. Gillette, T. R. Capo, J. L. Mate, and A. C. Baker. 2015b. Dynamic regulation of partner abundance mediates response of reef coral symbioses to environmental change. *Ecology* 96: 1411–1420.

Deng, C., C.-H. C. Cheng, H. Ye, X. He, and L. Chen. 2010. Evolution of an antifreeze protein by neofunctionalization under escape from adaptive conflict. *Proc. Natl. Acad. Sci. USA* 107: 21593–21598.

Deutsch C., A. Ferrel, B. Seibel B, H. O. Pörtner, and R. B. Huey. 2015. Climate change tightens a metabolic constraint on marine habitats. *Science* 348: 1132–1135.

Diaz, R. J., and R. Rosenberg. 2008. Spreading dead zones and consequences for marine ecosystems. *Science* 321: 926–929.

Dickson, A. 2010. Part I: Seawater carbonate chemistry. In U. Riebesell, V. N. Fabry, L. Hannson, and J.-P. Gattuso (eds.), *Guide to Best Practices for Ocean Acidification Research and Data Reporting*. Publications Office of the European Union, Luxembourg.

Dillon, M. E., G. Wang, and R. B. Huey. 2010. Global metabolic impacts of recent climate warming. *Nature* 467: 704–707.

Dixon, G. B., S. W. Davies, G. A. Aglyamova, E. Meyer, L. K. Bay, and M. V. Matz. 2015. Genomic determinants of coral heat tolerance across latitudes. *Science* 348: 1460–1462.

Doney, S. C., V. J. Fabry, R. A. Feely, and J. A. Kleypas. 2009. Ocean acidification: the other CO_2 problem. *Annu. Rev. Mar. Sci.* 1: 169–192.

Dong, Y.-W., L. P. Miller, J. G. Sanders, and G. N. Somero. 2008. Heat-shock protein 70 (Hsp70) expression in four limpets of the genus *Lottia*: interspecific variation in constitutive and inducible synthesis correlates with in situ exposure to heat stress. *Biol. Bull.* 215: 173–181.

Dowd, W. W., and G. N. Somero. 2013. Behavior and survival of *Mytilus* congeners following episodes of elevated body temperature in air and seawater. *J. Exp. Biol.* 216:502–514.

Dowd, W. W., F. A. King, and M. W. Denny. 2015. Thermal variation, thermal extremes and the physiological performance of individuals. *J. Exp. Biol.* 218: 1956–1967.

Fields, P. A., E. Rudomin, and G. N. Somero 2006. Temperature sensitivities of cytosolic malate dehydrogenases from native and invasive species of marine mussels (genus *Mytilus*): sequence-function linkages and correlations with biogeographic distribution. *J. Exp. Biol.* 209: 656–677.

Frieder, C. A., J. P. Gonzalez, E. E. Bockmon, M. O. Navarro, and L. I. Levin. 2014. Can variable pH and low oxygen moderate ocean acidification outcomes for mussel larvae? *Glob. Change Biol.* 20: 754–764.

Fuhrman, J. 2009. Microbial community structure and its functional implications. *Nature* 459: 193–199.

Gastaldo, R. A., S. L. Kamo, J. Neveling, J. W. Geissman, M. Bamford, and C. V. Looy. 2015. Is the vertebrate-defined Permian-Triassic boundary in the Karoo Basin, South Africa, the terrestrial expression of the end-Permian marine event? *Geology* 43: 939–942.

Gattuso, J. P., A. Magnan, R. Billé , W. W. L. Cheung , E. L. Howes, F. Joos, D. Allemand, L. Bopp, and 14 others. 2015. Contrasting futures for ocean and society from different anthropogenic CO_2 emissions scenarios. *Science* 349: doi:10.1126/science.aac4722

Gaylord, B., K. J. Kroeker, J. M. Sunday, K. M. Anderson, J. P. Barry, N. E. Brown, S. D. Connell, S. Dupont, and 12 others. 2015. Ocean acidification through the lens of ecological theory. *Ecology* 96: 3–15.

Geller, J. B. 1999. Decline of a native mussel masked by sibling species invasion. *Conserv. Biol.* 13: 661–664.

Geller, J. B., J. A. Darling, and J. T. Carlton. 2010. Genetic perspectives on marine biological invasions. *Annu. Rev. Mar. Sci.* 2: 367–393.

Gilbert, S., and D. Epel. 2015. *Ecological Developmental Biology.* Sinauer Associates, Sunderland, MA.

Gilman, S. E., M. C. Urban, J. Tewksbury, G. W. Gilchrist, and R. D. Holt. 2010. A framework for community interactions under climate change. *Trends Ecol. Evol.* 25: 325–331.

Gingerich, P. 2006. Environment and evolution through the Paleocene-Eocene thermal maximum. *Trends Ecol. Evol.* 21: 246–253.

Gleason, L. U., and R. S. Burton. 2013. Phenotypic evidence for local adaptation to heat stress in the marine snail *Chlorostoma* (formerly *Tegula*) *funebralis. J. Exp. Mar. Biol. Ecol.* 448: 360–366.

Gleason, L. U., and R. S. Burton. 2015. RNA-seq reveals regional differences in transcriptome response to heat stress in the marine snail *Chlorostoma funebralis. Mol. Ecol.* 24: 610–627.

Gobler, C. J., E. L. DePasquale, A. W. Griffith, and H. Baumann. 2014. Hypoxia and acidification have additive and synergistic negative effects on the growth, survival, and metamorphosis of early life stage bivalves. *PLoS One*: e83648.

Goldenfeld, N., and C. Woese. 2007. Biology's next revolution. *Nature* 445: 369.

Haigh, R., D. Ianson, C. A. Holt, H. E. Neate, and A. M. Edward. 2015. Effects of ocean acidification on temperate coastal marine ecosystems and fisheries in the northeast Pacific. *PLoS One* 10: e0117533.

Hamilton, T. J., A. Holcombe, and M. Tresguerres. 2014. CO_2-induced ocean acidification increases anxiety in rockfish via alteration of $GABA_A$ receptor functioning. *Proc. R. Soc. B* 281: 20132509.

Hardege, J. D., J. M. Rotchell, J. Terschak, and G. M. Greenway. 2011. Analytical challenges and the development of biomarkers to measure and to monitor the effects of ocean acidification. *Trends Anal. Chem.* 30: 1320–1326.

Harrison, P. M., and M. Gerstein. 2002. Studying genomes through the aeons: protein families, pseudogenes and proteome evolution. *J. Mol. Evol.* 318: 1155–1174.

Helmuth, B., C. D. G. Harley, P. M. Halpin, M. O'Donnell, G. E. Hofmann, and C. A. Blanchette. 2002. Climate change and latitudinal patterns of intertidal thermal stress. *Science* 298: 1015–1017.

Helmuth, B., B. D. Russell, S. D. Connell, Y. Dong, C. D. G. Harley, F. P. Lima, G. Sará, G. A. Williams, and N. Mieszkowska. 2014. Beyond long-term averages: making biological sense of a rapidly changing world. *Clim. Change Responses* 1: 6.

Hettinger, A., E. Sanford, T. M. Hill, J. D. Hosfelt, A. D. Russell, and B. Gaylord B. 2013. The influence of food supply on the response of Olympia oyster larvae to ocean acidification, *Biogeosciences* 10: 6629–6638.

Hilbish, T., P. Brannock, K. Jones, A. Smith, B. Bullock, and D. Wethey. 2010. Historical changes in the distributions of invasive and endemic marine invertebrates are contrary to global warming predictions: the effects of decadal climate oscillations. *J. Biogeogr.* 37: 423–431.

Hoegh-Guldberg, O., P. J. Mumby, A. J. Hooten, R. S. Steneck, P. Greenfield, E. Gomez, C. D. Harvell, P. F. Sale, and 9 others. 2007. Coral reefs under rapid climate change and ocean acidification. *Science* 318: 1737–1742.

Hofmann, G. E., and G. N. Somero. 1996. Interspecific variation in thermal denaturation of proteins in the congeneric mussels *Mytilus trossulus* and *M. galloprovincialis*: evidence from the heat-shock response and protein ubiquitination. *Mar. Biol.* 126: 65–75.

Hofmann, G. E., B. A. Buckley, S. Airaksinen, J. E. Keen, and G. N. Somero. 2000. Heat-shock protein expression is absent in the antarctic fish *Trematomus bernacchii* (family Nototheniidae). *J. Exp. Biol.* 203: 2331–2339.

Innan, H., and F. Kondrashov. 2010. The evolution of gene duplications: classifying and distinguishing between models. *Nat. Rev. Genet.* 11: 97–108.

IPCC. 2014. *Climate Change 2014: Synthesis Report. Contribution of Working Groups, I, II and II to the Fifth Assessment Report of the Intergovernmental Panel on Climate Change.* R. K. Pachauri et al. (eds.). IPCC, Geneva, Switzerland.

Jokiel, P. L., and S. L. Coles. 1990. Response of Hawaiian and other Indo-Pacific reef corals to elevated temperatures. *Coral Reefs* 8: 155–162.

Kawecki, T. J., and D. Ebert. 2004. Conceptual issues in local adaptation. *Ecol. Lett.* 7: 1225–1241.

Karasov, T., P. W. Messer, and D. A. Petrov. 2010. Evidence that adaptation in *Drosophila* is not limited by mutation at single sites. *PLoS Genet.* 6: e1000924.

Keeling, P. J., and J. D. Palmer. 2008. Horizontal gene transfer in eukaryotic evolution. *Nat. Rev. Genet.* 9: 605–618.

Keeling, R. F., A. Körtzinger, and N. Gruber. 2010. Ocean deoxygenation in a warming world. *Annu. Rev. Mar. Sci.* 9: 199–229.

Keller, A. A., V. Simon, F. Chan , W. W. Wakefield, M. E. Clarke, J. A. Barth, D. Kamikawa, and E. L. Fruh. 2010. Demersal fish and invertebrate biomass in relation to an offshore hypoxic zone along the US West Coast. *Fish. Oceanogr.* 19: 76–87.

Kellermann, V., J. Overgaard, A. A. Hoffmann, C. Fløjgaard, J. C. Svenning, and V. Loeschcke. 2012. Upper thermal limits of *Drosophila* are linked to species distributions and strongly constrained phylogenetically. *Proc. Natl. Acad. Sci. USA* 109: 16228–16233.

Kelly, M. W., E. Sanford, and R. K. Grosberg. 2012. Limited potential for adaptation to climate change in a broadly distributed marine crustacean. *Proc. R. Soc. B.* 279: 349–356.

Kleypas, J. 2015. Invisible barriers to dispersal. *Science* 348: 1086–1087.

Kristensen, T. N., J. Overgaard, J. Lassen, A. A. Hoffmann, and C. Sgrò. 2015. Low evolutionary potential for egg-to-adult viability in *Drosophila melanogaster* at high temperatures. *Evolution.* 69: 803–814.

Kroeker, K. J., R. L. Kordas, R. Crim, I. E. Hendriks, L. Ramajo, G. S. Singh, C. M. Duarte, and J.-P. Gattuso. 2013. Impacts of ocean acidification on marine organisms: quantifying sensitivities and interaction with warming. *Glob. Change Biol.* 19: 1884–1896.

Kroeker, K. J., B. Gaylord, T. M. Hill, J. D. Hosfelt, S. H. Miller, and E. Sanford. 2014. The role of temperature in determining species' vulnerability to ocean acidification: A case study using *Mytilus galloprovincialis. PLoS One* 9(7): e100353.

Kuo, E. S. L., and E. Sanford. 2009. Geographic variation in the upper thermal limits of an intertidal snail: implications for climate envelope models. *Mar. Ecol. Prog. Ser.* 388: 137–146.

Lande, R., and S. Shannon. 1996. The role of genetic variation in adaptation and population persistence in a changing environment. *Evolution* 50: 434–437.

Lockwood, B. L., and G. N. Somero. 2011a. Invasive and native blue mussels (genus *Mytilus*) on the California coast: The role of physiology in a biological invasion. *J. Exp. Mar. Biol. Ecol.* 400: 167–174.

Lockwood, B. L., and G. N. Somero. 2011b. Transcriptomic responses to salinity stress in invasive and native blue mussels (genus *Mytilus*). *Mol. Ecol.* 20: 517–529.

Lockwood, B. L., J. G. Sanders, and G. N. Somero. 2010. Transcriptomic responses to heat stress in invasive and native blue mussels (genus *Mytilus*): molecular correlates of invasive success. *J. Exp. Biol.* 213: 3548–3558.

López-García, P., Y. Zivanovic, P. Deschamps, and D. Moreira. 2015. Bacterial gene import and mesophilic adaptation in archaea. *Nat. Rev. Microbiol.* 13: 447–456.

Maneja, R. H., A. Y. Frommel, A. J. Geffen, A. Folkvord, U. Piatkowski, M. Y. Chang, and C. Clemmesen. 2013. Effects of ocean acidification on the calcification of otoliths of larval Atlantic cod *Gadus morhua. Mar. Ecol. Prog. Ser.* 477: 251–258.

McClatchie, S., R. Goericke, R. Cosgrove, G. Auad, and R. Vetter. 2010. Oxygen in the Southern California Bight: Multidecadal trends and implications for demersal fisheries. *Geophys. Res. Lett.* L19602.

McDonald, J. H., and R. K. Koehn. 1988. The mussels *Mytilus galloprovincialis* and *Mytilus trossulus* on the Pacific coast of North America. *Mar. Biol.* 99: 111–118.

McFall-Ngai, M. J. 2015. Giving microbes their due—animal life in a microbially dominant world. *J. Exp. Biol.* 218: 1968–1973.

McFall-Ngai, M. J., M. G. Hadfield, T. C. G., Bosch, H. V. Carey, T. Domazel-Lošo, A. E. Douglas, N. Dubilier, G. Eberl, and 18 others. 2013. Animals in a bacterial world: a new imperative for the life sciences. *Proc. Natl. Acad. Sci. USA* 110: 3229–3236.

McMichael, A. J., and K. B. G. Dear. 2010. Climate change: Heat, health, and longer horizons. *Proc. Natl. Acad. Sci. USA* 107: 9483–9484.

Melzner, F. P., K. Strange, K. Trubenback, I. Thomsen, U. Casties, S. Panknin, N. Gorb, and M. A. Gutowska. 2011. Food supply and seawater pCO_2 impact calcification and internal shell dissolution in the blue mussel *Mytilus edulis. PLoS One* 6: e24223.

Memmott, J., P. G. Craze, N. M. Waser, and M. V. Price. 2007. Global warming and the disruption of plant-pollinator interactions. *Ecol. Lett.* 10: 710–717.

Montgomery, J., and K. Clements. 2000. Disaptation and recovery in the evolution of Antarctic fishes. *Trends Ecol. Evol.* 15: 267–271.

Mooney, H. A., and E. E. Cleland. 2001. The evolutionary impact of invasive species. *Proc. Natl. Acad. Sci. USA* 98: 5446–5451.

Moran, N. A., and T. Jarvik. 2010. Lateral transfer of genes from fungi underlies carotenoid production in aphids. *Science* 328: 624–627.

Muir, P. R., C. C. Wallace, T. Done, and J. D. Aguirre. 2015. Limited scope for latitudinal extension of reef corals. *Science* 348: 1135–1138.

Munday, P. L., N. E. Crawley, and G. Nilsson. 2009. Interacting effects of elevated temperature and ocean acidification on the aerobic performance of coral reef fishes. *Mar. Ecol. Prog. Series* 388: 235–242.

Munday, P. L., M. I. McCormick, and G. E. Nilsson. 2012. Impact of global warming and rising CO_2 levels on coral reef fishes: what hope for the future? *J. Exp. Biol.* 215: 3865–3873.

Nagelkerken, I., and S. D. Connell. 2015. Global alteration of ocean ecosystem functioning due to increasing human CO_2 emissions. *Proc. Natl. Acad. Sci. USA* 112: 13272–13277.

Nilsson, G., D. L. Dixson, P. Domenici, M. I. McCormick, C. Sorensen, S. A. Watson, and P. I. Munday. 2012. Near-future carbon dioxide levels alter fish behaviour by interfering with neurotransmitter function. *Nat. Clim. Change* 2: 201–204.

O'Brien, K. M., I. A. Mueller, J. I. Orczewska, K. P. Dullen, and M. Ortego. 2014. Hearts of some Antarctic fishes lack mitochondrial creatine kinase. *Comp. Biochem. Physiol. A. Mol. Integr. Physiol.* 178: 30–36.

Oliver, T. A., and S. R. Palumbi. 2011. Do fluctuating temperature environments elevate coral thermal tolerance? *Coral Reefs* 30: 429–440.

Orr, J. C., V. J. Fabry, O. Aumont, L. Bopp, S.C. Doney, R.A. Feely, A. Gnanadesikan, N. Gruber, and 19 others. 2005.

Anthropogenic ocean acidification over the twenty-first century and its impact on calcifying organisms. *Nature* 437: 681-686.

Pal, J. S., and E. A. B. Eltahir. 2016. Future temperature in southwest Asia projected to exceed a threshold for human adaptability. *Nat. Clim. Change* 6: 197–200.

Palumbi, S. R. 2004. Marine reserves and ocean neighborhoods: the spatial scale of marine populations and their management. *Annu. Rev. Environ. Resour.* 29: 31–68.

Palumbi, S. R., D. J. Barshis, N. Taylor-Knowles, and R. A. Bay. 2014. Mechanisms of reef coral resistance to future climate change. *Science* 344: 895–898.

Pan, T.-C. F., S. L. Applebaum, and D. T. Manahan. 2015. Experimental acidification alters the allocation of metabolic energy. *Proc. Natl. Acad. Sci. USA* 112: 4696–4701.

Parmesan, C., and G. Yohe. 2003. A globally coherent fingerprint of climate change impacts across natural systems. *Nature* 421: 37–42.

Pearson, R. G., and T. P. Dawson. 2003. Predicting the impacts of climate change on the distribution of species: Are bioclimate envelope models useful? *Glob. Ecol. Biogeogr.* 12: 361–371.

Peck, L. S., S. A. Morley, J. Richard, and M. S. Clark. 2014. Acclimation and thermal tolerance in Antarctic marine ectotherms. *J. Exp. Biol.* 217: 16–22.

Pereira, R. J., F. S. Barreto, and R. S. Burton. 2014. Ecological novelty by hybridization: experimental evidence for increased thermal tolerance by transgressive segregation in *Tigriopus californicus*. *Evolution* 68: 204–215.

Perry, A. L., P. J. Low, J. R. Ellis, and J. D. Reynolds. 2005. Climate change and distribution shifts in marine fishes. *Science* 308: 1912–1915.

Pespeni, M. H., E. Sanford, B. Gaylord, T. M. Hill, J. D. Hosfelt, H. K. Jaris, M. LaVigne, E. A. Lenz, and 3 others. 2013a. Evolutionary change during experimental ocean acidification. *Proc. Natl. Acad. Sci. USA* 110: 6937–6942.

Pespeni, M. H., F. Chan, B. A. Menge, and S. R. Palumbi. 2013b. Signs of adaptation to local pH conditions across an environmental mosaic in the California current ecosystem. *Integr. Comp. Biol.* 53: 857–870.

Pilson, M. E. Q. 1998. *An Introduction to the Chemistry of the Sea.* Prentice Hall, Upper Saddle River, NJ.

Pinsky, M. L., B. Worm, M. J. Fogarty, J. L. Sarmiento, and S. A. Levin. 2013. Marine taxa track local climate velocities. *Science* 342: 1239–1242.

Pochon, X., and R. D. Gates. 2010. A new *Symbiodinium* clade (Dinophyceae) from soritid foraminifera in Hawai'i. *Mol. Phylogen. Evol.* 56: 492–497.

Pörtner, H. O. 2010. Oxygen- and capacity-limitation of thermal tolerance: a matrix for integrating climate-related stressor effects in marine ecosystems. *J. Exp. Biol.* 213: 881–893.

Pörtner, H. O., and A. P. Farrell. 2008. Physiology and climate change. *Science.* 322: 690–692.

Regen, C. T. 1913. The Antarctic fishes of the Scottish National Antarctic Expedition. *Trans. R. Soc. Edin.* 49: 229–292.

Reisert, J., and D. Restrepo. 2009. Molecular tuning of odorant receptors and its implication for odor signal processing. *Chem. Senses* 34: 535–545.

Rohner, N., D. F. Jarosz, J. E. Kowalko, M. Yoshizawa, W. R. Jeffrey, R. L. Borowsky, S. Lindquist, and C. J. Tabin. 2015. Cryptic variation in morphological evolution: Hsp90 as a capacitor for loss of eyes in cavefish. *Science* 342: 1372–1375.

Röhl, U., T. J. Bralower, R. D. Norris, and G. Wefer. 2000. New chronology for the late Paleocene thermal maximum and its environmental implications. *Geology* 28: 927–930.

Rosa R., and B. A. Seibel. 2008. Synergistic effects of climate-related variables suggest future physiological impairment in a top ocean predator. *Proc. Natl. Acad. Sci. USA* 105: 20776–20780.

Rowan, R. 2004. Thermal adaptation in reef coral symbionts. *Nature* 430: 742.

Ruddiman, W. F., E. C. Ellis, J. O Kaplan, and D. Q. Fuller. 2015. Defining the epoch we live in. *Science* 348: 6229–6230.

Sanford, E., and M. W. Kelly. 2011. Local adaptation in marine invertebrates. *Annu. Rev. Mar. Sci.* 3: 509–535.

Sanford, E., M. S. Roth, G. C. Johns, J. P. Wares, and G. N. Somero. 2003. Local selection and latitudinal variation in a marine predator-prey interaction. *Science.* 300: 1135–1137.

Savolainen, O., T. Pyhäjärvi, and T. Knürr. 2007. Gene flow and local adaptation in trees. *Annu. Rev. Ecol. Syst.* 38: 595–619.

Schär, C. 2015. Climate extremes: The worst heat waves to come. *Nat. Clim. Change* 6: 128–129.

Schneider, K., and B. Helmuth. 2007. Spatial variability in habitat temperature may drive patterns of selection between an invasive and native mussel species. *Mar. Ecol. Prog. Ser.* 339: 157–167.

Schönknecht, G., W.-H. Chen, C. M. Ternes, G. G. Barbier, R. P. Shrestha, M. Stanke, A. Bräutigam, B. J. Baker, and 11 others. 2013. Gene transfer from bacteria and archaea facilitated evolution of an extremophilic eukaryote. *Science* 339: 1207–1210.

Schoville, S. D., F. S. Barreto, G. W. Moy, A. Wolff, and R. S. Burton. 2012. Investigating the molecular basis of local adaptation to thermal stress: population differences in gene expression across the transcriptome of the copepod *Tigriopus californicus. BMC Evol. Biol.* 12: 170.

Seibel, B. A. 2011. Critical oxygen levels and metabolic suppression in oceanic oxygen minimum zones. *J. Exp. Biol.* 214: 326–336.

Seibel, B. A., N. Sören Häfker, K. Trübenbach, J. Zhand, S. N. Tessier, H.-O. Pörtner, R. Rosa, and K. B. Storey. 2014. Metabolic suppression during protracted exposure to hypoxia in the jumbo squid, *Dosidicus gigas*, living in an oxygen minimum zone. *J. Exp. Biol.* 217: 2555–2568.

Sherwood, S. C., and M. Huber. 2010. An adaptability limit to climate change due to heat stress. *Proc. Natl. Acad. Sci. USA* 107: 9552–9555.

Sibbick, J. 2007. *Ecology.* Cold Spring Harbor Laboratory Press, New York.

Slamovits, C. H., N. Okamoto, L. Burri, E. R. James, and P. J. Keeling. 2011. A bacterial proteorhodopsin proton pump in marine eukaryotes. *Nat. Commun.* 2: 183.

Somero, G. N. 2002. Thermal physiology and vertical zonation of intertidal animals: Optima, limits, and costs of living. *Integr. Comp. Biol.* 42: 780–789

Somero, G. N., J. Beers, F. Chan, T. Hill, T. Klinger, and L. Litvin. 2016. What changes in the carbonate system, oxygen, and temperature portend for the northeastern Pacific Ocean. *BioScience* 66: 14–26.

Somero, G. N., and A. L. DeVries. 1967. Temperature tolerance of some Antarctic fishes. *Science* 156: 257–258.

Stager, C. 2011. *Deep Future: the Next 100,000 Years of Life on Earth.* St. Martin's Press, New York.

Stenseng, E., C. Braby, and G. N. Somero 2005. Evolutionary and acclimation-induced variation in the thermal limits of heart function in congeneric marine snails (genus *Tegula*): Implications for vertical zonation. *Biol. Bull.* 208: 138–144.

Stillman, J. H. 2003. Acclimation capacity underlies susceptibility to climate change. *Science* 301: 65.

Stillman, J. H., and A. W. Paganini. 2015. Biochemical adaptation to ocean acidification. *J. Exp. Biol.* 218: 1946–1955.

Stillman, J. H., and G. N. Somero. 1996. Adaptation to temperature stress and aerial exposure in congeneric species of intertidal porcelain crabs (genus *Petrolisthes*): Correlation of physiology, biochemistry and morphology with vertical distribution. *J. Exp. Biol.* 199: 1845–1855.

Stillman, J. H., and G. N. Somero. 2000. A comparative analysis of the upper thermal tolerance limits of eastern Pacific porcelain crabs, genus *Petrolisthes*: Influences of latitude, vertical zonation, acclimation, and phylogeny. *Physiol. Biochem. Zool.* 73: 200–208.

Stramma, L., G. C. Johnson, J. Sprintall, and V. Mohrholz. 2008. Expanding oxygen minimum zones in the tropical oceans. *Science* 320: 655–658.

Stuart-Smith, R. D., G. J. Edgar, N. S. Barrett, S. J. Kininmonth, and A. E. Bates. 2015. Thermal biases and vulnerability to warming in the world's marine fauna. *Nature* 528: 88–92.

Sunday, J. M., A. E. Bates, and N. K. Dulvy. 2012. Thermal tolerance and the global redistribution of animals. *Nat. Clim. Change* 2: 686–690.

Sunday, J. M., R. N. Crim, C. D. G. Harley, and M. W. Hart. 2011. Quantifying rates of evolutionary adaptation in response to ocean acidification. *PLoS One* 6: e22881.

Tepolt, C. K., and S. R. Palumbi. 2015. Transcriptome sequencing reveals both neutral and adaptive genome dynamics in a marine invader. *Mol. Ecol.* 24: 4145–4158.

Tepolt, C. K., and G. N. Somero. 2014. Master of all trades: thermal acclimation and adaptation of cardiac function in a broadly distributed marine invasive species, the European green crab, *Carcinus maenas. J. Exp. Biol.* 217: 1129–1138.

Tewksbury, J. J., R. B. Huey, and C. A. Deutsch. 2008. Putting the heat on tropical animals. *Science* 320: 1296–1297.

Tomanek, L., and E. Sanford. 2003. Heat-shock protein 70 (Hsp70) as a biochemical stress indicator: an experimental field test in two congeneric intertidal gastropods (genus: *Tegula*). *Biol. Bull.* 205: 276–284.

Tomanek, L., and G. N. Somero. 1999. Evolutionary and acclimation-induced variation in the heat-shock responses of congeneric marine snails (genus *Tegula*) from different thermal habitats: implications for limits of thermotolerance and biogeography. *J. Exp. Biol.* 202: 2925–2936.

Tomanek, L., and M. Zuzow. 2010. The proteomic response of the mussel congeners *Mytilus galloprovincialis* and *M. trossulus* to acute heat stress: implications for thermal tolerance limits and metabolic costs of thermal stress. *J. Exp. Biol.* 213: 3905–3916.

Tomanek, L., M. Zuzow, L. Hitt, L. Serafini, and J. J. Valenzuela. 2012. Proteomics of hyposaline stress in blue mussel congeners (genus *Mytilus*): implications for biogeographic range limits in response to climate change. *J. Exp. Biol.* 215: 3905–3916.

Torres, J., M. D. Grigsby, and M. E. Clarke. 2012. Aerobic and anaerobic metabolism in oxygen minimum layer fishes: the role of alcohol dehydrogenase. *J. Exp. Biol.* 215: 1905–1914.

Trübenbach, K., T. Teixeira, M. Diniz, and R. Rosa. 2013. Hypoxia tolerance and antioxidant defense system of juvenile jumbo squids in oxygen minimum zones. *Deep-Sea Res. II* 95: 209–217.

Tseng, Y.-C., M. Y. Hu, M. Stumpp, L.-Y. Lin, F. Melzner, and P.-P. Hwang. 2013. CO_2-driven seawater acidification differentially affects development and molecular plasticity along life history of fish (*Oryzias latipes*). *Comp. Biochem. Physiol. A.* 165: 119–130.

Tuchman, B. 1978. *A Distant Mirror: The Calamitous 14th Century.* A.A. Knopf, New York.

Vaquer-Sunyer, R., and C. M. Duarte. 2008. Thresholds of hypoxia for marine biodiversity. *Proc. Natl. Acad. Sci. USA* 105: 15452–15457.

Verberk, W. C. E. P., D. T. Bilton, P. Calosi, and J. I. Spicer. 2011. Oxygen supply in aquatic ectotherms: Partial pressure and solubility together explain biodiversity and size patterns. *Ecology* 92: 1565–1572.

Waldbusser, G., B. Hales, C. J. Langdon, B. A. Haley, P. Schrader, E. L. Brunner, M. W. Gray, C. A. Miller, and 2 others. 2015. Ocean acidification has multiple modes of action on bivalve larvae. *PLoS One* 10: e0128376.

Watanabe, J. M. 1982. *Aspects of community organization in a kelp forest habitat: Factors influencing the bathymetric segregation of three species of herbivorous gastropods.* Ph.D. Diss., University of California, Berkeley.

Willett, C. S. 2010. Potential fitness trade-offs for thermal tolerance in the intertidal copepod *Tigriopus californicus. Evolution* 64: 2521–2534.

Zeebe, R. E., J. C. Zachos, K. Caldeira, and T. Tyrrell 2008. Oceans: carbon emissions and acidification. *Science* 321: 51–52.

Zeidberg, L. D., and B. H. Robison. 2007. Invasive range expansions by the Humboldt squid, *Dosidicus gigas* in the eastern North Pacific. *Proc. Natl. Acad. Sci. USA* 104: 12948–12950.

Zhaxybayeva, O., and W. F. Doolittle. 2011. Lateral gene transfer: a primer. *Curr. Biol.* 21: R242–R246.

Illustration Credits

Chapter 1 Page 4: Courtesy of the University of Washington. Page 20: © Images & Stories/Alamy Stock Photo. Page 22 *Gillichthys*: Courtesy of Dr. T.-H. Yang. Page 22 *Trematomus*: Photograph by G. N. Somero. Page 23: Courtesy of Dr. Luke Miller. Page 24: Courtesy of Dr. Gregory Jensen.

Chapter 2 2.41 *adult*: Courtesy of Dr. Jason Podrabsky. 2.43 *turtle*: © Alberthep/istock. Page 40 *reconstruction*: © Science Photo Library/Alamy Stock Photo. Page 40 *inset*: © iStock.com/Natasha Litova. Page 64: © 2015 MBARI. Page 73: © Nancy Nehring/Getty Images. Page 74: © Danny Laps/NiS/Getty Images. Page 88: Courtesy of Dr. Geoff Dilly. Page 89: © Gregory G. Dimijian, M.D./Science Source. Page 96: Photo by G. N. Somero.

Chapter 3 3.1 *tardigrade*: © Sciencefoto.De, Dr. Andre Kempe/Getty Images. 3.1 *beetle*: Courtesy of Oivind Toien. 3.1 *Trematomus*: Photograph by G. N. Somero. 3.1 *Gillichthys*: Courtesy of Dr. T.-H. Yang. 3.1 *Echinolittorina*: Courtesy of Dr. Yunwei Dong. 3.1 *vent*: Courtesy of the University of Washington. 3.1 *vent inset*: Courtesy of the Archivo Angels Tapias y Fabrice Confalonieri. 3.4: Courtesy of Dr. Peter Fields. 3.12: Courtesy of David McIntyre. 3.15A: Data from PDB 1T2K. D. Panne et al., 2004. *EMBO J.* 23: 4384. 3.15B: Data from PDB 1JUN. F. Junius et al., 1996. *J. Biol. Chem.* 271: 13663. 3.55A, B: From Ross, Pawlina, and Barnash, 2009. *Atlas of Descriptive Histology.* Sinauer Associates: Sunderland, MA. 3.59A *swordfish*: © Panya Kuanun/Shutterstock. Page 165 *Lottia*: Courtesy of Dr. J. M. Watanabe. Page 248: Courtesy of Dr. Jason Podrabsky. Page 282: *Sisyphus* (1548–49) by Titian, Prado Museum, Madrid. Page 294: © iStock.com/Valeriy Kirsanov. Page 321: Courtesy of the New York State Department of Environmental Conservation. Page 322: Courtesy of Oivind Toien.

Chapter 4 Box 4.3A: © Ammit/Alamy Stock Photo. Box 4.6: © Veronika Burmeister/Visuals Unlimited, Inc. Page 356 *Methanosarcina*: © Ralph Robinson/Visuals Unlimited, Inc. Page 356 *ponds*: © Kevin Kemmerer/ Getty Images. Page 367: © Peter Scoones/Getty Images. Page 370 *snailfish*: From Yancey et al. 2014. Page 370 *rattail fish*: © 2002 MBARI. Page 376: © Andrey Nekrasov/Getty Images. Page 379: Photograph by Dr. James Childress, University of California, Santa Barbara. Page 416: © Sciencefoto.De, Dr. Andre Kempe/ Getty Images.

Chapter 5 5.19B: Illustration by G. M. Woodward, from Regan, C. Tate, 1914. *Fishes.* London: British Museum. 5.20 *Trematomus*: Photograph by G. N. Somero. 5.20 *Gillichthys*: Courtesy of Dr. T.-H. Yang. Box 5.3B: Courtesy of Goddard SVS/NASA. Page 443: © 2003 MBARI. Page 448: Courtesy of Emma Kissling. Page 450 *Plotosus*: © Nobuo Matsumura/Alamy Stock Photo. Page 454: © David Liittschwager/Getty Images. Page 457: © iStock.com/arousa. Page 465 *Chlorostoma*: © Doug Sokell/Visuals Unlimited, Inc. Page 467: Courtesy of Dr. J. M. Watanabe. Page 469: © blickwinkel/Alamy Stock Photo. Page 478: © Nancy Nehring/Getty Images. Page 482 *Nucella*: © Randimal/ Shutterstock. Page 483: © NatalieJean/Shutterstock. Page 486: © Roberto Nistri/Alamy Stock Photo. Page 493 *top*: Courtesy of Gerald Schoenknecht. Page 493 *bottom*: Courtesy of Christine Oesterhelt. Page 495: © karelnoppe/Shutterstock.

Index

Page numbers in *italic* type indicate that the information will be found in an illustration.